VOLUME ONE HUNDRED AND THIRTY ONE

CURRENT TOPICS IN DEVELOPMENTAL BIOLOGY

Plant Development and Evolution

CURRENT TOPICS IN DEVELOPMENTAL BIOLOGY

"A meeting-ground for critical review and discussion of developmental processes"
A.A. Moscona and Alberto Monroy (Volume 1, 1966)

VOLUME ONE HUNDRED AND THIRTY ONE

CURRENT TOPICS IN DEVELOPMENTAL BIOLOGY

Plant Development and Evolution

Edited by

UELI GROSSNIKLAUS

Department of Plant and Microbial Biology & Zurich-Basel Plant Science Center, University of Zurich, Zurich, Switzerland

ELSEVIER

ACADEMIC PRESS

An imprint of Elsevier

Academic Press is an imprint of Elsevier
50 Hampshire Street, 5th Floor, Cambridge, MA 02139, United States
525 B Street, Suite 1650, San Diego, CA 92101, United States
The Boulevard, Langford Lane, Kidlington, Oxford OX5 1GB, United Kingdom
125 London Wall, London, EC2Y 5AS, United Kingdom

First edition 2019

Notices

Knowledge and best practice in this field are constantly changing. As new research and experience broaden our understanding, changes in research methods, professional practices, or medical treatment may become necessary.

Practitioners and researchers must always rely on their own experience and knowledge in evaluating and using any information, methods, compounds, or experiments described herein. In using such information or methods they should be mindful of their own safety and the safety of others, including parties for whom they have a professional responsibility.

To the fullest extent of the law, neither the Publisher nor the authors, contributors, or editors, assume any liability for any injury and/or damage to persons or property as a matter of products liability, negligence or otherwise, or from any use or operation of any methods, products, instructions, or ideas contained in the material herein.

ISBN: 978-0-12-809804-2
ISSN: 0070-2153

Publisher: Zoe Kruze
Acquisition Editor: Fiona Pattison
Editorial Project Manager: Shellie Bryant
Production Project Manager: Denny Mansingh
Cover Designer: Greg Harris

Typeset by SPi Global, India

Contents

Contributors

Alma Armenta-Medina
Laboratorio Nacional de Genómica para la Biodiversidad (Langebio), Unidad de Genómica Avanzada, Centro de Investigación y Estudios Avanzados del Instituto Politécnico Nacional (CINVESTAV-IPN), Irapuato, Guanajuato, México

Célia Baroux
Department of Plant and Microbial Biology & Zurich-Basel Plant Science Center, University of Zurich, Zurich, Switzerland

John L. Bowman
School of Biological Sciences, Monash University, Melbourne, VIC, Australia

Liam N. Briginshaw
School of Biological Sciences, Monash University, Melbourne, VIC, Australia

John W. Chandler
Department of Biology, Biocenter, University of Cologne, Cologne, Germany

Marta Cubría-Radío
Department of Plant Biotechnology and Bioinformatics, Ghent University; VIB Center for Plant Systems Biology, Ghent, Belgium

Mainak Das Gupta
Max Planck Institute for Plant Breeding Research, Cologne, Germany

Thomas Dresselhaus
Cell Biology and Plant Biochemistry, University of Regensburg, Regensburg, Germany

Stevie N. Florent
School of Biological Sciences, Monash University, Melbourne, VIC, Australia

Hiroo Fukuda
Department of Biological Sciences, Graduate School of Science, The University of Tokyo, Tokyo, Japan

Charles S. Gasser
Department of Molecular and Cellular Biology, University of California, Davis, Davis, CA, United States

C. Stewart Gillmor
Laboratorio Nacional de Genómica para la Biodiversidad (Langebio), Unidad de Genómica Avanzada, Centro de Investigación y Estudios Avanzados del Instituto Politécnico Nacional (CINVESTAV-IPN), Irapuato, Guanajuato, México

Rita Groß-Hardt
Centre for Biomolecular Interactions, University of Bremen, Bremen, Germany

Ueli Grossniklaus
Department of Plant and Microbial Biology & Zurich-Basel Plant Science Center, University of Zurich, Zurich, Switzerland

Dieter Hackenberg
Department of Genetics and Genome Biology, University of Leicester, Leicester, United Kingdom

Alexander Kirbis
Department of Systematic and Evolutionary Botany & Zurich-Basel Plant Science Center, University of Zurich, Zurich, Switzerland

Elena M. Kramer
Harvard University, Cambridge, MA, United States

Gloria León-Martínez
Group of Reproductive Development and Apomixis, UGA Laboratorio Nacional de Genómica para la Biodiversidad (Langebio), Cinvestav Irapuato, Irapuato, Mexico

June B. Nasrallah
Section of Plant Biology, School of Integrative Plant Science, Cornell University, Ithaca, NY, United States

Lachezar A. Nikolov
Max Planck Institute for Plant Breeding Research, Cologne, Germany

Moritz K. Nowack
Department of Plant Biotechnology and Bioinformatics, Ghent University; VIB Center for Plant Systems Biology, Ghent, Belgium

Kyoko Ohashi-Ito
Department of Biological Sciences, Graduate School of Science, The University of Tokyo, Tokyo, Japan

Lars Østergaard
Department of Crop Genetics, John Innes Centre, Norwich, United Kingdom

Joakim Palovaara
Centre for Biomolecular Interactions, University of Bremen, Bremen, Germany

John B. Reese
Department of Ecology and Evolutionary Biology, University of Tennessee, Knoxville, TN, United States

Adam Runions
Max Planck Institute for Plant Breeding Research, Cologne, Germany

Isil Erbasol Serbes
Centre for Biomolecular Interactions, University of Bremen, Bremen, Germany

Vinay Shekhar
Department of Plant and Microbial Biology & Zurich-Basel Plant Science Center, University of Zurich, Zurich, Switzerland

Bihai Shi
Laboratoire Reproduction et Developpement des Plantes, Univ Lyon, ENS de Lyon, UCB Lyon1, CNRS, INRA, Lyon, France

Sara Simonini
Department of Crop Genetics, John Innes Centre, Norwich, United Kingdom

Debra J. Skinner
Department of Molecular and Cellular Biology, University of California, Davis, Davis, CA, United States

Dorothee Stöckle
Department of Plant and Microbial Biology & Zurich-Basel Plant Science Center, University of Zurich, Zurich, Switzerland

Péter Szövényi
Department of Systematic and Evolutionary Botany & Zurich-Basel Plant Science Center, University of Zurich, Zurich, Switzerland

Martha Thellmann
Department of Plant and Microbial Biology & Zurich-Basel Plant Science Center, University of Zurich, Zurich, Switzerland

Bennett Thomson
Smurfit Institute of Genetics, Trinity College Dublin, Dublin, Ireland

Miltos Tsiantis
Max Planck Institute for Plant Breeding Research, Cologne, Germany

David Twell
Department of Genetics and Genome Biology, University of Leicester, Leicester, United Kingdom

Karina van der Linde
Department of Cell Biology and Plant Biochemistry, University of Regensburg, Regensburg, Germany

Joop E.M. Vermeer
Department of Plant and Microbial Biology & Zurich-Basel Plant Science Center, University of Zurich, Zurich, Switzerland

Teva Vernoux
Laboratoire Reproduction et Developpement des Plantes, Univ Lyon, ENS de Lyon, UCB Lyon1, CNRS, INRA, Lyon, France

Jean-Philippe Vielle-Calzada
Group of Reproductive Development and Apomixis, UGA Laboratorio Nacional de Genómica para la Biodiversidad (Langebio), Cinvestav Irapuato, Irapuato, Mexico

Virginia Walbot
Department of Biology, Stanford University, Stanford, CA, United States

Manuel Waller
Department of Systematic and Evolutionary Botany & Zurich-Basel Plant Science Center, University of Zurich, Zurich, Switzerland

Frank Wellmer
Smurfit Institute of Genetics, Trinity College Dublin, Dublin, Ireland

Wolfgang Werr
Department of Biology, Biocenter, University of Cologne, Cologne, Germany

Joseph H. Williams
Department of Ecology and Evolutionary Biology, University of Tennessee, Knoxville, TN, United States

Liang-zi Zhou
Cell Biology and Plant Biochemistry, University of Regensburg, Regensburg, Germany

Preface

Plants are the basis of life on earth as we know it and have not only shaped, to a large extent, diverse ecosystems but also provide food, feed, fibers, and fuel for human subsistence. Understanding how plants evolved and develop to fulfill their manifold functions is thus of great fundamental and applied interest.

It will soon be 10 years since the last volume of *Current Topics in Developmental Biology* focused on plant development. Since then, biology has seen a revolution in methods for the investigation of genes, transcripts, and proteins, allowing a wide variety of "omics" approaches. Also, the tools to investigate gene function using forward-genetic, reverse-genetic, and alternative methods, such as artificial microRNAs and RNA interference, have provided deep insights into the molecular processes that regulate plant development. The arsenal of the plant biologist has recently been extended by the CRISPR/Cas9 system, which allows functional studies also in non-model systems, providing new insights into developmental diversity and plasticity. Clearly, focusing efforts on a model system, in this case *Arabidopsis thaliana*, has greatly accelerated progress in understanding plant function. On the other hand, a tunnel vision focusing on a single species is bound to miss important aspects of development that are not found in this model system. Certainly, plant biology has profited a lot from the powerful genetics of *Zea mays* (maize), which started well over a hundred years ago, and the recent focus on *Oryza sativa* (rice) as one of the most important crop plants. Over the last years, the range of plant species used in fundamental research has greatly expanded, providing insights into diverse developmental processes.

Unraveling the molecular control of plant development has also provided the material to look at the evolution of gene regulatory networks, investigating how preexisting mechanisms were co-opted for new developmental programs or how novelties arose during land plant evolution. Such evo-devo studies have provided great insights into the evolution of the enormous diversity of plants that exists today. It is more than timely then to provide a broad overview of plant development and evolution, as it is only possible in a multichapter book bringing together experts in diverse field to share their views on specific plant developmental processes and their evolution. The chapters provide a wide diversity of perspectives on both vegetative and reproductive development, summarizing the deep functional and mechanistic insights that have been gained in many fields of development. The chapters focusing on evolutionary aspects provide an excellent view on the directions the evo-devo field will take in the future by being able to do functional studies in nonmodel systems that promise to provide invaluable insights into the evolution of plant development.

I was delighted that so many of my colleagues enthusiastically accepted to contribute to this volume of *Current Topics in Developmental Biology* and I am thankful for their detailed reviews and insightful discussions on a wide range of topics in plant development and evolution. No doubt the next decade will see enormous progress and further deepen our understanding of plant function. Nevertheless, I am convinced that the readers will find this volume interesting and timely and hope they will enjoy reading it.

Ueli Grossniklaus
Department of Plant and Microbial Biology & Zurich-Basel Plant Science Center, University of Zurich, Zurich, Switzerland

CHAPTER ONE

Evolution of the plant body plan

Péter Szövényi*, Manuel Waller, Alexander Kirbis
Department of Systematic and Evolutionary Botany & Zurich-Basel Plant Science Center, University of Zurich, Zurich, Switzerland
*Corresponding author: e-mail address: peter.szoevenyi@uzh.ch

Contents

Abstract

Land plants evolved about 470 million years ago or even earlier, in a biological crust-dominated terrestrial flora. The origin of land plants was probably one of the most significant events in Earth's history, which ultimately contributed to the greening of the terrestrial environment and opened up the way for the diversification of both plant and non-plant lineages. Fossil and phylogenetic evidence suggest that land plants have evolved from fresh-water charophycean algae, which were physiologically, genetically, and developmentally potentiated to make the transition to land. Since all land plants have biphasic life cycles, in contrast to the haplontic life cycle of Charophytes, the evolution of land plants was linked to the origin of a multicellular sporophytic phase. Land plants have evolved complex body plans in a way that overall complexity

Current Topics in Developmental Biology, Volume 131
ISSN 0070-2153
https://doi.org/10.1016/bs.ctdb.2018.11.005

increased toward the tip of the land plant tree of life. Early forms were unbranched, with terminal sporangia and simple rhizoid rooting structures but without vasculature and leaves. Later on, branched forms with lateral sporangia appeared and paved the route for the evolution for indeterminacy. Finally, leaves and roots evolved to enable efficient nutrient transport to support a large plant body. The fossil record also suggests that almost all plant organs, such as leaves and roots, evolved multiple times independently over the course of land plant evolution. In this review, we summarize the current knowledge on the evolution of the land plant body plan by combining evidence of the fossil record, phylogenetics, and developmental biology.

1. The significance of land plants

Land plants evolved about 470 million years ago, or potentially even earlier, in the Cambrian (515 Ma) to Early Ordovician (473 Ma) when the terrestrial flora was dominated by a crust-forming assemblage of algae lichens, bacteria, cyanobacteria, and fungi (Berbee, James, & Strullu-Derrien, 2017; Morris et al., 2018; Salamon et al., 2018). The origin of land plants was one of the most significant events in the history of the globe, which paved the road for diversification of non-plant lineages in the terrestrial environment (Bateman et al., 1998; Berner, 1997). Among others, this was enabled by an increasing oxygen concentration produced through photosynthesis, initially by cyanobacteria and then eukaryotic algae and land plants (Kenrick, Wellman, Schneider, & Edgecombe, 2012). The resulting ozone layer reduced the amount of UV irradiation, which was probably necessary for plants to leave the water (Rozema et al., 2002; Rozema, Blokker, Mayoral Fuertes, & Broekman, 2009). Furthermore, land plants have also significantly contributed to the weathering of rocks and the formation of soil by preventing transportation of sediments by wind and water (Lenton, Crouch, Johnson, Pires, & Dolan, 2012; McMahon & Davies, 2018; Perkins, 2018). Altogether, plants have tremendously contributed to the greening of the planet and to the rapid increase of terrestrial biodiversity.

2. Understanding evolution of the plant body plan

Three major types of information can be utilized to investigate the main trends in the evolution of the land plant body plan, and to uncover the underlying genetic mechanisms. Phylogenetic analyses of extant land plants can provide information on the ancestor-descendant relationships of major lineages, and thus a rough timeline of their evolutionary origin.

Nevertheless, available phylogenetic evidence is mostly based on nucleotide data and includes only extant plants, providing little information about the combined evolutionary history of extant and fossil forms.

Another line of evidence is provided by the fossil data. Unfortunately, fossil record of the first land plants is sparse, and their interpretation can be ambiguous (Boyce & Kevin Boyce, 2010; Kenrick, 2018; Tomescu, 2009). Ambiguity in the assessment of morphological characters and their potential homology with body plans of extant land plants make the evolutionary placement of fossils questionable and combined analysis of data on fossils and extant land plants is challenging (Bateman, 1996; Puttick et al., 2018; Salamon et al., 2018; Stewart & Rothwell, 1993; Seward, 2011). Despite this, careful comparative morphological and evolutionary analyses of the fossil record and extant land plants gave rise to major theories about the evolutionary origin of main modular units of the land plant body plan, providing testable hypotheses (Chomicki, Coiro, & Renner, 2017).

Comparative analysis of fossils and phylogenetic relationships of extant land plants is necessary to reveal evolutionary trends and formulate testable hypotheses, but do not allow experimental validation. Comparative developmental biology and genetics/genomics of extant representatives of land plants can be used to experimentally test hypotheses on the developmental and genetic mechanisms underlying the evolutionary trends in land plant body plan evolution (Boyce & Kevin Boyce, 2010; Rothwell, Wyatt, & Tomescu, 2014; Tomescu, Wyatt, Hasebe, & Rothwell, 2014).

In this review, we provide insights into the recent advancement surrounding the evolution of the land plant body plan by evaluating evidence from phylogenetics/phylogenomics, the fossil record, comparative genomics, and the developmental biology of plants. We do so by summarizing recent evidence in these scientific fields and their implications for each major innovation in the evolution of the land plant body plan (see Fig. 1).

3. Phylogenetic relationships of land plants and evolution of the land plant body plan

Phylogenetic analysis of extant members of green plants unambiguously suggests that the lineage of land plants (embryophytes) together with several lineages of streptophytic algae (charophycean algae) forms the monophyletic group of streptophytes (Becker & Marin, 2009; Gitzendanner, Soltis, Wong, Ruhfel, & Soltis, 2018; Liu, Cox, Wang, & Goffinet, 2014; Wickett et al., 2014; Wodniok et al., 2011). The streptophytes are

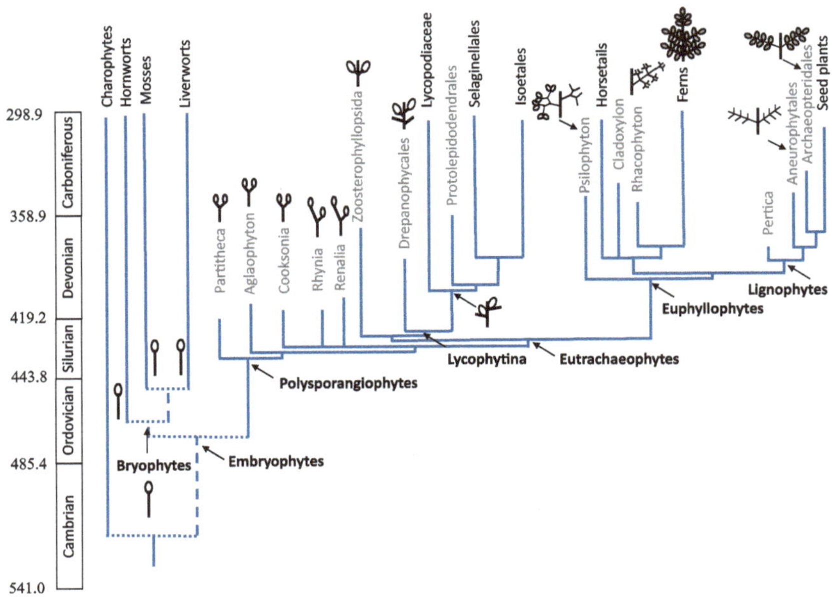

Fig. 1 Hypothetical phylogenetic relationship of extant and fossil land plants and Charophytes. The left bar shows the timescale of evolution. Names of extant groups are in black while extinct taxa/lineages are shown in gray. Dashed horizontal lines represent ambiguous phylogenetic relationships. Dotted vertical lines indicate the uncertainty in the timing of the origin of embryophytes and lineages of bryophytes. Current data using phylogenetic dating suggests that the deep splits in the tree may be considerably older than depicted here (Morris et al., 2018). Pictograms show the architectural features of major plant groups. Open elliptic structures refer to sporangia while gray filled structures depict leaves. *Redrawn, extended and modified from Kenrick, P., & Crane, P. (1997). The origin and early diversification of land plants: A Cladistic Study.* Science, 389(4), *33–39.*

further embedded into the large clade of the green lineage (green plants including green algae), the Viridiplantae. Three lineages of the paraphyletic grade of streptophyte algae, the Charophyceae, Coleochaetophyceae, and the Zygnematophyceae, are the closest relatives of land plants and share the presence of phragmoplast with land plants, a structure central to the formation of a new cell wall after cell division. Of the streptophyte algal lineages, the Zygnematophyceae appear to be the closest relative of land plants (Delwiche & Cooper, 2015; Timme, Bachvaroff, & Delwiche, 2012; Wickett et al., 2014; Wodniok et al., 2011). Extant Zygnematophyceae consist of unicellular and filamentous algae, occupying freshwater and terrestrial habitats, whose developmental features are very difficult to compare with the complexity of land plant body plans (Delwiche & Cooper, 2015).

Therefore, having the Zygnematophyceae as sister to land plants makes reconstruction of the character states of the shared common algal ancestor with land plants particularly challenging (de Vries & Archibald, 2018). Altogether, the sister relationship of Zygnematophyceae with land plants provides very little information on the potential order and extent of evolutionary transformations that led to the evolution of the complex body plans of early land plants.

Phylogenetic analysis of extant land plants (embryophytes) implies that land plants consist of the clade of vascular plants, which include the monophyletic groups of lycophytes, monilophytes, gymnosperms and angiosperms, and a group of three lineages, the mosses, liverworts, and hornworts, collectively referred to as bryophytes (Cox, Li, Foster, Martin Embley, & Civáň, 2014; Gitzendanner et al., 2018; Liu et al., 2014; Wickett et al., 2014; Wodniok et al., 2011). The phylogenetic inter-relationships of the bryophyte lineages and their relationship to vascular plants are still highly debated. For instance, earlier analyses, mainly based on molecular data, recovered all possible topologies between the three lineages of bryophytes and the rest of land plants. Either mosses, hornworts, or liverworts were reconstructed as sister to the rest of the land plants, and the three lineages of bryophytes were mainly resolved as a paraphyletic grade with some exceptions (Bremer, Humphries, Mishler, & Churchill, 1987; Chang & Graham, 2011; Cox et al., 2014; Finet, Timme, Delwiche, & Marlétaz, 2010; Fiz-Palacios, Schneider, Heinrichs, & Savolainen, 2011; Karol, 2001; Laurin-Lemay, Brinkmann, & Philippe, 2012; Mishler & Churchill, 1984; Qiu et al., 2006; Wodniok et al., 2011; Zhong, Liu, Yan, & Penny, 2013). Conversely, recent phylogenomic analyses implicate that mosses and liverworts are very likely monophyletic (Gitzendanner et al., 2018; Liu et al., 2014; Morris et al., 2018; Puttick et al., 2018; Wickett et al., 2014; Wodniok et al., 2011). Furthermore, some analyses give strong support to the monophyly of bryophytes revealing the deepest split between hornworts and a clade consisting of the liverworts and mosses (Cox et al., 2014; Gitzendanner et al., 2018; Liu et al., 2014; Wickett et al., 2014; Wodniok et al., 2011; Zhong et al., 2013).

4. Developmental patterns predating the origin of land plants

Recent evidence suggests that the genetic basis for cellular processes that allowed plants to colonize and successfully cope with the terrestrial environment were already present in the charophycean algae and, therefore, their evolution predated the origin of land plants (de Vries & Archibald, 2018;

de Vries, Curtis, Gould, & Archibald, 2018; Harholt, Moestrup, & Ulvskov, 2016; Selosse, Strullu-Derrien, Martin, Kamoun, & Kenrick, 2015). Similarly, it was found that most transcription factor families, key regulators of developmental processes thought to be specific to land plants, were already present in the charophycean algae (Catarino, Hetherington, Emms, Kelly, & Dolan, 2016; Wilhelmsson, Mühlich, Ullrich, & Rensing, 2017). This suggests that some developmental innovations relevant to land plant evolution may have been already present in the charophycean algae.

Charophycean algae are highly diverse in their development. For instance, some are unicellular, but others have attained multicellularity with thallose or filamentous forms (Delwiche & Cooper, 2015; Domozych, Popper, & Sørensen, 2016). They may or may not show apical growth, with some groups exhibiting complex apical growth and branching. Therefore, developmental patterns such as apical growth, branching, and multicellularity were present already in the charophycean algae. It is assumed that some of these developmental patterns, and likely the corresponding mechanisms, were retained through the evolution of early land plants. For instance, the thallose gametophytes of liverworts resemble those of the charophytes, and the underlying developmental mechanisms may have been present and retained from their common ancestor (Ligrone, Duckett, & Renzaglia, 2012a; Renzaglia, Duff, Nickrent, & Garbary, 2000). Similarly, the presence of both unicellular and multicellular forms in charophytes suggests that developmental processes for multicellularity might have been present in the common ancestor of charophytes and land plants and retained. Intriguingly, some experiments suggest that the multicellular-unicellular transition can be achieved in moss protonema by targeting genes involved in protein prenylation. Such mutants have unicellular cells undergoing unpolarized divisions and resemble those of unicellular algal cells (Antimisiaris & Running, 2014; Thole, Perroud, Quatrano, & Running, 2014). Furthermore, recent experiments with the alga *Chlamydomonas reinhardtii* suggest that multicellularity can be relatively easily attained under certain selection regimes or by altering the regulation of cell cycle genes (Boyd, Rosenzweig, & Herron, 2018; Hanschen et al., 2016; Herron, 2016; Niklas, 2014; Olson & Nedelcu, 2016; Ratcliff et al., 2013). The observation that the basic genetic tool kit of land plants is present in the genome of the charophyte *Klebsormidium flaccidum*, and that most key transcription factor families are present in various charophycean algae, further suggest that a set of developmental mechanisms and their genetic regulators were already present in the common ancestor of charophytes and land plants (Hori et al., 2014; Wilhelmsson et al., 2017).

Altogether, this implies that the common ancestor of land plants and charophytes was genetically and likely developmentally prepared to achieve the complex body plan of land plants through relatively few evolutionary steps. Nevertheless, except some studies on *MADS* box genes (Tanabe et al., 2005), information on the genetic regulation of specific developmental processes of charophytes is lacking, and future research will provide information about the potential homology of basic developmental mechanisms in land plants and charophytes.

5. Developmental innovations of land plants

Transition to the terrestrial environment brought some radical changes to the basic body plan of plants. The appearance of biphasic life cycles, sporopollenin and sporangia, three-dimensional growth and simple unbranched sporophytes appeared very early in land plant evolution.

5.1 Alternation of haploid (gametophyte) and diploid (sporophyte) generations

All charophycean algae had unicellular ancestral forms and a life cycle in which the haploid phase dominates, and the syngamy of gametes is immediately followed by meiosis of the zygote (Bowman, Sakakibara, Furumizu, & Dierschke, 2016; Niklas & Kutschera, 2009; Qiu, Yin-Long, Taylor, & McManus, 2012). Such life cycles are haplontic because all mitotic divisions occur in the haploid phase, and the diploid phase is only represented by a single cell, the zygote. The body plan of land plants radically broke this pattern by evolving multicellular alternating haploid (gametophytic) and diploid (sporophytic) phases. Relative dominance of the two phases has also changed during the course of land plant evolution, in such a way that the haploid phase dominated early in evolution, followed by the elaboration of the diploid and reduction of the haploid phase later in evolution (Bowman et al., 2016; Niklas & Kutschera, 2009; Qiu et al., 2012). Therefore, the origin of land plants is linked to the evolution of the multicellular sporophytic phase. Current phylogenetic evidence, consistently resolving charophycean algae with a purely haplontic life cycle as sister to all land plants, gives overwhelming support to this assertion. This finding is consistent with the antithetic theory proposing that land plants evolved from an algal ancestor with a haplontic life cycle with zygotic meiosis. In parallel, phylogenetic evidence rejects the homologous theory of Bower, assuming an algal ancestor with an isomorphic alternation of haploid and diploid generations. Altogether, current evidence implies that the multicellular sporophyte was an evolutionary innovation of land plants and originated by the intercalation into the ancestral haplontic life cycle of mitotic divisions in the zygote prior to meiosis (Bower, 1890; Haig, 2008).

Therefore, the major difference between the haplontic life cycles of charophytes and the biphasic life cycles of land plants is whether the zygote proceeds to meiosis without mitotic divisions or it proliferates mitotically prior to meiosis. Intensive research on this topic over the last years suggests that part of the genetic toolkit responsible for this developmental switch seems to be deeply rooted in the history of the green lineage (Bowman et al., 2016). Theory further suggests that the evolution of this toolkit may date back to the origin of green algal mating types, which likely evolved to impose a stringent control on the timing of developmental switches (Wang & Dohlman, 2005). The molecular toolkit controlling these critical developmental aspects of the biphasic life cycle of land plants seems to have evolved by co-opting the genetic network controlling meiosis in the zygote of simple unicellular algae, such as *C. reinhardtii* (Bowman et al., 2016; Lee, Lin, Joo, & Goodenough, 2008). In *C. reinhardtii*, heterodimerization of the BEL1-LIKE (BELL) homeodomain transcription factor GSP1 with the KNOTTED1-LIKE HOMEOBOX (KNOX) family transcription factor GSM1 is necessary to initiate zygotic gene expression and meiosis (Bowman et al., 2016; Lee et al., 2008). Homologs of both BELL and KNOX gene families are present in all land plants and they do heterodimerize (Floyd & Bowman, 2007; Frangedakis, Saint-Marcoux, Moody, Rabbinowitsch, & Langdale, 2017; Hay & Tsiantis, 2010; Horst et al., 2016). KNOX and BELL function is best studied in the moss *Physcomitrella patens*, in which a single BELL gene, *PpBELL1,* is sufficient and necessary to induce the sporophytic program. It is also known that KNOX genes went through a duplication prior to the origin of land plants, giving rise to Class 1 and Class 2 KNOX genes (Floyd & Bowman, 2007; Frangedakis et al., 2017; Furumizu, Alvarez, Sakakibara, & Bowman, 2015; Hay & Tsiantis, 2010). The Class 1 KNOX gene of *P. patens, MKN2,* is necessary for regulating sporophyte development in the moss, while the Class II KNOX genes appear to suppress the gametophytic program in the sporophytic phase (Sakakibara et al., 2013; Sakakibara, Nishiyama, Deguchi, & Hasebe, 2008; Sano et al., 2005). It is not yet clear with which KNOX protein PpBELL1 heterodimerizes with (Horst et al., 2016). Conversely, components of the Polycomb Repressive Complex 2 (PRC2), such as the homologs of the *Arabidopsis thaliana* proteins FERTILIZATION-INDEPENDENT ENDOSPERM (PpFIE) and CURLY LEAF (PpCLF), suppress the sporophytic program in the gametophytic phase (Mosquna et al., 2009; Okano et al., 2009; Pereman et al., 2016). Furthermore, transcriptomic data suggest early activation of meiosis-related genes in the apical cells of the sporophyte in the moss *P. patens* (Frank & Scanlon, 2015a, 2015b). Nevertheless, the link between the gene networks regulating the haploid-diploid switch and that of sporangial development is unknown. These findings suggest that an ancient regulatory network, whose evolution coincided with the origin of mating types in green algae, is deeply conserved across the green lineage. The putative ancestral function of this network was to enable a stringent control on the developmental switch between haploid and diploid programs (Bowman et al., 2016). This network was then used, and its regulatory role was extended, to govern various aspects of sporophytic development in land plants.

Although the core regulatory network controlling initiation of the sporophyte and gametophyte developmental programs is known, there is very little information available about the genetic program enabling multicellularity and three-dimensional growth in the sporophytic phase. It is possible that these developmental mechanisms were partly recruited from the gametophytic phase (Frank & Scanlon, 2015a, 2015b; Szovenyi, Rensing, Lang, Wray, & Shaw, 2010). Alternatively, they could have evolved de novo in land plants. It is known that zygotes of double mutants disrupting the FLORICAULA/LEAFY (FLO/LFY) homologs *PpLFY1* and *PpLFY2* arrest and are unable to divide mitotically (Tanahashi, Sumikawa, Kato, & Hasebe, 2005). Furthermore, *P. patens* zygotes lacking activity of the two WUSCHEL-RELATED HOMEOBOX 13-LIKE (*PpWOX13LA/B*) genes are unable to elongate and initiate the apical cell of the embryo (Sakakibara et al., 2014). Transcriptomic evidence from *P. patens* suggests that multicellularity and three-dimensional patterning may have been, at least partially, recruited from gametophytic programs (Frank et al., 2015; Frank & Scanlon, 2015b; Whitewoods et al., 2018). Nevertheless, many aspects of the evolution of the sporophytic phase are still unclear, and the interconnection of reproductive and proliferative programs is unknown.

Although, multiple lines of evidence support the origin of land plants from a charophycean ancestor, there is much ambiguity surrounding the evolutionary relationship of land plant lineages. The debate about the phylogenetic relationship of bryophyte lineages and their relationship with vascular plants has significant impact on how land plant body plan changes are interpreted. This includes the evolution of the biphasic life cycle and the evolution of other morphological traits, such as stomata, vascular tissues, etc.

For instance, studies suggesting a paraphyletic relationship of the three bryophyte lineages (Chang & Graham, 2011; Qiu et al., 2006, 2012) are compatible with a haploid-dominant ancestral land plant life cycle, which was retained in the paraphyletic grade of bryophytes but was followed by the dominance of the diploid phase in vascular plants (Bowman et al., 2016; Haig, 2008; Niklas & Kutschera, 2009; Qiu et al., 2012; Tomescu et al., 2014). In contrast, a monophyletic bryophyte clade may suggest that the haploid-dominant life cycle could have been a unique innovation of the bryophyte clade (Cox et al., 2014; Gitzendanner et al., 2018; Puttick et al., 2018; Wickett et al., 2014). According to this hypothesis, life cycle of the ancestral land plant could have been haploid-dominant, diploid-dominant, or equally dominant with close to isomorphic haploid and diploid phases. Furthermore, it is possible that the common ancestor of all land plants may have been more trachaeophyte-like, which may explain the origin of conducting tissues in mosses. The earliest fossil remains of land plants are currently interpreted as being stem trachaeophytes (Kenrick, 1994, 2018; Kenrick & Crane, 1997; Taylor, Kerp, & Hass, 2005). Nevertheless, it is possible that some of these fossils represent the common ancestor from which the monophyletic group of bryophytes evolved (Boyce & Kevin Boyce, 2010; Kenrick, 2018; Puttick et al., 2018). Therefore, there is considerable uncertainty surrounding the evolutionary origin of land plants, which makes reconstruction polarity of key characters in land plants challenging.

5.2 Evolution of three-dimensional growth in the haploid and diploid phases

Most charophycean algae grow in a planar form along a two-dimensional axis, while land plants evolved three-dimensional growth, enabled by the presence of a continuously rotating division plane in the stem cells (Delwiche & Cooper, 2015; Domozych et al., 2016; Langdale, 2008). The ability of flexibly changing division planes is missing from the algal relatives of land plants. The genetic mechanisms underlying this evolutionary transition can only be studied in land plants with a life cycle including both two- and three-dimensional growth patterns.

Importantly, some mosses exhibit a filamentous juvenile life cycle phase (protonemata) of the gametophyte, which is maintained by two-dimensional growth as seen in charophycean algae. After this stage, a change to three-dimensional growth may occur, giving rise to leafy shoots. This system is suitable to investigate the genetic mechanisms enabling the transition from two- to three-dimensional growth. Nevertheless, this approach assumes that the ontogeny of the moss *P. patens* recapitulates the evolutionary trajectory that occurred between algae and land plants (Harrison, Roeder, Meyerowitz, & Langdale, 2009). There are multiple genetic factors known to affect initiation of the three-dimensional growth pattern in *P. patens*. The *NO GAMETOPHORES1* (*PpNOG1*) gene regulates the transition to three-dimensional growth by inducing the degradation of proteins that likely repress the *P. patens* APB transcription factors, belonging to the AP2 class transcription factors homologous to *A. thaliana* AINTEGUMENTA, PLETHORA, and BABY BOOM (APB). The PpAPB transcription factors are necessary and sufficient to initiate three-dimensional buds (Aoyama et al., 2012; Moody, Kelly, Rabbinowitsch, & Langdale, 2018) The *PpNOG1* gene is likely also necessary in directing division planes in the apical cell, together with the *DEFECTIVE KERNEL1* homolog *PpDEK1*, to achieve proper three-dimensional growth (Demko et al., 2014; Olsen, Perroud, Johansen, & Demko, 2015; Perroud et al., 2014). *PpNOG1* and related genes occur only in land plants; therefore, it may be one of the key factors that regulate this innovation (Moody et al., 2018). Given that PpDEK1 and AP2 class transcription factors regulate similar processes in flowering plants and the moss, it is possible that this ancient network was already present in the common ancestor of land plants, and that its evolution coincided with the evolution of three-dimensional growth. Finally, homologs of the *CLAVATA* (*CLV*) pathway, crucial for meristem maintenance in the flowering plant shoot apical meristem, were recently shown to be critical to the correct orientation of division planes in the transition from two- to three-dimensional growth in *P. patens* (Whitewoods et al., 2018). The role of the *CLV* pathway in orienting division planes seems to be conserved between *A. thaliana* and *P. patens* and may represent the ancient function of this pathway. It is thus possible that the *CLV* pathway was recruited to regulate the proper orientation of division planes in two independent contexts: in the apical cell of the moss gametophyte and in the shoot apical meristem of flowering plants. Phenotypic effects of *PpDEK1, PpNOG1* and the moss homologs of the *CLV* pathway are overlapping, suggesting that they are likely members of a gene network governing the proper development of the gametophyte apical cell in *P. patens*. In line with these findings, transcriptomic evidence also suggests that the genetic mechanisms involved in the evolution of three-dimensional growth in gametophytes may have been partially recruited to support three-dimensional patterning in the sporophytic phase (Frank & Scanlon, 2015a).

5.3 Origin of spores, sporangia, and sporopollenin in land plants

All land plant spores are covered by sporopollenin, whereas algae only have heavy-walled zygotes that germinate via meiosis. Therefore, sporopollenin coated spores are an innovation specific to land plants. This assertion is in line with the finding that the earliest fossils with land plant affinities are the so-called cryptospores (470 million years old), whereas sporangial fragments are younger (450 million years old) (Edwards, Morris, Richardson, & Kenrick, 2014; Steemans et al., 2009). Cryptospores occur either in monads, diads, or in tetrads (see chapter "Evolution and co-option of developmental regulatory networks in early land plants" by Bowman et al., this issue). Their wall structure resembles that of some liverwort spores, such as the diads and tetrads of the genera *Haplomitrium* and *Sphaerocarpos*, respectively, but their affinity to extant bryophyte species is highly debated (Edwards, Richardson, Axe, & Davies, 2012; Gensel, 2008; Renzaglia et al., 2015). It is thought that sporopollenin-covered spores evolved earlier than sporophytes, likely via modification of the timing of meiosis and the deposition of sporopollenin, which significantly affected spore viability (Brown & Lemmon, 2011). Genetic pathways and their components are partially conserved across land plants, but conservation of the genetic components underlying the developmental process and its evolutionary trajectory is not known (Harrison, Alvey, & Henderson, 2010).

5.4 Origin of unbranched sporophyte forms

The earliest non-spore fossils assigned to land plants are believed to resemble present day vascular plants with bifurcating axes (Bowman, 2013; Boyce & Kevin Boyce, 2010; Kenrick, 2018; Kevin Boyce, 2005; Gensel, 2008; Langdale, 2008). Nevertheless, phylogeny and the earliest fossils of stem group polysporangiophytes suggest that early sporophytes may have been unbranched, with terminal sporangia such as in extant mosses. Furthermore, the earliest sporophytes are believed to be obligate matrotrophic (nutritionally supported by the gametophyte), with a well-developed photosynthetic gametophyte phase (Gensel, 2008; Kenrick, 2018; Qiu et al., 2012; Remy, Gensel, & Hass, 1993). Therefore, it is hypothesized that fossils of the earliest multicellular unbranched sporophyte forms are missing from the fossil record, as all known fossils show some bifurcation. Nevertheless, their small size and the lack of vascularization correspond to the organizational level of bryophyte sporophytes (Boyce & Kevin Boyce, 2010; Gensel, 2008;

Gerrienne et al., 2006; Remy et al., 1993; Taylor et al., 2005). It is assumed that these unbranched forms are retained in extant bryophytes, especially in the lineages of mosses and hornworts, which can be used to gain information about their developmental biology and evolution. Unfortunately, development of the sporophyte in the three lineages of bryophytes is strikingly different and, therefore, it is difficult to establish which features are ancestral to the group (Ligrone et al., 2012a; Ligrone, Duckett, & Renzaglia, 2012b) (Fig. 2).

Importantly, elongation of the liverwort sporophyte is almost exclusively due to cell expansion, and there is no apical or localized meristem activity with divisions occurring all over the sporophyte (Ligrone et al., 2012a; Renzaglia et al., 2000). In contrast, sporophyte growth is dominated by cell divisions in both hornworts and mosses, involving the activity of well-localized intercalary meristems and/or apical cells (Ligrone et al., 2012a; Villarreal & Renzaglia, 2015). Hornwort and moss sporophytes share the presence of a so-called multicellular proliferative zone within the sporophyte (Ligrone et al., 2012b). This proliferative zone produces cells upward, giving rise to the full body of the sporophyte in hornworts, whereas it produces cells downward in

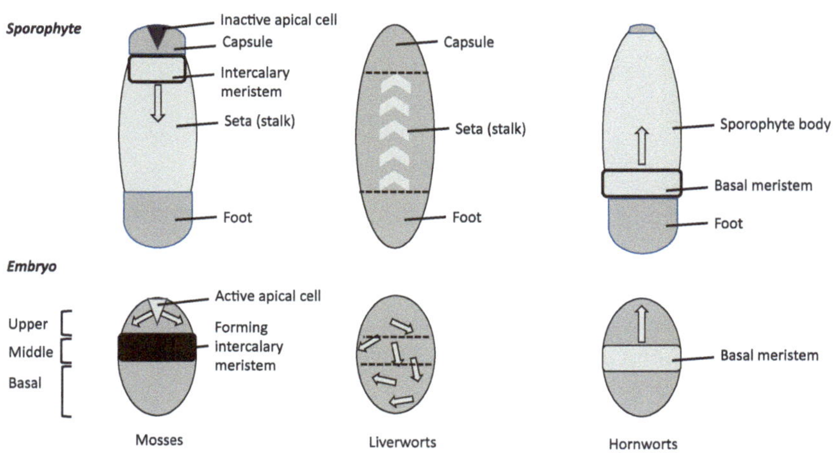

Fig. 2 Schematic representation of sporophyte development in the three lineages of bryophytes. The bottom row shows the three cell layers (basal, middle and upper) of the early embryo. The upper row depicts the developing sporophytes. Actively dividing cells are shown in gray and arrows refer to the direction of cell production. Inactive but preformed regions are shown in black. Randomly arranged arrows in the liverwort embryo symbolize cell divisions occurring across the whole embryo. Gray arrow heads in the developing sporophyte of liverworts depict cell elongation and a lack of cell divisions leading to the elongation of the seta.

mosses, giving rise to the stalk (seta) of the sporophyte (French & Paolillo, 1975b; Villarreal & Renzaglia, 2015). Developmental homology of these structures and their relation to the probably ancient uniaxial structures is unexplored. Nevertheless, historical experiments of moss sporophytes suggest that cytokinin may be involved in regulating activity of the intercalary meristem. Retarded growth of the sporophyte after decapitation can be alleviated by exogenous cytokinin application, which may represent a conserved mechanism with the angiosperms shoot apical meristem (SAM) (French & Paolillo, 1975a, 1975b; Ligrone et al., 2012b; Snipes et al., 2018; Truskina & Vernoux, 2018).

It is not yet clear, how developmental mechanisms of unbranched forms are linked to the origin of branched shoots of early vascular plants. There are two competing hypotheses explaining the evolutionary origin of early vascular shoot systems, the interpolation/sterilization and apical growth hypotheses (Bowman, 2013; Ligrone et al., 2012b; Tomescu et al., 2014). The apical growth hypothesis assumes that early shoots evolved from the extended meristematic activity of the transient apical cell of a moss-type embryo. In contrast, the interpolation/sterilization hypothesis suggests that the common ancestor of bryophyte and vascular plant sporophytes had apical reproductive activity, which went through a sterilization process to produce the axis of the shoot. Further comparative study on the unbranched sporophytes of bryophytes and that of vascular plants may clarify whether any of these hypotheses can be supported by experimental data.

5.5 Evolution of bifurcating axes

The earliest vascular plant remains had independent gametophyte and sporophyte generations, with bifurcating sporophyte axes that terminated in sporangia. This is in contrast to the unbranched and matrotroph sporophytes of extant mosses (Gensel, 2008; Kenrick, 2018; Remy et al., 1993; Taylor et al., 2005; Tomescu et al., 2014). Indeterminate sporophytes appear only much later in the fossil record, implying that evolution of bifurcation predates that of indeterminacy. Information on how branching evolved is provided by observations on fern and bryophyte sporophytes. In *P. patens* plants, deletion either of the TEOSINTE BRANCHED1/CYCLOIDEA/PROLIFERATING CELL FACTOR1 (TCP) transcription factor class gene *PpTCP5*, the TERMINAL EAR1-LIKE (TEL) gene *PpTEL*, or *PINFORMED1* (*PIN1*) homolog *PpPINB* gene will increase the proportion of bifurcating sporophytes compared to the wild type (Bennett et al., 2014; Ortiz-Ramírez et al., 2016; Vivancos et al., 2012). *PpFLO/LFY* mutants have a similar phenotype (Tanahashi et al., 2005). At present, it is unclear when exactly branching takes place and whether it occurs at the zygote or at the sporophyte-seta stage. It is also unclear whether the effect of these genes is achieved via similar developmental mechanisms. Observations on fern sporophytes show that branching is achieved by segregation and amplification of stem cells at the shoot apex (Harrison, Rezvani, & Langdale, 2007). Transcriptomic evidence suggests that genetic components necessary for this process are present in fern genomes and shoot apices, but the genetic mechanisms are still unclear (Evkaikina et al., 2017; Frank et al., 2015; Harrison, 2015). In contrast, gametophytic branching in mosses is regulated by the co-option of ancient hormonal effects involved in sporophyte branching in angiosperms, such as auxin, cytokinins, and strigolactons (Coudert, Bell, Edelin, & Harrison, 2017; Coudert et al., 2015). Polar auxin transport is essential for branching in flowering plants but in the moss bi-directional transport is required for normal branching (Coudert, 2017; Harrison, 2017).

5.6 Evolution of indeterminacy

The fossil record suggests that all indeterminate sporophyte axes have laterally arranged sporangia (Boyce & Kevin Boyce, 2010; Tomescu et al., 2014). Therefore, the evolution of indeterminacy of sporophytic axes and the lateral displacement of sporangia are linked, indicating spatially and temporally separated activity of reproductive and vegetative functions (Kenrick, 2018). Information on the molecular mechanisms underlying indeterminacy is exclusively coming from investigations on the determinant sporophyte of *P. patens*. It was shown that two components of the moss PRC2, encoded by *PpFIE* and *PpCLF*, are necessary to repress the meristematic activity of the sporophyte apical cell (Kenrick, 2018; Mosquna et al., 2009; Pereman et al., 2016). In mutants disrupting PRC2, the sporophytes produced branched structures by continuous proliferation. Class I KNOX genes are known to be responsible for the maintenance of the meristematic activity of the sporophyte apical cell (Sakakibara et al., 2008). That is, the interaction of PRC2 and Class I KNOX gene activity is key in the regulation of determinant and indeterminant growth. Class II KNOX genes are also expressed in and necessary for the development of the sporophyte by repressing the gametophytic program (Sakakibara et al., 2013). These observations imply that function of the PRC2 in repressing pluripotent sporophytic cells is conserved across land plants. Whether the antagonistic effect of Class I and II KNOX genes seen in flowering plants is also conserved in the development of the sporophyte of mosses is unknown (Furumizu et al., 2015). Nevertheless, it is possible that evolution of indeterminacy is partially achieved by the proper coregulation of Class I and Class II KNOX activities, which may connect reproductive and vegetative proliferative activities.

5.7 Evolution of meristems

Plant form and architecture are tightly linked to the activity of meristems. Indeterminate meristems have independently evolved in the gametophyte phase of bryophytes and ferns, and in the sporophyte of vascular plants. Gametophyte meristems considerably differ from those of sporophyte meristems, because the former are composed of a single stem cell, while the latter consist of a single stem cell, or a group of cells, overlaying several layers of proliferative cells (Ambrose & Vasco, 2016; Harrison et al., 2007, 2009; Langdale, 2008; Sanders, Darrah, & Langdale, 2011). Meristematic activity is not recognizable in the fossil record of early vascular plants. Furthermore, fossils at the divergence of vascular plants and bryophytes are missing, which makes it impossible to decide whether sporophytic indeterminant meristems have evolved once or multiple times independently. Therefore, all information is coming from observations made on extant taxa.

Among the extant representatives of the earliest diverged lineages, mosses, hornworts and some liverworts have multicellular proliferative sporophytic regions, which may be—to some extent—homologous to proliferative regions of the indeterminate meristematic regions of extant vascular plants (Bowman, 2013; Ligrone et al., 2012b; Tomescu et al., 2014). Importantly, hornwort basal meristems seem to be indeterminate, while moss intercalary meristems show only transient activity (Bowman, 2013; French & Paolillo, 1975b; Langdale, 2008; Ligrone et al., 2012a, 2012b; Villarreal & Renzaglia, 2015). It remains to be seen whether these structures are homologous to one another, and whether their proliferative activity shares common regulatory activity with that of the proliferative regions of vascular plant meristems.

The structure of indeterminate meristem activity of sporophyte shoots in lycophytes and ferns is highly variable, but it usually consists of a single apical cell, or groups of cells, overlaying a deeper layer of proliferative cells (Ambrose & Vasco, 2016; Frank et al., 2015; Vasco et al., 2016). Expression of key genes of the SAM in ferns and lycophytes suggests that their sporophytic meristems are multicellular structures, and their core regulatory mechanisms may be homologous to that of the flowering plant SAM (Evkaikina et al., 2017; Friedman, 2011). Nevertheless, transcriptomic evidences rather suggest that vascular plant meristems have followed divergent evolutionary trajectories, due either to lineage-specific changes or to independent evolutionary origins (Frank et al., 2015). Nevertheless, how an apical stem cell (or group of stem cells) evolved in the sporophyte stage is unknown, but the transient apical activity of moss sporophyte cells may represent the antecedent of these cells (Bowman, 2013; Tomescu et al., 2014). Transcriptomic and experimental evidence support the hypothesis that some regulatory mechanisms of the sporophytic SAM may have been recruited from those present in the apical cells of gametophytes (Frank et al., 2015; Whitewoods et al., 2018). Nevertheless, most experimental evidence suggest that this may have been fragmentary, because deletion of Class I KNOX and WOX transcription factor genes have no effect on the gametophyte, and Class III HD-Zip (C3HDZ) genes are not expressed in the gametophyte apical cell of the moss *P. patens* (Sakakibara et al., 2008, 2014; Yip, Floyd, Sakakibara, & Bowman, 2016).

5.8 Origin of leaves

Apical meristems of most plants have the function to reiterate the modules of leaves, which are optimized to capture light, and thus are the major places for photosynthesis. Leaves have evolved at least five times during the course of evolution: once in mosses, liverworts, lycophytes, ferns, and flowering plants (Boyce & Kevin Boyce, 2010; Rothwell et al., 2014; Tomescu, 2009). This is because early liverworts and ancestors of lycophytes and ferns were clearly leafless, supporting the non-homology of sporophytic and gametophytic leaves (Kevin Boyce, 2005; Tomescu, 2009; Vasco, Moran, & Ambrose, 2013). An independent origin of sporophytic and gametophytic leaves is also supported by the divergent mechanisms controlling their development (Plackett, Di Stilio, & Langdale, 2015).

The first lycophyte fossils were leafless, followed by the evolution of forms with vascularized outgrowths (Tomescu, 2009). Development of lycophyte leaves is best known from the extant genus *Selaginella*, and indicates that two epidermal cells initiate leaf development and further divisions establish dorsiventrality and inner tissue structure (Harrison et al., 2007). Nevertheless, development varies within lycophytes, and in Lycopodiales and Isoetaceae multiple cell layers are involved in the initiation of leaves (Plackett et al., 2015; Tomescu, 2009; Vasco et al., 2013).

In horsetails, the apical cells divide and will either attain leaf or intercalary meristem fate, of which the latter gives rise to the internodal tissues

(Renzaglia et al., 2000; Tomescu, Escapa, Rothwell, Elgorriaga, & Cúneo, 2017). The evolutionary origin of fern leaves is highly debated. All extant ferns are part of a monophyletic group of euphyllophytes, which is sister to the lycophytes. The fossil record suggests that the common ancestor of both lycophytes and euphyllophytes was leafless and, therefore, leaves in these two groups must have evolved independently (Galtier, 2010; Kenrick & Crane, 1997; Vasco et al., 2013). Fossil data and their phylogenetic placement further suggest that fern fronds and lignophyte fronds must have evolved independently. Nevertheless, it was also proposed that the petiole could be homologous across euphyllophyte leaves, but that evolution of the lamina occurred independently (Bierhorst, 1977; Rothwell, 1999; Tomescu, 2009). Homology of leaves within the monophyletic group of ferns is also debated due to the highly variable shape and developmental biology of fern leaves, and the lack of a proper fossil record, which would indicate how common ancestors of fern groups with widely varying leaf morphologies looked like (Rothwell, 1999; Rothwell & Nixon, 2006; Vasco et al., 2013).

Despite their diversity, fern leaves share some common features; they are determinate structures, have adaxial/abaxial identities, grow in a predefined phyllotaxis, and show a well-developed vasculature (Vasco et al., 2013). Fern fronds develop similar to leaves in the flowering plant SAM. Leaf primordia arise from the peripheral region of the SAM, and the spacing of primordia, and thus phyllotaxis, is regulated by inhibition from recently developed primordia. Fern fronds are usually produced by one apical cell, followed by ordered cell divisions at the tip of the developing structures. That is, fern fronds display acropetal tip grow, with an apical cell that is shared with bryophytes and lycophytes (Vasco et al., 2013). In contrast, division occurs across the whole developing organ and is coordinated by non-cell autonomous processes in angiosperms. Although cell-autonomous processes are typical for fate determination of fern fronds, hints for non-cell autonomous processes also exist, which would be a shared feature with flowering plant leaf primordia (Tomescu, 2009; Vasco et al., 2013).

There are three major theories proposed to explain the origin of lycophyte and euphyllophyte leaves from naked leafless ancestors. The most popular theory, the telome theory, was proposed by Zimmermann to explain how macrophylls may have evolved from bifurcating, leafless, shoot bearing ancestors (Zimmermann, 1951, 1965). There are three major processes that contributed to the origin of macrophylls: overtopping, planation, and webbing. Overtopping describes the mechanism by which some bifurcating shoots overgrow their neighbors. In the process of planation, slow growing axes attain a flat shape, forming the precursors of leaf venation. Finally, inter-shoot spaces are filled up with tissues during the process of webbing, which gives rise to the lamina of the leaf. The telome theory also provides an explanation for the origin of lycophyte microphylls, which are supposed to have evolved by reduction from existing lateral branch systems bearing dichotomous sterile and fertile telomes (Zimmermann, 1951). In contrast to the telome theory, the enation theory proposes that microphylls evolved from naked shoots by tissue emergences that later became vascularized (Bower, 1959). Finally, a third theory of microphyll evolution, the sterilization theory, proposes that microphylls are the direct derivatives of sterilized sporangia (Crane & Kenrick, 1997; Kenrick & Crane, 1997).

The leaf developmental network is relatively well-studied in angiosperms (Bar & Ori, 2014; Gonzalez, Vanhaeren, & Inzé, 2012; Ichihashi & Tsukaya, 2015; Runions, Tsiantis, & Prusinkiewicz, 2017; Townsley & Sinha, 2012). Outside angiosperms, studies focused on the C3HDZ network and the repression of Class I KNOX genes by the ASYMMETRIC LEAVES1/ROUGH SHEATH2/PHANTASTICA (ARP) family of MYB transcription factors, the KNOX-ARP network. Expression of KNOX and ARP family genes seem to be conserved across vascular plants, likely suggesting some level of homology in the leaf initiation process (Harrison et al., 2005; Vasco et al., 2013). Expression of members of the C3HDZ network is less clear. Early experiments showed that C3HDZ genes are expressed in leaf primordia but later disappear from the developing leaves in *Selaginella kraussiana* (Harrison et al., 2005). In contrast, in the fern *Osmunda regalis*, C3HDZ genes are expressed in an adaxial manner, consistent with what we know about their function in angiosperms. This difference was interpreted as a sign of non-homology of micro- and macrophylls or that their homology is restricted to the level of lateral organ formation. This early observation supports the notion that homology in leaf development of lycophytes and euphyllophytes exists at the branch system level, which is in line with the telome theory (Floyd & Bowman, 2007). According to this interpretation the KNOX-ARP network was co-opted from a shoot-specific program to regulate leaf development.

A considerably extended analysis of C3HDZ gene expression in ferns and lycophytes, however, put theories of leaf evolution into a new light. Vasco et al. (2016) showed that C3HDZ genes exhibit conserved expression patterns in euphyllophytes (ferns and seed plants), being expressed on the adaxial side of developing leaves. In contrast, C3HDZ genes are only expressed in early leaf primordia in lycophyte microphylls. Nevertheless, C3HDZ genes are also expressed during both euphyllophyte and lycophyte sporangial development. This suggests that C3HDZ genes have an ancestral role in sporangial development, which is shared across all vascular plants. Furthermore, because the common ancestors of lycophytes and euphyllophytes were leafless, this implies that shared sporangial developmental programs were independently co-opted in the development of micro- and macrophylls, while downstream processes may differ. These results also suggest that ancestral function of C3HDZ genes was not in meristem maintenance, but in initiating leaves and in sporangial development. Therefore, their regulatory function in leaf development is a result of independent recruitment in multiple lineages of vascular plants.

The findings described above put evolution of sporophyte leaves into another context. The fact that initiating leaves and initiating sporangia, all lateral organs, are both regulated by C3HDZ genes suggests that leaf development is homologous at the level of lateral organ initiation across lycophytes and ferns. Assuming that the ancestral role of C3HDZ genes is governing sporangial development, this implies that microphylls must have evolved from some existing structures. Therefore, these observations contradict the enation theory but rather give support to either the sterilization or the telome-reduction theory. Finally, the observation that the genetic network of leaf initiation appears to be deeply conserved across vascular plants implies that leaves evolved on plants that were able to generate lateral organs instead of terminal organs. This developmental program was recruited from the program of lateral organ production, and ultimately from sporangial programs, in line with the telome theory (Vasco et al., 2016).

5.9 Evolution of rooting systems

All land plants and some streptophyte algae develop outgrowth on the interface between the plant and soil surface at some point in their development. These structures can be of two types: rhizoid rooting systems and typical root systems (Jones & Dolan, 2012).

Rhizoids are either unicellular or multicellular outgrowths on the surface of plants to facilitate attachment to the surface and acquisition of nutrients and water (Box, 1986; Buck, Jonathan Shaw, & Goffinet, n.d.; Kobiyama & Crandall-Stotler, 1999; McConaha, 1939, 1941; Odu, 1978; Trachtenberg & Zamski, 1979; Vermeer, Escher, Portielje, & de Klein, 2003). Rhizoid-like rooting structures appeared first in plant evolution, and their evolutionary origin pre-dates that of land plants. Rhizoids first occur in the gametophytic and then in the sporophytic phase of the life cycle. In extant bryophytes and streptophyte algae, rhizoids are only present in the haploid dominant-phase of the life cycle (Delwiche & Cooper, 2015; Domozych et al., 2016; Lewis & McCourt, 2004). For instance, some streptophyte algae, e.g., of the genera *Chara* and *Spirogyra*, develop rhizoid outgrowths (Delwiche & Cooper, 2015; Domozych et al., 2016; Lewis & McCourt, 2004). The bryophytes, including the monophyletic lineages of mosses, liverworts and hornworts, all bear rhizoids but only in the dominant haploid phase of their life cycle (Buck et al., n.d.; Crandall-Stotler & Stotler, 2008; Renzaglia, Villarreal, Joel Duff, & Goffinet, 2008). The very first fossil records of vascular plant ancestors also indicate that they had no roots but had naked rhizomatous sporophyte shoots bearing unicellular rhizoids (Hetherington & Dolan, 2018a). The early vascular plant fossil records also indicate the presence of independent gametophyte and sporophyte phases, and rhizoids occurred on some of the gametophytes found (Kenrick, 1994, 2018; Kenrick & Crane, 1997; Taylor et al., 2005). Therefore, rhizoids were present both on the gametophyte and sporophyte phases, suggesting that developmental mechanisms of rhizoid formation may have first evolved in the gametophyte phase and were then co-opted or recruited for rhizoid formation in the sporophyte phase.

Besides the earliest vascular plants with rhizoid-like rooting structures, almost all extant and fossilized vascular plants evolved roots, a radially symmetric axis that grows from a root meristem covered by the root cap that produces root hairs on its surface (Raven & Edwards, 2001). Fossil evidence indicates that roots evolved multiple times independently, at least once in lycophytes and euphyllophytes, despite all vascular plant roots bearing single celled outgrowths, the root hairs (Gensel, Kotyk, & Basinger, 2001; Hetherington & Dolan, 2018b; Kenrick, 2013; Kenrick & Crane, 1997; Kenrick & Strullu-Derrien, 2014; Raven & Edwards, 2001). Root hairs are functionally analogous to rhizoids, enabling and facilitating nutrient and water uptake form the soil (Marschner, 2012). Evolutionary-developmental evidence suggests that the functionally similar structures of rhizoids and root hairs also have a deeply shared genetic regulatory mechanism.

The ROOT HAIR DEFECTIVE SIX-LIKE (RSL) Group VIII basic helix-loop-helix (bHLH) transcription factors promote rhizoid development in the liverwort *Marchantia polymorpha* and the moss *P. patens*, as well as root hair development in angiosperms (Pires & Dolan, 2010a; Pires et al., 2013; Tam, Catarino, & Dolan, 2015; Yi, Menand, Bell, & Dolan, 2010). Furthermore, RSL genes also promote and are necessary for the induction of apical outgrowth of unicellular or multicellular structures in liverworts and in mosses (Honkanen & Dolan, 2016). This suggests that conserved genetic regulatory mechanisms of rhizoid and root hair initiation and growth were co-opted from a shared ancestral land plant developmental module that regulated outgrowth from epidermal cells. Further experiments also show that expression of genes involved in tip growth of rhizoids and root hairs is highly conserved between flowering plants, liverworts, and mosses (Honkanen et al., 2016; Huang, Shi, Wang, Ryu, & Schiefelbein, 2017).

The *Lotus japonicus* ROOTHAIRLESS LIKE (LRL) Group XI bHLH transcription factors are also major regulators of root hair growth in flowering plants, with Class I and Class II genes promoting and inhibiting root hair growth, respectively (Pires & Dolan, 2010a; Yi et al., 2010). Duplication of these genes occurred in the vascular plant lineage (Pires & Dolan, 2010a, 2010b). LRL family genes are positive regulators of rhizoid growth both in mosses and liverworts, implying that this was the ancestral function of these genes (Pires et al., 2013; Proust et al., 2016; Tam et al., 2015). After duplication, Class I and Class II genes obtained divergent antagonistic functions (Honkanen & Dolan, 2016; Honkanen et al., 2016). The RSL-LRL regulatory network appears to have evolved increasing complexity via duplication and functional specialization (Pires et al., 2013). Deep homology of root hairs and rhizoids provides a good example of modularity, and how modularity can facilitate evolutionary innovations.

5.10 Roots

Roots differ from the earlier rhizoid rooting systems by (i) growing from a meristem that is protected by the root cap and (ii) showing gravitropism (Raven & Edwards, 2001). Because early fossils of vascular plants and ancestors of lycophytes and euphyllophytes were all rootless, roots have evolved independently in these two groups, and potentially even had multiple origins within the euphyllophyte clade (Boyce & Kevin Boyce, 2010; Friedman, Moore, & Purugganan, 2004; Kenrick & Crane, 1997; Kevin Boyce, 2005; Raven & Edwards, 2001). The very early fossil remains of lycophyte roots are the stigmarian root structures of lycophyte trees from the Carboniferous. Similarity of shoot and root systems in these fossils led to the hypothesis of the shoot origin of roots. Stigmarian roots are unique and consists of a bifurcating axis that is covered by bifurcating rootlets, arranged in a phyllotactic manner covered by root hairs (Hetherington, Berry, & Dolan, 2016). Intriguingly, this architecture only occurs in extant herbaceous quillworts (Isoetes) and represents a special innovation restricted to this clade (Hetherington et al., 2016).

Extant lycophytes have bifurcating roots covered with root hairs that either arise laterally during embryonic development or postembryonically on the so-called rhizophores (Foster & Gifford, 1959; Harvey-Gibson, 1902; Saxelby, 1908; Uphof, 1920). Whether rhizophores have shoot or root affinity is still debated; nevertheless, expression of Class I KNOX genes at the tip of rhizophores rather suggests a shoot affinity (Kawai, Tanabe, Soma, & Ito, 2010; Lu & Jernstedt, 1996). Very early euphyllophyte roots are from fossilized *Cladoxylons* that have bifurcating axes originating from the swollen base of the plant (Stein, Mannolini, Hernick, Landing, & Berry, 2007; Xu et al., 2017). Extant fern roots either initiate at the dorsal end of the embryo, or later at the base of the shoot or on rhizomes, and lateral roots develop from the internal endodermis layer (Eames, 1936). The progymnosperms had either bifurcating or laterally branching roots like in seed plants (Raven & Edwards, 2001).

It is suggested that euphyllophyte and probably also lycophyte roots evolved from modified shoot systems. This idea is supported by the fossil record and by some experimental evidence, including gene expression analyses and the effect of hormones on root development (de Vries et al., 2016; Kawai et al., 2010). Although the molecular mechanism of root development in angiosperms is very well understood, there have been very few experiments to test the functions of homologous genes in non-flowering plant lineages to reveal the evolution of rooting structures. Comparative transcriptomic analysis across seed plants and lycophytes revealed a conserved set of genes that is used throughout root development, despite the strikingly different morphology of lycophyte roots (Huang & Schiefelbein, 2015; Huang et al., 2017). This either suggests that a core gene set was co-opted by all lineages or, alternatively, a core regulatory network of genes was extended in various ways in lycophyte and euphyllophyte lineages to regulate root development. WOX genes are key in maintaining pluripotency in shoot and root meristems and group into three phylogenetic clades; the ancestral clades, the intermediate clade, and the WUSCHEL (WUS) clade. Members of the WUS clade occur in all leptosporangiate ferns and all seed plants but not in lycophytes and bryophytes, and their function in the maintenance of pluripotency in root development seems to be conserved across ferns and seed plants (Ge et al., 2016; Huang et al., 2017; Nardmann & Werr, 2012). Nevertheless, the way stem cell maintenance is achieved likely differs between ferns and seed plants. Both fern and gymnosperm WUS homologs could rescue the *wox5–1* mutant's stem cell defect in the *A. thaliana* root meristem, but only the gymnosperm WUS homolog was able to move to the cell adjacent to the quiescent center of the root meristem (Zhang, Jiao, Jiao, Zhao, & Zhu, 2017). Members of the intermediate-clade occur in lycophytes, ferns, and seed plants. These genes are unlikely to be used in root development of lycophytes but were likely recruited for adventitious root formation in the common ancestor of ferns and seed plants (Huang et al., 2017; Zhu et al., 2017). Further duplication in this clade led to two groups of genes, of which one was retained in adventitious root regulation while the other was recruited to function in primary root organogenesis (Huang et al., 2017). On the other hand, phylogenetic distribution and functional analysis of LATERAL ORGAN BOUNDARIES-DOMAIN/ASYMMETRIC LEAVES2-LIKE (LBD/ASL) domain proteins involved in root initiation in angiosperms suggest divergent regulatory mechanisms in root development of lycophytes and euphyllophytes (Coudert, Dievart, Droc, & Gantet, 2013; Zhang et al., 2017).

6. Conclusions and perspectives

Besides the putative conservation of developmental programs across land plants, our review also suggests emerging general evolutionary-developmental mechanisms of land plant body plan evolution. First of all, current evidence suggests that co-option of highly conserved gene networks in independently evolved organs with similar functions plays a significant role in the evolution of the body plan of land plants. Nevertheless, it remains to be seen why some networks were frequently co-opted while other regulatory networks have evolved de novo. Whether robustness and evolvability of regulatory circuits is a significant factor determining co-option or de novo evolution is unexplored (Boukhibar & Barkoulas, 2015). There are also multiple examples suggesting that gene duplications have significantly contributed to the evolution of complex regulatory circuits. Future studies should reveal the molecular mechanisms, working through *cis* or *trans*, via which new functions and regulatory connections/rewiring were achieved (Moriyama & Koshiba-Takeuchi, 2018). Finally, it appears that ancestors of land plants were genetically, physiologically, and developmentally potentiated to evolve more and more complex body plans. This may imply that relatively radical phenotypic changes could be achieved by few genetic changes. Therefore, evolution of the land plant body plan might have followed the punctuated equilibrium and not the gradual model of evolution (Kohsokabe & Kaneko, 2016). However, the relationship between the genotype and phenotype map of specific regulatory networks is unknown, currently hindering any advancement concerning this issue (Hong et al., 2018; Kohsokabe & Kaneko, 2016).

We are confident that future research combining the toolbox of rapidly developing high-throughput geno- and phenotyping technologies across major lineages of land plants will bring new insights into the developmental mechanisms that contributed to the evolution of the land plant body plan.

Acknowledgments

We are thankful for the financial support of the Swiss National Science Foundation (Grants 160004 and 131726), the Georges and Antoine Claraz Foundation (Switzerland), the US National Science Foundation, the Forschungskredit and the University Research Priority Program "Evolution in Action" of the University of Zurich.

References

Ambrose, B. A., & Vasco, A. (2016). Bringing the multicellular fern meristem into focus. *The New Phytologist*, *210*(3), 790–793.

Antimisiaris, M. F., & Running, M. P. (2014). Turning moss into algae. *Plant Signaling & Behavior*, *9*(7), e29314.

Aoyama, T., Hiwatashi, Y., Shigyo, M., Kofuji, R., Kubo, M., Ito, M., et al. (2012). AP2-type transcription factors determine stem cell identity in the moss *Physcomitrella patens*. *Development*, *139*(17), 3120–3129.

Bar, M., & Ori, N. (2014). Leaf development and morphogenesis. *Development*, *141*(22), 4219–4230.

Bateman, R. M. (1996). Nonfloral homoplasy and evolutionary scenarios in living and fossil land plants. In *Homoplasy* (pp. 91–130). Academic Press, Elsevier Inc.

Bateman, R. M., Crane, P. R., DiMichele, W. A., Kenrick, P. R., Rowe, N. P., Speck, T., et al. (1998). EARLY EVOLUTION OF LAND PLANTS: Phylogeny, physiology, and ecology of the primary terrestrial radiation. *Annual Review of Ecology and Systematics*, *29*(1), 263–292.

Becker, B., & Marin, B. (2009). Streptophyte algae and the origin of embryophytes. *Annals of Botany*, *103*(7), 999–1004.

Bennett, T. A., Liu, M. M., Aoyama, T., Bierfreund, N. M., Braun, M., Coudert, Y., et al. (2014). Plasma membrane-targeted PIN proteins drive shoot development in a moss. *Current Biology: CB*, *24*(23), 2776–2785.

Berbee, M. L., James, T. Y., & Strullu-Derrien, C. (2017). Early diverging fungi: Diversity and impact at the dawn of terrestrial life. *Annual Review of Microbiology*, *71*, 41–60.

Berner, R. A. (1997). Geochemistry and geophysics: The rise of plants and their effect on weathering and atmospheric CO2. *Science*, *276*(5312), 544–546.

Bierhorst, D. W. (1977). On the stem apex, leaf initiation and early leaf ontogeny in filicalean ferns. *American Journal of Botany*, *64*(2), 125.

Boukhibar, L. M., & Barkoulas, M. (2015). The developmental genetics of biological robustness. *Annals of Botany*, *117*(5), 699–707.

Bower, F. O. (1890). On antithetic as distinct from homologous alternation of generations in plants. *Annals of Botany*, *os-4*(3), 347–370.

Bower, F. O. (1959). *Primitive land plants, also known as the archegoniatae*. Jaipur: Arihanta Publishers.

Bowman, J. L. (2013). Walkabout on the long branches of plant evolution. *Current Opinion in Plant Biology*, *16*(1), 70–77.

Bowman, J. L., Sakakibara, K., Furumizu, C., & Dierschke, T. (2016). Evolution in the cycles of life. *Annual Review of Genetics*, *50*, 133–154.

Box, R. J. (1986). Quantitative short-term uptake of inorganic phosphate by the chara hispida rhizoid. *Plant, Cell & Environment*, *9*(6), 501–506.

Boyce, C. K., & Kevin Boyce, C. (2010). The evolution of plant development in a paleontological context. *Current Opinion in Plant Biology*, *13*(1), 102–107. https://doi.org/10.1016/j.pbi.2009.10.001.

Boyd, M., Rosenzweig, F., & Herron, M. D. (2018). Analysis of motility in multicellular *Chlamydomonas reinhardtii* evolved under predation. *PLoS One*, *13*(1), e0192184.

Bremer, K., Humphries, C. J., Mishler, B. D., & Churchill, S. P. (1987). On cladistic relationships in green plants. *Taxon*, *36*(2), 339.

Brown, R. C., & Lemmon, B. E. (2011). Spores before sporophytes: Hypothesizing the origin of sporogenesis at the algal-plant transition. *The New Phytologist*, *190*(4), 875–881.

Buck, W. R., Jonathan Shaw, A., & Goffinet, B. (n.d.). Morphology, anatomy, and classification of the bryophyta. In: A. Jonathan Shaw (Ed.), Bryophyte biology *Bryophyte biology* (pp. 55–138). Cambridge University Press.

Catarino, B., Hetherington, A. J., Emms, D. M., Kelly, S., & Dolan, L. (2016). The stepwise increase in the number of transcription factor families in the precambrian predated the diversification of plants on land. *Molecular Biology and Evolution*, *33*(11), 2815–2819.

Chang, Y., & Graham, S. W. (2011). Inferring the higher-order phylogeny of mosses (Bryophyta) and relatives using a large, multigene plastid data set. *American Journal of Botany*, *98*(5), 839–849.

Chomicki, G., Coiro, M., & Renner, S. S. (2017). Evolution and ecology of plant architecture: Integrating insights from the fossil record, extant morphology, developmental genetics and phylogenies. *Annals of Botany*, *120*(6), 855–891.

Coudert, Y. (2017). The evolution of branching in land plants: Between conservation and diversity. In L. Nuno de la Rosa & G. Müller (Eds.), *Evolutionary developmental biology* (pp. 1–17). Springer.

Coudert, Y., Bell, N. E., Edelin, C., & Harrison, C. J. (2017). Multiple innovations underpinned branching form diversification in mosses. *The New Phytologist*, *215*(2), 840–850.

Coudert, Y., Dievart, A., Droc, G., & Gantet, P. (2013). ASL/LBD phylogeny suggests that genetic mechanisms of root initiation downstream of auxin are distinct in lycophytes and euphyllophytes. *Molecular Biology and Evolution*, *30*(3), 569–572.

Coudert, Y., Palubicki, W., Ljung, K., Novak, O., Leyser, O., & Jill Harrison, C. (2015). Three ancient hormonal cues co-ordinate shoot branching in a moss. *eLife*, *4*, e06808. https://doi.org/10.7554/elife.06808.

Cox, C. J., Li, B., Foster, P. G., Martin Embley, T., & Civáň, P. (2014). Conflicting phylogenies for early land plants are caused by composition biases among synonymous substitutions. *Systematic Biology*, *63*(2), 272–279.

Crandall-Stotler, B., & Stotler, R. E. (2008). Morphology and classification of the marchantiophyta. In J. A. Shaw (Ed.), *Bryophyte biology* (pp. 21–70). Cambridge University Press.

Crane, P. R., & Kenrick, P. (1997). Diverted development of reproductive organs: A source of morphological innovation in land plants. *Plant Systematics and Evolution = Entwicklungsgeschichte Und Systematik Der Pflanzen*, *206*(1–4), 161–174.

de Vries, J., & Archibald, J. M. (2018). Plant evolution: Landmarks on the path to terrestrial life. *The New Phytologist*, *217*(4), 1428–1434.

de Vries, J., Curtis, B. A., Gould, S. B., & Archibald, J. M. (2018). Embryophyte stress signaling evolved in the algal progenitors of land plants. *Proceedings of the National Academy of Sciences of the United States of America*, *115*(15), E3471–E3480.

de Vries, J., Fischer, A. M., Roettger, M., Rommel, S., Schluepmann, H., Bräutigam, A., et al. (2016). Cytokinin-induced promotion of root meristem size in the fern azolla supports a shoot-like origin of euphyllophyte roots. *The New Phytologist*, *209*(2), 705–720.

Delwiche, C. F., & Cooper, E. D. (2015). The evolutionary origin of a terrestrial flora. *Current Biology: CB*, *25*(19), R899–R910.

Demko, V., Perroud, P.-F., Johansen, W., Delwiche, C. F., Cooper, E. D., Remme, P., et al. (2014). Genetic analysis of DEFECTIVE KERNEL1 loop function in three-dimensional body patterning in *Physcomitrella patens*. *Plant Physiology*, *166*(2), 903–919.

Domozych, D. S., Popper, Z. A., & Sørensen, I. (2016). Charophytes: Evolutionary giants and emerging model organisms. *Frontiers in Plant Science*, 7, 1470.

Eames, A. J. (1936). *Morphology of vascular plants: Lower groups (psilophytales to filicales)*. R. E. Krieger Pub. Co.

Edwards, D., Morris, J. L., Richardson, J. B., & Kenrick, P. (2014). Cryptospores and cryptophytes reveal hidden diversity in early land floras. *The New Phytologist*, *202*(1), 50–78.

Edwards, D., Richardson, J. B., Axe, L., & Davies, K. L. (2012). A new group of early devonian plants with valvate sporangia containing sculptured permanent dyads. *Botanical Journal of the Linnean Society*, *168*(3), 229–257.

Evkaikina, A. I., Berke, L., Romanova, M. A., Proux-Wéra, E., Ivanova, A. N., Rydin, C., et al. (2017). The Huperzia selago shoot tip transcriptome sheds new light on the evolution of leaves. *Genome Biology and Evolution*, *9*(9), 2444–2460.

Finet, C., Timme, R. E., Delwiche, C. F., & Marlétaz, F. (2010). Multigene phylogeny of the green lineage reveals the origin and diversification of land plants. *Current Biology: CB*, *20*(24), 2217–2222.

Fiz-Palacios, O., Schneider, H., Heinrichs, J., & Savolainen, V. (2011). Diversification of land plants: Insights from a family-level phylogenetic analysis. *BMC Evolutionary Biology*, *11*, 341.

Floyd, S. K., & Bowman, J. L. (2007). The ancestral developmental tool kit of land plants. *International Journal of Plant Sciences*, *168*(1), 1–35.

Foster, A. S., & Gifford, E. M. (1959). *Comparative morphology of vascular plants*. W. H. Freeman & Co.

Frangedakis, E., Saint-Marcoux, D., Moody, L. A., Rabbinowitsch, E., & Langdale, J. A. (2017). Nonreciprocal complementation of KNOX gene function in land plants. *The New Phytologist*, *216*(2), 591–604.

Frank, M. H., Edwards, M. B., Schultz, E. R., McKain, M. R., Fei, Z., Sørensen, I., et al. (2015). Dissecting the molecular signatures of apical cell-type shoot meristems from two ancient land plant lineages. *The New Phytologist*, *207*(3), 893–904.

Frank, M. H., & Scanlon, M. J. (2015a). Cell-specific transcriptomic analyses of three-dimensional shoot development in the moss *Physcomitrella patens*. *The Plant Journal: For Cell and Molecular Biology*, *83*(4), 743–751.

Frank, M. H., & Scanlon, M. J. (2015b). Transcriptomic evidence for the evolution of shoot meristem function in sporophyte-dominant land plants through concerted selection of ancestral gametophytic and sporophytic genetic programs. *Molecular Biology and Evolution*, *32*(2), 355–367.

French, J. C., & Paolillo, D. J. (1975a). Effect of exogenously supplied growth regulators on intercalary meristematic activity and capsule expansion in funaria. *The Bryologist*, *78*(4), 431.

French, J. C., & Paolillo, D. J. (1975b). Intercalary meristematic activity in the sporophyte of *Funaria* (musci). *American Journal of Botany*, *62*(1), 86–96.

Friedman, W. E. (2011). Plant genomics: Homoplasy heaven in a lycophyte genome. *Current Biology: CB*, *21*(14), R554–R556.

Friedman, W. E., Moore, R. C., & Purugganan, M. D. (2004). The evolution of plant development. *American Journal of Botany*, *91*(10), 1726–1741.

Furumizu, C., Alvarez, J. P., Sakakibara, K., & Bowman, J. L. (2015). Antagonistic roles for KNOX1 and KNOX2 genes in patterning the land plant body plan following an ancient gene duplication. *PLoS Genetics*, *11*(2), e1004980.

Galtier, J. (2010). The origins and early evolution of the megaphyllous leaf. *International Journal of Plant Sciences, 171*(6), 641–661.

Ge, Y., Liu, J., Zeng, M., He, J., Qin, P., Huang, H., et al. (2016). Identification of WOX family genes in *Selaginella kraussiana* for studies on stem cells and regeneration in lycophytes. *Frontiers in Plant Science, 7*, 93.

Gensel, P. G. (2008). The earliest land plants. *Annual Review of Ecology, Evolution, and Systematics, 39*(1), 459–477.

Gensel, P. G., Kotyk, M. E., & Basinger, J. F. (2001). 5. Morphology of above- and below-ground structures in early devonian (pragian–emsian) plants. In P. G. Gensel & D. Edwards (Eds.), *Plants invade the land* (pp. 83–102). Columbia University Press.

Gerrienne, P., Dilcher, D. L., Bergamaschi, S., Milagres, I., Pereira, E., & Rodrigues, M. A. C. (2006). An exceptional specimen of the early land plant cooksonia paranensis, and a hypothesis on the life cycle of the earliest eutracheophytes. *Review of Palaeobotany and Palynology, 142*(3–4), 123–130.

Gitzendanner, M. A., Soltis, P. S., Wong, G. K.-S., Ruhfel, B. R., & Soltis, D. E. (2018). Plastid phylogenomic analysis of green plants: A billion years of evolutionary history. *American Journal of Botany, 105*, 291–301. https://doi.org/10.1002/ajb2.1048.

Gonzalez, N., Vanhaeren, H., & Inzé, D. (2012). Leaf size control: Complex coordination of cell division and expansion. *Trends in Plant Science, 17*(6), 332–340.

Haig, D. (2008). Homologous versus antithetic alternation of generations and the origin of sporophytes. *The Botanical Review, 74*(3), 395–418.

Hanschen, E. R., Marriage, T. N., Ferris, P. J., Hamaji, T., Toyoda, A., Fujiyama, A., et al. (2016). The gonium pectorale genome demonstrates co-option of cell cycle regulation during the evolution of multicellularity. *Nature Communications, 7*, 11370.

Harholt, J., Moestrup, Ø., & Ulvskov, P. (2016). Why plants were terrestrial from the beginning. *Trends in Plant Science, 21*(2), 96–101.

Harrison, C. J. (2015). Shooting through time: New insights from transcriptomic data. *Trends in Plant Science, 20*(8), 468–470.

Harrison, C. J. (2017). Auxin transport in the evolution of branching forms. *The New Phytologist, 215*(2), 545–551.

Harrison, C. J., Alvey, E., & Henderson, I. R. (2010). Meiosis in flowering plants and other green organisms. *Journal of Experimental Botany, 61*(11), 2863–2875.

Harrison, C. J., Jill Harrison, C., Corley, S. B., Moylan, E. C., Alexander, D. L., Scotland, R. W., et al. (2005). Independent recruitment of a conserved developmental mechanism during leaf evolution. *Nature, 434*(7032), 509–514.

Harrison, C. J., Rezvani, M., & Langdale, J. A. (2007). Growth from two transient apical initials in the meristem of *Selaginella kraussiana. Development, 134*(5), 881–889.

Harrison, C. J., Roeder, A. H. K., Meyerowitz, E. M., & Langdale, J. A. (2009). Local cues and asymmetric cell divisions underpin body plan transitions in the moss *Physcomitrella patens. Current Biology: CB, 19*(6), 461–471.

Harvey-Gibson, R. J. (1902). Contributions towards a knowledge of the anatomy of the genus Selaginella. *Annals of Botany, os-16*(3), 449–466.

Hay, A., & Tsiantis, M. (2010). KNOX genes: Versatile regulators of plant development and diversity. *Development, 137*(19), 3153–3165.

Herron, M. D. (2016). Origins of multicellular complexity: Volvox and the volvocine algae. *Molecular Ecology, 25*(6), 1213–1223.

Hetherington, A. J., Berry, C. M., & Dolan, L. (2016). Networks of highly branched stigmarian rootlets developed on the first giant trees. *Proceedings of the National Academy of Sciences of the United States of America, 113*(24), 6695–6700.

Hetherington, A. J., & Dolan, L. (2018a). Bilaterally symmetric axes with rhizoids composed the rooting structure of the common ancestor of vascular plants. *Philosophical Transactions of the Royal Society of London. Series B, Biological Sciences, 373*(1739), 20170042. https://doi.org/10.1098/rstb.2017.0042.

Hetherington, A. J., & Dolan, L. (2018b). Stepwise and independent origins of roots among land plants. *Nature*, *561*(7722), 235–238.

Hong, L., Dumond, M., Zhu, M., Tsugawa, S., Li, C.-B., Boudaoud, A., et al. (2018). Heterogeneity and robustness in plant morphogenesis: From cells to organs. *Annual Review of Plant Biology*, *69*, 469–495. https://doi.org/10.1146/annurev-arplant-042817-040517.

Honkanen, S., & Dolan, L. (2016). Growth regulation in tip-growing cells that develop on the epidermis. *Current Opinion in Plant Biology*, *34*, 77–83.

Honkanen, S., Jones, V. A. S., Morieri, G., Champion, C., Hetherington, A. J., Kelly, S., et al. (2016). The mechanism forming the cell surface of tip-growing rooting cells is conserved among land plants. *Current Biology: CB*, *26*(23), 3238–3244.

Hori, K., Maruyama, F., Fujisawa, T., Togashi, T., Yamamoto, N., Seo, M., et al. (2014). Klebsormidium flaccidum genome reveals primary factors for plant terrestrial adaptation. *Nature Communications*, *5*, 3978.

Horst, N. A., Katz, A., Pereman, I., Decker, E. L., Ohad, N., & Reski, R. (2016). A single homeobox gene triggers phase transition, embryogenesis and asexual reproduction. *Nature Plants*, *2*, 15209.

Huang, L., & Schiefelbein, J. (2015). Conserved gene expression programs in developing roots from diverse plants. *The Plant Cell*, *27*(8), 2119–2132.

Huang, L., Shi, X., Wang, W., Ryu, K. H., & Schiefelbein, J. (2017). Diversification of root hair development genes in vascular plants. *Plant Physiology*, *174*(3), 1697–1712.

Ichihashi, Y., & Tsukaya, H. (2015). Behavior of leaf meristems and their modification. *Frontiers in Plant Science*, *6*, 1060.

Jones, V. A. S., & Dolan, L. (2012). The evolution of root hairs and rhizoids. *Annals of Botany*, *110*(2), 205–212.

Karol, K. G. (2001). The closest living relatives of land plants. *Science*, *294*(5550), 2351–2353.

Kawai, J., Tanabe, Y., Soma, S., & Ito, M. (2010). Class 1 KNOX gene expression supports the Selaginella rhizophore concept. *Journal of Plant Biology = Singmul Hakhoe Chi*, *53*(4), 268–274.

Kenrick, B. (1994). Alternation of generations in land plants: New phylogenetic and palaeobotanical evidence. *Biological Reviews of the Cambridge Philosophical Society*, *69*(3), 293–330.

Kenrick, P. (2013). The origin of roots. In A. Eshel & T. Beeckman (Eds.), *Plant roots: The hidden half* (4th ed.).

Kenrick, P. (2018). Changing expressions: A hypothesis for the origin of the vascular plant life cycle. *Philosophical Transactions of the Royal Society of London. Series B, Biological Sciences*, *373*(1739), pii: 20170149. https://doi.org/10.1098/rstb.2017.0149.

Kenrick, P., & Crane, P. (1997). The origin and early diversification of land plants: A cladistic study. *Science*, *389*(4), 33–39.

Kenrick, P., & Strullu-Derrien, C. (2014). The origin and early evolution of roots. *Plant Physiology*, *166*(2), 570–580.

Kenrick, P., Wellman, C. H., Schneider, H., & Edgecombe, G. D. (2012). A timeline for terrestrialization: Consequences for the carbon cycle in the palaeozoic. *Philosophical Transactions of the Royal Society of London. Series B, Biological Sciences*, *367*(1588), 519–536.

Kevin Boyce, C. (2005). The evolutionary history of roots and leaves. In N. M. Holbrook & M. Zwieniecki (Eds.), *Vascular transport in plants* (pp. 479–499). Academic Press.

Kobiyama, Y., & Crandall-Stotler, B. (1999). Studies of specialized pitted parenchyma cells of the liverwort conocephalum hill and their phylogenetic implications. *International Journal of Plant Sciences*, *160*(2), 351–370.

Kohsokabe, T., & Kaneko, K. (2016). Evolution-development congruence in pattern formation dynamics: Bifurcations in gene expression and regulation of networks structures. *Journal of Experimental Zoology. Part B, Molecular and Developmental Evolution*, *326*(1), 61–84.

Langdale, J. A. (2008). Evolution of developmental mechanisms in plants. *Current Opinion in Genetics & Development, 18*(4), 368–373.
Laurin-Lemay, S., Brinkmann, H., & Philippe, H. (2012). Origin of land plants revisited in the light of sequence contamination and missing data. *Current Biology: CB, 22*(15), R593–R594.
Lee, J.-H., Lin, H., Joo, S., & Goodenough, U. (2008). Early sexual origins of homeoprotein heterodimerization and evolution of the plant KNOX/BELL family. *Cell, 133*(5), 829–840.
Lenton, T. M., Crouch, M., Johnson, M., Pires, N., & Dolan, L. (2012). First plants cooled the ordovician. *Nature Geoscience, 5*(2), 86–89.
Lewis, L. A., & McCourt, R. M. (2004). Green algae and the origin of land plants. *American Journal of Botany, 91*(10), 1535–1556.
Ligrone, R., Duckett, J. G., & Renzaglia, K. S. (2012a). Major transitions in the evolution of early land plants: A bryological perspective. *Annals of Botany, 109*(5), 851–871.
Ligrone, R., Duckett, J. G., & Renzaglia, K. S. (2012b). The origin of the sporophyte shoot in land plants: A bryological perspective. *Annals of Botany, 110*(5), 935–941.
Liu, Y., Cox, C. J., Wang, W., & Goffinet, B. (2014). Mitochondrial phylogenomics of early land plants: Mitigating the effects of saturation, compositional heterogeneity, and codon-usage bias. *Systematic Biology, 63*(6), 862–878.
Lu, P., & Jernstedt, J. A. (1996). Rhizophore and root development in *Selaginella martensii*: Meristem transitions and identity. *International Journal of Plant Sciences, 157*(2), 180–194.
Marschner, H. (2012). *Marschner's mineral nutrition of higher plants*. Academic Press.
McConaha, M. (1939). Ventral surface specializations of *Conocephalum conicum*. *American Journal of Botany, 26*(6), 353–355.
McConaha, M. (1941). Ventral structures effecting capillarity in the marchantiales. *American Journal of Botany, 28*(4), 301.
McMahon, W. J., & Davies, N. S. (2018). Evolution of alluvial mudrock forced by early land plants. *Science, 359*(6379), 1022–1024.
Mishler, B. D., & Churchill, S. P. (1984). A cladistic approach to the phylogeny of the "bryophytes" *Brittonia, 36*(4), 406.
Moody, L. A., Kelly, S., Rabbinowitsch, E., & Langdale, J. A. (2018). Genetic regulation of the 2D to 3D growth transition in the moss *Physcomitrella patens*. *Current Biology: CB, 28*(3), 473–478.e5.
Moriyama, Y., & Koshiba-Takeuchi, K. (2018). Significance of whole-genome duplications on the emergence of evolutionary novelties. *Briefings in Functional Genomics, 17*, 329–338https://doi.org/10.1093/bfgp/ely007.
Morris, J. L., Puttick, M. N., Clark, J. W., Edwards, D., Kenrick, P., Pressel, S., et al. (2018). The timescale of early land plant evolution. *Proceedings of the National Academy of Sciences of the United States of America, 115*(10), E2274–E2283.
Mosquna, A., Katz, A., Decker, E. L., Rensing, S. A., Reski, R., & Ohad, N. (2009). Regulation of stem cell maintenance by the polycomb protein FIE has been conserved during land plant evolution. *Development, 136*(14), 2433–2444.
Nardmann, J., & Werr, W. (2012). The invention of WUS-like stem cell-promoting functions in plants predates leptosporangiate ferns. *Plant Molecular Biology, 78*(1–2), 123–134.
Niklas, K. J. (2014). The evolutionary-developmental origins of multicellularity. *American Journal of Botany, 101*(1), 6–25.
Niklas, K. J., & Kutschera, U. (2009). The evolution of the land plant life cycle. *The New Phytologist, 185*(1), 27–41.
Odu, E. A. (1978). The adaptive importance of moss rhizoids for attachment to the substratum. *Journal of Bryology, 10*(2), 163–181.
Okano, Y., Aono, N., Hiwatashi, Y., Murata, T., Nishiyama, T., Ishikawa, T., et al. (2009). A polycomb repressive complex 2 gene regulates apogamy and gives evolutionary insights into early land plant evolution. *Proceedings of the National Academy of Sciences of the United States of America, 106*(38), 16321–16326.

Olsen, O.-A., Perroud, P.-F., Johansen, W., & Demko, V. (2015). DEK1; missing piece in puzzle of plant development. *Trends in Plant Science*, *20*(2), 70–71.

Olson, B. J., & Nedelcu, A. M. (2016). Co-option during the evolution of multicellular and developmental complexity in the volvocine green algae. *Current Opinion in Genetics & Development*, *39*, 107–115.

Ortiz-Ramírez, C., Hernandez-Coronado, M., Thamm, A., Catarino, B., Wang, M., Dolan, L., et al. (2016). A transcriptome atlas of *Physcomitrella patens* provides insights into the evolution and development of land plants. *Molecular Plant*, *9*(2), 205–220.

Pereman, I., Mosquna, A., Katz, A., Wiedemann, G., Lang, D., Decker, E. L., et al. (2016). The polycomb group protein CLF emerges as a specific tri-methylase of H3K27 regulating gene expression and development in *Physcomitrella patens*. *Biochimica et Biophysica Acta*, *1859*(7), 860–870.

Perkins, S. (2018). Early plants slowed mud from flowing into sea. *Science*. https://doi.org/10.1126/science.aat4589.

Perroud, P.-F., Demko, V., Johansen, W., Wilson, R. C., Olsen, O.-A., & Quatrano, R. S. (2014). Defective kernel 1 (DEK1) is required for three-dimensional growth in *Physcomitrella patens*. *The New Phytologist*, *203*(3), 794–804.

Pires, N., & Dolan, L. (2010a). Early evolution of bHLH proteins in plants. *Plant Signaling & Behavior*, *5*(7), 911–912.

Pires, N., & Dolan, L. (2010b). Origin and diversification of basic-helix-loop-helix proteins in plants. *Molecular Biology and Evolution*, *27*(4), 862–874.

Pires, N. D., Yi, K., Breuninger, H., Catarino, B., Menand, B., & Dolan, L. (2013). Recruitment and remodeling of an ancient gene regulatory network during land plant evolution. *Proceedings of the National Academy of Sciences of the United States of America*, *110*(23), 9571–9576.

Plackett, A. R. G., Di Stilio, V. S., & Langdale, J. A. (2015). Ferns: The missing link in shoot evolution and development. *Frontiers in Plant Science*, *6*, 972.

Proust, H., Honkanen, S., Jones, V. A. S., Morieri, G., Prescott, H., Kelly, S., et al. (2016). RSL class I genes controlled the development of epidermal structures in the common ancestor of land plants. *Current Biology: CB*, *26*(1), 93–99.

Puttick, M. N., Morris, J. L., Williams, T. A., Cox, C. J., Edwards, D., Kenrick, P., et al. (2018). The interrelationships of land plants and the nature of the ancestral embryophyte. *Current Biology: CB*, *28*(5), 733–745.e2.

Qiu, Y.-L., Li, L., Wang, B., Chen, Z., Knoop, V., Groth-Malonek, M., et al. (2006). The deepest divergences in land plants inferred from phylogenomic evidence. *Proceedings of the National Academy of Sciences of the United States of America*, *103*(42), 15511–15516.

Qiu, Y.-L., Yin-Long, Q., Taylor, A. B., & McManus, H. A. (2012). Evolution of the life cycle in land plants. *Journal of Systematics and Evolution*, *50*(3), 171–194.

Ratcliff, W. C., Herron, M. D., Howell, K., Pentz, J. T., Rosenzweig, F., & Travisano, M. (2013). Experimental evolution of an alternating uni- and multicellular life cycle in *Chlamydomonas reinhardtii*. *Nature Communications*, *4*, 2742.

Raven, J. A., & Edwards, D. (2001). Roots: Evolutionary origins and biogeochemical significance. *Journal of Experimental Botany*, *52*(Spec Issue), 381–401.

Remy, W., Gensel, P. G., & Hass, H. (1993). The gametophyte generation of some early devonian land plants. *International Journal of Plant Sciences*, *154*(1), 35–58.

Renzaglia, K. S., Crandall-Stotler, B., Pressel, S., Duckett, J. G., Schuette, S., & Strother, P. K. (2015). Permanent spore dyads are not "a thing of the past": On their occurrence in the liverwort *Haplomitrium* (Haplomitriopsida). *Botanical Journal of the Linnean Society*, *179*(4), 658–669.

Renzaglia, K. S., Duff, R. J. T., Nickrent, D. L., & Garbary, D. J. (2000). Vegetative and reproductive innovations of early land plants: Implications for a unified phylogeny. *Philosophical Transactions of the Royal Society of London. Series B, Biological Sciences*, *355*(1398), 769–793.

Renzaglia, K. S., Villarreal, J. C., Joel Duff, R., & Goffinet, B. (2008). New insights into morphology, anatomy, and systematics of hornworts. In A. Jonathan Shaw (Ed.), *Bryophyte biology* (pp. 139–172). Cambridge University Press.

Rothwell, G. W. (1999). Fossils and ferns in the resolution of land plant phylogeny. *The Botanical Review*, *65*(3), 188–218.

Rothwell, G. W., & Nixon, K. C. (2006). How does the inclusion of fossil data change our conclusions about the phylogenetic history of euphyllophytes? *International Journal of Plant Sciences*, *167*(3), 737–749.

Rothwell, G. W., Wyatt, S. E., & Tomescu, A. M. F. (2014). Plant evolution at the interface of paleontology and developmental biology: An organism-centered paradigm. *American Journal of Botany*, *101*(6), 899–913.

Rozema, J., Björn, L. O., Bornman, J. F., Gaberscik, A., Häder, D.-P., Trost, T., et al. (2002). The role of UV-B radiation in aquatic and terrestrial ecosystems—An experimental and functional analysis of the evolution of UV-absorbing compounds. *Journal of Photochemistry and Photobiology. B, Biology*, *66*(1), 2–12.

Rozema, J., Blokker, P., Mayoral Fuertes, M. A., & Broekman, R. (2009). UV-B absorbing compounds in present-day and fossil pollen, spores, cuticles, seed coats and wood: Evaluation of a proxy for solar UV radiation. *Photochemical & Photobiological Sciences: Official Journal of the European Photochemistry Association and the European Society for Photobiology*, *8*(9), 1233–1243.

Runions, A., Tsiantis, M., & Prusinkiewicz, P. (2017). A common developmental program can produce diverse leaf shapes. *The New Phytologist*, *216*(2), 401–418.

Sakakibara, K., Ando, S., Yip, H. K., Tamada, Y., Hiwatashi, Y., Murata, T., et al. (2013). KNOX2 genes regulate the haploid-to-diploid morphological transition in land plants. *Science*, *339*(6123), 1067–1070.

Sakakibara, K., Nishiyama, T., Deguchi, H., & Hasebe, M. (2008). Class 1 KNOX genes are not involved in shoot development in the moss *Physcomitrella patens* but do function in sporophyte development. *Evolution & Development*, *10*(5), 555–566.

Sakakibara, K., Reisewitz, P., Aoyama, T., Friedrich, T., Ando, S., Sato, Y., et al. (2014). WOX13-like genes are required for reprogramming of leaf and protoplast cells into stem cells in the moss *Physcomitrella patens*. *Development*, *141*(8), 1660–1670.

Salamon, M. A., Gerrienne, P., Steemans, P., Gorzelak, P., Filipiak, P., Le Hérissé, A., et al. (2018). Putative late ordovician land plants. *The New Phytologist*, *218*, 1305–1309. https://doi.org/10.1111/nph.15091.

Sanders, H. L., Darrah, P. R., & Langdale, J. A. (2011). Sector analysis and predictive modelling reveal iterative shoot-like development in fern fronds. *Development*, *138*(14), 2925–2934.

Sano, R., Juárez, C. M., Hass, B., Sakakibara, K., Ito, M., Banks, J. A., et al. (2005). KNOX homeobox genes potentially have similar function in both diploid unicellular and multicellular meristems, but not in haploid meristems. *Evolution & Development*, 7(1), 69–78.

Saxelby, E. M. (1908). The origin of the roots in *Lycopodium selago*. *Annals of Botany*, *os-22*(1), 21–33.

Selosse, M.-A., Strullu-Derrien, C., Martin, F. M., Kamoun, S., & Kenrick, P. (2015). Plants, fungi and oomycetes: A 400-million year affair that shapes the biosphere. *The New Phytologist*, *206*(2), 501–506.

Seward, A. C. (2011). Difficulties and sources of error in the determination of fossil plants. In A. C. Seward (Ed.), *Fossil plants* (pp. 93–109). Cambridge University Press.

Snipes, S. A., Rodriguez, K., DeVries, A. E., Miyawaki, K. N., Perales, M., Xie, M., et al. (2018). Cytokinin stabilizes WUSCHEL by acting on the protein domains required for nuclear enrichment and transcription. *PLoS Genetics*, *14*(4), e1007351.

Steemans, P., Herisse, A. L., Melvin, J., Miller, M. A., Paris, F., Verniers, J., et al. (2009). Origin and radiation of the earliest vascular land plants. *Science*, *324*(5925), 353.

Stein, W. E., Mannolini, F., Hernick, L. V., Landing, E., & Berry, C. M. (2007). Giant cladoxylopsid trees resolve the enigma of the earth's earliest forest stumps at Gilboa. *Nature*, *446*(7138), 904–907.

Stewart, W. N., & Rothwell, G. W. (1993). *Paleobotany and the evolution of plants*. Cambridge University Press.

Szovenyi, P., Rensing, S. A., Lang, D., Wray, G. A., & Shaw, A. J. (2010). Generation-biased gene expression in a bryophyte model system. *Molecular Biology and Evolution*, *28*(1), 803–812.

Tam, T. H. Y., Catarino, B., & Dolan, L. (2015). Conserved regulatory mechanism controls the development of cells with rooting functions in land plants. *Proceedings of the National Academy of Sciences of the United States of America*, *112*(29), E3959–E3968.

Tanabe, Y., Hasebe, M., Sekimoto, H., Nishiyama, T., Kitani, M., Henschel, K., et al. (2005). Characterization of MADS-box genes in charophycean green algae and its implication for the evolution of MADS-box genes. *Proceedings of the National Academy of Sciences of the United States of America*, *102*(7), 2436–2441.

Tanahashi, T., Sumikawa, N., Kato, M., & Hasebe, M. (2005). Diversification of gene function: Homologs of the floral regulator FLO/LFY control the first zygotic cell division in the moss *Physcomitrella patens*. *Development*, *132*(7), 1727–1736.

Taylor, T. N., Kerp, H., & Hass, H. (2005). Life history biology of early land plants: Deciphering the gametophyte phase. *Proceedings of the National Academy of Sciences of the United States of America*, *102*(16), 5892–5897.

Thole, J. M., Perroud, P.-F., Quatrano, R. S., & Running, M. P. (2014). Prenylation is required for polar cell elongation, cell adhesion, and differentiation in Physcomitrella patens. *The Plant Journal: For Cell and Molecular Biology*, *78*(3), 441–451.

Timme, R. E., Bachvaroff, T. R., & Delwiche, C. F. (2012). Broad phylogenomic sampling and the sister lineage of land plants. *PLoS One*, *7*(1), e29696.

Tomescu, A. M. F. (2009). Megaphylls, microphylls and the evolution of leaf development. *Trends in Plant Science*, *14*(1), 5–12.

Tomescu, A. M. F., Escapa, I. H., Rothwell, G. W., Elgorriaga, A., & Cúneo, N. R. (2017). Developmental programmes in the evolution of Equisetum reproductive morphology: A hierarchical modularity hypothesis. *Annals of Botany*, *119*(4), 489–505.

Tomescu, A. M. F., Wyatt, S. E., Hasebe, M., & Rothwell, G. W. (2014). Early evolution of the vascular plant body plan—The missing mechanisms. *Current Opinion in Plant Biology*, *17*, 126–136.

Townsley, B. T., & Sinha, N. R. (2012). A new development: Evolving concepts in leaf ontogeny. *Annual Review of Plant Biology*, *63*, 535–562.

Trachtenberg, S., & Zamski, E. (1979). The apoplastic conduction of water in *Polytrichum juniperinum* willd. gametophytes. *The New Phytologist*, *83*(1), 49–52.

Truskina, J., & Vernoux, T. (2018). The growth of a stable stationary structure: Coordinating cell behavior and patterning at the shoot apical meristem. *Current Opinion in Plant Biology*, *41*, 83–88.

Uphof, J. C. T. (1920). Contributions towards a knowledge of the anatomy of the genus Selaginella. *Annals of Botany*, *os-34*(4), 493–517.

Vasco, A., Moran, R. C., & Ambrose, B. A. (2013). The evolution, morphology, and development of fern leaves. *Frontiers in Plant Science*, *4*, 345.

Vasco, A., Smalls, T. L., Graham, S. W., Cooper, E. D., Wong, G. K.-S., Stevenson, D. W., et al. (2016). Challenging the paradigms of leaf evolution: Class III HD-Zips in ferns and lycophytes. *The New Phytologist*, *212*(3), 745–758.

Vermeer, C. P., Escher, M., Portielje, R., & de Klein, J. J. M. (2003). Nitrogen uptake and translocation by chara. *Aquatic Botany*, *76*(3), 245–258.

Villarreal, J. C., & Renzaglia, K. S. (2015). The hornworts: Important advancements in early land plant evolution. *Journal of Bryology*, *37*(3), 157–170.

Vivancos, J., Spinner, L., Mazubert, C., Charlot, F., Paquet, N., Thareau, V., et al. (2012). The function of the RNA-binding protein TEL1 in moss reveals ancient regulatory mechanisms of shoot development. *Plant Molecular Biology*, *78*(4–5), 323–336.

Wang, Y., & Dohlman, H. G. (2005). Pheromone-regulated sumoylation of transcription factors that mediate the invasive to mating developmental switch in yeast. *The Journal of Biological Chemistry*, *281*(4), 1964–1969.

Whitewoods, C. D., Cammarata, J., Nemec Venza, Z., Sang, S., Crook, A. D., Aoyama, T., et al. (2018). CLAVATA was a genetic novelty for the morphological innovation of 3D growth in land plants. *Current Biology: CB*, *28*(15), 2365–2376.e5.

Wickett, N. J., Mirarab, S., Nguyen, N., Warnow, T., Carpenter, E., Matasci, N., et al. (2014). Phylotranscriptomic analysis of the origin and early diversification of land plants. *Proceedings of the National Academy of Sciences of the United States of America*, *111*(45), E4859–E4868.

Wilhelmsson, P. K. I., Mühlich, C., Ullrich, K. K., & Rensing, S. A. (2017). Comprehensive genome-wide classification reveals that many plant-specific transcription factors evolved in streptophyte algae. *Genome Biology and Evolution*, *9*(12), 3384–3397.

Wodniok, S., Brinkmann, H., Glöckner, G., Heidel, A. J., Philippe, H., Melkonian, M., et al. (2011). Origin of land plants: Do conjugating green algae hold the key? *BMC Evolutionary Biology*, *11*, 104.

Xu, H.-H., Berry, C. M., Stein, W. E., Wang, Y., Tang, P., & Fu, Q. (2017). Unique growth strategy in the earth's first trees revealed in silicified fossil trunks from China. *Proceedings of the National Academy of Sciences of the United States of America*, *114*(45), 12009–12014.

Yi, K., Menand, B., Bell, E., & Dolan, L. (2010). A basic helix-loop-helix transcription factor controls cell growth and size in root hairs. *Nature Genetics*, *42*(3), 264–267.

Yip, H. K., Floyd, S. K., Sakakibara, K., & Bowman, J. L. (2016). Class III HD-Zip activity coordinates leaf development in *Physcomitrella patens*. *Developmental Biology*, *419*(1), 184–197.

Zhang, Y., Jiao, Y., Jiao, H., Zhao, H., & Zhu, Y.-X. (2017). Two-step functional innovation of the stem-cell factors WUS/WOX5 during plant evolution. *Molecular Biology and Evolution*, *34*(3), 640–653.

Zhong, B., Liu, L., Yan, Z., & Penny, D. (2013). Origin of land plants using the multispecies coalescent model. *Trends in Plant Science*, *18*(9), 492–495.

Zhu, Y., Chen, L., Zhang, C., Hao, P., Jing, X., & Li, X. (2017). Global transcriptome analysis reveals extensive gene remodeling, alternative splicing and differential transcription profiles in non-seed vascular plant *Selaginella moellendorffii*. *BMC Genomics*, *18* (Suppl. 1), 1042.

Zimmermann, W. (1951). Main results of the "TELOME Theory" *Paleobotanist*, *1*, 456–470.

Zimmermann, W. (1965). Die telomtheorie. *Fortschritte der Evolutionsforschung*, *1*. G. Fischer, Stuttgart.

CHAPTER TWO

Evolution and co-option of developmental regulatory networks in early land plants

John L. Bowman[1], Liam N. Briginshaw, Stevie N. Florent
School of Biological Sciences, Monash University, Melbourne, VIC, Australia
[1]Corresponding author: e-mail address: john.bowman@monash.edu

Contents

Abstract

Land plants evolved from an ancestral alga from which they inherited developmental and physiological characters. A key innovation of land plants is a life cycle with an alternation of generations, with both haploid gametophyte and diploid sporophyte generations having complex multicellular bodies. The origins of the developmental genetic programs patterning these bodies, whether inherited from an algal ancestor or evolved *de novo*, and whether programs were co-opted between generations, are largely open questions. We first provide a framework for land plant evolution and co-option of developmental regulatory pathways and then examine two cases in more detail.

1. The algal origin of land plants

Land plants evolved from an ancestral charophycean algae with the order Zygnematales or a Zygnematales plus Coleochaetales clade being the extant sister group, and the Charales more distantly related (Finet, Timme,

Current Topics in Developmental Biology, Volume 131
ISSN 0070-2153
https://doi.org/10.1016/bs.ctdb.2018.10.001

Delwiche, & Marlétaz, 2012; Laurin-Lemay, Brinkmann, & Philippe, 2012; Timme, Bachvaroff, & Delwiche, 2012). However, the morphological complexity and physiology of the ancestral charophycean alga from which land plants evolved are not easily deduced due to the disparities among extant charophycean algae. For example, the Zygnematales contain both unbranched filamentous and unicellular species, while some Coleochaetales are parenchymous disc-shaped organisms in which the egg is maternally retained (reviewed in Delwiche & Cooper, 2015). The general consensus is that the ancestral alga was multicellular, with its morphological complexity a mixture of ancestral characters present in extant charophycean lineages along with derived characters. Alternatively, some have argued that the ancestral alga may have been single-celled (Stebbins & Hill, 1980). Likewise, given the semi-terrestrial nature of many extant charophycean algal species (reviewed in Delwiche & Cooper, 2015; Holzinger, 2016; Lewis & McCourt, 2004), it may be surmised that the ancestral land plant inherited at least some of the required physiological machinery, such as desiccation tolerance, from the ancestral alga (reviewed in Delwiche & Cooper, 2015; Harholt, Moestrup, & Peter, 2016). Consistent with these scenarios, many land plant gene families have deep roots in the charophycean algae (e.g., Bowman et al., 2017; Floyd, Zalewski, & Bowman, 2006; Hori et al., 2014; Ju et al., 2015; Tanabe et al., 2005).

A defining land plant feature is an alternation of generations, whereby both haploid gametophyte and diploid sporophyte develop complex multicellular bodies (Hofmeister, 1862), in contrast to the haplontic life cycles of multicellular charophycean algae in which only the gametophyte is multicellular. The evolution of a multicellular diploid body allowed the production of large numbers of progeny because a single fertilization event eventually leads to the formation of many spores, likely an adaptation to terrestrial environments where aqueous fertilization is limited by water availability (Bower, 1890). How much of the genetic program that patterns the charophyte gametophyte body plan was inherited to pattern land plant body plans is an open question. The relative simplicity of the charophyte gametophyte is based on filamentous growth, which in most taxa is decentralized, i.e., there is not a "meristem" with a pool of stem cells. In contrast, land plant gametophytes exhibit localized growth from a meristem with an apical cell possessing multiple cutting faces (Campbell, 1918; Watson, 1964). Both focal localization of "meristematic" activity and differentiation of cells with multiple division planes suggest an increase in cell-cell communication pathways, reflected in the *de novo* origin of both new phytohormone signaling pathways and peptide ligand-receptor signaling pathways in the ancestral land plant (Bowman et al., 2017; Hori et al., 2014). A second question is whether, or how much of, gametophyte developmental programs were co-opted to pattern the multicellular sporophyte. To address these questions, we must first have an understanding of the relationships of extant land plant and charophycean algal lineages, as well as the nature of their fossil record.

2. Early land plants

The resolution of basal land plant lineages remains enigmatic, with nearly all possible relationships being postulated with phylogenies reconstructed using molecular characters (e.g., Nishiyama et al., 2004; Puttick et al., 2018; Qiu et al., 2006; Wickett et al., 2014), and the two most plausible phylogenetic topologies resulting in very different hypotheses for ancestral land plant characteristics (Fig. 1). In one scenario, bryophytes (liverworts, mosses, and hornworts) are a paraphyletic grade, most often postulated as a moss + liverwort clade that is either sister to all other land plants, or alternatively, sister to vascular plants (Puttick et al., 2018; Wickett et al., 2014). In either scenario, the ancestral land plant was likely bryophyte-like, with a life cycle that was gametophyte dominant and antithetic in origin, i.e., the gametophyte and sporophyte generations were fundamentally different from their beginning (Bower, 1890). In contrast, the second scenario, bryophyte monophyly (Nishiyama et al., 2004; Puttick et al., 2018), opens up the unsettling possibility that the ancestral land plant possessed isomorphic generations [consistent with the homologous theory of the origin of the alternation of generations (Scott, 1895)], with bryophytes and tracheophytes (lycophytes, ferns, and seed plants) diverging in opposite generational dominance. In the case of bryophyte monophyly, much less can be inferred about the ancestral land plant based on extant lineages, placing more reliance on the fossil record.

2.1 Cryptospores and cryptophytes

Few identified macrofossils represent the early colonization of land, but cryptospores recovered from terrestrial or near offshore deposits dating from as early as the mid-Cambrian are plentiful (Fig. 1; Edwards, Duckett, & Richardson, 1995; Edwards et al., 2014; Rubinstein, Gerrienne, de la Puenta, Astini, & Steemans, 2010; Strother & Taylor, 2018; Strother, Traverse, & Vecoli, 2015; Taylor, 1995; Wellman, Osterloff, & Mohiuddin, 2003). Cryptospores are a non-phylogenetic assemblage of spores with sporopollenin

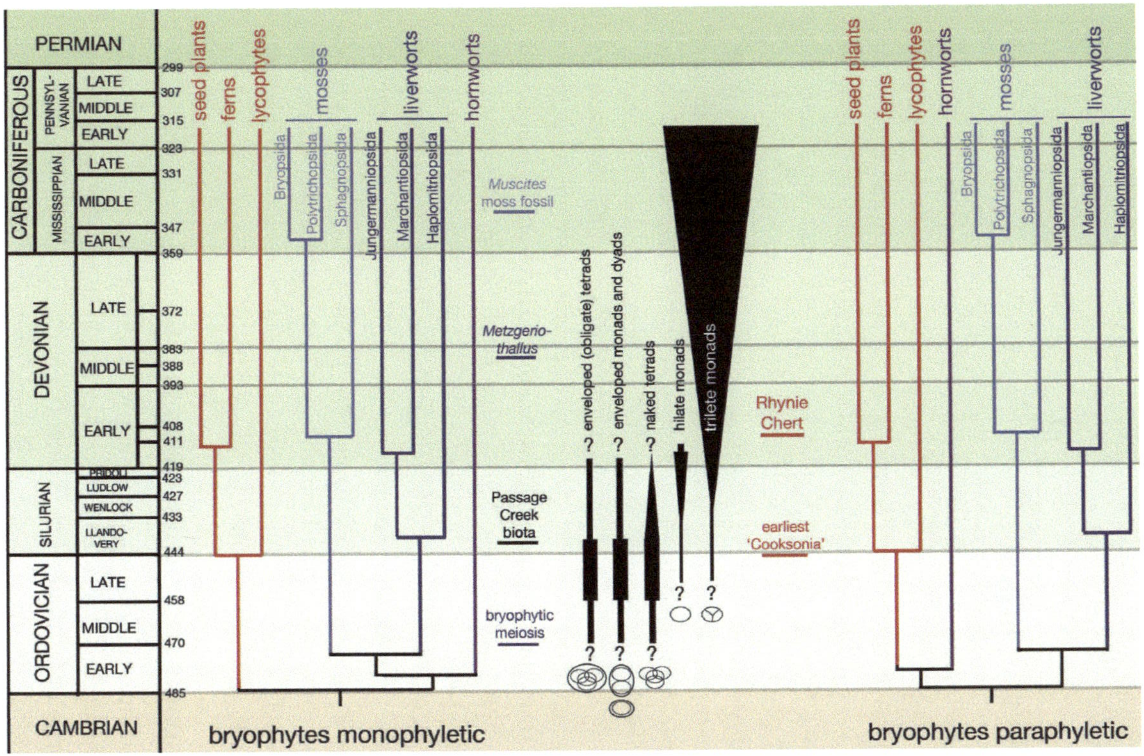

Fig. 1 Early land plant evolution. Contrasting phylogenies of land plants based on constraining molecular phylogenies with the known fossil record relative to the geological time scale (Morris et al., 2018; Puttick et al., 2018). The paraphyletic bryophyte tree places liverworts + mosses sister to other land plants (Puttick et al., 2018) instead of hornworts occupying the sister group position (Morris et al., 2018; Wickett et al., 2014). The monophyletic and paraphyletic bryophyte phylogenetic trees flank the fossil record of early land plants: cryptospores, black center, with spore types and time-courses shown (Edwards, Morris, Richardson, & Kenrick, 2014; Kenrick, Wellman, Schneider, & Edgecombe, 2012); thalloid fossils of unknown origin, black left; bryophyte fossils, blue left; protrachaeophytes, red right. The constrained phylogenies predict a Cambrian or Early Ordovician origin for land plants, with the three extant bryophyte lineages diverging in the Ordovician and the initial divergence of extant vascular plants in the Early Silurian. These times are significantly earlier than macrofossils assigned to the stem lineages of these groups, but overlap with cryptospore assemblages of unknown affinity.

containing walls distinct from trilete and monolete spores and pollen grains, but resembling land plant spores (Strother, 1991). Monolete spores have a single line at the position where they were separated from the spore mother cell whereas trilete spores have three radiating lines indicating derivation from a meiotic tetrad. Cryptospores occur as monads, dyads and tetrads and are presumably meiotic products, with the dyads and tetrads either being naked or obligate, the latter with the dyads or tetrads enclosed by an envelope (Edwards et al., 2014). Obligate tetrads might be advantageous in habitat colonization especially if species are dioecious (Gray, 1985). Among extant land plants obligate tetrads occur in liverwort taxa, e.g., *Sphaerocarpus terrestris, Haplomitrium gibbsiae, Riccia curtisii* (Austin, 1869; Douin, 1909; McAllister, 1916; Renzaglia et al., 2015; Strasburger, 1909), with ultrastructural and geochemical analyses indicating some Silurian cryptospore walls are chemically similar to those of extant liverwort spores (Steemans, Lepot, Marshall, Le Herisse, & Javaux, 2010). An association of cryptospores with bryophytic organisms is suggested by some Lower Devonian bryophyte-like plants that possess *in situ* cryptospores (Edwards, Wellman, & Axe, 1999); however, some putative protracheophyte fossils, such as Devonian Cooksonioid-species also produced dyad and tetrad spores, and thus they could be characteristic of multiple early land plant lineages (Edwards et al., 2014).

Cambrian and Lower Ordovician cryptospores exhibit morphological affinity with streptophyte algal lineages rather than embryophytes, with sporopollenin in the presumed wall of the zygote that serves as the dispersal or overwintering entity (Strother & Taylor, 2018). However, a dramatic shift in morphology occurs in the Middle Ordovician (c. 470 Ma), with cryptospore assemblages acquiring embryophyte characteristics indicating establishment of a bryophytic style of meiosis by this time (Strother & Taylor, 2018). The geographic distribution of these Middle Ordovician cryptospores might suggest a Gondwanan origin for land plants (Rubinstein et al., 2010; Wellman, 2010). The morphological transition to embryophytic cryptospores corresponds to a heterochronic shift whereby sporopollenin deposition is developmentally delayed until after meiosis, such that it is found in the walls of the meiospores rather than the zygote (Strother & Taylor, 2018). This is consistent with the sporopollenin-transfer hypothesis for the evolution of embryophytic meiosis and spores (Graham, 1993).

Subsequent to the Middle Ordovician, distinct cryptospore assemblages occur at different times in the fossil record (Edwards et al., 2014; Gray, 1985; Wellman & Gray, 2000). The relatively stable composition of cryptospores, of envelope-enclosed (obligate) or naked tetrads, dyads and monads, and their geographically widespread presence from the Middle Ordovician to the Early Silurian, has been taken as sign of evolutionary stasis for 40–50 Myr. In late Early Silurian deposits, the abundance of enveloped spores declines and naked varieties dominate the assemblages. From the Middle-Upper Silurian, increasingly complex spore morphologies are seen and while tetrads are still present, they are not obligate and by the Early Devonian cryptospore assemblages decline in abundance, being replaced by trilete spores marked with lines reflecting meiotic tetrads. The shift in cryptospore assemblages has been taken as evidence of diversification of early land plant lineages whose abundance may have been diminished when their habitats were abruptly dominated by the evolution of the tracheophyte lineage during the Silurian. Unfortunately, the nature of the cryptophytes, the plants producing cryptospores, is largely unknown. Consistent with the early cryptospore record, molecule-based phylogenies constrained with fossil calibrations suggest a rapid diversification of extant land plant lineages, with all bryophyte lineages and a tracheophyte stem group evolving as early as the Ordovician (Morris et al., 2018).

2.2 Macrofossil record

The earliest land plant macrofossils, such as *Cooksonia* (Late Silurian to Devonian) and those found in the Rhynie Chert (Early Devonian; c. 410 Ma) have traditionally been classified as prototracheophytes (Fig. 1; Kenrick & Crane, 1997). The exceptional preservation of the Rhynie Chert flora allows identification of gametophyte (presence of either antheridia or archegonia) and sporophyte (presence of spore capsules) generations of individual species, and has led to the idea that prototracheophytes possessed isomorphic generations, albeit with sporophytes much larger than gametophytes (Gerrienne & Gonez, 2011; Kenrick, 2018; Kenrick & Crane, 1997). Both generations were erect, dichotomously branching, leafless axes that bore either spore capsules or gametangia at their tips. In contrast, the early Devonian *Cooksonia paranensis* has been postulated as having a heteromorphic alternation of generations, with a small thalloid gametophyte accompanying an erect, dichotomously branching, leafless sporophyte (Gerrienne et al., 2006), but this interpretation is tempered by the lack of diagnostic characters, rendering the designation of sporophyte versus gametophyte generations ambiguous (Kenrick, 2018). Earlier *Cooksonia*-like fossils (c. 445–425 Ma) also suffer the same lack of clarity; however, it has been speculated that polysporangiate *Cooksonia* may have existed in the late Silurian (Salamon et al., 2018). Some unclassified late Silurian-Early Devonian fossils (e.g., *Sporogonites, Tortilicaulis*, Nematophytes, Spongiophytes) could represent land plants, but their affinities are uncertain (reviewed in Taylor, Taylor, & Krings, 2009). However, given the hypothesized earlier origin of land plants, further examination of terrestrial Silurian and Ordovician deposits harboring probable land plant fossils might be more promising (Salamon et al., 2018; Tomescu, Pratt, Rothwell, Strother, & Nadon, 2009).

Regardless of whether bryophytes are mono- or paraphyletic, each of the three lineages has a long isolated evolutionary history and each is united by a suite of distinct morphological and anatomical characters (reviewed in Campbell, 1918; Watson, 1964). Phylogenies indicate rapid and ancient divergences between the bryophyte lineages implying that Late Silurian macrofossils of each should be present, but they either have not been found or are not recognized as such. For example, the thalloid terrestrial wetland communities of the Early Silurian Passage Creek biota provide evidence of land plants of undetermined affinity and could represent bryophyte fossils (Tomescu & Rothwell, 2006). The earliest known fossil definitively assigned to the liverwort lineage, the middle Devonian *Metzgeriothallus sharonae* (Fig. 1), is a simple thalloid gametophyte with multistratose costa and unistratose wings and an apparently typical liverwort sporophyte (Hernick, Landing, & Bartowski, 2008) that might not appear out of place within extant liverwort diversity. Consistent with their proposed antiquity, fossils ascribed to all three major liverwort lineages are found by the Early Permian (reviewed in Oostendorp, 1987; Tomescu, Bomfleur, Bippus, & Savoretti, 2018). In contrast, the first definitive moss fossils occur 40 million years later in the Early Carboniferous [although *Sphagnum*-like fossils from the Ordovician were recently reported (Cardona-Correa et al., 2016)], and the first known hornwort fossils over 200 million years later in the Early Cretaceous (Drinnan & Chambers, 1986; Hübers & Kepr, 2012). Given the proposed rapid divergences of the bryophyte lineages, where are the missing fossils? One possibility is that the earliest fossils of the moss and hornwort lineages are morphologically disparate from extant species, and perhaps at least some presently unclassified Silurian-Ordovician fossils represent early members of these lineages prior to their morphological diversification, thus providing a possible avenue for reconciliation of the fossil record with molecular phylogenies.

3. The ancestral land plant

In terms of visualizing the ancestral land plant phenotype, where do the molecular phylogenies and fossil record lead us? Again, there are two very different scenarios depending upon whether bryophytes are monophyletic versus paraphyletic. First, consider paraphyletic bryophytes, as it is the simpler case with its results also applicable to the ancestral bryophyte in the case of monophyletic bryophytes. Some plesiomorphic (ancestral state) characters may be surmised based on the morphology of extant bryophytes. For instance, the life cycle is projected to be heteromorphic, with a dominant gametophyte generation producing archegonia and antheridia during sexual reproduction. All bryophyte lineages have gametophytic rhizoids as rooting structures, with unicellular rhizoids, present in extant liverworts and hornworts, likely ancestral (Ligrone, Duckett, & Renzaglia, 2012a). Likewise, the ancestral bryophyte also likely possessed mucilage cells. Rhizoids may have been inherited from an ancestral charophycean alga as some lineages of extant charophytes produce rhizoid-like structures. Mucilage cells, however, are likely a land plant adaptation to the terrestrial environment. The nature of the ancestral gametophyte body plan is enigmatic, with dorsi-ventral thalloid forms (Campbell, 1891; Cavers, 1910; Mishler & Churchill, 1985), erect leafy axes (Evans, 1939; Harris, 1938; Kashyap, 1919; Wettstein, 1908) and leafless axes (Ligrone et al., 2012a) all proposed. While the precise form is contested, all share a shoot apical meristem (SAM) containing a single apical cell with multiple division planes, a character predicted to be present in the ancestral land plant, and therefore representing a land plant innovation.

The monosporangiate sporophyte generation in extant bryophytes consists of a foot embedded in the maternal gametophyte through which nutrient transfer occurs, a seta that may elevate the sporophyte, and a sporangium, or capsule, in which a subset of cells undergo meiosis to produce haploid spores. These characters are plesiomorphic. In contrast, the presence and location of sporophyte shoot meristems vary between bryophyte lineages—liverworts largely lack a localized meristem, mosses possess both a SAM and an intercalary seta meristem, and hornworts grow from a basal intercalary meristem. The relationships between bryophyte shoot meristems and indeterminate tracheophyte SAMs have been debated (see Albert, 1999; Bowman, 2013; Kato & Akiyama, 2005; Ligrone, Duckett, & Renzaglia, 2012b; Mishler & Churchill, 1984; Tomescu, Wyatt, Hasebe, & Rothwell, 2014).

If bryophytes are monophyletic, much less can be predicted about the ancestral land plant. For instance, it becomes possible that the ancestral land plant possessed "isomorphic" generations, as found is some Rhynie Chert plants, and even opens the prospect of a polysporangiate ancestor. In this scenario, all bryophyte gametophyte body plans could be highly derived, as vascular plant gametophyte body plans are now perceived. However, at a minimum, the ancestral gametophyte body would have possessed a SAM with basal rhizoid-like rooting structures.

Given these two contrasting scenarios, refinement of both the fossil record and molecular-based phylogenies is required. For the latter, inclusion of additional extant basal bryophyte and tracheophyte lineages, including analyses focused on genomic characters may (or may not) provide resolution of extant land plant relationships. Considering the fossil record, the 40–50 million year interval from the earliest cryptospores to the first macrofossils is key—is the unique cryptospore assemblage during this time indicative of an equally unique body plan? If so, can they be assigned to any extant land plant stem group lineage? Intuitively, one might envisage the first land plants to be prostrate rather than erect, but this, as with most other attributes of the ancestral land plant, remains an open question.

4. Co-option and novelty in developmental innovation

A question in the evolution of developmental innovations is whether pre-existing genetic programs were co-opted for directing an organism's development or whether the evolution of novel genetic programs was required. Given the framework presented for land plant evolution from a charophycean alga ancestor, we can begin to address this question at two evolutionary transitions—first, what land plant developmental genetic programs were inherited from the alga ancestor, and second, were developmental genetic programs originally active in the land plant gametophyte co-opted to pattern the sporophyte?

4.1 Of rhizoids and root hairs

The differentiation of both bryophyte rhizoids and angiosperm root hairs is controlled by orthologous basic helix-loop-helix (bHLH) VIIIc subfamily genes, referred to as *ROOT HAIR DEFECTIVE SIX-LIKE* (*RSL*) Class I genes (Pires & Dolan, 2010). In the moss *Physcomitrella patens*, RSL genes (Pp*RSL*) direct the development of multicellular gametophytic rhizoids (Jang, Yi, Pires, Menand, & Dolan, 2011; Menand et al., 2007), while in the liverwort *Marchantia polymorpha* a single RSL Class I gene (Mp*RSL1*) promotes the formation of unicellular gametophytic rhizoids (Proust et al., 2016). Therefore, development of rooting structures with analogous functions in anchorage and nutrient acquisition is directed by orthologous genes in both gametophyte and sporophyte generations in bryophytes and angiosperms, respectively (Kenrick, 2018; Menand et al., 2007; Pires & Dolan, 2010). Does this reflect a deep homology (Shubin, Tabin, & Carroll, 2009) or a more complex relationship?

Mp*RSL1* has roles in the formation of slime papillae and gemmae, which also both originate from individual epidermal cells in the gametophyte, suggesting a broader role in the development of structures derived from epidermal outgrowths (Proust et al., 2016). Pp*RSL* genes also act in caulonemal cell formation, which in some cases form as outgrowths at the junction of filamentous protonemal cells (Menand et al., 2007). These observations suggest that bHLH-VIIIc members originally functioned more broadly, acting gametophytically in the land plant ancestor to control structures originating from individual epidermal cells (Proust et al., 2016). Mp*RSL1* is also expressed in the sporophyte (Bowman et al., 2017), although its function there is unexplored. Thus, the role of bHLH-VIIIc genes in vascular plant sporophyte root hairs could have been derived from a co-option from its gametophytic role in promoting epidermal outgrowths, e.g., rhizoids. Alternatively, since some Rhynie Chert sporophytes produced rhizoids, their presence may be merely a relic of the ancestral sporophyte if the ancestral land plant possessed isomorphic generations.

Some extant charophycean algae (e.g., Coleochaetales and Charales) have rhizoid-like anchoring cells, suggesting land plant rhizoids may have been inherited from an algal ancestor. The ancestral land plant apparently had two bHLH-VIIIc genes, with *Arabidopsis thaliana* genes from both clades involved in root hair differentiation (Pires & Dolan, 2010; Yi, Menand, Bell, & Dolan, 2010), suggesting a role in epidermal outgrowth formation and differentiation may have been an ancestral function for both land plant genes. At least some charophycean algae have a gene orthologous to the two land plant bHLH-VIIIc clades (Bowman et al., 2017). Phylogenetic analysis of potential charophycean algal bHLH-VIIIc orthologs revealed representatives in the Coleochaetales, but not the Zygnematales or Charales (Fig. 2A), raising the possibility that epidermal rhizoid-like outgrowths in *Coleochaete* (Fig. 2B) could be controlled by an orthologous genetic machinery. The inferred loss of bHLH-VIIc in Zygnematales might be due to its reduced morphology and lifecycle, but the lack of orthologs in Charales suggests that its rhizoid-like structures may have a different genetic basis.

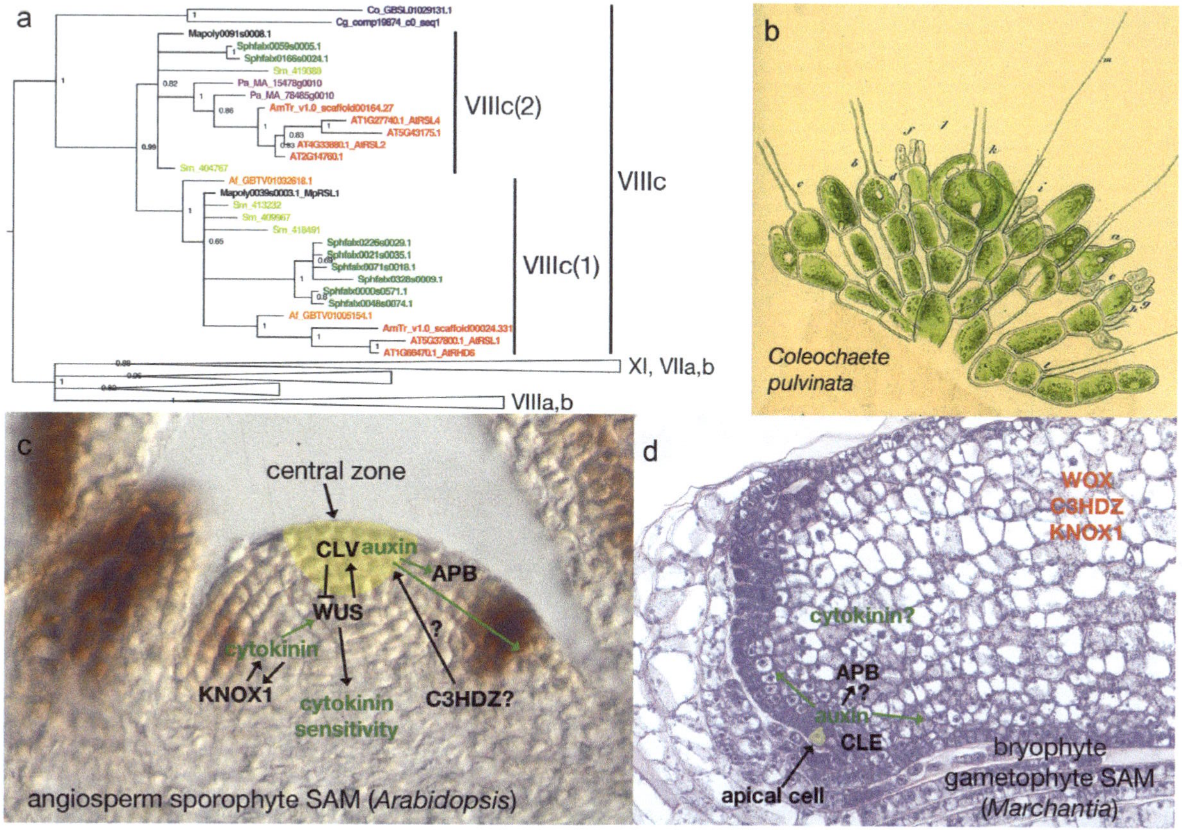

Fig. 2 Rhizoids and SAMs. (A) Rooted phylogram of bHLH-VIIIc genes in Streptophytes was generated using Bayesian analysis (5,000,000 runs, 50,000 burnin, 75 amino acid characters) of 104 aligned bHLH Streptophyte sequences. Values are posterior probabilities; clade names based on previous studies (Pires & Dolan, 2010). AmTr, *Amborella trichopoda (red)*; Pa, *Picea abies (purple)*; Af, *Azolla filiculoides (orange)*; Sm, *Selaginella moellendorffi (light green)*; Sphfalx, *Sphagnum fallax (dark green)*; Mapoly, *Marchantia polymorpha (black)*; Cg, *Chaetosphaeridium globosum* (blue); Co, *Coleochaete orbicularis* (blue). (B). *Coleochaete pulvinata* with rhizoid-like outgrowths (Pringsheim, 1860). (C and D) Angiosperm sporophyte SAM and bryophyte gametophyte SAM; central zone and apical cell (yellow); genetic pathways (black), hormones (green), and their known interactions (arrows) are shown, with genetic pathways not active in red.

4.2 The shoot apical meristem

SAMs differ in structure in different taxa (Fig. 2C and D). Tracheophytes harbor large sporophytic SAMs exhibiting a zonation whereby a central zone, consisting of either a single cell or, in the case of seed plants, several cells, constitutes a pool of pluripotent stem cells that are supplied, via cell division and displacement, to an outer peripheral zone where tissue differentiation occurs. Bryophyte gametophyte SAMs are characterized by a single apical cell acting in a manner analogous to the central zone of a vascular plant sporophyte meristem. Apical cell derivatives divide in stereotypical patterns producing a merophyte, with all cells of the merophyte differentiating to form a defined portion of the gametyphyte body.

As with most land plant developmental biology, the SAM of angiosperms is the most well characterized, with the roles of several gene classes acting in the formation and maintenance of the sporophyte SAM firmly established (Fig. 2C; Barton, 2010). Class I KNOX (KNOX1) genes act in SAM formation and maintenance, in part by increasing cytokinin production (reviewed in Hay & Tsiantis, 2010). Central zone size is regulated by a negative feedback loop between the WUSCHEL transcription factor, which also appears to increase cytokinin production and sensitivity, and a CLAVATA (CLV) signaling system consisting of a secreted ligand (CLV3) of the CLE family and its receptor(s) (CLV1, CLV2) (reviewed in Somssich, Je, Simon, & Jackson, 2016). *AINTEGUMENTA/PLETHORA/BABYBOOM* (APB) activity has been linked to SAM function, where these genes act in both specifying stem cell fate and differentiation of their derivatives (reviewed in Scheres & Krizek, 2018). Class III HD-Zip genes (C3HDZ) also play a role in the establishment and maintenance of the sporophyte SAM, although pleiotropic effects on shoot development have obscured their precise roles (Emery et al., 2003; Prigge et al., 2005). Both C3HDZ and APB gene functions have been intimately associated with auxin (Ilegems et al., 2010; Scheres & Krizek, 2018), which is produced in the SAM and whose transport is crucial for organogenesis at the SAM periphery (reviewed in Truskina & Vernoux, 2018; Wang & Jiao, 2018). While all functional genetic experiments have focused on angiosperms, expression patterns of KNOX1 genes in ferns and lycophytes support a role in the sporophyte SAM in these lineages (Harrison et al., 2005; Sano et al., 2005). In contrast, C3HDZ expression in fern and lycophyte sporophyte SAMs is variable (Floyd & Bowman, 2006; Prigge & Clark, 2006; Vasco et al., 2016).

How much can be extrapolated to gametophyte SAMs of more basal land plants? At present, data are limited primarily to two species and are largely non-overlapping in nature. KNOX1 genes are not expressed in the fern gametophyte SAM (Sano et al., 2005) and functional analyses in *P. patens* confirmed KNOX1 activity has no role in the moss gametophyte SAM, although it functions in proliferation in sporophyte meristems (Fig. 2D; Sakakibara, Nishiyama, Deguchi, & Hasebe, 2008). This result is not surprising in light of an ancestral Viridiplantae role for KNOX genes initiating the zygotic gene developmental program post-fertilization (Bowman, Sakakibara, Furumizu, & Dierschke, 2016; Lee, Lin, Joo, & Goodenough, 2008). As suggested by observations on lycophytes and ferns, no role was found for C3HDZ genes in the gametophyte SAM of *P. patens* (Yip, Floyd, Sakakibara, & Bowman, 2016). The CLE signaling pathway is a land plant innovation (Bowman et al., 2017) and acts to regulate cell proliferation and cell division planes in gametophyte SAMs of *P. patens*, roles shared with CLE activity in sporophyte SAMs of *A. thaliana* (Whitewoods et al., 2018). The *P. patens* CLE peptides act through orthologous CLV receptors, indicating that the signaling module is conserved (Whitewoods et al., 2018). In contrast, WOX gene activity was not found to play any role in the *P. patens* gametophyte SAM (Sakakibara et al., 2014), such that CLE signaling influencing gametophyte meristem architecture must be acting via different downstream targets as compared to sporophyte meristems of angiosperms. As in sporophyte SAMs, the liverwort *M. polymorpha* gametophyte SAM acts as a source of auxin that is required for patterning most aspects of the gametophyte body (Eklund et al., 2015; Flores-Sandoval, Eklund, & Bowman, 2015; Kato et al., 2015). APB genes were found to be indispensable for the formation of apical cells of the *P. patens* gametophyte SAM (Aoyama et al., 2012) and could provide a genetic component linking auxin to pluripotency in SAMs.

In the simplest co-option scenario, a gametophyte SAM genetic program would come under control of KNOX regulation such that it would be expressed in the sporophyte. However, the SAM as an auxin source and APB activity are the only presently known shared features of gametophyte and sporophyte SAMs, with the roles of CLV signaling and cytokinins in bryophytes as yet untested. The involvement of auxin is consistent with the hypothesis that the origin of the auxin transcriptional response in the ancestral land plant was instrumental in the evolution of focal regions of cell division, i.e., meristems (Flores-Sandoval et al., 2018). However, the paucity of known overlaps in control mechanisms suggests that either we do not have a fundamental understanding of SAM function, perhaps due to angiosperm sporophyte SAMs being highly derived and bryophyte SAMs largely unexplored, or alternatively, extant gametophyte and sporophyte SAMs have followed different evolutionary trajectories.

5. Conclusions

Two characters considered here, rhizoids and the SAM, provide contrasting examples. The genetic program for rhizoid development (or epidermal projections) appears to be controlled by orthologous genes in both generations and could have been co-opted from the gametophytic generation to the sporophytic generation in the ancestral land plant, provided the orthologous genes in charophycean algae perform a similar function. In contrast, the genetic programs for sporophyte and gametophyte SAMs as presently known differ significantly. This provides support for the antithetic hypothesis, but it may also reflect the derived nature of the angiosperm sporophyte SAM and the lack of knowledge of gametophyte SAM function. In summary, a better understanding of early land plants and/or the relationships among extant land plants is required for ascertaining most ancestral or derived characters, and this in turn is required for a broader consideration of antithetic versus homologous interpretations of the alternation of generations in land plants.

Acknowledgments

We thank Mihai Tomescu, Paul Kenrick, David Smyth, Tom Dierschke, and Tom Fisher for constructive comments and criticisms of the ideas presented here and we take full responsibility for any errors and we apologize to authors of literature we failed to cite due to space constraints. Bowman lab supported by the Australian Research Council (DP160100892, DP170100049 to J.L.B.).

References

Albert, V. A. (1999). Shoot apical meristems and floral patterning: An evolutionary perspective. *Trends in Plant Science*, *4*, 84–86.

Aoyama, T., Hiwatashi, Y., Shigyo, M., Kofuji, R., Kubo, M., Ito, M., et al. (2012). AP2-type transcription factors determine stem cell identity in the moss *Physcomitrella patens*. *Development*, *139*, 3120–3129.

Austin, C. F. (1869). Characters of some new Hepaticae (mostly North American), together with notes on a few imperfectly described species. *Proceedings of the Academy of Natural Sciences of Philadelphia*, *1869*, 219–234.

Barton, M. K. (2010). Twenty years on: The inner workings of the shoot apical meristem, a developmental dynamo. *Developmental Biology*, *341*, 95–113.

Bower, F. O. (1890). On antithetic as distinct from homologous alternation of generations in plants. *Annals of Botany*, *4*, 347–370.

Bowman, J. L. (2013). Walkabout on the long branches of plant evolution. *Current Opinion in Plant Biology*, *16*, 70–77.

Bowman, J. L., Kohchi, T., Yamato, K. T., Jenkins, J., Shu, S., Ishizaki, K., et al. (2017). Insights into land plant evolution garnered from the *Marchantia polymorpha* genome. *Cell*, *171*, 287–304.

Bowman, J. L., Sakakibara, K., Furumizu, C., & Dierschke, T. (2016). Evolution in the cycles of life. *Annual Review of Genetics*, *50*, 133–154.

Campbell, D. H. (1891). On the relationships of the Archegoniata. *Botanical Gazette, 16*, 323–333.

Campbell, D. H. (1918). *Mosses and ferns*. New York: The Macmillan Company, 708.

Cardona-Correa, C., Piotrowski, M. J., Knack, J. J., Kodner, R. E., Geary, D. H., & Graham, L. E. (2016). Peat moss-like vegetative remains from ordovician carbonates. *International Journal of Plant Sciences, 177*, 523–538.

Cavers, F. (1910). The inter-relationships of the Bryophyta. *New Phytologist, 9*, 341–353. 81–112, 157–186, 193–234, 269–304.

Delwiche, C. F., & Cooper, E. D. (2015). The evolutionary origin of a terrestrial Flora. *Current Biology, 25*, R899–R910.

Douin, C. (1909). Nouvelles observations sur *Sphaerocarpus*. *Revue Bryologique et Lichénologique, 36*, 37–41.

Drinnan, A. N., & Chambers, T. C. (1986). Flora of the lower cretaceous koonwarra fossil bed (Korumburra group), South Gippsland, Victoria. *Memoirs of the Association of Australasian Palaeontologists, 3*, 1–77.

Edwards, D., Duckett, J. G., & Richardson, J. B. (1995). Hepatic characters in the earliest land plants. *Nature, 374*, 635–636.

Edwards, D., Morris, J. L., Richardson, J. B., & Kenrick, P. (2014). Cryptospores and cryptophytes reveal hidden diversity in early land floras. *New Phytologist, 202*, 50–78.

Edwards, D., Wellman, C. H., & Axe, L. (1999). Tetrads in sporangia and spore masses from the Upper Silurian and Lower Devonian of the Welsh Borderland. *Botanical Journal of the Linnean Society, 130*, 111–156.

Eklund, D. M., Ishizaki, K., Flores-Sandoval, E., Kikuchi, S., Takebayashi, Y., Tsukamoto, S., et al. (2015). Auxin produced by the indole-3-pyruvate pathway regulates development and gemmae dormancy in the liverwort *Marchantia polymorpha*. *The Plant Cell, 27*, 1650–1669.

Emery, J. F., Floyd, S. K., Alvarez, J., Eshed, Y., Hawker, N. P., Izhaki, A., et al. (2003). Radial patterning of *Arabidopsis* shoots by class III HD-Zip and KANADI genes. *Current Biology, 13*, 1768–1774.

Evans, A. W. (1939). The classification of the Hepaticae. *Botanical Review, 5*, 49–96.

Finet, C., Timme, R. E., Delwiche, C. F., & Marlétaz, F. (2012). Multigene phylogeny of the green lineage reveals the origin and diversification of land plants. *Current Biology, 22*, 1456–1457.

Flores-Sandoval, E., Eklund, D. M., & Bowman, J. L. (2015). A simple auxin transcriptional response system regulates multiple morphogenetic processes in the liverwort *Marchantia polymorpha*. *PLoS Genetics, 11*, e1005207.

Flores-Sandoval, E., Eklund, D. M., Hong, S.-F., Alvarez, J. P., Fisher, T. J., Lampugnani, E. R., et al. (2018). Class C ARFs evolved before the origin of land plants and antagonize differentiation and developmental transitions in *Marchantia polymorpha*. *New Phytologist, 218*, 1612–1630.

Floyd, S. K., & Bowman, J. L. (2006). Distinct developmental mechanisms reflect the independent origins of leaves in vascular plants. *Current Biology, 16*, 1911–1917.

Floyd, S. K., Zalewski, C. S., & Bowman, J. L. (2006). Evolution of class III Homeodomain leucine zipper genes in streptophytes. *Genetics, 173*, 373–388.

Gerrienne, P., Dilcher, D. L., Bergamaschi, S., Milagres, I., Pereira, E., & Rodrigues, M. A. C. (2006). An exceptional specimen of the early land plant *Cooksonia paranensis*, and a hypothesis on the life cycle of the earliest eutracheophytes. *Review of Palaeobotany and Palynology, 142*, 123–130.

Gerrienne, P., & Gonez, P. (2011). Early evolution of life cycles in embryophytes: A focus on the fossil evidence of gametophyte/sporophyte size and morphological complexity. *Journal of Systematics and Evolution, 49*, 1–16.

Graham, L. E. (1993). *Origin of land plants*. New York: John Wiley and Sons Inc.

Gray, J. (1985). The microfossil record of early land plants—Advances in understanding of early Terrestrialization, 1970–1984. *Philosophical Transactions of the Royal Society of London. Series B, Biological Sciences, 309*, 167–195.

Harholt, J., Moestrup, Ø., & Peter, U. (2016). Why plants were terrestrial from the beginning. *Trends in Plant Science, 21*, 96–101.

Harris, T. M. (1938). *The British rhaetic flora*. London: Trustees of the British Museum, 84.

Harrison, C. J., Corley, S. B., Moylan, E. C., Alexander, D. L., Scotland, R. W., & Langdale, J. A. (2005). Independent recruitment of a conserved developmental mechanism during leaf evolution. *Nature, 434*, 509–514.

Hay, A., & Tsiantis, M. (2010). KNOX genes: Versatile regulators of plant development and diversity. *Development, 137*, 3153–3165.

Hernick, L. V., Landing, E., & Bartowski, K. E. (2008). Earth's oldest liverworts—*Metzgeriothallus sharonae* sp nov from the middle Devonian (Givetian) of eastern New York, USA. *Review of Palaeobotany and Palynology, 148*, 154–162.

Hofmeister, W. F. B. (1862). *On the germination, development, and fructification of the higher Cryptogamia, and on the fructification of the Coniferae*. London: Ray Society, 506.

Holzinger, A. (2016). Abiotic stress tolerance of charophyte green algae: New challenges for omics techniques. *Frontiers in Plant Science, 7*, 678.

Hori, K., Maruyama, F., Fujisawa, T., Togashi, T., Yamamoto, N., Seo, M., et al. (2014). *Klebsormidium flaccidum* genome reveals primary factors for plant terrestrial adaptation. *Nature Communications, 5*, 4978.

Hübers, M., & Kepr, H. (2012). Oldest known mosses discovered in Mississippian (late Visean) strata of Germany. *Geology, 40*, 755–758.

Ilegems, M., Douet, V., Meylan-Bettex, M., Uyttewaal, M., Brand, L., Bowman, J. L., et al. (2010). Interplay of auxin, KANADI and class III HD-ZIP transcription factors in vascular tissue formation. *Development, 137*, 975–984.

Jang, G., Yi, K. K., Pires, N. D., Menand, B., & Dolan, L. (2011). RSL genes are sufficient for rhizoid system development in early diverging land plants. *Development, 138*, 2273–2281.

Ju, C., Van de Poel, B., Cooper, E. D., Theirer, J. H., Gibbons, T. R., Delwiche, C. F., et al. (2015). Conservation of ethylene as a plant hormone over 450 million years of evolution. *Nature Plants, 1*, 14004.

Kashyap, S. R. (1919). The relationships of liverworts especially in the light of some recently discovered Himalayan forms. *Proceedings of the Asiatic Society of Bengal, 15*, 152–166.

Kato, M., & Akiyama, H. (2005). Interpolation hypothesis for the origin of the vegetative sporophyte of land plants. *Taxon, 52*, 443–450.

Kato, H., Ishizaki, K., Kouno, M., Shirakawa, M., Bowman, J. L., Nishihama, R., et al. (2015). Auxin-mediated transcriptional system with a minimal set of components is critical for morphogenesis through the life cycle in *Marchantia polymorpha. PLoS Genetics, 11*, e1005084.

Kenrick, P. (2018). Changing expressions: A hypothesis for the origin of the vascular plant life cycle. *Philosophical Transactions of the Royal Society of London. Series B, Biological Sciences, 373*, 20170149.

Kenrick, P., & Crane, P. R. (1997). *The origin and early diversification of land plants: A cladistic study*. Washington: Smithsonian Institution Press, 441.

Kenrick, P., Wellman, C. H., Schneider, H., & Edgecombe, G. D. (2012). A timeline for terrestrialization: Consequences for the carbon cycle in the Palaeozoic. *Philosophical Transactions of the Royal Society of London. Series B, Biological Sciences, 367*, 519–536.

Laurin-Lemay, S., Brinkmann, H., & Philippe, H. (2012). Origin of land plants revisited in the light of sequence contamination and missing data. *Current Biology, 22*, R593–R594.

Lee, J.-H., Lin, H., Joo, S., & Goodenough, U. (2008). Early sexual origins of Homeoprotein Heterodimerization and evolution of the plant KNOX/BELL family. *Cell, 133*, 829–840.

Lewis, L. A., & McCourt, R. M. (2004). Green algae and the origin of land plants. *American Journal of Botany, 91*, 1535–1556.
Ligrone, R., Duckett, J. G., & Renzaglia, K. S. (2012a). Major transitions in the evolution of early land plants: A bryological perspective. *Annals of Botany, 109*, 851–871.
Ligrone, R., Duckett, J. G., & Renzaglia, K. S. (2012b). The origin of the sporophyte shoot in land plants: A bryological perspective. *Annals of Botany, 110*, 935–941.
McAllister, F. (1916). The morphology of *Thallocarpus curtisii*. *Bulletin of the Torrey Botanical Club, 43*, 117–126.
Menand, B., Yi, K. K., Jouannic, S., Hoffmann, L., Ryan, E., Linstead, P., et al. (2007). An ancient mechanism controls the development of cells with a rooting function in land plants. *Science, 316*, 1477–1480.
Mishler, B. D., & Churchill, S. P. (1984). A cladistic approach to the phylogeny of the bryophytes. *Brittonia, 36*, 406–424.
Mishler, B. D., & Churchill, S. P. (1985). Transition to a land flora: Phylogenetic relationships of the green algae and bryophytes. *Cladistics, 1*, 305–328.
Morris, J. L., Puttick, M. N., Clark, J. W., Edwards, D., Kenrick, P., Pressel, S., et al. (2018). The timescale of early land plant evolution. *Proceedings of the National Academy of Sciences of the United States of America, 115*, E2274–E2283.
Nishiyama, T., Wolf, P. G., Kugita, M., Sinclair, R. B., Sugita, M., Sugiura, C., et al. (2004). Chloroplast phylogeny indicates that bryophytes are monophyletic. *Molecular Biology and Evolution, 21*, 1813–1819.
Oostendorp, C. (1987). The bryophytes of paleozoic and the mesozoic. In J.-P. Frahm & S. R. Gradstein (Eds.), *Bryophytorum Bibliotheca: Vol. 34* (pp. 1–110). J. Cramer; Berlin/Stuttgart.
Pires, N., & Dolan, L. (2010). Origin and diversification of basic-helix-loop-helix proteins in plants. *Molecular Biology and Evolution, 27*, 862–874.
Prigge, M. J., & Clark, S. E. (2006). Evolution of the class III HD-Zip gene family in land plants. *Evolution & Development, 8*, 350–361.
Prigge, M. J., Otsuga, D., Alonso, J. M., Ecker, J. R., Drews, G. N., & Clark, S. E. (2005). Class III homeodomain-leucine zipper gene family members have overlapping, antagonistic, and distinct roles in *Arabidopsis* development. *Plant Cell, 17*, 61–76.
Pringsheim, N. (1860). Beiträge zur Morphologie und Systematik der Algen III Die Coleochaeteen. *Jahrbucher fur Wissenschaftliche Botanik, 2*, 1–38.
Proust, H., Honkanen, S., Jones, V. A. S., Morieri, G., Prescott, H., Kelly, S., et al. (2016). RSL class I genes controlled the development of epidermal structures in the common ancestor of land plants. *Current Biology, 26*, 93–99.
Puttick, M. N., Morris, J. L., Williams, T. A., Cox, C. J., Edwards, D., Kenrick, P., et al. (2018). The interrelationships of land plants and the nature of the ancestral embryophyte. *Current Biology, 28*, 733–745.
Qiu, Y. L., Li, L. B., Wang, B., Chen, Z. D., Knoop, V., Groth-Malonek, M., et al. (2006). The deepest divergences in land plants inferred from phylogenomic evidence. *Proceedings of the National Academy of Sciences of the United States of America, 103*, 15511–15516.
Renzaglia, K. S., Crandall-Stotler, B., Pressel, S., Duckett, J. G., Schuette, S., & Strother, P. K. (2015). Permanent spore dyads are not 'a thing of the past'—On their occurrence in the liverwort *Haplomitrium* (Haplomitriopsida). *Botanical Journal of the Linnean Society, 179*, 658–669.
Rubinstein, C. V., Gerrienne, P., de la Puenta, G. S., Astini, R. A., & Steemans, P. (2010). Early middle Ordovician evidence for land plants in Argentina (eastern Gondwana). *New Phytologist, 188*, 365–369.
Sakakibara, K., Nishiyama, T., Deguchi, H., & Hasebe, M. (2008). Class 1 KNOX genes are not involved in shoot development in the moss *Physcomitrella patens* but do function in sporophyte development. *Evolution & Development, 10*, 555–566.

Sakakibara, K., Reisewitz, P., Aoyama, T., Friedrich, T., Ando, S., Sato, Y., et al. (2014). WOX13-like genes are required for reprogramming of leaf and protoplast cells into stem cells in the moss *Physcomitrella patens*. *Development*, *141*, 1660–1670.

Salamon, M. A., Gerrienne, P., Steemans, P., Gorzelak, P., Filipiak, P., Hérisse, A. L., et al. (2018). Putative late Ordovician land plants. *New Phytologist*, *218*, 1305–1309.

Sano, R., Juarez, C. M., Hass, B., Sakakibara, K., Ito, M., Banks, J. A., et al. (2005). KNOX homeobox genes potentially have similar function in both diploid unicellular and multicellular meristems, but not in haploid meristems. *Evolution & Development*, 7, 69–78.

Scheres, B., & Krizek, B. A. (2018). Coordination of growth in root and shoot apices by AIL/PLT transcription factors. *Current Opinion in Plant Biology*, *41*, 95–101.

Scott, D. H. (1895). Nathanael Pringsheim. *Nature*, *51*, 399–402.

Shubin, N., Tabin, C., & Carroll, S. (2009). Deep homology and the origins of evolutionary novelty. *Nature*, *457*, 818–823.

Somssich, M., Je, B., Simon, R., & Jackson, D. (2016). Clavata-Wuschel signaling in the shoot meristem. *Development*, *143*, 3238–3248.

Stebbins, G. L., & Hill, G. J. C. (1980). Did multicellular plants invade the land? *The American Naturalist*, *115*, 342–353.

Steemans, P., Lepot, K., Marshall, C. P., Le Herisse, A., & Javaux, E. J. (2010). FTIR characterisation of the chemical composition of Silurian miospores (cryptospores and trilete spores) from Gotland, Sweden. *Review of Palaeobotany and Palynology*, *162*, 577–590.

Strasburger, E. (1909). *Histologische Beiträge Heft VII. Zeitpunkt der Bestimmung des Geschlechts, Apogamie, Pathenogenesis und Reduktionsteilung*. Jena: Gustav Fischer.

Strother, P. K. (1991). A classification schema for the cryptospores. *Palynology*, *15*, 219–236.

Strother, P. K., & Taylor, W. A. (2018). The evolutionary origin of the plant spore in relation to the antithetic origin of the plant sporophyte. In M. Krings, C. J. Harper, N. R. Cúneo, & G. W. Rothwell (Eds.), *Transformative paleobotany: Papers to commemorate the life and legacy of Thomas N. Taylor* (pp. 3–20). Amsterdam, Netherlands: Elsevier.

Strother, P. K., Traverse, A., & Vecoli, M. (2015). Cryptospores from the Hanadir Shale Member of the Qasim Formation, Ordovician (Darriwilian) of Saudi Arabia—Taxonomy and systematics. *Review of Palaeobotany and Palynology*, *212*, 97–110.

Tanabe, Y., Hasebe, M., Sekimoto, H., Nishiyama, T., Kitani, M., Henschel, K., et al. (2005). Characterization of MADS-box genes in charophycean green algae and its implication for the evolution of MADS-box genes. *Proceedings of the National Academy of Sciences of the United States of America*, *102*, 2436–2441.

Taylor, W. A. (1995). Spores in earliest land plants. *Nature*, *373*, 391–392.

Taylor, T., Taylor, E., & Krings, M. (2009). *Paleobotany: The biology and evolution of fossil plants*. New York: Academic Press, Elsevier.

Timme, R. E., Bachvaroff, T. R., & Delwiche, C. F. (2012). Broad Phylogenomic sampling and the sister lineage of land plants. *PLoS One*, 7, e29696.

Tomescu, A. M. F., Bomfleur, B., Bippus, A. C., & Savoretti, M. A. (2018). Why are bryophytes so rare in the fossil record? A spotlight on taphonomy and fossil preservation. In M. Krings, C. J. Harper, N. R. Cúneo, & G. W. Rothwell (Eds.), *Transformative paleobotany: Papers to commemorate the life and legacy of Thomas N. Taylor* (pp. 375–416). Amsterdam, Netherlands: Elsevier.

Tomescu, A. M. F., Pratt, L. M., Rothwell, G. W., Strother, P. K., & Nadon, G. C. (2009). Carbon isotopes support the presence of extensive land floras pre-dating the origin of vascular plants. *Palaeogeography, Palaeoclimatology, Palaeoecology*, *283*, 46–59.

Tomescu, A. M. F., & Rothwell, G. W. (2006). Wetlands before tracheophytes—Thalloid terrestrial communities of the early Silurian Passage Creek biota (Virginia). *Geological Society of America Special Papers*, *399*, 41–56.

Tomescu, A. M. F., Wyatt, S. E., Hasebe, M., & Rothwell, G. W. (2014). Early evolution of the vascular plant body plan—The missing mechanisms. *Current Opinion in Plant Biology*, *17*, 126–136.

Truskina, J., & Vernoux, T. (2018). The growth of a stable stationary structure: Coordinating cell behavior and patterning at the shoot apical meristem. *Current Opinion in Plant Biology, 41*, 83–88.

Vasco, A., Smalls, T. L., Graham, S. W., Cooper, E. D., Wong, G. K. S., Stevenson, D. W., et al. (2016). Challenging the paradigms of leaf evolution: Class III HD-Zips in ferns and lycophytes. *New Phytologist, 212*, 745–758.

Wang, Y., & Jiao, Y. L. (2018). Auxin and above-ground meristems. *Journal of Experimental Botany, 69*, 147–154.

Watson, E. V. (1964). *The structure and life of bryophytes*. London: Hutchinson & Co, 211.

Wellman, C. H. (2010). The invasion of the land by plants: When and where? *New Phytologist, 188*, 306–309.

Wellman, C. H., & Gray, J. (2000). The microfossil record of early land plants. *Philosophical Transactions of the Royal Society of London. Series B, Biological Sciences, 355*, 717–731.

Wellman, C. H., Osterloff, P. L., & Mohiuddin, U. (2003). Hepatic characters in the earliest land plants. *Nature, 425*, 282–285.

Wettstein, R. V. (1908). *Handbuch der systematischen Botanik*. Leipzig: F. Deuticke, 577.

Whitewoods, C. D., Cammarata, J., Venza, Z. N., Sang, S., Crook, A. D., Aoyama, T., et al. (2018). CLAVATA was a genetic novelty for the morphological innovation of 3D growth in land plants. *Current Biology, 28*, 2365–2376.

Wickett, N. J., Mirarab, S., Nguyen, N., Warnow, T., Carpenter, E., Matasci, N., et al. (2014). Phylotranscriptomic analysis of the origin and early diversification of land plants. *Proceedings of the National Academy of Sciences of the United States of America, 111*, E4859–E4868.

Yi, K., Menand, B., Bell, E., & Dolan, L. (2010). A basic helix-loop-helix transcription factor controls cell growth and size in root hairs. *Nature Genetics, 42*, 264–267.

Yip, H. K., Floyd, S. K., Sakakibara, K., & Bowman, J. L. (2016). Class III HD-Zip activity coordinates leaf development in *Physcomitrella patens*. *Developmental Biology, 419*, 184–197.

CHAPTER THREE

The role of plant root systems in evolutionary adaptation

Vinay Shekhar[†], Dorothee Stöckle[†], Martha Thellmann[†], Joop E.M. Vermeer*

Department of Plant and Microbial Biology & Zurich-Basel Plant Science Center, University of Zurich, Zurich, Switzerland

*Corresponding author: e-mail address: joop.vermeer@botinst.uzh.ch

Contents

Abstract

Roots provide a means to plants for gathering belowground resources. They are plastic and can adapt to ever-changing environmental cues. The plasticity of the roots comes from their ability to branch out by developing lateral and/or adventitious roots. In this chapter, we make an attempt to document the diversity in plant root systems and understand their role in evolutionary adaptation. After a brief introduction to different root systems, such as homorhizic and allorhizic ones, the relationship of plant roots with their surroundings, i.e., the rhizosphere and its effect on adaptation, will be discussed.

[†] Equal contribution.

Current Topics in Developmental Biology, Volume 131
ISSN 0070-2153
https://doi.org/10.1016/bs.ctdb.2018.11.011

Despite the difficulty to conclusively construct a timeline of evolution of plant root systems, documented facts from previous publications are examined and an effort has been made to delve into how rooting structures in plants adapted to prevailing conditions by bringing about endogenous changes vis-à-vis evolutionary development and exogenous changes to their surroundings.

1. Introduction

Plant evolution is usually characterized as descriptive biology with the aim to explain how the present species diversity arose over a geological time frame. Studying morphological traits is a prime feature of paleobiology and still leads to fundamental insights (Hetherington, Berry, & Dolan, 2016). However, paleontologists are highly dependent on the fossil record which is often very scarce, particularly for root specimens that preserve very poorly (Raven & Edwards, 2001). With the advent of next generation sequencing techniques and the massive drop in sequencing costs, it becomes increasingly feasible to answer questions that bear relevance to evolutionary biology. For example, due to an increase in the number of fully sequenced plant genomes, it is possible to refine the phylogenetic tree, as recently been exemplified by Puttick et al. (2018). Their analysis of bryophyte transcriptomes suggested monophyly for the bryophytes (hornworts, liverworts, and mosses, Fig. 2), a hitherto unanticipated finding (Rensing, 2018). Comparative analysis on a genome scale will also enable researchers to discover whole-genome duplications, events known to be major drivers of plant speciation (Li et al., 2015). Moreover, population-scale approaches such as genome-wide association studies (GWAS) offer the possibility to connect quantitative traits (plant height, yield or resistance) with the underlying molecular framework. Engineering plant root systems provides a means to react to rapidly altered environmental conditions and changes in soil characteristics of different habitats (Li et al., 2017; Uga et al., 2013). In this review, evolution of plant root systems is in the spotlight as an example of the above-mentioned advances in technology. In the first part, the plant root and root systems will be introduced to set the stage for the following complex details needed to appreciate the "news" from the fossil record. Subsequently, we look at root systems across the phylogenetic tree and provide the latest insights into their root system architecture (RSA) and adaptive capacity.

2. Roots, root systems, and root biotic colonization

2.1 What is a root?

Roots are one of the earliest adaptations that enabled plants to colonize land (Kenrick & Crane, 1997). They perform essential functions necessary for the survival and development of plants, such as anchoring them to a substrate, providing mechanical support, storing photoassimilates, and absorbing water and nutrients (Bellini, Pacurar, & Perrone, 2014; Fitter, 2002; Olsen & Kemper, 1968).

In seed plants, the development of the primary root (PR) begins with the elongation of the radicle from the embryo during germination. The root apical meristem (RAM) is already defined during embryogenesis (Dolan et al., 1993; Scheres et al., 1994; Van Norman, Liberman, & Benfey, 2013) and comprises a quiescent center (QC) surrounded by stem cells (initials) for the various root tissue types (Fig. 1A).

The initials in the proximal meristem divide to form vascular, ground, and epidermal tissues, which undergo cell elongation and thus lead to the growth of the radially symmetric root. Despite its simplicity, the root of the model plant *Arabidopsis thaliana* is comprised of all canonical tissue types: overlying the central stele (xylem and phloem) are single cell layers each of pericycle, endodermis, cortex, and epidermis (Fig. 1A; Grierson, Nielsen, Ketelaarc, & Schiefelbein, 2014; Scheres, Benfey, & Dolan, 2002; Van Norman et al., 2013). Certain epidermal cells produce root hairs—unicellular enations that increase the surface area for absorption of nutrients and water.

The distal meristem gives rise to the columella/root cap cells (Scheres et al., 1994), which not only lubricate and protect the root, but also perceive chemical cues (Ferrieri et al., 2017; Huang et al., 2018; Kutschera-Mitter, Barmicheva, & Sobotik, 1998; Newcombe & Rhodes, 1904; Raven & Edwards, 2001). A similar structure is not present at the shoot apical meristem. The embryonic primary root grows vertically downward along the gravity vector, displaying strong gravitropism (Darwin, 1880; Su, Gibbs, Jancewicz, & Masson, 2017). Therefore, the plasticity of the root system is mostly dependent on the post-embryonically formed lateral (root-borne) and adventitious (shoot-borne) roots (Bellini et al., 2014; Malamy & Benfey, 1997). Lateral root (LR) and adventitious root (AR) programs are similar but different, and thus a closer look is needed to distinguish them.

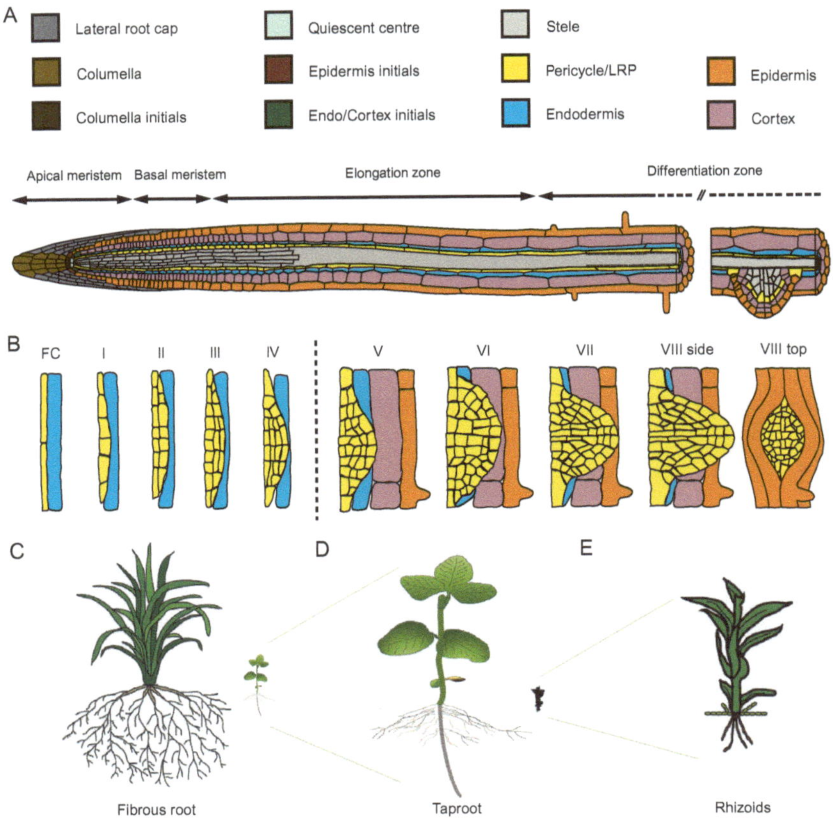

Fig. 1 Root and LR tissue organization leading to different RSA. (A) Schematic representation of the *A. thaliana* root from the root apical meristem (RAM), via the basal meristem and elongation zone to the differentiation zone, which harbors LR primordia and root hairs. (B) Progression of LR founder cells through the eight stages of LR development; transition of the endodermis occurs between stages IV and V; cortex and epidermis remain turgid as seen in VIII top view, (C–E) types of root system architectures: (C) Fibrous root system of a rice plant. (D) Soy bean seedling with a tap root. (E) *Physcomitrella patens* with single-celled rhizoids. *Images sources: Reused and modified with permission from Panel A: Peret, B. (2017). Primary and lateral root.ai.* Figshare. *Retrieved from https://doi.org/10.6084/m9.figshare.5143987.v4. Panel B: Stoeckle, D., Thellmann, M., & Vermeer, J. E. (2018). Breakout—Lateral root emergence in* Arabidopsis thaliana. Current Opinion in Plant Biology, 41, *67–72. Panel C: Sparks, E. (2017a). Cartoon of a rice plant.* Figshare. *Retrieved from https://doi.org/10.6084/m9.figshare.4688329.v1. Panel D: Tamang, B. (2018). Soybean Plant Graphical Illustration.* Figshare. *Retrieved from https://doi.org/10.6084/m9.figshare.5769882.v3. Panel E: Sparks, E. (2017b). Physcomitrella patens (moss).* Figshare. *Retrieved from https://doi.org/10.6084/m9.figshare.4688395.v1.*

2.2 Lateral and adventitious roots

Since most insights into LR and AR development are gained from *A. thaliana*, we focus our description of the processes to this example. This dicot provides an excellent system to study roots due to a simple and predictable arrangement of cells in radial layers (Dolan et al., 1993).

LR formation starts at the basal meristem (Fig. 1A) with an auxin-dependent pre-initiation event called "priming," which grants xylem pole pericycle (XPP) cells a LR founder cell identity (De Smet et al., 2007). Upon activation by the plant hormone auxin, the LR founder cells start differentiating by increasing their cell volume; concomitant migration of the nuclei precedes an asymmetric anticlinal cell division (Casimiro et al., 2001; De Rybel et al., 2010; Dubrovsky, Rost, Colón-Carmona, & Doerner, 2001; Vermeer et al., 2014). This marks the initiation of LRs and takes place in the elongation zone (Bellini et al., 2014; Casero, Casimiro, & Lloret, 1995). The origin of ARs is not as clear and differs across species. Both programs rely on the action of the auxin regulatory gene *SOLITARY ROOT* (*SLR*; *AUX/IAA*), as the gain-of-function mutant *slr* lacks both secondary root types. In contrast, *monopteros* mutants (lacking a functional *AUXIN RESPONSE FACTOR5/ARF5* gene) cannot make LRs but are able to go through the AR program. The few molecular studies available for AR formation suggest that distinct players [*ARF17*, *ARGONAUT1* (AGO1)] act in the AR program, since the LR program of *arf17* or *ago1* mutants is not hampered (Bellini et al., 2014; Gutierrez et al., 2009; Przemeck, Mattsson, Hardtke, Sung, & Berleth, 1996; Sorin et al., 2005).

After initiation, the differentiating cells, now forming the LR primordium (LRP), undergo several rounds of anticlinal and periclinal divisions in a stereotypical pattern to form a dome-shaped primordium. According to Malamy and Benfey (1997), the LR process can be divided into eight stages (Fig. 1B), with stages 1–4 taking place before growth through the endodermis and stages 5–7 when the LRP breaches the cortex and epidermis (Malamy & Benfey, 1997; Stoeckle, Thellmann, & Vermeer, 2018). Further proliferation and beginning cell elongation give rise to a stage 8, fully-emerged LR (von Wangenheim et al., 2016).

Auxin plays a major role in this process and in addition to LR priming and initiation, it is also involved in signaling the endodermal cells overlying the LRP to accommodate the growing LRP through controlled volume loss (Vermeer et al., 2014). Degradation of cell adhesions in tissues overlying the endodermis is also mediated by auxin, thus allowing the growing LRP to traverse cortex and epidermis (Swarup et al., 2008). LR research in other species is less advanced. It seems that many regulators are conserved and perform similar functions, but also differences have been observed. In *Zeas mays* (maize) and *Oryza sativa* (rice), for example, the endodermis takes part in LRP formation and constitutes the new epidermal layer (Bellini et al., 2014). It also needs to be tested whether the mechanisms identified in *A. thaliana* also hold true in plant species that have a larger root system with more cell layers that need to be traversed.

Adventitious rooting lacks the deep understanding at the molecular level that we have for LRs. This is mostly because *A. thaliana* is producing ARs only under adverse conditions (etiolated hypocotyls, de-rooted seedlings) and research in non-model species lags behind (Steffens & Rasmussen, 2016). However, shoot-borne rooting is a very important horticultural and agronomical trait, and thus there is a body of knowledge about the impact of different treatments on their potential to induce ARs from a variety of species and tissues [*Fragaria* spp. (strawberry), *Malus pumila* (apple), *Vitis vinifera* (grapevine), *Humulus lupulus* (hops); stolon, leave, stem] (Verstraeten, Schotte, & Geelen, 2014). The unifying theme is that there exists a huge variability in treatments that promote rooting in one species do not do so in others. Significantly, many hormones influence AR and LR in opposite directions. For example, the volatile plant hormone ethylene has been shown to promote AR formation, whereas it inhibits LR formation (Lewis et al., 2011; Muday, Rahman, & Binder, 2012; Negi, Ivanchenko, & Muday, 2008; Negi, Sukumar, Liu, Cohen, & Muday, 2010; Ruzicka et al., 2007; Vandenbussche, Vriezen, & Straeten, 2007; Verstraeten et al., 2014). Most hormone application acts through modulating auxin fluxes and concentrations, which also acts as a key regulator in AR formation. Regardless, emergence of ARs seems to be inhibited rather than promoted by exogenous auxin application (Bellini et al., 2014).

As exemplified in rice and *Solanum lycopersicum* (tomato), the formation of ARs is a way to react to environmental changes like flooding (Mcnamara & Mitchell, 1991). Long periods of hypoxia to the root system are detrimental, and thus being able to produce new roots from stems above the water level is a vital trait and contributes to the plasticity and flexibility of the RSA.

2.3 Tap roots, fibrous roots, and root system architecture

Some of the most astonishing root systems that can be observed above ground are AR-based, such as mangrove stilt roots or aerial roots of *Hedera helix* (ivy) and epiphytic orchids. Less well known is that adult root systems of monocots (monocotyledons—embryo with one cotyledon), such as maize and rice, only use ARs. Even in dicots (eudicotyledons—embryo with two cotyledon), such as tomato, *Vigna radiata* (mung bean) and *Helianthus annuus* (sun flower), ARs substantially add to the RSA.

RSA is defined by the amount, shape and organization of PR, LRs, and ARs (Smith & De Smet, 2012). In eudicots, the primary root tends to dominate, giving rise to LRs and forming an allorhizic root system, commonly known as the tap root system. Monocots have an ephemeral primary root, which allows other types of roots, such as seminal and adventitious roots derived from shoots, to dominate the root system. This is a homorhizic root system, commonly known as the fibrous root system (Bellini et al., 2014; Seago & Fernando, 2013).

During their life cycle, plants have to survive in the soil environment with a nutrient availability that is mostly heterogenous in space and time. An important role in the adaptive responses of plants to their ever-changing environment is thus their ability to shape the root system in order to efficiently extract nutrients and water from the soil. It is well known that excess or deficiency of different nutrients induces changes in RSA and root morphology, which can to some extent be attributed to specific nutrients (Giehl & von Wiren, 2014).

Mineral elements have complex interactions with themselves and other soil constituents, and can be carried out of the rooted soil volume by water. For a plant to be productive and have competitive success, it is important to be able to efficiently acquire these elements. In addition to nutrient status, very strong gradients in temperature, oxygen, water availability, pH, and bulk density occur with soil depth over a scale of centimeters (Lynch, 1995).

In order to get additional insights into how plants can modify their RSA to nutrient availability, the use of functional-structural plant models, such as *SimRoot*, can be very powerful. Using this approach, the three-dimensional development of a maize RSA was simulated to predict the optimal root branching density for phosphorus or nitrogen uptake. This revealed that a high branching density with short LRs was optimal for phosphorus acquisition and sparse branching with long LRs was optimal for nitrogen acquisition. This experiment showed that there is no such thing as an optimal branching density because it depends heavily on the nitrate-to-phosphorus ratio in the soil environment. This means there is a RSA tradeoff in most soils: plants may have different optimal branching behavior in different soil domains (Postma, Dathe, & Lynch, 2014).

In recent years, several studies have explored how the root system has adapted to local growth conditions. Most of these studies have been performed using *A. thaliana*, where the advent of cheap, large-scale, high-resolution genotyping and sequencing technologies has made the sequencing of many plant strains feasible (Alonso-Blanco et al., 2016). This has enabled GWASs to identify alleles responsible for a large variety of quantitative traits (Weigel, 2012) by measuring them in many different accessions and subsequently associating them with single nucleotide polymorphisms (SNPs). As these accessions have been collected from different geographical locations, the observed changes in RSA and general growth reveal how these traits might have played a role in the adaptation to their growth environment. GWASs have been successfully used to identify regulators of root growth under normal, different nutrient and hormone treatments, and stress conditions (Julkowska et al., 2017; Meijon, Satbhai, Tsuchimatsu, & Busch, 2014; Ogura & Busch, 2015; Ristova et al., 2016; Rosas et al., 2013; Slovak et al., 2014). In addition, it was reported that natural *brevis radix* (*brx*) loss-of-function alleles confer adaptation to acidic soils in *A. thaliana* (Gujas, Alonso-Blanco, & Hardtke, 2012). Although screening of different accessions has been mostly used in *A. thaliana*, this approach was shown to also have a big potential for understanding the role of RSA in the adaptation of rice to drought (Uga et al., 2013). With the rapidly evolving sequencing technologies, we expect that GWAS with more plant species (including crops), and comparisons between wild and domesticated species, will yield many new insights into how the evolution of RSA has contributed to adaptation. As most of the studies have relied on plate-based analyses of growth and RSA phenotypes, it is clear that we need to grow and phenotype roots of plants growing under near physiological conditions in order to gain better insights into how RSA can be exploited for successful adaptation to different environmental conditions. Recent advancements in X-ray micro-computed tomography and bioluminescence-based systems like the Growth and Luminescence Observatory of roots (GLO-roots) will be important to get a more complete picture of how dynamic RSA plays a role in the adaptive responses of plants (Mairhofer, Sturrock, Bennett, Mooney, & Pridmore, 2015; Rellán-Álvarez et al., 2015).

2.4 Rhizomes and rhizoids

Flowering plants like hops, *Zingiber officinale* (ginger), *Curcuma longa* (turmeric), and *Iris* spp., as well as non-flowering plants like horsetails, ferns and lycophytes, have a belowground structure called rhizome, which is commonly confused to be a part of the root system. Although rhizomes are anatomically not a root, they do fulfill root functions. In order to increase the surface area, rhizomes mostly bear single-celled rhizoids, a structure that anatomically and molecularly closely resembles root hairs (Jones & Dolan, 2012). However different in origin and shape, rhizomes and true roots are all subterranean and as such, intimately connected and influenced by the rhizosphere with its microorganisms. Their influence on root systems is a major driving force of plant evolution with large geochemical implications (Pieterse, de Jonge, & Berendsen, 2016; Schwartz et al., 2013).

2.5 Mycorrhizae and Rhizobia

Plants have been accompanied by fungi, bacteria, viruses, and protista over millions of years, and these interactions are thought to have accelerated the adaptation to land (Bonfante & Genre, 2010).

Endophytic soil bacteria enter the plant root, colonize them and provide beneficial functions, such as promotion of plant growth, health, and systemic resistance (Ryan, Germaine, Franks, Ryan, & Dowling, 2008; Shahzad, Khalid, Arshad, Tahir, & Mahmood, 2010). They often secrete growth hormones like auxin, cytokinin and gibberellic acid, or reduce ethylene production and thus connect plant growth promotion with changes in RSA (Lugtenberg & Kamilova, 2009; Ortíz-Castro, Contreras-Cornejo, Macías-Rodríguez, & López-Bucio, 2009; Schwartz et al., 2013). The interaction between *Rhizobium* spp. and legumes is a prominent and the oldest known example of a symbiosis between bacteria and plants (Hellriegel & Wilfarth, 1888; Lavin, Herendeen, & Wojciechowski, 2005). In exchange for carbon, these nitrogen-fixing bacteria provide plant-accessible ammonium made from atmospheric N_2. This process happens in so-called nodules. Nitrogen-fixing symbioses can be found among the Rosid 1 clade (including legumes; Soltis et al., 2000). Many agriculturally relevant plants, such as *Glycine* max (soy bean), *Cicer arietinum* (chickpea), *Pisum sativum* (pea), and *Trifolium* spp. (clover), interact with *Rhizobia* spp. As they have maintained this symbiosis for an extended period of time, it must have provided an evolutionary advantage in the adaptation to their environment. We anticipate many novel insights into plant-bacteria interactions as the systematic exploration of the root microbiome by high-throughput sequencing techniques advances.

Symbiosis between plants and fungi evolved over 400 million years ago, with arbuscular mycorrhizal fungi (AMF) being the oldest and most abundant (Field et al., 2015). The success of these connections is evident by the estimated 90% of land plants that interact with AMF (Bonfante & Genre, 2010). To colonize the land, symbiosis was an essential means to improve nutrient uptake (phosphorus and nitrogen) in exchange for carbon (Bidartondo et al., 2011; Chiu, Choi, & Paszkowski, 2018; Kiers et al., 2011; Martin, Uroz, & Barker, 2017; Pirozynski & Malloch, 1975; Wang et al., 2010). AMF can influence the RSA of several plant species through an increase or decrease of root branching, and they can even overcome LR mutant phenotypes (reviewed in Gutjahr & Parniske, 2013). During evolution, the *Brassica* lineage (including *A. thaliana*) lost the ability to interact with AMF (Bonfante & Genre, 2010). However, recent studies revealed that the fungus *Colletotrichum tofieldiae* can support the phosphorus uptake of *A. thaliana* roots under phosphorus starvation conditions (Hiruma et al., 2016).

3. From first roots to angiosperm root diversity

3.1 Devonian rooting structures

Multicellular, non-vascular plants that can be described as being similar to extant bryophytes are accepted as the first colonizers of land close to lake shorelines during the mid to late Ordovician of the Paleozoic era (ca. 450 Mya; Fig. 2; Kenrick, Wellman, Schneider, & Edgecombe, 2012; Steemans et al., 2009), although a recent study has placed the event at a much earlier, middle Cambrian–Early Ordovician period (ca. 525 Mya; Morris et al., 2018). It is difficult to conclude what kind of roots these first land plants had because there are barely any plant body fossils for study from this period.

Initially, rooting structures developed to help anchor the plants, which moved from water to land. The first rooting structures were simple rhizoid-based rooting systems (RBRSs) that grew mostly horizontally on the surface and, when subterraneous, shallow on moist lowlands (Algeo, Berner, Maynard, & Scheckler, 1995; Jones & Dolan, 2012). The earliest fossil evidence of RBRS comes from Rhynie Chert in Scotland from the Devonian period (410 Mya; Dolan & Hetherington, 2017). Their role in evolutionary adaptation was to anchor the thallus to the substrate and assist in taking up nutrients (Kenrick & Strullu-Derrien, 2014). The liverwort *Marchantia polymorpha*, an extant bryophyte, is generally used as a model for the earliest terrestrial plants that gave rise to RBRSs from rhizoid bearing axes of their gametophyte and sporophyte generation (Graham, Wilcox, Cook, & Gensel, 2004; Menand et al., 2007; Shimamura, 2016). *M. polymorpha* also

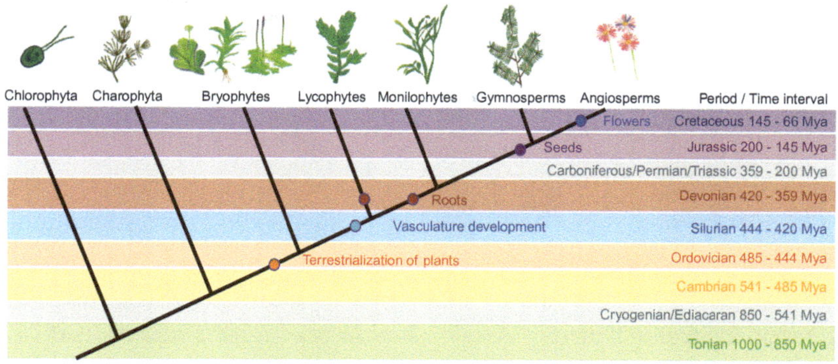

Fig. 2 Key events and innovations that shaped plant evolution. Land colonization (orange dot) happened at the transition from the Cambrian to Ordovician period. Vascular tissues (cyan dot) date back to the Silurian. Evolution of roots (red dots) happened independently in lycophytes and the ancestor of all euphyllophytes, and is set in the Devonian period. More than 200 million years later, first seed plants (violet dot) occur in the Jurassic fossil record. During the Cretaceous period, flowers arise in the angiosperm lineage. *Copyright of the image belongs to Eftychis Frangedakis. Modified with permission.*

exhibits two distinct rooting structures: tuberculate rhizoids scattered underneath the thallus of the gametophyte, mainly for anchoring the plant to its substrate, and smooth clustered rhizoids in the sporophyte, mainly for absorption of nutrients (Graham et al., 2004). Unlike *M. polymorpha*, the early land plants bore only one type of smooth rhizoids. They were filamentous and unbranched (Kenrick & Strullu-Derrien, 2014). When comparative studies were conducted on rhizoids of *M. polymorpha* and the moss *Physcomitrella patens* with root hairs of *A. thaliana*, it was found that a highly conserved molecular network controls the development of these cell types in all three species (Honkanen et al., 2017; Menand et al., 2007).

One of the commonly observed functions of roots in extant terrestrial plants is the ability to provide mechanical support. Shallow growing rhizoids could barely be expected to provide mechanical support. Instead, early vascular plants possessed subterranean rhizomatous axes which bore rhizoids (Kenrick & Strullu-Derrien, 2014). The study of the fossils of *Drephenophycus*, an early vascular plant, has revealed the importance of rhizomatous growth as an early precursor of deep-rooted plants. These belowground rhizomes were found to grow extensively and form complex subterranean networks. They could produce dense vegetation and protect the substrate surface from erosion (Xue et al., 2016). These rhizomes are similar in their clonal growth to some species of extant clubmosses and ferns of the genera *Lycopodium* and *Pteridium*, respectively, nicely tying in with the fact that some basal members of lycophyte and euphyllophyte clades still retain the feature of rhizomatous growth (Kenrick, 2013; Kenrick & Strullu-Derrien, 2014; Xue et al., 2016).

Owing to the lack of reliable fossils for a period of nearly 40 million years, the transition from RBRS to first true roots is poorly understood; hence, this is a highly debated and controversial topic. Depending on the methods of investigation, few theories exist on the evolution of roots. It was proposed that roots evolved separately in different clades of early vascular plants (Kenrick & Strullu-Derrien, 2014). One theory suggests that roots evolved from rhizoid-bearing axes, while another suggests the possibility of leaves turning into roots (Bierhorst, 1971; Fujinami et al., 2017; Rothwell & Erwin, 1985). It has also been postulated that roots arose as de novo organs and not from any other plant part (Matsunaga & Tomescu, 2016). Fujinami et al. (2017) studied the diversity of RAMs in different species of extant lycophytes, which are believed to have diversified at the same time that roots evolved, reflecting the multiple origins of the root.

3.2 Paraphyletic origin of true roots

As plants started to get bigger, the rooting system also got more complex. This could be seen in lycophytes, the first vascular plants. Contrary to vascular plants from higher divisions, lycophytes have leaves with a single vascular bundle called microphylls instead of leaves with a more complex network of vasculature called megaphylls (Doyle, 2013; Kenrick, 2013). Organisms in this order are the club-, fir- and spike mosses that are commonly confused with ferns. Species belonging to the extinct genus *Zosterophyllum* were one of the first known vascular plants to bear rooting structures intermediate to rhizoids and true roots. Studies of fossils have revealed that these plants had a belowground rhizome that bore roots in dense tufts (Hao, Xue, Guo, & Wang, 2010). The differentiating factor between early roots in lycophytes and the rhizoids in extinct bryophytes is that this was the first instance where structures similar to aerial stems grew belowground in a distinct opposite orientation to the aerial stems. However, they were narrower and shorter without any microphylls or leaves and bifurcating irregularly, suggesting that they developed to have a specific rooting function (Gensel & Edwards, 2001). The meristems frequently found on rhizomes and erect stems that gave rise to these early roots were dormant and could develop into either shoots or roots (Kenrick, 2013). Fossil records of the cottonwood canyon lycophyte, an early basal lycophyte, exhibit regular branching of this belowground stem (K-branching) and suggest that LRs developed from these branches (Matsunaga & Tomescu, 2016). However, these were not true roots. While advancing our knowledge, studies like this especially highlight the parts of early root evolution that are still unknown. Nevertheless, based on developmental differences, there is a consensus in the community that roots have paraphyletic origin, i.e., they evolved independently at least twice: once at the base of the lycophytes and once in the euphyllophyte clades (Fig. 2; Kenrick & Crane, 1997; Kenrick & Strullu-Derrien, 2014; Raven & Edwards, 2001). Prominently, lycophyte roots branch apically by meristematic division, while euphyllophyte roots originate endogenously from cells of the pericycle or endodermis (Friedman, Moore, & Purugganan, 2004).

Another theory proposes further independent evolution of roots in euphyllophytes based on their origin during embryogenesis: in eudicots, the root is formed very early from the hypophysis of the eight-cell stage embryo while in monocots, it forms from the ground meristem of the globular stage embryo (Bennett & Scheres, 2010).

3.3 The oldest root meristem

A major determinant to distinguish and categorize root fossils is the structure of the RAM. Lycophyte and fern RAMs are very distinct from those of the later euphyllophyte clades as they only have a single active meristematic apical cell of an inverted-prism structure. By dividing asymmetrically in three directions, the apical cell forms the tissue that later acquires radial symmetry. Euphyllophytes have RAMs of the above-mentioned structure (Fig. 1A), with angiosperms having rather small organizing centers (few initials) while many proliferating cells surround the QC of gymnosperms.

It is extremely challenging to find these small (<2 mm) organs fossilized. However, a recent study has accomplished this (Hetherington, Dubrovsky, & Dolan, 2016). In the Oxford natural history collection, they found two specimens in a carboniferous rock that date back 320 million years and show a remarkable state of preservation. Moreover, they found one RAM to be fossilized in an actively growing state (no differentiated vascular tissue close to the QC). This allowed them to perform modern lineage tracing analysis and led to the classification of the RAM belonging to the gymnosperm clade. By employing the "Körper-Kappe" theory, developed by the Swiss botanist Otto Schüepp (Schüepp, 1917), they reasoned that *Radix carbonica* is to be treated as a separate species (Hetherington, Dubrovsky, et al., 2016).

3.4 Angiosperm root system plasticity

Among all the vascular plants currently inhabiting the earth, angiosperms form the most diverse group with 416 families and nearly 300,000 species (Christenhusz & Byng, 2016). The name literally means encased seeds, which happens to be a characteristic feature of angiosperms, differentiating them from gymnosperms. It is hard to formulate a common characteristic root system for angiosperms as the species range from aquatic lilies to woody magnolias. Additionally, most of these species are hardly studied with respect to their root systems, the exceptions being monocots and dicots. Most of the world's economically important plant species belong to these two divisions and are, hence, well studied.

3.4.1 Monocots

Many economically important species, including members of the *Orchidaceae* (orchids), *Poaceae* (true grasses), *Arecaceae* (palms), *Musaceae* (bananas, plantains), *Zingiberaceae* (ginger), *Liliaceae* (lilies, tulips), and *Amaryllidaceae* (onions, garlic, daffodils) are monocots. Monocots exhibit one of the two major types of root systems seen in angiosperms, the fibrous or homorhizic root system. In this type of root system, the embryonic root that arises from the seed is only important during germination and early stages of seedling development. Later, post-embryonic roots that emerge from the shoot, take over the rooting functions (Bellini et al., 2014).

Unlike LRs, which always arise from pericycle cells, post-embryonic roots in monocots arise from different tissues in the shoot, and hence from different cell types. However, they always develop from cells neighboring vascular tissues, such as hypocotyl pericycle cells, phloem/xylem parenchyma cells, interfascicular cambium cells, or cells close/adjacent to the vascular cylinder of the stem. All embryonic and post-embryonic roots in monocots form LRs in order to branch out (Bellini et al., 2014).

Orchids represent 50% of the monocots and are an interesting case study with respect to the role of RSA in evolutionary adaptation. Orchids can be epiphytic or terrestrial. Epiphytic orchids produce aerial roots that anchor the plant to trees and, with an adaptive spongy epidermis called velamen radicum, absorb nutrients from the humid air. In contrast to the terrestrial orchids, epiphytic orchids do not make LRs. It was recently reported that two genes, *AGAMOUS LIKE12* (*AGL12*) and *ARABIDOPSIS NITRATE REGULATED1* (*ANR1*) could have played a major role in the segregation of root phenotypes in epiphytic and terrestrial orchids as they were either absent or showed reduced expression in epiphytes (Zhang et al., 2017). Studies in *A. thaliana* have shown that *AGL12* is involved in root cell differentiation, and *ANR1* is a key regulator of LR development in response to external nitrate supply (Par˘enicová et al., 2003; Zhang & Forde, 1998).

Beside the orchids, cereals represent another large group of the monocots and belong to the *Poaceae*. Maize, rice, and *Triticum aestivum* (wheat), the top three staple crops by global annual production, are all cereals (Gennari, 2015). They have complex and widely distinct RSAs and can bear different types of roots: in addition to PRs, they can form seminal roots (SRs), crown roots (CR), and brace roots (BR) (Feldmann, 1994). The development of these roots even within one species is firmly dependent on external conditions and varies between subspecies. There are also differences in total root mass per plant, number of root branches per unit length, lateral spread, and number of SRs (Feldmann, 1994; Pagès & Kervella, 2018).

Photosynthesis type also seems to have an effect on root branching and root biomass (Atkinson et al., 2016). Among all monocots, the root system of cereals is the best described and studied because of their obvious economic importance. However, this knowledge tells us little about evolutionary adaptation of RSA in monocots due to human intervention in breeding modern varieties of cereals. Maize, rice, and wheat have all been domesticated by humans for thousands of years and, until recently, little to no consideration was given to root phenotypes during selection. In fact, cereals have been directly selected for higher grain yield, which led to indirect selection of weaker and smaller roots (Gaudin, McClymont, Soliman, & Raizada, 2014; Schmidt, Bowles, & Gaudin, 2016). A comparative study of two cultivated varieties of maize, sweet and popping corn, and teosinte (*Z. mays* subsp. *parviglumis*), the ancestor of modern maize, has revealed that the shoot-to-root ratio was significantly higher while root thickness was significantly lower in cultivated varieties compared to teosinte (Szoboszlay et al., 2015). Another study investigated the effect of altered dosage of an allele of *Teosinte branched1* (*Tb1*), a major domestication gene in maize that is known to repress shoot branching on roots. It was reported that decreased expression of *Tb1* had a significant impact on RSA, with increased branching via an increase in the number of CRs and LRs (Gaudin et al., 2014). RSA in domesticated cereals has also been significantly altered by agricultural practices in recent times due to the increased use of fertilizers and irrigation. A comparative study of 16 varieties of cultivated maize developed in different time periods over the last 100 years was conducted to understand the effect of nitrogen levels in soil on RSA. BRs and CRs grew less steep and LRs grew longer in newer compared to older varieties, suggesting adaptation to better nutrient foraging and reduced water foraging (York, Galindo-Castañeda, Schussler, & Lynch, 2015). A similar result was observed in a comparison of the upland rainfed rice variety Azucena and lowland irrigated rice variety IR64. Azucena was found to have deeply rooted PRs and CRs, while IR64 was shallow rooted but highly branched, suggesting the tradeoffs made by the cultivated varieties depend on the conditions they were selected in (Clark et al., 2011). Rebouillat et al. (2009) described a remarkable adaptation to survive in waterlogged conditions, where LRs in rice form aerenchyma cells in the outer cortical layers to enhance the diffusion of oxygen. Increased awareness of the effect of indirect breeding on root phenotypes is leading to more efforts to create new varieties of cereals by introgression of genes that have been lost or altered during domestication (Dixit et al., 2012; Gaudin, McClymont, & Raizada, 2011; Meister, Rajani, Ruzicka, & Schachtman, 2014).

Wheat has been cultivated for over 10,000 years. During its domestication some traits were lost that were in fact important for survival (Zohary, 2004). A recent study describes an increase of SR number in tetra- and hexaploid wheat, but also a reduced number of SRs in tetraploid *Triticum dicoccum* (wild emmer) (Golan, Hendel, Méndez Espitia, Schwartz, & Peleg, 2018). Furthermore, this study revealed that the reduced number of the SRs is due to suppressed primordia in the wild form, and that this lower number is necessary to recover from temporal water stress. In general, deeper rooting varieties have an advantage in periods of drought (Manschadi, Manske, & Vlek, 2013). Temperature changes, as well as root pruning in combination with a carbohydrate-rich nutrition, result in a different behavior of LR formation (Bingham & Stevenson, 1993).

One possibility to assess the effects of domestication on RSA evolution is to contrast the above finding with the model grass *Brachypodium distachyon*, which also belongs to the *Poaceae* and thus is a close relative of wheat and barley (Kellogg, 2001; Vogel, Garvin, Leong, & Hayden, 2006). *B. distachyon* is a wild species that was never used in breeding (Girin et al., 2014). At present, the research community has access to over 200 natural accessions and many of them have already been sequenced (Gordon et al., 2017). In recent years, researchers started to investigate the RSA of *B. distachyon* and its natural variations (Ingram et al., 2012; Pacheco-Villalobos & Hardtke, 2012; Pacheco-Villalobos, Sankar, Ljung, & Hardtke, 2013). It was shown that the root system of different *B. distachyon* accessions displays different responses to phosphorous and nitrogen starvation conditions (Ingram et al., 2012).

3.4.2 Dicots

Dicots form a paraphyletic group, with basal angiosperms diverging earlier than the monocots did (APG IV, 2016; Stuessy, 2010). As mentioned earlier, the simple morphology of the root of *A. thaliana*, the vast genetic tools available in this model, and its accessibility for microscopy has made this species very suitable to study root development. The *A. thaliana* root system has a taproot with several orders of LRs. In recent years, a very detailed picture has emerged about the molecular and genetic pathways underlying LR development in this species, which we will, however, not discuss this in this chapter. For more information regarding the underlying developmental mechanisms we refer the reader to some excellent recent reviews (Otvos & Benkova, 2017; Stoeckle et al., 2018; Vilches-Barro & Maizel, 2015). One of the challenges for the coming years is to scrutinize to which extent the molecular mechanisms identified in *A. thaliana* are conserved in other plant species. For instance, in the legume *Medicago truncatula*, LRs also initiate from the xylem pole pericycle and, in general, follow a similar developmental program. However, the overlying endodermis and cortex of *M. truncatula* will reactivate the cell cycle and will become an integral part of the newly formed LR (Bensmihen, 2015; Herrbach, Rembliere, Gough, & Bensmihen, 2014). This is also observed in other plants, such as tomato (Ivanchenko, Coffeen, Lomax, & Dubrovsky, 2006) and several cereals (Orman-Ligeza et al., 2013), suggesting that the system of *A. thaliana* might not serve as a stereotypical LR model. Interestingly, *M. truncatula* can also engage in a symbiosis with bacterial *Rhizobia* spp., leading to the formation of a new lateral organ, the root nodule. It has been suggested that this process has exploited the LR developmental program (Bensmihen, 2015). Interestingly, non-legume actinorhizal plants and *Parasponia* spp. form a different type of nodule that also initiates from the pericycle and has a structure much more similar to LRs, as these nodules have a central vasculature instead of the peripheral vasculature observed in legume nodules (Pawlowski & Bisseling, 1996).

In order to address how RSA could have played a role in the adaptation to the growth environment, it is important to compare RSA of plants grown in different conditions and to identify the mechanisms that are casual for these changes. It has been reported that there is a striking natural variation between different *A. thaliana* accessions, and that it is very likely that these changes occurred during the adaptation to their growth environment (Horton et al., 2012). Indeed, there is also substantial variation between the RSA in these different accessions (Meijon et al., 2014; Rosas et al., 2013). Preliminary results also suggest that this is the case for *B. distachyon* (Pacheco-Villalobos & Hardtke, 2012).

4. Geochemical consequences of root evolution

With the advent of roots in the Devonian period, plants have become increasingly successful to colonize land. Due to their dependence on rock-derived minerals like phosphorous, potassium, calcium, magnesium and iron, plants possess the mechanism to rapidly weather rocks to release these minerals. This process was accelerated by the evolution of embryophytes and their new rooting structures (Raven & Edwards, 2001). When early vascular plants started to colonize land in the late Ordovician period, they caused a massive drawdown of atmospheric CO_2, which was rapidly forming stable compounds with the minerals that had been exposed to the atmosphere. This led to a temporary but substantial drop in atmospheric CO_2 levels, plunging the earth into an ice age and causing mass extinction of marine organisms (Algeo et al., 1995; Lenton, Crouch, Johnson, Pires, & Dolan, 2012). In the late Devonian, plants innovated arborescence (tree-like stature). Plants belonging to the *Archeopteris* genus dominated the earth during this period, and they grew up to 30 m in height. Increase in aboveground stature meant that the belowground biomass needed to reciprocate: roots increased in thickness and grew deeper into the earth. New geoclimate models also hint at a striking contribution of arbuscularization of roots to accelerate the change in geochemical conditions on the early planet earth (Mills, Batterman, & Field, 2018).

5. Conclusions

In conclusion, we like to revisit a statement made by Hetherington, Berry, et al. (2016), which was also reiterated and emphasized by Tomescu (2016) due to its importance for all the current and future scientists who are trying to reach conclusions about plant evolution: "extant anatomical diversity represents only a subset of all the diversity that ever existed". Sampling only living specimens will thus not help us to improve and complete the bigger picture. Therefore, we need to embrace the possibilities the fossil record might hold, and look forward to the exploration of treasures from yet unexplored natural historic collections.

Acknowledgments

Work in the Vermeer laboratory is supported by the Swiss National Science Foundation (PP00P3_157524 and 316030_164086), the Netherlands Organization for Scientific Research (NWO 864.13.008), and the University of Zurich.

References

Algeo, T., Berner, R. A., Maynard, J. B., & Scheckler, S. (1995). Late Devonian oceanic anoxic events and biotic crises: "Rooted" in the evolution of vascular plants. *GSA Today*, *5*(45), 64–66.

Alonso-Blanco, C., Andrade, J., Becker, C., Bemm, F., Bergelson, J., Borgwardt, K. M. M., et al. (2016). 1,135 genomes reveal the global pattern of polymorphism in *Arabidopsis thaliana*. *Cell*, *166*(2), 481–491.

APG IV. (2016). An update of the angiosperm phylogeny group classification for the orders and families of flowering plants: APG IV. *Botanical Journal of the Linnean Society*, *141*(4), 399–436.

Atkinson, R. R. L., Mockford, E. J., Bennett, C., Christin, P. A., Spriggs, E. L., Freckleton, R. P., et al. (2016). C4 photosynthesis boosts growth by altering physiology, allocation and size. *Nature Plants*, *2*(5), 1–5.

Bellini, C., Pacurar, D. I., & Perrone, I. (2014). Adventitious roots and lateral roots: Similarities and differences. *Annual Review of Plant Biology*, *65*(1), 639–666.

Bennett, T., & Scheres, B. (2010). Root development—Two meristems for the Price of one? *Current Topics in Developmental Biology*, *91*, 67–102.

Bensmihen, S. (2015). Hormonal control of lateral root and nodule development in legumes. *Plants (Basel)*, *4*(3), 523–547.

Bidartondo, M. I., Read, D. J., Trappe, J. M., Merckx, V., Ligrone, R., & Duckett, J. G. (2011). The dawn of symbiosis between plants and fungi. *Biology Letters*, 7(4), 574–577.

Bierhorst, D. W. (1971). *Morphology of vascular plants*. Macmillan.

Bingham, I. J., & Stevenson, E. A. (1993). Control of root growth: Effects of carbohydrates on the extension, branching and rate of respiration of different fractions of wheat roots. *Physiologia Plantarum*, *88*(1), 149–158.

Bonfante, P., & Genre, A. (2010). Mechanisms underlying beneficial plant–fungus interactions in mycorrhizal symbiosis. *Nature Communications*, *1*, 48.

Casero, P. J., Casimiro, I., & Lloret, P. G. (1995). Lateral root initiation by asymmetrical transverse divisions of pericycle cells in four plant species: *Raphanus sativus*, *Helianthus annuus*, *Zea mays*, and *Daucus carota*. *Protoplasma*, *188*(1–2), 49–58.

Casimiro, I., Marchant, A., Bhalerao, R. P., Beeckman, T., Dhooge, S., Swarup, R., et al. (2001). Auxin transport promotes arabidopsis lateral root initiation. *The Plant Cell*, *13*(4), 843.

Chiu, C. H., Choi, J., & Paszkowski, U. (2018). Independent signalling cues underpin arbuscular mycorrhizal symbiosis and large lateral root induction in rice. *New Phytologist*, *217*(2), 552–557.

Christenhusz, M. J. M., & Byng, J. W. (2016). The number of known plants species in the world and its annual increase. *Phytotaxa*, *261*(3), 201–217.

Clark, R. T., MacCurdy, R. B., Jung, J. K., Shaff, J. E., McCouch, S. R., Aneshansley, D. J., et al. (2011). Three-dimensional root phenotyping with a novel imaging and software platform. *Plant Physiology*, *156*(2), 455–465.

Darwin, C. (1880). The circumnutating movements of seedling plants. In C. Darwin (Ed.), *The power of movement in plants* (1st ed., pp. 10–66). Cambridge: Cambridge University Press.

De Rybel, B., Vassileva, V., Parizot, B., Demeulenaere, M., Grunewald, W., Audenaert, D., et al. (2010). A novel Aux/IAA28 signaling cascade activates GATA23-dependent specification of lateral root founder cell identity. *Current Biology*, *20*(19), 1697–1706.

De Smet, I., Tetsumura, T., De Rybel, B., Frey, N. F. d., Laplaze, L., Casimiro, I., et al. (2007). Auxin-dependent regulation of lateral root positioning in the basal meristem of *Arabidopsis*. *Development*, *134*(4), 681–690.

Dixit, S., Swamy, B. P. M., Vikram, P., Ahmed, H. U., Sta Cruz, M. T., Amante, M., et al. (2012). Fine mapping of QTLs for rice grain yield under drought reveals sub-QTLs conferring a response to variable drought severities. *Theoretical and Applied Genetics*, *125*(1), 155–169.

Dolan, L., & Hetherington, A. J. (2017). Bilaterally symmetric axes with rhizoids composed the rooting structure of the common ancestor of vascular plants. *Philosophical Transactions of the Royal Society B: Biological Sciences*, *373*(1), 20170042.

Dolan, L., Janmaat, K., Willemsen, V., Linstead, P., Poethig, S., Roberts, K., et al. (1993). Cellular organisation of the *Arabidopsis thaliana* root. *Development*, *119*(1), 71–84.

Doyle, J. A. (2013). Phylogenetic analyses and morphological innovations in land plants. In B. A. Ambrose & M. Purugganan (Eds.), *Vol. 45. Annual plant reviews* (1st ed., pp. 1–50). Blackwell Publishing Ltd.

Dubrovsky, J. G., Rost, T. L., Colón-Carmona, A., & Doerner, P. (2001). Early primordium morphogenesis during lateral root initiation in *Arabidopsis thaliana*. *Planta*, *214*(1), 30–36.

Feldmann, L. (1994). The maize root. In M. Freeling & V. Walbot (Eds.), *The maize handbook* (1st ed., pp. 29–37). US: Springer.

Ferrieri, A. P., Machado, R. A. R., Arce, C. C. M., Kessler, D., Baldwin, I. T., & Erb, M. (2017). Localized micronutrient patches induce lateral root foraging and chemotropism in *Nicotiana attenuata*. *Journal of Integrative Plant Biology*, *59*(10), 759–771.

Field, K. J., Leake, J. R., Tille, S., Allinson, K. E., Rimington, W. R., Bidartondo, M. I., et al. (2015). From mycoheterotrophy to mutualism: Mycorrhizal specificity and functioning in *Ophioglossum vulgatum* sporophytes. *New Phytologist*, *205*(4), 1492–1502.

Fitter, A. (2002). Characteristics and functions of root systems. In Y. Waisel, A. Eshel, & U. Kafkafi (Eds.), *Plant roots: The hidden half* (3rd ed). Marcel Dekker Inc. p. 2.1-2.30.

Friedman, W. E., Moore, R. C., & Purugganan, M. D. (2004). The evolution of plant development. *American Journal of Botany*, *91*(10), 1726–1741.

Fujinami, R., Yamada, T., Nakajima, A., Takagi, S., Idogawa, A., Kawakami, E., et al. (2017). Root apical meristem diversity in extant lycophytes and implications for root origins. *New Phytologist*, *215*(3), 1210–1220.

Gaudin, A. C. M., McClymont, S. A., & Raizada, M. N. (2011). The nitrogen adaptation strategy of the wild teosinte ancestor of modern maize, *Zea mays* subsp. *parviglumis*. *Crop Science*, *51*(6), 2780–2795.

Gaudin, A. C. M., McClymont, S. A., Soliman, S. S. M., & Raizada, M. N. (2014). The effect of altered dosage of a mutant allele of *Teosinte branched 1 (tb1-ref)* on the root system of modern maize. *BMC Genetics*, *15*, 1–14.

Gennari, P. (2015). *FAO statistical pocketbook 2015—World food and agriculture*. Rome: FAO.

Gensel, P. G., & Edwards, D. (2001). *Plants invade the land: Evolutionary and environmental perspectives*. Columbia University Press.

Giehl, R. F., & von Wiren, N. (2014). Root nutrient foraging. *Plant Physiology*, *166*(2), 509–517.

Girin, T., David, L. C., Chardin, C., Sibout, R., Krapp, A., Ferrario-Méry, S., et al. (2014). *Brachypodium*: A promising hub between model species and cereals. *Journal of Experimental Botany*, *65*(19), 5683–5696.

Golan, G., Hendel, E., Méndez Espitia, G. E., Schwartz, N., & Peleg, Z. (2018). Activation of seminal root primordia during wheat domestication reveals underlying mechanisms of plant resilience. *Plant, Cell and Environment*, *41*(4), 755–766.

Gordon, S. P., Contreras-Moreira, B., Woods, D. P., Des Marais, D. L., Burgess, D., Shu, S., et al. (2017). Extensive gene content variation in the *Brachypodium distachyon* pan-genome correlates with population structure. *Nature Communications*, *8*(2184), 1–13.

Graham, L. E., Wilcox, L. W., Cook, M. E., & Gensel, P. G. (2004). Resistant tissues of modern marchantioid liverworts resemble enigmatic early Paleozoic microfossils. *Proceedings of the National Academy of Sciences of the United States of America*, *101*(30), 11025–11029.

Grierson, C., Nielsen, E., Ketelaarc, T., & Schiefelbein, J. (2014). Root hairs. *The Arabidopsis Book*, *12*, e0172.

Gujas, B., Alonso-Blanco, C., & Hardtke, C. S. (2012). Natural *Arabidopsis brx* loss-of-function alleles confer root adaptation to acidic soil. *Current Biology*, *22*(20), 1962–1968.

Gutierrez, L., Bussell, J. D., Pacurar, D. I., Schwambach, J., Pacurar, M., & Bellini, C. (2009). Phenotypic plasticity of adventitious rooting in *Arabidopsis* is controlled by complex regulation of AUXIN RESPONSE FACTOR transcripts and MicroRNA abundance. *The Plant Cell, 21*(10), 3119–3132.

Gutjahr, C., & Parniske, M. (2013). Cell and developmental biology of arbuscular mycorrhiza Symbiosis. *Annual Review of Cell and Developmental Biology, 29*(1), 593–617.

Hao, S., Xue, J., Guo, D., & Wang, D. (2010). Earliest rooting system and root: Shoot ratio from a new *Zosterophyllum* plant. *New Phytologist, 185*(2009), 217–225.

Hellriegel, H., & Wilfarth, H. (1888). *Untersuchungen über die Stickstoffnahrung der Gramineen und Leguminosen.* Berlin: Buchdruckerei der "Post", Kayssler & Co.

Herrbach, V., Rembliere, C., Gough, C., & Bensmihen, S. (2014). Lateral root formation and patterning in *Medicago truncatula. Journal of Plant Physiology, 171*(3–4), 301–310.

Hetherington, A. J., Berry, C. M., & Dolan, L. (2016). Networks of highly branched stigmarian rootlets developed on the first giant trees. *Proceedings of the National Academy of Sciences of the United States of America, 113*(24), 6695–6700.

Hetherington, A. J., Dubrovsky, J. G., & Dolan, L. (2016). Unique cellular organization in the oldest root meristem. *Current Biology, 26*(12), 1629–1633.

Hiruma, K., Gerlach, N., Sacristan, S., Nakano, R. T., Hacquard, S., Kracher, B., et al. (2016). Root endophyte *Colletotrichum tofieldiae* confers plant fitness benefits that are phosphate status dependent. *Cell, 165*(2), 464–474.

Honkanen, S., Jones, V. A. S., Morieri, G., Champion, C., Hetherington, A. J., Kelly, S., et al. (2017). The mechanism forming the cell surface of tip-growing rooting cells is conserved among land plants. *Current Biology, 27*(20), 3224.

Horton, M. W., Hancock, A. M., Huang, Y. S., Toomajian, C., Atwell, S., Auton, A., et al. (2012). Genome-wide patterns of genetic variation in worldwide *Arabidopsis thaliana* accessions from the RegMap panel. *Nature Genetics, 44*(2), 212–216.

Huang, G., Liang, W., Sturrock, C. J., Pandey, B. K., Giri, J., Mairhofer, S., et al. (2018). Rice actin binding protein RMD controls crown root angle in response to external phosphate. *Nature Communications, 9*(1), 2346.

Ingram, P. A., Zhu, J., Shariff, A., Davis, I. W., Benfey, P. N., & Elich, T. (2012). High-throughput imaging and analysis of root system architecture in *Brachypodium distachyon* under differential nutrient availability. *Philosophical Transactions of the Royal Society of London. Series B, Biological Sciences, 367*(1595), 1559–1569.

Ivanchenko, M. G., Coffeen, W. C., Lomax, T. L., & Dubrovsky, J. G. (2006). Mutations in the *Diageotropica (Dgt)* gene uncouple patterned cell division during lateral root initiation from proliferative cell division in the pericycle. *Plant Journal, 46*(3), 436–447.

Jones, V. A. S., & Dolan, L. (2012). The evolution of root hairs and rhizoids. *Annals of Botany, 110*(2), 205–212.

Julkowska, M. M., Koevoets, I. T., Mol, S., Hoefsloot, H., Feron, R., Tester, M. A., et al. (2017). Genetic components of root architecture remodeling in response to salt stress. *Plant Cell, 29*(12), 3198–3213.

Kellogg, E. A. (2001). Evolutionary history of the grasses. *Plant Physiology, 125*(3), 1198–1205.

Kenrick, P. (2013). The origin of roots. In A. Eshel & T. Beeckman (Eds.), *Plant roots: The hidden half* (4th ed.). Taylor & Francis Group. p. 1.1-1.14.

Kenrick, P., & Crane, P. R. (1997). The origin and early evolution of plants on land. *Nature, 389*, 33–39.

Kenrick, P., & Strullu-Derrien, C. (2014). The origin and early evolution of roots. *Plant Physiology, 166*(2), 570–580.

Kenrick, P., Wellman, C. H., Schneider, H., & Edgecombe, G. D. (2012). A timeline for terrestrialization: Consequences for the carbon cycle in the Palaeozoic. *Philosophical Transactions of the Royal Society of London. Series B, Biological Sciences, 367*(1588), 519–536.

Kiers, E. T., Duhamel, M., Beesetty, Y., Mensah, J. A., Franken, O., Verbruggen, E., et al. (2011). Reciprocal rewards stabilize cooperation in the mycorrhizal symbiosis. *Science, 333*(6044), 880–882.
Kutschera-Mitter, L., Barmicheva, K. M., & Sobotik, M. (1998). The importance of root-cap mucilage for plant and soil. In J. E. Box (Ed.), *Root demographics and their efficiencies in sustainable agriculture, grasslands and forest ecosystems. Developments in plant and soil sciences* (pp. 673–683). Dordrecht: Springer Netherlands.
Lavin, M., Herendeen, P. S., & Wojciechowski, M. F. (2005). Evolutionary rates analysis of leguminosae implicates a rapid diversification of lineages during the tertiary. *Systematic Biology, 54*(4), 575–594.
Lenton, T. M., Crouch, M., Johnson, M., Pires, N., & Dolan, L. (2012). First plants cooled the Ordovician. *Nature Geoscience, 5*(2), 86–89.
Lewis, D. R., Ramirez, M. V., Miller, N. D., Vallabhaneni, P., Ray, W. K., Helm, R. F., et al. (2011). Auxin and ethylene induce Flavonol accumulation through distinct transcriptional networks. *Plant Physiology, 156*(1), 144–164.
Li, Z., Baniaga, A. E., Sessa, E. B., Scascitelli, M., Graham, S. W., Rieseberg, L. H., et al. (2015). Early genome duplications in conifers and other seed plants. *Science Advances, 1*(10), 1–8.
Li, X. K., Guo, Z. L., Lv, Y., Cen, X., Ding, X. P., Wu, H., et al. (2017). Genetic control of the root system in rice under normal and drought stress conditions by genome-wide association study. *PLoS Genetics, 13*(7), 24.
Lugtenberg, B., & Kamilova, F. (2009). Plant-growth-promoting Rhizobacteria. *Annual Review of Microbiology, 63*, 541–556.
Lynch, J. (1995). Root architecture and plant productivity. *Plant Physiology, 109*(1), 7–13.
Mairhofer, S., Sturrock, C. J., Bennett, M. J., Mooney, S. J., & Pridmore, T. P. (2015). Extracting multiple interacting root systems using X-ray microcomputed tomography. *The Plant Journal: For Cell and Molecular Biology, 84*(5), 1034–1043.
Malamy, J. E., & Benfey, P. N. (1997). Organization and cell differentiation in lateral roots of *Arabidopsis thaliana. Development, 124*(1), 33–44.
Manschadi, A. M., Manske, G. G. B., & Vlek, P. L. G. (2013). Root architecture and resource acquisition: Wheat as a model plant. In A. Eshel & T. Beeckman (Eds.), *Plant roots: The hidden half* (4th ed.). CRC Press. pp. 22-1-22-18.
Martin, F. M., Uroz, S., & Barker, D. G. (2017). Ancestral alliances: Plant mutualistic symbioses with fungi and bacteria. *Science, 356*(6340).
Matsunaga, K. K. S., & Tomescu, A. M. F. (2016). Root evolution at the base of the lycophyte clade: Insights from an early Devonian lycophyte. *Annals of Botany, 117*(4), 585–598.
Mcnamara, S. T., & Mitchell, C. A. (1991). Roles of auxin and ethylene in adventitious root-formation by a flood-resistant tomato genotype. *Hortscience, 26*(1), 57–58.
Meijon, M., Satbhai, S. B., Tsuchimatsu, T., & Busch, W. (2014). Genome-wide association study using cellular traits identifies a new regulator of root development in *Arabidopsis. Nature Genetics, 46*(1), 77–81.
Meister, R., Rajani, M. S., Ruzicka, D., & Schachtman, D. P. (2014). Challenges of modifying root traits in crops for agriculture. *Trends in Plant Science, 19*(12), 779–788.
Menand, B., Yi, K., Jouannic, S., Hoffmann, L., Ryan, E., Linstead, P., et al. (2007). An ancient mechanism controls the development of cells with a rooting function in land plants. *Science, 316*(13), 1477–1481.
Mills, B. J. W., Batterman, S. A., & Field, K. J. (2018). Nutrient acquisition by symbiotic fungi governs Palaeozoic climate transition. *Philosophical Transactions of the Royal Society of London. Series B, Biological Sciences, 373*(1739).
Morris, J. L., Puttick, M. N., Clark, J. W., Edwards, D., Kenrick, P., Pressel, S., et al. (2018). The timescale of early land plant evolution. *Proceedings of the National Academy of Sciences of the United States of America, 115*(10), E2274–E2283.

Muday, G. K., Rahman, A., & Binder, B. M. (2012). Auxin and ethylene: Collaborators or competitors? *Trends in Plant Science*, *17*(4), 181–195.
Negi, S., Ivanchenko, M. G., & Muday, G. K. (2008). Ethylene regulates lateral root formation and auxin transport in *Arabidopsis thaliana*. *The Plant Journal*, *55*(2), 175–187.
Negi, S., Sukumar, P., Liu, X., Cohen, J. D., & Muday, G. K. (2010). Genetic dissection of the role of ethylene in regulating auxin-dependent lateral and adventitious root formation in tomato. *The Plant Journal*, *61*(1), 3–15.
Newcombe, F. C., & Rhodes, A. L. (1904). Chemotropism of roots. *Botanical Gazette*, *37*(1), 22–35.
Ogura, T., & Busch, W. (2015). From phenotypes to causal sequences: Using genome wide association studies to dissect the sequence basis for variation of plant development. *Current Opinion in Plant Biology*, *23*, 98–108.
Olsen, S. R., & Kemper, W. D. (1968). Movement of nutrients to plant roots. In A. G. Norman (Ed.), *Vol. 20. Advances in agronomy* (pp. 91–151).
Orman-Ligeza, B., Parizot, B., Gantet, P. P., Beeckman, T., Bennett, M. J., & Draye, X. (2013). Post-embryonic root organogenesis in cereals: Branching out from model plants. *Trends in Plant Science*, *18*(8), 1360–1385.
Ortíz-Castro, R., Contreras-Cornejo, H. A., Macías-Rodríguez, L., & López-Bucio, J. (2009). The role of microbial signals in plant growth and development. *Plant Signaling & Behavior*, *4*(8), 701–712.
Otvos, K., & Benkova, E. (2017). Spatiotemporal mechanisms of root branching. *Current Opinion in Genetics & Development*, *45*, 82–89.
Pacheco-Villalobos, D., & Hardtke, C. S. (2012). Natural genetic variation of root system architecture from *Arabidopsis* to *Brachypodium*: Towards adaptive value. *Philosophical Transactions of the Royal Society of London. Series B, Biological Sciences*, *367*(1595), 1552–1558.
Pacheco-Villalobos, D., Sankar, M., Ljung, K., & Hardtke, C. S. (2013). Disturbed local auxin homeostasis enhances cellular anisotropy and reveals alternative wiring of auxin-ethylene crosstalk in *Brachypodium distachyon* seminal roots. *PLoS Genetics*, *9*(6), e1003564.
Pagès, L., & Kervella, J. (2018). Seeking stable traits to characterize the root system architecture. Study on 60 species located at two sites *in natura*. *Annals of Botany*, *122*(1), 107–115.
Parˇenicová, L., de Folter, S., Kieffer, M., Horner, D. S., Favalli, C., Busscher, J., et al. (2003). Molecular and phylogenetic analyses of the complete MADS-box transcription factor family in Arabidopsis. *The Plant Cell*, *15*(7), 1538–1551.
Pawlowski, K., & Bisseling, T. (1996). Rhizobial and Actinorhizal symbioses: What are the shared features? *The Plant Cell*, *8*(10), 1899–1913.
Pieterse, C. M. J., de Jonge, R., & Berendsen, R. L. (2016). The soil-borne supremacy. *Trends in Plant Science*, *21*(3), 171–173.
Pirozynski, K. A., & Malloch, D. W. (1975). The origin of land plants: A matter of mycotrophism. *Bio Systems*, *6*(3), 153–164.
Postma, J. A., Dathe, A., & Lynch, J. P. (2014). The optimal lateral root branching density for maize depends on nitrogen and phosphorus availability. *Plant Physiology*, *166*(2), 590–602.
Przemeck, G. K. H., Mattsson, J., Hardtke, C. S., Sung, Z. R., & Berleth, T. (1996). Studies on the role of the *Arabidopsis* gene *MONOPTEROS* in vascular development and plant cell axialization. *Planta*, *200*(2), 229–237.
Puttick, M. N., Morris, J. L., Williams, T. A., Cox, C. J., Edwards, D., Kenrick, P., et al. (2018). The interrelationships of land plants and the nature of the ancestral Embryophyte. *Current Biology*, *28*(5), 733–745.
Raven, J. A., & Edwards, D. (2001). Roots: Evolutionary origins and biogeochemical significance. *Journal of Experimental Botany*, *52*, 381–401. Roots Special Issue.

Rebouillat, J., Dievart, A., Verdeil, J. L., Escoute, J., Giese, G., Breitler, J. C., et al. (2009). Molecular genetics of rice root development. *Rice*, *2*(1), 15–34.

Rellán-Álvarez, R., Lobet, G., Lindner, H., Pradier, P.-L., Sebastian, J., Yee, M.-C., et al. (2015). GLO-roots: An imaging platform enabling multidimensional characterization of soil-grown root systems. *eLife*, *4*, e07597.

Rensing, S. A. (2018). Plant evolution: Phylogenetic relationships between the earliest land plants. *Current Biology*, *28*(5), R210–R213.

Ristova, D., Carre, C., Pervent, M., Medici, A., Kim, G. J., Scalia, D., et al. (2016). Combinatorial interaction network of transcriptomic and phenotypic responses to nitrogen and hormones in the *Arabidopsis thaliana* root. *Science Signaling*, *9*(451), rs13.

Rosas, U., Cibrian-Jaramillo, A., Ristova, D., Banta, J. A., Gifford, M. L., Fan, A. H., et al. (2013). Integration of responses within and across Arabidopsis natural accessions uncovers loci controlling root systems architecture. *Proceedings of the National Academy of Sciences of the United States of America*, *110*(37), 15133–15138.

Rothwell, G. W., & Erwin, D. M. (1985). The rhizomorph apex of Paurodendron; implications for homologies among the rooting organs of Lycopsida. *American Journal of Botany*, *72*(1), 86–98.

Ruzicka, K., Ljung, K., Vanneste, S., Podhorska, R., Beeckman, T., Friml, J., et al. (2007). Ethylene regulates root growth through effects on auxin biosynthesis and transport-dependent auxin distribution. *The Plant Cell*, *19*(7), 2197–2212.

Ryan, R. P., Germaine, K., Franks, A., Ryan, D. J., & Dowling, D. N. (2008). Bacterial endophytes: Recent developments and applications. *FEMS Microbiology Letters*, *278*(1), 1–9.

Scheres, B., Benfey, P., & Dolan, L. (2002). Root development. *The Arabidopsis Book*, *1*, e0101.

Scheres, B., Wolkenfelt, H., Willemsen, V., Terlouw, M., Lawson, E., Dean, C., et al. (1994). Embryonic origin of the *Arabidopsis* primary root and root meristem initials. *Development*, *120*, 2475–2487.

Schmidt, J. E., Bowles, T. M., & Gaudin, A. C. M. (2016). Using ancient traits to convert soil health into crop yield: Impact of selection on maize root and rhizosphere function. *Frontiers in Plant Science*, 7(373), 1–11.

Schüepp, O. (1917). Untersuchungen über Wachstum und Formwechsel von Vegetationspunkten. *Jahrbücher Für Wissenschaftliche Botanik*, *57*, 17–79.

Schwartz, R. A., Ortiz, I., Maymon, M., Herbold, W. C., Fujishige, A. N., Vijanderan, A. J., et al. (2013). *Bacillus simplex*—A little known PGPB with anti-fungal activity—Alters pea legume root architecture and nodule morphology when Coinoculated with *Rhizobium leguminosarum* bv. Viciae. *Agronomy*, *3*(4), 595–620.

Seago, J. L., & Fernando, D. D. (2013). Anatomical aspects of angiosperm root evolution. *Annals of Botany*, *112*(2), 223–238.

Shahzad, S. M., Khalid, A., Arshad, M., Tahir, J., & Mahmood, T. (2010). Improving nodulation, growth and yield of *Cicer arietinum* L. through bacterial ACC-deaminase induced changes in root architecture. *European Journal of Soil Biology*, *46*(5), 342–347.

Shimamura, M. (2016). *Marchantia polymorpha*: Taxonomy, phylogeny and morphology of a model system. *Plant and Cell Physiology*, *57*(2), 230–256.

Slovak, R., Goschl, C., Su, X., Shimotani, K., Shiina, T., & Busch, W. (2014). A scalable open-source pipeline for large-scale root phenotyping of *Arabidopsis*. *The Plant Cell*, *26*(6), 2390–2403.

Smith, S., & De Smet, I. (2012). Root system architecture: Insights from *Arabidopsis* and cereal crops. *Philosophical Transactions of the Royal Society of London. Series B, Biological Sciences*, *367*(1595), 1441–1452.

Soltis, D. E., Soltis, P. S., Chase, M. W., Mort, M. E., Albach, D. C., Zanis, M., et al. (2000). Angiosperm phylogeny inferred from 18S rDNA, rbcL, and atpB sequences. *Botanical Journal of the Linnean Society*, *133*(4), 381–461.

Sorin, C., Bussell, J. D., Camus, I., Ljung, K., Kowalczyk, M., Geiss, G., et al. (2005). Auxin and light control of adventitious rooting in *Arabidopsis*. *The Plant Cell, 17*(5), 1343–1359.

Steemans, P., Hérissé, A. L., Melvin, J., Miller, M. A., Paris, F., Verniers, J., et al. (2009). Origin and radiation of the earliest vascular land plants. *Science, 324*(5925), 353.

Steffens, B., & Rasmussen, A. (2016). The physiology of adventitious roots. *Plant Physiology, 170*(2), 603–617.

Stoeckle, D., Thellmann, M., & Vermeer, J. E. (2018). Breakout—Lateral root emergence in *Arabidopsis thaliana*. *Current Opinion in Plant Biology, 41*, 67–72.

Stuessy, T. F. (2010). Paraphyly and the origin and classification of angiosperms. *Taxon, 59*(3), 689–693.

Su, S. H., Gibbs, N. M., Jancewicz, A. L., & Masson, P. H. (2017). Molecular mechanisms of root gravitropism. *Current Biology, 27*(17), R964–R972.

Swarup, K., Benková, E., Swarup, R., Casimiro, I., Péret, B., Yang, Y., et al. (2008). The auxin influx carrier LAX3 promotes lateral root emergence. *Nature Cell Biology, 10*(8), 946–954.

Szoboszlay, M., Lambers, J., Chappell, J., Kupper, J. V., Moe, L. A., & McNear, D. H. (2015). Comparison of root system architecture and rhizosphere microbial communities of Balsas teosinte and domesticated corn cultivars. *Soil Biology and Biochemistry, 80*, 34–44.

Tomescu, A. M. F. (2016). Development: Paleobotany at the high table of evo–devo. *Current Biology, 26*(12), R505–R508.

Uga, Y., Sugimoto, K., Ogawa, S., Rane, J., Ishitani, M., Hara, N., et al. (2013). Control of root system architecture by *DEEPER ROOTING 1* increases rice yield under drought conditions. *Nature Genetics, 45*(9), 1097–1102.

Van Norman, J. M., Liberman, L. M., & Benfey, P. N. (2013). Arabidopsis root. In A. Eshel & T. Beeckman (Eds.), *Plant roots: The hidden half.* (4th ed.). Taylor and Francis Group. 2-1–2-17.

Vandenbussche, F., Vriezen, W. H., & Straeten, D. V. D. (2007). Ethylene biosynthesis and signaling: A puzzle yet to be completed. In P. Hedden & S. G. Thomas (Eds.), *Annual plant reviews Volume 24: Plant hormone signaling*.

Vermeer, J. E. M., von Wangenheim, D., Barberon, M., Lee, Y., Stelzer, E. H. K., Maizel, A., et al. (2014). A spatial accommodation by neighboring cells is required for organ initiation in *Arabidopsis*. *Science, 343*(6167), 178–183.

Verstraeten, I., Schotte, S., & Geelen, D. (2014). Hypocotyl adventitious root organogenesis differs from lateral root development. *Frontiers in Plant Science, 5*(495), 1–13.

Vilches-Barro, A., & Maizel, A. (2015). Talking through walls: Mechanisms of lateral root emergence in *Arabidopsis thaliana*. *Current Opinion in Plant Biology, 23*, 31–38.

Vogel, J. P., Garvin, D. F., Leong, O. M., & Hayden, D. M. (2006). *Agrobacterium*-mediated transformation and inbred line development in the model grass *Brachypodium distachyon*. *Plant Cell, Tissue and Organ Culture, 84*(2), 199–211.

von Wangenheim, D., Fangerau, J., Schmitz, A., Smith, R. S., Leitte, H., Stelzer, E. H., et al. (2016). Rules and self-organizing properties of post-embryonic plant organ cell division patterns. *Current Biology, 26*(4), 439–449.

Wang, D., Griffitts, J., Starker, C., Fedorova, E., Limpens, E., Ivanov, S., et al. (2010). A nodule-specific protein secretory pathway required for nitrogen-fixing symbiosis. *Science, 327*(5969), 1126–1129.

Weigel, D. (2012). Natural variation in Arabidopsis: From molecular genetics to ecological genomics. *Plant Physiology, 158*(1), 2–22.

Xue, J., Deng, Z., Huang, P., Huang, K., Benton, M. J., Cui, Y., et al. (2016). Belowground rhizomes in paleosols: The hidden half of an early devonian vascular plant. *Proceedings of the National Academy of Sciences of the United States of America, 113*(34), 9451–9456.

York, L. M., Galindo-Castañeda, T., Schussler, J. R., & Lynch, J. P. (2015). Evolution of US maize (*Zea mays* L.) root architectural and anatomical phenes over the past 100 years corresponds to increased tolerance of nitrogen stress. *Journal of Experimental Botany, 66*(8), 2347–2358.

Zhang, H., & Forde, B. G. (1998). An *Arabidopsis* MADS box gene that controls nutrient-induced changes in root architecture. *Science*, *279*(5349), 407–409.
Zhang, G. Q., Liu, K. W., Li, Z., Lohaus, R., Hsiao, Y. Y., Niu, S. C., et al. (2017). The Apostasia genome and the evolution of orchids. *Nature*, *549*(7672), 379–383.
Zohary, D. (2004). Unconscious selection and the evolution of domesticated plants. *Economic Botany*, *58*(1), 5–10.

CHAPTER FOUR

Patterning at the shoot apical meristem and phyllotaxis

Bihai Shi, Teva Vernoux*
Laboratoire Reproduction et Developpement des Plantes, Univ Lyon, ENS de Lyon, UCB Lyon1, CNRS, INRA, Lyon, France
*Corresponding author: e-mail address: teva.vernoux@ens-lyon.fr

Contents

Abstract

The shoot apical meristem (SAM) generates all above-ground organs throughout the life of plants. The development and maintenance of the SAM are crucial for building the plant architecture. The spatiotemporal patterning of lateral organs (leaves and flowers), called phyllotaxis, is one of the best-characterized self-organizing systems and has long been proposed to be driven by inhibitory fields generated by the existing organs and blocking new initiations in their vicinity. Recent years have seen impressive progress in our understanding of the molecular mechanisms controlling SAM function, and on how these mechanisms act in phyllotactic patterning. In this chapter, we first review the regulation of SAM stem cell activity and discuss how feedback signals

Current Topics in Developmental Biology, Volume 131
ISSN 0070-2153
https://doi.org/10.1016/bs.ctdb.2018.10.003

coming from the differentiated organs affect stem cell homeostasis. Then we highlight experimental and theoretical works that have revealed the chemical and biophysical factors acting in the regulation of phyllotaxis. Finally, we summarize the important roles of SAM geometry in phyllotaxis.

1. Introduction

In contrast to animals, in which organogenesis occurs mainly during embryogenesis, plants continuously produce new organs during their post-embryonic development. This unique developmental ability is due to the existence of stem cell-containing tissues located at the growing apices of plants, called meristems. The shoot apical meristem (SAM), located at the shoot apex, and the root apical meristem (RAM), located at the root tip, give birth to all the aerial and underground parts of a plant, respectively. In several basal plants, the shoot apex is made of a single triangular cell (Golub & Wetmore, 1948). But in higher plants, like *Arabidopsis thaliana*, *Zea mays* (maize), and *Oryza sativa* (rice), the SAM is multicellular, with distinct cell layers and domains.

Both the SAM and the RAM are established during embryogenesis. Once environmental conditions, such as temperature and humidity, are suitable for germination, the meristems are activated to initiate organogenesis (although for some species like maize, organogenesis by the SAM is already active during embryogenesis), which is repetitive and modular. The modules are called phytomers (Hay & Tsiantis, 2010; Teichmann & Muhr, 2015). During the vegetative phase of development, each phytomer consists of a compressed internode, a leaf on the node, and an axillary meristem in the leaf axil. When plants reach the floral transition stage, internodes start elongating, and leaves are narrower, and their axillary meristems develop into branches. Later in the reproductive phase, phytomers bear flowers, with axillary leaves often being absent. Meristem activity is not only controlled by molecular and hormonal networks in the stem cells and surrounding regions but also by feedback regulation through long-distance signals from the differentiated cells in the previously formed organs. We will discuss these regulatory mechanisms below.

Phyllotaxis (from ancient Greek: *phýllon* refers to "leaf" and *táxis* means "arrangement") refers to the spatial organization of lateral structures (leaves, branches, flowers, etc.) along the stem. Phyllotaxis has been attracting the attention of mathematicians, physicists, and biologists for several centuries (for review, Adler, 1997; Galvan-Ampudia, Chaumeret, Godin, & Vernoux, 2016;

Kuhlemeier, 2017; Traas, 2013). In nature, there are different types of phyllotactic patterns, including alternate (distichous), opposite (decussate), whorled, and spiral. Among these, spiral phyllotaxis is the most frequent (Jean & Barabe, 1998), with an average divergence angle between single consecutive organ primordia close to 137.5°, the so-called "golden angle" defined from the Fibonacci series. The primary determinant of phyllotaxis is the spatiotemporal pattern of organogenesis in the meristem (Fig. 1A). During development, these patterns sometimes shift (Bartlett & Thompson, 2014). For example, in *A. thaliana*, the cotyledons and the first pair of true leaves are initiated in a nearly opposite decussate pattern, while all the following leaves and flowers initiate following a spiral pattern and later, during flower formation, floral organs are arranged in whorled patterns.

In contrast to the extensive modeling work available, it is only in the early 21st century that the molecular mechanisms underlying phyllotactic patterning in the SAM started to be uncovered. In this chapter, we will discuss our current knowledge of these mechanisms. We will first present a brief overview of the development of the SAM, of its organization and of the regulation

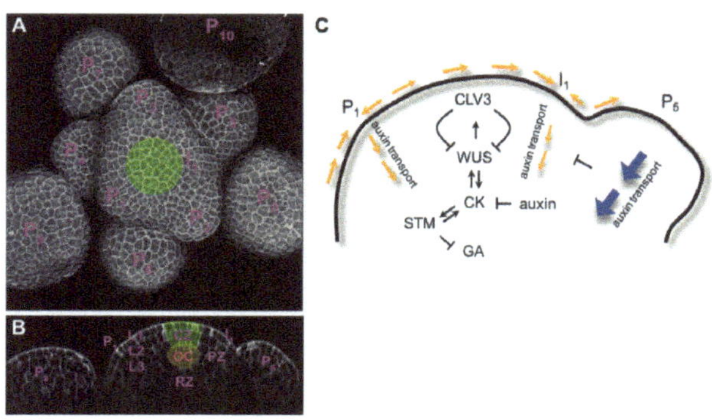

Fig. 1 The inflorescence shoot apical meristem of *A. thaliana* and its regulatory network controlling stem cell activity. (A) Top view of the SAM showing the center zone (CZ) and sequentially initiated primordia, with P_1 referring to the youngest primordium, P_2 the second youngest one, and so on; I_1 refers the oldest incipient primordium. (B) Side view of the SAM indicating the different structural organizations: three specific cell layers (L1, L2, and L3) and three different functional zones, the CZ, the peripheral zone (PZ), and the rib zone (RZ) that harbor the organizing center (OC). (C) Conceptual summary of the regulation of stem cell activity in the SAM. The interaction between CLV and WUS forms a negative regulatory feedback loop in the meristem center, with cytokinin (CK) generating positive loops with WUS and STM to maintain SAM development, auxin inhibiting the action of CK, and gibberellins (GA) mediating the coordination of STM regulation. Feedback from the differentiated organs regulates stem cell activity through auxin transport modulation (see Shi et al., 2018 for details). Bars = 20 μm.

of stem cell activity. We will then summarize the history of research on phyllotaxis and discuss the multiple regulatory levels involved in the regulation of phyllotaxis. We will notably highlight the role of the phytohormones auxin and cytokinin, of biomechanical signals, as well as of SAM geometry and of post-meristematic growth.

2. Development of the shoot apical meristem

2.1 A brief overview of the organization and genetic regulation of shoot apical meristem activity

Numerous reviews have been written on the genes regulating the activity of the SAM (Barton, 2010; Gaillochet, Daum, & Lohmann, 2015; Galli & Gallavotti, 2016; Soyars, James, & Nimchuk, 2016; Truskina & Vernoux, 2018; Wang, Smith, & Li, 2018). Here, we will only briefly summarize the current knowledge. SAMs of dicots, such as the one of *A. thaliana*, contain three well-defined cell layers (Fig. 1B). Whereas the two outer layers (L1 and L2) contain cells with anticlinal cell division planes (new cell wall being perpendicular to the meristem surface), cells from the third layer (L3) divide in all directions. During development, the L1 forms the epidermis, while the L2 and L3 form the subepidermal and vascular tissues of aerial tissues. In monocots, such as maize and rice, the SAM consists only of two layers with the L1 producing similarly the epidermis and the L2 building the internal tissues.

The SAM can also be divided into different functional domains based on their gene expression patterns and signaling activities. The central zone (CZ) harbors the stem cells that divide slowly. The homeostasis of this population of cells is essential to maintain a stable meristem size. With division in the CZ, some of the daughter cells are moved outward from the CZ to the peripheral zone (PZ), where new organ primordia (leaves and flowers) are initiated. Underneath the CZ is the rib zone (RZ) that is responsible for the development of the internal tissues of the stem. In addition, there is a group of cells located in the upper part of the RZ, referred to as the organizing center (OC) that establishes a stem cell niche and regulates the stem cell population above it (Fig. 1C). First identified in *A. thaliana*, the transcription factor gene *WUSCHEL* (*WUS*) is expressed in the OC (Mayer et al., 1998). The WUS protein moves upward into the CZ through plasmodesmata (Daum, Medzihradszky, Suzaki, & Lohmann, 2014), activating in the stem cells the expression of the *CLAVATA3* (*CLV3*) gene, which is the founding member of the CLV3/EMBRYO SURROUNDING

REGION (CLE) peptide family and is expressed specifically in the CZ (Clark, Running, & Meyerowitz, 1995; Opsahl-Ferstad, Deunff, Dumas, & Rogowsky, 1997). Reciprocally, when secreted from the CZ, the CLV3 peptide is perceived by membrane receptors, notably the receptor-like kinase CLV1, in the OC to repress *WUS* activity (Clark, Williams, & Meyerowitz, 1997; Ottoline Leyser & Furner, 1992). This interaction between *CLV3* and *WUS* establishes a negative feedback regulatory loop that contributes to stem cell homeostasis in the SAM. Other independent regulators of stem cell niche homeostasis have been identified more recently, providing robustness to the organization of the SAM (Mandel et al., 2016; Schuster et al., 2014; Zhou et al., 2015). Numerous regulators of organogenesis are also known (reviewed in Besnard, Vernoux, & Hamant, 2011; Sassi & Vernoux, 2013; Sluis & Hake, 2015; Wang & Jiao, 2018), and we will discuss later how auxin acts as the master regulator of the organogenesis program.

2.2 Hormonal regulation of stem cell activity

Plant hormones have long been known to have an indispensable role in the regulation of stem cell activity and in self-organization of the SAM (Fig. 1C). First, cytokinin (CK) is thought to be important for SAM development because it can prevent cell differentiation and promote cell division of stem cells. Supporting this idea, a two-component sensor (TCS) that reports CK activity *in planta*, detects a maximum of CK signaling activity in the OC, overlapping with the expression of *WUS* (Gordon, Chickarmane, Ohno, & Meyerowitz, 2009; Muller & Sheen, 2008). This is consistent with the previous observation that the WUS protein inhibits transcription of two genes encoding CK signaling repressors, *ARABIDOPSIS RESPONSE REGULATOR7* (*ARR7*) and *ARR15* (Leibfried et al., 2005; Zhao et al., 2010), indicating that WUS positively regulates CK response sensitivity. Gordon and co-workers also reported that the CK signaling pathway, in turn, activates *WUS* levels and regulates stem cell number in the meristem (Gordon et al., 2009). Additionally, it was shown that a gene encoding a nucleotide sugar transporter, *REPRESSOR OF CYTOKININ DEFICIENCY1* (*ROCK1*), fine-tunes CK activity through modification of CK-degrading enzymes, the CK oxidases/dehydrogenases (CKXs), and thus regulates meristem activity (Niemann et al., 2015). On the other hand, the Class I *Knotted1-like homeobox* (KNOX1) genes, including *SHOOTMERISTEMLESS* (*STM*) in *A. thaliana* and *KNOTTED1* in maize, were shown to upregulate CK biosynthesis and consequently the CK response in the SAM (Jasinski et al., 2005; Kerstetter, Laudencia-Chingcuanco, Smith, & Hake, 1997; Yanai et al., 2005). The KNOX1 gene family encodes a group of transcription factors and is essential for meristem development and maintenance (Hay & Tsiantis, 2010). An activating enzyme in CK biosynthesis of rice, encoded by the *LONELY GUY* (*LOG*) gene, is required for meristem maintenance, with mRNA expression levels of KNOX1 family genes in *log* mutants being dramatically decreased (Kurakawa et al., 2007). Thus, KNOX1 family genes and CK may act in a positive regulatory loop during SAM development and maintenance.

In contrast to CK, auxin has a low activity level in the OC of the SAM, as indicated by the auxin signaling sensors DR5 and DII-VENUS, and negatively regulates the stem cell population (Benkova et al., 2003; Brunoud et al., 2012; Shi et al., 2018). However, at the same time, periodic auxin maxima have been shown to be driving organ initiation in the SAM (Benkova et al., 2003; Heisler et al., 2005; Reinhardt et al., 2003). We will discuss this aspect more thoroughly in the phyllotaxis section below. Within the SAM, it is reported that auxin negatively regulates the CK signaling inhibitors *ARR7* and *ARR15* through the action of MONOPTEROS/AUXIN RESPONSE FACTOR5 (MP/ARF5) transcription factor, suggesting a cross-talk between CK and auxin in the regulation of stem cell activity and SAM development (Zhao et al., 2010).

Besides CK and auxin, gibberellins (GAs), a group of plant hormones that promote cell elongation and cell division and thus play an essential role in plant growth, are found to be regulated by KNOX genes in the SAM. In *A. thaliana*, GA levels are coordinated by STM activity in the SAM through the inhibition of expression of the GA biosynthesis gene *GA20-oxidase1*, which is highly expressed in organ primordia (Jasinski et al., 2005). However, how GAs affect SAM maintenance remains poorly understood.

2.3 Feedback from differentiated cells

In animals, stem cell proliferation is controlled not only by signals from the stem cell niche but also by those from differentiating progeny cells (Lander, Gokoffski, Wan, Nie, & Calof, 2009; Rue & Martinez Arias, 2015). In the SAM, it has long been hypothesized that signals from lateral organs (leaves, flowers, and branches) have a feedback on meristem homeostasis (for details, see Barton, 2010). For example, the two *A. thaliana* YABBY genes *FILAMENTOUS FLOWER* (*FIL*) and *YABBY3* (*YAB3*) might non-autonomously regulate SAM activity through two putative signals originating from the meristem-primordia boundaries. Meristem size in *fil-8 yab3-2* double mutants is dramatically increased with enlarged *WUS* and *CLV3* expression domains (Goldshmidt, Alvarez, Bowman, & Eshed, 2008). It has also recently been reported that a maize CLE family gene, *ZmFCP1*, encoding a peptide secreted from leaf primordia, suppresses *WUS* expression and thus stem cell activity in the meristem (Je et al., 2016). More recently, combining biological experiments and mathematical modeling, it was found that polar auxin transport from lateral organs inhibits polar auxin transport from the SAM (including leaf primordia P_1 and incipient primordia I_1), keeping a constant low auxin level in the SAM, thereby maintaining a robust SAM stem cell activity and thus normal meristem growth and development (Shi et al., 2018). Based on this work, it was proposed that the lateral organs feedback on the growth of the SAM through an auxin transport switch mechanism (Fig. 1C). In summary, all these studies suggest that both signals emanating from the meristem-primordia boundaries and/or long-distance hormone transportation from lateral organs are responsible for a feedback regulation on stem cell proliferation and meristem development, which might balance the self-maintenance and organogenesis activities of the meristem.

3. A few words on the history of phyllotaxis research: The development of the inhibitory field concept

Phyllotaxis is a striking example of self-organized patterning in plants and, more generally, in developmental biology. Due to the unique mathematical regularities of phyllotaxis, phyllotaxis research has long been interdisciplinary and started at least as early as the 14th century (Adler, 1997). Numerous models have been proposed to explain phyllotactic patterning, and among them, the one based on the inhibitory field hypothesis is now well accepted. This hypothesis proposes that each formed organ primordium generates an inhibitory field (the inhibition could be explained non-exclusively by geometry, chemical, or mechanical signals) that prevents initiation of new primordia in its vicinity (Douady & Couder, 1996; Hofmeister, 1868; Snow & Snow, 1962). Therefore, as the SAM grows, incipient primordia can be initiated at sites where the inhibition is the lowest (Fig. 2).

As early as 1830, mathematicians discovered that in spiral phyllotaxis, the number of contact parastichies (spirals connecting two closest neighboring organs) in clockwise and counter-clockwise orientations, respectively, corresponds to two consecutive numbers of the Fibonacci series, where the first

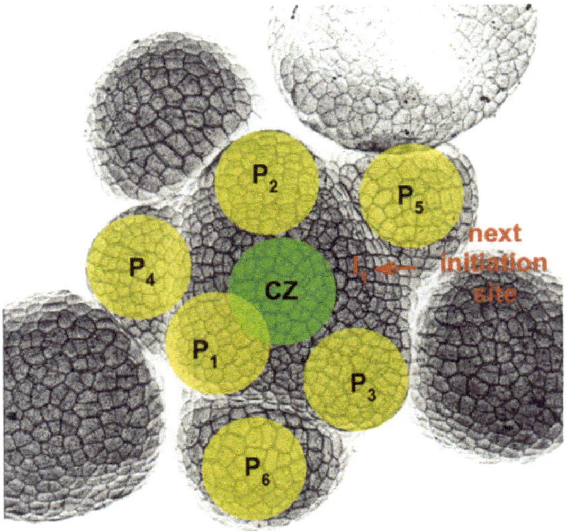

Fig. 2 Conceptual model of the inhibitory field hypothesis. Outside of the central zone (CZ), primordia initiated at the periphery produce inhibitory fields (yellow-filled circles) with the next incipient primordium (I_1) emerging at the site where the inhibitory effect is lowest.

numbers are 1 and 2, and then each number is the sum of the two previous ones. The golden angle (137.5°) is mathematically related to the Fibonacci sequence as it corresponds to the division of a circle (360°) by the value of the limit of the ratio between two consecutive numbers in the Fibonacci sequence, a ratio also called the "golden ratio." This is one of the reasons why spiral phyllotaxis has received much attention from both theoretical and experimental scientists.

One of the first models to explain phyllotaxis was put forward in the 19th century by Hofmeister (1868). He proposed that new primordia appear periodically at the apex boundary in the largest available gap left by previously formed primordia. Then, a chemical theory was put forward by Schoute (1913), stating that the placement of incipient primordium is determined by a chemical inhibitor secreted by the previously formed primordia that prevents a new primordium from emerging too close. Richard favored Schoute's hypothesis and called it a "field theory" (Richards, 1948, 1951). This inhibitory field concept also found support from microsurgical experiments conducted on meristems of *Lupinus albus* (Snow & Snow, 1932). Snows found that the position where a new leaf primordium is initiated is influenced by the previously formed primordia adjacent to the site of initiation, and thus, they proposed that "each new leaf is determined in the first space on the growing apical cone that attains a necessary minimum size and minimum distance below the tip" (Snow & Snow, 1962). This microsurgery results were recently re-explored and further confirmed using laser-ablation approaches (Reinhardt, Frenz, Mandel, & Kuhlemeier, 2005). Wardlaw (1949) also conducted microsurgery experiments but in shoot meristems of ferns and favored an explanation of inhibition due to physical stresses. Numerical simulations were used to analyze the system dynamics emerging from inhibitory fields. In their seminal work, Douady and Couder demonstrated that an inhibitory field hypothesis can recapitulate a wide variety of phyllotactic patterns by only adjusting one geometric parameter, corresponding to the ratio between the radius of lateral inhibitory fields and that of the central zone (Douady & Couder, 1996).

Taken together, the existing evidence suggests that inhibitory fields that may arise from either chemical or physical signals or even both could be necessary and sufficient for self-organization of phyllotactic patterning in the SAM. In the following sections, we will discuss the role of chemical (auxin and cytokinin) and physical signals in phyllotaxis (Fig. 3).

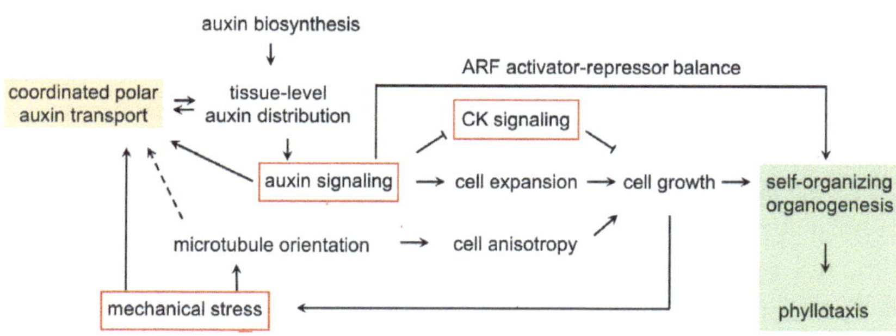

Fig. 3 Graphical summary of the regulation of chemical (auxin, CK) and mechanical signals on phyllotaxis in the SAM. First, a positive feedback loop between auxin and PIN1 polarity self-organizes the auxin distribution in the SAM. Self-organization of organogenesis driving phyllotaxis is further dependent on spatial differences in auxin signaling capacities. The signaling capacities also depend on the local balance between ARF activators and repressors. Regulation of auxin biosynthesis also contributes to the distribution of auxin. Second, CK signaling acts downstream auxin signaling in the temporal regulation of phyllotaxis. Third, polar auxin transport, primarily mediated by PIN1 in the SAM, responds directly to mechanical stress mediated by microtubule orientation. Microtubule organization also controls cell wall anisotropy, thus determining the direction of cell growth. Together with auxin that controls cell wall expansion and growth rate, this mechanism regulates cell growth. Cell growth, in turn, generates a feedback on mechanical stress, thus channeling organogenesis.

4. Phyllotaxis and chemical signals

4.1 Coordinated polar auxin transport: Tissue level auxin distribution allowing for self-organizing organ initiation

Studies over the past 20 years have identified auxin as a major actor in the molecular mechanisms of organ initiation and phyllotaxis. In contrast to that of other phytohormones, transport of auxin in plants is mainly directional, i.e., polar, involving the action of specific carriers. Among these carriers, the PIN-FORMED1 (PIN1) protein is the major regulator of polar auxin transport in the SAM. It is localized asymmetrically on plasma membranes and is thought to direct the auxin flux (Petrasek et al., 2006; Wisniewska et al., 2006). Mutants affecting *PIN1* develop almost no floral buds on the inflorescence meristem, resulting in a "pin-shaped" inflorescence, a phenotype that can be mimicked by treatment with the auxin transport inhibitor *N*-1-naphthylphthalamic acid (NPA) (Galweiler et al., 1998; Okada, 1991).

These results were the first to indicate that polar auxin transport is required for organ initiation. It was reported later that exogenous auxin treatment could rescue organogenesis of *pin1* inflorescences in *A. thaliana* or of NPA-induced pin-shaped vegetative SAMs in tomato (Reinhardt, 2000; Vernoux, Kronenberger, Grandjean, Laufs, & Traas, 2000), suggesting that auxin maxima regulate organ initiation. In addition, immunodetection and advanced live-imaging methods showed that cellular PIN1 polarities create a convergence pattern toward incipient primordia initiation sites, suggesting that polar auxin transport mediated by PIN1 forms auxin maxima in the meristem, and activates organ initiation (Heisler et al., 2005; Reinhardt et al., 2003). Using the biosensors DII-VENUS and its derivative R2D2 allowed the visualization of auxin levels in the SAM, confirming auxin accumulation at the sites where primordia initiate (Bhatia et al., 2016; Vernoux et al., 2011). Using both computer- and experiment-based analysis, the AUXIN1/LIKE AUXIN1 (AUX/LAX) family of auxin influx carriers has also been proposed to contribute significantly to auxin distribution dynamics, and thus to stabilize phyllotactic patterning (Bainbridge et al., 2008; Kramer, 2004; Reinhardt et al., 2003). Importantly, besides controlling auxin accumulation, the convergence patterns of PIN1 polarities at incipient primordia have also been proposed to deplete auxin around the primordia (Heisler et al., 2005; Reinhardt et al., 2003), a prediction that was confirmed using the DII-VENUS biosensor (Vernoux et al., 2011). Taken together, these studies indicate that depletion of auxin, an activator of organogenesis, in the vicinity of primordia by auxin efflux transporters can generate a chemical inhibitory field.

But how is PIN1 polarity spatiotemporally regulated at the tissue level to get a coordinated auxin flux driving a self-organizing organogenesis pattern? We have extensive knowledge on the control of PIN1 polarity at the cellular level in *A. thaliana* (Friml, 2010), but our understanding at the tissue level is still limited. Different models have been proposed based on experimental observations. The concentration-based model postulates that PIN1 in the L1 of the SAM can be polarized against the auxin gradient in the tissue. For a given cell, polarity is established toward the neighboring cell that contains the highest auxin concentration, thereby resulting in auxin accumulation at the PIN1 convergence points (Jonsson, Heisler, Shapiro, Meyerowitz, & Mjolsness, 2006; Smith et al., 2006). The flux-based hypothesis, originally proposed to explain mid-vein formation (Sachs, 1969, 1981), also allows the simulation of realistic self-organization of the PIN1 network in the SAM. This hypothesis proposes that PIN1 localization at the membrane is proportional to the auxin flux through the membrane, thus stabilizing the auxin efflux (Abley, Sauret-Güeto, Marée, & Coen, 2016; Stoma et al., 2008). But, this model requires a transient auxin decrease during the formation of PIN1 convergence sites, which has not been observed to date.

A dual-polarization hypothesis has also been put forward and proposes that an up-the-gradient mechanism regulates PIN1 polarities in the epidermal cells, and that simultaneously a with-the-flux polarization acts in the internal tissues, allowing PIN1 localization to shift to the basal side of cells and to initiate venation (Bayer et al., 2009). The existence of different mechanisms of polarization is possibly supported by the analysis of a sister clade of *PIN1*, *Sister-of-PIN1* (*SoPIN1*), present in all flowering species excluding the Brassicaceae (and thus *A. thaliana*), and of two duplicated forms of *PIN1* in grasses, *PIN1a* and *PIN1b* (O'Connor et al., 2014). The spatiotemporal patterns of polarities of SoPIN1, PIN1a and PIN1b proteins in the SAM are compatible with SoPIN1 being polarized in an up-the-gradient manner in the epidermal and with PIN1a and PIN1b being localized in the inner cells as predicted by the with-the-flux mechanism (O'Connor et al., 2017, 2014).

While the assumptions are different, all models have in common the long-standing idea that auxin has a positive feedback on its transport. Taken together, they thus indicate that a regulatory loop between auxin, PIN1 protein expression, and PIN1 localization is likely responsible for the self-organization of cellular PIN1 polarities and thus of organ initiation in the

SAM. However, all models have significant limitations and further biological data and alternative mathematical models are required in the future to fully understand how polar auxin transport self-organizes the auxin distribution in the SAM (for discussion, see van Berkel, de Boer, Scheres, & ten Tusscher, 2013).

4.2 Auxin signaling: Downstream regulation of auxin distribution

Auxin signaling has also been reported more recently to play a role in phyllotaxis together with the spatial auxin distribution. Central to auxin signaling is the interaction between the AUXIN/INDOLE ACETIC ACID (AUX/IAA) transcriptional repressors and the AUXIN RESPONSE FACTOR (ARF) transcription factors. As cellular auxin levels increase, auxin promotes the association of AUX/IAA proteins with the TRANSPORT INHIBITOR RESPONSE1 (TIR1) F-box protein [or another one of three related AUXIN-RELATED F-BOX (AFB) proteins], which is part of an SCF E3 ubiquitin ligase complex. The AFBs act as co-receptors of auxin and their association with the SCF complex induces the proteolytic degradation of AUX/IAA proteins and thus releases the ARFs from the AUX/IAA repressors, allowing the ARFs to regulate transcriptional responses (Calderon Villalobos et al., 2012; Dharmasiri, Dharmasiri, & Estelle, 2005; Kepinski & Leyser, 2005; Tan et al., 2007). Some AUX/IAA family genes are also targets of ARFs, allowing for a negative feedback loop.

There are 29 AUX/IAAs and 23 ARFs in *A. thaliana*, with most ARFs being regulatory repressors and only 5 of them (ARF5 to ARF8 and ARF19) being activators (Guilfoyle & Hagen, 2007). Using in situ hybridization, Vernoux et al. (2011) have drawn a spatial regulatory map of auxin signaling in the SAM by analyzing the expression of most TIR1/AFB, AUX/IAA, and ARF family genes. This analysis showed a general trend with a low level of expression of ARF and Aux/IAA genes in the CZ and a high expression level in the PZ. Thus, there is large-scale co-expression of ARF activators and repressors in the SAM. A mathematical model of auxin signaling, based on a large-scale analysis of the AUX/IAA-ARF interaction networks, predicted that (1) the differential expression patterns between ARF activators and repressors produce a differential sensitivity to auxin between the CZ (low) and the PZ (high) and (2) the balance between ARF activators and repressors provides a buffering capacity toward input fluctuations, thereby stabilizing regulatory responses (Vernoux et al., 2011). The auxin signaling input sensor DII-Venus, and the output signaling reporter DR5::Venus, could be used to confirm these two predictions. These results likely explain why exogenous auxin treatment cannot induce organ initiation in the CZ of the *pin* mutant (Reinhardt, 2000) and that organogenesis occurs only in the PZ. Also, it suggests that auxin signaling plays a role in the robustness of phyllotaxis.

More evidence on the role of the auxin signaling pathway comes from studies of the MP/ARF5 transcription factor, one of the ARF activators in *A. thaliana*. The *mp* mutant shows a *pin*-like inflorescence meristem phenotype, but unlike in the *pin1* mutant, exogenous auxin application is not able to rescue its flower initiation defects (Hardtke & Berleth, 1998; Reinhardt et al., 2003; Zhao et al., 2010). MP/ARF5 was further shown to regulate floral organ initiation through the direct activation of key regulators of flower development: *LEAFY* (*LFY*), *AINTEGUMENTA* (*ANT*), *AINTEGUMENTA-LIKE6/PLETHORA3* (*AIL6/PLT3*), and *FIL*, and the cytokinin response factors *ARR7*, *TARGET OF MONOPTEROS3/CYTOKININ RESPONSE FACTOR2* (Weijers & Wagner, 2016; Yamaguchi et al., 2013). This allows for a better understanding of the molecular mechanisms by which auxin directly controls organogenesis. More recently, work combining live-imaging and molecular genetics proposed that MP is necessary in the SAM to stabilize the PIN1 convergence site before primordia outgrowth (Bhatia et al., 2016). Auxin signaling would thus be directly involved in the mechanisms of PIN1 polarization, thus playing a central role alongside polar auxin transport in the regulation of phyllotaxis. Finally, a recent study found that auxin transcriptional responses are restricted to the boundary between domains of expression of dorsoventral polarity regulators, i.e., *REVOLUTA* and *KANADI* genes (Caggiano et al., 2017), suggesting that organ polarity genes can also be essential for phyllotaxis by regulating auxin signaling capacities.

4.3 Regulation of auxin biosynthesis: Upstream regulation of auxin distribution

As mentioned in Section 1, *A. thaliana* plants change their leaf phyllotactic pattern from decussate to spiral after the initiation of the first pair of true leaves; however, this shift is dramatically delayed in the *plt3 plt5 plt7* triple mutant (Prasad et al., 2011). *PLT* genes encode members of the AP2-domain transcriptional factor family. The *PLT3*, *PLT5*, and *PLT7* genes are found expressed both in the CZ and the PZ (Prasad et al., 2011). Furthermore, genetics and functional analysis revealed that the PLT proteins regulate the expression of the auxin biosynthesis genes *YUC1* and *YUC4*

(Pinon, Prasad, Grigg, Sanchez-Perez, & Scheres, 2013). It was also found that three INDETERMINATE DOMAIN (IDD) transcription factors, IDD14, IDD15, and IDD16, promote auxin accumulation by directly regulating the expression of *PIN1* and the two auxin biosynthesis genes *YUC5* and *TRYPTOPHAN AMINOTRANSFERASE of ARABIDOPSIS* (Cui et al., 2013). Thus, these results indicate a role for PLT- or IDD-dependent auxin biosynthesis in the *A. thaliana* meristem. In maize, it was reported that two auxin biosynthesis genes, *VANISHING TASSEL2* and *SPARSE INFLORESCENCE1*, act in the same pathway to regulate vegetative and reproductive meristem development and phyllotaxis, with their double mutant exhibiting a barren inflorescence meristem without lateral organs or functional spikelets (Gallavotti, Yang, Schmidt, & Jackson, 2008; Phillips et al., 2011). Taken together, all these results suggest that auxin biosynthesis is also required for phyllotaxis.

The above-mentioned findings are consistent with earlier work, which revealed that mutations affecting *YUC1* and *YUC4* enhance the *pin1* mutant phenotype, and that disruption of *AUX1* in the *yuc1 yuc2 yuc4 yuc6* quadruple mutant phenocopies the *yuc1 yuc4 pin1* triple mutant (Cheng, Dai, & Zhao, 2007). These results indicate that auxin biosynthesis and polar auxin transport contribute synergistically to phyllotaxis. However, we are still missing a clear view of the sites of auxin biosynthesis in the SAM, and auxin biosynthesis remains to be incorporated into auxin transport-based phyllotaxis models to get a more comprehensive understanding of self-organized patterning in the SAM.

4.4 Cytokinin: A secondary inhibitory field acting on the plastochrone

Given that auxin and CK signaling cross-talk during SAM development (see above), one might ask whether there is a role for CK in organ initiation and phyllotaxis? It was reported that CK, along with auxin, is required for organ formation (Yoshida, Mandel, & Kuhlemeier, 2011). In this section, we will discuss how cytokinin is involved in phyllotaxis.

A. thaliana mutants affecting the CK signaling inhibitor *ARABIDOPSIS HISTIDINE PHOSPHOTRANSFER PROTEIN6* (*AHP6*) have a phenotype of increased organ co-initiation that results in permutations of the order of siliques along the stem, while the divergence angle between organ initiation sites in the SAM is unchanged (Besnard, Rozier, & Vernoux, 2014). Further observations found that *AHP6* expression in organ primordia is directly activated by auxin through MP and appears specifically one plastochron later than expression of the DR5 reporter (Besnard et al., 2014). More interestingly, it was discovered that the AHP6 protein can move out of the primordia where it is transcribed, establishing a differential of accumulation between successive incipient primordia: AHP6 is low in I_1 (Fig. 2), while it accumulates in I_2 where the next primordium is expected to form after I_1. Blocking AHP6 protein movement by fusing three copies of Venus to the AHP6 protein was not able to rescue *ahp6* mutants in contrast to fusions with a single copy of Venus that can move. This finding supports a scenario where low levels of AHP6 at I_1 allow for high CK signaling and organ initiation, while high levels of AHP6 at I_2 non-cell autonomously repress CK signaling and organ initiation, thereby avoiding organ co-initiation and stabilizing the plastochron (Besnard et al., 2014). Thus, in addition to the primary auxin inhibitory fields that specify the sites of organ initiation, AHP6, acting downstream of auxin, produces a CK-based secondary inhibitory field that increases the robustness of the timing of phyllotaxis patterning (Besnard et al., 2014).

These findings also indicate that AHP6 filters out noise in phyllotaxis, as co-initiation results in permutations of organ order along the stem, and thus in changes of the relative angles on the stem. Phyllotaxis disorder resulting from noise in plastochron initiation has been shown to be common in plants (Refahi et al., 2016). A modeling approach was used to further explore the origin of co-initiation. More precisely, the authors revisited the deterministic model of phyllotaxis developed by Douady and Couder (1996) and developed a stochastic model of primordia initiation at the SAM combining cell-autonomous decisions and stochastic signal perception by cells (Refahi et al., 2016). This model accurately recapitulates the classical phyllotactic patterns and, in addition, produces realistic permutation disorders as a result of stochasticity in signal perception by cells. The model thus predicts that perception and processing of the signals (among which auxin) driving phyllotaxis have stochastic properties (a prediction that remains to be proven), which can induce the co-initiation of primordia. Moreover, the CK-based fields allow to limit the propagation of noise from the cellular level to the tissue scale and thus its impact on the construction of the stem.

5. Mechanical signals: How does physics affect phyllotaxis?

While the role of chemical signals is now well established, the role of mechanical signals in phyllotaxis is still elusive; but accumulating evidence indicates that they also play a role in patterning the SAM. Using atomic force microscopy imaging and quantitative modeling, several groups have found that the SAM shows locally distinct mechanical properties. In the L1, cell walls of the CZ are much stiffer than those of the PZ (Kierzkowski et al., 2012; Milani et al., 2011, 2014), and incipient primordia and growing primordia have an increased cell wall elasticity (Peaucelle et al., 2011). These findings suggest that organ initiation in the SAM is correlated with a change in mechanical properties (Robinson et al., 2013).

Plant cells are under turgor pressure and glued together by cell walls that contain cellulose and hemicellulose fibrils embedded in a matrix of pectin and proteins (Keegstra, 2010; Wolf, Hématy, & Höfte, 2012). Organogenesis is a process involving coordinated local cell division and directional cell growth (Cosgrove, 2005; Sablowski & Carnier Dornelas, 2014). Cell growth is, in turn, based on irreversible changes in cell wall expansion (i.e., growth rate) and cell wall anisotropy (i.e., growth direction) (Hamant & Traas, 2010).

Irreversible cell expansion induced by auxin during growth is linked to both acidic conditions in the cell wall and the activity of cell wall remodeling enzymes. First, according to the acid growth hypothesis, an auxin-induced decrease of apoplastic pH increases the activity of expansion proteins (which can release the cross-links between polysaccharides complexes), thereby resulting in cell wall loosening (Arsuffi & Braybrook, 2018; Cosgrove, 2005; Murray, Jones, Godin, & Traas, 2012; Nakayama et al., 2012). Second, pectin demethylesterification by PECTIN METHYLESTERASES (PMEs) mediates local auxin-dependent cell wall loosening and thereby organogenesis (Braybrook & Peaucelle, 2013; Peaucelle et al., 2011, 2008). The homeodomain transcription factor BELLRINGER (BLR) was found to be required for the establishment of normal phyllotaxis by restricting PME5 activity to organ primordia (Peaucelle et al., 2011). Thus, during organogenesis, auxin probably triggers cell expansion through modifications of the mechanical properties of the cell wall. Adding to the complexity, other signals, such as the phytohormones GA and brassinosteroid (BR) as well as an active regulation of turgor pressure, are also likely involved, but their precise impact on cell walls and cell expansion remains to be established.

Microtubules are important for the determination of cell wall anisotropy, which is established via the spatial organization of cellulose microfibrils in the cell wall. The cellulose microfibrils determine the direction of growth and their orientation is guided by microtubules through controlling the insertion and movement of cellulose synthase complexes in the plasma membrane (Bringmann et al., 2012; Lei, Li, Du, Bashline, & Gu, 2013; Li, Lei, Somerville, & Gu, 2012). It has been reported that cortical microtubule orientations align parallel to the predicted mechanical stress (Hamant et al., 2008). Laser-ablation and pharmacological treatments weakening the cell wall further support a scenario where microtubules sense the mechanical stress on the SAM surface, and align accordingly to guide anisotropic cellulose deposition, thus counteracting the mechanical stress (Hamant et al., 2008). This stress-driven microtubule reorientation relies on the action of the microtubule-severing protein KATANIN (Uyttewaal et al., 2012).

In addition to microtubule arrays, mechanical stress has been proposed as a possible factor orienting cellular PIN1 polarity (Heisler et al., 2010; Nakayama et al., 2012). In the *A. thaliana* SAM, PIN1 localizes to anticlinal cell walls parallel to the main microtubule orientation (Heisler et al., 2010). Using mathematical modeling, it was shown that mechanical signals can mediate the feedback loop between auxin and the polar auxin transport system, thereby influencing phyllotaxis (Heisler et al., 2010). Further support comes from experiments in tomato where mechanical stress in the cell walls, mediated through plasma membrane properties, would influence PIN1 levels in each cell and the proportion of PIN1 localized to specific areas of the plasma membrane (Nakayama et al., 2012). Auxin was later shown to act on the organization of the cortical microtubule arrays during organ initiation to directly reduce the mechanical anisotropy of cells, thereby promoting organ initiation (together with the effects on cell wall mechanical properties) (Sassi et al., 2014). Intuitively, considering the role of MP in auxin signaling and phyllotaxis (see Section 4.2), the positive feedback loop between auxin, PIN1, and microtubule orientation is proposed to act through a localized auxin response mediated by MP activity (reviewed in Bhatia & Heisler, 2018).

6. The SAM geometry: Phyllotaxis diversity

The inhibitory field models and notably the work from Douady and Couder (1996) predict that the geometry of the SAM should determine phyllotactic patterning through potentially changing the ratio between the radius of inhibitory fields and that of the CZ of the SAM. Consistent with this idea, the analysis of all major lineages of land plants revealed that the level of phyllotactic diversity correlates with that of SAM organization (Gola & Banasiak, 2016).

Several pieces of experimental work highlight this link between SAM geometry and phyllotaxis, and identified some regulatory genes mediating it. First, mutants of maize in the cytokinin-inducible two-component response regulator *ABERRANT PHYLLOTAXY1* (*ABPH1*) display an opposite (decussate) rather than a normal alternative (distichous) phyllotaxis as in wild-type maize and other grasses. The role of *ABPH1* in phyllotaxis regulation has been shown to correlate with an enlargement of the meristem, thought to result from changes in cytokine signaling, whereas the size of lateral organs remains unchanged (Giulini, Wang, & Jackson, 2004; Jackson & Hake, 1999). Second, a dominant maize mutant, *abph2*, has the same meristematic and phyllotactic phenotype as *abph1*. The molecular origin of the mutation is a transposition of a glutaredoxin gene, *MALE STERILE CONVERTED ANTHER1* (*MSCA1*), to another site in the genome. MSAC1 directly interacts with the bZIP transcription factor *FASCIATED EAR4* (FEA4) that regulates SAM size, suggesting that MSAC1 modifies phyllotaxis through a FEA4-mediated meristem size modification (Pautler et al., 2015; Yang et al., 2015). Similarly, a rice mutant, *decussate* (*dec*), was reported to present an enlarged meristem and an alternate-to-decussate shift in phyllotaxis. Several type A response regulator genes that act in the cytokinin signaling pathway were found to be down-regulated in the *dec* mutant, suggesting that a change in cytokinin signaling is also involved in this case (Itoh, Hibara, Kojima, Sakakibara, & Nagato, 2012). Taken together, all these findings indicate that changes in SAM size can trigger a transition in phyllotactic patterning. Future discoveries on factors influencing SAM geometry will certainly allow for a better understanding of how changes in SAM geometry during development mediate phyllotactic transitions.

7. Conclusions

In summary, the SAM, from which all above-ground organs are derived, is a self-organizing structure implicating a multilayered regulation of stem cell homeostasis and multiple factors (among which hormones and biomechanical forces) that determine phyllotactic patterning.

Since the molecular cloning of the first genes regulating the development and function of the SAM in maize (*KNOTTED1* in 1989) and *A. thaliana* (*AGAMOUS* in 1990), the past 30 years have seen a tremendous expansion of our knowledge of the underlying molecular mechanisms. Systems biology and multidisciplinary experimental approaches, such as advanced live-imaging, quantitative analysis, and computational modeling, provided a dynamic view of SAM development and phyllotaxis. Further insights into the molecular mechanisms underlying the self-organization of this unique developmental system will certainly come in the future with the use of tools such as CRISPR-Cas gene editing and single cell genomics.

Novel experimental tools will allow the answering of many remaining questions on how hormones like auxin, cytokinin, and others establish cross-talks, and on the structure and properties of the gene networks influenced by these dynamic signals. How auxin distribution is coordinately established by the polar auxin transport system is also a key question; addressing it will certainly require an in depth analysis of the cellular polarization mechanisms in the SAM to understand its emerging properties. Furthermore, although growth is a crucial parameter in inhibitory field models that drive the dynamics of self-organization, global growth regulation in the SAM has largely been overlooked. Thus, it will be interesting to explore the contribution of growth to phyllotactic patterning. Finally, it would be interesting to know how cells sense mechanical signals, as this will likely be needed to clarify the role of these signals in meristem homeostasis and phyllotaxis.

Acknowledgments

We apologize to all colleagues whose original work could not be cited due to space constraints. The work of the authors is funded by the French National Research Agency.

References

Abley, K., Sauret-Güeto, S., Marée, A. F., & Coen, E. (2016). Formation of polarity convergences underlying shoot outgrowths. *eLife*, *5*, e18165. https://doi.org/10.7554/eLife.18165.

Adler, I. (1997). A history of the study of phyllotaxis. *Annals of Botany*, *80*(3), 231–244. https://doi.org/10.1006/anbo.1997.0422.

Arsuffi, G., & Braybrook, S. A. (2018). Acid growth: An ongoing trip. *Journal of Experimental Botany*, *69*(2), 137–146. https://doi.org/10.1093/jxb/erx390.

Bainbridge, K., Guyomarc'h, S., Bayer, E., Swarup, R., Bennett, M., Mandel, T., et al. (2008). Auxin influx carriers stabilize phyllotactic patterning. *Genes & Development*, *22*(6), 810–823. https://doi.org/10.1101/gad.462608.

Bartlett, M. E., & Thompson, B. (2014). Meristem identity and phyllotaxis in inflorescence development. *Frontiers in Plant Science*, *5*, 508–518. https://doi.org/10.3389/fpls.2014.00508.

Barton, M. K. (2010). Twenty years on: The inner workings of the shoot apical meristem, a developmental dynamo. *Developmental Biology*, *341*(1), 95–113. https://doi.org/10.1016/j.ydbio.2009.11.029.

Bayer, E. M., Smith, R. S., Mandel, T., Nakayama, N., Sauer, M., Prusinkiewicz, P., et al. (2009). Integration of transport-based models for phyllotaxis and midvein formation. *Genes & Development*, *23*(3), 373–384. https://doi.org/10.1101/gad.497009.

Benkova, E., Michniewicz, M., Sauer, M., Teichmann, T., Seifertová, D., Jürgens, G., et al. (2003). Local efflux-dependent auxin gradients as a common module for plant organ formation. *Cell*, *115*(5), 591–602. https://doi.org/10.1016/S0092-8674(03)00924-3.

Besnard, F., Rozier, F., & Vernoux, T. (2014). The AHP6 cytokinin signaling inhibitor mediates an auxin-cytokinin crosstalk that regulates the timing of organ initiation at the shoot apical meristem. *Plant Signaling & Behavior*, *9*(4), e28788.

Besnard, F., Vernoux, T., & Hamant, O. (2011). Organogenesis from stem cells *in planta*: Multiple feedback loops integrating molecular and mechanical signals. *Cellular and Molecular Life Sciences: CMLS*, *68*(17), 2885–2906. https://doi.org/10.1007/s00018-011-0732-4.

Bhatia, N., Bozorg, B., Larsson, A., Ohno, C., Jönsson, H., & Heisler, M. G. (2016). Auxin acts through *MONOPTEROS* to regulate plant cell polarity and pattern phyllotaxis. *Current Biology*, *26*(23), 3202–3208. https://doi.org/10.1016/j.cub.2016.09.044.

Bhatia, N., & Heisler, M. G. (2018). Self-organizing periodicity in development: Organ positioning in plants. *Development*, *145*(3), dev149336. https://doi.org/10.1242/dev.149336.

Braybrook, S. A., & Peaucelle, A. (2013). Mechano-chemical aspects of organ formation in *Arabidopsis thaliana*: The relationship between auxin and pectin. *PLoS One*, *8*(3), e57813. https://doi.org/10.1371/journal.pone.0057813.

Bringmann, M., Landrein, B., Schudoma, C., Hamant, O., Hauser, M.-T., & Persson, S. (2012). Cracking the elusive alignment hypothesis: The microtubule-cellulose synthase nexus unraveled. *Trends in Plant Science*, *17*(11), 666–674. https://doi.org/10.1016/j.tplants.2012.06.003.

Brunoud, G., Wells, D. M., Oliva, M., Larrieu, A., Mirabet, V., Burrow, A. H., et al. (2012). A novel sensor to map auxin response and distribution at high spatio-temporal resolution. *Nature*, *482*(7383), 103–106. https://doi.org/10.1038/nature10791.

Caggiano, M. P., Yu, X., Bhatia, N., Larsson, A., Ram, H., Ohno, C. K., et al. (2017). Cell type boundaries organize plant development. *eLife*, *6*, e27421. https://doi.org/10.7554/eLife.27421.

Calderon Villalobos, L. I. A., Lee, S., De Oliveira, C., Ivetac, A., Brandt, W., Armitage, L., et al. (2012). A combinatorial TIR1/AFB-Aux/IAA co-receptor system for differential sensing of auxin. *Nature Chemical Biology*, *8*(5), 477–485. https://doi.org/10.1038/nchembio.926.

Cheng, Y., Dai, X., & Zhao, Y. (2007). Auxin synthesized by the YUCCA flavin monooxygenases is essential for embryogenesis and leaf formation in *Arabidopsis*. *The Plant Cell*, *19*(8), 2430–2439. https://doi.org/10.1105/tpc.107.053009.

Clark, S. E., Running, M. P., & Meyerowitz, E. M. (1995). *CLAVATA3* is a specific regulator of shoot and floral meristem development affecting the same processes as *CLAVATA1*. *Development*, *121*(7), 2057. Retrieved from: http://dev.biologists.org/content/121/7/2057.abstract.

Clark, S. E., Williams, R. W., & Meyerowitz, E. M. (1997). The *CLAVATA1* gene encodes a putative receptor kinase that controls shoot and floral meristem size in *Arabidopsis*. *Cell*, *89*(4), 575–585.

Cosgrove, D. J. (2005). Growth of the plant cell wall. *Nature Reviews Molecular Cell Biology*, *6*(11), 850–861. https://doi.org/10.1038/nrm1746.

Cui, D., Zhao, J., Jing, Y., Fan, M., Liu, J., Wang, Z., et al. (2013). The *Arabidopsis* IDD14, IDD15, and IDD16 cooperatively regulate lateral organ morphogenesis and gravitropism by promoting auxin biosynthesis and transport. *PLoS Genetics*, *9*(9), e1003759. https://doi.org/10.1371/journal.pgen.1003759.

Daum, G., Medzihradszky, A., Suzaki, T., & Lohmann, J. U. (2014). A mechanistic framework for noncell autonomous stem cell induction in *Arabidopsis*. *Proceedings of the National Academy of Sciences of the United States of America*, *111*(40), 14619–14624. https://doi.org/10.1073/pnas.1406446111.

Dharmasiri, N., Dharmasiri, S., & Estelle, M. (2005). The F-box protein TIR1 is an auxin receptor. *Nature*, *435*(7041), 441–445. https://doi.org/10.1038/nature03543.

Douady, S., & Couder, Y. (1996). Phyllotaxis as a dynamical self organizing process part II: The spontaneous formation of a periodicity and the coexistence of spiral and whorled patterns. *Journal of Theoretical Biology, 178*(3), 275–294. https://doi.org/10.1006/jtbi.1996.0025.

Friml, J. (2010). Subcellular trafficking of PIN auxin efflux carriers in auxin transport. *European Journal of Cell Biology, 89*(2–3), 231–235. https://doi.org/10.1016/j.ejcb.2009.11.003.

Gaillochet, C., Daum, G., & Lohmann, J. U. (2015). O cell, where art thou? The mechanisms of shoot meristem patterning. *Current Opinion in Plant Biology, 23*, 91–97. https://doi.org/10.1016/j.pbi.2014.11.002.

Gallavotti, A., Yang, Y., Schmidt, R. J., & Jackson, D. (2008). The relationship between auxin transport and maize branching. *Plant Physiology, 147*(4), 1913–1923. https://doi.org/10.1104/pp.108.121541.

Galli, M., & Gallavotti, A. (2016). Expanding the regulatory network for meristem size in plants. *Trends in Genetics, 32*(6), 372–383. https://doi.org/10.1016/j.tig.2016.04.001.

Galvan-Ampudia, C. S., Chaumeret, A. M., Godin, C., & Vernoux, T. (2016). Phyllotaxis: From patterns of organogenesis at the meristem to shoot architecture: Organogenesis at the meristem and shoot architecture. *Wiley Interdisciplinary Reviews: Developmental Biology, 5*(4), 460–473. https://doi.org/10.1002/wdev.231.

Galweiler, L., Guan, C., Müller, A., Wisman, E., Mendgen, K., Yephremov, A., et al. (1998). Regulation of polar auxin transport by *AtPIN1* in *Arabidopsis* vascular tissue. *Science (New York, N.Y.), 282*(5397), 2226–2230.

Giulini, A., Wang, J., & Jackson, D. (2004). Control of phyllotaxy by the cytokinin-inducible response regulator homologue *ABPHYL1*. *Nature, 430*(7003), 1031–1034. https://doi.org/10.1038/nature02778.

Gola, E. M., & Banasiak, A. (2016). Diversity of phyllotaxis in land plants in reference to the shoot apical meristem structure. *Acta Societatis Botanicorum Poloniae, 85*(4), 3529–3549. https://doi.org/10.5586/asbp.3529.

Goldshmidt, A., Alvarez, J. P., Bowman, J. L., & Eshed, Y. (2008). Signals derived from *YABBY* gene activities in organ primordia regulate growth and partitioning of *Arabidopsis* shoot apical meristems. *The Plant Cell, 20*(5), 1217–1230. https://doi.org/10.1105/tpc.107.057877.

Golub, S. J., & Wetmore, R. H. (1948). Studies of development in the vegetative shoot of equisetum arvense L. I. The shoot apex. *American Journal of Botany, 35*(10), 755.

Gordon, S. P., Chickarmane, V. S., Ohno, C., & Meyerowitz, E. M. (2009). Multiple feedback loops through cytokinin signaling control stem cell number within the *Arabidopsis* shoot meristem. *Proceedings of the National Academy of Sciences of the United States of America, 106*(38), 16529–16534. https://doi.org/10.1073/pnas.0908122106.

Guilfoyle, T. J., & Hagen, G. (2007). Auxin response factors. *Current Opinion in Plant Biology, 10*(5), 453–460. https://doi.org/10.1016/j.pbi.2007.08.014.

Hamant, O., Heisler, M. G., Jönsson, H., Krupinski, P., Uyttewaal, M., Bokov, P., et al. (2008). Developmental patterning by mechanical signals in *Arabidopsis*. *Science, 322*(5908), 1650–1655. https://doi.org/10.1126/science.1165594.

Hamant, O., & Traas, J. (2010). The mechanics behind plant development. *The New Phytologist, 185*(2), 369–385. https://doi.org/10.1111/j.1469-8137.2009.03100.x.

Hardtke, C. S., & Berleth, T. (1998). The *Arabidopsis* gene *MONOPTEROS* encodes a transcription factor mediating embryo axis formation and vascular development. *The EMBO Journal, 17*(5), 1405–1411. https://doi.org/10.1093/emboj/17.5.1405.

Hay, A., & Tsiantis, M. (2010). KNOX genes: Versatile regulators of plant development and diversity. *Development, 137*(19), 3153–3165. https://doi.org/10.1242/dev.030049.

Heisler, M. G., Hamant, O., Krupinski, P., Uyttewaal, M., Ohno, C., Jönsson, H., et al. (2010). Alignment between PIN1 polarity and microtubule orientation in the shoot apical meristem reveals a tight coupling between morphogenesis and auxin transport. *PLoS Biology, 8*(10), e1000516. https://doi.org/10.1371/journal.pbio.1000516.

Heisler, M. G., Ohno, C., Das, P., Sieber, P., Reddy, G. V., Long, J. A., et al. (2005). Patterns of auxin transport and gene expression during primordium development revealed by live imaging of the *Arabidopsis* inflorescence meristem. *Current Biology, 15*(21), 1899–1911. https://doi.org/10.1016/j.cub.2005.09.052.

Hofmeister. (1868). Allgemeine morphologie der gewachse. In A. de Bary, T. H. Irmisch, & J. Sachs (Eds.), *Handbuch der Physiologischen Botanik* (pp. 405–664). Leipzig: Engelmann.

Itoh, J., Hibara, K., Kojima, M., Sakakibara, H., & Nagato, Y. (2012). Rice *DECUSSATE* controls phyllotaxy by affecting the cytokinin signaling pathway. *The Plant Journal, 72*(6), 869–881. https://doi.org/10.1111/j.1365-313x.2012.05123.x.

Jackson, D., & Hake, S. (1999). Control of phyllotaxy in maize by the *abphyl1* gene. *Development (Cambridge, England), 126*(2), 315–323.

Jasinski, S., Piazza, P., Craft, J., Hay, A., Woolley, L., Rieu, I., et al. (2005). KNOX action in *Arabidopsis* is mediated by coordinate regulation of cytokinin and gibberellin activities. *Current Biology, 15*(17), 1560–1565. https://doi.org/10.1016/j.cub.2005.07.023.

Je, B. I., Gruel, J., Lee, Y. K., Bommert, P., Arevalo, E. D., Eveland, A. L., et al. (2016). Signaling from maize organ primordia via *FASCIATED EAR3* regulates stem cell proliferation and yield traits. *Nature Genetics, 48*(7), 785–791. https://doi.org/10.1038/ng.3567.

Jean, R. V., & Barabe, D. (1998). *Symmetry in plants*. Singapore, River Edge, NJ: World Scientific. Retrieved from: https://site.ebrary.com/id/10698923.

Jonsson, H., Heisler, M. G., Shapiro, B. E., Meyerowitz, E. M., & Mjolsness, E. (2006). An auxin-driven polarized transport model for phyllotaxis. *Proceedings of the National Academy of Sciences of the United States of America, 103*(5), 1633–1638. https://doi.org/10.1073/pnas.0509839103.

Keegstra, K. (2010). Plant cell walls. *Plant Physiology, 154*(2), 483–486. https://doi.org/10.1104/pp.110.161240.

Kepinski, S., & Leyser, O. (2005). The *Arabidopsis* F-box protein TIR1 is an auxin receptor. *Nature, 435*(7041), 446–451. https://doi.org/10.1038/nature03542.

Kerstetter, R. A., Laudencia-Chingcuanco, D., Smith, L. G., & Hake, S. (1997). Loss-of-function mutations in the maize homeobox gene, *knotted1*, are defective in shoot meristem maintenance. *Development (Cambridge, England), 124*(16), 3045–3054.

Kierzkowski, D., Nakayama, N., Routier-Kierzkowska, A.-L., Weber, A., Bayer, E., Schorderet, M., et al. (2012). Elastic domains regulate growth and organogenesis in the plant shoot apical meristem. *Science (New York, N.Y.), 335*(6072), 1096–1099. https://doi.org/10.1126/science.1213100.

Kramer, E. M. (2004). PIN and AUX/LAX proteins: Their role in auxin accumulation. *Trends in Plant Science, 9*(12), 578–582. https://doi.org/10.1016/j.tplants.2004.10.010.

Kuhlemeier, C. (2017). Phyllotaxis. *Current Biology, 27*(17), R882–R887. https://doi.org/10.1016/j.cub.2017.05.069.

Kurakawa, T., Ueda, N., Maekawa, M., Kobayashi, K., Kojima, M., Nagato, Y., et al. (2007). Direct control of shoot meristem activity by a cytokinin-activating enzyme. *Nature, 445*(7128), 652–655. https://doi.org/10.1038/nature05504.

Lander, A. D., Gokoffski, K. K., Wan, F. Y. M., Nie, Q., & Calof, A. L. (2009). Cell lineages and the logic of proliferative control. *PLoS Biology*, 7(1), e1000015. https://doi.org/10.1371/journal.pbio.1000015.

Lei, L., Li, S., Du, J., Bashline, L., & Gu, Y. (2013). *CELLULOSE SYNTHASE INTERACTIVE3* regulates cellulose biosynthesis in both a microtubule-dependent and microtubule-independent manner in *Arabidopsis*. *The Plant Cell, 25*(12), 4912–4923. https://doi.org/10.1105/tpc.113.116715.

Leibfried, A., To, J. P. C., Busch, W., Stehling, S., Kehle, A., Demar, M., et al. (2005). WUSCHEL controls meristem function by direct regulation of cytokinin-inducible response regulators. *Nature, 438*(7071), 1172–1175. https://doi.org/10.1038/nature04270.

Li, S., Lei, L., Somerville, C. R., & Gu, Y. (2012). Cellulose synthase interactive protein 1 (CSI1) links microtubules and cellulose synthase complexes. *Proceedings of the National Academy of Sciences of the United States of America*, *109*(1), 185–190. https://doi.org/10.1073/pnas.1118560109.

Mandel, T., Candela, H., Landau, U., Asis, L., Zelinger, E., Carles, C. C., et al. (2016). Differential regulation of meristem size, morphology and organization by the ERECTA, CLAVATA and class III HD-ZIP pathways. *Development*, *143*(9), 1612–1622. https://doi.org/10.1242/dev.129973.

Mayer, K. F., Schoof, H., Haecker, A., Lenhard, M., Jürgens, G., & Laux, T. (1998). Role of *WUSCHEL* in regulating stem cell fate in the *Arabidopsis* shoot meristem. *Cell*, *95*(6), 805–815.

Milani, P., Gholamirad, M., Traas, J., Arnéodo, A., Boudaoud, A., Argoul, F., et al. (2011). *In vivo* analysis of local wall stiffness at the shoot apical meristem in *Arabidopsis* using atomic force microscopy. *The Plant Journal: for Cell and Molecular Biology*, *67*(6), 1116–1123. https://doi.org/10.1111/j.1365-313X.2011.04649.x.

Milani, P., Mirabet, V., Cellier, C., Rozier, F., Hamant, O., Das, P., et al. (2014). Matching patterns of gene expression to mechanical stiffness at cell resolution through quantitative tandem epifluorescence and nanoindentation. *Plant Physiology*, *165*(4), 1399–1408. https://doi.org/10.1104/pp.114.237115.

Muller, B., & Sheen, J. (2008). Cytokinin and auxin interaction in root stem-cell specification during early embryogenesis. *Nature*, *453*(7198), 1094–1097. https://doi.org/10.1038/nature06943.

Murray, J. A. H., Jones, A., Godin, C., & Traas, J. (2012). Systems analysis of shoot apical meristem growth and development: Integrating hormonal and mechanical signaling. *The Plant Cell*, *24*(10), 3907–3919. https://doi.org/10.1105/tpc.112.102194.

Nakayama, N., Smith, R. S., Mandel, T., Robinson, S., Kimura, S., Boudaoud, A., et al. (2012). Mechanical regulation of auxin-mediated growth. *Current Biology*, *22*(16), 1468–1476. https://doi.org/10.1016/j.cub.2012.06.050.

Niemann, M. C. E., Bartrina, I., Ashikov, A., Weber, H., Novák, O., Spíchal, L., et al. (2015). *Arabidopsis* ROCK1 transports UDP-GlcNAc/UDP-GalNAc and regulates ER protein quality control and cytokinin activity. *Proceedings of the National Academy of Sciences of the United States of America*, *112*(1), 291–296. https://doi.org/10.1073/pnas.1419050112.

O'Connor, D. L., Elton, S., Ticchiarelli, F., Hsia, M. M., Vogel, J. P., & Leyser, O. (2017). Cross-species functional diversity within the PIN auxin efflux protein family. *eLife*, *6*, e31804. https://doi.org/10.7554/eLife.31804.

O'Connor, D. L., Runions, A., Sluis, A., Bragg, J., Vogel, J. P., Prusinkiewicz, P., et al. (2014). A division in PIN-mediated auxin patterning during organ initiation in grasses. *PLoS Computational Biology*, *10*(1), e1003447. https://doi.org/10.1371/journal.pcbi.1003447.

Okada, K. (1991). Requirement of the auxin polar transport system in early stages of *Arabidopsis* floral bud formation. *The Plant Cell*, *3*(7), 677–684. https://doi.org/10.1105/tpc.3.7.677.

Opsahl-Ferstad, H.-G., Deunff, E. L., Dumas, C., & Rogowsky, P. M. (1997). *ZmEsr*, a novel endosperm-specific gene expressed in a restricted region around the maize embryo. *The Plant Journal*, *12*(1), 235–246. https://doi.org/10.1046/j.1365-313X.1997.12010235.x.

Ottoline Leyser, H. M., & Furner, I. J. (1992). Characterisation of three shoot apical meristem mutants of *Arabidopsis thaliana*. *Development*, *116*(2), 397. http://dev.biologists.org/content/116/2/397.

Pautler, M., Eveland, A. L., LaRue, T., Yang, F., Weeks, R., Lunde, C., et al. (2015). *FASCIATED EAR4* encodes a bZIP transcription factor that regulates shoot meristem size in maize. *The Plant Cell*, *27*(1), 104–120. https://doi.org/10.1105/tpc.114.132506.

Peaucelle, A., Braybrook, S. A., Le Guillou, L., Bron, E., Kuhlemeier, C., & Höfte, H. (2011). Pectin-induced changes in cell wall mechanics underlie organ initiation in *Arabidopsis*. *Current Biology, 21*(20), 1720–1726. https://doi.org/10.1016/j.cub.2011.08.057.

Peaucelle, A., Louvet, R., Johansen, J. N., Höfte, H., Laufs, P., Pelloux, J., et al. (2008). *Arabidopsis* phyllotaxis is controlled by the methyl-esterification status of cell-wall pectins. *Current Biology, 18*(24), 1943–1948. https://doi.org/10.1016/j.cub.2008.10.065.

Petrasek, J., Mravec, J., Bouchard, R., Blakeslee, J. J., Abas, M., Seifertová, D., et al. (2006). PIN proteins perform a rate-limiting function in cellular auxin efflux. *Science (New York, N.Y.), 312*(5775), 914–918. https://doi.org/10.1126/science.1123542.

Phillips, K. A., Skirpan, A. L., Liu, X., Christensen, A., Slewinski, T. L., Hudson, C., et al. (2011). *Vanishing tassel2* encodes a grass-specific tryptophan aminotransferase required for vegetative and reproductive development in maize. *The Plant Cell, 23*(2), 550–566. https://doi.org/10.1105/tpc.110.075267.

Pinon, V., Prasad, K., Grigg, S. P., Sanchez-Perez, G. F., & Scheres, B. (2013). Local auxin biosynthesis regulation by PLETHORA transcription factors controls phyllotaxis in *Arabidopsis*. *Proceedings of the National Academy of Sciences of the United States of America, 110*(3), 1107–1112. https://doi.org/10.1073/pnas.1213497110.

Prasad, K., Grigg, S. P., Barkoulas, M., Yadav, R. K., Sanchez-Perez, G. F., Pinon, V., et al. (2011). *Arabidopsis* PLETHORA transcription factors control phyllotaxis. *Current Biology, 21*(13), 1123–1128. https://doi.org/10.1016/j.cub.2011.05.009.

Refahi, Y., Brunoud, G., Farcot, E., Jean-Marie, A., Pulkkinen, M., Vernoux, T., et al. (2016). A stochastic multicellular model identifies biological watermarks from disorders in self-organized patterns of phyllotaxis. *eLife, 5*, e14093. https://doi.org/10.7554/eLife.14093.

Reinhardt, D. (2000). Auxin regulates the initiation and radial position of plant lateral organs. *The Plant Cell, 12*(4), 507–518. https://doi.org/10.1105/tpc.12.4.507.

Reinhardt, D., Frenz, M., Mandel, T., & Kuhlemeier, C. (2005). Microsurgical and laser ablation analysis of leaf positioning and dorsoventral patterning in tomato. *Development (Cambridge, England), 132*(1), 15–26. https://doi.org/10.1242/dev.01544.

Reinhardt, D., Pesce, E.-R., Stieger, P., Mandel, T., Baltensperger, K., Bennett, M., et al. (2003). Regulation of phyllotaxis by polar auxin transport. *Nature, 426*(6964), 255–260. https://doi.org/10.1038/nature02081.

Richards, F. J. (1948). The geometry of phyllotaxis and its origin. *Symposium of the Society for Experimental Biology, 2*, 217–245.

Richards, F. J. (1951). Phyllotaxis: Its quantitative expression and relation to growth in the apex. *Philosophical Transactions of the Royal Society B: Biological Sciences, 235*(629), 509–564. https://doi.org/10.1098/rstb.1951.0007.

Robinson, S., Burian, A., Couturier, E., Landrein, B., Louveaux, M., Neumann, E. D., et al. (2013). Mechanical control of morphogenesis at the shoot apex. *Journal of Experimental Botany, 64*(15), 4729–4744. https://doi.org/10.1093/jxb/ert199.

Rue, P., & Martinez Arias, A. (2015). Cell dynamics and gene expression control in tissue homeostasis and development. *Molecular Systems Biology, 11*(2), 792. https://doi.org/10.15252/msb.20145549.

Sablowski, R., & Carnier Dornelas, M. (2014). Interplay between cell growth and cell cycle in plants. *Journal of Experimental Botany, 65*(10), 2703–2714. https://doi.org/10.1093/jxb/ert354.

Sachs, T. (1969). Polarity and the induction of organized vascular tissues. *Annals of Botany, 33*(2), 263–275. https://doi.org/10.1093/oxfordjournals.aob.a084281.

Sachs, T. (1981). The control of the patterned differentiation of vascular tissues. *Advances in Botanical Research: Vol. 9* (pp. 151–262). Elsevier. https://doi.org/10.1016/S0065-2296(08)60351-1.

Sassi, M., Ali, O., Boudon, F., Cloarec, G., Abad, U., Cellier, C., et al. (2014). An auxin-mediated shift toward growth isotropy promotes organ formation at the shoot meristem in *Arabidopsis*. *Current Biology, 24*(19), 2335–2342. https://doi.org/10.1016/j.cub.2014.08.036.

Sassi, M., & Vernoux, T. (2013). Auxin and self-organization at the shoot apical meristem. *Journal of Experimental Botany, 64*(9), 2579–2592. https://doi.org/10.1093/jxb/ert101.

Schoute. (1913). Beitrage zur blattstellunglehre. I. Die theorie. *Recueilde Travaux Botaniques Neerlandais, 10*, 153–339.

Schuster, C., Gaillochet, C., Medzihradszky, A., Busch, W., Daum, G., Krebs, M., et al. (2014). A regulatory framework for shoot stem cell control integrating metabolic, transcriptional, and phytohormone signals. *Developmental Cell, 28*(4), 438–449. https://doi.org/10.1016/j.devcel.2014.01.013.

Shi, B., Guo, X., Wang, Y., Xiong, Y., Wang, J., Hayashi, K., et al. (2018). Feedback from lateral organs controls shoot apical meristem growth by modulating auxin transport. *Developmental Cell, 44*(2), 204–216.e6. https://doi.org/10.1016/j.devcel.2017.12.021.

Sluis, A., & Hake, S. (2015). Organogenesis in plants: Initiation and elaboration of leaves. *Trends in Genetics, 31*(6), 300–306. https://doi.org/10.1016/j.tig.2015.04.004.

Smith, R. S., Guyomarc'h, S., Mandel, T., Reinhardt, D., Kuhlemeier, C., & Prusinkiewicz, P. (2006). A plausible model of phyllotaxis. *Proceedings of the National Academy of Sciences of the United States of America, 103*(5), 1301–1306. https://doi.org/10.1073/pnas.0510457103.

Snow, M., & Snow, R. (1932). Experiments on phyllotaxis. I. The effect of isolating a primordium. *Philosophical Transactions of the Royal Society B: Biological Sciences, 221*(474–482), 1–43. https://doi.org/10.1098/rstb.1932.0001.

Snow, M., & Snow, R. (1962). A theory of the regulation of phyllotaxis based on *Lupinus albus*. *Philosophical Transactions of the Royal Society B: Biological Sciences, 244*(717), 483–513. https://doi.org/10.1098/rstb.1962.0003.

Soyars, C. L., James, S. R., & Nimchuk, Z. L. (2016). Ready, aim, shoot: Stem cell regulation of the shoot apical meristem. *Current Opinion in Plant Biology, 29*, 163–168. https://doi.org/10.1016/j.pbi.2015.12.002.

Stoma, S., Lucas, M., Chopard, J., Schaedel, M., Traas, J., & Godin, C. (2008). Flux-based transport enhancement as a plausible unifying mechanism for auxin transport in meristem development. *PLoS Computational Biology, 4*(10), e1000207. https://doi.org/10.1371/journal.pcbi.1000207.

Tan, X., Calderon-Villalobos, L. I. A., Sharon, M., Zheng, C., Robinson, C. V., Estelle, M., et al. (2007). Mechanism of auxin perception by the TIR1 ubiquitin ligase. *Nature, 446*(7136), 640–645. https://doi.org/10.1038/nature05731.

Teichmann, T., & Muhr, M. (2015). Shaping plant architecture. *Frontiers in Plant Science, 6*. https://doi.org/10.3389/fpls.2015.00233.

Traas, J. (2013). Phyllotaxis. *Development, 140*(2), 249–253. https://doi.org/10.1242/dev.074740.

Truskina, J., & Vernoux, T. (2018). The growth of a stable stationary structure: Coordinating cell behavior and patterning at the shoot apical meristem. *Current Opinion in Plant Biology, 41*, 83–88. https://doi.org/10.1016/j.pbi.2017.09.011.

Uyttewaal, M., Burian, A., Alim, K., Landrein, B., Borowska-Wykręt, D., Dedieu, A., et al. (2012). Mechanical stress acts via katanin to amplify differences in growth rate between adjacent cells in *Arabidopsis*. *Cell, 149*(2), 439–451. https://doi.org/10.1016/j.cell.2012.02.048.

van Berkel, K., de Boer, R. J., Scheres, B., & ten Tusscher, K. (2013). Polar auxin transport: Models and mechanisms. *Development (Cambridge, England), 140*(11), 2253–2268. https://doi.org/10.1242/dev.079111.

Vernoux, T., Brunoud, G., Farcot, E., Morin, V., Van den Daele, H., Legrand, J., et al. (2011). The auxin signalling network translates dynamic input into robust patterning at the shoot apex. *Molecular Systems Biology, 7*(1), 508. https://doi.org/10.1038/msb.2011.39.

Vernoux, T., Kronenberger, J., Grandjean, O., Laufs, P., & Traas, J. (2000). PIN-FORMED 1 regulates cell fate at the periphery of the shoot apical meristem. *Development*, *127*(23), 5157–5165.

Wang, Y., & Jiao, Y. (2018). Auxin and above-ground meristems. *Journal of Experimental Botany*, *69*(2), 147–154. https://doi.org/10.1093/jxb/erx299.

Wang, B., Smith, S. M., & Li, J. (2018). Genetic regulation of shoot architecture. *Annual Review of Plant Biology*, *69*(1). https://doi.org/10.1146/annurev-arplant-042817-040422.

Wardlaw, C. W. (1949). Phyllotaxis and organogenesis in ferns. *Nature*, *164*(4161), 167–169. https://doi.org/10.1038/164167a0.

Weijers, D., & Wagner, D. (2016). Transcriptional responses to the auxin hormone. *Annual Review of Plant Biology*, *67*, 539–574. https://doi.org/10.1146/annurev-arplant-043015-112122.

Wisniewska, J., Xu, J., Seifertová, D., Brewer, P. B., Ruzicka, K., Blilou, I., et al. (2006). Polar PIN localization directs auxin flow in plants. *Science*, *312*(5775), 883. https://doi.org/10.1126/science.1121356.

Wolf, S., Hématy, K., & Höfte, H. (2012). Growth control and cell wall signaling in plants. *Annual Review of Plant Biology*, *63*, 381–407. https://doi.org/10.1146/annurev-arplant-042811-105449.

Yamaguchi, N., Wu, M.-F., Winter, C. M., Berns, M. C., Nole-Wilson, S., Yamaguchi, A., et al. (2013). A molecular framework for auxin-mediated initiation of flower primordia. *Developmental Cell*, *24*(3), 271–282. https://doi.org/10.1016/j.devcel.2012.12.017.

Yanai, O., Shani, E., Dolezal, K., Tarkowski, P., Sablowski, R., Sandberg, G., et al. (2005). *Arabidopsis* KNOXI proteins activate cytokinin biosynthesis. *Current Biology*, *15*(17), 1566–1571. https://doi.org/10.1016/j.cub.2005.07.060.

Yang, F., Bui, H. T., Pautler, M., Llaca, V., Johnston, R., Lee, B., et al. (2015). A maize glutaredoxin gene, *Abphyl2*, regulates shoot meristem size and phyllotaxy. *The Plant Cell*, *27*(1), 121–131. https://doi.org/10.1105/tpc.114.130393.

Yoshida, S., Mandel, T., & Kuhlemeier, C. (2011). Stem cell activation by light guides plant organogenesis. *Genes & Development*, *25*(13), 1439–1450. https://doi.org/10.1101/gad.631211.

Zhao, Z., Andersen, S. U., Ljung, K., Dolezal, K., Miotk, A., Schultheiss, S. J., et al. (2010). Hormonal control of the shoot stem-cell niche. *Nature*, *465*(7301), 1089–1092. https://doi.org/10.1038/nature09126.

Zhou, Y., Liu, X., Engstrom, E. M., Nimchuk, Z. L., Pruneda-Paz, J. L., Tarr, P. T., et al. (2015). Control of plant stem cell function by conserved interacting transcriptional regulators. *Nature*, *517*(7534), 377–380. https://doi.org/10.1038/nature13853.

Further reading

Anderson, T. R., & Slotkin, T. A. (1975). Maturation of the adrenal medulla—IV. Effects of morphine. *Biochemical Pharmacology*, *24*(16), 1469–1474.

Crowell, E. F., Timpano, H., Desprez, T., Franssen-Verheijen, T., Emons, A.-M., Hofte, H., et al. (2011). Differential regulation of cellulose orientation at the inner and outer face of epidermal cells in the *Arabidopsis* hypocotyl. *The Plant Cell*, *23*(7), 2592–2605. https://doi.org/10.1105/tpc.111.087338.

Gutierrez, R., Lindeboom, J. J., Paredez, A. R., Emons, A. M. C., & Ehrhardt, D. W. (2009). *Arabidopsis* cortical microtubules position cellulose synthase delivery to the plasma membrane and interact with cellulose synthase trafficking compartments. *Nature Cell Biology*, *11*(7), 797–806. https://doi.org/10.1038/ncb1886.

Hu, C., Zhu, Y., Cui, Y., Cheng, K., Liang, W., Wei, Z., et al. (2018). A group of receptor kinases are essential for *CLAVATA* signalling to maintain stem cell homeostasis. *Nature Plants*, *4*(4), 205–211. https://doi.org/10.1038/s41477-018-0123-z.

Makar, A. B., McMartin, K. E., Palese, M., & Tephly, T. R. (1975). Formate assay in body fluids: Application in methanol poisoning. *Biochemical Medicine*, *13*(2), 117–126.

Mandel, T., Moreau, F., Kutsher, Y., Fletcher, J. C., Carles, C. C., & Williams, L. E. (2014). The ERECTA receptor kinase regulates *Arabidopsis* shoot apical meristem size, phyllotaxy and floral meristem identity. *Development*, *141*(4), 830–841. https://doi.org/10.1242/dev.104687.

Paredez, A. R., Somerville, C. R., & Ehrhardt, D. W. (2006). Visualization of cellulose synthase demonstrates functional association with microtubules. *Science (New York, N.Y.)*, *312*(5779), 1491–1495. https://doi.org/10.1126/science.1126551.

Rayle, D. L., & Cleland, R. E. (1992). The acid growth theory of auxin-induced cell elongation is alive and well. *Plant Physiology*, *99*(4), 1271–1274.

Schoof, H., Lenhard, M., Haecker, A., Mayer, K. F., Jürgens, G., & Laux, T. (2000). The stem cell population of *Arabidopsis* shoot meristems is maintained by a regulatory loop between the *CLAVATA* and *WUSCHEL* genes. *Cell*, *100*(6), 635–644. https://doi.org/10.1016/S0092-8674(00)80700-X.

Shapiro, B. E., Tobin, C., Mjolsness, E., & Meyerowitz, E. M. (2015). Analysis of cell division patterns in the *Arabidopsis* shoot apical meristem. *Proceedings of the National Academy of Sciences of the United States of America*, *112*(15), 4815–4820. https://doi.org/10.1073/pnas.1502588112.

Somssich, M., Je, B. I., Simon, R., & Jackson, D. (2016). CLAVATA-WUSCHEL signaling in the shoot meristem. *Development*, *143*(18), 3238–3248. https://doi.org/10.1242/dev.133645.

CHAPTER FIVE

Leaf development and evolution

Lachezar A. Nikolov[†], **Adam Runions**[†], **Mainak Das Gupta**[†], **Miltos Tsiantis***
Max Planck Institute for Plant Breeding Research, Cologne, Germany
*Corresponding author: e-mail address: tsiantis@mpipz.mpg.de

Contents

Abstract

Plant leaves are differentiated organs that arise sequentially from a population of pluripotent stem cells at the shoot apical meristem (SAM). There is substantial diversity in leaf shape, much of which depends on the size and arrangement of outgrowths at the leaf margin. These outgrowths are generated by a patterning mechanism similar to the phyllotactic processes producing organs at the SAM, which involves the transcription factors CUP-SHAPED COTYLEDON and the phytohormone auxin. In the leaf, this patterning mechanism creates sequential protrusions and indentations along the margin. The size, shape, and distribution of these protrusions also depend on the overall growth of the leaf lamina. Globally, growth is regulated by a complex genetic network

[†] These authors contributed equally and the order was determined by drawing lots.

Current Topics in Developmental Biology, Volume 131
ISSN 0070-2153
https://doi.org/10.1016/bs.ctdb.2018.11.006

controlling the distribution of cell proliferation and the timing of differentiation. Evolutionary changes in margin form arise from changes in two different classes of homeobox genes that modify the outcome of marginal patterning in diverse ways, and are under intense investigation.

1. Introduction

Leaves are the main photosynthetic organs in plants and exhibit a staggering diversity of forms. This variation provides an excellent model for understanding how biological shapes are generated and how they evolved (Bar & Ori, 2015; Rebeiz & Tsiantis, 2017). Much of the diversity in leaf shape reflects variation of leaf margins (Fig. 1A). These can be entire or variously elaborated to form shallow protrusions called serrations, or deeper protrusions called lobes. In the case of dissected leaves (also referred to as compound), the leaf is fully divided into individual leaflets. The leaflets can themselves bear serrations, lobes, or even additional leaflets, in the latter case leading to self-similar forms. Here we review genetic pathways influencing seed plant leaf development with emphasis on the margin, as well as current insights into the evolutionary diversification of these pathways that resulted in morphological diversity.

2. Leaf initiation and marginal patterning

2.1 Leaf initiation

Leaves emerge from a pool of pluripotent cells located at the tips of growing shoots, termed the shoot apical meristem (SAM). They grow from small populations of founder cells of approximately 100–200 cells in *Zea mays* (maize) (Poethig, 1984) and 15 cells in *Arabidopsis thaliana* (Irish & Sussex, 1992) into a differentiated organ. Studies in simple leaf model systems indicate that founder cell delimitation involves two mechanisms (Fig. 1B). The first mechanism depends on the mutual repression of *ASYMMETRIC LEAVES1 (AS1)/ROUGH SHEATH2/PHANTASTICA* (ARP) MYB-domain proteins and Class I *KNOTTED1*-like homeobox (KNOXI) proteins, which confines the meristem-promoting activity of KNOXI to the SAM, and ARPs and their AS2 cofactors to leaves (Byrne et al., 2000; Lin, Shuai, & Springer, 2003; Timmermans, Hudson, Becraft, & Nelson, 1999; Tsiantis, Schneeberger, Golz, Freeling, & Langdale, 1999). The second mechanism involves localized delivery of the phytohormone auxin to sites of leaf

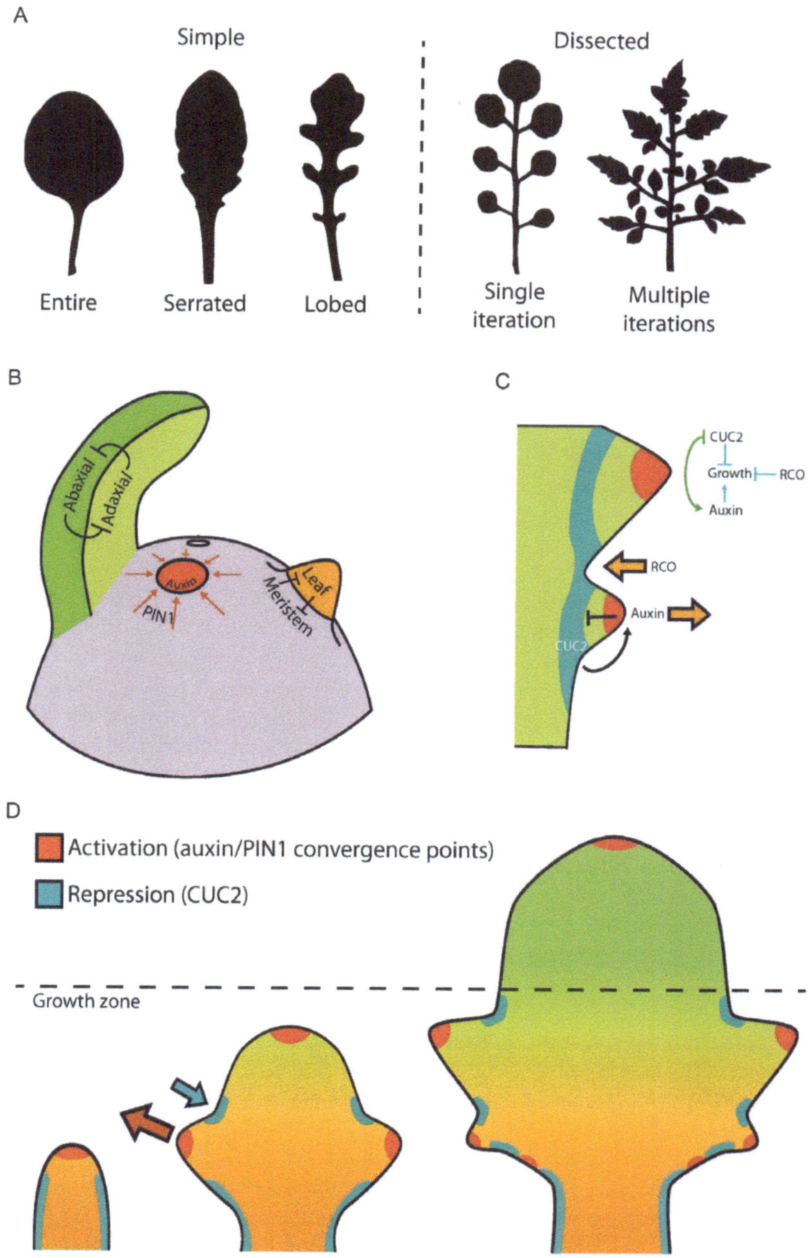

Fig. 1 See legend on next page.

inception by the PIN-FORMED1 (PIN1) auxin efflux transporter, producing auxin maxima. These two mechanisms are likely interconnected, as PIN1 and auxin signaling are also required for KNOXI exclusion from *A. thaliana* leaf primordia (Hay, Barkoulas, & Tsiantis, 2006; Heisler et al., 2005). Notably, PIN1-mediated auxin transport at the apex not only contributes to KNOXI repression but enables the patterned arrangement of leaf primordia at the apex, a process known as phyllotaxis. Central to the organization of this pattern is a feedback loop between auxin and PIN1, which allows auxin to promote its own transport by influencing the subcellular localization of PIN1 (Bhatia et al., 2016; Jonsson, Heisler, Shapiro, Meyerowitz, & Mjolsness, 2006; Reinhardt et al., 2003; Smith et al., 2006). This mechanism causes PIN1 to point toward the auxin maxima, resulting in PIN1 convergence points (Jonsson et al., 2006; Smith et al., 2006).

Fig. 1 Leaf morphology and ontogeny. (A) Outlines of leaves with simple (*Arabidopsis thaliana* and *Arabidopsis lyrata*), and dissected (*Cardamine hirsuta* and *Solanum lycopersicum*) margins (outer boundary). Leaves with an undivided blade are simple and can have margins that are smooth (entire, first leaf of *A. thaliana*), punctuated with protrusions, termed serrations, when small (adult leaf of *A. thaliana*), and lobes when large (*A. lyrata*). Leaves with divided blades are dissected, also termed compound, and have leaflets (small leaf-like forms) arranged either along a single stalk (singly dissected) or hierarchically (iteratively dissected). (B) Early stages of leaf development. Directional transport of auxin by PIN1 (red arrows) accumulates auxin at discrete sites in the periphery of the SAM, creating an auxin maximum that marks the site of leaf initiation (red disk). Establishment of leaf identity also involves the antagonism between factors promoting leaf identity (orange) and meristem identity (purple). Blade outgrowth depends on the establishment of abaxial/adaxial polarity, to which antagonism between adaxial (light green) and abaxial (dark green) identity factors contribute. The white disk marks the tip of the SAM, the pluripotent cell population that gives rise to lateral organs, including leaves. (C) Marginal pattern generation emerges from a feedback between auxin (red) and *CUC2* (cyan), where *CUC2* enables auxin accumulation by the auxin efflux carrier PIN1; in turn, auxin represses *CUC2* producing an interspersed pattern of auxin maxima and *CUC2* expression along the leaf margin. Auxin enhances outgrowth of the margin, while *CUC2* represses outgrowth. Protrusion shape is sculpted by additional growth regulators that locally modulate growth. For example, *RCO* inhibits growth in indentations, producing more dissected forms. (D) The three interacting components generating leaf shape in a developing *A. thaliana* leaf (developmental age increases from left to right): Component 1, background growth establishes a growth zone with a basipetal growth gradient (orange indicates growth intensity; dashed line marks the end of the proximal growth zone); component 2, the marginal pattern generator producing an interspersed pattern of (component 3) local growth activation (red) and repression (cyan). The marginal pattern generator introduces new growth activator foci as space is generated by growth. *Figure adapted with permission from Runions, A., & Tsiantis, M. (2017). The shape of things to come: From typology to predictive models for leaf diversity.* American Journal of Botany. 104, *1437–1441.*

2.2 Establishment of the leaf blade and margin

Development of leaf shape from the founder cells is a gradual process. Starting as a small bump at the shoot apex, the leaf flattens as it increases in size. The lateral growth giving rise to a flat blade and the margin depends on delimitation of the upper (adaxial) and lower (abaxial) sides of the leaf relative to the central axis defined by the meristem (Fig. 1B). Abaxial and adaxial identities are elaborated through the action of a multi-layered regulatory network involving transcription factors and small noncoding RNAs, which delimit the activity of adaxial and meristem-promoting Class III HD-ZIP (C3HDZ) homeodomain transcription factors as well as abaxial determinants, such as genes of the KANADI family, to their respective domains (reviewed in Kalve, De Vos, & Beemster, 2014; Lenhard, 2017; Merelo, Paredes, Heisler, & Wenkel, 2017; Skopelitis, Benkovics, Husbands, & Timmermans, 2017). While details of this process are outside the scope of this chapter, a key point is that the adaxial-abaxial boundary appears to define a competence zone in the leaf margin where protrusions develop (Caggiano et al., 2017). These protrusions characterize the leaf morphology of different species as they give rise to serrations, lobes, or leaflets in the mature leaf. Differences in the arrangement and shape of outgrowths along the leaf margin are a major contributor to leaf form diversity, both between and within species (Cartolano et al., 2015; Chitwood et al., 2014; Runions, Tsiantis, & Prusinkiewicz, 2017). In recent years, significant progress has been made in understanding the molecular processes regulating margin development, and how these processes are integrated into the global progression of leaf maturation. Studies in model plants indicate that these processes are largely conserved (Bar & Ori, 2015), and have common features with those patterning organs in the SAM (see chapter "Patterning at the shoot apical meristem and phyllotaxis" by Shi and Vernoux, this issue). Similar to the initiation of organs, leaf margin outgrowths initiate at auxin maxima coinciding with PIN1 convergence points (Barkoulas, Hay, Kougioumoutzi, & Tsiantis, 2008; Bilsborough et al., 2011; Hay et al., 2006; Koenig, Bayer, Kang, Kuhlemeier, & Sinha, 2009; Scarpella, Marcos, Friml, & Berleth, 2006). Recent work in *A. thaliana* indicates that leaf margin founder cells may originate from the peripheral zone where organs are initiated (Caggiano et al., 2017). Therefore, marginal patterning may represent a continuation of the processes organizing lateral organs in a different developmental context. In spite of the strong parallels between margin development and organ initiation, there are differences between these two processes. After organ initiation, identity factors play a key role in determining the nature and shape of the resulting outgrowth (e.g., leaf, sepal, petal, or stamen) (Bowman, Smyth, & Meyerowitz, 2012; Tsiantis & Hay, 2003). Currently, there is no evidence that identity factors marking protrusions of different size and shape exist. Instead, the emerging picture for the leaf outgrowths is that they are shaped by a multitude of molecular factors that act together to modulate growth at protrusions or indentations.

2.3 Molecular regulation of the margin

Marginal patterning has been most extensively studied in *A. thaliana*. In this species, serrations are organized in the proliferative, fast growing region near the leaf base (Biot et al., 2016; Donnelly, Bonetta, Tsukaya, Dengler, & Dengler, 1999) (Fig. 1C). In *A. thaliana*, epidermal PIN1 expression is necessary for the formation of auxin maxima and the patterning of serrations (Bilsborough et al., 2011). However, patterning also depends on the transcription factor CUP-SHAPED COTYLEDON2 (CUC2), which is expressed in the indentations (sinuses) between serrations during marginal patterning (Bilsborough et al., 2011; Nikovics et al., 2006). Together, auxin maxima and *CUC2* expression form an alternating pattern along the proximal margin of young leaves that is required for the normal patterning of serrations (Bilsborough et al., 2011; Hay et al., 2006; Nikovics et al., 2006). Auxin maxima are not observed in *cuc2* mutants, and broader *CUC2* expression throughout the leaf epidermis results in smooth margins. Auxin contributes to the establishment of this alternating pattern, both transcriptionally and post transcriptionally via activation of the microRNA *miR164A*, which represses *CUC2* (Bilsborough et al., 2011; Nikovics et al., 2006). Thus, auxin restricts *CUC2* expression to the sinus and inhibits growth in an expression-dependent manner to increase sinus depth (Nikovics et al., 2006). Consequently, *CUC2* may play a dual-role in sculpting the leaf margin by promoting the organization of growth-enhancing auxin maxima while inhibiting growth in its expression domain (Bilsborough et al., 2011; Kawamura, Horiguchi, & Tsukaya, 2010; Nikovics et al., 2006). Additional factors control the final shape of protrusions by modulating auxin gradients generated by marginal patterning (Kasprzewska et al., 2015), auxin response (Tameshige et al., 2016), or by accentuating growth repression near the leaf base in the case of growth repressors like *REDUCED COMPLEXITY* (*RCO*) (Vlad et al., 2014).

An integrated picture of serration development is provided by computational models (Bilsborough et al., 2011) (Fig. 1D). In this model, the leaf grows primarily at the base with *CUC2* expressed along its flanks.

Within the *CUC2* expression domain, a feedback between auxin and PIN1 organizes auxin maxima. By assuming that PIN1 proteins polarize toward neighboring cells with high auxin concentrations (i.e., up-the-gradient) (Jonsson et al., 2006; Smith et al., 2006), initially small auxin concentration differences are enhanced, leading to the emergence of discrete auxin maxima and PIN1 convergence points. Transport toward the maxima locally depletes auxin, inhibiting auxin maxima formation in regions around the convergence points. CUC2 is subsequently eliminated at the auxin maxima, which stabilizes its position along the margin by disabling the feedback between auxin and PIN1. Together, the outlined interactions of auxin, PIN1, and CUC2 provide a mechanism that organizes additional maxima in *CUC2* expressing regions, as space is made available due to growth. The model assumes a strong basipetal gradient of growth and proliferation (consistent with experimental observations, Section 3), causing maxima to be organized sequentially at the leaf base. Such a model reproduces the observed pattern of interspersed auxin activity and *CUC2* expression, as well as the asymmetric shape of serrations. It also highlights how the global pattern of proliferation and maturation broadly constrains the operation of marginal patterning. As growth influences patterning by defining the available space within which peaks of activators—in this case auxin—can form, alterations in the growth domain can alter the order of maxima initiation. For example, broadening the growth domain causes maxima to form in an intercalated pattern (i.e., between pre-existing maxima) instead of sequentially, as it happens when available space is only present at the organ base. This is consistent with mutants where differentiation is delayed or abolished, which have a dramatic overproduction of serrations (Alvarez, Furumizu, Efroni, Eshed, & Bowman, 2016; Bresso, Chorostecki, Rodriguez, Palatnik, & Schommer, 2018). In other species, correlations between growth pattern and the location of protrusion emergence have likewise been observed, with outgrowths appearing with a regular spacing in proliferative regions (Ikeuchi, Igarashi, Okada, & Tsukaya, 2014; Nakamasu, Nakayama, Nakayama, Suematsu, & Kimura, 2014; Nakamasu, Suematsu, & Kimura, 2017).

Geometric models integrating marginal patterning, local regulation of growth at protrusions and sinuses, and gradients of growth and differentiation indicate that interactions between these components can explain aspects of leaf diversity (Runions et al., 2017). In these models, protrusions are introduced based on available space along the margin and initiate new growth directions coinciding with the direction of veins. This changes the growth distribution of the leaf, guiding the introduction of subsequent protrusions in a self-organizing process. Varying parameters of the model reproduces important aspects of leaf diversity. For instance, increasing growth inhibition between protrusions can produce transitions between serrated, lobed, and dissected leaves, and varying the size of the region where marginal patterning occurs changes protrusion number and arrangement. This highlights the idea that leaf diversity may largely result from the interplay of local marginal patterning and growth regulation within more global patterns of growth and differentiation.

3. Background leaf growth and cellular behavior during leaf development

As margin development is integrally connected with the global growth of the leaf, the interplay of marginal and broad growth patterns is instrumental in determining final morphology. Other than local genetic control at the margin, it is therefore essential to understand how development of the leaf is broadly coordinated. This requires (i) exploring how growth is distributed during development and how this distribution is controlled to produce shape, and (ii) direct investigation of the cellular parameters of growth and the possible relationship of cell proliferation and differentiation to leaf form.

3.1 Distribution of leaf growth: Models and measurements

Many leaves, including those of *A. thaliana*, are spoon shaped. This geometry indicates that tissue growth is not uniform throughout the organ; otherwise, leaves would be disk-like. How growth is distributed to produce particular geometries poses a challenge, because plant tissues are mechanically constrained in their growth. Neighboring cells are stuck together, which produces discrepancies between specified growth (i.e., how a region would grow if unconstrained by neighboring regions) and resultant growth (i.e., how the region actually grows) (Kennaway, Coen, Green, & Bangham, 2011). Computational approaches coupled with tissue and cell level growth measurements based on time lapse or clonal sectors (Kuchen et al., 2012; Poethig, 1984; Poethig & Sussex, 1985; Vlad et al., 2014) are instrumental in efforts to connect growth and form. Although many leaves have a gradient to their growth (basipetal in *A. thaliana*), growth measurements indicate that during later stages, growth is largely the same in all directions (isotropic) (Alim, Armon, Shraiman, & Boudaoud, 2016;

Mitchison, 2016; Richards & Kavanagh, 1943). However, during the early stages of leaf development in *A. thaliana*, when leaf shape is established, growth has a preferred direction (anisotropic) toward the leaf tip (Kuchen et al., 2012). Physically based leaf models, where growth was specified locally (i.e., at each point in the tissue) and the resulting change in form was determined by finding the resulting mechanical equilibrium, can explain the overall development of organ shape in leaves without serrations (e.g., the first leaf of *A. thaliana*) (Kuchen et al., 2012). This work shows that overall growth and shape may be accounted for by an organ-wide coordinate system, polarized toward the leaf tip that is carried along (i.e., deformed) with the leaf as it grows. In this model, growth rates are faster at the leaf base, and have an anisotropic component oriented toward the leaf tip that are highest in the mid-vein. By varying growth anisotropy at the leaf tip and the rate of lateral growth, this model can account for a broad range of simple leaf shapes. A global polarity field, as posited by this model, is consistent with recent analysis of ectopic expression of the cell polarity marker BREAKING OF ASYMMETRY IN THE STOMATAL LINEAGE (BASL) in *A. thaliana* leaves (Mansfield et al., 2018). When expressed ectopically throughout the epidermis, BASL has a coordinated polarity toward the leaf base that is broadly aligned with that predicted by this model. Nonetheless, a key question is the biological nature of the signal that polarizes growth in the primordium, as BASL expression in the leaf epidermis is normally restricted to the stomatal lineage (Dong, MacAlister, & Bergmann, 2009). Candidates include the PIN family of auxin transporter proteins, which polarize toward the leaf tip during initiation. Additional support for this hypothesis is provided by studies of petal development, where PIN polarities diverge away from the base and mirror growth directions at the tips of developing petals, which are spoon shaped but lack a pointy tip (Sauret-Güeto, Schiessl, Bangham, Sablowski, & Coen, 2013).

Such approaches, linking modeling and biological experimentation, are essential for exploring processes that are difficult to conceptualize using basic intuitions (Chickarmane et al., 2010; Prusinkiewicz & Runions, 2012). For organ form, the differences between resultant and specified growth mentioned above provide such an example. However, in evaluating models, it is essential to explore why they generate specific forms, and gage whether they are compatible with our understanding of biology (Coen & Kennaway, 2018). This is an important research direction that promises to provide invaluable insights into the emergence of organ geometry.

3.2 Cellular parameters of leaf growth

While measurements and modeling of growth are critical for conceptualizing development of the overall geometry of the leaf, cells are the building blocks of organs. Thus, it is necessary to understand the control of cellular behavior during leaf development, and how this contributes to growth. This development is characterized by an early proliferative phase, followed by an expansive or differentiation phase (Fig. 2), which together determine final organ size (Horiguchi, Ferjani, Fujikura, & Tsukaya, 2006). During the proliferative phase, the average size of proliferating cells remains approximately constant. Subsequently, cells undergo a significant increase in vacuole size, ploidy, and overall size. The duration of cell proliferation influences the number of cells, which can later differentiate and grow. Premature cessation of proliferation or onset of differentiation can lead to the formation of smaller leaves. Simple ink-spot experiments and earlier clonal analysis studies tracking surface

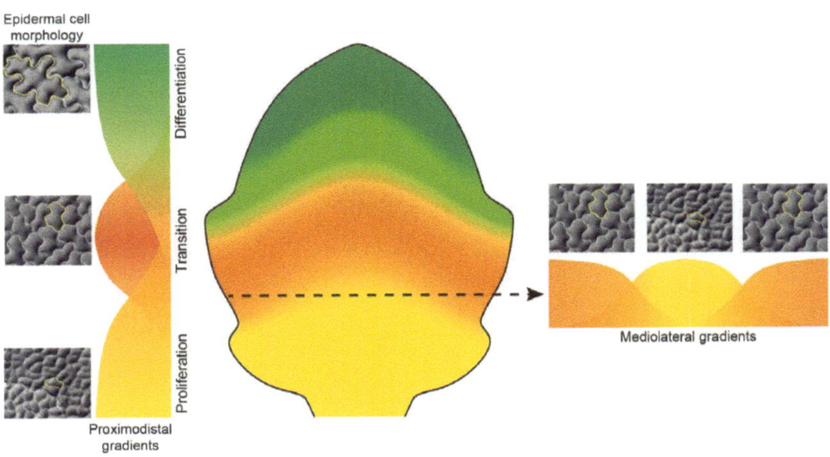

Fig. 2 Medio-lateral and proximo-distal gradients in a growing leaf. Cells proliferate throughout the organ in young primordia. In many species, including *A. thaliana*, an arrest front (green) appears at the distal tip of the leaf and progresses proximally toward the base. The arrest front coincides with cellular differentiation, which is reflected in the morphology of the epidermal cells. Differentiation and proliferation zones are bridged by a transition zone, where proliferating cells progressively acquire a differentiated state. This transition zone coincides with the expression of heterochronic regulators, such as Class II TCPs. As a result, the Class II TCP genes have a dynamic expression pattern, interfacing the distal differentiation zone and the proximal proliferation zone, where expression is maximal in the center of the transition zone. The transition zone (and corresponding Class II TCP expression front) has a convex shape, indicating that cells near the margin differentiate before those in the medial region of the leaf (dashed black line). This produces a mild medio-lateral gradient in cell differentiation and Class II TCP expression at this interface.

growth demonstrate that regions of sustained proliferation have a greater contribution to the final area of the mature leaf (Fig. 3) (Das Gupta & Nath, 2015; Poethig & Sussex, 1985). Nevertheless, a large correlative analysis of 200 *A. thaliana* genotypes suggests that growth itself may be the primary driver of division and endoreduplication, and thus determine organ size (Massonnet et al., 2011). On one hand, this is consistent with the similar final organ size of leaves exhibiting reduced proliferation and larger cells, and increased proliferation and smaller cells, a phenomenon known as compensation (Horiguchi & Tsukaya, 2011; Tsukaya, 2008). On the other hand, experiments on petal growth in *A. thaliana* suggest that cell size and proliferation can affect lateral organ growth independently, without "compensating" for each other (Randall, Sornay, Dewitte, & Murray, 2015). These results might be reconciled by a conceptual model positing a defined temporal growth window for organ development, which is divided into proliferative and expansive phases (Randall et al., 2015). Changing the partitioning of the growth window between proliferation and expansion leads to apparent compensation. In contrast, varying the total duration of the growth window would alter organ size.

Fig. 3 Growth gradients in a tobacco leaf. To track growth distributions during development, ink spots were placed on a young tobacco leaf (left) and followed until the leaf reached maturity (right). Differences in the distribution of ink spots between the two stages show that proximal regions of the leaf, which proliferate longer, have a greater contribution to final surface area (red arrow indicates a corresponding spot at both time-points). The faster progression of growth arrest along the margin causes sequences of spots forming a line along the straight medio-lateral axis to become concave in the mature leaf (dashed yellow lines).

As the timing of the differentiation phase plays an important role in regulating the final size of the organ, there is considerable regulation of the spatiotemporal regulation of this process. For example, model species like *A. thaliana*, *Antirrhinum majus*, and maize have a strong proximo-distal gradient of cell proliferation arrest, which starts near the distal end of the leaf and proceeds proximally (Donnelly et al., 1999; Nath, Crawford, Carpenter, & Coen, 2003; Nelissen et al., 2012). This progressing arrest front divides the leaf into three zones—a proximal proliferative zone, a distal differentiating zone, and an interfacing transition zone (Fig. 2). These zones are also reflected in cell size and shape, with small polygonal cells in the proliferating base, a gradient of growing cells in the transition zone, and large puzzle shaped cells near the leaf tip. The proximo-distal arrest gradient coincides with the growth gradient through the leaf, which is highest at the base and decreases gradually toward the distal tip (Andriankaja et al., 2012; Kuchen et al., 2012).

The proximo-distal gradient of growth has a medio-lateral component. Together these gradients produce a transition zone with convex shape as the arrest front progresses faster along the leaf margins compared to the medial part of the leaf. The importance of these gradients is evident from mutants with altered leaf morphologies, which result from perturbations in the gradient of growth or the timing of exit from proliferation. Effects of perturbation in growth gradients are discussed in the following sections.

3.3 Cell proliferation and leaf size

The molecular mechanisms underlying the proximo-distal gradient of cell proliferation and differentiation involve opposing gradients of transcription factors, microRNAs (*miRs*) and plant hormones. The proliferative domain in *A. thaliana* leaves is marked by the largely overlapping expression domains of *CYCLIN B1;1* and GROWTH-REGULATING FACTOR (GRF) family genes, exhibiting gradients with highest expression near the leaf base (Debernardi, Rodriguez, Mecchia, & Palatnik, 2012; Rodriguez et al., 2010). GRFs act as positive regulators of cell proliferation and their loss-of-function mutants have smaller leaves with reduced cell number (Horiguchi, Kim, & Tsukaya, 2005; Kim & Kende, 2004). The expression of *GRF* genes is controlled by regulatory *miR396*, with *miR396* expression decreasing gradually from the tip toward the base (Debernardi et al., 2012). Overexpression of *miR396* reduces *GRF* gene expression, leading to a significant reduction in leaf size and cell number (Rodriguez et al., 2010;

Wang, Gu, et al., 2011). Overexpression of GRF family members or expression of *miR396*-resistant versions of *GRF* genes leads to larger leaves with more cells (Debernardi et al., 2012; Nelissen et al., 2012). An antagonistic expression pattern of *miR396* and GRF genes is also observed in species with different patterns of leaf growth. For example, the leaves of *Dillenia indica*, unlike *A. thaliana* and maize, differentiate from the base and show highest expression of *miR396* near the base and highest expression of *GRF2* near the tip (Das Gupta & Nath, 2015). A broader survey of species with different distributions of leaf growth and proliferation (e.g., highest near the tip, or uniform throughout the leaf) reinforce the strong correlation between gradients of growth and GRF gene expression.

The function of GRFs is positively regulated by a group of cofactors, GRF INTERACTING FACTORS (GIFs), that exhibit a GRF-like expression gradient. GIFs promote cell proliferation and their loss-of-function mutants display smaller leaves with reduced cell number. The GIFs regulate GRF functions by recruiting chromatin-remodeling factors to the GRF regulatory regions. Analysis of the GRF-GIF interactions along the proximo-distal axis of maize leaves revealed that the molecular associations between the GRFs and the GIFs vary along this axis (Nelissen et al., 2015). For example, the GIFs form a complex with seven of the GRFs in the proliferative zone, but only two of the GRFs in the transition zone. As such, GRF function is dynamically regulated at the transition zone by transcriptional and post-transcriptional mechanisms to ensure the precise exit of cells from the proliferation to the differentiation zone (Nelissen et al., 2015).

Alterations in the size of the proliferation zone in the proximo-distal axis can affect leaf size, without significantly affecting the overall shape of the organ (Das Gupta & Nath, 2015; Tsukaya, 2002). GRFs, GIFs, and *miR396* alter leaf size by affecting cell number (Nelissen et al., 2012; Rodriguez et al., 2010; Rodriguez, Schommer, & Palatnik, 2016). In contrast, factors that coordinate the timing of cell proliferation and differentiation in both the proximal-distal and medio-lateral axes can significantly affect leaf shape. This is due, in part, to their influence on marginal patterning, as sustained growth and proliferation at the margin can prolong marginal patterning and fuel the outgrowth of marginal protrusions. One such class of regulators are the *CINCINNATA* (*CIN*)-like TCP (*TEOSINTE BRANCHED, CYCLOIDEA* and *PCF1/2*) family genes (also known as Class II TCPs) that promote differentiation (Martin-Trillo & Cubas, 2010).

The transition zone features dynamic gene expression and hormone activity. In maize leaves, it is superimposed with a peak of bioactive gibberellins (GA) that regulate the size of the proximal proliferative zone (Nelissen et al., 2012). The dynamic peak of GA results from the combined activities of GA biosynthetic genes in the proximal proliferation zone and GA catabolic genes in the distal expansion zone. GA-deficient maize plants produce smaller leaves, while plants with higher GA production make bigger leaves. Alteration of GA activity influences leaf size by changing the size of the proliferation zone, which correlates positively with GA levels. The maize leaf also shows highest levels of bioactive auxin and cytokinin (CK) in the proliferative zone (Nelissen et al., 2012).

3.4 Timing of differentiation and leaf shape

The Class II TCPs are plant-specific transcription factors that regulate organ growth (Martin-Trillo & Cubas, 2010). They are regulated by the conserved *miR319* that restricts TCP function to the proliferating zone of developing leaves (Ori et al., 2007; Palatnik et al., 2003). Mutant analysis in several plant species suggests that Class II TCPs promote differentiation to regulate leaf shape and size (Martin-Trillo & Cubas, 2010; Narumi et al., 2011). Expression of *CIN* in *A. majus* has a dynamic pattern along the proximo-distal axis, and a mild medio-lateral expression gradient, which together mirror the convex arrest front progressing toward the leaf base (Fig. 2) (Das Gupta, Aggarwal, & Nath, 2014; Nath et al., 2003). In the *cin* mutant, progression of the arrest front slows down, allowing excess cell proliferation and leading to the formation of larger leaves. Interestingly, the shape of the convex arrest front in wild-type plants changes to strongly concave in the *cin* mutant, prolonging growth especially along the leaf margins and resulting in dramatic buckling and crinkling along the margin (Nath et al., 2003). The extreme phenotypic effects also indicate the importance of genes that harmonize the exit from proliferation along multiple tissue axes. *CIN* appears to coordinate the differentiation of cells in the two axes by simultaneously regulating multiple phytohormone-related genes. For example, the auxin signaling gene *INDOLE-3-ACETIC ACID INDUCIBLE3* and cytokinin receptor gene *HISTIDINE KINASE4* are directly bound by CIN and show medio-lateral and proximo-distal expression gradients (Das Gupta et al., 2014). TCP function in tomato is also partly mediated by the GA signaling pathway (Yanai, Shani, Russ, & Ori, 2011), and the dynamic peak of GA in the maize leaf is reminiscent of the TCP peak in *A. majus*, indicating a link between TCP and GA functions (Nelissen et al., 2012).

A strong TCP dependent effect on leaf shape is observed in other model species where marginal patterning contributes to leaf form, such as *A. thaliana* and *Solanum lycopersicum* (tomato). *A. thaliana* has five Class II TCP genes (*TCP2–5* and *TCP10*) that are targeted by *miR319*. Gain-of-function in *miR319* leads to simultaneous downregulation of all five TCPs, resulting in crinkly leaves (Palatnik et al., 2003). Additionally, loss-of-function mutants affecting TCP genes have delayed differentiation along the leaf margins, resulting in deep serrations (Bresso et al., 2018; Palatnik et al., 2003). The loss of TCP function in the compound leaves of tomato leads to the formation of highly dissected leaves due to a prolonged proliferative phase (Ori et al., 2007). The effects of mutants affecting TCP genes on leaf shape are more pronounced when combined with other genes that redundantly control leaf margin differentiation. In *A. thaliana*, a combination of higher order loss-of-function mutants affecting genes of the TCP and NGATHA (NGA) family exhibits highly elaborated crinkly leaf margins with deep lobes, as a result of abolished marginal differentiation and continuously sustained proliferation (Alvarez et al., 2016). Intriguingly, these effects do not appear to result from altered expression of KNOXI and CUC proteins that promote formation of marginal outgrowths. Rather, they involve WUSCHEL-LIKE HOMEOBOX (WOX) proteins that promote cell proliferation in different contexts in the plant, including the leaf margin, where they are required for lateral outgrowth and development of a flat lamina (Nardmann, Ji, Werr, & Scanlon, 2004). Current genetic evidence suggests that WOX family members act downstream of adaxial/abaxial patterning, and may promote blade outgrowth by helping to organize an intermediate domain between the adaxial and abaxial sides of the leaf that encompasses the margin (Nakata et al., 2012). These complex interactions illustrate that an important task for the future will be to understand the mechanistic interrelationships and downstream effectors of these different pathways influencing leaf margin development.

4. The genetic basis for the evolution of leaf shape

Having discussed the main processes underlying leaf development, it becomes possible to ask what are the specific developmental and molecular changes that caused natural variation in leaf form at different evolutionary scales. The mustard family (Brassicaceae) is an excellent model that harbors a representative sample of the leaf diversity of flowering plants (Nikolov & Tsiantis, 2017). This diversity can be apparent within a single individual, both throughout development and under different environmental conditions. It can also be seen among populations of a single species, and between species at different phylogenetic scales. Several examples of the aforementioned diversity follow. Many species exhibit age-related variation in leaf morphology. For example, *A. thaliana* forms rounder and smoother juvenile leaves and more elongated adult leaves, featuring pronounced serrations (Poethig, 2013). Leaflet number in *Cardamine hirsuta* increases with age (Cartolano et al., 2015; Rubio-Somoza et al., 2014). A single individual of the aquatic mustard *Rorippa aquatica* can form different leaves, depending on submergence or temperature, with complex leaves developing at lower temperatures (Nakayama et al., 2014). *C. hirsuta* exhibits variation in leaf shape among different accessions that may reflect local adaptation (Cartolano et al., 2015). Within the genus *Cardamine*, leaves vary from simple (*Cardamine alpina*), lobed (*Cardamine resedifolia*), palmately dissected tri- and pentafoliate (*Cardamine pentaphyllos*) to pinnately dissected (*C. hirsuta*) into individual leaflets (Hay et al., 2014). Leaf shape variation is extensive at the family level (Nikolov & Tsiantis, 2017). Comparative studies have started to elucidate the molecular mechanisms that give rise to leaf shape diversity during development (Chitwood & Sinha, 2016). Among the many genes that can potentially alter leaf shape, only a few were repeatedly employed during evolution, depending on the phylogenetic scale in question. This convergence at the molecular level also reflects a broader developmental convergence, where independently derived differences in growth and differentiation at the leaf margin give rise to similar leaf forms. This phenomenon was also captured by morphologists with the concept of the marginal blastozone (Efroni, Eshed, & Lifschitz, 2010; Hagemann & Gleissberg, 1996), which was used to denote the ability of the leaf margin to generate outgrowths of different types. It has been speculated that such convergences reflect the deep evolutionary origin of leaves from branched shoots, for example, by Arber (2012), who emphasized the developmental similarities of shoots, leaves, and leaflets. Molecular evidence now helps make those debates more concrete and, in this section, we focus on cases where genetic evidence has provided insights into the causal molecular changes underlying diversity.

4.1 Macroevolutionary changes

Forward genetic screens and the candidate gene approach based on species with contrasting leaf morphologies have revealed genetic components underlying the divergence in leaf shape at the macroevolutionary scale. KNOXI genes were the first regulators discovered to control leaf diversification in a number of distantly related species (Bar & Ori, 2015; Bharathan et al., 2002; Hareven, Gutfinger, Parnis, Eshed, & Lifschitz, 1996; Hay & Tsiantis, 2006). KNOXI genes are typically expressed in the SAM of species with simple leaves, like *A. thaliana*, but this expression extends into developing leaves to promote leaflet initiation in distantly related species with complex leaves, like tomato and *C. hirsuta* (Hay & Tsiantis, 2006). In eudicots, KNOX genes have been identified as important regulators of leaf development and complexity (Hake et al., 2004; Tsuda & Hake, 2015), where KNOXI gene expression in leaves broadly correlates with the formation of complex leaf primordia (Bharathan et al., 2002).

While the understanding of how KNOXI proteins promote leaflet formation is fragmentary, there is substantial evidence that this function in leaves involves the reactivation of SAM-maintenance pathways. In part, these effects involve the direct regulation of genes influencing hormone homeostasis that establish a low GA—high CK regime, by inhibiting GA activity and promoting CK activity (Bolduc et al., 2012; Hay et al., 2002; Jasinski et al., 2005; Sakamoto, Kamiya, Ueguchi-Tanaka, Iwahori, & Matsuoka, 2001; Yanai et al., 2005). These interactions may underlie the ability of KNOXI genes to prolong marginal patterning and delay differentiation (Bar & Ori, 2015; Hay & Tsiantis, 2010). The effect of KNOXI genes is antagonistic to TCP and NGA activities (Hareven et al., 1996; Ori et al., 2007). Crosstalk exists among these modules as KNOXI proteins exert their effects within a TCP-dependent competence window (Ori et al., 2007; Shleizer-Burko, Burko, Ben-Herzel, & Ori, 2011). Additionally, KNOXI expression promotes leaflet outgrowth by reinforcing auxin-dependent cell polarization via the PIN1 auxin efflux transporter, which supports leaflet formation and outgrowth (Barkoulas et al., 2008). The action of KNOXI genes on margin patterning may, at least in part, involve mutual reinforcement with CUC proteins (Blein et al., 2008). Conversely, KNOXI action in leaflets is also subject to negative regulation by Class II KNOX genes (Furumizu, Alvarez, Sakakibara, & Bowman, 2015). These and related findings (Ben-Gera et al., 2016) further underscore the fact that multiple processes acting in parallel influence growth and differentiation in the leaf margin, and highlight the need for a precise understanding of how each one controls the direction, amount, and duration of growth to sculpt leaf form.

When gene expression differences between species are found to contribute to macroevolutionary differences in form, the question arises of what underlies the mechanistic basis for these divergent expression patterns. This can be addressed by swapping regulatory and coding sequences in the context of interspecific gene transfers (Nikolov & Tsiantis, 2015). Such experiments showed that *cis*-regulatory divergence underlies the difference between *A. thaliana*, where KNOXI genes are not expressed in leaves, and *C. hirsuta* and *A. lyrata*, where KNOXI genes are expressed in leaves (Hay & Tsiantis, 2006; Piazza et al., 2010). Comparative analysis of *A. thaliana* and *C. hirsuta* mutants, combined with interspecific gene transfers, also revealed that the KNOXI genes *SHOOTMERISTEMLESS* (*STM*) and *BREVIPEDICELLUS* (*BP*) have different potency to modify shape, which correlates inversely with the degree of pleiotropy of these genes (Rast-Somssich et al., 2015). Notably, *BP* regulatory sequences were more potent to elicit an increase in leaf complexity in *A. thaliana* than those of *STM*, although *bp* mutants show less severe developmental defects compared to *stm* mutants. This finding suggests that one way to alter shape involves genes with demonstrated functional capacity and low pleiotropy that can be readily integrated into, or excluded from, existing genetic networks to contribute to divergence in morphology.

Comparative genetic studies in a phylogenetic framework demonstrated that the absence of KNOXI expression in *A. thaliana* leaves is derived. In *C. hirsuta* leaves, *BP* expression is negatively regulated by the MYB transcription factor AS1 (Rast-Somssich et al., 2015), and a regulatory module comprising *miR164A* and its target *CUC2* (Hay & Tsiantis, 2016; Nikovics et al., 2006; Rast-Somssich et al., 2015). This creates crosstalk between components of the leaf gene regulatory network that does not occur in *A. thaliana*. Thus, *cis*-regulatory changes not only partition *BP* activity between the SAM and the leaf of *A. thaliana* and *C. hirsuta*, but also result in different genetic interactions between common components of the leaf gene regulatory networks of both species. These findings highlight the importance of comparative studies in multiple species to fully understand the logic of organ development.

Another pathway influencing leaf complexity is defined by *REDUCED COMPLEXITY* (*RCO*), which encodes for a growth repressor. *RCO* contributes to leaf shape changes in a number of mustards, including *C. hirsuta* and *A. thaliana* (Vlad et al., 2014), and *Capsella* species (Sicard et al., 2014). *RCO* arose from the tandem gene duplication that also gave rise to *LATE MERISTEM IDENTITY1* (*LMI1*), a gene first identified as a floral regulator (Saddic et al., 2006). Coupled coding and regulatory evolution of *RCO* after the duplication resulted in a novel expression domain at the base of the leaf and reduced protein stability, which likely mitigated detrimental pleiotropic effects (Vlad et al., 2014; Vuolo et al., 2016) (Fig. 4). Furthermore, *RCO* gene loss likely contributed to leaf simplification in *A. thaliana*. Introducing *RCO* back into *A. thaliana* as a transgene results in complex leaves, demonstrating that despite having simple leaves, much of the ancestral regulatory network contributing to leaflet complexity has remained intact in this species. *RCO* evolution provides a clear example for the role of gene duplication coupled with regulatory and protein sequence change and loss to modify morphology. Polymorphism patterns at the *RCO* locus display signatures for positive selection in its regulatory and coding sequences, indicating that *RCO*-mediated changes in leaf form may have been of adaptive significance (Vuolo et al., 2016). This conclusion is reinforced by experiments of plant physiological performance, such as enhanced CO_2 fixation, and correlation between seed biomass and levels of *RCO* activity in transgenic plants (Vuolo et al., 2016). Similar observations are reported in *Gossypium hirsutum* (cotton), where okra-leaf genotypes, which exhibit high leaf complexity as a result of altered *LMI1* expression (Andres et al., 2017; Chang et al., 2016), show an increase in photosynthetic rates (Wells, Meredith, & Williford, 1986). An open question is why changes in RCO and leaf complexity would alter leaf physiology. Possibilities include changes in gas exchange due to reduced air boundary layer thickness as in other complex leaves (Royer & Wilf, 2006), altered light capture, owing to reduced self-shading (Niklas, 1988), and hydraulics, owing to altered vasculature organization in a complex blade (Dengler & Kang, 2001; Sack et al., 2012).

To understand how this newly discovered pathway regulating leaf complexity operates, it will be important to identify downstream effectors via which *RCO* and *LMI1* regulate cell growth and consequently shape. A first step in this direction was the recent isolation of the mitotic blocker *WEE1* as a direct target of LMI1 in leaves (Vuolo et al., 2018). This finding was made after discovering that *LMI1* is active in stipules, vestigial outgrowths at the flanks of *A. thaliana* and *C. hirsuta* leaves, where it represses their growth by causing an early transition to endoreduplication, through *WEE1* activation. Notably, endoreduplication is often associated with

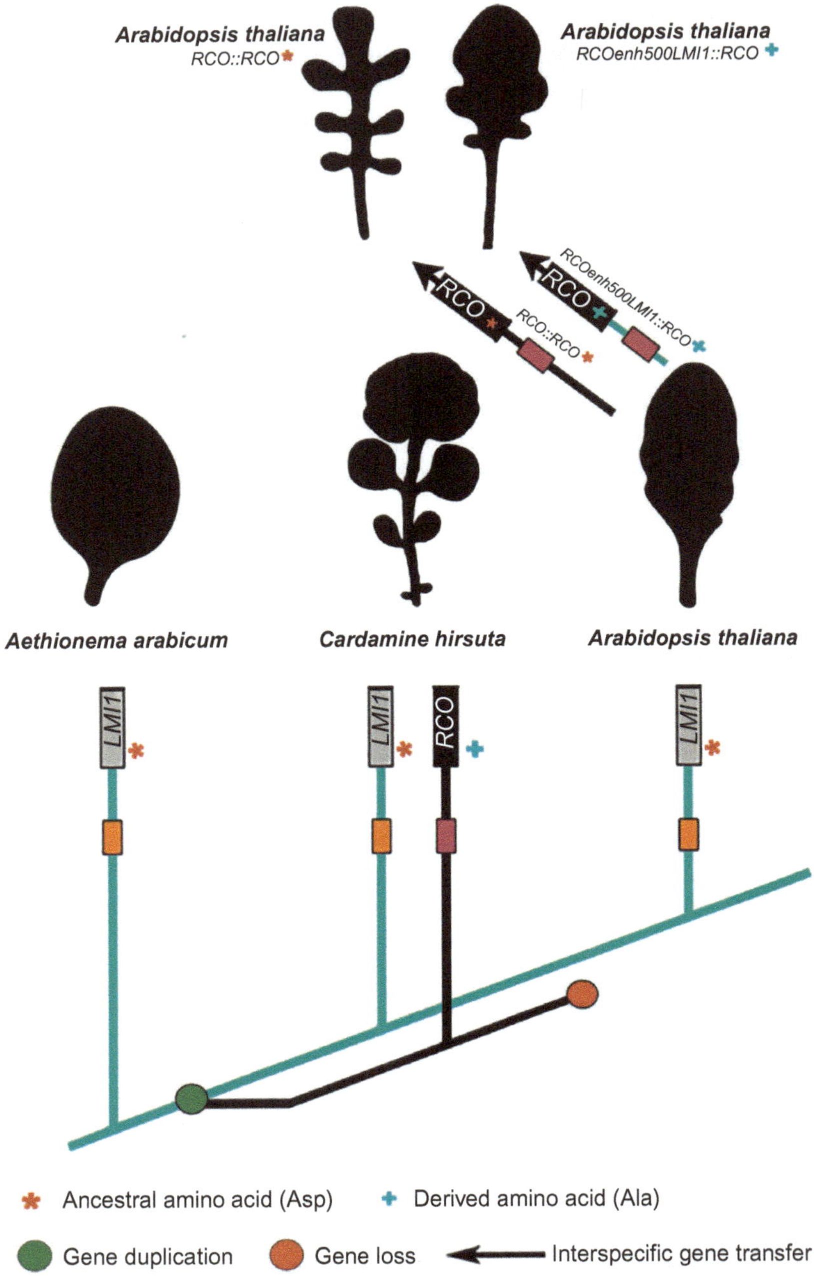

Fig. 4 See legend on opposite page.

increased cell and organ size, as is the case for fleshy fruits (Chevalier et al., 2011). However, mathematical modeling indicates that the consequence of endoreduplication on tissue growth in a proliferating cell population depends on the timing of activation (Vuolo et al., 2018). Early onset of endoreduplication, as induced in stipules by *LMI1*, eliminates mitotic divisions and limits the pool of cells available for future tissue growth. As a consequence, final stipule size is reduced despite an early increase in cell growth. Conversely, in *lmi1* mutants, stipules grow out as small ectopic leaves, simple in *A. thaliana* and complex in *C. hirsuta*, indicating that stipules can be considered transformed leaves. Accordingly, *Tendrilless* (*Tl*; pea *LMI1*) is not expressed in the large leafy stipules of pea plants, but is active in the upper part of the leaf, where it promotes formation of bladeless climbing structures called tendrils (Hofer et al., 2009; Vuolo et al., 2018). It is thus possible that the concurrent stipule overgrowth and transformation of leaflets to tendrils that occurs in some legumes, as noted by Darwin (1865), depends on this recently discovered LMI1 endoreduplication module. It will be important to explore how broadly this *LMI1* endoreduplication module is used to regulate plant development, and whether it also mediates *RCO* function in different taxa.

In addition to the diverse morphologies produced by variation of the leaf margin, leaves can develop into specialized adaptive morphologies that involve extensive 3D deformation of the blade. This is exemplified by carnivorous plants that develop specialized lobes or pitchers to trap and digest small animals. Understanding how these forms develop provides an exciting

Fig. 4 Enhancer and coding sequence evolution at the *RCO* locus shaped leaf diversity in crucifer plants. A single copy *LMI1-like* homeobox gene is present in the genome of *Aethionema arabicum*, the sister lineage to the rest of Brassicaceae, which features simple leaves. The gene is expressed distally in the leaf under the control of a ~500 bp enhancer element (orange). A gene duplication event (green circle) following the divergence of *Aethionema* gave rise to two paralogous gene lineages. One of the paralogs, *LMI1*, has retained the ancestral expression pattern. The other paralog, *RCO*, is expressed proximally in the leaf, following divergence of the enhancer element (purple). *RCO* was lost from the genome of *A. thaliana* (red circle). The *RCO* enhancer co-evolved with a protein coding change, decreasing the stability and potency of the protein (D48A; blue cross). Introduction of *RCO* or a chimeric *LMI1* gene harboring the *RCO* enhancer into *A. thaliana* increases leaf complexity, partially reversing the effect of *RCO* loss. Introducing a mutated RCO protein bearing the ancestral amino acid at position 48 (aspartic acid) into *A. thaliana* produces more severely dissected leaves. This suggests that coupled regulatory and coding sequence evolution created the potential for morphological novelty by limiting pleiotropic effects.

opportunity to unravel the origin of complex tissue geometries, and we refer readers to recent studies and reviews for more information (Fukushima et al., 2017, 2015; Whitewoods & Coen, 2017).

4.2 Microevolutionary changes, heterochrony, and heterophylly

The presented studies provide important insights into the genetic differences that underlie variation between reproductively isolated species. However, variation emerges at the population level, and it is important to evaluate whether pathways that allow variation between and within species rely on the same genetic components to achieve similar phenotypic output. These approaches can also help understand the ecological significance of variation in leaf shape. Quantitative trait locus (QTL) analysis is a powerful way to identify the genetic components underlying leaf morphological differences. Such an approach, based on a cross between two naturally occurring strains of *C. hirsuta* that differ in leaflet number and thus leaf complexity, identified a number of QTLs that contribute to phenotypic variation (Cartolano et al., 2015). One of the larger effect QTLs reflects allelic variation in the expression of *FLOWERING LOCUS C* (*FLC*), a floral repressor that can also influence age-dependent variation in leaf development (Willmann & Poethig, 2011). The common genetic control results in early flowering plants, progressing to adult shapes with more leaflets faster than their slower-aging counterparts. This mechanism helps coordinate leaf development with reproductive timing and may link vegetative growth to reproduction to optimize resource allocation to seeds. Such optimization might be particularly important for annual plants like *C. hirsuta*, where flowering time can rapidly diversify. Examples where early flowering genotypes developed leaves with increased complexity were already noted in the 1940s in cotton and other species (Ashby, 1948). Thus, current molecular investigations provide opportunities to interpret these earlier findings and test their generality. For example, genes controlling age-dependent variation in leaf form were found to associate with strong heteroblastic changes in diverse species like *Gevuina avellana* and acacia, where there is a striking variation in leaf form (Ostria-Gallardo et al., 2016; Wang, Park, et al., 2011), and heteroblastic differences also contribute to leaf shape variation in *Vitis vinifera* (grapevine) (Chitwood et al., 2016). An important area of research in the future is to understand the degree to which this variation in leaf morphology results from neutral processes as opposed to local adaptation.

Although the adaptive significance of different leaf margin morphologies is a still matter of debate, environmentally controlled variation in leaf shape (heterophylly) may have a strong adaptive component (Wells & Pigliucci, 2000). It is a classic example of phenotypic plasticity, where a single individual develops leaves with different morphologies under different environmental conditions. Heterophylly is particularly pronounced in plants growing at the border of terrestrial and aquatic environments that often withstand partial submergence (Nakayama, Sinha, & Kimura, 2017). One such example is *R. aquatica*, which develops pinnately compound leaves when submerged and simpler leaves in terrestrial conditions (Nakayama et al., 2014). The difference in morphology correlates with the expression of KNOXI family genes, and subsequently with the accumulation of GA in the developing complex leaves. Exogenous application of GA can trigger complex leaf development, indicating the sufficiency of this plant hormone to alter shape. Similar pharmacological approach demonstrated that a variety of hormonal cues likely regulate heterophylly in other species, including abscisic acid in *Potamogeton nodosus* and *Ludwigia arcuata*, and ethylene in *L. arcuata* (Nakayama et al., 2017). *Capsella rubella* shows phenotypic plasticity as a response to temperature that correlates with the level of *RCO* expression, providing a possible mechanistic link between leaf development and the environment (Sicard et al., 2014). However, as *C. rubella* arose in the context of a strong bottleneck (Guo et al., 2009), where effects of genetic drift were likely substantial, the adaptive significance of this finding may be difficult to investigate.

5. Conclusions

These examples highlight the substantive advances that have been made in understanding leaf development and diversity. However, despite this progress, many pieces remain missing in the puzzle of attaining a predictive knowledge of leaf shape generation and evolution. Nevertheless, recent work (Bassel & Smith, 2016; Coen, Kennaway, & Whitewoods, 2017; Vlad et al., 2014) also highlights paths forward to solving these problems. A major challenge for the future is to build a physically based computational framework at the cellular level, which allows global consideration of the action of the multiple regulatory hierarchies that generate diverse leaf forms. Doing this successfully will require advanced imaging to identify the links between spatially and temporally distributed actions of the genetic networks shaping form and growth, as well as an understanding of how critical developmental regulators act at the subcellular level to regulate growth. The latter task will involve dissecting signaling pathways driving growth, using more refined molecular biology, biochemistry and biomechanics methodologies, including single cell methods. This combination of approaches exploited in a comparative and phylogenetically informed framework will no doubt lead to substantial advances on how biological forms emerge and diversify. Additionally, mechanistic work on diversity will stimulate informed physiological and ecological studies, which should help elucidate the evolutionary forces that shape leaf diversity and its ecophysiological significance.

Acknowledgments

We thank Mike Ambrose (JIC) for providing pea germplasm. A.R. gratefully acknowledges support from the European Commission through a Marie Skłodowska-Curie fellowship (IF-2015 703886), and L.N. support by a Humboldt-Bayer Research Fellowship. Relevant work in the Tsiantis laboratory is supported by DFG FOR (Forschergruppe) 2581 and a core grant by the Max Planck Society.

References

Alim, K., Armon, S., Shraiman, B. I., & Boudaoud, A. (2016). Leaf growth is conformal. *Physical Biology*, *13*, 05LT01.

Alvarez, J. P., Furumizu, C., Efroni, I., Eshed, Y., & Bowman, J. L. (2016). Active suppression of a leaf meristem orchestrates determinate leaf growth. *eLife*, *5*, e15023.

Andres, R. J., Coneva, V., Frank, M. H., Tuttle, J. R., Smayoa, L. F., Han, S. W., et al. (2017). Modifications to a LATE MERISTEM IDENTITY1 gene are responsible for the major leaf shapes of Upland cotton (Gossypium hirsutum L.). *Proceedings of the National Academy of Sciences of the United States of America*, *114*, E57–E66.

Andriankaja, M., Dhondt, S., De Bodt, S., Vanhaeren, H., Coppens, F., De Milde, L., et al. (2012). Exit from proliferation during leaf development in *Arabidopsis thaliana*: A not-so-gradual process. *Developmental Cell*, *22*, 64–78.

Arber, A. (2012). *The natural philosophy of plant form*. Cambridge University Press.

Ashby, E. (1948). Studies in the morphogenesis of leaves. *New Phytologist*, *47*, 153–176.

Bar, M., & Ori, N. (2015). Compound leaf development in model plant species. *Current Opinion in Plant Biology*, *23*, 61–69.

Barkoulas, M., Hay, A., Kougioumoutzi, E., & Tsiantis, M. (2008). A developmental framework for dissected leaf formation in the *Arabidopsis* relative *Cardamine hirsuta*. *Nature Genetics*, *40*, 1136–1141.

Bassel, G. W., & Smith, R. S. (2016). Quantifying morphogenesis in plants in 4D. *Current Opinion in Plant Biology*, *29*, 87–94.

Ben-Gera, H., Dafna, A., Alvarez, J. P., Bar, M., Mauerer, M., & Ori, N. (2016). Auxin-mediated lamina growth in tomato leaves is restricted by two parallel mechanisms. *The Plant Journal*, *86*, 443–457.

Bharathan, G., Goliber, T. E., Moore, C., Kessler, S., Pham, T., & Sinha, N. R. (2002). Homologies in leaf form inferred from *KNOXI* gene expression during development. *Science*, *296*, 1858–1860.

Bhatia, N., Bozorg, B., Larsson, A., Ohno, C., Jönsson, H., & Heisler, M. G. (2016). Auxin acts through MONOPTEROS to regulate plant cell polarity and pattern phyllotaxis. *Current Biology, 26*, 3202–3208.

Bilsborough, G. D., Runions, A., Barkoulas, M., Jenkins, H. W., Hasson, A., Galinha, C., et al. (2011). Model for the regulation of *Arabidopsis thaliana* leaf margin development. *Proceedings of the National Academy of Sciences of the United States of America, 108*, 3424–3429.

Biot, E., Cortizo, M., Burguet, J., Kiss, A., Oughou, M., Maugarny-Cales, A., et al. (2016). Multiscale quantification of morphodynamics: MorphoLeaf software for 2D shape analysis. *Development, 143*, 3417–3428.

Blein, T., Pulido, A., Vialette-Guiraud, A., Nikovics, K., Morin, H., Hay, A., et al. (2008). A conserved molecular framework for compound leaf development. *Science, 322*, 1835–1839.

Bolduc, N., Yilmaz, A., Mejia-Guerra, M. K., Morohashi, K., O'Connor, D., Grotewold, E., et al. (2012). Unraveling the KNOTTED1 regulatory network in maize meristems. *Genes & Development, 26*, 1685–1690.

Bowman, J. L., Smyth, D. R., & Meyerowitz, E. M. (2012). The ABC model of flower development: Then and now. *Development, 139*, 4095–4098.

Bresso, E. G., Chorostecki, U., Rodriguez, R. E., Palatnik, J. F., & Schommer, C. (2018). Spatial control of gene expression by miR319-regulated TCP transcription factors in leaf development. *Plant Physiology, 176*, 1694–1708.

Byrne, M. E., Barley, R., Curtis, M., Arroyo, J. M., Dunham, M., Hudson, A., et al. (2000). *Asymmetric leaves1* mediates leaf patterning and stem cell function in *Arabidopsis*. *Nature, 408*, 967.

Caggiano, M. P., Yu, X., Bhatia, N., Larsson, A., Ram, H., Ohno, C. K., et al. (2017). Cell type boundaries organize plant development. *eLife, 6*, e27421.

Cartolano, M., Pieper, B., Lempe, J., Tattersall, A., Huijser, P., Tresch, A., et al. (2015). Heterochrony underpins natural variation in *Cardamine hirsuta* leaf form. *Proceedings of the National Academy of Sciences of the United States of America, 112*, 10539–10544.

Chang, L., Fang, L., Zhu, Y., Wu, H., Zhang, Z., Liu, C., et al. (2016). Insights into interspecific hybridization events in allotetraploid cotton formation from characterization of a gene-regulating leaf shape. *Genetics, 204*, 799–806.

Chevalier, C., Nafati, M., Mathieu-Rivet, E., Bourdon, M., Frangne, N., Cheniclet, C., et al. (2011). Elucidating the functional role of endoreduplication in tomato fruit development. *Annals of Botany, 107*, 1159–1169.

Chickarmane, V., Roeder, A. H., Tarr, P. T., Cunha, A., Tobin, C., & Meyerowitz, E. M. (2010). Computational morphodynamics: A modeling framework to understand plant growth. *Annual Review of Plant Biology, 61*, 65–87.

Chitwood, D. H., Klein, L. L., O'Hanlon, R., Chacko, S., Greg, M., Kitchen, C., et al. (2016). Latent developmental and evolutionary shapes embedded within the grapevine leaf. *New Phytologist, 210*, 343–355.

Chitwood, D. H., Ranjan, A., Kumar, R., Ichihashi, Y., Zumstein, K., Headland, L. R., et al. (2014). Resolving distinct genetic regulators of tomato leaf shape within a heteroblastic and ontogenetic context. *Plant Cell, 26*, 3616–3629.

Chitwood, D. H., & Sinha, N. R. (2016). Evolutionary and environmental forces sculpting leaf development. *Current Biology, 26*, R297–R306.

Coen, E., & Kennaway, R. (2018). Early shaping of a leaf. *Nature Plants, 4*, 618.

Coen, E., Kennaway, R., & Whitewoods, C. (2017). On genes and form. *Development, 144*, 4203–4213.

Darwin, C. (1865). On the movements and habits of climbing plants. *Journal of the Linnean Society of London, Botany, 9*, 1–118.

Das Gupta, M., Aggarwal, P., & Nath, U. (2014). *CINCINNATA* in *Antirrhinum majus* directly modulates genes involved in cytokinin and auxin signaling. *New Phytologist*, *204*, 901–912.

Das Gupta, M., & Nath, U. (2015). Divergence in patterns of leaf growth polarity is associated with the expression divergence of miR396. *Plant Cell*, *27*, 2785–2799.

Debernardi, J. M., Rodriguez, R. E., Mecchia, M. A., & Palatnik, J. F. (2012). Functional specialization of the plant miR396 regulatory network through distinct microRNA-target interactions. *PLoS Genetics*, *8*, e1002419.

Dengler, N., & Kang, J. (2001). Vascular patterning and leaf shape. *Current Opinion in Plant Biology*, *4*, 50–56.

Dong, J., MacAlister, C. A., & Bergmann, D. C. (2009). BASL controls asymmetric cell division in *Arabidopsis*. *Cell*, *137*, 1320–1330.

Donnelly, P. M., Bonetta, D., Tsukaya, H., Dengler, R. E., & Dengler, N. G. (1999). Cell cycling and cell enlargement in developing leaves of *Arabidopsis*. *Developmental Biology*, *215*, 407–419.

Efroni, I., Eshed, Y., & Lifschitz, E. (2010). Morphogenesis of simple and compound leaves: A critical review. *The Plant Cell*, *22*, 1019–1032. https://doi.org/10.1105/tpc.109.073601.

Fukushima, K., Fang, X., Alvarez-Ponce, D., Cai, H., Carretero-Paulet, L., Chen, C., et al. (2017). Genome of the pitcher plant *Cephalotus* reveals genetic changes associated with carnivory. *Nature Ecology & Evolution*, *1*, 0059.

Fukushima, K., Fujita, H., Yamaguchi, T., Kawaguchi, M., Tsukaya, H., & Hasebe, M. (2015). Oriented cell division shapes carnivorous pitcher leaves of *Sarracenia purpurea*. *Nature Communications*, *6*, 6450.

Furumizu, C., Alvarez, J. P., Sakakibara, K., & Bowman, J. L. (2015). Antagonistic roles for *KNOX1* and *KNOX2* genes in patterning the land plant body plan following an ancient gene duplication. *PLoS Genetics*, *11*, e1004980.

Guo, Y.-L., Bechsgaard, J. S., Slotte, T., Neuffer, B., Lascoux, M., Weigel, D., et al. (2009). Recent speciation of *Capsella rubella* from *Capsella grandiflora*, associated with loss of self-incompatibility and an extreme bottleneck. *Proceedings of the National Academy of Sciences of the United States of America*, *106*, 5246–5251.

Hagemann, W., & Gleissberg, S. (1996). Organogenetic capacity of leaves: The significance of marginal blastozones in angiosperms. *Plant Systematics and Evolution*, *199*, 121–152.

Hake, S., Smith, H. M., Holtan, H., Magnani, E., Mele, G., & Ramirez, J. (2004). The role of *KNOX* genes in plant development. *Annual Review of Cell and Developmental Biology*, *20*, 125–151.

Hareven, D., Gutfinger, T., Parnis, A., Eshed, Y., & Lifschitz, E. (1996). The making of a compound leaf: Genetic manipulation of leaf architecture in tomato. *Cell*, *84*, 735–744.

Hay, A., Barkoulas, M., & Tsiantis, M. (2006). *ASYMMETRIC LEAVES1* and auxin activities converge to repress *BREVIPEDICELLUS* expression and promote leaf development in Arabidopsis. *Development*, *133*, 3955–3961.

Hay, A., Kaur, H., Phillips, A., Hedden, P., Hake, S., & Tsiantis, M. (2002). The gibberellin pathway mediates KNOTTED1-type homeobox function in plants with different body plans. *Current Biology*, *12*, 1557–1565.

Hay, A. S., Pieper, B., Cooke, E., Mandáková, T., Cartolano, M., Tattersall, A. D., et al. (2014). *Cardamine hirsuta*: A versatile genetic system for comparative studies. *The Plant Journal*, *78*, 1–15.

Hay, A., & Tsiantis, M. (2006). The genetic basis for differences in leaf form between *Arabidopsis thaliana* and its wild relative *Cardamine hirsuta*. *Nature Genetics*, *38*, 942–947.

Hay, A., & Tsiantis, M. (2010). KNOX genes: Versatile regulators of plant development and diversity. *Development*, *137*, 3153–3165.

Hay, A., & Tsiantis, M. (2016). *Cardamine hirsuta*: A comparative view. *Current Opinion in Genetics & Development*, *39*, 1–7.

Heisler, M. G., Ohno, C., Das, P., Sieber, P., Reddy, G. V., Long, J. A., et al. (2005). Patterns of auxin transport and gene expression during primordium development revealed by live imaging of the *Arabidopsis* inflorescence meristem. *Current Biology*, *15*, 1899–1911.

Hofer, J., Turner, L., Moreau, C., Ambrose, M., Isaac, P., Butcher, S., et al. (2009). Tendril-less regulates tendril formation in pea leaves. *Plant Cell*, *21*, 420–428.

Horiguchi, G., Ferjani, A., Fujikura, U., & Tsukaya, H. (2006). Coordination of cell proliferation and cell expansion in the control of leaf size in *Arabidopsis thaliana*. *Journal of Plant Research*, *119*, 37–42.

Horiguchi, G., Kim, G. T., & Tsukaya, H. (2005). The transcription factor AtGRF5 and the transcription coactivator AN3 regulate cell proliferation in leaf primordia of *Arabidopsis thaliana*. *The Plant Journal*, *43*, 68–78.

Horiguchi, G., & Tsukaya, H. (2011). Organ size regulation in plants: Insights from compensation. *Frontiers in Plant Science*, *2*, 24.

Ikeuchi, M., Igarashi, H., Okada, K., & Tsukaya, H. (2014). Acropetal leaflet initiation of *Eschscholzia californica* is achieved by constant spacing of leaflets and differential growth of leaf. *Planta*, *240*, 125–135.

Irish, V. F., & Sussex, I. M. (1992). A fate map of the *Arabidopsis* embryonic shoot apical meristem. *Development*, *115*, 745–753.

Jasinski, S., Piazza, P., Craft, J., Hay, A., Woolley, L., Rieu, I., et al. (2005). KNOX action in *Arabidopsis* is mediated by coordinate regulation of cytokinin and gibberellin activities. *Current Biology*, *15*, 1560–1565.

Jonsson, H., Heisler, M. G., Shapiro, B. E., Meyerowitz, E. M., & Mjolsness, E. (2006). An auxin-driven polarized transport model for phyllotaxis. *Proceedings of the National Academy of Sciences*, *103*, 1633–1638.

Kalve, S., De Vos, D., & Beemster, G. T. S. (2014). Leaf development: A cellular perspective. *Frontiers in Plant Science*, *5*, 362.

Kasprzewska, A., Carter, R., Swarup, R., Bennett, M., Monk, N., Hobbs, J. K., et al. (2015). Auxin influx importers modulate serration along the leaf margin. *The Plant Journal*, *83*, 705–718.

Kawamura, E., Horiguchi, G., & Tsukaya, H. (2010). Mechanisms of leaf tooth formation in *Arabidopsis*. *The Plant Journal*, *62*, 429–441.

Kennaway, R., Coen, E., Green, A., & Bangham, A. (2011). Generation of diverse biological forms through combinatorial interactions between tissue polarity and growth. *PLoS Computional Biology*, 7, e1002071.

Kim, J. H., & Kende, H. (2004). A transcriptional coactivator, AtGIF1, is involved in regulating leaf growth and morphology in *Arabidopsis*. *Proceedings of the National Academy of Sciences of the United States of America*, *101*, 13374–13379.

Koenig, D., Bayer, E., Kang, J., Kuhlemeier, C., & Sinha, N. (2009). Auxin patterns *Solanum lycopersicum* leaf morphogenesis. *Development*, *136*, 2997–3006.

Kuchen, E. E., Fox, S., de Reuille, P. B., Kennaway, R., Bensmihen, S., Avondo, J., et al. (2012). Generation of leaf shape through early patterns of growth and tissue polarity. *Science*, *335*, 1092–1096.

Lenhard, M. (2017). Plant development: Keeping on the straight and narrow and flat. *Current Biology*, *27*, R1277–R1280.

Lin, W.-C., Shuai, B., & Springer, P. S. (2003). The Arabidopsis LATERAL ORGAN BOUNDARIES-domain gene *ASYMMETRIC LEAVES2* functions in the repression of *KNOX* gene expression and in adaxial-abaxial patterning. *The Plant Cell*, *15*, 2241–2252.

Mansfield, C., Newman, J. L., Olsson, T. S. G., Hartley, M., Chan, J., & Coen, E. (2018). Ectopic BASL reveals tissue cell polarity throughout leaf development in Arabidopsis Thaliana. *Current Biology*, *28*, 2638–2646.

Martin-Trillo, M., & Cubas, P. (2010). *TCP* genes: A family snapshot ten years later. *Trends in Plant Science*, *15*, 31–39.

Massonnet, C., Tisné, S., Radziejwoski, A., Vile, D., De Veylder, L., Dauzat, M., et al. (2011). New insights into the control of endoreduplication: Endoreduplication could be driven by organ growth in *Arabidopsis* leaves. *Plant Physiology*, *157*, 2044–2055.

Merelo, P., Paredes, E. B., Heisler, M. G., & Wenkel, S. (2017). The shady side of leaf development: The role of the REVOLUTA/KANADI1 module in leaf patterning and auxin-mediated growth promotion. *Current Opinion in Plant Biology*, *35*, 111–116.

Mitchison, G. (2016). Conformal growth of *Arabidopsis* leaves. *Journal of Theoretical Biology*, *408*, 155–166.

Nakamasu, A., Nakayama, H., Nakayama, N., Suematsu, N. J., & Kimura, S. (2014). A developmental model for branching morphogenesis of lake cress compound leaf. *PLoS One*, *9*, e111615.

Nakamasu, A., Suematsu, N. J., & Kimura, S. (2017). Asymmetries in leaf branch are associated with differential speeds along growth axes: A theoretical prediction. *Developmental Dynamics*, *246*, 981–991.

Nakata, M., Matsumoto, N., Tsugeki, R., Rikirsch, E., Laux, T., & Okada, K. (2012). Roles of the middle domain-specific *WUSCHEL-RELATED HOMEOBOX* genes in early development of leaves in *Arabidopsis*. *The Plant Cell*, *24*, 519–535.

Nakayama, H., Nakayama, N., Seiki, S., Kojima, M., Sakakibara, H., Sinha, N., et al. (2014). Regulation of the KNOX-GA gene module induces heterophyllic alteration in North American lake cress. *The Plant Cell*, *26*, 4733–4748.

Nakayama, H., Sinha, N. R., & Kimura, S. (2017). How do plants and phytohormones accomplish heterophylly, leaf phenotypic plasticity, in response to environmental cues. *Frontiers in Plant Science*, *8*, 1717.

Nardmann, J., Ji, J., Werr, W., & Scanlon, M. J. (2004). The maize duplicate genes *narrow sheath1* and *narrow sheath2* encode a conserved homeobox gene function in a lateral domain of shoot apical meristems. *Development*, *131*, 2827–2839.

Narumi, T., Aida, R., Koyama, T., Yamaguchi, H., Sasaki, K., Shikata, M., et al. (2011). *Arabidopsis* chimeric TCP3 repressor produces novel floral traits in *Torenia fournieri* and *Chrysanthemum morifolium*. *Plant Biotechnology*, *28*, 131–140.

Nath, U., Crawford, B. C., Carpenter, R., & Coen, E. (2003). Genetic control of surface curvature. *Science*, *299*, 1404–1407.

Nelissen, H., Eeckhout, D., Demuynck, K., Persiau, G., Walton, A., van Bel, M., et al. (2015). Dynamic changes in ANGUSTIFOLIA3 complex composition reveal a growth regulatory mechanism in the maize leaf. *Plant Cell*, *27*, 1605–1619.

Nelissen, H., Rymen, B., Jikumaru, Y., Demuynck, K., Van Lijsebettens, M., Kamiya, Y., et al. (2012). A local maximum in gibberellin levels regulates maize leaf growth by spatial control of cell division. *Current Biology*, *22*, 1183–1187.

Niklas, K. J. (1988). Biophysical limitations on plant form and evolution. In L. D. Gottlieb & S. K. Jain (Eds.), *Plant evolutionary biology* (pp. 185–220). Dordrecht, the Netherlands: Springer.

Nikolov, L. A., & Tsiantis, M. (2015). Interspecies gene transfer as a method for understanding the genetic basis for evolutionary change: Progress, pitfalls, and prospects. *Frontiers in Plant Science*, *6*, 1135.

Nikolov, L. A., & Tsiantis, M. (2017). Using mustard genomes to explore the genetic basis of evolutionary change. *Current Opinion in Plant Biology*, *36*, 119–128.

Nikovics, K., Blein, T., Peaucelle, A., Ishida, T., Morin, H., Aida, M., et al. (2006). The balance between the *MIR164A* and *CUC2* genes controls leaf margin serration in *Arabidopsis*. *Plant Cell*, *18*, 2929–2945.

Ori, N., Cohen, A. R., Etzioni, A., Brand, A., Yanai, O., Shleizer, S., et al. (2007). Regulation of *LANCEOLATE* by *miR319* is required for compound-leaf development in tomato. *Nature Genetics*, *39*, 787–791.

Ostria-Gallardo, E., Ranjan, A., Chitwood, D. H., Kumar, R., Townsley, B. T., Ichihashi, Y., et al. (2016). Transcriptomic analysis suggests a key role for *SQUAMOSA PROMOTER BINDING PROTEIN LIKE*, *NAC* and *YUCCA* genes in the heteroblastic development of the temperate rainforest tree *Gevuina avellana* (*Proteaceae*). *New Phytologist*, *210*, 694–708.

Palatnik, J. F., Allen, E., Wu, X., Schommer, C., Schwab, R., Carrington, J. C., et al. (2003). Control of leaf morphogenesis by microRNAs. *Nature*, *425*, 257–263.

Piazza, P., Bailey, C. D., Cartolano, M., Krieger, J., Cao, J., Ossowski, S., et al. (2010). *Arabidopsis thaliana* leaf form evolved via loss of KNOX expression in leaves in association with a selective sweep. *Current Biology*, *20*, 2223–2228.

Poethig, S. (1984). Cellular parameters of leaf morphogenesis in maize and tobacco. In R. A. White & W. C. Dickison (Eds.), *Contemporary problems in plant anatomy* (pp. 235–259). Elsevier.

Poethig, R. S. (2013). Vegetative phase change and shoot maturation in plants. In A. E. Rougvie & M. B. O'Connor (Eds.), *Current topics in developmental biology* (pp. 125–152). Elsevier.

Poethig, R. S., & Sussex, I. M. (1985). The developmental morphology and growth dynamics of the tobacco leaf. *Planta*, *165*, 158–169.

Prusinkiewicz, P., & Runions, A. (2012). Computational models of plant development and form. *New Phytologist*, *193*, 549–569.

Randall, R. S., Sornay, E., Dewitte, W., & Murray, J. A. (2015). *AINTEGUMENTA* and the D-type cyclin CYCD3;1 independently contribute to petal size control in *Arabidopsis*: Evidence for organ size compensation being an emergent rather than a determined property. *Journal of Experimental Botany*, *66*, 3991–4000.

Rast-Somssich, M. I., Broholm, S., Jenkins, H., Canales, C., Vlad, D., Kwantes, M., et al. (2015). Alternate wiring of a KNOXI genetic network underlies differences in leaf development of *A. thaliana* and *C. hirsuta*. *Genes & Development*, *29*, 2391–2404.

Rebeiz, M., & Tsiantis, M. (2017). Enhancer evolution and the origins of morphological novelty. *Current Opinion in Genetics & Development*, *45*, 115–123.

Reinhardt, D., Pesce, E. R., Stieger, P., Mandel, T., Baltensperger, K., Bennett, M., et al. (2003). Regulation of phyllotaxis by polar auxin transport. *Nature*, *426*, 255–260.

Richards, O. W., & Kavanagh, A. J. (1943). The analysis of the relative growth gradients and changing form of growing organisms: Illustrated by the tobacco leaf. *The American Naturalist*, *77*, 385–399.

Rodriguez, R. E., Mecchia, M. A., Debernardi, J. M., Schommer, C., Weigel, D., & Palatnik, J. F. (2010). Control of cell proliferation in *Arabidopsis thaliana* by microRNA miR396. *Development*, *137*, 103–112.

Rodriguez, R. E., Schommer, C., & Palatnik, J. F. (2016). Control of cell proliferation by microRNAs in plants. *Current Opinion in Plant Biology*, *34*, 68–76.

Royer, D. L., & Wilf, P. (2006). Why do toothed leaves correlate with cold climates? Gas exchange at leaf margins provides new insights into a classic Paleotemperature proxy. *International Journal of Plant Sciences*, *167*, 11–18.

Rubio-Somoza, I., Zhou, C., Confraria, A., Martinho, C., von Born, P., Baena-Gonzalez, E., et al. (2014). Temporal control of leaf complexity by miRNA-regulated licensing of protein complexes. *Current Biology*, *24*, 2714–2719.

Runions, A., Tsiantis, M., & Prusinkiewicz, P. (2017). A common developmental program can produce diverse leaf shapes. *New Phytologist*, *216*, 401–418.
Sack, L., Scoffoni, C., McKown, A. D., Frole, K., Rawls, M., Havran, J. C., et al. (2012). Developmentally based scaling of leaf venation architecture explains global ecological patterns. *Nature Communications*, *3*, 837.
Saddic, L. A., Huvermann, B., Bezhani, S., Su, Y., Winter, C. M., Kwon, C. S., et al. (2006). The LEAFY target LMI1 is a meristem identity regulator and acts together with LEAFY to regulate expression of *CAULIFLOWER*. *Development*, *133*, 1673–1682.
Sakamoto, T., Kamiya, N., Ueguchi-Tanaka, M., Iwahori, S., & Matsuoka, M. (2001). KNOX homeodomain protein directly suppresses the expression of a gibberellin biosynthetic gene in the tobacco shoot apical meristem. *Genes & Development*, *15*, 581–590.
Sauret-Güeto, S., Schiessl, K., Bangham, A., Sablowski, R., & Coen, E. (2013). JAGGED controls *Arabidopsis* petal growth and shape by interacting with a divergent polarity field. *PLoS Biology*, *11*, e1001550.
Scarpella, E., Marcos, D., Friml, J., & Berleth, T. (2006). Control of leaf vascular patterning by polar auxin transport. *Genes & Development*, *20*, 1015–1027.
Shleizer-Burko, S., Burko, Y., Ben-Herzel, O., & Ori, N. (2011). Dynamic growth program regulated by LANCEOLATE enables flexible leaf patterning. *Development*, *138*, 695–704.
Sicard, A., Thamm, A., Marona, C., Lee, Y. W., Wahl, V., Stinchcombe, J. R., et al. (2014). Repeated evolutionary changes of leaf morphology caused by mutations to a homeobox gene. *Current Biology*, *24*, 1880–1886.
Skopelitis, D. S., Benkovics, A. H., Husbands, A. Y., & Timmermans, M. C. P. (2017). Boundary formation through a direct threshold-based readout of mobile small RNA gradients. *Developmental Cell*, *43*, 265–273, (e266).
Smith, R. S., Guyomarc'h, S., Mandel, T., Reinhardt, D., Kuhlemeier, C., & Prusinkiewicz, P. (2006). A plausible model of phyllotaxis. *Proceedings of the National Academy of Sciences of the United States of America*, *103*, 1301–1306.
Tameshige, T., Okamoto, S., Lee, J. S., Aida, M., Tasaka, M., Torii, K. U., et al. (2016). A secreted peptide and its receptors shape the auxin response pattern and leaf margin morphogenesis. *Current Biology*, *26*, 2478–2485.
Timmermans, M., Hudson, A., Becraft, P. W., & Nelson, T. (1999). ROUGH SHEATH2: A Myb protein that represses *knox* homeobox genes in maize lateral organ primordia. *Science*, *284*, 151–153.
Tsiantis, M., & Hay, A. (2003). Comparative plant development: The time of the leaf? *Nature Reviews. Genetics*, *4*, 169.
Tsiantis, M., Schneeberger, R., Golz, J. F., Freeling, M., & Langdale, J. A. (1999). The maize *rough sheath2* gene and leaf development programs in monocot and dicot plants. *Science*, *284*, 154–156.
Tsuda, K., & Hake, S. (2015). Diverse functions of KNOX transcription factors in the diploid body plan of plants. *Current Opinion in Plant Biology*, *27*, 91–96.
Tsukaya, H. (2002). Leaf development. *The Arabidopsis Book*, *1*, e0072.
Tsukaya, H. (2008). Controlling size in multicellular organs: Focus on the leaf. *PLoS Biology*, *6*, e174.
Vlad, D., Kierzkowski, D., Rast, M. I., Vuolo, F., Dello Ioio, R., Galinha, C., et al. (2014). Leaf shape evolution through duplication, regulatory diversification, and loss of a homeobox gene. *Science*, *343*, 780–783.
Vuolo, F., Kierzkowski, D., Runions, A., Hajheidari, M., Mentink, R. A., Das Gupta, M., et al. (2018). LMI1 homeodomain protein regulates organ proportions by spatial modulation of endoreduplication. *Genes & Development*, *32*, 1361–1366.
Vuolo, F., Mentink, R. A., Hajheidari, M., Bailey, C. D., Filatov, D. A., & Tsiantis, M. (2016). Coupled enhancer and coding sequence evolution of a homeobox gene shaped leaf diversity. *Genes & Development*, *30*, 2370–2375.

Wang, L., Gu, X., Xu, D., Wang, W., Wang, H., Zeng, M., et al. (2011). miR396-targeted AtGRF transcription factors are required for coordination of cell division and differentiation during leaf development in *Arabidopsis*. *Journal of Experimental Botany*, *62*, 761–773.

Wang, J.-W., Park, M. Y., Wang, L.-J., Koo, Y., Chen, X.-Y., Weigel, D., et al. (2011). miRNA control of vegetative phase change in trees. *PLoS Genetics*, 7, e1002012.

Wells, R., Meredith, W. R., & Williford, J. R. (1986). Canopy photosynthesis and its relationship to plant productivity in near-isogenic cotton lines differing in leaf morphology. *Plant Physiology*, *82*, 635–640.

Wells, C. L., & Pigliucci, M. (2000). Adaptive phenotypic plasticity: The case of heterophylly in aquatic plants. *Perspectives in Plant Ecology, Evolution, and Systematics*, *3*, 1–18.

Whitewoods, C. D., & Coen, E. (2017). Growth and development of three-dimensional plant form. *Current Biology*, *27*, R910–R918.

Willmann, M. R., & Poethig, R. S. (2011). The effect of the floral repressor *FLC* on the timing and progression of vegetative phase change in *Arabidopsis*. *Development*, *138*, 677–685.

Yanai, O., Shani, E., Dolezal, K., Tarkowski, P., Sablowski, R., Sandberg, G., et al. (2005). Arabidopsis KNOXI proteins activate cytokinin biosynthesis. *Current Biology*, *15*, 1566–1571.

Yanai, O., Shani, E., Russ, D., & Ori, N. (2011). Gibberellin partly mediates LANCEOLATE activity in tomato. *The Plant Journal*, *68*, 571–582.

CHAPTER SIX

Vascular tissue development in plants

Hiroo Fukuda*, Kyoko Ohashi-Ito
Department of Biological Sciences, Graduate School of Science, The University of Tokyo, Tokyo, Japan
*Corresponding author: e-mail address: fukuda@bs.s.u-tokyo.ac.jp

Contents

Abstract

The plant vasculature is a sophisticated system that has greatly contributed to the evolution of land plants over the past few hundred million years. The formation of the vascular system is a well-organized plant developmental process, but it is also flexible in response to environmental changes. Provascular cells arise after asymmetric cell division in early embryos and differentiate into various vascular cells, including procambial cells, which function as vascular stem cells. Mutual regulation by auxin and cytokinin is essential for vascular pattern formation, in which the xylem, phloem, and procambium are arranged in a species-specific manner. The hierarchical expression of different transcription factors contributes to the sequential development of vascular tissues. These transcription factors sometimes form feedback loops involving plant hormones. Non-cell autonomous signals, including hormones, peptides, and miRNAs, function in the organization of vascular tissue development.

Current Topics in Developmental Biology, Volume 131
ISSN 0070-2153
https://doi.org/10.1016/bs.ctdb.2018.10.005

1. Introduction

During evolution, plants have acquired a vascular system, which has greatly contributed to the success of land plants. The vascular system allows plants to rapidly transport water and nutrients over long distances between roots, growing deeply in soil, and shoots, growing upward toward the sun. Importantly, it also provides mechanical support for the plant body. The vascular system consists of xylem, phloem, and the intervening procambium, which produces cambial, xylem, and phloem cells. During secondary growth, the cambium continues to give rise to xylem and phloem cells. The evolution of secondary growth has enabled the emergence of large trees and long-lived plants. Procambial and cambial cells are referred to as vascular stem cells based on their functions (Hirakawa et al., 2008).

All vascular tissues in the plant axis (root and hypocotyl) are derived from only four initial cells in the globular embryo (Scheres et al., 1995). These cells divide and differentiate, giving rise to provascular tissue during embryogenesis. In the root, vascular initial cells next to the quiescent center continue to give rise to vascular cells, and procambial cells subsequently divide to increase their population. In lateral organs, such as lateral roots and buds, newly formed lateral organs form provascular tissues, which connect to the existing vasculature in the plant axis. Recent studies revealed various molecular events related to vascular tissue development during embryogenesis and postembryogenesis (Furuta, Helmann, & Helariutta, 2014). In particular, there have been important advances in our understanding of key transcription factors and signaling molecules that regulate vascular cell fate (Heo, Blob, & Helariutta, 2017; Vaughan-Hirsch, Goodall, & Bishopp, 2018). In this review, we focus on recent advances in our understanding of the regulation of vascular fate in view of the functions of key transcription factors and signaling molecules.

2. Initiation of vascular system development

2.1 Organization of vascular cells in the embryo

The lower tier of the early globular embryo (see Armenta-Medina & Gillmor, 2019) consists of three cell layers: precursors of epidermal, ground, and vascular cells (Fig. 1). The provascular cell layer is located in the center of the embryo and includes four cells, which subsequently undergo asymmetric cell divisions to create an additional cell layer consisting of four

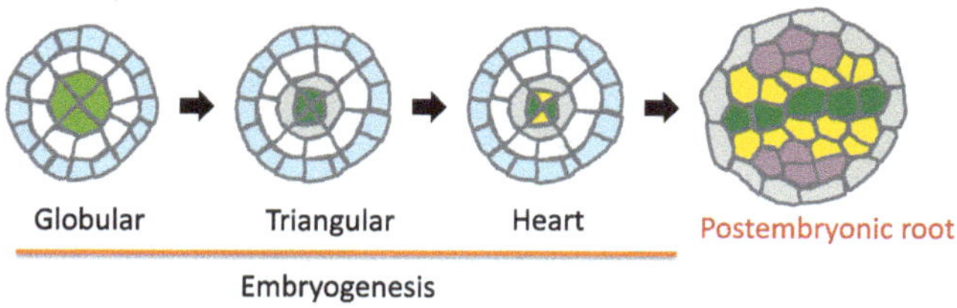

Fig. 1 Vascular cell organization in the embryo and root. Provascular cells (light green) are located in the center of the globular embryo. Asymmetric cell divisions give rise to vascular initial cells (green) in the center of the triangular embryo. Two vascular initial cells become xylem precursor cells, and based on this pattern, subsequent cell divisions lead to the formation of a pattern in the root vasculature. The illustration of a post-embryonic root only shows the vascular region of the root.

cells (Fig. 1) (Scheres et al., 1995; Yoshida et al., 2014). All vascular cells in the root and hypocotyl are derived from the four newly emerging cells in the triangular embryo (vascular initial cells). Therefore, there appear to be two stages in the determination of vascular cell identity: provascular cell identity before the early globular stage and vascular initial cell identity after the late globular stage.

The ground and provascular cell layers originate from the same cell layer. Recent cell type-specific transcriptome analysis showed that vascular cell marker genes, e.g. *SHORTROOT* (*SHR*), begin to be expressed earlier than ground cell marker genes prior to early globular embryo formation. This finding suggests that provascular cell identity is determined in early embryogenesis and earlier than ground cell identity (Palovaara, Saiga, & Wendrich, 2017). The regulation of the planes of cell division (a process critical for the establishment of vascular initial cells) and the regulation of vascular cell identity appear to be closely related. A strong auxin response is observed in provascular cells in the globular embryo (Moller et al., 2017). Initially, the four vascular initial cells are also highly responsive to auxin, and the auxin responses of two of the four cells gradually become stronger (De Rybel et al., 2014). The selection of these two cells is thought to be based on their positions relative to the cotyledon primordia, which function as sources of auxin (De Rybel et al., 2014).

Detailed observations of auxin-responsive mutants indicated that auxin plays a crucial role in asymmetric cell division and vascular cell identity in the embryo (Yoshida et al., 2014). Indeed, the loss-of-function of the auxin response transcription factor MONOPTEROS/AUXIN RESPONSE FACTOR5 (MP/ARF5) causes severe defects in asymmetric cell divisions in the lower tier of the globular embryo, leading to the loss of vascular cells (Hardtke & Berleth, 1998). The molecular mechanism underlying the MP/ARF5-dependent establishment of vascular initial cells is now becoming clear. *TARGET OF MONOPTEROS5* (*TMO5*), a direct target gene of MP, encodes a basic helix-loop-helix (bHLH) transcription factor that is specifically expressed in provascular cells (De Rybel et al., 2013; Schlereth et al., 2010). *TMO5* is first expressed in all four vascular initial cells, but soon its expression is restricted to two cells facing each other, i.e. the xylem initial cells that form the xylem axis (De Rybel et al., 2014). *TMO5* continues to be expressed only in the xylem axis (De Rybel et al., 2013). Thus, xylem axis identity is correlated with the auxin response. TMO5 belongs to a small subfamily including TMO5, TMO5-LIKE1 (T5L1), T5L2, and T5L3, which function as heterodimers with the bHLH transcription factor LONESOME HIGHWAY (LHW), and its homologs (De Rybel et al., 2013; Ohashi-Ito & Bergmann, 2007). Embryos of the loss-of-function *lhw* and *tmo5 t5l1* mutants have almost the same phenotype, with cell divisions giving rise to vascular initial cells being abnormal (De Rybel et al., 2013; Ohashi-Ito, Matsukawa, & Fukuda, 2013). Moreover, a quadruple mutant of *tmo5* and its close *t5l1-3* homologs produces a root consisting of three cell layers lacking the layer of vascular initial cells (De Rybel et al., 2013), suggesting that the *MP-TMO5/LHW* pathway regulates the production of vascular initial cells. However, how TMO5/LHW and related heterodimers control asymmetric cell division remains a mystery.

2.2 Pattern formation in vascular tissues

The postembryonic root vasculature has a conserved vascular cell arrangement. In *Arabidopsis thaliana*, a xylem axis is present in the center of the vascular region, two phloem poles are located symmetrically with respect to the xylem axis, and procambial cells lie between the xylem and phloem (Fig. 1). The formation of this pattern begins soon after vascular initial cells are established in the embryo and is also regulated by TMO5/LHW complexes. *TMO5* continues to be expressed only in the xylem axis, not only during embryogenesis, but also in the postembryonic root (De Rybel et al., 2013).

TMO5/LHW also plays an important role in the production of procambial cells. As reduced numbers of vascular cells are commonly observed in cytokinin biosynthesis and response mutants as well as *lhw* and *tmo5 t5l1* mutants (Mahonen, Bishopp, et al., 2006; Mahonen et al., 2001; Tokunaga et al., 2012), the regulation of cytokinin signaling through TMO5/LHW is thought to dominate this process. TMO5/LHW upregulates cytokinin biosynthesis genes, such as *LONELY GUY3* (*LOG3*) and *LOG4*, in the xylem axis (De Rybel et al., 2014; Ohashi-Ito et al., 2014). Cytokinin, in turn, triggers cell divisions in cells adjacent to xylem axis cells, and these dividing cells become procambial cells. Thus, procambial cell initiation is organized by xylem axis cells in a non-cell autonomous manner. This relationship between the xylem axis and procambial cells is reminiscent of the "organizer and stem cell" mechanisms that function in the shoot and root apical meristems; it is likely that the xylem axis functions as an organizer of vascular stem cells.

Mutual regulation of auxin and cytokinin signaling is essential for vascular pattern formation. The strong auxin response in the vasculature is restricted to the xylem axis. This restriction is built upon the mutual regulation between auxin and cytokinin signaling (Bishopp et al., 2011; De Rybel et al., 2014). The auxin- and cytokinin-responsive regions are clearly separated and are located in the xylem axis and procambial cells, respectively. Although cytokinin is initially produced in the xylem axis, as mentioned above, no cytokinin response occurs in the xylem axis; in this region, auxin induces the expression of *ARABIDOPSIS HISTIDINE PHOSPHOTRANSFER PROTEIN6* (*AHP6*), encoding a negative regulator of cytokinin responses, via multiple pathways (Mahonen, Higuchi, et al., 2006; Ohashi-Ito et al., 2014). *AHP6* suppresses the cytokinin response in protoxylem precursor cells but is not present in metaxylem precursor cells. Thus, another factor that suppresses the cytokinin response in metaxylem must be identified. All vascular cells in roots of cytokinin receptor mutants, such as *woodenleg* (*wol*) and *ahk2 ahk3 ahk4* affecting multiple copies of the ARABIDOPSIS HISTIDINE KINASE (AHK) family, differentiate into protoxylem vessels (Mahonen, Bishopp, et al., 2006; Mahonen et al., 2001). This finding suggests that the cytokinin response is required to restrict the auxin response to the xylem axis. Cytokinin regulates auxin transport and suppresses auxin responses in procambial cells in postembryonic roots (Bishopp et al., 2011). This mechanism also appears to function in embryos. In addition, modeling studies support the notion that the mutual regulation between auxin and cytokinin signaling is essential for pattern formation in the vascular tissue (De Rybel et al., 2014; el-Showk et al., 2015; Mellor et al., 2017; Muraro et al., 2014), although the source of this cytokinin is a matter of debate. New cytokinin transporters, the PURINE PERMEASEs, were recently identified (Zürcher, Liu, di Donato, Geisler, & Müller, 2016). Further analyses, including studies exploring cytokinin transporters, should shed light on this issue.

A recent study suggested that the boundary between the procambium and the xylem axis is determined by mobile transcription factors known as AT-HOOK MOTIF NUCLEAR LOCALIZED PROTEINs (AHLs) (Zhou, Wang, Lee, & Lee, 2013). In *ahl3* and *ahl4* single mutants, both ectopic protoxylem vessels and ectopic metaxylem vessels form in the procambial region adjacent to the xylem axis, suggesting that the xylem axis domain is enlarged (Zhou et al., 2013).

3. Maintenance of vascular stem cells

In multicellular organisms, stem cells are maintained by cell divisions that renew the stem cell and give rise to a daughter cell that differentiates into specific cells types. Vascular tissues continuously form due to the presence of procambial and cambial cells, which function as vascular stem cells. The maintenance of vascular stem cells is regulated by *TRACHEARY ELEMENT DIFFERENTIATION INHIBITORY FACTOR (TDIF)-TDIF RECEPTOR/PHLOEM INTERCALATED WITH XYLEM* (*TDR/PXY*) signaling (Fig. 2). TDIF, a 12 amino-acid peptide with two hydroxylated proline residues, is encoded by *CLE41* and *CLE44* in *A. thaliana* (Ito et al., 2006). The receptor of TDIF is TDR, a Class XI LEUCINE-RICH

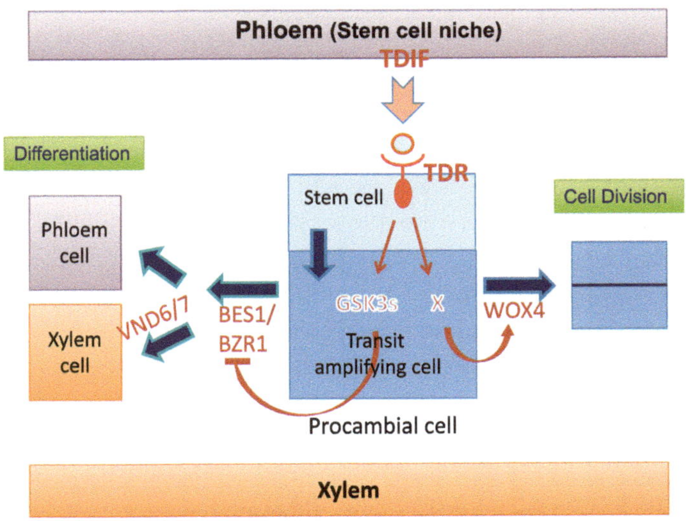

Fig. 2 Regulation of procambial cell fate by the TDIF-TDR signaling pathway. TDIF is produced in phloem cells, secreted from phloem cells, and perceived by TDR in procambial cells. TDIF signaling induces the differentiation of stem cells into transit amplifying cells during the procambial stage. In transit amplifying cells, the TDIF-WOX4 and TDIF-GSK3/SKs-BES1/BZR1 signaling pathways regulate the promotion of cell division and the inhibition of the differentiation into xylem and phloem cells, respectively.

REPEAT RECEPTOR-LIKE KINASE (LRR-RK), that consists of an extracellular LRR domain and a cytoplasmic kinase domain (Hirakawa et al., 2008). Recent X-ray analyses uncovered structural features of the TDIF-TDR complex. The extracellular domain of TDR forms a twisted right-handed superhelix composed of 22 LRRs and the N-terminus, which recognizes TDIF via its inner concave surface (Morita et al., 2016; Zhang, Lin, Han, Qu, & Chai, 2016). SOMATIC EMBRYOGENESIS RECEPTOR KINASE (SERK) family proteins function as co-receptors of TDR (Zhang, Lin, Han, Wang, et al., 2016). In the TDIF-TDR-SERK2 complex, TDIF mediates interactions between TDR and SERK2, with its C-terminal side sandwiched between the LRRs of TDR and SERK2. Similarly, CLE42, the closest homolog of TDIF, induces an interaction between PXY-LIKE2 (PXL2) and SERK2 (Mou et al., 2017). The addition of SERK2 does not alter the basic structure of the TDIF-TDR complex (Zhang, Lin, Han, Wang, et al., 2016), suggesting that TDIF functions as a "glue" that joins TDR and a SERK, which might phosphorylate a cytoplasmic downstream factor.

TDIF is produced in (and secreted from) phloem cells (Fig. 2). The secreted TDIF is perceived by TDR on the plasma membrane of procambial cells (Hirakawa et al., 2008). *TDIF-TDR* signaling functions in the maintenance of procambial cell populations through the regulation of two processes: (i) the promotion of procambial cell division and (ii) the suppression of the differentiation of procambial cells into tracheary elements (Hirakawa, Kondo, & Fukuda, 2010). *WUSCHEL-LIKE HOMEOBOX4* (*WOX4*) is a key target of the *TDIF-T*DR signaling pathway in terms of cell proliferation (Fig. 2) (Hirakawa et al., 2010). *WOX4* is preferentially expressed in the procambium and cambium, and its expression level is upregulated upon application of TDIF in a *TDR*-dependent manner. Genetic analyses showed that *WOX4* is required to promote the proliferation of procambial/cambial cells but not for repressing their commitment to xylem differentiation in response to the *TDIF* signal. This signal also controls the orientation of cell divisions in procambial cells (Etchells & Turner, 2010; Fisher & Turner, 2007). Indeed, cambium-specific expression of the aspen receptor kinase PttPXY and its peptide ligand PttCLE41, which are functional orthologs of TDR and TDIF, generated trees with a twofold increase in the rate of wood formation, increased height and producing larger leaves than the control (Etchells, Mishra, Kumar, Campbell, & Turner, 2015). These results indicate that the *TDIF-TDR* pathway regulates secondary growth in woody plants.

Kondo and coworkers (2014) found that BRASSINOSTEROID-INSENSITIVE2 (BIN2), a GLYCOGEN SYNTHASE KINASE3/SHAGGY-LIKE KINASE protein (GSK3/SK), binds to the cytoplasmic kinase domain of TDR (Fig. 2) (Kondo et al., 2014). Extracellular TDIF induces the release of BIN2 from TDR in the cytoplasm, which activates BIN2. The *A. thaliana* genome encodes 10 GSK3/SK family members, which are classified into four subgroups (SK I-IV) (Saidi, Hearn, & Coates, 2012). Of these, SK I and SK II subgroup members interact with TDR; these interactions are regulated in a TDIF-dependent fashion (Kondo et al., 2014). These GSK3/SKs redundantly mediate TDIF-TDR signaling, leading to the inhibition of xylem differentiation. Indeed, the inhibition of GSK3/SK kinase activity by the chemical compound bikinin promotes tracheary element differentiation (Kondo et al., 2014). Based on this finding, a novel system for in vitro tracheary element differentiation, VISUAL (Vascular cell Induction culture System Using Arabidopsis Leaves) was established (Kondo et al., 2016; Kondo, Fujita, Sugiyama, & Fukuda, 2015). In VISUAL, mesophyll cells in cultured cotyledons and leaf segments differentiate at high frequency and semi-synchronously into tracheary elements in the presence of bikinin, auxin, and cytokinin. *BRI1-EMS-SUPRESSOR1* (*BES1*) functions downstream of the GSK3/SKs, as *bes1* loss-of-function mutants exhibit reduced ectopic xylem differentiation in response to bikinin (Kondo et al., 2014). VISUAL using *brassinazole resistant 1-1 (bzr1-1)* and *bes1-1 bzr1-2* mutants revealed that not only *BES1* but also *BZR1*, the closest homolog of *BES1*, acts redundantly in promoting tracheary element differentiation downstream of the GSK3/SKs (Fig. 2) (Saito, Kondo, & Fukuda, 2018). Further analyses indicated that *BES1* and *BZR1* promote phloem differentiation, as well as xylem differentiation, from vascular stem cells (Saito et al., 2018). This finding is consistent with that of a previous report, showing that both *bes1-D* and *bzr1-D* partially rescued the phenotype of the *octopus* mutant, in which root protophloem differentiation is prone to be suppressed (Anne et al., 2015).

In multicellular organisms, stem cells often give rise to transit amplifying cells, which rapidly divide and have restricted differentiation potential. The transit amplifying cells then differentiate into specific cell types (Singh & Bhalla, 2006). Therefore, TDIF signaling induces the differentiation of stem cells into transit amplifying cells during the procambial stage. In transit amplifying cells, the TDIF-WOX4 and TDIF-GSK3/SKs-BES1/BZR1 signaling pathways regulate the promotion of cell division and the inhibition of differentiation into xylem and phloem cells, respectively (Fig. 2). This idea is supported by the finding that cell cycle-related genes, as well as procambium-, early xylem- and early phloem-specific genes, are highly expressed in *bes1* (Saito et al., 2018).

4. Xylem cell differentiation

4.1 Establishment of xylem cell fate

Higher-order mutations of TMO5 or LHW family genes lead to impaired xylem vessel differentiation (De Rybel et al., 2013; Ohashi-Ito, Oguchi, Kojima, Sakakibara, & Fukuda, 2013). Conversely, the overexpression of *LHW* and *T5L1* leads to ectopic tracheary element differentiation (Katayama et al., 2015). These findings suggest that *TMO5/LHW* confers xylem cell fate in a cell autonomous manner.

Moreover, several NAC-domain [for NO APICAL MERISTEM (NAM), ARABIDOPSIS TRANSCRIPTION ACTIVATION FACTOR (ATAF), CUP-SHAPED COTYLEDON (CUC)] transcription factors play crucial roles in vascular development. VASCULAR-RELATED NAC-DOMAIN1-7 (VND1-7) proteins regulate xylem differentiation in *A. thaliana* (Fig. 3). Overexpression of *VND6* and *VND7* induces ectopic tracheary element formation in various tissues of *A. thaliana*, as well as of other plants such as poplar (Kubo et al., 2005). A recent study indicates that other genes of the VND family, including *VND1*, *VND2*, and *VND3*, also play a role in inducing ectopic tracheary element formation (Endo et al., 2015; Zhou, Zhong, & Ye, 2014). Interestingly, *VND1*, *VND2*, and

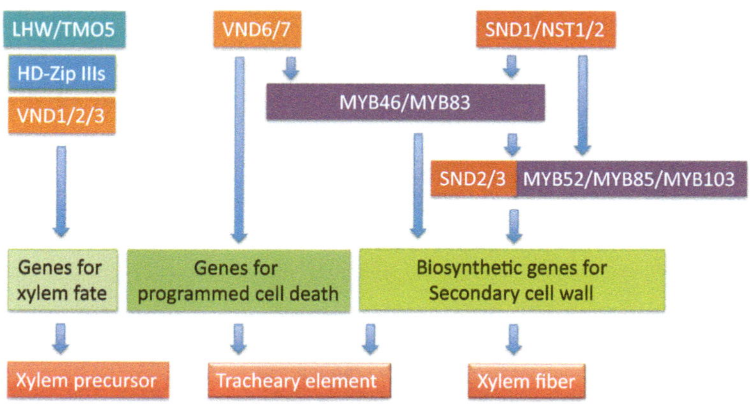

Fig. 3 Orchestrated gene expression by master transcription factors. The VNDs and SNDs/NSTs transcription factors lead to differentiation into different xylem cell types through hierarchical expression of gene expression.

VND3 are expressed in xylem precursor cells in the root apical meristem of *A. thaliana*, while *VND4-VND7* are expressed in xylem precursor cells in the root elongation and differentiation zones (Kubo et al., 2005). In xylogenic cultures, *VND1-VND3* are expressed during very early stages of vascular differentiation together with procambial marker genes (Kondo et al., 2016; Saito et al., 2018; Tan et al., 2018). In *vnd1 vnd2 vnd3* mutant seedlings, the differentiation of tracheary elements in the secondary veins of cotyledons is inhibited in the dark but not in the light (Tan et al., 2018). Together with the observation that VND1-VND3 bind to the promoter of *VND7* (Endo et al., 2015), these results suggest that *VND1-VND3* are expressed in cells that are fated to form xylem cells and may function in the establishment of xylem precursor cells.

4.2 Determination of xylem differentiation

The xylem consists of tracheary elements, xylem parenchyma cells, and xylem fiber cells. Tracheary elements are dead, hollow cells with patterned cell walls comprising xylem vessels and tracheids, which function as conductive hollow tubes for water and nutrient transport throughout the plant body. Xylem fiber cells, with evenly thickened secondary cell walls, provide mechanical support to the plant body. VND6 and VND7 are master regulators of tracheary element differentiation (Fig. 3). The overexpression of *VND6* and *VND7* forces parenchyma cells to differentiate into tracheary elements. Indeed, 80–90% of undifferentiated culture cells differentiate into tracheary elements in response to *VND6* overexpression (Oda, Iida, Kondo, & Fukuda, 2010), and most mesophyll cells differentiate into tracheary elements in response to *VND7* overexpression (Yamaguchi, Mitsuda, Ohtani, Ohme-Takagi, & Demura, 2011). Other VND genes can also induce tracheary element formation, but at a much lower efficiency (Endo et al., 2015; Zhou et al., 2014). Therefore, *VND6* and *VND7* are determinants of tracheary element differentiation, whereas *VND1-VND5* might induce tracheary element differentiation indirectly, for example, by inducing the expression of *VND6* and *VND7* (Endo et al., 2015). VND6 and VND7 bind to a tracheary element-specific *cis* element (TERE) in genes that function in secondary cell wall formation and programmed cell death (PCD) to form rigid, hollow tubes (Ohashi-Ito, Oda, & Fukuda, 2010; Pyo, Demura, & Fukuda, 2007; Yamaguchi et al., 2011). At the same time, VND6 and VND7 directly upregulate genes encoding transcription factors such as MYB46 and MYB83, which function as master regulators of secondary cell wall formation involving all three major secondary wall components: cellulose, lignin, and xylan (Fig. 3) (McCarthy, Zhong, & Ye, 2009; Zhong, Richardson, & Ye, 2007). This orchestrated gene expression contributes to the rapid biosynthesis of the wide variety of components required for secondary cell wall formation or PCD (see Cubría-Radío & Nowack, 2019).

Since VND6 and VND7 function as final determinants of cell differentiation, their production and stability must be precisely regulated. VND-INTERACTING2 (VNI2) functions as a transcriptional repressor of VND7 to spatiotemporally regulate VND7 activity (Yamaguchi et al., 2010). VND7 and LATERAL ORGAN BOUNDARIES-DOMAIN15 (LBD15) may form a positive feedback loop by mutually promoting each other's expression (Ohashi-Ito, Iwamoto, & Fukuda, 2018). This positive feedback loop leads to an exponential increase in *VND7* levels, which contributes to the rapid progression of secondary wall thickening and PCD. LBD18/ASYMMETRIC LEAVES2-LIKE20 (ASL20) and LBD30/ASL19 might also be involved in the positive feedback regulation of *VND7* expression (Soyano, Thitamadee, Machida, & Chua, 2008).

NAC SECONDARY WALL THICKENING PROMOTING FACTOR1 (NST1), NST2, and SECONDARY WALL-ASSOCIATED NAC DOMAIN PROTEIN1 (SND1), which belong to the subclade that is most similar to the VND family, are master regulators of xylem fiber cell differentiation (Fig. 3) (Ko, Yang, Park, Lerouxel, & Han, 2007; Mitsuda et al., 2007; Mitsuda, Seki, Shinozaki, & Ohme-Takagi, 2005; Zhong, Demura, & Ye, 2006). Indeed, the overexpression of *SND1*, *NST1*, and *NST2* can induce the ectopic deposition of secondary cell walls, whereas *snd1 nst1* mutant plants exhibit reduced xylem fiber formation (Mitsuda et al., 2007). SND1 (as well as VND6, as mentioned above) directly upregulates *MYB46* and *MYB83*, encoding master regulators of secondary cell wall formation (Ohashi-Ito et al., 2010; Zhong, Lee, & Ye, 2010; Zhong et al., 2007). On the other hand, whereas VND6 initiates tracheary element-specific PCD by directly inducing the expression of PCD-related genes, such as protease and nuclease genes, SND1 preferentially upregulates genes encoding lignin monomer biosynthetic enzymes (Ohashi-Ito et al., 2010). Thus, members of the SND subclade function as master regulators of xylem fiber differentiation by orchestrating gene expression. Orthologs of VND and NST/SND genes have been identified in various vascular plants (Zhong et al., 2010), suggesting that they function as general regulators of vascular tissue development in vascular plants. However, the moss

Physcomitrella patens, a non-vascular plant, also contains VND-related NACs with roles in cell wall thickening. Overexpressing the VND-related NAC gene *PpVNS* from *P. patens* promotes the expression of genes supporting water-conducting cell formation, leading to ectopic cell wall thickening and PCD in both *P. patens* and *A. thaliana* (Xu et al., 2014). This finding points to the conservation of the transcriptional regulation and cellular functions of water-conducting cells between mosses and angiosperms.

4.3 Regulation of protoxylem and metaxylem differentiation

Xylem vessels include protoxylem and metaxylem vessels. Protoxylem vessels, which have annular or spiral secondary cell walls and sometimes collapse during growth, form in elongating regions of organs. The secondary cell walls of metaxylem vessels have a reticulate or pitted pattern, which provides rigidity to the cells, and therefore these vessels form later than protoxylem in regions where elongation has stopped. *A. thaliana* roots contain three files of metaxylem cells located between two files of protoxylem. These observations have prompted the following question: How is protoxylem and metaxylem cell formation regulated spatiotemporally?

Class III HD-ZIP (C3HDZ) transcription factors, including CORONA (CNA), PHABULOSA (PHB), REVOLUTA (REV), PHAVOLUTA (PHV), and ATHB8, function in various developmental processes (Ramachandran, Carlsbecker, & Etchell, 2017; Roodbarkelari & Groot, 2017). Studies of various combinations of loss-of-function as well as gain-of-function mutants indicate that high and low levels of C3HDZ transcription factors result in the formation of metaxylem and protoxylem, respectively (Carlsbecker et al., 2010; Miyashima, Koi, Hashimoto, & Nakajima, 2011). In quadruple loss-of-function mutants of the C3HDZ family, protoxylem cells develop throughout the xylem axis, while in gain-of-function *phb-d* mutants, whose transcripts are resistant to RNA interference by *miR165/166*, metaxylem cells develop in peripheral xylem positions. Therefore, C3HDZ transcription factors act redundantly in the switching of metaxylem and protoxylem cells. Indeed, during root vascular tissue development, SHORT ROOT (SHR) proteins move from the stele to the endodermis, where they induce the expression of the GRAS family transcription factor gene *SCARECROW* (*SCR*) and, together with SCR, upregulate the expression of target genes, including *miR165/166* genes (Cui et al., 2007; Gallagher & Benfey, 2009; Levesque et al., 2006; Nakajima, Sena, Nawy, & Benfey, 2001). The mature *miR165/166* microRNAs move back from the endodermis to the pericycle and protoxylem vessel poles in the stele, where they degrade the transcripts of C3HDZ genes including *PHB* (Carlsbecker et al., 2010), allowing protoxylem differentiation to take place.

The role of C3HDZ proteins in protoxylem/metaxylem switching might involve the regulation of auxin signaling. The *indole acetic acid20-1* (*iaa20-2*) *iaa30-1* double mutant exhibits a high frequency of double protoxylem strands, instead of a single protoxylem strand at the peripheral xylem axis. Conversely, elevated *IAA30* expression reduces protoxylem formation while enhancing metaxylem formation (Müller et al., 2015). PHB directly binds to the promoters of both *MONOPTEROS* (*MP*) and *IAA20*, and the altered activity of *MP* and *IAA20* results in the formation of distorted xylem (Müller et al., 2015). Therefore, C3HDZ transcription factors appear to mediate a feed-forward loop formed by *MP* and *IAA20* that focuses and stabilizes the auxin response, leading to the patterned differentiation of protoxylem and metaxylem cells. Indeed, *trp2* mutants defective in the TRYPTOPHAN SYNTHASE BETA-SUBUNIT 1, which is involved in tryptophan-mediated auxin biosynthesis, have a defect in metaxylem vessel formation but not in protoxylem vessel formation, suggesting that auxin biosynthesis is involved in metaxylem vessel formation (Ursache et al., 2014).

Cytokinin is another factor that functions in the patterned differentiation of protoxylem and metaxylem tissue. Triple loss-of-function *arabidopsis response regulator1-3* (*arr1-3*) *arr10-5 arr12-1* mutants, disrupting type-B response regulators, exhibit an aborted primary root with protoxylem but no metaxylem (Argyros et al., 2008). Similarly, in mutants of cytokinin receptors, such as *wol* and *ahk2 ahk3 ahk4*, all stele cells differentiate into protoxylem cells (Mahonen, Bishopp, et al., 2006; Mahonen et al., 2001). Thus, cytokinin is required for metaxylem formation but inhibits protoxylem formation. Various signaling molecules regulate xylem patterning via cytokinin signaling. For instance, a loss-of-function mutant of *FUMONISIN B1-RESISTANT12* (*FBR12*) produces extra protoxylem vessels due to reduced cytokinin sensitivity (Ren et al., 2013). CLE10 inhibits the formation of protoxylem vessels but not metaxylem vessels by suppressing the expression of type-A ARR genes, including *ARR5* and *ARR6* (Kondo, Hirakawa, & Fukuda, 2011). In *Oryza sativa* (rice), however, the CLE peptide FLORAL ORGAN NUMBER 2-LIKE CLE PEPTIDE2 (FCP2) negatively regulates metaxylem cell differentiation (Chu et al., 2013; Kamiya, Nagasaki, Morikami, Sato, & Matsuoka, 2003). Therefore, the role of *CLE* signaling in regulating root xylem cell development may not be conserved between *A. thaliana* and rice.

A loss-of-function mutant of *ACAULIS5* (*ACL5*) in *A. thaliana* exhibits excess protoxylem differentiation in various organs (Muñiz et al., 2008). *ACL5* encodes thermospermine synthase, which promotes the production of SUPPRESSOR OF ALC5-51-LIKE (SACL) proteins of the bHLH family (Imai et al., 2006; Katayama et al., 2015; Vera-Sirera et al., 2015). SACL proteins interact with LHW to interfere with the functioning of LHW/TMO5 dimers during xylem differentiation. *ACL5* is preferentially expressed in metaxylem precursors, suggesting that the spatial regulation of *ACL5* might be crucial for metaxylem differentiation.

5. Conclusions and perspectives

We described our current understanding of vascular formation. In brief, provascular cells develop in early embryos after asymmetric cell divisions. Transcription factors involved in auxin and auxin responses play crucial roles in asymmetric cell division and vascular cell identity in the embryo. The interacting phytohormones auxin and cytokinin regulate the formation of vascular patterns in which xylem, phloem, and procambium precursor cells are arranged in a species-specific manner. Two bHLH transcription factors, TMO5 and LHW, mediate this mutual hormonal regulation. During postembryonic stages, procambial cells behave as vascular stem cells. The maintenance of procambial cells is regulated via the non-cell autonomous signaling of the CLE peptide TDIF and its receptor, TDR/PXY. NAC domain transcription factors are major players in xylem differentiation: VND1-VND3 initiate xylem fate, and various other VNDs and SNDs/NSTs execute the differentiation of tracheary elements and xylem fibers, respectively. C3HDZ transcription factors, as well as cytokinin levels, control differentiation into either protoxylem or metaxylem. These findings indicate that the sequential, organized development of the vascular system requires the hierarchical expression of transcription factors and spatiotemporally regulated non-cell autonomous signals, including plant hormones, peptides, microRNAs, and so on.

In this chapter, we focused on xylem cell differentiation, as it represents a final step in vascular cell differentiation. However, procambial cells also differentiate into sieve elements and companion cells. To understand the pluripotency of procambial cells, xylem differentiation should be considered together with phloem differentiation. Fortunately, recent advances in our understanding of phloem differentiation have uncovered a number of components essential for this process (Anne & Hardtke, 2017; Blob, Heo, & Helariutta, 2018; Rodriguez-Villalon, 2016), such that we are now ready to compare xylem with phloem differentiation. However, due to space limitations, we did not discuss these new findings. Nevertheless, we would like to point out the following. As BES1 is involved in the differentiation of procambial cells not only into xylem cells but also into phloem cells, we expect that xylem and phloem differentiations share a common initiation mechanism. On the other hand, we still know little about the very early stage of vascular development. Efroni and coworkers succeeded in exploring the fates of single regenerating cells produced after the removal of the root meristem using single cell RNA sequencing (Efroni et al., 2016). This approach is promising for elucidating the early developmental program in vascular cells, particularly in roots or via VISUAL.

Acknowledgments

This work was supported in part by Grants-in-Aid from the Ministry of Education, Culture, Sports, Science and Technology of Japan [15H05958 to H.F., 25113004 to K.O.-I.], from the Japan Society for the Promotion of Science [16H06377 to H.F., 15K07098 to K.O.-I.], and from the Naito Foundation to H.F.

References

Anne, P., Azzopardi, M., Gissot, L., Beaubiat, S., Hématy, K., & Palauqui, J. C. (2015). OCTOPUS negatively regulates BIN2 to control phloem differentiation in *Arabidopsis thaliana*. *Current Biology*, *25*, 2584–2590.

Anne, P., & Hardtke, C. S. (2017). Phloem function and development—Biophysics meets genetics. *Current Opinion in Plant Biology*, *43*, 22–28.

Argyros, R. D., Mathews, D. E., Chiang, Y.-H., Palmer, C. M., Thibault, D. M., Etheridge, N., et al. (2008). Type B response regulators of Arabidopsis play key roles in cytokinin signaling and plant development. *Plant Cell*, *20*, 2102–2116.

Armenta-Medina, A., & Gillmor, C. S. (2019). Genetic, molecular and parent-of-origin regulation of early embryogenesis in flowering plants. In U. Grossniklaus (Ed.), *Current topics in developmental biology*: *Vol. 131*. Plant development and evolution (pp. 497–543). Elsevier.

Bishopp, A., Help, H., El-Showk, S., Weijers, D., Scheres, B., Friml, J., et al. (2011). A mutually inhibitory interaction between auxin and cytokinin specifies vascular pattern in roots. *Current Biology*, *21*, 917–926.

Blob, B., Heo, J. O., & Helariutta, Y. (2018). Phloem differentiation: An integrative model for cell specification. *Journal of Plant Research*, *131*, 31–36.

Carlsbecker, A., Lee, J. Y., Roberts, C. J., Dettmer, J., Lehesranta, S., Zhou, J., et al. (2010). Cell signalling by microRNA165/6 directs gene dose-dependent root cell fate. *Nature*, *465*, 316–321.

Chu, H., Liang, W., Li, J., Hong, F., Wu, Y., Wang, L., et al. (2013). A CLE-WOX signalling module regulates root meristem maintenance and vascular tissue development in rice. *Journal of Experimental Botany*, *64*, 5359–5369.

Cubría-Radío, M., & Nowack, M. K. (2019). Transcriptional networks orchestrating programmed cell death during plant development. In U. Grossniklaus (Ed.), *Current topics in developmental biology*: *Vol. 131*. Plant development and evolution (pp. 161–184). Elsevier.

Cui, H., Levesque, M. P., Vernoux, T., Jung, J. W., Paquette, A. J., Gallagher, K. L., et al. (2007). An evolutionarily conserved mechanism delimiting SHR movement defines a single layer of endodermis in plants. *Science*, *316*, 421–425.

De Rybel, B., Adibi, M., Breda, A. S., Wendrich, J. R., Smit, M. E., Novak, O., et al. (2014). Plant development. Integration of growth and patterning during vascular tissue formation in Arabidopsis. *Science*, *345*, 1255215.

De Rybel, B., Moller, B., Yoshida, S., Grabowicz, I., de Reuille, P. B., Boeren, S., et al. (2013). A bHLH complex controls embryonic vascular tissue establishment and indeterminate growth in Arabidopsis. *Developmental Cell*, *24*, 426–437.

Efroni, I., Mello, A., Nawy, T., Ip, P. L., Rahni, R., DelRose, N., et al. (2016). Root regeneration triggers an embryo-like sequence guided by hormonal interactions. *Cell*, *165*, 1721–1733.

el-Showk, S., Help-Rinta-Rahko, H., Blomster, T., Siligato, R., Maree, A. F., Mahonen, A. P., et al. (2015). Parsimonious model of vascular patterning links transverse hormone fluxes to lateral root initiation: Auxin leads the way, while cytokinin levels out. *PLoS Computational Biology*, *11*, e1004450.

Endo, H., Yamaguchi, M., Tamura, T., Nakano, Y., Nishikubo, N., Yoneda, A., et al. (2015). Multiple classes of transcription factors regulate the expression of VASCULAR-RELATED NAC-DOMAIN7, a master switch of xylem vessel differentiation. *Plant & Cell Physiology*, *56*, 242–254.

Etchells, J. P., Mishra, L. S., Kumar, M., Campbell, L., & Turner, S. R. (2015). Wood formation in trees is increased by manipulating PXY-regulated cell division. *Current Biology*, *25*, 1050–1055.

Etchells, J. P., & Turner, S. R. (2010). The PXY-CLE41 receptor ligand pair defines a multifunctional pathway that controls the rate and orientation of vascular cell division. *Development*, *137*, 767–774.

Fisher, K., & Turner, S. (2007). PXY, a receptor-like kinase essential for maintaining polarity during plant vascular-tissue development. *Current Biology*, *17*, 1061–1066.

Furuta, K. M., Helmann, E., & Helariutta, Y. (2014). Molecular control of cell specification and cell differentiation during procambial development. *Annual Review of Plant Biology*, *65*, 607–638.

Gallagher, K. L., & Benfey, P. N. (2009). Both the conserved GRAS domain and nuclear localization are required for SHORT-ROOT movement. *The Plant Journal*, *57*, 785–797.

Hardtke, C. S., & Berleth, T. (1998). The Arabidopsis gene *MONOPTEROS* encodes a transcription factor mediating embryo axis formation and vascular development. *The EMBO Journal*, *17*, 1405–1411.

Heo, J. O., Blob, B., & Helariutta, Y. (2017). Differentiation of conductive cells: A matter of life and death. *Current Opinion in Plant Biology*, *35*, 23–29.

Hirakawa, Y., Kondo, Y., & Fukuda, H. (2010). TDIF peptide signaling regulates vascular stem cell proliferation via the WOX4 homeobox gene in Arabidopsis. *Plant Cell*, *22*, 2618–2629.

Hirakawa, Y., Shinohar,a, H., Kondo, Y., Inoue, A., Nakanomyo, I., Ogawa, M., et al. (2008). Non-cell-autonomous control of vascular stem cell fate by a CLE peptide/receptor system. *Proceedings of the National Academy of Sciences of the United States of America*, *105*, 15208–15213.

Imai, A., Hanzawa, Y., Komura, M., Yamamoto, K., Komeda, Y., & Takahashi, T. (2006). The dwarf phenotype of the Arabidopsis acl5 mutant is suppressed by a mutation in an upstream ORF of a bHLH gene. *Development*, *133*, 3575–3585.

Ito, Y., Nakanomyo, I., Motose, H., Iwamoto, K., Sawa, S., Dohmae, N., et al. (2006). Dodeca-CLE peptides as suppressors of plant stem cell differentiation. *Science*, *313*, 842–845.

Kamiya, N., Nagasaki, H., Morikami, A., Sato, Y., & Matsuoka, M. (2003). Isolation and characterization of a rice WUSCHEL-type homeobox gene that is specifically expressed in the central cells of a quiescent center in the root apical meristem. *The Plant Journal*, *35*, 429–441.

Katayama, H., Iwamoto, K., Kariya, Y., Asakawa, T., Kan, T., Fukuda, H., et al. (2015). A negative feedback loop controlling bHLH complexes is involved in vascular cell division and differentiation in the root apical meristem. *Current Biology*, *25*, 3144–3150.

Ko, J. H., Yang, S. H., Park, A. H., Lerouxel, O., & Han, K. H. (2007). ANAC012, a member of the plant-specific NAC transcription factor family, negatively regulates xylary fiber development in *Arabidopsis thaliana*. *The Plant Journal*, *50*, 1035–1048.

Kondo, Y., Fujita, T., Sugiyama, M., & Fukuda, H. (2015). A novel system for xylem cell differentiation in *Arabidopsis thaliana*. *Molecular Plant*, *8*, 612–621.

Kondo, Y., Hirakawa, Y., & Fukuda, H. (2011). CLE peptides can negatively regulate protoxylem vessel formation via cytokinin signaling. *Plant & Cell Physiology*, *52*, 37–48.

Kondo, Y., Ito, T., Nakagami, H., Hirakawa, Y., Saito, M., Tamaki, T., et al. (2014). Plant GSK3 proteins regulate xylem cell differentiation downstream of TDIF–TDR signalling. *Nature Communications*, *5*, 3504.

Kondo, Y., Nurani, A. M., Saito, C., Ichihashi, Y., Saito, M., Yamazaki, K., et al. (2016). Vascular cell induction culture system using Arabidopsis leaves (VISUAL) reveals the sequential differentiation of sieve element-like cells. *Plant Cell*, *28*, 1250–1262.

Kubo, M., Udagawa, M., Nishikubo, N., Horiguchi, G., Yamaguchi, M., Ito, J., et al. (2005). Transcription switches for protoxylem and metaxylem vessel formation. *Genes & Development*, *19*, 1855–1860.

Levesque, M. P., Vernoux, T., Busch, W., Cui, H., Wang, J. Y., Blilou, I., et al. (2006). Whole-genome analysis of the SHORT-ROOT developmental pathway in Arabidopsis. *PLoS Biology*, *4*, e143.

Mahonen, A. P., Bishopp, A., Higuchi, M., Nieminen, K., Kinoshita, K., Tormakangas, K., et al. (2006). Cytokinin signaling and its inhibitor AHP6 regulate cell fate during vascular development. *Science*, *311*, 94–98.

Mahonen, A. P., Bonke, M., Kauppinen, L., Riikonen, M., Benfey, P. N., & Helariutta, Y. (2001). A novel two-component hybrid molecule regulates vascular morphogenesis of the Arabidopsis root. *Genes & Development*, *14*, 2938–2943.

Mahonen, A. P., Higuchi, M., Tormakangas, K., Miyawaki, K., Pischke, M. S., Sussman, M. R., et al. (2006). Cytokinins regulate a bidirectional phosphorelay network in Arabidopsis. *Current Biology*, *16*, 1116–1122.

McCarthy, R. L., Zhong, R., & Ye, Z. H. (2009). MYB83 is a direct target of SND1 and acts redundantly with MYB46 in the regulation of secondary cell wall biosynthesis in Arabidopsis. *Plant & Cell Physiology*, *50*, 1950–1964.

Mellor, N., Adibi, M., El-Showk, S., De Rybel, B., King, J., Mahonen, A. P., et al. (2017). Theoretical approaches to understanding root vascular patterning: A consensus between recent models. *Journal of Experimental Botany*, *68*, 5–16.

Mitsuda, N., Iwase, A., Yamamoto, H., Yoshida, M., Seki, M., Shinozaki, K., et al. (2007). NAC transcription factors, NST1 and NST3, are key regulators of the formation of secondary walls in woody tissues of Arabidopsis. *Plant Cell*, *19*, 270–280.

Mitsuda, N., Seki, M., Shinozaki, K., & Ohme-Takagi, M. (2005). The NAC transcription factors NST1 and NST2 of Arabidopsis regulate secondary wall thickenings and are required for anther dehiscence. *Plant Cell*, *17*, 2993–3006.

Miyashima, S., Koi, S., Hashimoto, T., & Nakajima, K. (2011). Non-cell-autonomous microRNA165 acts in a dose-dependent manner to regulate multiple differentiation status in the Arabidopsis root. *Development*, *138*, 2303–2313.

Moller, B. K., Ten Hove, C. A., Xiang, D., Williams, N., Lopez, L. G., Yoshida, S., et al. (2017). Auxin response cell-autonomously controls ground tissue initiation in the early Arabidopsis embryo. *Proceedings of the National Academy of Sciences of the United States of America*, *114*, E2533–E2539.

Morita, J., Kato, K., Nakane, T., Kondo, Y., Fukuda, H., Nishimasu, H., et al. (2016). Crystal structure of the plant receptor-like kinase TDR in complex with the TDIF peptide. *Nature Communications*, 7, 12383.

Mou, S., Zhang, X., Han, Z., Wang, J., Gong, X., & Chai, J. (2017). CLE42 binding induces PXL2 interaction with SERK2. *Protein & Cell*, *8*, 612–617.

Müller, C. J., Valdés, A. E., Wang, G., Ramachandran, P., Beste, L., Uddenberg, D., et al. (2015). PHABULOSA mediates an auxin signaling loop to regulate vascular patterning in Arabidopsis. *Plant Physiology*, *170*, 956–970.

Muñiz, L., Minguet, E. G., Singh, S. K., Pesquet, E., Vera-Sirera, F., Moreau-Courtois, C. L., et al. (2008). ACAULIS5 controls Arabidopsis xylem specification through the prevention of premature cell death. *Development*, *135*, 2573–2582.

Muraro, D., Mellor, N., Pound, M. P., Help, H., Lucas, M., Chopard, J., et al. (2014). Integration of hormonal signaling networks and mobile microRNAs is required for vascular patterning in Arabidopsis roots. *Proceedings of the National Academy of Sciences of the United States of America*, *111*, 857–862.

Nakajima, K., Sena, G., Nawy, T., & Benfey, P. N. (2001). Intercellular movement of the putative transcription factor SHR in root patterning. *Nature*, *413*, 307–311.

Oda, Y., Iida, Y., Kondo, Y., & Fukuda, H. (2010). Wood cell-wall structure requires local 2D-microtubule disassembly by a novel plasma membrane-anchored protein. *Current Biology*, *20*, 1197–1202.

Ohashi-Ito, K., & Bergmann, D. C. (2007). Regulation of the Arabidopsis root vascular initial population by LONESOME HIGHWAY. *Development*, *134*, 2959–2968.

Ohashi-Ito, K., Iwamoto, K., & Fukuda, H. (2018). LOB DOMAIN-CONTAINING PROTEIN 15 positively regulates expression of *VND7*, a master regulator of tracheary elements. *Plant & Cell Physiology*, *59*, 989–996.

Ohashi-Ito, K., Matsukawa, M., & Fukuda, H. (2013). An atypical bHLH transcription factor regulates early xylem development downstream of auxin. *Plant & Cell Physiology*, *54*, 398–405.

Ohashi-Ito, K., Oda, Y., & Fukuda, H. (2010). Arabidopsis VASCULAR-RELATED NAC-DOMAIN6 directly regulates the genes that govern programmed cell death and secondary wall formation during xylem differentiation. *Plant Cell*, *22*, 3461–3473.

Ohashi-Ito, K., Oguchi, M., Kojima, M., Sakakibara, H., & Fukuda, H. (2013). Auxin-associated initiation of vascular cell differentiation by LONESOME HIGHWAY. *Development*, *140*, 765–769.

Ohashi-Ito, K., Saegusa, M., Iwamoto, K., Oda, Y., Katayama, H., & Kojima, M. (2014). A bHLH complex activates vascular cell division via cytokinin action in root apical meristem. *Current Biology*, *24*, 2053–2058.

Palovaara, J., Saiga, S., Wendrich, J. R., van 't Wout Hofland, N., van Schayck, J. P., Hater, F., et al. (2017). Transcriptome dynamics revealed by a gene expression atlas of the early Arabidopsis embryo. *Nature Plants*, *3*, 894–904.

Pyo, H., Demura, T., & Fukuda, H. (2007). TERE; a novel cis-element responsible for a coordinated expression of genes related to programmed cell death and secondary wall formation during differentiation of tracheary elements. *The Plant Journal*, *51*, 955–965.

Ramachandran, P., Carlsbecker, A., & Etchell, J. P. (2017). Class III HD-ZIPs govern vascular cell fate: An HD view on patterning and differentiation. *Journal of Experimental Botany*, *68*, 55–69.
Ren, B., Chen, Q., Hong, S., Zhao, W., Feng, J., Feng, H., et al. (2013). The Arabidopsis eukaryotic translation initiation factor eIF5A-2 regulates root protoxylem development by modulating cytokinin signaling. *The Plant Cell*, *25*, 3841–3857.
Rodriguez-Villalon, A. (2016). Wiring a plant: Genetic networks for phloem formation in *Arabidopsis thaliana* roots. *The New Phytologist*, *210*, 45–50.
Roodbarkelari, F., & Groot, E. P. (2017). Regulatory function of homeodomain-leucine zipper (HD-ZIP) family proteins during embryogenesis. *The New Phytologist*, *213*, 98–104.
Saidi, Y., Hearn, T. J., & Coates, J. C. (2012). Function and evolution of 'green' GSK3/Shaggy-like kinases. *Trends in Plant Science*, *17*, 39–46.
Saito, M., Kondo, Y., & Fukuda, H. (2018). BES1 and BZR1 redundantly promote phloem and xylem differentiation. *Plant & Cell Physiology*, *59*, 590–600.
Scheres, B., Laurenzio, L. D., Willemsen, V., Hauser, M. T., Janmaat, K., Weisbeek, P., et al. (1995). Mutations affecting the radial organisation of the Arabidopsis root display specific defects throughout the embryonic axis. *Development*, *121*, 53–62.
Schlereth, A., Moller, B., Liu, W., Kientz, M., Flipse, J., Rademacher, E. H., et al. (2010). MONOPTEROS controls embryonic root initiation by regulating a mobile transcription factor. *Nature*, *464*, 913–916.
Singh, M. B., & Bhalla, P. L. (2006). Plant stem cells carve their own niche. *Trends in Plant Science*, *11*, 241–246.
Soyano, T., Thitamadee, S., Machida, Y., & Chua, N. H. (2008). ASYMMETRIC LEAVES2-LIKE19/LATERAL ORGAN BOUNDARIES DOMAIN30 and ASL20/LBD18 regulate tracheary element differentiation in Arabidopsis. *Plant Cell*, *20*, 3359–3373.
Tan, T. T., Endo, H., Sano, R., Kurata, T., Yamaguchi, M., Ohtani, M., et al. (2018). Transcription factors VND1-VND3 contribute to cotyledon xylem vessel formation. *Plant Physiology*, *176*, 773–789.
Tokunaga, H., Kojima, M., Kuroha, T., Ishida, T., Sugimoto, K., Kiba, T., et al. (2012). Arabidopsis lonely guy (LOG) multiple mutants reveal a central role of the LOG-dependent pathway in cytokinin activation. *The Plant Journal*, *69*, 355–365.
Ursache, R., Miyashima, S., Chen, Q., Vatén, A., Nakajima, K., Carlsbecker, A., et al. (2014). Tryptophan-dependent auxin biosynthesis is required for HD-ZIP III-mediated xylem patterning. *Development*, *141*, 1250–1259.
Vaughan-Hirsch, J., Goodall, B., & Bishopp, A. (2018). North, East, South, West: Mapping vascular tissues onto the Arabidopsis root. *Current Opinion in Plant Biology*, *41*, 16–22.
Vera-Sirera, F., De Rybel, B., Úrbez, C., Kouklas, E., Pesquera, M., Álvarez-Mahecha, J. C., et al. (2015). A bHLH-based feedback loop restricts vascular cell proliferation in plants. *Developmental Cell*, *35*, 432–443.
Xu, B., Ohtani, M., Yamaguchi, M., Toyooka, K., Wakazaki, M., Sato, M., et al. (2014). Contribution of NAC transcription factors to plant adaptation to land. *Science*, *343*, 1505–1508.
Yamaguchi, M., Mitsuda, N., Ohtani, M., Ohme-Takagi, M., & Demura, T. (2011). VASCULAR-RELATED NAC-DOMAIN7 directly regulates the expression of a broad range of genes for xylem vessel formation. *The Plant Journal*, *66*, 579–590.
Yamaguchi, M., Ohtani, M., Mitsuda, N., Kubo, M., Ohme-Takagi, M., Fukuda, H., et al. (2010). VND-INTERACTING2, a NAC domain transcription factor, negatively regulates xylem vessel formation in Arabidopsis. *Plant Cell*, *22*, 1249–1263.

Yoshida, S., de Reuille, P. B., Lane, B., Bassel, G. W., Prusinkiewicz, P., Smith, R. S., et al. (2014). Genetic control of plant development by overriding a geometric division rule. *Developmental Cell, 29*, 75–87.
Zhang, H., Lin, X., Han, Z., Qu, L. J., & Chai, J. (2016). Crystal structure of PXY-TDIF complex reveals a conserved recognition mechanism among CLE peptide-receptor pairs. *Cell Research, 26*, 543–555.
Zhang, H., Lin, X., Han, Z., Wang, J., Qu, L. J., & Chai, J. (2016). SERK family receptor-like kinases function as co-receptors with PXY for plant vascular development. *Molecular Plant, 9*, 1406–1414.
Zhong, R., Demura, T., & Ye, Z. H. (2006). SND1, a NAC domain transcription factor, is a key regulator of secondary wall synthesis in fibers of Arabidopsis. *Plant Cell, 18*, 3158–3170.
Zhong, R., Lee, C., & Ye, Z. H. (2010). Evolutionary conservation of the transcriptional network regulating secondary cell wall biosynthesis. *Trends in Plant Science, 15*, 625–632.
Zhong, R., Richardson, E. A., & Ye, Z. H. (2007). The MYB46 transcription factor is a direct target of SND1 and regulates secondary wall biosynthesis in Arabidopsis. *Plant Cell, 19*, 2776–2792.
Zhou, J., Wang, X., Lee, J. Y., & Lee, J. Y. (2013). Cell-to-cell movement of two interacting AT-hook factors in Arabidopsis root vascular tissue patterning. *The Plant Cell, 25*, 187–201.
Zhou, J., Zhong, R., & Ye, Z. H. (2014). Arabidopsis NAC domain proteins, VND1 to VND5, are transcriptional regulators of secondary wall biosynthesis in vessels. *PLoS One, 22*, e105726.
Zürcher, E., Liu, J., di Donato, M., Geisler, M., & Müller, B. (2016). Plant development regulated by cytokinin sinks. *Science, 353*, 1027–1030.

CHAPTER SEVEN

Transcriptional networks orchestrating programmed cell death during plant development

Marta Cubría-Radío[a,b], Moritz K. Nowack[a,b,*]
[a]Department of Plant Biotechnology and Bioinformatics, Ghent University, Ghent, Belgium
[b]VIB Center for Plant Systems Biology, Ghent, Belgium
*Corresponding author: e-mail address: moritz.nowack@vib.be

Contents

Abstract

Transcriptional gene regulation is a fundamental biological principle in the development of eukaryotes. It does control not only cell proliferation, specification, and differentiation, but also cell death processes as an integral feature of an organism's developmental program. As in animals, developmentally regulated cell death in plants occurs in numerous contexts and is of vital importance for plant vegetative and reproductive development. In comparison with the information available on the molecular regulation of programmed cell death (PCD) in animals, however, our knowledge on plant PCD still remains scarce.

Here, we discuss the functions of different classes of transcription factors that have been implicated in the control of developmentally regulated cell death. Though doubtlessly representing but a first layer of PCD regulation, information on PCD-regulating transcription factors and their targets represents a promising strategy to understand the complex machinery that ensures the precise and failsafe execution of PCD processes in plant development.

Current Topics in Developmental Biology, Volume 131
ISSN 0070-2153
https://doi.org/10.1016/bs.ctdb.2018.10.006

1. Introduction

Many aspects of eukaryote development are determined by differential gene expression, which is in turn controlled by complex gene regulatory networks. Protein levels are regulated at multiple steps from gene transcription over de novo protein translation to protein degradation. Transcription factors (TFs) are defined as proteins that show sequence-specific DNA binding and are capable of activating and/or repressing gene transcription. In this context, TFs control differential gene expression and thus can act as central switches to coordinate cellular differentiation and development by modulating expression of individual or groups of genes (Riechmann et al., 2000).

Just as cell division, growth, and differentiation, also cell death is a fundamental biological process of cellular life. A multitude of causes can bring about cellular death in eukaryotes. Instantaneous and catastrophic death of cells by severe physical, chemical, or mechanical insults is referred to as "accidental cell death," while "regulated cell death" (RCD) describes different forms of a genetically encoded, actively controlled demise of cells (Galluzzi et al., 2018). Historically, metazoan cell death processes have been divided according to morphological characteristics into type I or apoptosis, type II or autophagy, and type III or necrosis. To date, a dozen major RCD routines are recognized, which can display morphological features ranging from fully necrotic to fully apoptotic and from pro-inflammatory to anti-inflammatory. These different RCD processes can be triggered by diverse kinds of stresses which are too severe or prolonged for an adaptive response. However, RCD occurs also in the absence of stresses as an inherent part of regular physiological or developmental programs. These cases of cell death are generally referred to as "programmed cell death" (PCD) (Galluzzi et al., 2018).

Different forms of PCD have crucial functions in the development and the physiology of animals and plants. In comparison to metazoans, plants have a set of unique cellular components. The presence of cell walls, chloroplasts and large central vacuoles necessitates and offers independent solutions for the regulation and execution of plant PCD (Van Durme & Nowack, 2016). Furthermore, central apoptotic regulators, including those homologous to Bcl-2 family proteins or caspases, are not encoded in plant genomes, suggesting that the genetics of plant PCD differs substantially from classical animal PCD modes (van Doorn, 2011). As for metazoan PCD, there have been attempts to classify plant PCD into different categories, e.g., based on morphology (van Doorn, 2011) or biological context (Daneva, Gao, Van Durme, & Nowack, 2016). Morphologically, vacuolar cell death has been described as depending on a combination of autophagy-like processes and the release of lytic enzymes from the disintegrating central vacuole. In contrast, plant necrosis has been characterized by the absence of vacuolar cell death features, early rupture of the plasma membrane, and protoplast shrinkage (van Doorn, 2011). Regarding the biological context, environmental PCD (ePCD) triggered by biotic and abiotic stresses has been contrasted with developmentally controlled PCD (dPCD) that occurs in the absence of environmental perturbation (Daneva et al., 2016). Of note, the transcriptional regulation of several genes has been found to be positively associated with different cases of dPCD, whereas the same genes have not been found to be regulated in ePCD contexts (Olvera-Carrillo et al., 2015).

In comparison with the information available about the molecular regulation of animal PCD, our knowledge of plant PCD still remains fragmented. In animals, evidence suggests that the core of the PCD machinery is largely regulated post-translationally, i.e., on the level of protein localization, activation, or destruction (Fuchs & Steller, 2015). Nevertheless, especially in the context of cancer and development, also the regulation of gene transcription has been shown to be crucial for animal PCD initiation and execution. The tumor suppressor p53, for instance, is a TF regulating the expression of over 500 genes, among them many pro-apoptotic regulators (Aubrey, Kelly, Janic, Herold, & Strasser, 2017). Mutations in p53 are commonly found in human cancer cells, particularly in the DNA-binding domain (Olivier et al., 2002), documenting the importance of transcriptional control of PCD in the prevention of cancer. Similarly, the tumor-suppressing action of the retinoblastoma protein pRB occurs largely via the TF E2F1, which controls the transcription of pro-apoptotic genes (Polager & Ginsberg, 2009). In *Drosophila melanogaster*, the *reaper*, *head involution defective* and *grim* genes (encoding the inhibitor-of-apoptosis antagonizing RHG proteins) are transcriptionally regulated during fly embryogenesis. Whereas deletion of *RHG* genes blocks apoptosis, inducible expression is sufficient to cause wide-spread ectopic apoptosis (White, Tahaoglu, & Steller, 1996). Finally, the pro-apoptotic gene *EGG-LAYING ABNORMAL-1* (*EGL-1*) is transcriptionally controlled, repressed by CELL DEATH SPECIFICATION-1 (CES-1) and activated by a basic HELIX-LOOP-HELIX (bHLH) TF complex, to regulate apoptosis in *Caenorhabditis elegans* neuronal cell death (Thellmann, Hatzold, & Conradt, 2003).

Also in plants, the regulation of gene transcription has been identified to be an important regulatory principle for different types of PCD. In this chapter, we will give an overview on the involvement of major plant TF classes in different developmentally controlled PCD processes.

2. NAC transcription factor family

Transcription factors belonging to the NAC [for NO APICAL MERISTEM (NAM), ARABIDOPSIS TRANSCRIPTION ACTIVATION FACTOR (ATAF), CUP-SHAPED COTYLEDON (CUC)] family comprise one of the largest groups of plant-specific transcriptional regulators. NAC TFs are involved in a multitude of biological processes, from plant development to stress responses (Fan et al., 2018). Interestingly, NAC TFs have been implicated in different types of cell death, including ePCD and dPCD processes. In *Brassica napus* (rapeseed), several NAC TFs modulate the reaction to environmental stresses by promoting accumulation of reactive oxygen species and a hypersensitive response like ePCD (Chen et al., 2017; Yan et al., 2018). Also in dPCD, NAC TFs play a prominent role, as will be discussed in the following paragraphs.

The evolution of efficient long-distance water and solute transport systems was crucial for the successful colonization of terrestrial habitats by plants (Ohtani, Akiyoshi, Takenaka, Sano, & Demura, 2017). PCD and PCD-like processes play an important role during the development of plant vascular tissues, particularly in the xylem tracheary elements and the phloem sieve elements (see recent review by Heo, Blob, & Helariutta, 2017). The xylem in vascular plants is the transport network for water and extends from the root to the shoot tissues. In order to achieve efficient long-distance transport through cell lumina, conductive xylem cells undergo a PCD process characterized by total cytoplasmic clearance and preservation of the reinforced secondary cell wall (SCW) (Escamez & Tuominen, 2014). Over the last decades, the developmental program of xylem development has been elucidated in quite some detail, including xylem specification, patterning, differentiation, and PCD.

Many aspects of xylem differentiation have been discovered in xylogenic cell cultures of *Zinnia elegans* and *Arabidopsis thaliana*, in which xylem differentiation can be induced by hormones (Demura et al., 2002; Kubo et al., 2005). Downstream of hormonal signaling, numerous TFs have been implicated in the different processes of xylem differentiation, including PCD as its final step (Fig. 1) (Fukuda, 2004). In *A. thaliana*, a microarray analysis

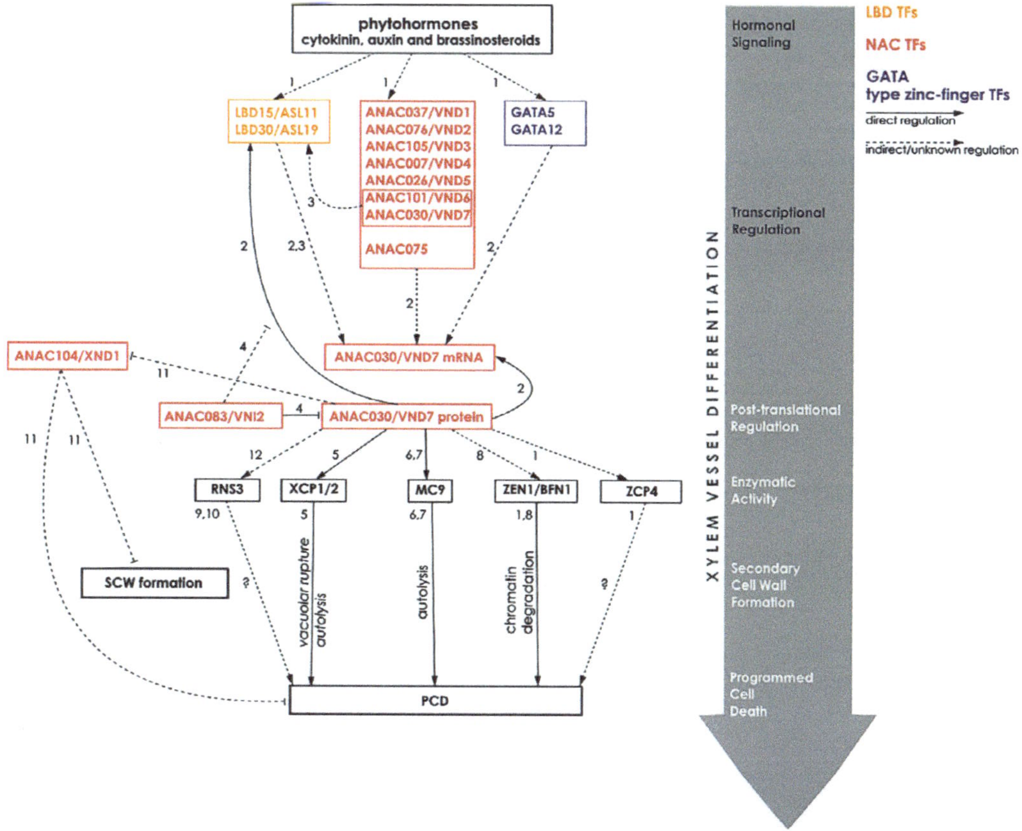

Fig. 1 Transcriptional regulation of xylem vessel differentiation and PCD. LBD TFs are indicated in yellow; NAC TFs in magenta; GATA type zinc-finger TFs in blue. Non-TF proteins are indicated in gray. Solid lines indicate direct regulation; dotted lines indicate indirect regulation. Arrow-headed lines represent positive regulation, and bar-headed lines negative regulation of targets. Corresponding references: 1: Fukuda (2004); 2: Endo et al. (2015); 3: Soyano, Thitamadee, Machida, and Chua (2008); 4: Yamaguchi et al. (2010); 5: Avci, Petzold, Ismail, Beers, and Haigler (2008); 6: Bollhöner et al. (2013); 7: Escamez et al. (2016); 8: Ito and Fukuda (2002); 9: Bariola et al. (1994); 10: Olvera-Carrillo et al. (2015); 11: Zhao, Avci, Grant, Haigler, and Beers (2008); 12: Zhong, Lee, and Ye (2010).

revealed that 1705 genes were differentially expressed upon induction of xylem formation (Kubo et al., 2005). Among the upregulated genes, there were several NAC TFs with similarity to the xylem-expressed Zinnia TF Z567, and the seven members of the corresponding NAC clade in *A. thaliana* were designated *VASCULAR-RELATED NAC-DOMAIN1* (*VND1*) to *VND7* (Kubo et al., 2005). While recessive loss-of-function approaches failed to produce specific phenotypes, overexpression of *VND6* and *VND7* caused ectopic formation of xylem-like cells, which resembled protoxylem in the case of *VND7* and metaxylem in the case of *VND6*. Accordingly, *pVND6::YFP-NLS* reporters were expressed in the differentiating metaxylem, whereas *pVND7::YFP-NLS* reporters showed expression in the protoxylem. Ultimately, overexpression of both *VND6* and *VND7* leads to ectopic cell death, suggesting that these TFs control both SCW formation and PCD (Kubo et al., 2005). Since the initial discovery of the VNDs, many other elements have been identified (Fig. 1). Of over 80 TFs expressed during xylem vessel differentiation, 14 were found to activate the expression of *VND7* in transient expression assays, including *VND1-6* and *VND7* itself. These putative *VND7* regulators were found bind to a specific motif in the *cis*-regulatory region of *VND7* (Endo et al., 2015). Inducible overexpression of *VND1-5* and *ANAC075* triggered ectopic differentiation of xylem vessels. Surprisingly, however, whereas VND7 targets were upregulated after overexpression of these TFs, *VND7* mRNA levels themselves were not markedly raised, indicating a complex regulation of VND7 activity *in planta* (Endo et al., 2015). Next to NAC TFs, also LATERAL ORGAN BOUNDARIES DOMAIN/ASYMMETRIC LEAVES2-LIKE (LBD/ASL) and GATA-type zinc-finger TF family members were identified as putative *VND7* regulators (Fig. 1) (Endo et al., 2015).

More recently, *VND1*, *VND2*, and *VND3* have been identified to contribute to xylem vessel differentiation in cotyledons independent of *VND6* and *VND7* (Tan et al., 2018). The formation of ectopic xylem in detached hormone-treated cotyledons was found to depend on the presence of functional *VND1*, *VND2*, and *VND3*. Furthermore, a *vnd1 vnd2 vnd3* triple mutant showed inhibited xylem vessel formation of secondary veins in cotyledons under dark conditions (Tan et al., 2018). These findings suggest that *VND1* to *VND3* have specific functions and might be linking light conditions to vessel formation (Tan et al., 2018).

Next to TFs promoting vessel formation, also differentiation-inhibiting NAC TFs have been identified in controlling correct xylem differentiation. For instance, VND-INTERACTING2 (VNI2)/ANAC083 represses xylem vessel differentiation and PCD through interacting with VND7, thereby repressing VND7 target genes (Yamaguchi et al., 2010). The presence of a C-terminal PEST degradation motif suggests that VNI2 is post-translationally controlled to fine-tune VND7 activity, and C-terminally truncated VNI2 dominantly inhibited the development of xylem vessels (Yamaguchi et al., 2010). The xylem-expressed *XYLEM NAC DOMAIN1* (*XND1*)/*ANAC104* has also been shown to negatively regulate xylem differentiation and PCD (Fig. 1). *XND1* overexpression generates dwarf plants in which viable xylem parenchyma cells lacking fortified SCWs replaced functional xylem vessels (Zhao et al., 2008). Interestingly, XND1 has recently been shown to interact with the cell cycle and differentiation regulator RETINOBLASTOMA-RELATED (RBR) via highly conserved C-terminal motifs. Deletion of these motifs severely compromises the effects of *XND1* overexpression, possibly linking *RBR*-controlled processes with xylem differentiation and PCD control (Zhao et al., 2017).

Taken together, NAC TFs play a prominent role in the developmental program of xylem formation, controlling early decisions as well as the late differentiation steps of SCW formation and PCD (Fig. 1). Among the downstream targets of this transcriptional network are several enzymes that have been implicated in aspects of PCD execution (Fig. 1). *XYLEM CYSTEINE PROTEASE1* (*XCP1*) and *XCP2*, for instance, have been shown to contribute to cellular autolysis before PCD (micro-autolysis) and after PCD (macro-autolysis). As a consequence, the *xcp1* and *xcp1 xcp2* mutants show a delay in cytoplasmic clearance of dying xylem vessels (Fig. 1) (Avci et al., 2008). Similarly, *METACASPASE9* (*MC9*) has been shown to participate in the cytoplasmic clearing after tonoplast rupture (Fig. 1) (Bollhöner et al., 2013; Escamez et al., 2016). The cysteine protease ZCP4 has been shown to specifically function during xylem vessel PCD in Zinnia (Fukuda, 2004). It degrades the chromatin structure inside the nucleus, helping a rapid nuclear degradation triggered by vacuole rupture during PCD in *Z. elegans* (Obara, Kuriyama, & Fukuda, 2001). Next to these proteases, also other PCD-associated hydrolytic enzymes have been identified as being regulated by VND7, including *RIBONUCLEASE3* (*RNS3*), *BIFUNCTIONAL NUCLEASE1* (*BFN1*), and *PUTATIVE ASPARTIC PROTEINASE3* (*PASPA3*) (Zhong et al., 2010), although to date the specific functions of most of them remain unclear.

In contrast to the xylem vessels, the mature phloem sieve elements remain alive. However, to optimize nutrient transport, also sieve elements reinforce their cell walls and get cleared of most organelles, including the nucleus, in a process that is reminiscent of cellular clearance in PCD (Geldner, 2014). Acting downstream of the MYB TF ALTERED PHLOEM DEVELOPMENT (APL), ANAC045 and ANAC086 have been shown to redundantly regulate the phloem enucleation process by controlling the expression of *NEN4*, encoding a phloem-specific nuclease (Furuta et al., 2014).

Outside the vascular bundle, dPCD processes also play important roles in plant development. In the root cap of *A. thaliana*, for instance, PCD processes have been shown to regulate organ size. The root cap consists of two tissues, the columella root cap at the very tip of the root, and the lateral root cap (LRC), which flanks the root meristem (Kumpf & Nowack, 2015). Although there is continuous root cap cell production by specific stem cells, root cap organ size does not increase accordingly, implying a disposal of mature cells to compensate for the production of new cells. While columella cells get shed off the root in regular intervals as packets of living cells, expression of numerous dPCD-associated genes in the LRC, including *RNS3*, *BFN1*, *PASPA3*, and *MC9*, hinted to a PCD process occurring in the root cap (Olvera-Carrillo et al., 2015). Upstream of *PASPA3* and *BFN1*, the NAC TF SOMBRERO (SMB/ANAC033) has been identified as a master regulator of root cap PCD (Fendrych et al., 2014). *SMB* was first implicated in root cap maturation based on the observation that *smb* mutants show prolonged root cap division activity and delayed root cap differentiation (Bennett et al., 2010; Willemsen et al., 2008). This delay of maturation also affects expression of PCD-associated genes and causes a delay of PCD execution, which in *smb* mutants occurs further up in the root elongation zone as compared to the wild type. In this zone, *smb* root cap cells die in an aberrant way without post-mortem corpse clearance, which leads to the root proper being covered in unprocessed LRC cell corpses (Fendrych et al., 2014). Recently, two additional NAC TFs, ANAC087 and ANAC046, have been identified to regulate distinct aspects of root cap PCD (Huysmans et al., 2018). *ANAC087* is expressed in the entire root cap, whereas *ANAC046* expression is restricted to the columella root cap. *ANAC087* indirectly promotes expression of *BFN1* to control nuclear degradation after root cap cell death in the root elongation zone. ANAC046, however, is a direct regulator of *BFN1* and other PCD-associated genes and, redundantly with *ANAC087*, controls PCD onset in the columella root cap during and after shedding of these cells into the rhizosphere (Huysmans et al., 2018). Downstream of these NAC

TFs, *BFN1* has been implicated in chromatin degradation in the root cap, as a *bfn1* loss-of-function mutant shows a delay of nuclear degradation (Fendrych et al., 2014). The *Z. elegans BFN1* homolog, *ZINNIA ENDONUCLEASE1* (*ZEN1*), has been shown to have a similar loss-of-function phenotype in xylogenic cell cultures of *Z. elegans* (Ito & Fukuda, 2002), although to date, no *bfn1* phenotype has been described in xylem vessels of *A thaliana*.

NAC TFs have also been implicated in senescence processes and age-induced PCD (Daneva et al., 2016). Ethylene-regulated TFs control the onset of leaf senescence (Koyama, 2014). ETHYLENE-INSENSITIVE3 (EIN3) directly activates expression of *ORESARA1 (ORE1)/ANAC092/AtNAC2*, encoding a central positive regulator of leaf senescence (Kim et al., 2014). Also, the NAC TF gene *AtNAP* is under the direct control of EIN3, whereas the *ORE1* paralog *ORS1* (Balazadeh et al., 2011) and the *ANAC019* and *ANAC055* genes are activated during leaf senescence in an EIN3-independent manner (Hickman et al., 2013). In all cases, NAC TFs positively regulate leaf senescence (Koyama, 2014). More recently, it has been shown that the *BnaANAC087* gene is induced by multiple stress and hormone treatments and is highly expressed in senescent leaves (Yan et al., 2018).

Senescence of the ephemeral floral organs has also been shown to be controlled by ethylene and NAC TFs (Shibuya, 2018). Floral organ senescence can be initiated by successful pollination, but also flowers remaining unpollinated will eventually start to senesce in a species-specific aging process (Rogers, 2013). During the age-induced senescence of unpollinated flowers, ORE1 and a novel flower-specific NAC TF, KIRA1 (KIR1/ANAC074), have been identified as partially redundantly operating regulators of age-induced stigma degeneration (Gao et al., 2018). In dominant-negative single mutants and *ore1 kir1* double mutants, the stigma life span is substantially prolonged. Interestingly, the stigma's function supporting pollen tube growth toward the ovules is not extended accordingly, suggesting that *ORE1* and *KIR1* control only the PCD aspect during floral senescence but not the time span of stigma receptivity. Nevertheless, misexpression of *KIR1* was sufficient to cause ectopic cell death in tobacco and *A. thaliana*, and early expression of *KIR1* in the stigma caused precocious cell death and loss of floral receptivity, suggesting that NAC-controlled PCD is sufficient to restrict floral receptivity in time (Gao et al., 2018).

3. MYB transcription factor family

The first TF cloned from plants was COLORED1 (C1) of *Zea mays* (maize), a MYB TF with high structural homology to the vertebrate proto-oncogene *c-MYB* (*my*elo*b*lastosis) (Dubos et al., 2010). In contrast to the plant-specific NACs, MYB TFs are represented in all eukaryotes. Yet, just like NAC TFs, MYB TFs occur in large gene families and regulate a great diversity of processes in development, differentiation, metabolism, and responses to environmental stresses and PCD processes (Ambawat, Sharma, Yadav, & Yadav, 2013).

A prominent role for *MYB* genes is controlling aspects of dPCD during tapetum development. The tapetal cell layer borders the anther locules and is crucial for microsporogenesis and pollen maturation. In particular, tapetal PCD plays a vital secretory role in the development of pollen grains by contributing metabolites and enzymes that are important for pollen wall synthesis and pollen coat deposition (Liu & Fan, 2013). Manipulating tapetal PCD invariably affects male fertility, indicating its importance for pollen development and coat formation (Gómez, Talle, & Wilson, 2015). The most upstream TF in tapetal development is GAMYB, a MYB TF important in gibberellic acid (GA) signaling in cereal seeds and during anther development (Fig. 2A and B) (Aya et al., 2009). In *Oryza sativa* (rice), GA-deficient and GA-insensitive mutants exhibit similar defects in tapetal development and PCD as do *gamyb* mutants (Aya et al., 2009). Interestingly, *GAMYB* function is also conserved in dicots, for instance, in *A. thaliana* (Plackett, Thomas, Wilson, & Hedden, 2011). GAMYBs are post-transcriptionally repressed by the microRNA *mirR159*, which is GA-regulated in *A. thaliana*, but GA-independent in rice (Fig. 2A and B) (Alonso-Peral et al., 2010; He et al., 2015).

Downstream of GAMYB, a network of MYB, bHLH, and PLANT HOMEODOMAIN (PHD)-finger TFs coordinates tapetum development. Among the MYB TFs in *A. thaliana* and rice, *TAPETAL DEVELOPMENT AND FUNCTION1* (*TDF1/MYB35*) and *MALE STERILE188* (*MS188/MYB80*) occupy a central position in the tapetum control network together with bHLH TFs (Gu et al., 2014) (Fig. 2). The *A. thaliana tdf1* mutant develops overproliferating and enlarged tapetal cells that fail to maintain tapetum fate and typical degeneration patterns (Zhu et al., 2008). Rice *tdf1* mutants show a similar loss-of-function phenotype as *A. thaliana*

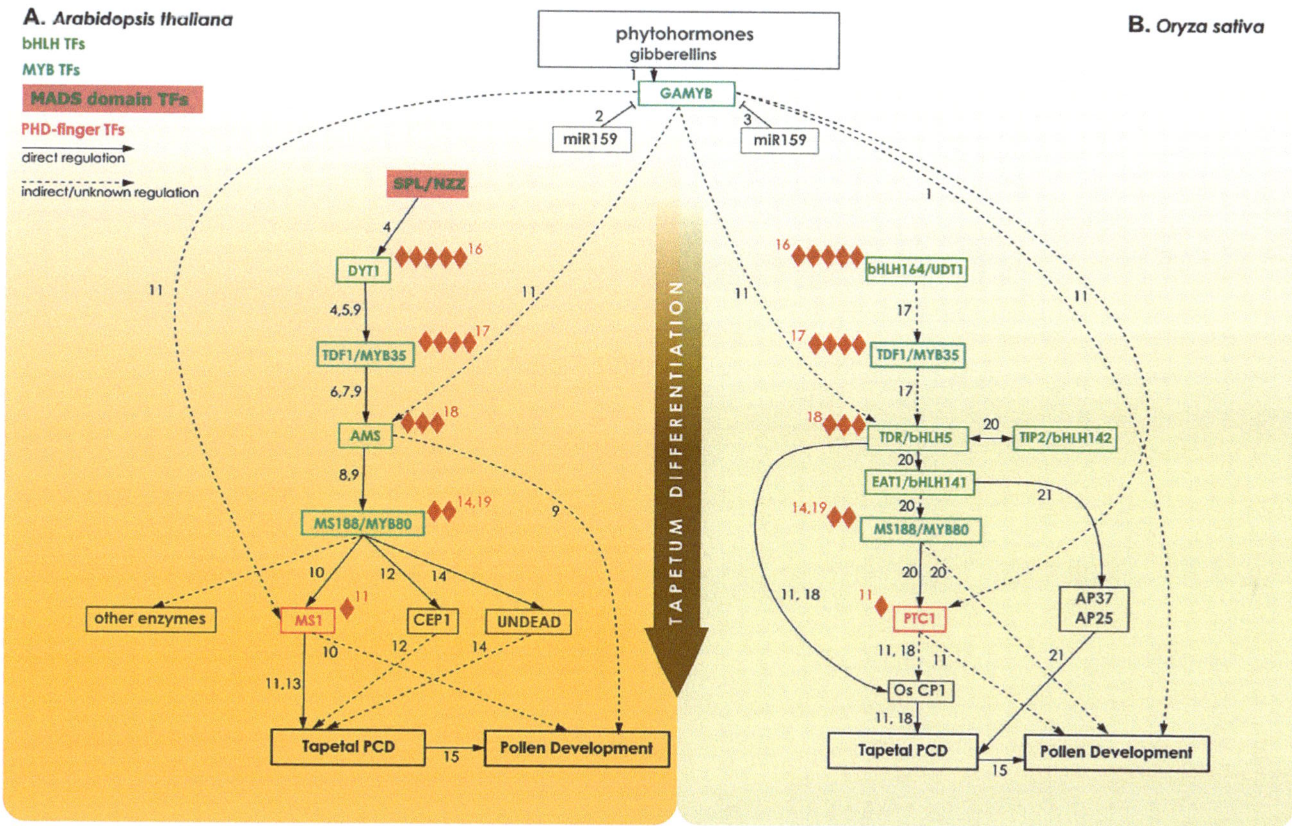

Fig. 2 The MYB/bHLH network: transcriptional regulation of tapetal differentiation and PCD in *A. thaliana* (A) and in *O. sativa* (rice) (B). MYB TFs are indicated in turquoise, bHLH TFs in green, MADS-domain TFs in green-squared pink, and PHD-finger TFs in magenta. Non-TF proteins are indicated in gray. Solid lines indicate direct regulation; dotted lines indirect or unknown nature of regulation. Arrow-headed lines represent positive regulation, and bar-headed lines negative regulation of targets. Red rhombi indicate homology between *A. thaliana* and *O. sativa*. Corresponding references: 1: Aya et al. (2009); 2: Alonso-Peral et al. (2010); 3: He, Wang, Wang, and Yang (2015); 4: Zhang et al. (2006); 5: Feng et al. (2012); 6: Xu et al. (2010); 7: Zhu et al. (2008); 8: Sorensen et al. (2003); 9: Li, Xue, Zhu, and Yang (2017); 10: Zhang et al. (2007); 11: Li et al. (2011); 12: Zhang et al. (2014); 13: Vizcay-Barrena and Wilson (2006); 14: Phan, Iacuone, Li, and Parish (2011); 15: Gómez et al. (2015); 16: Jung et al. (2005); 17: Cai et al. (2015); 18: Li et al. (2006); 19: Phan, Li, and Parish (2012); 20: Ko et al. (2017); 21: Niu et al. (2013).

mutants, and rice *TDF1* is able to cross-complement the *tdf1* mutant in *A. thaliana* (Cai et al., 2015). Further downstream, *MS188/MYB80* plays a central role in controlling the timing of tapetal PCD. The *ms188/myb80* mutants show premature tapetal degeneration, characterized by an early onset of vacuolation and chromatin degradation in tapetal cells, as shown with a terminal deoxynucleotidyl transferase Dutp nick end labeling (TUNEL) assay (Phan et al., 2011). Interestingly, *MYB80* homologs in the crop plants rice, rapeseed, *Triticum aestivum* (wheat) and *Gossypium hirsutum* (cotton) have similar expression patterns and functions as *MS188/MYB80* in *A. thaliana* (Phan et al., 2012). Moreover, the crop MYB80 homologs could restore the wild-type phenotype in the *A. thaliana ms188/myb80* mutant. These data show that the regulation of tapetal PCD has a high degree of evolutionary conservation (Fig. 2A and B).

Downstream of MS188/MYB80, TFs including the PHD-finger TF MALE STERILITY1 (MS1) as well as the protease UNDEAD (Vizcay-Barrena & Wilson, 2006) have been identified (Fig. 2A). UNDEAD is an aspartic protease that prevents the early onset of PCD, since *undead* knock-down lines show precocious PCD similar to *ms188/myb80* mutants. As UNDEAD shows a mitochondrial target sequence, it might act by hydrolyzing (a) still unknown apoptosis-inducing target protein(s) in mitochondria (Phan et al., 2011).

CYSTEINE ENDOPEPTIDASE1 (*CEP1*), encoding a papain-like cysteine protease promoting tapetal PCD, is indirectly regulated by *MS188/MYB80* (Fig. 2A). *CEP1* overexpression causes premature tapetal PCD and male sterility, whereas *cep1* mutants show an interrupted tapetum PCD associated with a decrease of pollen fertility (Zhang et al., 2014). These results demonstrate the importance of the precise timing of tapetal PCD for full male fertility. MS188/MYB80 also regulates the expression of *RESPIRATORY-BRUST OXIDASE HOMOLOG E* (*RBOHE*), implicating the transcriptional control of reactive oxygen species production in tapetal PCD (Xie, Wan, Li, & Zhang, 2014).

Besides controlling PCD processes in the tapetum, MYB TFs are also playing important roles in cell death events that occur during plant fertilization. When a growing pollen tube reaches the female gametophyte, both the pollen tube cell and the receptive synergid cell undergo coordinated disintegration to facilitate gamete fusion (Dresselhaus, Sprunck, & Wessel, 2016). The three pollen tube-expressed, redundantly acting MYB TFs MYB97, MYB101 and MYB120 have been shown to regulate this process on the male side. Triple mutant pollen tubes do not show the rupture of the pollen tube cell to release the sperm cells, and fail to trigger synergid cell degeneration (Leydon et al., 2013; Liang et al., 2013). Of several pollen tube-expressed target genes (Leydon et al., 2013), a secreted cysteine-rich peptide might be involved in the communication processes required for successful fertilization. Cell-to-cell signaling, depending on the receptor-like kinase FERONIA and specific calcium signatures in both synergid and pollen tube cells, has been shown to orchestrate the synchronized degeneration of synergid and pollen tube cells (Ngo, Vogler, Lituiev, Nestorova, & Grossniklaus, 2014; see also Zhou & Dresselhaus, 2019).

4. bHLH transcription factor family

The bHLH TFs are present in all eukaryotes (Riechmann et al., 2000). In *A. thaliana*, the bHLH family represents the second largest class of TFs comprising more than 150 genes (Feller, Machemer, Braun, & Grotewold, 2011). In plants, bHLH TFs have prominent roles in developmental patterning processes, fruit dehiscence, secondary metabolite biosynthesis, hormone signaling, and stress responses (Feller et al., 2011; Goossens, Mertens, & Goossens, 2017; Ohashi-Ito & Fukuda, 2016; Ran, Shen, Liu, & Wang, 2013). They often interact with MYB TFs in heteromeric protein complexes to fine-tune these processes (Feller et al., 2011). Although MYB and bHLH TFs have been implicated in dPCD regulation, no such interaction has been shown so far in this particular context.

Alongside MYB TFs, several bHLH TFs regulate tapetal differentiation and PCD (Fig. 2A). The bHLH TF-encoding gene *DISFUNCTIONAL TAPETUM1* (*DYT1*) is strongly expressed in the tapetum and developing microspores (Zhang et al., 2006) and is in turn controlled by the MADS-domain TF SPOROCYTELESS/NOZZLE (Zhang et al., 2006). Interestingly, a functional *pDYT1::DYT1-GFP* fusion protein was localized exclusively to the tapetum and could not be detected in microspores, suggesting that the critical function of *DYT1* takes place in the tapetum (Gu et al., 2014). Loss of *DYT1* function results in a precocious vacuolation of tapetum cells, the downregulation of many tapetum-expressed genes, and male sterility (Zhang et al., 2006). Likewise, *A. thaliana* plants expressing a dominant-repressive *DYT1-SRDX* version display impaired male fertility, suggesting that *DYT1* acts as a transcriptional activator (Feng et al., 2012). DYT1 was found to directly promote expression of *TDF1/MYB35* by binding to its promotor region in vivo (Fig. 2A). Expression of *TDF1* under the control of the *DYT1* promoter is able to restore the expression of downstream TFs including *MS188/MYB80* and the PHD-finger TF-encoding gene *MS1* (Fig. 2A). In this *trans*-complementation line, tapetum development was partially restored, although male fertility was not (Gu et al., 2014). Thus, *DYT1* is indirectly required for the expression of *MS188/MYB80* (Feng et al., 2012), the bHLH TF-encoding gene *ABORTED MICROSPORES* (Galluzzi et al., 2018; Sorensen et al., 2003), and *MS1* (Vizcay-Barrena & Wilson, 2006), indicating that *DYT1* is acting upstream of *TDF1/MYB35*, *AMS*, *MS188/MYB80*, and *MS1* (Fig. 2A). AMS is a regulator of tapetal PCD and, as such, important for pollen wall formation, as explained in the previous section (Fig. 2A) (Xu et al., 2010). *AMS* is expressed in the tapetum of *A. thaliana*, and loss of *AMS* function causes premature tapetum cell death (Fig. 2A) (Sorensen et al., 2003). AMS probably regulates lipid transfer proteins, which are involved in pollen coat formation, and E3 ubiquitin ligases to degrade particular target proteins (Li et al., 2017). Consistent with the model of a transcriptional cascade, *AMS* and *MS1* are expressed in the tapetum later than *DYT1* during anther development (Fig. 2A) (Li et al., 2017).

In the tapetum of rice, several bHLH TFs have been functionally characterized, some of them acting homologously to *A. thaliana* tapetum bHLH TFs (Fig. 2B). *TAPETUM DEGENERATION RETARDATION* (*TDR*)/*bHLH5* is a putative *AMS* ortholog, although male sterility in rice *tdr* mutants has been associated with delayed rather than precocious tapetal degeneration (Li et al., 2006). *DYT1* has a putative homolog in rice, *UNDEVELOPED TAPETUM1 (UDT1/bHLH164)*, which causes aberrant tapetum differentiation and male sterility when mutated (Jung et al., 2005). Additionally, *TDR INTERACTING PROTEIN2* (*TIP2/bHLH142*) expression has been found to be essential for timely tapetal PCD and pollen development in rice. *TIP2* deletion causes male sterility by blocking tapetal PCD (Ko et al., 2017). Interestingly, *TIP2* misexpression under a ubiquitous promoter triggers premature tapetum PCD, which equally leads to male sterility (Ko et al., 2017), again highlighting the importance of the accurate control of tapetum PCD for male fertility. TIP2 has been shown to interact with TDR to modulate the expression of yet another bHLH TF, ETERNAL TAPETUM1 (EAT1)/DTD/bHLH141 (Ko et al., 2017). Similar to *tip2* mutants, the *eat1* mutant exhibits delayed tapetal PCD. Notably, the EAT1 TF has been shown to directly regulate the expression of two aspartic proteases, *OsAP37* and *OsAP25*, which in turn are sufficient to execute PCD when misexpressed (Niu et al., 2013). In addition, bHLH142/TIP2 cooperates with TDR through protein-protein interaction to modulate the transcriptional activity of *EAT1* (Fig. 2B) (Ko et al., 2017). Taken together, both in rice and *A. thaliana*, there is a largely conserved transcriptional network to control tapetum differentiation and PCD (Cai et al., 2015).

Similar to the tapetum surrounding and supporting developing microspores, the endosperm tissue in the developing seed supports the growing embryo. During early seed development in *A. thaliana*, the endosperm proliferates vigorously, but later ceases growth and subsequently gets almost completely degraded, providing space and nutrients to the expanding embryo (Ingram, 2010). *RETARDED GROWTH OF EMBRYO1* (*RGE1*)/ *ZHOUPI* (*ZOU*) encodes an endosperm-specific bHLH TF, and *rge1/zou* loss-of-function mutants develop an endosperm that fails to degenerate after cellularization, persisting throughout mid-seed development. As a consequence, embryo growth is severely impeded and mature seeds show a collapsed seed phenotype at maturity (Kondou et al., 2008; Yang et al., 2008). Interestingly, the dwarfed embryos are viable and can develop into healthy plants if germination is supported by the addition of sucrose (Yang et al., 2008). Loss-of-function of the related *INDUCER OF CBP EXPRESSION1* (*ICE1*) causes a similar endosperm persistence phenotype as *rge1/zou* (Denay et al., 2014). *RGE1/ZOU* and *ICE1* are co-expressed in the endosperm and interact in yeast via their bHLH domains, which is important for their role in endosperm breakdown (Denay et al., 2014). Despite the fact that maize has a largely persistent endosperm, knock-down of the single maize *ZOU* homolog, *ZmZOU*, causes a slightly reduced embryo size. Interestingly, not only the embryo-surrounding region, but also the embryonic suspensor show retarded breakdown, suggesting partly conserved and partly divergent roles of *ZOU* in regulating tissue degeneration in monocots and dicots (Grimault et al., 2015).

Endosperm degradation in *A. thaliana* is characterized by expression of dPCD-associated genes (Olvera-Carrillo et al., 2015), and of these at least *PASPA3* and *BFN1* have been shown to be strongly downregulated in *rge1/zou* mutants (Fourquin et al., 2016). Quantification of the mechanical properties of developing seeds revealed an increased apparent stiffness, indicating an increased turgor of the persistent endosperm (Fourquin et al., 2016). Furthermore, endosperm cell wall breakdown was reduced, suggesting that *RGE1/ZOU* primarily promotes cell wall degradation. Cell wall breakdown would allow the growing embryo to exert mechanical stress on endosperm cells (Fourquin et al., 2016). The questions as to whether this mechanical stress simply causes a passive physical disruption of endosperm cells, thus representing a form of accidental cell death, or whether embryo growth and development provide mechanical or other cues to cause an active endosperm PCD process remain to be elucidated.

5. Other transcription factor families involved in PCD

Besides TFs of the large NAC, MYB, and bHLH classes, members of several other TF families have been implicated in dPCD control.

During xylem vessel differentiation, TFs of the LBD/ASL, GATA, and class III HD-Zip families are acting as upstream regulators of *VND7* (Fig. 1) (Soyano et al., 2008). LBD15/ASL11, LBD30/ASL19, GATA5, and GATA12 are able to induce *VND7* promoter activity in a dual luciferase (LUC) assay (Endo et al., 2015). In *35S::LBD15-SRDX* expressing cotyledons, vein formation was incomplete causing a discontinuity of the xylem network (Ohashi-Ito, Iwamoto, & Fukuda, 2018). Inducible *LBD30* expression triggered trans-differentiation of cells from nonvascular tissues into xylem vessel-like cells (Soyano et al., 2008). Similar cells were formed in transgenic plants overexpressing *GATA12*, but not *GATA5* (Endo et al., 2015). Still, *pGATA5::YFP-NLS* and *pGATA12::YFP-NLS* signals were overlapping with those of *pVND7::YFP-NLS* in the differentiation zone of the root stele (Endo et al., 2015).

As mentioned in Sections 3 and 4, *MS1* encodes a PHD-finger TF and is regulated by MS188/MYB80 in the *A. thaliana* tapetal PCD network (Fig. 2A) (Vizcay-Barrena & Wilson, 2006). *PTC1*, the rice homolog of *A. thaliana MS1* (Li et al., 2011), is also involved in tapetal PCD (Fig. 2B). The PTC1 TF controls tapetal PCD and pollen development acting downstream of GAMYB and bHLH142 (Fig. 2B) (Li et al., 2011). The rice *ptc1* mutant shows phenotypic similarity to *A. thaliana ms1* lacking tapetal DNA fragmentation, delayed tapetal degeneration, abnormal pollen wall formation, and aborted pollen development (Li et al., 2011). PTC1 together with TDR regulates the expression of the PCD-promoting *OsCP1* gene, encoding a cysteine protease involved in protein degradation (Li et al., 2011, 2006). The *oscp1* mutant confers pollen degradation as a result of functional defects in the tapetum (Lee, Jung, An, & Chung, 2004).

During fertilization, both the pollen tube and the receptive synergid degenerate. While *MYB97*, *MYB101*, and *MYB120* regulate this process on the pollen tube side, a heterodimer of the REPRODUCTIVE MERISTEM (REM) TFs VALKYRIE (Galluzzi et al., 2018) and VERDANDI (VDD) has been implicated in controlling these cell death processes on the female side in the receptive synergid cell (Mendes et al., 2016). Although *vdd-1* and *val* loss-of-function ovules are still able to attract pollen tubes, their synergid cells do not degenerate, and the attracted pollen tube does not rupture to release its sperm cells (Mendes et al., 2016).

After fertilization, the nucellus degenerates in wild-type seeds of *A. thaliana*, likely representing yet another dPCD process (Xu et al., 2016). The *A. thaliana* MADS-domain TF-encoding gene *AGAMOUS-LIKE62* (*AGL62*) is expressed specifically in the endosperm and initiates PCD signaling in the nucellus. *AGL62* acts through inhibiting the function of *Polycomb* group proteins, which repress nucellus degeneration before fertilization (Xu et al., 2016). Further downstream in this pathway, another two MADS-domain TFs, TRANSPARENT TESTA16 (TT16) and GORDITA (GOA), redundantly promote nucellus PCD. In the *tt16-1* mutant, and even more so in the *tt16 goa* double mutant, nucellar degradation is delayed (Xu et al., 2016). In rice, the MADS-domain TF MADS29 has been shown to promote nucellus PCD, possibly by regulating the expression of cysteine proteases (Yin & Xue, 2012).

6. Conclusions

Complex transcriptional networks coordinate cellular specification and differentiation. For several plant cell types, including xylem vessels, root cap cells, the tapetum layer, or the endosperm, degenerative processes represent the ultimate step of cellular differentiation. Not surprisingly, cell death events have to be tightly controlled in both the spatial and the temporal dimension, as demonstrated by severe developmental defects in mutants affected in PCD processes.

In the developmental context, different forms of PCD are intricately interconnected with earlier steps of the cellular differentiation process. Interestingly, to date there are only few TFs known to exclusively regulate cell death; rather, most TFs regulate several features of cellular differentiation with PCD as its final stage.

Although our understanding of transcriptional PCD regulation remains fragmented, evidence so far suggests that there is a transcriptionally controlled core machinery of developmental PCD in plants. Intriguingly, there seems to be a gradient from tissue-specific transcriptional regulation to process-specific gene expression patterns (Fig. 3). Most TFs identified to date are tissue-specific, for instance VNDs in the xylem, SMB in the root cap, or DYT1 in the tapetum. These TFs likely act as master differentiation regulators of specific cell types, integrating developmental signals to ensure their proper cellular differentiation, of which dPCD represents the ultimate step. Downstream of these key TFs exist complex gene regulatory networks and TF cascades. Some of the TFs in these networks are less tissue-specific and are commonly expressed in several cell types preparing for dPCD.

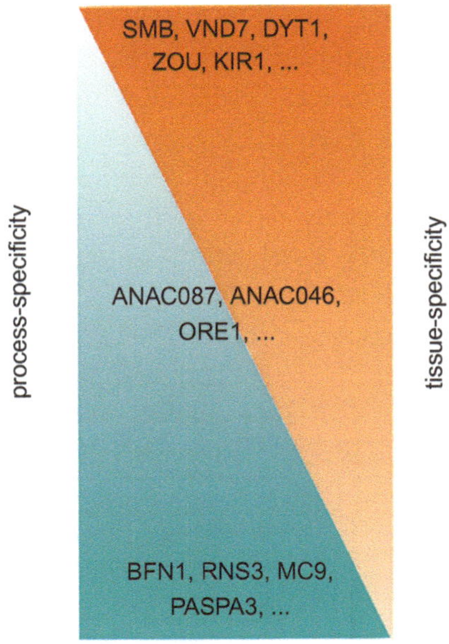

Fig. 3 PCD gene regulatory networks show a trade-off between tissue- and process-specificity. Many upstream TFs are tissue- or cell type-specific (e.g., SMB in the root cap; VND7 in protoxylem; DYT1 in tapetum differentiation; RGE1/ZOU in endosperm degeneration; KIR1 in the degenerating floral stigma) while a few intermediate TFs are commonly regulated in several cell types preparing for PCD (e.g., *ANAC087* and *ANAC046* are commonly expressed in xylem and root cap cells; *ORE1* is expressed in stigma and senescent leaves). PCD-associated downstream target genes (e.g., *BFN1*, *RNS3*, *PASPA3*, *MC9*, and others) are not tissue-specific, but commonly expressed in many cell types or tissues undergoing dPCD.

Although we still know few such intermediate TFs, recent examples include ANAC087 and ANAC046, which are expressed in both the root cap and xylem vessels, and ORE1 that regulates gene expression in both senescent leaves and floral organs. Finally, a plethora of genes, most of them encoding hydrolytic enzymes like BFN1, RNS3 and PASPA3, has been found to be commonly expressed prior to almost all dPCD processes, independent of the tissue context (Fig. 3). These dPCD-associated genes thus tend to be not specific for a certain cell type, but rather for the particular process of dPCD.

Dedicated approaches combining both genetics and network modeling approaches are required to unravel the complex transcriptional networks that ensure the failsafe integration of PCD processes in the differentiation programs of specific plant cells, tissues, or organs. Although transcriptional modulation undoubtedly represents only one facet of PCD control in the developmental context, knowledge about the TFs and their target genes will be an important way forward for a comprehensive understanding of PCD regulation in plants.

Acknowledgments

The authors want to thank members of the VIB Programmed Cell Death laboratory for critical feedback on this manuscript, and Annick Bleys for help in preparing it. We gratefully acknowledge funding by the ERC StG PROCELLDEATH (Project Number: 639234) to M.K.N.

References

Alonso-Peral, M. M., Li, J., Li, Y., Allen, R. S., Schnippenkoetter, W., Ohms, S., et al. (2010). The microRNA159-regulated *GAMYB-like* genes inhibit growth and promote programmed cell death in Arabidopsis. *Plant Physiology*, *154*, 757–771.

Ambawat, S., Sharma, P., Yadav, N. R., & Yadav, R. C. (2013). MYB transcription factor genes as regulators for plant responses: An overview. *Physiology and Molecular Biology of Plants*, *19*, 307–321.

Aubrey, B. J., Kelly, G. L., Janic, A., Herold, M. J., & Strasser, A. (2017). How does p53 induce apoptosis and how does this relate to p53-mediated tumour suppression? *Cell Death and Differentiation*, *25*, 104–113.

Avci, U., Petzold, H. E., Ismail, I. O., Beers, E. P., & Haigler, C. H. (2008). Cysteine proteases XCP1 and XCP2 aid micro-autolysis within the intact central vacuole during xylogenesis in Arabidopsis roots. *Plant Journal*, *56*, 303–315.

Aya, K., Ueguchi-Tanaka, M., Kondo, M., Hamada, K., Yano, K., Nishimura, M., et al. (2009). Gibberellin modulates anther development in rice via the transcriptional regulation of GAMYB. *Plant Cell*, *21*, 1453–1472.

Balazadeh, S., Kwasniewski, M., Caldana, C., Mehrnia, M., Zanor, M. I., Xue, G.-P., et al. (2011). ORS1, an H_2O_2-responsive NAC transcription factor, controls senescence in *Arabidopsis thaliana*. *Molecular Plant*, *4*, 346–360.

Bariola, P. A., Howard, C. J., Taylor, C. B., Verburg, M. T., Jaglan, V. D., & Green, P. J. (1994). The *Arabidopsis* ribonuclease gene *RNS1* is tightly controlled in response to phosphate limitation. *Plant Journal*, *6*, 673–685.

Bennett, T., van den Toorn, A., Sanchez-Perez, G. F., Campilho, A., Willemsen, V., Snel, B., et al. (2010). SOMBRERO, BEARSKIN1, and BEARSKIN2 regulate root cap maturation in *Arabidopsis*. *Plant Cell*, *22*, 640–654.

Bollhöner, B., Zhang, B., Stael, S., Denancé, N., Overmyer, K., Goffner, D., et al. (2013). *Post mortem* function of AtMC9 in xylem vessel elements. *New Phytologist*, *200*, 498–510.

Cai, C.-F., Zhu, J., Lou, Y., Guo, Z.-L., Xiong, S.-X., Wang, K., et al. (2015). The functional analysis of *OsTDF1* reveals a conserved genetic pathway for tapetal development between rice and Arabidopsis. *Science Bulletin*, *60*, 1073–1082.

Chen, Q., Niu, F., Yan, J., Chen, B., Wu, F., Guo, X., et al. (2017). Oilseed rape NAC56 transcription factor modulates reactive oxygen species accumulation and hypersensitive response-like cell death. *Physiologia Plantarum*, *160*, 209–221.

Daneva, A., Gao, Z., Van Durme, M., & Nowack, M. K. (2016). Functions and regulation of programmed cell death in plant development. *Annual Review of Cell and Developmental Biology*, *32*, 441–468.

Demura, T., Tashiro, G., Horiguchi, G., Kishimoto, N., Kubo, M., Matsuoka, N., et al. (2002). Visualization by comprehensive microarray analysis of gene expression programs during transdifferentiation of mesophyll cells into xylem cells. *Proceedings of the National Academy of Sciences of the United States of America, 99*, 15794–15799.

Denay, G., Creff, A., Moussu, S., Wagnon, P., Thévenin, J., Gérentes, M.-F., et al. (2014). Endosperm breakdown in *Arabidopsis* requires heterodimers of the basic helix-loop-helix proteins ZHOUPI and INDUCER OF CBP EXPRESSION 1. *Development, 141*, 1222–1227.

Dresselhaus, T., Sprunck, S., & Wessel, G. M. (2016). Fertilization mechanisms in flowering plants. *Current Biology, 26*, R125–R139.

Dubos, C., Stracke, R., Grotewold, E., Weisshaar, B., Martin, C., & Lepiniec, L. (2010). MYB transcription factors in *Arabidopsis*. *Trends in Plant Science, 15*, 573–581.

Endo, H., Yamaguchi, M., Tamura, T., Nakano, Y., Nishikubo, N., Yoneda, A., et al. (2015). Multiple classes of transcription factors regulate the expression of *VASCULAR-RELATED NAC-DOMAIN7*, a master switch of xylem vessel differentiation. *Plant & Cell Physiology, 56*, 242–254.

Escamez, S., André, D., Zhang, B., Bollhöner, B., Pesquet, E., & Tuominen, H. (2016). METACASPASE9 modulates autophagy to confine cell death to the target cells during *Arabidopsis* vascular xylem differentiation. *Biology Open, 5*, 122–129.

Escamez, S., & Tuominen, H. (2014). Programmes of cell death and autolysis in tracheary elements: When a suicidal cell arranges its own corpse removal. *Journal of Experimental Botany, 65*, 1313–1321.

Fan, K., Li, F., Chen, J., Li, Z., Lin, W., Cai, S., et al. (2018). Asymmetric evolution and expansion of the NAC transcription factor in polyploidized cotton. *Frontiers in Plant Science, 9*, 47.

Feller, A., Machemer, K., Braun, E. L., & Grotewold, E. (2011). Evolutionary and comparative analysis of MYB and bHLH plant transcription factors. *Plant Journal, 66*, 94–116.

Fendrych, M., Van Hautegem, T., Van Durme, M., Olvera-Carrillo, Y., Huysmans, M., Karimi, M., et al. (2014). Programmed cell death controlled by ANAC033/SOMBRERO determines root cap organ size in *Arabidopsis*. *Current Biology, 24*, 931–940.

Feng, B., Lu, D., Ma, X., Peng, Y., Sun, Y., Ning, G., et al. (2012). Regulation of the Arabidopsis anther transcriptome by DYT1 for pollen development. *Plant Journal, 72*, 612–624.

Fourquin, C., Beauzamy, L., Chamot, S., Creff, A., Goodrich, J., Boudaoud, A., et al. (2016). Mechanical stress mediated by both endosperm softening and embryo growth underlies endosperm elimination in *Arabidopsis* seeds. *Development, 143*, 3300–3305.

Fuchs, Y., & Steller, H. (2015). Live to die another way: Modes of programmed cell death and the signals emanating from dying cells. *Nature Reviews Molecular Cell Biology, 16*, 329–344.

Fukuda, H. (2004). Signals that control plant vascular cell differentiation. *Nature Reviews Molecular Cell Biology, 5*, 379–391.

Furuta, K. M., Yadav, S. R., Lehesranta, S., Belevich, I., Miyashima, S., Heo, J.-O., et al. (2014). *Arabidopsis* NAC45/86 direct sieve element morphogenesis culminating in enucleation. *Science, 345*, 933–937.

Galluzzi, L., Vitale, I., Aaronson, S. A., Abrams, J. M., Adam, D., Agostinis, P., et al. (2018). Molecular mechanisms of cell death: Recommendations of the Nomenclature Committee on Cell Death 2018. *Cell Death and Differentiation, 25*, 486–541.

Gao, Z., Daneva, A., Salanenka, Y., Van Durme, M., Huysmans, M., Lin, Z., et al. (2018). KIRA1 and ORESARA1 terminate flower receptivity by promoting cell death in the stigma of *Arabidopsis*. *Nature Plants, 4*, 365–375.

Geldner, N. (2014). Making phloem—A near-death experience. *Science, 345*, 875–876.

Gómez, J. F., Talle, B., & Wilson, Z. A. (2015). Anther and pollen development: A conserved developmental pathway. *Journal of Integrative Plant Biology, 57*, 876–891.

Goossens, J., Mertens, J., & Goossens, A. (2017). Role and functioning of bHLH transcription factors in jasmonate signalling. *Journal of Experimental Botany*, *68*, 1333–1347.
Grimault, A., Gendrot, G., Chamot, S., Widiez, T., Rabillé, H., Gérentes, M.-F., et al. (2015). ZmZHOUPI, an endosperm-specific basic helix–loop–helix transcription factor involved in maize seed development. *The Plant Journal*, *84*, 574–586.
Gu, J.-N., Zhu, J., Yu, Y., Teng, X.-D., Lou, Y., Xu, X.-F., et al. (2014). DYT1 directly regulates the expression of *TDF1* for tapetum development and pollen wall formation in Arabidopsis. *Plant Journal*, *80*, 1005–1013.
He, D., Wang, Q., Wang, K., & Yang, P. (2015). Genome-wide dissection of the microRNA expression profile in rice embryo during early stages of seed germination. *PLoS One*, *10*, e0145424.
Heo, J.-O., Blob, B., & Helariutta, Y. (2017). Differentiation of conductive cells: A matter of life and death. *Current Opinion in Plant Biology*, *35*, 23–29.
Hickman, R., Hill, C., Penfold, C. A., Breeze, E., Bowden, L., Moore, J. D., et al. (2013). A local regulatory network around three NAC transcription factors in stress responses and senescence in Arabidopsis leaves. *Plant Journal*, *75*, 26–39.
Huysmans, M., Skorzinski, N., Buono, R. A., Cubría-Radío, M., De Winter, F., Parizot, B., et al. (2018). ANAC087 and ANAC046 control distinct aspects of programmed cell death in the Arabidopsis columella and lateral root cap. *Plant Cell*, *30*(9), 2197–2213. https://doi.org/10.1105/tpc.18.00293.
Ingram, G. C. (2010). Family life at close quarters: Communication and constraint in angiosperm seed development. *Protoplasma*, *247*, 195–214.
Ito, J., & Fukuda, H. (2002). ZEN1 is a key enzyme in the degradation of nuclear DNA during programmed cell death of tracheary elements. *Plant Cell*, *14*, 3201–3211.
Jung, K.-H., Han, M.-J., Lee, Y.-S., Kim, Y.-W., Hwang, I., Kim, M.-J., et al. (2005). Rice *Undeveloped Tapetum1* is a major regulator of early tapetum development. *Plant Cell*, *17*, 2705–2722.
Kim, H. J., Hong, S. H., Kim, Y. W., Lee, I. H., Jun, J. H., Phee, B.-K., et al. (2014). Gene regulatory cascade of senescence-associated NAC transcription factors activated by ETHYLENE-INSENSITIVE2-mediated leaf senescence signalling in *Arabidopsis*. *Journal of Experimental Botany*, *65*, 4023–4036.
Ko, S.-S., Li, M.-J., Lin, Y.-J., Hsing, H.-X., Yang, T.-T., Chen, T.-K., et al. (2017). Tightly controlled expression of *bHLH142* is essential for timely tapetal programmed cell death and pollen development in rice. *Frontiers in Plant Science*, *8*, 1258.
Kondou, Y., Nakazawa, M., Kawashima, M., Ichikawa, T., Yoshizumi, T., Suzuki, K., et al. (2008). RETARDED GROWTH OF EMBRYO1, a new basic helix-loop-helix protein, expresses in endosperm to control embryo growth. *Plant Physiology*, *147*, 1924–1935.
Koyama, T. (2014). The roles of ethylene and transcription factors in the regulation of onset of leaf senescence. *Frontiers in Plant Science*, *5*, 650.
Kubo, M., Udagawa, M., Nishikubo, N., Horiguchi, G., Yamaguchi, M., Ito, J., et al. (2005). Transcription switches for protoxylem and metaxylem vessel formation. *Genes & Development*, *19*, 1855–1860.
Kumpf, R. P., & Nowack, M. K. (2015). The root cap: A short story of life and death. *Journal of Experimental Botany*, *66*, 5651–5662.
Lee, S., Jung, K.-H., An, G. H., & Chung, Y.-Y. (2004). Isolation and characterization of a rice cysteine protease gene, *OSCP1*, using T-DNA gene-trap system. *Plant Molecular Biology*, *54*, 755–765.
Leydon, A. R., Beale, K. M., Woroniecka, K., Castner, E., Chen, J., Horgan, C., et al. (2013). Three MYB transcription factors control pollen tube differentiation required for sperm release. *Current Biology*, *23*, 1209–1214.

Li, D.-D., Xue, J.-S., Zhu, J., & Yang, Z.-N. (2017). Gene regulatory network for tapetum development in *Arabidopsis thaliana*. *Frontiers in Plant Science*, *8*, 1559.

Li, H., Yuan, Z., Vizcay-Barrena, G., Yang, C., Liang, W., Zong, J., et al. (2011). *PERSISTENT TAPETAL CELL1* encodes a PHD-finger protein that is required for tapetal cell death and pollen development in rice. *Plant Physiology*, *156*, 615–630.

Li, N., Zhang, D.-S., Liu, H.-S., Yin, C.-S., Li, X.-x., Liang, W.-q., et al. (2006). The rice *tapetum degeneration retardation* gene is required for tapetum degradation and anther development. *Plant Cell*, *18*, 2999–3014.

Liang, Y., Tan, Z.-M., Zhu, L., Niu, Q.-K., Zhou, J.-J., Li, M., et al. (2013). MYB97, MYB101 and MYB120 function as male factors that control pollen tube-synergid interaction in *Arabidopsis thaliana* fertilization. *PLoS Genetics*, *9*, e1003933.

Liu, L., & Fan, X. D. (2013). Tapetum: Regulation and role in sporopollenin biosynthesis in *Arabidopsis*. *Plant Molecular Biology*, *83*, 165–175.

Mendes, M. A., Guerra, R. F., Castelnovo, B., Silva-Velazquez, Y., Morandini, P., Manrique, S., et al. (2016). Live and let die: A REM complex promotes fertilization through synergid cell death in *Arabidopsis*. *Development*, *143*, 2780–2790.

Ngo, Q. A., Vogler, H., Lituiev, D. S., Nestorova, A., & Grossniklaus, U. (2014). A calcium dialog mediated by the *FERONIA* signal transduction pathway controls plant sperm delivery. *Developmental Cell*, *29*, 491–500.

Niu, N., Liang, W., Yang, X., Jin, W., Wilson, Z. A., Hu, J., et al. (2013). EAT1 promotes tapetal cell death by regulating aspartic proteases during male reproductive development in rice. *Nature Communications*, *4*, 1445.

Obara, K., Kuriyama, H., & Fukuda, H. (2001). Direct evidence of active and rapid nuclear degradation triggered by vacuole rupture during programmed cell death in Zinnia. *Plant Physiology*, *125*, 615–626.

Ohashi-Ito, K., & Fukuda, H. (2016). Functional mechanism of bHLH complexes during early vascular development. *Current Opinion in Plant Biology*, *33*, 42–47.

Ohashi-Ito, K., Iwamoto, K., & Fukuda, H. (2018). LOB DOMAIN-CONTAINING PROTEIN 15 positively regulates expression of *VND7*, a master regulator of tracheary elements. *Plant & Cell Physiology*, *59*, 989–996.

Ohtani, M., Akiyoshi, N., Takenaka, Y., Sano, R., & Demura, T. (2017). Evolution of plant conducting cells: Perspectives from key regulators of vascular cell differentiation. *Journal of Experimental Botany*, *68*, 17–26.

Olivier, M., Eeles, R., Hollstein, M., Khan, M. A., Harris, C. C., & Hainaut, P. (2002). The IARC TP53 database: New online mutation analysis and recommendations to users. *Human Mutation*, *19*, 607–614.

Olvera-Carrillo, Y., Van Bel, M., Van Hautegem, T., Fendrych, M., Huysmans, M., Simaskova, M., et al. (2015). A conserved core of programmed cell death indicator genes discriminates developmentally and environmentally induced programmed cell death in plants. *Plant Physiology*, *169*, 2684–2699.

Phan, H. A., Iacuone, S., Li, S. F., & Parish, R. W. (2011). The MYB80 transcription factor is required for pollen development and the regulation of tapetal programmed cell death in *Arabidopsis thaliana*. *Plant Cell*, *23*, 2209–2224.

Phan, H. A., Li, S. F., & Parish, R. W. (2012). MYB80, a regulator of tapetal and pollen development, is functionally conserved in crops. *Plant Molecular Biology Reporter*, *78*, 171–183.

Plackett, A. R. G., Thomas, S. G., Wilson, Z. A., & Hedden, P. (2011). Gibberellin control of stamen development: A fertile field. *Trends in Plant Science*, *16*, 568–578.

Polager, S., & Ginsberg, D. (2009). p53 and E2f: Partners in life and death. *Nature Reviews Cancer*, *9*, 738–748.

Ran, J.-H., Shen, T.-T., Liu, W.-J., & Wang, X.-Q. (2013). Evolution of the bHLH genes involved in stomatal development: Implications for the expansion of developmental complexity of stomata in land plants. *PLoS One*, *8*, e78997.

Riechmann, J. L., Heard, J., Martin, G., Reuber, L., Jiang, C.-Z., Keddie, J., et al. (2000). *Arabidopsis* transcription factors: Genome-wide comparative analysis among eukaryotes. *Science, 290*, 2105–2110.

Rogers, H. J. (2013). From models to ornamentals: How is flower senescence regulated? *Plant Molecular Biology, 82*, 563–574.

Shibuya, K. (2018). Molecular aspects of flower senescence and strategies to improve flower longevity. *Breeding Science, 68*, 99–108.

Sorensen, A.-M., Kröber, S., Unte, U. S., Huijser, P., Dekker, K., & Saedler, H. (2003). The Arabidopsis *ABORTED MICROSPORES* (*AMS*) gene encodes a MYC class transcription factor. *Plant Journal, 33*, 413–423.

Soyano, T., Thitamadee, S., Machida, Y., & Chua, N.-H. (2008). *ASYMMETRIC LEAVES2-LIKE19/LATERAL ORGAN BOUNDARIES DOMAIN30* and *ASL20/LBD18* regulate tracheary element differentiation in *Arabidopsis*. *Plant Cell, 20*, 3359–3373.

Tan, T. T., Endo, H., Sano, R., Kurata, T., Yamaguchi, M., Ohtani, M., et al. (2018). Transcription factors VND1-VND3 contribute to cotyledon xylem vessel formation. *Plant Physiology, 176*, 773–789.

Thellmann, M., Hatzold, J., & Conradt, B. (2003). The Snail-like CES-1 protein of *C. elegans* can block the expression of the BH3-only cell-death activator gene *egl-1* by antagonizing the function of bHLH proteins. *Development, 130*, 4057–4071.

van Doorn, W. G. (2011). Classes of programmed cell death in plants, compared to those in animals. *Journal of Experimental Botany, 62*, 4749–4761.

Van Durme, M., & Nowack, M. K. (2016). Mechanisms of developmentally controlled cell death in plants. *Current Opinion in Plant Biology, 29*, 29–37.

Vizcay-Barrena, G., & Wilson, Z. A. (2006). Altered tapetal PCD and pollen wall development in the *Arabidopsis ms1* mutant. *Journal of Experimental Botany, 57*, 2709–2717.

White, K., Tahaoglu, E., & Steller, H. (1996). Cell killing by the *Drosophila* gene *reaper*. *Science, 271*, 805–807.

Willemsen, V., Bauch, M., Bennett, T., Campilho, A., Wolkenfelt, H., Xu, J., et al. (2008). The NAC domain transcription factors FEZ and SOMBRERO control the orientation of cell division plane in *Arabidopsis* root stem cells. *Developmental Cell, 15*, 913–922.

Xie, H.-T., Wan, Z.-Y., Li, S., & Zhang, Y. (2014). Spatiotemporal production of reactive oxygen species by NADPH oxidase is critical for tapetal programmed cell death and pollen development in *Arabidopsis*. *Plant Cell, 26*, 2007–2023.

Xu, W., Fiume, E., Coen, O., Pechoux, C., Lepiniec, L., & Magnani, E. (2016). Endosperm and nucellus develop antagonistically in Arabidopsis seeds. *Plant Cell, 28*, 1343–1360.

Xu, J., Yang, C., Yuan, Z., Zhang, D., Gondwe, M. Y., Ding, Z., et al. (2010). The *ABORTED MICROSPORES* regulatory network is required for postmeiotic male reproductive development in *Arabidopsis thaliana*. *Plant Cell, 22*, 91–107.

Yamaguchi, M., Ohtani, M., Mitsuda, N., Kubo, M., Ohme-Takagi, M., Fukuda, H., et al. (2010). VND-INTERACTING2, a NAC domain transcription factor, negatively regulates xylem vessel formation in *Arabidopsis*. *Plant Cell, 22*, 1249–1263.

Yan, J., Tong, T., Li, X., Chen, Q., Dai, M., Niu, F., et al. (2018). A novel NAC-type transcription factor, NAC87, from oilseed rape modulates reactive oxygen species accumulation and cell death. *Plant & Cell Physiology, 59*, 290–303.

Yang, S., Johnston, N., Talideh, E., Mitchell, S., Jeffree, C., Goodrich, J., et al. (2008). The endosperm-specific *ZHOUPI* gene of *Arabidopsis thaliana* regulates endosperm breakdown and embryonic epidermal development. *Development, 135*, 3501–3509.

Yin, L.-L., & Xue, H.-W. (2012). The *MADS29* transcription factor regulates the degradation of the nucellus and the nucellar projection during rice seed development. *Plant Cell, 24*, 1049–1065.

Zhang, D., Liu, D., Lv, X., Wang, Y., Xun, Z., Liu, Z., et al. (2014). The cysteine protease CEP1, a key executor involved in tapetal programmed cell death, regulates pollen development in *Arabidopsis*. *Plant Cell*, *26*, 2939–2961.

Zhang, W., Sun, Y., Timofejeva, L., Chen, C., Grossniklaus, U., & Ma, H. (2006). Regulation of *Arabidopsis* tapetum development and function by *DYSFUNCTIONAL TAPETUM1* (*DYT1*) encoding a putative bHLH transcription factor. *Development*, *133*, 3085–3095.

Zhang, Z.-B., Zhu, J., Gao, J.-F., Wang, C., Li, H., Li, H., et al. (2007). Transcription factor *AtMYB103* is required for anther development by regulating tapetum development, callose dissolution and exine formation in Arabidopsis. *Plant Journal*, *52*, 528–538.

Zhao, C., Avci, U., Grant, E. H., Haigler, C. H., & Beers, E. P. (2008). XND1, a member of the NAC domain family in *Arabidopsis thaliana*, negatively regulates lignocellulose synthesis and programmed cell death in xylem. *Plant Journal*, *53*, 425–436.

Zhao, C., Lasses, T., Bako, L., Kong, D., Zhao, B., Chanda, B., et al. (2017). XYLEM NAC DOMAIN1, an angiosperm NAC transcription factor, inhibits xylem differentiation through conserved motifs that interact with RETINOBLASTOMA-RELATED. *New Phytologist*, *216*, 76–89.

Zhong, R., Lee, C., & Ye, Z.-H. (2010). Global analysis of direct targets of secondary wall NAC master switches in *Arabidopsis*. *Molecular Plant*, *3*, 1087–1103.

Zhou, L.-Z., & Dresselhaus, T. (2019). Friend or foe: Signaling mechanisms during double fertilization in flowering seed plants. In U. Grossniklaus (Ed.), *Current topics in developmental biology: Vol. 131*. Plant development and evolution (pp. 453–496). Cambridge, United States: Academic Press.

Zhu, J., Chen, H., Li, H., Gao, J.-F., Jiang, H., Wang, C., et al. (2008). *Defective in Tapetal development and function 1* is essential for anther development and tapetal function for microspore maturation in Arabidopsis. *Plant Journal*, *55*, 266–277.

CHAPTER EIGHT

Molecular regulation of flower development

Bennett Thomson, Frank Wellmer*
Smurfit Institute of Genetics, Trinity College Dublin, Dublin, Ireland
*Corresponding author: e-mail address: wellmerf@tcd.ie

Contents

Abstract

Over the past three decades, several hundred genes with important regulatory functions during reproductive development in angiosperms have been identified. While we do not yet know, in most cases, how these genes and their products act, fundamental insights into the molecular mechanisms underlying the formation of flowers have been obtained in recent years. These advances were made possible to a large extent by studying the functions of master regulators of flower development through a multitude of experimental approaches, ranging from basic genetic analysis to genome-wide surveys. Based on the results of this work, several models for the molecular control of flower formation have been proposed, which have been tested and largely validated. These models have guided and informed research in the field, and facilitated recent efforts to delineate the composition and architecture of the gene regulatory networks underlying flower development. In this chapter, we aim to describe the current state of flowering research with a focus on recent progress in the field. We also discuss open questions that we believe need to be addressed in the future to further our understanding of the regulatory mechanisms that control floral morphogenesis and evolution.

Current Topics in Developmental Biology, Volume 131
ISSN 0070-2153
https://doi.org/10.1016/bs.ctdb.2018.11.007

1. Introduction

Three decades of research into the molecular mechanisms underlying the formation of flowers have yielded a rich harvest. The results of genetic analyses and transcriptomics studies have led to the identification of several hundred genes with known or likely regulatory functions in various aspects of flower development. Many of these genes encode transcription factors and transcriptional co-regulators (for a sample list, see, e.g., Wils & Kaufmann, 2017), microRNAs (miRNAs) (Nag & Jack, 2010), or proteins involved in the epigenetic control of gene expression (Vachon, Engelhorn, & Carles, 2018). The molecular mechanisms, through which these regulators act, are in many cases not yet understood. Nevertheless, much progress has been made in this area in recent years due to major advances in technology, facilitating the use of methods such as genome-wide studies of transcription factor-binding sites (Kaufmann, Muino, et al., 2010), proteomic analyses (Smaczniak, Li, et al., 2012), and various new imaging techniques for flowering research (Prunet, Jack, & Meyerowitz, 2016). The results of comparative analyses suggest that the functions of floral regulators are often (but not always) conserved across distinct angiosperms. Despite the considerable progress in the flowering field over recent years, we are far from a systems-wide understanding of the gene regulatory networks underlying this key developmental process. Furthermore, many open questions remain, especially in relation to floral evolution and the genetic diversity that forms the basis for the dazzling array of floral morphologies found among extant angiosperms.

In this chapter, we will attempt to provide an overview of ongoing research activities in the field and summarize recent progress in the following areas: floral meristem (FM) formation, specification, and termination; floral patterning; and floral organ specification and morphogenesis. We will also discuss open questions related to these topics that we think need to be addressed in the coming years to further advance our understanding of flower development. Given the immense growth of the field, which has seen the introduction of new models for flowering research in recent years (e.g., Damerval & Becker, 2017; Wang et al., 2015), our discussion will inevitably be vastly incomplete, and we apologize in advance to those colleagues whose work we could not discuss here due to space constraints or the specific topics we have chosen for this chapter.

2. Formation and specification of floral meristems

Flowers are formed by FMs, which are initiated in specific positions of the inflorescence meristem (IM) (Smyth, Bowman, & Meyerowitz, 1990). The emergence of FMs is preceded by the formation of local maxima of the phytohormone auxin through polar auxin transport (Bhatia & Heisler, 2018; Heisler et al., 2005) (see Fig. 1A for an overview of the genes and

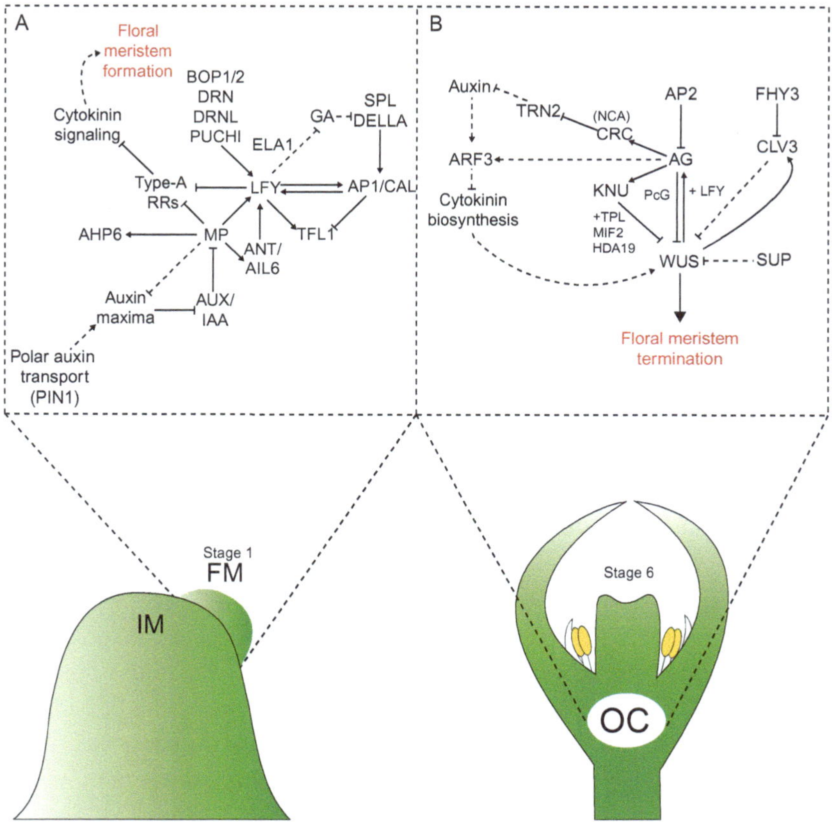

Fig. 1 Genetic control of floral meristem initiation and termination. Schematics representing the gene regulatory networks underlying the formation of the floral meristem (A) from stage 1 (Smyth et al., 1990), and the cessation of stem cell activity at stage 6 (B) within the organizing center (OC). Refer to the main text for detailed descriptions of both events. Solid lines represent direct regulation of target genes and dashed lines represent indirect regulation. IM, inflorescence meristem; FM, floral meristem; PcG, polycomb group proteins; NCA, non-cell autonomous activity; RRs, response regulators.

processes involved in FM emergence and specification). In the model plant *Arabidopsis thaliana*, this process depends on the activity of the auxin efflux carrier PIN-FORMED1 (PIN1) (Galweiler et al., 1998), as demonstrated by *pin1* mutants that do not form flowers but that can be rescued by exogenous auxin treatments (Reinhardt et al., 2003). The auxin maxima in the outer cell layers of the IM lead to the local activation of the AUXIN RESPONSE FACTOR (ARF) MONOPTEROS (MP; also known as ARF5) (Bhatia et al., 2016; Bhatia & Heisler, 2018). MP has been shown to recruit the SWI/SNF ATPase chromatin remodelers SPLAYED (SYD) and BRAHMA (BRM) to target gene promoters and to increase chromatin accessibility (Wu et al., 2015). Among the genes activated through this mechanism are the transcription factor-encoding genes *AINTEGUMENTA* (*ANT*) and *AINTEGUMENTA-LIKE6* (*AIL6*) (Yamaguchi et al., 2013), which promote floral growth (Krizek, 2009). MP also activates expression of *ARABIDOPSIS HISTIDINE PHOSPHOTRANSFER PROTEIN6* (*AHP6*) (Besnard et al., 2014), which encodes a negative regulator of the signaling pathway for the phytohormone cytokinin (Mahonen et al., 2006). AHP6 has been shown to suppresses, through intercellular movement, meristematic cell activity in the tissue surrounding floral primordia and thus to assist in establishing phyllotaxis, the regular pattern of primordia emergence (Besnard et al., 2014).

The control of auxin transport and/or synthesis in FMs has also been linked to the activities of the related DORNRÖSCHEN (DRN), DORNRÖSCHEN-LIKE (DRNL), and PUCHI transcription factors that have been shown to control FM identity and floral organ initiation in a partially redundant manner together with the BLADE-ON-PETIOLE1 (BOP1) and BOP2 proteins (Chandler & Werr, 2017), which are components of an E3 ubiquitin ligase complex (Zhang et al., 2017). These proteins promote the expression of *LEAFY* (*LFY*), a master regulatory gene for FM identity specification (Weigel, Alvarez, Smyth, Yanofsky, & Meyerowitz, 1992). Importantly, *LFY* expression is also activated in an auxin-dependent manner by MP and ANT/AIL6 (Yamaguchi, Jeong, Nole-Wilson, Krizek, & Wagner, 2016; Yamaguchi et al., 2013), and by several flowering time pathways including components of florigen, which mediates flowering in response to seasonal changes in photoperiod (Abe et al., 2005; Wigge et al., 2005).

The plant-specific LFY transcription factor, whose structure has been elucidated in atomic detail (Hames et al., 2008; Sayou et al., 2016) and whose target genes have been identified on a genome-wide scale (Moyroud et al., 2011; Winter et al., 2011), has a multitude of known functions during the earliest stages of flower development. Notably, it downregulates, together or in parallel with MP (Zhao et al., 2010), the expression of type-A response regulators (Chahtane et al., 2013), which negatively regulate cytokinin responses. The resulting local increase in cytokinin signaling is then thought to lead to the establishment of a stem cell population and an organizing center in emerging floral primordia and thus to a fully functional FM (reviewed in Denay, Chahtane, Tichtinsky, & Parcy, 2017). *LFY* has additional links to hormone signaling pathways: it controls auxin responses and hence appears to reinforce the auxin-dependent initiation of floral primordia (Li et al., 2013). It also reduces gibberellin (GA) levels in the FM via up-regulation of the cytochrome P450 GA-catabolic enzyme EUI-LIKE P450 A1 (ELA1). This depletion of GA leads to an accumulation of the GA-sensitive DELLA proteins, which then act together with SQUAMOSA PROMOTER BINDING PROTEIN-LIKE (SPL) transcription factors to activate another key FM identity gene, *APETALA1* (*AP1*) (Yamaguchi et al., 2014, 2009). *AP1* and the paralogous and partially redundant *CAULIFLOWER* (*CAL*) gene, which are also directly activated by LFY (Wagner, Sablowski, & Meyerowitz, 1999; William et al., 2004), encode MADS-domain transcription factors (Kempin, Savidge, & Yanofsky, 1995; Mandel, Gustafson-Brown, Savidge, & Yanofsky, 1992), and thus belong to a family of regulators that plays many essential roles in the reproductive development of plants (Gramzow & Theissen, 2010; Theissen, Melzer, & Rumpler, 2016). The AP1 (and probably also CAL) transcription factor directly promotes *LFY* expression as part of a positive feedback loop to ensure a rapid and strong activation of these FM identity genes (Kaufmann, Wellmer, et al., 2010). The combined activities of these regulators, as well as of several other MADS-domain transcription factors, such as AGAMOUS-LIKE24 (AGL24) and SHORT VEGETATIVE PHASE (SVP) (reviewed in Denay et al., 2017), lead to the specification of FM identity. This is accomplished by the (partial) suppression of the genetic program for shoot development and the activation of a battery of genes with flower-specific expression (Kaufmann, Wellmer, et al., 2010). It appears that this process is mediated to a large extent by AP1/CAL and LFY acting together on a shared set of target genes (Winter, Yamaguchi, Wu, & Wagner, 2015). However, recent evidence suggests that these transcription factors also have independent and even partially antagonistic activities. Notably, it has been shown that the shoot identity gene *TERMINAL FLOWER1*

(*TFL1*) (Bradley, Ratcliffe, Vincent, Carpenter, & Coen, 1997), which needs to be suppressed so that flower formation can commence (Baumann et al., 2015), is repressed by AP1 but activated by LFY (Denay et al., 2017; Goslin et al., 2017; Serrano-Mislata et al., 2017). It has been proposed that this antagonistic activity of AP1/CAL and LFY may ensure that flower formation commences only when the levels of these regulators are sufficiently high to enable the stable running of the developmental program for flower formation (Goslin et al., 2017).

3. Floral patterning

FM identity regulators such as LFY and AP1/CAL play essential roles in the patterning of FMs because they control the expression of genes needed for the specification of floral organ primordia, which are initiated either in a spiral (especially in basal angiosperms) or a whorled arrangement. These floral organ identity genes were discovered through the characterization of homeotic mutants that exhibited floral organ transformations (e.g., Bowman, Smyth, & Meyerowitz, 1989). Analyses of the genetic interactions between the different mutants led to the formulation of the ABC model, which is based on the notion that the floral organ identity genes act in a combinatorial manner to control floral organ specification (Coen & Meyerowitz, 1991). Over the past 30 years, the characterization and functional study of these genes, especially in *Antirrhinum majus* (snapdragon) and *A. thaliana*, but also in several other species such as *Petunia hybrida* (petunia) and *Oryza sativa* (rice), led to fundamental insights into the regulatory processes underlying flower development. The history and the main findings of this seminal work are well documented in the recent literature (Bowman, Smyth, & Meyerowitz, 2012; Causier, Schwarz-Sommer, & Davies, 2010; O'Maoileidigh, Graciet, & Wellmer, 2014; Pajoro, Biewers, et al., 2014; Stewart, Graciet, & Wellmer, 2016) and we will therefore only briefly introduce fundamental concepts in this review.

The original ABC model posits that A function genes specify sepals, and together with B function genes, petals, while B and C function activities together lead to stamen development, and C function alone controls carpel formation (Coen & Meyerowitz, 1991) (Fig. 2). An additional principle of the model is that A and C function genes act in an antagonistic manner. This model was later extended to the ABCDE model, with D function genes controlling ovule development and E function genes being involved in the specification of all types of floral organs (Krizek & Fletcher, 2005)

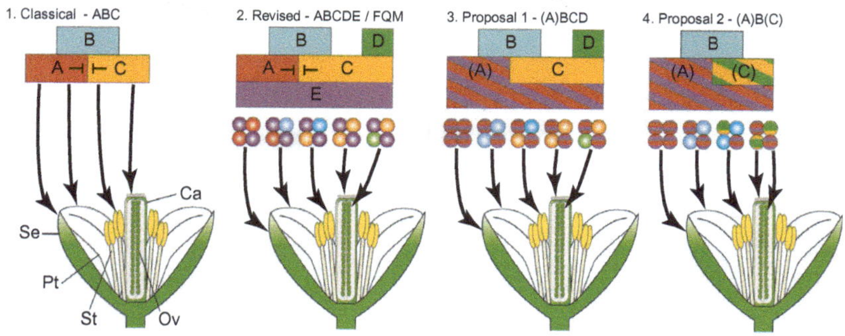

Fig. 2 Proposed models for flower development in *Arabidopsis thaliana*. Models describing the combinatorial activities of different classes of homeotic genes, which function in specific whorls of the flower to specify floral organ identity. Refer to the main texts for details of each of the models. Arrows signify the organs/whorls in which the different classes of genes are active. For models 2–4, tetrameric complexes are colored in accordance with the classes to which the proteins belong. For models 3 and 4, "(A)" represents the consolidation of (A) and (E) class genes into a single class and "(C)" represents the consolidation of (C) and (D) class genes into a single class. FQM, floral quartet model; Se, sepals; Pt, petals; St, stamens; Ca, carpel; Ov, ovules.

(Fig. 2). Because A function mutants resembling those found in *A. thaliana* could not be identified in most other species, and A and E function genes were found to be also involved in specifying FMs, it was proposed that their combined activities (as well as the activities of other regulators that control FM fate) define the ground state of a floral organ, the sepal (Causier et al., 2010). The resulting (A)BCD model (with (A) incorporating both A and E function; Fig. 2) is more widely applicable across diverse angiosperms to explain floral organ identities (e.g., Wu et al., 2017). Recently, this model was modified further into an (A)B(C) model (with (C) being C+D; Fig. 2) based on the notion that C and D genes are closely related and are likely derived from genes that had combined C and D function activities (Theissen et al., 2016). In agreement with this idea, C and D function genes in petunia have been found to have largely overlapping activities (Heijmans et al., 2012), while in *A. thaliana* these gene activities are more functionally diverse (Pinyopich et al., 2003). Thus, the flowering field has gone almost full circle from ABC to (A)B(C) over the course of almost 30 years.

The expression of the floral organ identity genes, which all encode MIKC-type MADS-domain transcription factors (with an N-terminal DNA-binding MADS (M) domain, an intervening (I) domain, a keratin-like (K) domain, and a C-terminal (C) domain as common domain structure; Gramzow & Theissen, 2010), occurs mainly in the organs they help to specify (summarized in O'Maoileidigh et al., 2014). As mentioned above, FM identity regulators play key roles in the activation of these genes. However, other regulatory inputs, both positive and negative, are required to precisely establish the spatiotemporal expression domains of the floral organ identity genes (reviewed in Krizek & Fletcher, 2005; O'Maoileidigh et al., 2014). For example, activation of the C function gene *AGAMOUS* (*AG*) in the center of FMs in *A. thaliana* depends on the combined activities of LFY and the stem cell regulator WUSCHEL (WUS) (Lenhard, Bohnert, Jurgens, & Laux, 2001; Lohmann et al., 2001). At the same time, *AG* expression is suppressed in the outer regions of the flower by the APETALA2 (AP2) transcription factor (Jofuku, den Boer, Van Montagu, & Okamuro, 1994), which was originally proposed to confer A function. Indeed, AP2 appears to mediate the antagonism between A and C function genes (as well as controlling the expression of other floral homeotic genes) (Dinh et al., 2012; Krogan, Hogan, & Long, 2012; Wollmann, Mica, Todesco, Long, & Weigel, 2010), but may otherwise not be directly involved in floral organ specification. Notably, both *AP2* and *AG* are not only controlled at the transcriptional, but also at the post-transcriptional level, highlighting the complexity of the regulatory mechanisms involved—*AP2* is negatively regulated by the *miR172* miRNA (Chen, 2004; Wollmann et al., 2010), and *AG* pre-mRNA processing is controlled by the activity of several RNA binding proteins (Cheng, Kato, Wang, Li, & Chen, 2003; Rodriguez-Cazorla et al., 2015).

While most aspects of the (A)B(C) model appear to be fairly well conserved across angiosperms (Causier & Davies, 2014; Causier et al., 2010; Theissen et al., 2016), the mechanism underlying the antagonism between A and C function genes may be more variable. This is suggested by results from petunia and snapdragon where it has been shown that C function gene activity is suppressed in the outer floral whorls by homologous miRNAs of the *miR169* family that target NF-YA transcription factor-encoding genes (Cartolano et al., 2007). In petunia, it has recently been demonstrated that the *BLIND* miRNA acts in parallel with the transcription factor AP2B/BLIND ENHANCER (BEN) (Morel et al., 2017). Notably, while BEN is a transcription factor of the AP2 family, it belongs to a subgroup that, in *A. thaliana*, is involved in the control of flowering time, not floral patterning.

In addition to establishing the expression domains of the floral homeotic genes, boundaries must be set during floral patterning so as to demarcate individual whorls and organ primordia. This is in part accomplished by genes that are more generally involved in boundary formation throughout plant development such as the *CUP-SHAPED COTYLEDON* (*CUC*) genes (Aida, Ishida, Fukaki, Fujisawa, & Tasaka, 1997; Vroemen, Mordhorst, Albrecht, Kwaaitaal, & de Vries, 2003), but also through genes with flower-specific activity. An example for the latter is *SUPERMAN* (*SUP*), which encodes a C2H2 zinc-finger transcription factor (Sakai, Medrano, & Meyerowitz, 1995) that controls the boundary between the third and fourth whorls, and possibly represses the B function genes *APETALA3* (*AP3*) and *PISTILLATA* (*PI*) in the center of the FM (Prunet, Yang, Das, Meyerowitz, & Jack, 2017; Sakai, Krizek, Jacobsen, & Meyerowitz, 2000). Similarly, the related RABBIT EARS (RBE) transcription factor, which is involved in petal formation (Huang & Irish, 2015; Takeda, Matsumoto, & Okada, 2004), sets the boundary between the second and third whorls by repressing *AG* (Krizek, Lewis, & Fletcher, 2006). How exactly these regulators exert their functions is not yet fully known, but it has been shown that SUP acts in part by controlling phytohormone response pathways (Nibau, Di Stilio, Wu, & Cheung, 2011; Xu et al., 2018), while RBE regulates the expression of *miR164* miRNAs (Huang, Lopez-Giraldez, Townsend, & Irish, 2012), which are known to target *CUC* boundary genes in flowers (Baker, Sieber, Wellmer, & Meyerowitz, 2005; Sieber, Wellmer, Gheyselinck, Riechmann, & Meyerowitz, 2007).

4. Floral organ specification

How do the floral homeotic transcription factors control the gene regulatory programs needed for the formation of the different floral organ types? According to the floral quartet model (Theissen, 2001), transcription factor complexes composed of four floral MADS-domain proteins (with the identity of the proteins predicted by the ABCDE model) associate with target gene promoters, with the two dimers of the tetrameric complex binding to proximate CArG-box sequences (CArG: C-A-rich-G; consensus: $5'$-CC(A/T)$_6$GG-$3'$) to induce the looping of chromatin. This model is now well supported by experimental evidence. Importantly, it has been shown through proteomic analysis that the different complexes predicted by the floral quartet model exist in planta (Smaczniak, Immink, et al., 2012), but their exact stoichiometry is not yet known. Furthermore, the induction of DNA looping by floral quartets has been demonstrated both in vitro and in vivo (Melzer, Verelst, & Theissen, 2009; Mendes et al., 2013), but it is unclear whether it is an absolute prerequisite for floral quartet activity.

Another unanswered question is to what extent tetrameric complexes with different compositions are active in a given floral whorl. For example, in the third whorl where stamens form, the quartet model predicts a complex composed of B, C and E function factors. However, it is also possible that in the same cells complexes assemble that are only composed of C and E factors, as predicted for the fourth whorl. Likewise, MADS-domain protein dimers (rather than tetramers) may carry out some functions required for floral organ specification. Testing these different possibilities will likely be important for a full understanding of the activities of the floral homeotic protein complexes but will also be technically challenging, as any experiments to address this problem would arguably require a specific control over dimer and/or tetramer formation.

Recently, Theissen and co-workers proposed a model for how the floral quartets may interact with chromatin to regulate the expression of their target genes (Theissen et al., 2016). This nucleosome mimicry model posits that floral quartets are structurally similar to nucleosomes (or nucleosome halves) and may be able to evict and replace nucleosomes in the promoters of target genes where CArG-boxes are present. DNA looping induced by the floral quartets could then resemble chromatin wrapping around nucleosomes. After being incorporated into chromatin, the floral quartets would interact, for example, with basal transcription factors needed to induce transcription. While this model is currently not well supported by experimental evidence, it is intriguing and will likely inspire future research in this area.

Gel-filtration experiments have shown that complexes containing the E function protein SEPALLATA3 (SEP3) are much larger than what would be expected from a tetrameric MADS-domain protein complex (Smaczniak, Immink, et al., 2012), implying the association with additional proteins or, alternatively, the existence of higher-order complexes containing multiple MADS-domain proteins. In agreement with the former possibility, proteomic and other studies have identified proteins that appear to be associated with the floral quartets (Simonini et al., 2012; Smaczniak, Immink, et al., 2012), including non-MADS-domain transcription factors such as ARFs (Ripoll et al., 2015), as well as proteins involved in the epigenetic control of gene expression such as the histone H3 lysine 27 demethylase RELATIVE OF EARLY FLOWERING6 (REF6) (Smaczniak, Immink, et al., 2012). While the significance of these interactions is not well understood at present, the idea of higher-order transcription factor complexes is intriguing and may be the reason why the floral homeotic factors have fairly distinct sets of target genes despite similar genome-wide binding patterns (see below).

Likewise, the association of the floral quartet complexes with epigenetic modifiers could explain why binding of these complexes to DNA appears to be often followed by increases in chromatin accessibility, suggesting that the floral quartets may function as pioneer transcription factors (Pajoro, Madrigal, et al., 2014). Thus, several attractive hypotheses have been put forward in recent years to explain the activities of these regulatory complexes, but more work will be required to fully understand, at the molecular level, their modes of activity.

Target genes of the floral homeotic factors in *A. thaliana* and other species have been identified through a combination of genome-wide binding analyses and transcriptomics studies (Gomez-Mena, de Folter, Costa, Angenent, & Sablowski, 2005; Kaufmann et al., 2009; Kaufmann, Wellmer, et al., 2010; O'Maoileidigh et al., 2013; Pajoro, Madrigal, et al., 2014; Wellmer, Riechmann, Alves-Ferreira, & Meyerowitz, 2004; Wuest et al., 2012). This work showed that they are bifunctional with gene activation being possibly more prevalent than gene repression. As observed for other transcription factors in plants and animals, the number of binding sites for each of the transcription factors (several thousand) appears to exceed the number of genes that are under transcriptional control (several hundred) by an order of magnitude. There also seems to be a large overlap in the binding sites for the different floral homeotic proteins (O'Maoileidigh et al., 2013), in agreement with the finding that the different floral organ identity complexes have partially overlapping binding specificities (Jetha, Theissen, & Melzer, 2014; Khanday et al., 2016; Smaczniak, Muino, Chen, Angenent, & Kaufmann, 2017; Yan, Chen, & Kaufmann, 2016). As expected for developmental master regulators, the target genes of the floral homeotic factors are heavily enriched for genes encoding other regulatory proteins and miRNAs. Many of these regulators have known functions in processes such as hormone signaling and cell cycle control (summarized in Pajoro, Biewers, et al., 2014), suggesting that the floral homeotic factors bring about floral organ specification in part by controlling general developmental pathways of the plant.

5. Floral organ development

Floral organ development has been studied extensively and the results of this work are summarized in several recent reviews (e.g., Gomez, Talle, & Wilson, 2015; Huang & Irish, 2016; Marsch-Martinez & de Folter, 2016; Moubayidin & Ostergaard, 2017; Song, Qi, Huang, & Xie, 2013). Here we will take a broader view at this process and specifically discuss how the floral homeotic factors control the gene networks that regulate their development. The results of transcriptomics studies (e.g., in *A. thaliana*: Jiao & Meyerowitz, 2010; Ryan et al., 2015; Schmid et al., 2005; Wellmer, Alves-Ferreira, Dubois, Riechmann, & Meyerowitz, 2006; Wellmer et al., 2004; Zhang et al., 2005) suggest that these networks are exceptionally complex, with several thousand genes exhibiting specific expression during floral organ development and the vast majority of these genes being active in the reproductive organs. While many of the genes that act during early floral stages are under direct control of the floral homeotic factors, the influence of these master regulators appears reduced as flowers mature (Ryan et al., 2015; Wuest et al., 2012). In fact, once activated, the floral homeotic genes remain expressed throughout most of floral organ development. In agreement with this prolonged expression, it has been shown through stage-specific perturbation experiments that they control diverse processes during floral organ morphogenesis, such as the promotion of growth and sporogenesis (Ito, Ng, Lim, Yu, & Meyerowitz, 2007; O'Maoileidigh et al., 2013; Sablowski, 2015; Wuest et al., 2012). Furthermore, there is evidence that in some cases, their activities are required at different floral stages for the development of a given organ type (summarized in Wellmer, Graciet, & Riechmann, 2014).

How do the floral homeotic factors control floral organ development at the molecular level? From the available data, it appears that their sets of target genes substantially change as flower development progresses, such that they contain more genes involved in differentiation processes, and fewer with regulatory functions, as flowers become more mature (Ryan et al., 2015). Another angle to look at these target genes is to consider whether they are specifically expressed in flowers or are also active elsewhere in the plant. One of the earliest, and still most striking examples for the former, is *NOZZLE/SPOROCYTELESS* (*NZZ/SPL*) (Balasubramanian & Schneitz, 2000; Yang, Ye, Xu, & Sundaresan, 1999), which is activated by AG (Ito et al., 2004) and controls male and female sporogenesis, but also plays a role in conferring stamen identity (Liu et al., 2009). Another example for a gene with flower-specific expression that is activated by AG (Gomez-Mena et al., 2005; O'Maoileidigh et al., 2013) is *CRABS CLAW* (*CRC*), which plays essential roles in carpel and nectary development (Bowman & Smyth, 1999).

In addition to triggering the expression of many regulators with flower-specific activities, the floral homeotic factors modify, as briefly mentioned above, developmental pathways with more general functions in the plant. An important example is the hormone-dependent control of growth. For instance, it has been shown that AG controls jasmonate-dependent stamen growth by regulating the expression of the *DEFECTIVE IN ANTHER DEHISCENCE1* (*DAD1*) gene (Ito et al., 2007). AG also appears to control cytokinin biosynthesis and/or response (O'Maoileidigh, Stewart, Zheng, Coupland, & Wellmer, 2018; O'Maoileidigh et al., 2013) as well as the response to auxin (Zhang et al., 2018), and thus phytohormone pathways that have been shown to be particularly important for the formation of the reproductive floral organs (Marsch-Martinez & de Folter, 2016; Song et al., 2013). In addition, the floral homeotic factors regulate the expression of genes involved in growth regulation. A particularly well-studied gene in this context encodes the C2H2 zinc-finger transcription factor JAGGED (JAG), which controls lateral organ growth (Dinneny, Yadegari, Fischer, Yanofsky, & Weigel, 2004; Ohno, Reddy, Heisler, & Meyerowitz, 2004). *JAG* and its paralog *NUBBIN* (*NUB*) (Dinneny, Weigel, & Yanofsky, 2006) influence floral organ growth especially in the apical (distal) region. Molecular and quantitative analyses showed that JAG promotes anisotropic growth (Schiessl, Kausika, Southam, Bush, & Sablowski, 2012) and acts in part by suppressing genes coding for KIP RELATED PROTEIN (KIP) cell cycle inhibitors (Schiessl, Muino, & Sablowski, 2014).

The observation that many genes regulated by the floral homeotic factors have broader functions during plant development, and that many of those genes exhibit peak expression in leaves (O'Maoileidigh et al., 2013), renewed efforts to understand how leaves are transformed into floral organs. This principle was first recognized by von Goethe in the late 18th century (von Goethe, 1790) and was validated 200 years later through elegant genetic experiments in which all floral quartets were rendered inactive simultaneously, leading to the formation of leaves in place of floral organs (Bowman, Drews, & Meyerowitz, 1991; Ditta, Pinyopich, Robles, Pelaz, & Yanofsky, 2004). Also, ectopic expression of combinations of the floral homeotic factors results in the transformation of leaves into floral organs (Goto, Kyozuka, & Bowman, 2001; Honma & Goto, 2001; Pelaz, Tapia-Lopez, Alvarez-Buylla, & Yanofsky, 2001). In further support of Goethe's hypothesis, it has been shown that floral organs share common features with leaves, such as determinacy control via suppression of leaf meristematic activity (Alvarez, Furumizu, Efroni, Eshed, & Bowman, 2016). From these observations, it appears that parts of the genetic program for leaf development remain active in floral organs while others may be modified or even completely suppressed. An example for the latter is the suppression of trichomes (a hallmark of leaf development) on reproductive organs in *A. thaliana*, which has been shown to depend on the combined activities of B, C, and D function factors (O'Maoileidigh et al., 2018, 2013). These regulators seem to repress activators of trichome initiation such as *GLABRA1* (*GL1*) and also contribute to the up-regulation of repressors of trichome formation during flower development. Furthermore, it has been shown that *AG* limits cytokinin responses and/or synthesis, which are known to promote trichome initiation, in gynoecia (O'Maoileidigh et al., 2018). Thus, it appears that the floral homeotic factors suppress trichome formation on reproductive floral organs by modifying the activities of several modules of the gene regulatory network that controls trichome initiation throughout plant development.

In summary, the floral homeotic factors have at least two broad activities during floral organ development: (i) they activate other regulatory genes with flower-specific expression that carry out important functions in the control of floral morphogenesis and (ii), they modulate or suppress the activities of genes and pathways that are active elsewhere in the plant, including components of the genetic program for leaf development.

6. Floral meristem termination

In contrast to IMs, FMs are determinate structures and cease their activity when the formation of all floral organs has been initiated (at floral stage 6 in *A. thaliana*). Central to the termination of FM activity is the suppression of the main regulator of stem cell identity, *WUS* (Laux, Mayer, Berger, & Jurgens, 1996), which encodes a homeodomain transcription factor (Mayer et al., 1998). *WUS* expression is regulated in FMs by several different transcription factors (Fig. 1B). These multiple inputs likely ensure a precise temporal control of stem cell termination and thus prevent overgrowth of the FM. The first (and arguably the most important) regulator of *WUS* identified in FMs is the floral homeotic protein AG (Lenhard et al., 2001; Lohmann et al., 2001), which is expressed in stamen and carpel primordia in the center of the FM (where the stem cells reside) from stage 3 onward (Ito et al., 2004; Yanofsky et al., 1990). AG was shown to bind to the *WUS* promoter and to recruit a *Polycomb* group (PcG) complex, leading to a gradual decline in *WUS* expression through the deposition of repressive chromatin marks (Liu et al., 2011).

In addition to this direct effect, AG also controls floral stem cell activity indirectly. Key to this mechanism is the *KNUCKLES* (*KNU*) gene, which encodes a C2H2 zinc-finger transcription factor (Payne, Johnson, & Koltunow, 2004). *KNU* is a direct target gene of AG and is activated in the center of FMs at stage 6 when the termination of floral stem cell activity occurs (Sun, Xu, Ng, & Ito, 2009). The precise timing of *KNU* activation has been shown to depend on the AG-dependent eviction of the PcG complex from the *KNU* promoter, which leads to the gradual removal of repressive chromatin marks (Sun et al., 2014). It was demonstrated recently that KNU in both *A. thaliana* and *Solanum lycopersicum* (tomato) is part of a complex that binds directly to the *WUS* locus to silence its expression (Bollier et al., 2018). In *A. thaliana*, this complex contains the MINI ZINC FINGER2 (MIF2) protein, the TOPLESS (TPL) co-repressor protein, and HISTONE DEACETYLASE19 (HDA19).

In addition to *KNU*, AG acts indirectly on floral stem cells through activation of the above-mentioned *CRC* gene (Gomez-Mena et al., 2005; O'Maoileidigh et al., 2013). While *CRC* transcription also commences at stage 6 (Bowman & Smyth, 1999), its expression domain does not encompass the FM center as in the case of *KNU*. Instead it is confined to the abaxial side of carpel primordia (Bowman & Smyth, 1999). Thus, the CRC transcription factor must act non-cell autonomously on floral stem cells. It has recently been shown that this is mediated at least in part through an auxin-dependent mechanism: CRC represses expression of the *TORNADO2* (*TRN2*) gene, which controls auxin homeostasis in the FM (Yamaguchi, Huang, Xu, Tanoi, & Ito, 2017).

Another link between auxin and FM termination that has been recently described involves the auxin-dependent transcription factor ARF3 [also known as ETTIN (ETT)]. It has been shown that ARF3/ETT represses cytokinin biosynthesis genes to reduce cytokinin levels in FMs and that it may act synergistically with AG in this process. The reduced level of cytokinin is then thought to affect cell division activities in the FM (Zhang et al., 2018). Additional regulators that help to bring about FM determinacy include the above-mentioned SUP transcription factor (Prunet et al., 2017; Xu et al., 2018), the Class III HD-ZIP transcription factor-coding genes *PHABULOSA* (*PHB*) and *PHAVOLUTA* (*PHV*), which are targets of the *miR165/166* miRNAs (Ji et al., 2011), as well as FAR-RED ELONGATED HYPOCOTYL3 (FHY3), which was shown to repress the stem cell regulatory gene *CLAVATA3* (*CLV3*) (Li et al., 2016).

While all the regulators discussed above promote FM termination, it has recently been shown that the AP2 transcription factor supports floral stem cell maintenance by counteracting the repressive activity that AG has on *WUS* (Huang et al., 2017; Ji et al., 2011). This finding is in agreement with the known role of AP2 in stem cell maintenance in the shoot apical meristem

(Wurschum, Gross-Hardt, & Laux, 2006). Thus, the antagonism between A and C function genes proposed by the ABC model (see above) is not limited to floral organ identity specification but extends to the control of FM determinacy.

7. Perspectives and future challenges

In the sections above, we have aimed to describe the current state of research on flower development and highlighted some of the most recent findings in the field. Despite the considerable progress that has been made in all areas, we are far from a complete understanding of the molecular mechanisms that regulate the development of flowers. We believe that among the most pressing open questions that need to be addressed in the coming years are the following (for a more comprehensive discussion of questions, see Scutt & Vandenbussche, 2014; Wellmer, Bowman, et al., 2014):

1. How do floral homeotic transcription factors regulate their target genes and how do they achieve specificity in genomes that contain thousands of potential binding sites? As described above, there is evidence that the formation of higher-order regulatory complexes and differences in DNA-binding activities may be essential determinants of this specificity. Therefore, a detailed analysis of the biochemical and biophysical properties of the different complexes will likely be necessary to address this question. Given the substantial technical advances in recent years, for example in the field of proteomics, it appears likely that rapid progress in answering this question will be made in the near future.
2. How do the target genes of the floral organ identity factors mediate floral organ morphogenesis? As described above, several thousand genes with flower-specific expression have been identified in *A. thaliana* and other species. The analysis of these genes through reverse genetics approaches has not always led to substantial insights, possibly due to high levels of functional redundancy. With the advent of new genome-editing techniques that allow the targeting of multiple loci simultaneously (Peterson et al., 2016), this work should be much more feasible and fruitful in the future. Nevertheless, methods of systems biology as well as mathematical approaches (Alvarez-Buylla et al., 2007; Davila-Velderrain, Martinez-Garcia, & Alvarez-Buylla, 2016; La Rota et al., 2011) will arguably be required to generate models that describe, integrate and predict the regulatory mechanisms underlying floral morphogenesis.

3. What is the genetic basis for the large variations in floral morphologies found among angiosperms? Comparative genomics combined with new models for research on flower development will undoubtedly be necessary to make significant progress in this area, but there are already impressive case studies to understand the genetic basis of floral shape variation. For example, a recent analysis of the boundary gene *CUPULIFORMIS* (*CUP*) in snapdragon has shown that variations in boundary gene expression likely played an important role in determining complex floral shapes (Rebocho, Kennaway, Bangham, & Coen, 2017). In another example, it has been demonstrated that co-option of a gene network composed of organ polarity regulators is involved in determining stamen filament morphology (de Almeida et al., 2014).

The substantial advances in our understanding of arguably all aspects of flower development have not only allowed unprecedented insights into the reproductive biology of higher plants, but have also opened up new and promising avenues for crop improvement. These approaches, which can be much more targeted now than in previous decades by virtue of the introduction of highly specific genome-editing techniques, and the availability of genome sequence data for most crop plants, are informed and underpinned by the ever-expanding knowledge of plant biology (Alfred, Dangl, Kamoun, & McCouch, 2014; Ronald, 2014). Thus, through plant translational research, we can now target, and attempt to optimize, specific traits to improve crop yields. Because flowers produce most of the food people and their livestock consume, floral regulators are especially promising targets for crop improvement. In fact, it is now known that genes encoding MADS-domain proteins have been targets for selection during crop domestication, and it has therefore been proposed that future efforts for crop improvement should specifically focus on floral regulatory genes to bring about, for example, beneficial changes to inflorescence architecture (Park et al., 2014; Schilling, Pan, Kennedy, & Melzer, 2018). Thus, the progress made in the field of flowering research over the past decade has not only provided fundamental insights into plant reproduction, but has given us the knowledge base needed to address some of the most important societal challenges of the 21st century.

Acknowledgments

We thank the members of our laboratory for discussions. Work on flower development in our group was funded by grants from Science Foundation Ireland and the Irish Research Council.

References

Abe, M., Kobayashi, Y., Yamamoto, S., Daimon, Y., Yamaguchi, A., Ikeda, Y., et al. (2005). FD, a bZIP protein mediating signals from the floral pathway integrator FT at the shoot apex. *Science, 309*, 1052–1056.

Aida, M., Ishida, T., Fukaki, H., Fujisawa, H., & Tasaka, M. (1997). Genes involved in organ separation in *Arabidopsis*: An analysis of the *cup-shaped cotyledon* mutant. *Plant Cell, 9*, 841–857.

Alfred, J., Dangl, J. L., Kamoun, S., & McCouch, S. R. (2014). New horizons for plant translational research. *PLoS Biology, 12*, e1001880.

Alvarez, J. P., Furumizu, C., Efroni, I., Eshed, Y., & Bowman, J. L. (2016). Active suppression of a leaf meristem orchestrates determinate leaf growth. *eLife, 5*, e15023.

Alvarez-Buylla, E. R., Benitez, M., Davila, E. B., Chaos, A., Espinosa-Soto, C., & Padilla-Longoria, P. (2007). Gene regulatory network models for plant development. *Current Opinion in Plant Biology, 10*, 83–91.

Baker, C. C., Sieber, P., Wellmer, F., & Meyerowitz, E. M. (2005). The *early extra petals1* mutant uncovers a role for microRNA *miR164c* in regulating petal number in *Arabidopsis*. *Current Biology, 15*, 303–315.

Balasubramanian, S., & Schneitz, K. (2000). NOZZLE regulates proximal-distal pattern formation, cell proliferation and early sporogenesis during ovule development in *Arabidopsis thaliana*. *Development, 127*, 4227–4238.

Baumann, K., Venail, J., Berbel, A., Domenech, M. J., Money, T., Conti, L., et al. (2015). Changing the spatial pattern of *TFL1* expression reveals its key role in the shoot meristem in controlling *Arabidopsis* flowering architecture. *Journal of Experimental Botany, 66*, 4769–4780.

Besnard, F., Refahi, Y., Morin, V., Marteaux, B., Brunoud, G., Chambrier, P., et al. (2014). Cytokinin signalling inhibitory fields provide robustness to phyllotaxis. *Nature, 505*, 417–421.

Bhatia, N., Bozorg, B., Larsson, A., Ohno, C., Jonsson, H., & Heisler, M. G. (2016). Auxin acts through MONOPTEROS to regulate plant cell polarity and pattern phyllotaxis. *Current Biology, 26*, 3202–3208.

Bhatia, N., & Heisler, M. G. (2018). Self-organizing periodicity in development: Organ positioning in plants. *Development, 145*, dev149336.

Bollier, N., Sicard, A., Leblond, J., Latrasse, D., Gonzalez, N., Gevaudant, F., et al. (2018). At-MINI ZINC FINGER2 and Sl-INHIBITOR OF MERISTEM ACTIVITY, a conserved missing link in the regulation of floral meristem termination in *Arabidopsis* and tomato. *Plant Cell, 30*, 83–100.

Bowman, J. L., Drews, G. N., & Meyerowitz, E. M. (1991). Expression of the *Arabidopsis* floral homeotic gene *AGAMOUS* is restricted to specific cell types late in flower development. *Plant Cell, 3*, 749–758.

Bowman, J. L., & Smyth, D. R. (1999). *CRABS CLAW*, a gene that regulates carpel and nectary development in *Arabidopsis*, encodes a novel protein with zinc finger and helix-loop-helix domains. *Development, 126*, 2387–2396.

Bowman, J. L., Smyth, D. R., & Meyerowitz, E. M. (1989). Genes directing flower development in *Arabidopsis*. *Plant Cell, 1*, 37–52.

Bowman, J. L., Smyth, D. R., & Meyerowitz, E. M. (2012). The ABC model of flower development: Then and now. *Development, 139*, 4095–4098.

Bradley, D., Ratcliffe, O., Vincent, C., Carpenter, R., & Coen, E. (1997). Inflorescence commitment and architecture in *Arabidopsis*. *Science, 275*, 80–83.

Cartolano, M., Castillo, R., Efremova, N., Kuckenberg, M., Zethof, J., Gerats, T., et al. (2007). A conserved microRNA module exerts homeotic control over *Petunia hybrida* and *Antirrhinum majus* floral organ identity. *Nature Genetics, 39*, 901–905.

Causier, B., & Davies, B. (2014). Flower development in the asterid lineage. *Methods in Molecular Biology, 1110*, 35–55.
Causier, B., Schwarz-Sommer, Z., & Davies, B. (2010). Floral organ identity: 20 Years of ABCs. *Seminars in Cell and Developmental Biology, 21*, 73–79.
Chahtane, H., Vachon, G., Le Masson, M., Thevenon, E., Perigon, S., Mihajlovic, N., et al. (2013). A variant of LEAFY reveals its capacity to stimulate meristem development by inducing *RAX1*. *Plant Journal, 74*, 678–689.
Chandler, J. W., & Werr, W. (2017). DORNROSCHEN, DORNROSCHEN-LIKE, and PUCHI redundantly control floral meristem identity and organ initiation in *Arabidopsis*. *Journal of Experimental Botany, 68*, 3457–3472.
Chen, X. (2004). A microRNA as a translational repressor of APETALA2 in *Arabidopsis* flower development. *Science, 303*, 2022–2025.
Cheng, Y., Kato, N., Wang, W., Li, J., & Chen, X. (2003). Two RNA binding proteins, HEN4 and HUA1, act in the processing of *AGAMOUS* pre-mRNA in *Arabidopsis thaliana*. *Developmental Cell, 4*, 53–66.
Coen, E. S., & Meyerowitz, E. M. (1991). The war of the whorls: Genetic interactions controlling flower development. *Nature, 353*, 31–37.
Damerval, C., & Becker, A. (2017). Genetics of flower development in *Ranunculales*—A new, basal eudicot model order for studying flower evolution. *New Phytologist, 216*, 361–366.
Davila-Velderrain, J., Martinez-Garcia, J. C., & Alvarez-Buylla, E. R. (2016). Dynamic network modelling to understand flowering transition and floral patterning. *Journal of Experimental Botany, 67*, 2565–2572.
de Almeida, A. M., Yockteng, R., Schnable, J., Alvarez-Buylla, E. R., Freeling, M., & Specht, C. D. (2014). Co-option of the polarity gene network shapes filament morphology in angiosperms. *Scientific Reports, 4*, 6194.
Denay, G., Chahtane, H., Tichtinsky, G., & Parcy, F. (2017). A flower is born: An update on *Arabidopsis* floral meristem formation. *Current Opinion in Plant Biology, 35*, 15–22.
Dinh, T. T., Girke, T., Liu, X., Yant, L., Schmid, M., & Chen, X. (2012). The floral homeotic protein APETALA2 recognizes and acts through an AT-rich sequence element. *Development, 139*, 1978–1986.
Dinneny, J. R., Weigel, D., & Yanofsky, M. F. (2006). NUBBIN and JAGGED define stamen and carpel shape in *Arabidopsis*. *Development, 133*, 1645–1655.
Dinneny, J. R., Yadegari, R., Fischer, R. L., Yanofsky, M. F., & Weigel, D. (2004). The role of JAGGED in shaping lateral organs. *Development, 131*, 1101–1110.
Ditta, G., Pinyopich, A., Robles, P., Pelaz, S., & Yanofsky, M. F. (2004). The *SEP4* gene of *Arabidopsis thaliana* functions in floral organ and meristem identity. *Current Biology, 14*, 1935–1940.
Galweiler, L., Guan, C., Muller, A., Wisman, E., Mendgen, K., Yephremov, A., et al. (1998). Regulation of polar auxin transport by AtPIN1 in *Arabidopsis* vascular tissue. *Science, 282*, 2226–2230.
Gomez, J. F., Talle, B., & Wilson, Z. A. (2015). Anther and pollen development: A conserved developmental pathway. *Journal of Integrative Plant Biology, 57*, 876–891.
Gomez-Mena, C., de Folter, S., Costa, M. M., Angenent, G. C., & Sablowski, R. (2005). Transcriptional program controlled by the floral homeotic gene *AGAMOUS* during early organogenesis. *Development, 132*, 429–438.
Goslin, K., Zheng, B., Serrano-Mislata, A., Rae, L., Ryan, P. T., Kwasniewska, K., et al. (2017). Transcription factor interplay between LEAFY and APETALA1/CAULIFLOWER during floral initiation. *Plant Physiology, 174*, 1097–1109.
Goto, K., Kyozuka, J., & Bowman, J. L. (2001). Turning floral organs into leaves, leaves into floral organs. *Current Opinion in Genetics & Development, 11*, 449–456.

Gramzow, L., & Theissen, G. (2010). A hitchhiker's guide to the MADS world of plants. *Genome Biology*, *11*, 214.

Hames, C., Ptchelkine, D., Grimm, C., Thevenon, E., Moyroud, E., Gerard, F., et al. (2008). Structural basis for LEAFY floral switch function and similarity with helix-turn-helix proteins. *EMBO Journal*, *27*, 2628–2637.

Heijmans, K., Ament, K., Rijpkema, A. S., Zethof, J., Wolters-Arts, M., Gerats, T., et al. (2012). Redefining C and D in the petunia ABC. *Plant Cell*, *24*, 2305–2317.

Heisler, M. G., Ohno, C., Das, P., Sieber, P., Reddy, G. V., Long, J. A., et al. (2005). Patterns of auxin transport and gene expression during primordium development revealed by live imaging of the *Arabidopsis* inflorescence meristem. *Current Biology*, *15*, 1899–1911.

Honma, T., & Goto, K. (2001). Complexes of MADS-box proteins are sufficient to convert leaves into floral organs. *Nature*, *409*, 525–529.

Huang, T., & Irish, V. F. (2015). Temporal control of plant organ growth by TCP transcription factors. *Current Biology*, *25*, 1765–1770.

Huang, T., & Irish, V. F. (2016). Gene networks controlling petal organogenesis. *Journal of Experimental Botany*, *67*, 61–68.

Huang, T., Lopez-Giraldez, F., Townsend, J. P., & Irish, V. F. (2012). RBE controls *microRNA164* expression to effect floral organogenesis. *Development*, *139*, 2161–2169.

Huang, Z., Shi, T., Zheng, B., Yumul, R. E., Liu, X., You, C., et al. (2017). APETALA2 antagonizes the transcriptional activity of AGAMOUS in regulating floral stem cells in *Arabidopsis thaliana*. *New Phytologist*, *215*, 1197–1209.

Ito, T., Ng, K. H., Lim, T. S., Yu, H., & Meyerowitz, E. M. (2007). The homeotic protein AGAMOUS controls late stamen development by regulating a jasmonate biosynthetic gene in *Arabidopsis*. *Plant Cell*, *19*, 3516–3529.

Ito, T., Wellmer, F., Yu, H., Das, P., Ito, N., Alves-Ferreira, M., et al. (2004). The homeotic protein AGAMOUS controls microsporogenesis by regulation of *SPOROCYTELESS*. *Nature*, *430*, 356–360.

Jetha, K., Theissen, G., & Melzer, R. (2014). *Arabidopsis* SEPALLATA proteins differ in cooperative DNA-binding during the formation of floral quartet-like complexes. *Nucleic Acids Research*, *42*, 10927–10942.

Ji, L., Liu, X., Yan, J., Wang, W., Yumul, R. E., Kim, Y. J., et al. (2011). ARGONAUTE10 and ARGONAUTE1 regulate the termination of floral stem cells through two microRNAs in *Arabidopsis*. *PLoS Genetics*, 7, e1001358.

Jiao, Y., & Meyerowitz, E. M. (2010). Cell-type specific analysis of translating RNAs in developing flowers reveals new levels of control. *Molecular Systems Biology*, *6*, 419.

Jofuku, K. D., den Boer, B. G., Van Montagu, M., & Okamuro, J. K. (1994). Control of *Arabidopsis* flower and seed development by the homeotic gene *APETALA2*. *Plant Cell*, *6*, 1211–1225.

Kaufmann, K., Muino, J. M., Jauregui, R., Airoldi, C. A., Smaczniak, C., Krajewski, P., et al. (2009). Target genes of the MADS transcription factor SEPALLATA3: Integration of developmental and hormonal pathways in the *Arabidopsis* flower. *PLoS Biology*, 7, e1000090.

Kaufmann, K., Muino, J. M., Osteras, M., Farinelli, L., Krajewski, P., & Angenent, G. C. (2010). Chromatin immunoprecipitation (ChIP) of plant transcription factors followed by sequencing (ChIP-SEQ) or hybridization to whole genome arrays (ChIP-CHIP). *Nature Protocols*, *5*, 457–472.

Kaufmann, K., Wellmer, F., Muino, J. M., Ferrier, T., Wuest, S. E., Kumar, V., et al. (2010). Orchestration of floral initiation by APETALA1. *Science*, *328*, 85–89.

Kempin, S. A., Savidge, B., & Yanofsky, M. F. (1995). Molecular basis of the *cauliflower* phenotype in *Arabidopsis*. *Science*, *267*, 522–525.

Khanday, I., Das, S., Chongloi, G. L., Bansal, M., Grossniklaus, U., & Vijayraghavan, U. (2016). Genome-wide targets regulated by the OsMADS1 transcription factor reveals its DNA recognition properties. *Plant Physiology*, *172*, 372–388.

Krizek, B. (2009). AINTEGUMENTA and AINTEGUMENTA-LIKE6 act redundantly to regulate *Arabidopsis* floral growth and patterning. *Plant Physiology*, *150*, 1916–1929.

Krizek, B. A., & Fletcher, J. C. (2005). Molecular mechanisms of flower development: An armchair guide. *Nature Reviews. Genetics*, *6*, 688–698.

Krizek, B. A., Lewis, M. W., & Fletcher, J. C. (2006). RABBIT EARS is a second-whorl repressor of *AGAMOUS* that maintains spatial boundaries in *Arabidopsis* flowers. *Plant Journal*, *45*, 369–383.

Krogan, N. T., Hogan, K., & Long, J. A. (2012). APETALA2 negatively regulates multiple floral organ identity genes in *Arabidopsis* by recruiting the co-repressor TOPLESS and the histone deacetylase HDA19. *Development*, *139*, 4180–4190.

La Rota, C., Chopard, J., Das, P., Paindavoine, S., Rozier, F., Farcot, E., et al. (2011). A data-driven integrative model of sepal primordium polarity in *Arabidopsis*. *Plant Cell*, *23*, 4318–4333.

Laux, T., Mayer, K. F., Berger, J., & Jurgens, G. (1996). The *WUSCHEL* gene is required for shoot and floral meristem integrity in *Arabidopsis*. *Development*, *122*, 87–96.

Lenhard, M., Bohnert, A., Jurgens, G., & Laux, T. (2001). Termination of stem cell maintenance in *Arabidopsis* floral meristems by interactions between WUSCHEL and AGAMOUS. *Cell*, *105*, 805–814.

Li, D., Fu, X., Guo, L., Huang, Z., Li, Y., Liu, Y., et al. (2016). *FAR-RED ELONGATED HYPOCOTYL3* activates *SEPALLATA2* but inhibits *CLAVATA3* to regulate meristem determinacy and maintenance in *Arabidopsis*. *Proceedings of the National Academy of Sciences of the United States of America*, *113*, 9375–9380.

Li, W., Zhou, Y., Liu, X., Yu, P., Cohen, J. D., & Meyerowitz, E. M. (2013). LEAFY controls auxin response pathways in floral primordium formation. *Science Signaling*, *6*, ra23.

Liu, X., Huang, J., Parameswaran, S., Ito, T., Seubert, B., Auer, M., et al. (2009). The *SPOROCYTELESS/NOZZLE* gene is involved in controlling stamen identity in *Arabidopsis*. *Plant Physiology*, *151*, 1401–1411.

Liu, X., Kim, Y. J., Muller, R., Yumul, R. E., Liu, C., Pan, Y., et al. (2011). AGAMOUS terminates floral stem cell maintenance in *Arabidopsis* by directly repressing *WUSCHEL* through recruitment of Polycomb Group proteins. *Plant Cell*, *23*, 3654–3670.

Lohmann, J. U., Hong, R. L., Hobe, M., Busch, M. A., Parcy, F., Simon, R., et al. (2001). A molecular link between stem cell regulation and floral patterning in *Arabidopsis*. *Cell*, *105*, 793–803.

Mahonen, A. P., Bishopp, A., Higuchi, M., Nieminen, K. M., Kinoshita, K., Tormakangas, K., et al. (2006). Cytokinin signaling and its inhibitor AHP6 regulate cell fate during vascular development. *Science*, *311*, 94–98.

Mandel, M. A., Gustafson-Brown, C., Savidge, B., & Yanofsky, M. F. (1992). Molecular characterization of the *Arabidopsis* floral homeotic gene *APETALA1*. *Nature*, *360*, 273–277.

Marsch-Martinez, N., & de Folter, S. (2016). Hormonal control of the development of the gynoecium. *Current Opinion in Plant Biology*, *29*, 104–114.

Mayer, K. F., Schoof, H., Haecker, A., Lenhard, M., Jurgens, G., & Laux, T. (1998). Role of WUSCHEL in regulating stem cell fate in the *Arabidopsis* shoot meristem. *Cell*, *95*, 805–815.

Melzer, R., Verelst, W., & Theissen, G. (2009). The class E floral homeotic protein SEPALLATA3 is sufficient to loop DNA in 'floral quartet'-like complexes in vitro. *Nucleic Acids Research*, *37*, 144–157.

Mendes, M. A., Guerra, R. F., Berns, M. C., Manzo, C., Masiero, S., Finzi, L., et al. (2013). MADS domain transcription factors mediate short-range DNA looping that is essential for target gene expression in *Arabidopsis*. *Plant Cell*, *25*, 2560–2572.

Morel, P., Heijmans, K., Rozier, F., Zethof, J., Chamot, S., Bento, S. R., et al. (2017). Divergence of the floral A-function between an asterid and a rosid species. *Plant Cell*, *29*, 1605–1621.

Moubayidin, L., & Ostergaard, L. (2017). Gynoecium formation: An intimate and complicated relationship. *Current Opinion in Genetics & Development*, *45*, 15–21.

Moyroud, E., Minguet, E. G., Ott, F., Yant, L., Pose, D., Monniaux, M., et al. (2011). Prediction of regulatory interactions from genome sequences using a biophysical model for the *Arabidopsis* LEAFY transcription factor. *Plant Cell*, *23*, 1293–1306.

Nag, A., & Jack, T. (2010). Sculpting the flower; the role of microRNAs in flower development. *Current Topics in Developmental Biology*, *91*, 349–378.

Nibau, C., Di Stilio, V. S., Wu, H. M., & Cheung, A. Y. (2011). *Arabidopsis* and tobacco *SUPERMAN* regulate hormone signalling and mediate cell proliferation and differentiation. *Journal of Experimental Botany*, *62*, 949–961.

Ohno, C. K., Reddy, G. V., Heisler, M. G., & Meyerowitz, E. M. (2004). The *Arabidopsis JAGGED* gene encodes a zinc finger protein that promotes leaf tissue development. *Development*, *131*, 1111–1122.

O'Maoileidigh, D. S., Graciet, E., & Wellmer, F. (2014). Genetic control of *Arabidopsis* flower development. In F. Fornara (Ed.), *Advances in botanical research* (pp. 159–190): Academic Press.

O'Maoileidigh, D. S., Stewart, D., Zheng, B., Coupland, G., & Wellmer, F. (2018). Floral homeotic proteins modulate the genetic program for leaf development to suppress trichome formation in flowers. *Development*, *145*, dev157784.

O'Maoileidigh, D. S., Wuest, S. E., Rae, L., Raganelli, A., Ryan, P. T., Kwasniewska, K., et al. (2013). Control of reproductive floral organ identity specification in *Arabidopsis* by the C function regulator AGAMOUS. *Plant Cell*, *25*, 2482–2503.

Pajoro, A., Biewers, S., Dougali, E., Leal Valentim, F., Mendes, M. A., Porri, A., et al. (2014). The (r)evolution of gene regulatory networks controlling *Arabidopsis* plant reproduction: A two-decade history. *Journal of Experimental Botany*, *65*, 4731–4745.

Pajoro, A., Madrigal, P., Muino, J. M., Matus, J. T., Jin, J., Mecchia, M. A., et al. (2014). Dynamics of chromatin accessibility and gene regulation by MADS-domain transcription factors in flower development. *Genome Biology*, *15*, R41.

Park, S. J., Jiang, K., Tal, L., Yichie, Y., Gar, O., Zamir, D., et al. (2014). Optimization of crop productivity in tomato using induced mutations in the florigen pathway. *Nature Genetics*, *46*, 1337–1342.

Payne, T., Johnson, S. D., & Koltunow, A. M. (2004). *KNUCKLES* (*KNU*) encodes a C2H2 zinc-finger protein that regulates development of basal pattern elements of the *Arabidopsis* gynoecium. *Development*, *131*, 3737–3749.

Pelaz, S., Tapia-Lopez, R., Alvarez-Buylla, E. R., & Yanofsky, M. F. (2001). Conversion of leaves into petals in Arabidopsis. *Current Biology*, *11*, 182–184.

Peterson, B. A., Haak, D. C., Nishimura, M. T., Teixeira, P. J., James, S. R., Dangl, J. L., et al. (2016). Genome-wide assessment of efficiency and specificity in CRISPR/Cas9 mediated multiple site targeting in Arabidopsis. *PLoS One*, *11*, e0162169.

Pinyopich, A., Ditta, G. S., Savidge, B., Liljegren, S. J., Baumann, E., Wisman, E., et al. (2003). Assessing the redundancy of MADS-box genes during carpel and ovule development. *Nature*, *424*, 85–88.

Prunet, N., Jack, T. P., & Meyerowitz, E. M. (2016). Live confocal imaging of *Arabidopsis* flower buds. *Developmental Biology*, *419*, 114–120.

Prunet, N., Yang, W., Das, P., Meyerowitz, E. M., & Jack, T. P. (2017). SUPERMAN prevents class B gene expression and promotes stem cell termination in the fourth whorl of *Arabidopsis thaliana* flowers. *Proceedings of the National Academy of Sciences of the United States of America, 114*, 7166–7171.

Rebocho, A. B., Kennaway, J. R., Bangham, J. A., & Coen, E. (2017). Formation and shaping of the *Antirrhinum* flower through modulation of the *CUP* boundary gene. *Current Biology, 27*, 2610–2622.

Reinhardt, D., Pesce, E. R., Stieger, P., Mandel, T., Baltensperger, K., Bennett, M., et al. (2003). Regulation of phyllotaxis by polar auxin transport. *Nature, 426*, 255–260.

Ripoll, J. J., Bailey, L. J., Mai, Q.-A., Wu, S. L., Hon, C. T., Chapman, E. J., et al. (2015). microRNA regulation of fruit growth. *Nature Plants, 1*, 15036.

Rodriguez-Cazorla, E., Ripoll, J. J., Andujar, A., Bailey, L. J., Martinez-Laborda, A., Yanofsky, M. F., et al. (2015). K-homology nuclear ribonucleoproteins regulate floral organ identity and determinacy in *Arabidopsis*. *PLoS Genetics, 11*, e1004983.

Ronald, P. C. (2014). Lab to farm: Applying research on plant genetics and genomics to crop improvement. *PLoS Biology, 12*, e1001878.

Ryan, P. T., O'Maoileidigh, D. S., Drost, H.-G., Kwasniewska, K., Gabel, A., Grosse, I., et al. (2015). Patterns of gene expression during Arabidopsis flower development from the time of initiation to maturation. *BMC Genomics, 16*, 488.

Sablowski, R. (2015). Control of patterning, growth, and differentiation by floral organ identity genes. *Journal of Experimental Botany, 66*, 1065–1073.

Sakai, H., Krizek, B. A., Jacobsen, S. E., & Meyerowitz, E. M. (2000). Regulation of *SUP* expression identifies multiple regulators involved in *Arabidopsis* floral meristem development. *Plant Cell, 12*, 1607–1618.

Sakai, H., Medrano, L. J., & Meyerowitz, E. M. (1995). Role of *SUPERMAN* in maintaining *Arabidopsis* floral whorl boundaries. *Nature, 378*, 199–203.

Sayou, C., Nanao, M. H., Jamin, M., Pose, D., Thevenon, E., Gregoire, L., et al. (2016). A SAM oligomerization domain shapes the genomic binding landscape of the LEAFY transcription factor. *Nature Communications, 7*, 11222.

Schiessl, K., Kausika, S., Southam, P., Bush, M., & Sablowski, R. (2012). JAGGED controls growth anisotropy and coordination between cell size and cell cycle during plant organogenesis. *Current Biology, 22*, 1739–1746.

Schiessl, K., Muino, J. M., & Sablowski, R. (2014). Arabidopsis JAGGED links floral organ patterning to tissue growth by repressing kip-related cell cycle inhibitors. *Proceedings of the National Academy of Sciences of the United States of America, 111*, 2830–2835.

Schilling, S., Pan, S., Kennedy, A., & Melzer, R. (2018). MADS-box genes and crop domestication: The jack of all traits. *Journal of Experimental Botany, 69*, 1447–1469.

Schmid, M., Davison, T. S., Henz, S. R., Pape, U. J., Demar, M., Vingron, M., et al. (2005). A gene expression map of Arabidopsis thaliana development. *Nature Genetics, 37*, 501–506.

Scutt, C. P., & Vandenbussche, M. (2014). Current trends and future directions in flower development research. *Annals of Botany, 114*, 1399–1406.

Serrano-Mislata, A., Goslin, K., Zheng, B., Rae, L., Wellmer, F., Graciet, E., et al. (2017). Regulatory interplay between LEAFY, APETALA1/CAULIFLOWER and TERMINAL FLOWER1: New insights into an old relationship. *Plant Signaling & Behavior, 12*, e1370164.

Sieber, P., Wellmer, F., Gheyselinck, J., Riechmann, J. L., & Meyerowitz, E. M. (2007). Redundancy and specialization among plant microRNAs: Role of the *MIR164* family in developmental robustness. *Development, 134*, 1051–1060.

Simonini, S., Roig-Villanova, I., Gregis, V., Colombo, B., Colombo, L., & Kater, M. M. (2012). Basic pentacysteine proteins mediate MADS domain complex binding to the DNA for tissue-specific expression of target genes in *Arabidopsis*. *Plant Cell, 24*, 4163–4172.

Smaczniak, C., Immink, R. G., Muino, J. M., Blanvillain, R., Busscher, M., Busscher-Lange, J., et al. (2012). Characterization of MADS-domain transcription factor complexes in *Arabidopsis* flower development. *Proceedings of the National Academy of Sciences of the United States of America, 109*, 1560–1565.

Smaczniak, C., Li, N., Boeren, S., America, T., van Dongen, W., Goerdayal, S. S., et al. (2012). Proteomics-based identification of low-abundance signaling and regulatory protein complexes in native plant tissues. *Nature Protocols*, 7, 2144–2158.

Smaczniak, C., Muino, J. M., Chen, D., Angenent, G. C., & Kaufmann, K. (2017). Differences in DNA binding specificity of floral homeotic protein complexes predict organ-specific target genes. *Plant Cell, 29*, 1822–1835.

Smyth, D. R., Bowman, J. L., & Meyerowitz, E. M. (1990). Early flower development in *Arabidopsis. Plant Cell, 2*, 755–767.

Song, S., Qi, T., Huang, H., & Xie, D. (2013). Regulation of stamen development by coordinated actions of jasmonate, auxin, and gibberellin in *Arabidopsis. Molecular Plant, 6*, 1065–1073.

Stewart, D., Graciet, E., & Wellmer, F. (2016). Molecular and regulatory mechanisms controlling floral organ development. *FEBS Journal, 283*, 1823–1830.

Sun, B., Looi, L. S., Guo, S., He, Z., Gan, E. S., Huang, J., et al. (2014). Timing mechanism dependent on cell division is invoked by Polycomb eviction in plant stem cells. *Science, 343*, 1248559.

Sun, B., Xu, Y., Ng, K. H., & Ito, T. (2009). A timing mechanism for stem cell maintenance and differentiation in the *Arabidopsis* floral meristem. *Genes & Development, 23*, 1791–1804.

Takeda, S., Matsumoto, N., & Okada, K. (2004). *RABBIT EARS*, encoding a SUPERMAN-like zinc finger protein, regulates petal development in *Arabidopsis thaliana. Development, 131*, 425–434.

Theissen, G. (2001). Development of floral organ identity: Stories from the MADS house. *Current Opinion in Plant Biology, 4*, 75–85.

Theissen, G., Melzer, R., & Rumpler, F. (2016). MADS-domain transcription factors and the floral quartet model of flower development: Linking plant development and evolution. *Development, 143*, 3259–3271.

Vachon, G., Engelhorn, J., & Carles, C. C. (2018). Interactions between transcription factors and chromatin regulators in the control of flower development. *Journal of Experimental Botany, 69*, 2461–2471.

von Goethe, J. W. (1790). *Versuch die Metamorphose der Pflanzen zu erklären*. Gotha, Germany: Ettinger.

Vroemen, C. W., Mordhorst, A. P., Albrecht, C., Kwaaitaal, M. A., & de Vries, S. C. (2003). The *CUP-SHAPED COTYLEDON3* gene is required for boundary and shoot meristem formation in *Arabidopsis. Plant Cell, 15*, 1563–1577.

Wagner, D., Sablowski, R. W., & Meyerowitz, E. M. (1999). Transcriptional activation of *APETALA1* by LEAFY. *Science, 285*, 582–584.

Wang, P., Liao, H., Zhang, W., Yu, X., Zhang, R., Shan, H., et al. (2015). Flexibility in the structure of spiral flowers and its underlying mechanisms. *Nature Plants, 2*, 15188.

Weigel, D., Alvarez, J., Smyth, D. R., Yanofsky, M. F., & Meyerowitz, E. M. (1992). LEAFY controls floral meristem identity in *Arabidopsis. Cell, 69*, 843–859.

Wellmer, F., Alves-Ferreira, M., Dubois, A., Riechmann, J. L., & Meyerowitz, E. M. (2006). Genome-wide analysis of gene expression during early *Arabidopsis* flower development. *PLoS Genetics, 2*, e117.

Wellmer, F., Bowman, J. L., Davies, B., Ferrandiz, C., Fletcher, J. C., Franks, R. G., et al. (2014). Flower development: Open questions and future directions. *Methods in Molecular Biology, 1110*, 103–124.

Wellmer, F., Graciet, E., & Riechmann, J. L. (2014). Specification of floral organs in Arabidopsis. *Journal of Experimental Botany, 65*, 1–9.

Wellmer, F., Riechmann, J. L., Alves-Ferreira, M., & Meyerowitz, E. M. (2004). Genome-wide analysis of spatial gene expression in *Arabidopsis* flowers. *Plant Cell, 16,* 1314–1326.

Wigge, P. A., Kim, M. C., Jaeger, K. E., Busch, W., Schmid, M., Lohmann, J. U., et al. (2005). Integration of spatial and temporal information during floral induction in *Arabidopsis*. *Science, 309,* 1056–1059.

William, D. A., Su, Y., Smith, M. R., Lu, M., Baldwin, D. A., & Wagner, D. (2004). Genomic identification of direct target genes of LEAFY. *Proceedings of the National Academy of Sciences of the United States of America, 101,* 1775–1780.

Wils, C. R., & Kaufmann, K. (2017). Gene-regulatory networks controlling inflorescence and flower development in Arabidopsis thaliana. *Biochimica et Biophysica Acta, 1860,* 95–105.

Winter, C. M., Austin, R. S., Blanvillain-Baufume, S., Reback, M. A., Monniaux, M., Wu, M. F., et al. (2011). LEAFY target genes reveal floral regulatory logic, *cis* motifs, and a link to biotic stimulus response. *Developmental Cell, 20,* 430–443.

Winter, C. M., Yamaguchi, N., Wu, M. F., & Wagner, D. (2015). Transcriptional programs regulated by both LEAFY and APETALA1 at the time of flower formation. *Physiologia Plantarum, 155,* 55–73.

Wollmann, H., Mica, E., Todesco, M., Long, J. A., & Weigel, D. (2010). On reconciling the interactions between APETALA2, miR172 and AGAMOUS with the ABC model of flower development. *Development, 137,* 3633–3642.

Wu, F., Shi, X., Lin, X., Liu, Y., Chong, K., Theissen, G., et al. (2017). The ABCs of flower development: Mutational analysis of *AP1/FUL*-like genes in rice provides evidence for a homeotic (A)-function in grasses. *Plant Journal, 89,* 310–324.

Wu, M. F., Yamaguchi, N., Xiao, J., Bargmann, B., Estelle, M., Sang, Y., et al. (2015). Auxin-regulated chromatin switch directs acquisition of flower primordium founder fate. *eLife, 4,* e09269.

Wuest, S. E., O'Maoileidigh, D. S., Rae, L., Kwasniewska, K., Raganelli, A., Hanczaryk, K., et al. (2012). Molecular basis for the specification of floral organs by APETALA3 and PISTILLATA. *Proceedings of the National Academy of Sciences of the United States of America, 109,* 13452–13457.

Wurschum, T., Gross-Hardt, R., & Laux, T. (2006). APETALA2 regulates the stem cell niche in the *Arabidopsis* shoot meristem. *Plant Cell, 18,* 295–307.

Xu, Y., Prunet, N., Gan, E. S., Wang, Y., Stewart, D., Wellmer, F., et al. (2018). SUPERMAN regulates floral whorl boundaries through control of auxin biosynthesis. *EMBO Journal, 37,* e97499.

Yamaguchi, N., Huang, J., Xu, Y., Tanoi, K., & Ito, T. (2017). Fine-tuning of auxin homeostasis governs the transition from floral stem cell maintenance to gynoecium formation. *Nature Communications, 8,* 1125.

Yamaguchi, N., Jeong, C. W., Nole-Wilson, S., Krizek, B. A., & Wagner, D. (2016). AINTEGUMENTA and AINTEGUMENTA-LIKE6/PLETHORA3 induce *LEAFY* expression in response to auxin to promote the onset of flower formation in *Arabidopsis*. *Plant Physiology, 170,* 283–293.

Yamaguchi, N., Winter, C. M., Wu, M. F., Kanno, Y., Yamaguchi, A., Seo, M., et al. (2014). Gibberellin acts positively then negatively to control onset of flower formation in *Arabidopsis*. *Science, 344,* 638–641.

Yamaguchi, N., Wu, M. F., Winter, C. M., Berns, M. C., Nole-Wilson, S., Yamaguchi, A., et al. (2013). A molecular framework for auxin-mediated initiation of flower primordia. *Developmental Cell, 24,* 271–282.

Yamaguchi, A., Wu, M. F., Yang, L., Wu, G., Poethig, R. S., & Wagner, D. (2009). The microRNA-regulated SBP-Box transcription factor SPL3 is a direct upstream activator of *LEAFY, FRUITFULL,* and *APETALA1*. *Developmental Cell, 17,* 268–278.

Yan, W., Chen, D., & Kaufmann, K. (2016). Molecular mechanisms of floral organ specification by MADS domain proteins. *Current Opinion in Plant Biology, 29*, 154–162.

Yang, W. C., Ye, D., Xu, J., & Sundaresan, V. (1999). The *SPOROCYTELESS* gene of *Arabidopsis* is required for initiation of sporogenesis and encodes a novel nuclear protein. *Genes & Development, 13*, 2108–2117.

Yanofsky, M. F., Ma, H., Bowman, J. L., Drews, G. N., Feldmann, K. A., & Meyerowitz, E. M. (1990). The protein encoded by the *Arabidopsis* homeotic gene *AGAMOUS* resembles transcription factors. *Nature, 346*, 35–39.

Zhang, X., Feng, B., Zhang, Q., Zhang, D., Altman, N., & Ma, H. (2005). Genome-wide expression profiling and identification of gene activities during early flower development in *Arabidopsis*. *Plant Molecular Biology, 58*, 401–419.

Zhang, B., Holmlund, M., Lorrain, S., Norberg, M., Bako, L., Fankhauser, C., et al. (2017). BLADE-ON-PETIOLE proteins act in an E3 ubiquitin ligase complex to regulate PHYTOCHROME INTERACTING FACTOR 4 abundance. *eLife, 6*, e26759.

Zhang, K., Wang, R., Zi, H., Li, Y., Cao, X., Li, D., et al. (2018). AUXIN RESPONSE FACTOR3 regulates floral meristem determinacy by repressing cytokinin biosynthesis and signaling. *Plant Cell, 30*, 324–346.

Zhao, Z., Andersen, S. U., Ljung, K., Dolezal, K., Miotk, A., Schultheiss, S. J., et al. (2010). Hormonal control of the shoot stem-cell niche. *Nature, 465*, 1089–1092.

CHAPTER NINE

Plus ça change, plus c'est la même chose: The developmental evolution of flowers

Elena M. Kramer*
Harvard University, Cambridge, MA, United States
*Corresponding author: e-mail address: ekramer@oeb.harvard.edu

Contents

Abstract

The study of floral developmental evolution may be as old as the study of botany itself, but its history as a modern field largely dates to the characterization of the ABC model of floral organ identity. As we have come to understand more about the wide range of genetic programs that influence floral development, the potential for comparative studies has increased, leading to a steady expansion of research questions. This review attempts to take a broad view of these many questions, highlighting classic and recent findings that are shaping the way we think about floral evolution. Conservation of genetic pathways is a common theme, but there are often differences, whether due to drift or co-option, that have the potential to surprise. On the whole, the field of floral evo-devo is in transition, as we apply new approaches that allow us to move away from our reliance on candidate genes and, potentially, discover genuinely new developmental mechanisms.

Current Topics in Developmental Biology, Volume 131
ISSN 0070-2153
https://doi.org/10.1016/bs.ctdb.2018.11.015

1. Introduction

The study of plant biology and diversity has always been based on comparative development and morphology, as best evidenced by Goethe's observation "Alles ist Blatt" ("all is leaf"; Goethe, 1790), which encompasses both the completely serial nature of plant development and the critical role of homeosis—the transformation of identity—in the evolution of plant form. However, the modern study of plant evolutionary developmental biology (evo-devo) largely started with the elucidation of the ABC model of floral organ identity and the initial observations made in the seminal "war of the whorls" paper regarding the evolutionary implications of this homeotic program (Coen & Meyerowitz, 1991). Coen and Meyerowitz (1991) noted that comparative studies of *Arabidopsis thaliana* and *Antirrhinum majus* would allow us to understand "whether the evolutionary changes that lead to the very different flower forms of the two plants are due to changes in the homeotic genes themselves, changes in their regulators, or changes in the downstream genes that they regulate." In many ways, this observation delineated the first decade of floral evo-devo, which largely focused on how modifications of the ABC model led to changes in floral form. There were several reasons for this. First, much to our good luck, it turned out that the genetic components of the ABC model are extraordinarily easy to identify in other taxa, and second, many interesting changes in floral form do relate to differences in floral organ identity. This approach defines one major theme in floral evo-devo, which depends on highly conserved genetic pathways that can be easily studied using candidate genes approaches—essentially, the experimental "low-hanging fruit." The candidate gene approach has yielded impressive and, in some cases, surprising results regarding the mechanisms that shape floral morphological evolution. Moving beyond these candidate gene approaches has proven more challenging, but recent advances in next generation sequencing, reverse genetics, and evolutionary genetics have the potential to help us answer the more challenging questions in floral evo-devo.

2. Highly conserved (or convergent) pathways

2.1 Floral organ identity

As originally defined, the ABC model posits that three classes of gene activity act in a combinatorial fashion to determine the identity of the four major types of floral organs: A class alone promoting sepal identity; A+B class, petal identity; B+C class, stamen identity, and C alone, carpel identity (Fig. 1A; see chapter "Molecular regulation of flower development" by Thomson & Wellmer, this issue). Almost all of the genes participating in this program are members of the MADS-domain superfamily of transcription factors and are now known to function in protein complexes,

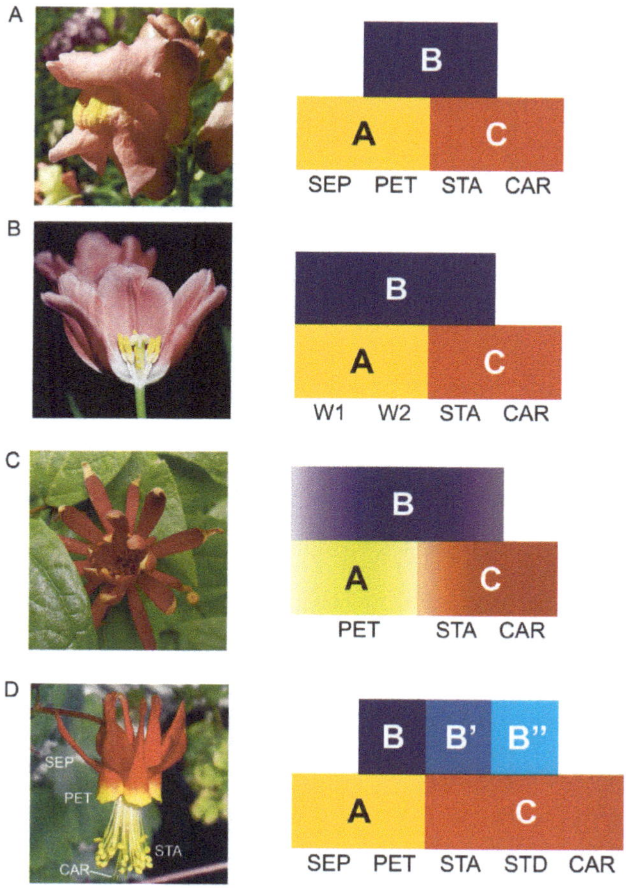

Fig. 1 (A) The classic ABC model of floral organ identity was first described in models such as snapdragon (left) and defines a mechanism for establishing the identity of sepals (SEP), petals (PET), stamens (STA), and carpels (CAR). See text for discussion of the challenges regarding A function. (B) The entirely petaloid perianth of many monocots, such as tulip (left), appears to be due to expression of the B class genes in both the first (W1) and second (W2) whorls. (C) In magnolid dicots such as species in the genus *Calycanthus* (left), intermediate organ types that grade between petaloid, stamenoid, and carpeloid identities are associated with gradients of identity gene expression across the floral meristem. (D) Species in the lower eudicot genus *Aquilegia* has five types of floral organs, the typical four plus stamindoes (STD). These are programmed by duplicate copies of the B class genes, here annotated as B, B′ and B″, which have experienced sub- and neofunctionalization (see text). *Photos in (A)–(C), public domain images from Wikimedia. Photos in (D) and (E) M. Kramer.*

explaining the combinatorial nature of the model (Theissen & Saedler, 2001). This homeotic program also gives us a straightforward mechanism to explain a specific class of dramatic transitions in floral morphology, namely, those that relate to transformations in organ identity. As can be observed in any trip to the local garden center, homeotic mutations abound in all lineages of the flowering plants and, in some cases, these changes have occurred in an evolutionary context as well as a horticultural one (reviewed in Theissen, 2010). The prevalence of such homeosis underscores the broad conservation of the ABC program, particularly the core identity programs controlling stamen and carpel identity (reviewed in Litt & Kramer, 2010). When we study homologs of the B class genes *APETALA3* (*AP3*) and *PISTILLATA* (*PI*), or the C class gene *AGAMOUS* (*AG*), there are sometimes differences in the details of their function, typically due to independent patterns of gene duplication (e.g., Causier et al., 2005; Davies et al., 1999), but the basic B + C = stamen and C = carpel code is highly conserved, consistent with the homology of stamens and carpels across the angiosperms.

Moving outside the reproductive organs to the sterile floral parts, collectively referred to as the perianth, things quickly become more complicated. Considering petal identity, I first have to distinguish between "petals," which are sterile, typically showy organs positioned in the second whorl of the flower, versus "petaloidy," which also refers to the condition of being sterile and showy but can occur in any whorl of the flower, even in extrafloral organs. It is very common for petaloid organs, especially in the second whorl of the flower, to express homologs of the B class genes and depend on these loci for their identity (reviewed in Litt & Kramer, 2010). Even taxa with undifferentiated petaloid organs, such as lilies, tulips, or magnolias, appear to rely on B gene homologs to promote the identity of their petaloid organs, regardless of their position in the flower or their phyllotaxy (Fig. 1B; Kanno, 2016; Kim, Koh, Yoo, et al., 2005). Likewise, magnoliid dicots whose floral organs gradually transition from one identity to another, express their B and C class genes in gradients across the floral primordia such that their chimeric organs are correlated with varying dosages of B and C gene expression (Fig. 1C; Kim, Koh, Ma, et al., 2005). At the same time, however, it is often difficult to say with certainty whether this similarity reflects common inheritance of a petal identity program and/or homology of the petaloid organs themselves. Both petals in the second whorl and petaloid organs in general are thought to have evolved many times independently in the angiosperms (Takhtajan, 1991), and it is possible that B class homologs have been recruited to control identity in these many instances (Baum & Whitlock, 1999).

Probably the best understood example of independent evolution of petaloid organs is the core eudicot Caryophyllales lineage (Brockington et al., 2009, 2012). Brockington and colleagues clearly demonstrated, using phylogenetic ancestral character state reconstruction, that petaloid organs were lost early in the diversification of the group but then subsequently regained in several lineages (Brockington et al., 2009). They further used detailed gene expression studies to investigate separate instances of the reappearance of petaloid organs (Brockington et al., 2012). In one case, the outer stamens have become petaloid and are found to express B gene homologs, but they also transiently express C homologs. This emphasizes the fact that stamens are likely to have been sterilized many times independently across the angiosperms and many, if not most, of these derived sterile organs will still express B class genes. Therefore, B gene expression alone cannot be taken as evidence of a commonly inherited identity program. However, in another lineage of Caryophyllales, the first whorl organs evolved petaloid traits but they completely lack B homolog expression. This result demonstrates that petaloid organs can evolve without the aid of B gene homologs, a pattern that has also been observed in petaloid sepals and bracts of diverse other lineages (Feng, Liu, Yu, Franks, & Xiang, 2012; Geuten et al., 2006; Jaramillo & Kramer, 2004; Park, Ishikawa, Yoshida, Kanno, & Kameya, 2003). In the lower eudicot genus *Aquilegia*, there is a similar example of petaloid sepals that do not rely on B gene function for their identity (Kramer et al., 2007), but these organs do use late B gene expression to activate anthocyanin production (Sharma & Kramer, 2017). This is a relatively rare instance of transference of attractive function to the first whorl sepals that is combined with a non-homeotic recruitment of the B class genes to promote aspects of petaloidy but not identity.

Just as the Caryophyllales are a clear example of independently derived petaloid organs, the buttercup family, Ranunculaceae, have become a surprising example of petal conservation. In this family, second whorl petals were always thought to have evolved many times independently by sterilization of outer stamens (Kosuge, 1994). However, reconstruction of the trait using an unbiased coding and modern phylogeny finds that the presence of petals is likely to have been ancestral (Zhang, Guo, et al., 2013). Furthermore, it turns out that early duplications of the *AP3* lineage in the last common ancestor of the broader Ranunculales produced three paralogous lineages, one of which, termed *AP3-3*, experienced subfunctionalization to specifically control petal identity (Fig. 1D; Sharma, Guo, Kong, & Kramer, 2011; Zhang, Guo, et al., 2013; Zhang, Steinmann, Nikolov, Kramer, & Davis, 2013). This petal-specific expression is widely conserved across both the Ranunculaceae and Berberidaceae sister families and has been functionally verified using reverse genetics and stable mutants in multiple Ranunculaceae genera (Goncalves et al., 2013; Sharma et al., 2011; Zhang, Guo, et al., 2013; Zhang, Steinmann, et al., 2013). Moreover, apetalous members of the family have lost expression of their *AP3-3* orthologs, which are often observed to be experiencing pseudogenization in the genome (Zhang, Guo, et al., 2013; Zhang, Steinmann, et al., 2013). Collectively, these data strongly suggest that the petals of the Ranunculaceae, and likely the Berberidaceae, are, in fact, homologous and share process homology via a commonly inherited, subfunctionalized petal identity program that was turned off many times independently. Thus, there is no hard and fast rule. Most petaloid organs appear to use B gene homologs to promote their identity, and this is likely due to a commonly inherited genetic program across large angiosperm clades, but there are many exceptions where petaloid organs have been lost and regained using alternative or homoplastic (convergently evolved) genetic mechanisms.

Up to now, I have been studiously avoiding any discussion of A function, which appears to be much more evolutionarily labile than B or C function. Considering the ABC model as it was originally defined in *A. thaliana*, we see that there were two components of A function: (1) sepal and petal identity and (2) repression of C function in the outer whorls of the flower (termed a cadastral function) (Bowman, Smyth, & Meyerowitz, 1991). In *A. thaliana*, the MADS-box gene *APETALA1* (*AP1*) was thought to function primarily in the first role, while the AP2-domain transcription factor *APETALA2* (*AP2*) controlled the second (Bowman, Alvarez, Weigel, Meyerowitz, & Smyth, 1993; Drews, Bowman, & Meyerowitz, 1991). However, comparative studies have revealed that *AP1* represents a core eudicot-specific paralogous lineage (Litt & Irish, 2003), whose function is not strictly required for sepal and petal identity, even in *A. thaliana* (reviewed in Litt, 2007). In the core eudicots, orthologs of *AP1* generally rarely contribute to sepal/petal identity, and outside the core eudicots, representatives of the ancestral *AP1/FRUITFULL* (*FUL*) lineage primarily function in leaf and fruit development rather than sepal and petal identity (Pabon-Mora, Sharma, Holappa, Kramer, & Litt, 2013). However, that does not mean that there are no other sepal and petal identity genes. For example, a comprehensive study of floral MADS-box gene function in the lower eudicot *Nigella damascena* found that another gene, *NdAGL6*, is essential for proper sepal and petal identity (Wang et al., 2015). Similarly, broad comparative expression studies in representatives of the magnoliid dicots found perianth-specific expression of the *AGL6* homologs, which may be consistent with a function in the identity of these organs (Kim, Koh, Yoo, et al., 2005). This complexity reminds us that we should try to avoid conflating aspects of genetic function, like sepal and petal identity, with the homology of a particular gene lineage. In the case of B and C gene function, the general conservation of their functional roles allows us to routinely talk about "B genes" as a short-hand for "genetic homologs of *AP3* and *PI* that play conserved roles in petal and stamen identity." With A class genes, this assumption cannot be made. All angiosperm genomes examined so far contain homologs of *AP1*, but, most likely, these do not function in sepal and petal identity, although other genes may contribute to the identity aspect of "A" function.

The cadastral function of *AP2* is equally complex. In *A. thaliana*, we know that *AP2* is regulated both at the transcriptional level and by the microRNA *miR172*, such that AP2 protein is only produced in the first and second whorl of the flower (Chen, 2004; Wollmann, Mica, Todesco, Long, & Weigel, 2010). The resultant AP2 protein acts as a transcriptional repressor by interacting with the co-factor TOPLESS to recruit histone deacetylase to target loci, including the C gene *AG* (Krogan, Hogan, & Long, 2012). This explains the loss of sepal and petal identity in *ap2* mutants, which is due to expansion of *AG* expression across the floral meristem (Drews et al., 1991). Orthologs of *AP2* have been studied in a number of other core eudicots, but none of them appear to have similar functions in regulating *AG* expression, although they do contribute to aspects of petal morphogenesis (reviewed in Litt, 2007). Recently, a tour de force genetic study in *Petunia hybrida* (petunia) revealed the evolutionary intricacy of cadastral regulation in this model species (Morel et al., 2017). Morel et al. (2017) used a combination of forward and reverse genetics to tease apart the functions of five different *AP2* family members: three that are close homologs of *AP2* (termed *PhROB1-3*) and two that are representatives of a different branch of the AP2 transcription factor family (termed *PhBEN* and *PhBOB*). It turns out that the closest homologs of *AP2*, the *PhROB1-3* loci, do contribute to normal floral organ development, but as cadastral loci they play only minor roles in repressing B gene expression in the sepals. In contrast, the gene *PhBEN* makes substantial contributions to limiting both the B and C gene expression domains, while its close paralog *PhBOB* functions in second whorl development. These diverse functions act in parallel to a completely different genetic pathway, defined by the miRNA *miR169* and its NF-YA transcription factor targets, to restrict C gene expression to the inner two whorls in petunia as well as other Asterids (Cartolano et al., 2007). Again, we see that this aspect of "A" function has experienced considerable developmental system drift (sensu True & Haag, 2001) during the diversification of the angiosperms. In almost every flower, there must be genetic mechanisms that restrict C function to the inner whorls, but the exact genetic players cannot be predicted with great confidence.

Finally, comparative studies of the ABC model allow us to consider the mechanisms underlying the evolution of novel floral organ identities. Although the majority of flowers have the four canonical organ types (or sometimes fewer in cases of unisexual flowers or perianth loss), it is surprisingly common for flowers to have five or more types of floral organs. Note that I am not referring to flowers with chimeric organs arranged in transformational grades, but rather five or more discrete types of floral organs. Such flowers require us to ask how the otherwise conserved ABC program can be modified to allow the intercalation of new organ types into the context of the flower. Such novel organs fall into two broad classes: those that are derived from a subset of previously present organs, or those that appear to have arisen de novo. The vast majority of novel organs are derived from previously existing floral parts, most commonly the stamens. These structures, termed staminodes, have arisen in dozens of different angiosperm families and range in morphological complexity (Walker-Larsen & Harder, 2000).

Staminodes have been best studied at the molecular level in two divergent angiosperm lineages, the monocot order Zingiberales, commonly referred to as the gingers, and the Ranunculaceae genus *Aquilegia*. Across the order Zingiberales, the transformation of particular fertile stamens into petaloid organs has occurred to varying degrees in multiple families (Specht, Yockteng, Almeida, Kirchoff, & Kress, 2012). This morphological transition would appear to require restriction of C gene expression but, in fact, studies suggest that the petaloid staminodes typically still express *AG* homologs (Morioka, Yockteng, Almeida, & Specht, 2015; Tian, Zou, Miao, Ning, & Liao, 2018). At the same time, there are other dramatic expression differences in the genes that control abaxial/adaxial development, consistent with the laminar expansion of the staminodes compared to the narrow filaments of the fertile stamens (Morioka et al., 2015; Tian et al., 2018). Although we still have no functional data in the gingers, it appears that pathways acting largely in parallel to the organ identity genes are responsible for staminode development. The situation in the genus *Aquilegia* is more phylogenetically restricted, such that only *Aquilegia* and its two closest sister genera produce a whorl of broad, sterile staminodes positioned between the fertile stamens and carpels (Kramer et al., 2007). Here, MADS-box gene duplication has been found to play a role, involving two of the three Ranunculaceae *AP3* paralogs mentioned above.

In addition to the petal-specific *AP3-3*, species of *Aquilegia* have two other *AP3* homologs, termed *AqAP3-1* and *AqAP3-2* (Kramer, Di Stilio, & Schluter, 2003). Functional studies have found that *AP3-2* in *Aquilegia coerulea* has evolved to specifically confer stamen identity while *AP3-1* has experienced neofunctionalization to promote the identity of the novel staminodes (Fig. 1D; Sharma & Kramer, 2013). These results suggest that comparatively old gene duplications can still give rise to neofunctionalization, in this case a new organ identity, long after the original gene duplication occurred.

One last example of a modified pre-existing organ is the labellum in members of the Orchidaceae (orchids). This case involves both floral zygomorphy (bilateral symmetry) and a novel organ identity. The labellum represents a modification of the petal whorl in which one petal has become fused with an adjacent stamen and then evolved a dramatically different morphology relative to the other two petals, thereby contributing to the bilateral symmetry of the flower. Interestingly, this case also involves duplication of the B gene *AP3*. Four ancient paralogous *AP3* lineages that arose before the diversification of the orchid family have evolved sub- and neofunctionalized roles that allow them to promote the identity of three discretely different types of petaloid organs—the orchid sepals, petals, and labellum—as well as the male reproductive tissues (Hsu et al., 2015; Mondragon-Palomino & Theissen, 2008). Although there appears to be a convergence between *A. coerulea* and orchids, in that they both use duplicated B genes to promote novel organ identity, it is important to note that duplications are rampant among the floral MADS-box genes across the angiosperms (Airoldi & Davies, 2012). Most instances of duplication are not associated with dramatic changes in floral morphology and, instead, appear to result in subfunctionalization or degrees of maintained redundancy (reviewed in Litt & Kramer, 2010), which may improve developmental robustness. Furthermore, in several separate lineages of flowering plants, there are instances of genuinely de novo organs that do not follow the *Aquilegia*/orchid model at all. Perhaps the most common example of this phenomenon is the corona, which is well known in the genera *Passiflora* and *Narcissus*, and the Aponcynaceae (milkweed family). These structures are typically positioned between the stamens and the petals and are quite late-arising, long after the other floral organs have initiated (Endress, 2016; Hemingway, Christensen, & Malcomber, 2011; Waters et al., 2013). The molecular basis of identity in the corona or corona-like structures has been explored in *Passiflora caerulea*, *Narcissus bulbocodium*, and all three genera of the Rafflesiaceae. Interestingly, they all share B and C gene homolog expression (Hemingway et al., 2011; Nikolov et al., 2013; Waters et al., 2013), but it remains entirely unclear how their initiation and distinct morphology are controlled. Clearly, there are still aspects of floral organ identity that are yet to be explored.

2.2 Floral symmetry

I have defined the experimental "low-hanging fruit" of plant evo-devo as questions that can easily be studied using candidate gene approaches. The ABC model clearly fits this definition but the second major developmental program that I have placed under this heading would not have been an obvious choice before the advent of molecular approaches. Floral bilateral symmetry, or zygomorphy, has arisen many times independently from radial symmetry (actinomorphy) across the flowering plants, often in association with major radiations and the evolution of specific pollinator relationships (reviewed in Citerne, Jabbour, Nadot, & Damerval, 2010; Hileman, 2014). Given this homoplastic pattern of evolution, it would be reasonable to assume that many different genetic mechanisms might be at play but, instead, researchers in floral evo-devo have found that in almost every instance examined, homologs of the same gene family have been independently recruited to control the development of this feature.

The genetic basis of zygomorphy was elucidated in *A. majus* (snapdragon) thanks to a series of insertional mutants that resulted in partial to complete radial transformation of floral organ morphology (Carpenter & Coen, 1990; Luo et al., 1999; Luo, Carpenter, Vincent, Copsey, & Coen, 1996). The two loci, called *CYCLOIDEA* (*CYC*) and *DICHOTOMA* (*DICH*), were identified as recently duplicated members of what is now known as the TEOSINTE BRANCHED1/CYCLOIDEA/PROLIFERATING CELL NUCLEAR ANTIGEN FACTOR1 (TCP) transcription factor family. These genes contribute to several aspects of floral symmetry, including the differential growth of dorsal and lateral petals, the abortion of the dorsal stamens, and the establishment of proper number of organs per whorl (termed merosity). Interestingly, the effect of *CYC/DICH* on lateral petals appears to be non-cell autonomous since the loci are not expressed in those organs (Luo et al., 1996). Deeper elucidation of the genetic program mediating snapdragon floral symmetry has explained this phenomenon. CYC/DICH was found to activate expression of a gene encoding a truncated MYB-domain protein, RADIALIS (RAD), that is thought to move non-cell autonomously within the early floral meristem (Corley, Carpenter, Copsey, & Coen, 2005). Under this model, a dorsal-ventral gradient of RAD protein then competes with a full-length MYB-domain protein, DIVARICATA (DIV), which promotes ventral organ identity (Fig. 2; Galego & Almeida, 2002). The negative effect of RAD on DIV function appears to be due to their mutual competition for other co-factors, the so-called DRIF proteins (Raimundo et al., 2013). Thus, in the dorsal area of the flower, full repression of DIV activity by RAD allows CYC/DICH to promote dorsal identity, while in the ventral area, unrestricted DIV promotes the alternate ventral morphology (Corley et al., 2005; Galego & Almeida, 2002; Raimundo et al., 2013). Given that the petals of *cyc dich div* triple mutants appear to all have lateral morphology (Almeida, Rocheta, & Galego, 1997), it may be that this developmental program is essentially the default condition of snapdragon petal identity, which is only active when RAD and DIV activities cancel each other out in lateral primordia.

Our appreciation of the conservation and convergence of *CYC/DICH* gene function has developed in a stepwise fashion. First, studies in maize discovered that a gene with sequence similarity to *CYC*, called *TEOSINTE BRANCHED1*, was responsible for regulating axillary meristem activity of lateral branches (Doebley, Stec, & Hubbard, 1997). Although this finding was not directly related to zygomorphy, it led to the definition of the TCP gene family and established a general function theme for its members as regulators of cell division (Cubas, Lauter, Doebley, & Coen, 1999). Comparative studies in other members of the Lamiales beyond *A. majus* found that *CYC*-like loci were also responsible for a radial mutant of *Linaria vulgaris* (Cubas, Vincent, & Coen, 1999) and implicated in shifts of floral symmetry

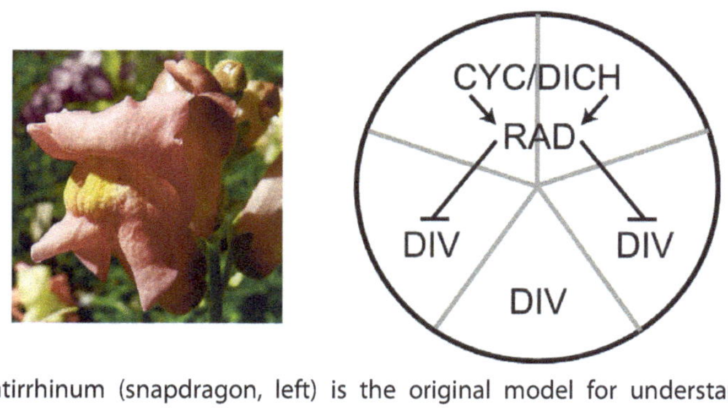

Fig. 2 Antirrhinum (snapdragon, left) is the original model for understanding the genetic program controlling floral zygomorphy (right). The paralogous TCP genes *CYCLOIDEA (CYC)* and *DICHOTOMA (DICH)* are expressed on the dorsal side of the developing floral meristem where they activate the MYB gene RADIALIS (RAD). RAD acts non-cell autonomously to oppose the activity of the MYB protein DIVARICATA (DIV).

in the genus *Mohavea* (Hileman, Kramer, & Baum, 2003). These findings, in themselves, were not overly surprising since the evolution of zygomorphy in the asterid Lamiales appears to have occurred quite early in their diversification, before the last common ancestor of the genera *Antirrhinum*, *Linaria*, and *Mohavea* (Hileman, 2014). However, in rapid succession, subsequent work in multiple divergent core eudicot lineages demonstrated that *CYC* homologs were responsible for independently derived zygomorphy in the rosid legumes (*Lotus japonicus*) and crucifers (*Iberis amara*), as well as Asteraceae (*Gerbera hybrida*) (Broholm et al., 2008; Busch & Zachgo, 2007; Feng et al., 2006). These systems are especially critical because the findings were based on functional data, including forward mutagenesis. There are some common themes across these taxa, such as duplications of the *CYC* homologs, a role in regulating cell proliferation, and combined functions in floral symmetry as well as the internal symmetry of individual organs, but there are also differences. The Asteraceae are a particularly complex case with as many as six distinct paralogous lineages (Chen, Shen, Guo, & Rao, 2018) that contribute to floret identity within the context of the capitulum inflorescence as well as the zygomorphy of individual florets (Broholm et al., 2008; Chapman et al., 2012; Juntheikki-Palovaara et al., 2014; Kim et al., 2008). Moreover, several of these paralogs function on the ventral side of the flower, rather than the dorsal. Recently, it was found that *RAD*- and *DIV*-like MYB genes also contribute to ray floret petal development in *Senecio vulgaris*, but their regulatory relationships to the *CYC*-like genes are still being teased apart (Garces, Spencer, & Kim, 2016). In contrast, regulatory connections between *CYC* and *RAD* homologs do not appear to exist in at least some rosid lineages (Costa, Fox, Hanna, Baxter, & Coen, 2005).

Comparative expression studies have expanded this list of examples by demonstrating that *CYC* homologs are consistently expressed in the dorsal petals of other zygomorphic eudicot lineages, including the Mapighiales and Dipsacales (reviewed in Hileman, 2014). Interestingly, several of these lineages offer the opportunity to investigate how changes in *CYC* expression are correlated with either shifts in zygomorphy or the complete reversion to actinomorphy. As a general rule, shifts in the axis of bilateral symmetry are perfectly reflected in changes in the expression of *CYC*, but the complete loss of zygomorphy can occur by numerous different pathways (reviewed in Hileman, 2014). These options include the loss of all *CYC* expression, ectopic radialization of *CYC* expression, and apparent instances in which actinomorphy is due to changes downstream of *CYC*. This variety is even observed across independent instances in the same family (Zhang, Guo, et al., 2013;

Zhang, Steinmann, et al., 2013). The role of *CYC* homologs in zygomorphic monocots is somewhat less clear but differential expression across the dorsal/ventral axis has been observed in some gingers, members of the Commelinaceae, and *Oryza sativa* (rice) (reviewed in Hileman, 2014). At the same time, *CYC*-like genes have not been implicated in the notably zygomorphic orchids (De Paolo, Gaudio, & Aceto, 2015). The one point of consistency across the zygomorphic rice, orchids, and Commelinaceae is that, unlike core eudicots, these monocot flowers have zygomorphic expression of their organ identity genes (Hsu et al., 2015; Preston & Hileman, 2012; Yamaguchi et al., 2006). Therefore, the program controlling floral symmetry in these lineages functions genetically upstream of organ identity, rather than in parallel to it.

Within the context of all this data, the surprising lacuna is a detailed understanding of *CYC* function in plants with actinomorphic flowers. We know that members of this gene family consistently influence cell division, both positively and negatively (Martin-Trillo & Cubas, 2010), but the *CYC* ortholog in *A. thaliana*, *TCP1*, has no loss-of-function phenotype. An activation-tagged allele of *TCP1* effects brassinosteroid biosynthesis (An, Guo, Gou, & Li, 2011; Guo et al., 2010), and the expression of *TCP1* is quite intriguing, being transiently expressed on the dorsal side of lateral floral meristems at very early developmental stages (Cubas, Coen, & Martinez Zapater, 2001). This has suggested a model in which the early dorsal expression of *CYC*-like genes has been developmentally extended many times independently, potentially via acquisition of positive auto-regulation (Yang et al., 2012). However, we have very little data on the expression of these loci in actinomorphic lineages, so assumptions about the ancestral functions of these genes are premature. Functional studies in other actinomorphic models, like the members of the Solanaceae, would help us understand why these genes have been so prone to independent co-option for convergent developmental roles.

3. Floral size

Floral organ identity and symmetry represent only a small fraction of the developmental processes that contribute to the wide range of angiosperm diversity. Another critical aspect is floral size, which incorporates three distinct developmental processes: the final size of the organs themselves, merosity, and the number of total whorls in the flower. Forward genetic studies, primarily in *A. thaliana*, have identified a broad set of potential candidate genes for these processes (Krizek & Anderson, 2013;

Moyroud & Glover, 2017), but the diversity of potential pathways is too daunting to take a gene-by-gene candidate approach. Instead, the application of evolutionary genetics in a wide array of new model systems is starting to provide insight into the mechanisms underlying variation in these traits.

3.1 Organ size

Our perception of floral size is often more a matter of final organ size, which, in turn, is controlled by the interaction between cell number and cell size. Given that ~95% of final organ size in plants is generated by cell expansion, control of cell size may seem like a good candidate for variation in final organ size, but it is not the only player. One powerful model system for exploring changes in floral organ size is the Brassiceae genus *Capsella*, in which a large-flowered outcrossing species *C. grandiflora* has recently given rise to a small-flowered selfing species, *C. rubella*, on a very recent time scale (reviewed in Sicard et al., 2011). A quantitative trait locus (QTL) analysis of *C. grandiflora* × *C. rubella* revealed six different loci that are responsible for their fivefold difference in petal size, which is primarily due to differences in cell number rather than cell size (Sicard et al., 2011). Two of the genes underlying these QTL have been characterized in detail. First, it was shown that the *C. rubella* allele of the F-box gene *STERILE APETALA* has acquired mutations in a petal-specific enhancer, such that the expression of the gene is reduced by 60% in those organs (Sicard et al., 2016). This change in expression reduces petal size by shortening the length of the cell proliferation period, thereby reducing the number of cells in the petal, particularly the distal blade. The second locus to be characterized tells quite a different story. In this case, the gene in question is a cytochrome P450 homologous to *CYP724A1* (Fujikura et al., 2018). The *C. rubella* allele of this locus has experienced two single nucleotide polymorphisms that result in increased splicing efficiency. This increased efficiency produces more functional protein that leads to higher-than-optimal levels of brassinosteroid (BR) synthesis. In *C. rubella*, this high BR level inhibits petal growth but also has pleiotropic effects on plant height, dry weight, and leaf size. Thus, in the case of the first QTL, we see a very specific effect on gene expression that minimizes pleiotropy, whereas the second QTL results in a surprisingly pleiotropic impact. Perhaps more importantly, these loci would never have been selected as candidates out of the dozens of genes that are known to impact cell proliferation, revealing the power of evolutionary genetic approaches.

Pollination biology strategies that differ between or even within species are also a theme in several other model systems. An example of within species variation is heterostyly, a phenomenon in which different individuals of the same species vary in terms of their stamen and style lengths (reviewed in Kappel, Huu, & Lenhard, 2017). In classic distyly, some individuals have short stamens and long styles, while others have long stamens and short styles. The production of these contrasting morphologies, which promote out-crossing, has evolved many times independently across the angiosperms. The genetic basis of the pattern has been the subject of study for over a century and is now yielding up its secrets thanks to evolutionary genetic and genomic approaches. Perhaps the best understood system right now is that of the genus *Primula*, in which the so-called *S* locus has been fully sequenced (Huu et al., 2016; Li et al., 2016, 2015; Nowak et al., 2015). The locus consists of a 280 kb insertion that is present in the hemizygous condition in S-morph (short style) individuals but completely absent in the L-morph (long style). The insertion interval contains five predicted genes, two of which have been directly connected to differences in style or stamen length. The first is a cytochrome P450 that is homologous to the *CYP734A50* lineage, which functions in BR homeostasis, in this case promoting hormone degradation (Huu et al., 2016; Nowak et al., 2015). When the gene is present with the S locus, high expression of the *CYP734A50* paralog decreases BR levels and results in less cell expansion and shorter styles, while its absence increases BR levels and produces longer styles. The second well-characterized component is a duplicate copy of the B class organ identity gene *GLOBOSA* (*GLO*) (Nowak et al., 2015). The exact mechanism of action is yet to be fully determined but it appears that when this extra *GLO* paralog is present, the increase in overall *GLO* expression results in higher anther position due to elongation of the corolla tube, to which the stamen filaments are fused. This result is rather surprising since such an effect on petal or stamen elaboration has not previously been described in transgenic lines with B gene over-expression (Krizek & Meyerowitz, 1996).

Several other model genera are being used to explore the genetic basis of natural variation in floral organ length or size by taking advantage of shifts in pollinator syndromes between inter-fertile species, which are often associated with changes in organ length and/or width, such as in the genera *Petunia*, *Penstemon* or *Mimulus*. In *Petunia axillaris* and *P. integrifolia*, differences in the length of the corolla tube depend on changes in both cell number and size (Stuurman et al., 2004), while differences affecting the stamen filaments and style between *P. axillaris* and *P. exserta* are primarily due to cell number

(Hermann, Klahre, Venail, Brandenburg, & Kuhlemeier, 2015). Although the causative loci underlying these phenotypes are yet to be identified, the identification of large effect QTL that impact multiple traits, including organ length as well as color and scent, reveal the existence of genomic "speciation islands" that can rapidly impact pollination biology and, therefore, genetic isolation (Hermann et al., 2013). A very similar pattern of large affect QTL controlling multiple pollinator-related traits has been observed in the *Penstemon neomexicanus* × *barbatus* system (Wessinger, Hileman, & Rausher, 2014). Members of the genus *Mimulus* are well-established evolutionary genetic model systems with an available genome sequence (Hellsten et al., 2013); they are routinely used for QTL and association mapping studies, as well as forward mutagenesis. All of these approaches are being applied to the question of floral organ size. For instance, QTL analysis of a cross between the small-flowered *M. laciniatus* and the large-flowered *M. guttatus* revealed few QTL of large effect, including one highly pleotropic locus that contains a homolog of the cell proliferation regulator *TCP4* (Ferris, Barnett, Blackman, & Willis, 2017). Another promising approach in *M. lewisii* is traditional forward mutagenesis, which has identified some surprising players in corolla tube width. One recent study recovered a semi-dominant mutant in *M. lewisii* that results in a decrease in corolla tube width due to a change in the final aspect ratio of the cells (Ding et al., 2017). This mutant turned out to be a dominant-negative allele of an actin locus, which results in cells that are narrower but do not change in length. Although it remains to be seen whether this kind of mutation could underlie natural variation in the trait, the difference in corolla tube width has been shown to affect pollinator visitation. Taken together, these studies in diverse systems suggest that numerous genetic pathways can influence organ size, which further underscores the importance of unbiased approaches that escape our dependence on candidate genes.

3.2 Organ number

An even more challenging question is how organ number may evolve in natural systems. The number of floral organs is due to several distinct developmental processes, including merosity and the number of whorls themselves. In some taxa, mechanisms such as ring primordia give rise to indeterminate organ numbers. Overall, we have a clear model for how primordium initiation is controlled by auxin trafficking (reviewed in Braybrook & Kuhlemeier, 2010) and at least a theoretical understanding of how transitions in organ initiation patterns (phyllotaxy) and merosity could occur (e.g., Smith & Bayer, 2009). Forward mutagenesis studies in several major model systems have revealed a wide array of loci that could potentially impact organ number (reviewed in Moyroud & Glover, 2017; Smyth, 2018), but many of these genes are also highly pleiotropic and may, therefore, be constrained in their evolutionary potential. Identifying the actual loci controlling evolution in floral organ number faces the additional challenge that traits such as merosity are often highly stable across large phylogenetic lineages (reviewed in Smyth, 2018), eliminating the option of using inter-species mapping. Finally, comparative transcriptomics may be unlikely to reveal dramatic differences in gene expression patterns since the same mechanisms of organ initiation are likely to be at work whether a flower has five organs per whorl or four. One intriguing and promising system is *Cardamine hirsuta*, in which petal number varies within the species (Monniaux, Pieper, & Hay, 2016). Initial results from QTL mapping reveal a complex system, in which many loci of small effect are responsible for variation in petal number (Monniaux et al., 2017). This, in turn, appears to reflect a release of developmental canalization, resulting in high variation in response to age and environmental conditions. Another system with unique potential is the Papaveraceae (poppy family) member *Eschscholzia californica*, whose indeterminate stamen numbers develop from a ring primordium (Becker, Gleissberg, & Smyth, 2005). While these models can be used to address control of merosity, a separate question is how the floral meristem decides when to stop making whorls, which is typically marked by the transition to carpel identity. Again, this process is well understood in *A. thaliana* (reviewed in Smyth, 2018) but little is known as to how conserved the program is across lineages. Here, natural variation may hold some promise since differences in whorl number, especially in the stamens, are seen among closely related species, including in evo-devo models such as *Aquilegia brevistyla* and *A. jonseii* (Munz, 1946).

4. Organ elaboration

Although the genetic programs controlling floral organ identity, floral symmetry, and primordium initiation largely function in parallel to one another at early stages of development (reviewed in Kramer, 2005), once organ elaboration begins, they interact in a synergistic fashion to yield the final morphology of each floral organ. Additional players in this process include the developmental pathways that commonly influence all lateral organ development, particularly adaxial/abaxial identity and the control of cell proliferation versus expansion during final stages of organ maturation. It seems likely that these many intersecting programs hold considerable potential for altering the final morphology of floral organs, but the genetic basis of this variation is just beginning to be explored.

One obvious place to start in exploring organ elaboration is with the fusion of floral organs, both within and between whorls, which is widely distributed across the angiosperms (Endress, 1994). Two of the most notable types of fusion are sympetaly and syncarpy. In sympetaly, the petals become fused into a corolla tube, which is often associated with specific pollinator interactions and is diagnostic of several large groups of eudicots. Likewise, syncarpy is one of the defining characteristics of the eudicots but it has also evolved independently in the monocots as well as smaller clades like the poppy family. Our understanding of how initiating primordia maintain their separation would seem to suggest an obvious hypothesis for how organ fusion occurs. Whorl-specific down-regulation of the *CUP-SHAPED COTYLEDON* (*CUC*) genes, which normally promote organ separation, is the simplest potential mechanism (Specht & Howarth, 2015; Zhong & Preston, 2015), but actual studies of sympetalous taxa suggest greater complexity. In petunia, some members of the CUC family have actually been found to promote organ fusion, rather than prevent it (Zhong, Powell, & Preston, 2016). Another sympetalous model is *M. lewisii*, which is amenable to standard forward mutagenesis. In this system, a mutation resulting in a partially unfused corolla was surprisingly found to correspond to a component of the pathway responsible for the processing of *trans*-acting siRNAs (tasiRNAs) (Ding et al., 2018). This, in turn, implicated members of the AUXIN RESPONSE FACTOR family, which are targeted by tasiRNAs. Studies of auxin signaling during corolla tube development suggest that the tissue responsible for joining adjacent petal primordia is not simply due to a suppression of the *CUC* genes, but to the independent recruitment of inter-primordial proliferation, almost like a late arising primordial ring.

Fused carpels are likewise the subject of considerable study but have proven to be challenging. All major genetic models have either syncarpous or monocarpic gynoecia and, although many mutants result in unfused carpels, which are typically due to defects in either organ identity or the differentiation of specific tissue types (reviewed in Pfannebecker, Lange, Rupp, & Becker, 2017b). In other words, none of these mutations are true reversions to apocarpy. Much like merosity and sympetaly, this is a trait that rarely varies among closely related species and is generally due to relatively ancient evolutionary events, which adds to the challenge of studying it. Luckily, the combination of careful studies of gene lineage evolution, functional tests in diverse lineages, and the use of systems like *E. californica* that have independently evolved syncarpy hold considerable promise for teasing apart this challenging question (e.g., Finet et al., 2016; Fourquin, Vinauger-Douard, Chambrier, Berne-Dedieu, & Scutt, 2007; Pfannebecker, Lange, Rupp, & Becker, 2017a; Pfannebecker et al., 2017b; Vialette-Guiraud et al., 2011).

Moving beyond features such as fusion, we should also consider three-dimensional organ shape. It is actually quite difficult, developmentally speaking, to make a plant organ flat (reviewed in Johnson & Lenhard, 2011). Floral organs, which are not optimized for photosynthesis but, rather, pollinator interactions, often achieve complex three-dimensional forms that similarly mirror a high degree of coordination between cell division and expansion patterns (reviewed in Whitewoods & Coen, 2017). While the floral organs of *A. thaliana* are rather unremarkable in this regard, the petals of *A. majus* are strikingly elaborate, integrating patterns of fusion and dramatic curvature (Crawford, Nath, Carpenter, & Coen, 2004). This model has been extensively exploited for several purposes. First, the existence of an active transposon in *A. majus* has made sector analysis a particularly powerful tool for generating models of tissue growth, both in wild-type and mutant backgrounds (e.g., Cui, Copsey, Green, Bangham, & Coen, 2010; Sauret-Gueto, Schiessl, Bangham, Sablowski, & Coen, 2013). Traditional mutant analyses have also been very informative. For instance, a *CUC* homolog in *A. majus* called *CULPIFORMIS* (*CUP*) has been found to play multiple roles during the development of the corolla (Rebocho, Kennaway, Bangham, & Coen, 2017). At earlier stages in development, *CUP* does appear to play a role in the generation of the corolla tube, being down-regulated in the region where the tube arises. During later stages, *CUP* expression becomes localized to specific domains along the proximal-distal and adaxial-abaxial axes of the organs. These domains reflect region-specific modulation of cell division patterns, which contribute to the curvature of the petal surface and yield the distinct shape of the snapdragon corolla. Further, the expression patterns differ between dorsal, lateral and ventral petals, suggesting that *CUP* expression integrates information from both the organ identity and floral symmetry pathways. This model is likely to be generalizable to organ sculpting in other organs where the additive effect of multiple parallel genetic programs results in localized patterns of cell division or expansion, ultimately producing laminar curvature.

One particularly dramatic example of petal curvature is the three-dimensional nectar spur, which has evolved many times independently across the angiosperms (Hodges & Arnold, 1995). This hollow tube is generally terminated with a nectary that rewards pollinators with the appropriate length of tongue and has been associated with adaptive radiations in many of the lineages in which it has evolved. Two evo-devo model systems are allowing the exploration of separate instances of the derivation of these fascinating structures. In the core eudicots, the genus *Linaria* is a relatively close relative of *A. majus* that shows considerable variation in spur length across its species (Cullen, Fernández-Mazuecos, & Glover, 2018). As it happens, previous work in *A. majus* pointed to a potential candidate gene for spur development. Several insertion mutants were found that resulted in the production of a spur-like outgrowth in the corolla tube (Golz, Keck, & Hudson, 2002). These insertions were traced back to two homologs of the KNOX gene family (Golz et al., 2002), which normally promote the maintenance of an undifferentiated state in apical meristems (reviewed in Hake et al., 2004). The insertion events actually caused ectopic expression of these KNOX genes, leading to the production of spur-like structures (Golz et al., 2002). Studies of these same genes in *L. vulgaris* revealed that they are, likewise, expressed in developing floral organs, suggesting that they could be critical to natural spur development (Box, Dodsworth, Rudall, Bateman, & Glover, 2011). Another question is what developmental mechanisms are responsible for variation in spur length. Comparative analysis of diverse species revealed that cell division is primarily responsible for the differences (Cullen et al., 2018). In particular, the long-spurred *L. becerrae* shows a dramatic early increase in cell number while the short-spurred *L. clementei* shows a much slower accumulation of cells over a longer period. This is followed by a similar degree of cell expansion to achieve a similar final shape. In parallel to these studies, work in multiple species of *Aquilegia* has revealed a completely different story. It does not appear that these spurs rely on KNOX gene expression but, rather, are the product of later localized cell divisions (Puzey, Gerbode, Hodges, Mahadevan, & Kramer, 2012; Yant, Collani, Puzey, Levy, & Kramer, 2015), much like the folding of petals in *A. majus*. In addition, variation in spur length closely correlates with final cell length, suggesting that this is the dominant developmental mechanism driving their differences (Puzey et al., 2012). Of course, such differences should not be surprising given their independent evolution and they reflect the many complementary pathways that evolutionary tinkering may take.

5. Conclusions and future perspectives

Considering the breadth of the last 20 years of plant evo-devo studies, there are themes of both stasis and change. There are numerous examples of conservation as well as convergence, often to surprising and yet to be understood degrees. Interesting future questions include: Why is A function so genetically labile compared to B and C? Why and how have *CYC* homologs been convergently recruited to promote zygomorphy so many times? Can we ultimately achieve a synthesis that helps us understand the collaborative interaction of all the various genetic programs influencing floral organ development? Of course, this impression of stasis may reflect the fact that we have largely been driven by candidate gene approaches, which will be biased to find conservation and convergence. Plant developmental evolution is itself experiencing an evolutionary process as we outgrow this dependence and start exploring more complicated research questions. Although we will always be drawn to the loci that are well known from established models, perhaps the greatest potential for novel discoveries will come from taking unbiased paths. The application of next generation sequencing to both genome and transcriptome characterization, together with CRISPR/Cas methods for functional studies, clearly represents the next era for evolutionary developmental biology, both in plants and animals.

References

Airoldi, C. A., & Davies, B. (2012). Gene duplication and the evolution of plant MADS-box transcription factors. *Journal of Genetics and Genomics*, *39*(4), 157–165.

Almeida, J., Rocheta, M., & Galego, L. (1997). Genetic control of flower shape in *Antirrhinum majus*. *Development*, *124*, 1387–1392.

An, J., Guo, Z., Gou, X., & Li, J. (2011). TCP1 positively regulates the expression of *DWF4* in *Arabidopsis thaliana*. *Plant Signaling & Behavior*, *6*(8), 1117–1118.

Baum, D. A., & Whitlock, B. A. (1999). Genetic clues to petal evolution. *Current Biology*, *9*, 525–527.

Becker, A., Gleissberg, S., & Smyth, D. R. (2005). Floral and vegetative morphogenesis in California poppy (*Eschscholzia californica* Cham). *International Journal of Plant Sciences*, *166*(4), 537–555.

Bowman, J. L., Alvarez, J., Weigel, D., Meyerowitz, E. M., & Smyth, D. R. (1993). Control of flower development in *Arabidopsis thaliana* by *APETALA1* and interacting genes. *Development*, *119*, 721–743.

Bowman, J. L., Smyth, D. R., & Meyerowitz, E. M. (1991). Genetic interactions among floral homeotic genes of *Arabidopsis*. *Development*, *112*, 1–20.

Box, M. S., Dodsworth, S., Rudall, P. J., Bateman, R. M., & Glover, B. J. (2011). Characterization of *Linaria* KNOX genes suggests a role in petal-spur development. *Plant Journal*, *68*(4), 703–714.

Braybrook, S. A., & Kuhlemeier, C. (2010). How a plant builds leaves. *Plant Cell*, *22*(4), 1006–1018.
Brockington, S. F., Alexandre, R., Ramdial, J., Moore, M. J., Crawley, S., Dhingra, A., et al. (2009). Phylogeny of the Caryophyllales senu lato: Revisiting hypotheses on pollination biology and perianth differentiation in the core Caryophyllales. *International Journal of Plant Sciences*, *170*(5), 627–643.
Brockington, S. F., Rudall, P. J., Frohlich, M. W., Oppenheimer, D. G., Soltis, P. S., & Soltis, D. E. (2012). 'Living stones' reveal alternative petal identity programs within the core eudicots. *Plant Journal*, *69*, 193–203.
Broholm, S. K., Tahtiharju, S., Laitinen, R. A. E., Albert, V. A., Teeri, T. H., & Elomaa, P. (2008). A TCP domain transcription factor controls flower type specification along the radial axis of the *Gerbera* (Asteraceae) inflorescence. *Proceedings of the National Academy of Sciences of the United States of America*, *105*(26), 9117–9122.
Busch, A., & Zachgo, S. (2007). Control of corolla monosymmetry in the Brassicaceae Iberis amara. *Proceedings of the National Academy of Sciences of the United States of America*, *104*(42), 16714–16719.
Carpenter, R., & Coen, E. S. (1990). Floral homeotic mutations produced by transposon-mutagenesis in *Antirrhinum majus*. *Genes and Development*, *4*, 1483–1493.
Cartolano, M., Castillo, R., Efremova, N., Kuckenberg, M., Zethof, J., Gerats, T., et al. (2007). A conserved microRNA module exerts homeotic control over *Petunia hybrida* and *Antirrhinum majus* floral organ identity. *Nature Genetics*, *39*(7), 901–905.
Causier, B., Castillo, R., Zhou, J. L., Ingram, R., Xue, Y. B., Schwarz-Sommer, Z., et al. (2005). Evolution in action: Following function in duplicated floral homeotic genes. *Current Biology*, *15*(16), 1508–1512.
Chapman, M. A., Tang, S. X., Draeger, D., Nambeesan, S., Shaffer, H., Barb, J. G., et al. (2012). Genetic analysis of floral symmetry in van Gogh's sunflowers reveals independent recruitment of *CYCLOIDEA* genes in the Asteraceae. *PLoS Genetics*, *8*(3), e1002628.
Chen, X. (2004). A microRNA as a translational repressor of *APETALA2* in *Arabidopsis* flower development. *Science*, *303*(5666), 2022–2025.
Chen, J., Shen, C. Z., Guo, Y. P., & Rao, G. Y. (2018). Patterning the Asteraceae Capitulum: Duplications and differential expression of the flower symmetry *CYC2*-like genes. *Frontiers in Plant Science*, *9*, 551.
Citerne, H., Jabbour, F., Nadot, S., & Damerval, C. (2010). The evolution of floral symmetry. *Advances in Botanical Research*, *54*(54), 85–137.
Coen, E. S., & Meyerowitz, E. M. (1991). The war of the whorls: Genetic interactions controlling flower development. *Nature*, *353*, 31–37.
Corley, S. B., Carpenter, R., Copsey, L., & Coen, E. (2005). Floral asymmetry involves an interplay between TCP and MYB transcription factors in *Antirrhinum*. *Proceedings of the National Academy of Sciences of the United States of America*, *102*(14), 5068–5073.
Costa, M. M. R., Fox, S., Hanna, A. I., Baxter, C., & Coen, E. (2005). Evolution of regulatory interactions controlling floral asymmetry. *Development*, *132*(22), 5093–5101.
Crawford, B. C. W., Nath, U., Carpenter, R., & Coen, E. (2004). *CINCINNATA* controls both cell differentiation and growth in petal lobes and leaves of *Antirrhinum*. *Plant Physiology*, *135*, 244–253.
Cubas, P., Coen, E., & Martinez Zapater, J. M. (2001). Ancient asymmetries in the evolution of flowers. *Current Biology*, *11*, 1050–1052.
Cubas, P., Lauter, N., Doebley, J., & Coen, E. (1999). The TCP domain: A motif found in proteins regulating plant growth and development. *Plant Journal*, *18*(2), 215–222.
Cubas, P., Vincent, C., & Coen, E. (1999). An epigenetic mutation responsible for natural variation in floral symmetry. *Nature*, *401*, 157–161.
Cui, M.-L., Copsey, L., Green, A. A., Bangham, J. A., & Coen, E. (2010). Quantitative control of organ shape by combinatorial gene activity. *PLoS Biology*, *8*(11), e1000538.

Cullen, E., Fernández-Mazuecos, M., & Glover, B. J. (2018). Evolution of nectar spur length in a clade of *Linaria* reflects changes in cell division rather than in cell expansion. *Annals of Botany, 122*, 801–809.

Davies, B., Motte, P., Keck, E., Saedler, H., Sommer, H., & Schwarz-Sommer, Z. (1999). *PLENA* and *FARINELLI*: Redundancy and regulatory interactions between two *Antirrhinum* MADS-box factors controlling flower development. *The EMBO Journal, 18*(14), 4023–4034.

De Paolo, S., Gaudio, L., & Aceto, S. (2015). Analysis of the TCP genes expressed in the inflorescence of the orchid *Orchis italica*. *Scientific Reports, 5*, 16265.

Ding, B., Lin, Q., Gurung, V., Sagawa, J. M., Stanley, L. E., Strobel, M., et al. (2018). Developmental genetics of corolla tube formation: Role of the tasiRNA-ARF pathway. *bioRxiv,* 253112. https://doi.org/10.1101/253112.

Ding, B. Q., Mou, F. J., Sun, W., Chen, S. L., Peng, F., Bradshaw, H. D., et al. (2017). A dominant-negative actin mutation alters corolla tube width and pollinator visitation in *Mimulus lewisii*. *New Phytologist, 213*(4), 1936–1944.

Doebley, J., Stec, A., & Hubbard, L. (1997). The evolution of apical dominance in maize. *Nature, 386*(6624), 485–488.

Drews, G. N., Bowman, J. L., & Meyerowitz, E. M. (1991). Negative regulation of the *Arabidopsis* homeotic gene *AGAMOUS* by the APETELA2 product. *Cell, 65*, 991–1002.

Endress, P. K. (1994). *Diversity and evolutionary biology of tropical flowers.* Cambridge: Cambridge University Press.

Endress, P. K. (2016). Development and evolution of extreme synorganization in angiosperm flowers and diversity: A comparison of Apocynaceae and Orchidaceae. *Annals of Botany, 117*(5), 749–767.

Feng, C.-M., Liu, X., Yu, Y., Franks, R. G., & Xiang, Q.-Y. (2012). Evolution of bract development and B-class MADS box gene expression in petaloid bracts of *Cornus* s. l. (Cornaceae). *New Phytologist, 196*, 631–643.

Feng, X. Z., Zhao, Z., Tian, Z. X., Xu, S. L., Luo, Y. H., Cai, Z. G., et al. (2006). Control of petal shape and floral zygomorphy in *Lotus japonicus*. *Proceedings of the National Academy of Sciences of the United States of America, 103*(13), 4970–4975.

Ferris, K. G., Barnett, L. L., Blackman, B. K., & Willis, J. H. (2017). The genetic architecture of local adaptation and reproductive isolation in sympatry within the *Mimulus guttatus* species complex. *Molecular Ecology, 26*(1), 208–224.

Finet, C., Floyd, S. K., Conway, S. J., Zhong, B. J., Scutt, C. P., & Bowmanb, J. L. (2016). Evolution of the YABBY gene family in seed plants. *Evolution & Development, 18*(2), 116–126.

Fourquin, C., Vinauger-Douard, M., Chambrier, P., Berne-Dedieu, A., & Scutt, C. P. (2007). Functional conservation between *CRABS CLAW* orthologues from widely diverged angiosperms. *Annals of Botany, 100*(3), 651–657.

Fujikura, U., Jing, R. C., Hanada, A., Takebayashi, Y., Sakakibara, H., Yamaguchi, S., et al. (2018). Variation in splicing efficiency underlies morphological evolution in Capsella. *Developmental Cell, 44*(2), 192.

Galego, L., & Almeida, J. (2002). Role of *DIVARICATA* in the control of dorsoventral asymmetry in *Antirrhinum* flowers. *Genes & Development, 16*, 880–891.

Garces, H. M. P., Spencer, V. M. R., & Kim, M. (2016). Control of floret symmetry by *RAY3*, *SvDIV1B*, and *SvRAD* in the Capitulum of *Senecio vulgaris*. *Plant Physiology, 171*(3), 2055–2068.

Geuten, K., Becker, A., Kaufmann, K., Caris, P., Janssens, S., Viaene, T., et al. (2006). Petaloidy and petal identity MADS-box genes in the balsaminoid genera *Impatiens* and *Marcgravia*. *Plant Journal, 47*(4), 501–518.

Goethe, J. W. V. (1790). *Versuch die Metamorphose der Pflanzen zu erklaren.* Gotha, Germany: C. W. Ettinger.

Golz, J. F., Keck, E. J., & Hudson, A. (2002). Spontaneous mutations in KNOX genes give rise to a novel floral structure in *Antirrhinum*. *Current Biology*, *12*, 515–522.

Goncalves, B., Nougue, O., Jabbour, F., Ridel, C., Morin, H., Laufs, P., et al. (2013). An *APETALA3* homolog controls both petal identity and floral meristem patterning in *Nigella damascena* L. (Ranunculaceae). *Plant Journal*, *76*(2), 223–235.

Guo, Z., Fujioka, S., Blancaflor, E. B., Miao, M. Z., Gou, X., & Li, J. (2010). TCP1 modulates brassinosteroid biosynthesis by regulating the expression of the key biosynthetic gene *DWARF4* in *Arabidopsis thaliana*. *Plant Cell*, *22*(4), 1161–1173.

Hake, S., Smith, H. M. S., Holtan, H., Magnani, E., Mele, G., & Ramirez, J. (2004). The role of knox genes in plant development. *Annual Review of Cell and Developmental Biology*, *20*, 125–151.

Hellsten, U., Wright, K. M., Jenkins, J., Shu, S. Q., Yuan, Y. W., Wessler, S. R., et al. (2013). Fine-scale variation in meiotic recombination in *Mimulus* inferred from population shotgun sequencing. *Proceedings of the National Academy of Sciences of the United States of America*, *110*(48), 19478–19482.

Hemingway, C. A., Christensen, A. R., & Malcomber, S. T. (2011). B- and C-class gene expression during corona development of the blue passionflower (*Passiflora caerulea*, Passifloraceae). *American Journal of Botany*, *98*(6), 923–934.

Hermann, K., Klahre, U., Moser, M., Sheehan, H., Mandel, T., & Kuhlemeier, C. (2013). Tight genetic linkage of prezygotic barrier loci creates a multifunctional speciation Island in *Petunia*. *Current Biology*, *23*(10), 873–877.

Hermann, K., Klahre, U., Venail, J., Brandenburg, A., & Kuhlemeier, C. (2015). The genetics of reproductive organ morphology in two *Petunia* species with contrasting pollination syndromes. *Planta*, *241*(5), 1241–1254.

Hileman, L. C. (2014). Trends in flower symmetry evolution revealed through phylogenetic and developmental genetic advances. *Philosophical Transactions of the Royal Society, B: Biological Sciences*, *369*(1648).

Hileman, L. C., Kramer, E. M., & Baum, D. A. (2003). Differential regulation of symmetry genes and the evolution of floral morphologies. *Proceedings of the National Academy of Sciences of the United States of America*, *100*(22), 12814–12819.

Hodges, S. A., & Arnold, M. L. (1995). Spurring plant diversification: Are floral nectar spurs a key innovation? *Philosophical Transactions of the Royal Society, B: Biological Sciences*, *262*(1365), 343–348.

Hsu, H. F., Hsu, W.-H., Yi, L., Mao, W. T., Yang, J. Y., Li, J. Y., et al. (2015). Model for perianth formation in orchids. *Nature Plants*, *1*, 15046.

Huu, C. N., Kappel, C., Keller, B., Sicard, A., Takebayashi, Y., Breuninger, H., et al. (2016). Presence versus absence of *CYP734A50* underlies the style-length dimorphism in primroses. *Elife*, *5*, e17956.

Jaramillo, M. A., & Kramer, E. M. (2004). *APETALA3* and *PISTILLATA* homologs exhibit novel expression patterns in the unique perianth in *Aristolochia* (Aristolochiaceae). *Evolution & Development*, *6*(6), 449–458.

Johnson, K., & Lenhard, M. (2011). Genetic control of plant organ growth. *New Phytologist*, *191*(2), 319–333.

Juntheikki-Palovaara, I., Tahtiharju, S., Lan, T. Y., Broholm, S. K., Rijpkema, A. S., Ruonala, R., et al. (2014). Functional diversification of duplicated *CYC2* clade genes in regulation of inflorescence development in *Gerbera hybrida* (Asteraceae). *Plant Journal*, *79*(5), 783–796.

Kanno, A. (2016). Molecular mechanism regulating floral architecture in monocotyledonous ornamental plants. *Horticulture Journal*, *85*(1), 8–22.

Kappel, C., Huu, C. N., & Lenhard, M. (2017). A short story gets longer: Recent insights into the molecular basis of heterostyly. *Journal of Experimental Botany*, *68*(21 – 22), 5719–5730.

Kim, M., Cui, M. L., Cubas, P., Gillies, A., Lee, K., Chapman, M. A., et al. (2008). Regulatory genes control a key morphological and ecological trait transferred between species. *Science*, *322*(5904), 1116–1119.

Kim, S., Koh, J., Ma, H., Hu, Y., Endress, P. K., Hauser, B. A., et al. (2005). Sequence and expression studies of A-, B-, and E-class MADS-box homologues in *Eupomatia* (Eupomatiaceae): Support for the bracteate origin of the calyptra. *International Journal of Plant Sciences*, *166*(2), 185–198.

Kim, S., Koh, J., Yoo, M. J., Kong, H. Z., Hu, Y., Ma, H., et al. (2005). Expression of floral MADS-box genes in basal angiosperms: Implications for the evolution of floral regulators. *Plant Journal*, *43*(5), 724–744.

Kosuge, K. (1994). Petal evolution in *Ranunculaceae*. *Plant Systematics and Evolution (Suppl)*, *8*, 185–191.

Kramer, E. M. (2005). Floral architecture: Regulation and diversity of floral shape and pattern. In T. CGN (Ed.), *Plant architecture and its manipulation* (pp. 120–147). Oxford, UK: Blackwell Publishing.

Kramer, E. M., Di Stilio, V. S., & Schluter, P. (2003). Complex patterns of gene duplication in the *APETALA3* and *PISTILLATA* lineages of the Ranunculaceae. *International Journal of Plant Sciences*, *164*(1), 1–11.

Kramer, E. M., Holappa, L., Gould, B., Jaramillo, M. A., Setnikov, D., & Santiago, P. (2007). Elaboration of B gene function to include the identity of novel floral organs in the lower eudicot *Aquilegia* (Ranunculaceae). *Plant Cell*, *19*(3), 750–766.

Krizek, B. A., & Anderson, J. T. (2013). Control of flower size. *Journal of Experimental Botany*, *64*(6), 1427–1437.

Krizek, B. A., & Meyerowitz, E. M. (1996). The *Arabidopsis* homeotic genes *APETALA3* and *PISTILLATA* are sufficient to provide the B class organ identity function. *Development*, *122*, 11–22.

Krogan, N. T., Hogan, K., & Long, J. A. (2012). APETALA2 negatively regulates multiple floral organ identity genes in *Arabidopsis* by recruiting the co-repressor TOPLESS and the histone deacetylase HDA19. *Development*, *139*, 4180–4190.

Li, J. H., Cocker, J. M., Wright, J., Webster, M. A., McMullan, M., Dyer, S., et al. (2016). Genetic architecture and evolution of the S locus supergene in *Primula vulgaris*. *Nature Plants*, *2*(12), 16188.

Li, J. H., Webster, M. A., Wright, J., Cocker, J. M., Smith, M. C., Badakshi, F., et al. (2015). Integration of genetic and physical maps of the *Primula vulgaris* S locus and localization by chromosome in situ hybridization. *New Phytologist*, *208*(1), 137–148.

Litt, A. (2007). An evaluation of A-function: Evidence from the *APETALA1* and *APETALA2* gene lineages. *International Journal of Plant Sciences*, *168*(1), 73–91.

Litt, A., & Irish, V. F. (2003). Duplication and diversification in the *APETALA1/FRUITFULL* floral homeotic gene lineage: Implications for the evolution of floral development. *Genetics*, *165*, 821–833.

Litt, A., & Kramer, E. M. (2010). The ABC model and the diversification of floral organ identity. *Seminars in Cell & Developmental Biology*, *21*(1), 129–137.

Luo, D., Carpenter, R., Copsey, L., Vincent, C., Clark, J., & Coen, E. (1999). Control of organ asymmetry in flowers of *Antirrhinum*. *Cell*, *99*, 367–376.

Luo, D., Carpenter, R., Vincent, C., Copsey, L., & Coen, E. (1996). Origin of floral asymmetry in *Antirrhinum*. *Nature*, *383*, 794–799.

Martin-Trillo, M., & Cubas, P. (2010). TCP genes: A family snapshot ten years later. *Trends in Plant Science*, *15*(1), 31–39.

Mondragon-Palomino, M., & Theissen, G. (2008). MADS about the evolution of orchid flowers. *Trends in Plant Science*, *13*(2), 51–59.

Monniaux, M., McKim, S. M., Cartolano, M., Thevenon, E., Parcy, F., Tsiantis, M., et al. (2017). Conservation vs divergence in LEAFY and APETALA1 functions between *Arabidopsis thaliana* and *Cardamine hirsuta*. *New Phytologist*, *216*(2), 549–561.

Monniaux, M., Pieper, B., & Hay, A. (2016). Stochastic variation in *Cardamine hirsuta* petal number. *Annals of Botany*, *117*(5), 881–887.

Morel, P., Heijmans, K., Rozier, F., Zethof, J., Chamot, S., Bento, S. R., et al. (2017). Divergence of the floral A-function between an Asterid and a Rosid species. *Plant Cell*, *29*(7), 1605.

Morioka, K., Yockteng, R., Almeida, A. M. R., & Specht, C. D. (2015). Loss of YABBY2-like gene expression may underlie the evolution of the laminar style in *Canna* and contribute to floral morphological diversity in the Zingiberales. *Frontiers in Plant Science*, *6*, 1106.

Moyroud, E., & Glover, B. J. (2017). The evolution of diverse floral morphologies. *Current Biology*, *27*(17), R941–R951.

Munz, P. A. (1946). *Aquilegia*: The cultivated and wild columbines. *Gentes Herbarium*, *7*, 1–150.

Nikolov, L. A., Endress, P. K., Sugumaran, M., Sasirat, S., Vessabutr, S., Kramer, E. M., et al. (2013). Developmental origins of the world's largest flowers, Rafflesiaceae. *Proceedings of the National Academy of Sciences of the United States of America*, *110*, 18578–18583.

Nowak, M. D., Russo, G., Schlapbach, R., Huu, C. N., Lenhard, M., & Conti, E. (2015). The draft genome of *Primula veris* yields insights into the molecular basis of heterostyly. *Genome Biology*, *16*, 12.

Pabon-Mora, N., Sharma, B., Holappa, L. D., Kramer, E. M., & Litt, A. (2013). The *Aquilegia FRUITFULL*-like genes play key roles in leaf morphogenesis and inflorescence development. *Plant Journal*, *74*, 197–212.

Park, J.-H., Ishikawa, Y., Yoshida, R., Kanno, A., & Kameya, T. (2003). Expression of *AODEF*, a B-functional MADS-box gene, in stamens and inner tepals of the dioecious species *Asparagus officinalis* L. *Plant Molecular Biology*, *51*, 867–875.

Pfannebecker, K. C., Lange, M., Rupp, O., & Becker, A. (2017a). An evolutionary framework for carpel developmental control genes. *Molecular Biology and Evolution*, *34*(2), 330–348.

Pfannebecker, K. C., Lange, M., Rupp, O., & Becker, A. (2017b). Seed plant-specific gene lineages involved in carpel development. *Molecular Biology and Evolution*, *34*(4), 925–942.

Preston, J. C., & Hileman, L. C. (2012). Parallel evolution of TCP and B-class genes in Commelinaceae flower bilateral symmetry. *EvoDevo*, *3*, 6.

Puzey, J. R., Gerbode, S., Hodges, S. A., Mahadevan, L., & Kramer, E. M. (2012). Evolution of spur-length diversity in *Aquilegia* petals is achieved solely through cell-shape anisotropy. *Proceedings of the Royal Society of London, Series B: Biological Sciences*, *279*(1733), 1640–1645.

Raimundo, J., Sobral, R., Bailey, P., Azevedo, H., Galego, L., Almeida, J., et al. (2013). A subcellular tug of war involving three MYB-like proteins underlies a molecular antagonism in *Antirrhinum* flower asymmetry. *Plant Journal*, *75*(4), 527–538.

Rebocho, A. B., Kennaway, J. R., Bangham, J. A., & Coen, E. (2017). Formation and shaping of the Antirrhinum flower through modulation of the *CUP* boundary gene. *Current Biology*, *27*(17), 2610 – 2622.e3.

Sauret-Gueto, S., Schiessl, K., Bangham, A., Sablowski, R., & Coen, E. (2013). *JAGGED* controls *Arabidopsis* petal growth and shape by interacting with a divergent polarity field. *PLoS Biology*, *11*(4), 16.

Sharma, B., Guo, C., Kong, H., & Kramer, E. M. (2011). Petal-specific subfunctionalization of an *APETALA3* paralog in the Ranunculales and its implications for petal evolution. *New Phytologist*, *190*, 870–883.

Sharma, B., & Kramer, E. M. (2013). Sub- and neofunctionalization of *APETALA3* paralogs have contributed to the evolution of novel floral organ identity in *Aquilegia* (columbine, Ranunculaceae). *New Phytologist*, *197*, 949–957.

Sharma, B., & Kramer, E. M. (2017). *Aquilegia* B gene homologs promote petaloidy of the sepals and maintenance of the C domain boundary. *EvoDevo, 8*, 22.
Sicard, A., Kappel, C., Lee, Y. W., Wozniak, N. J., Marona, C., Stinchcombe, J. R., et al. (2016). Standing genetic variation in a tissue-specific enhancer underlies selfing-syndrome evolution in *Capsella. Proceedings of the National Academy of Sciences of the United States of America, 113*(48), 13911–13916.
Sicard, A., Stacey, N., Hermann, K., Dessoly, J., Neuffer, N., Baurle, I., et al. (2011). Genetics, evolution, and adaptive significance of the Selfing syndrome in the genus *Capsella. Plant Cell, 23*, 3156–3171.
Smith, R. S., & Bayer, E. M. (2009). Auxin transport-feedback models of patterning in plants. *Plant Cell and Environment, 32*(9), 1258–1271.
Smyth, D. R. (2018). Evolution and genetic control of the floral ground plan. *New Phytologist, 220*(1), 70–86.
Specht, C. D., & Howarth, D. G. (2015). Adaptation in flower form: A comparative evodevo approach. *New Phytologist, 206*(1), 74–90.
Specht, C. D., Yockteng, R., Almeida, A. M., Kirchoff, B. K., & Kress, W. J. (2012). Homoplasy, pollination, and emerging complexity during the evolution of floral development in the tropical gingers (Zingiberales). *Botanical Review, 78*(4), 440–462.
Stuurman, J., Hoballah, M. E., Broger, L., Moore, J., Basten, C., & Kuhlemeier, C. (2004). Dissection of floral pollination syndromes in petunia. *Genetics, 168*(3), 1585–1599.
Takhtajan, A. (1991). *Evolutionary trends in flowering plants.* New York: Columbia University Press.
Theissen, G. (2010). Homeosis of the angiosperm flower: Studies on three candidate cases of saltational evolution. *Palaeodiversity, 3*(Suppl), 131–139.
Theissen, G., & Saedler, H. (2001). Floral quartets. *Nature, 409*, 469–471.
Tian, X. Y., Zou, P., Miao, M. Z., Ning, Z. L., & Liao, J. P. (2018). RNA-Seq analysis reveals the distinctive adaxial-abaxial polarity in the asymmetric one-theca stamen of *Canna indica. Molecular Genetics and Genomics, 293*(2), 391–400.
True, J. R., & Haag, E. S. (2001). Developmental system drift and flexibility in evolutionary trajectories. *Evolution & Development, 3*(2), 109–119.
Vialette-Guiraud, A. C. M., Adam, H., Finet, C., Jasinski, S., Jouannic, S., & Scutt, C. P. (2011). Insights from ANA-grade angiosperms into the early evolution of *CUP-SHAPED COTYLEDON* genes. *Annals of Botany, 107*(9), 1511–1519.
Walker-Larsen, J., & Harder, L. D. (2000). The evolution of staminodes in angiosperms: Patterns of stamen reduction, loss, and functional re-invention. *American Journal of Botany, 87*(10), 1367–1384.
Wang, P., Liao, H., Zhang, W., Yu, X., Zhang, R., Shan, H., et al. (2015). Flexiiblity in the structure of spiral flowers and its underlying mechanisms. *Nature Plants, 2*, 15188.
Waters, M. T., Tiley, A. M. M., Kramer, E. M., Meerow, A. W., Langdale, J. A., & Scotland, R. W. (2013). The corona of the daffodil *Narcissus bulbocodium* shares stamen-like identity and is distinct from the orthodox floral whorls. *Plant Journal, 74*(4), 615–625.
Wessinger, C. A., Hileman, L. C., & Rausher, M. D. (2014). Identification of major quantitative trait loci underlying floral pollination syndrome divergence in Penstemon. *Philosophical Transactions of the Royal Society, B: Biological Sciences, 369*, 20130349. https://doi.org//10.1098/rstb.2013.0349.
Whitewoods, C. D., & Coen, E. (2017). Growth and development of three-dimensional plant form. *Current Biology, 27*(17), R910–R918.
Wollmann, H., Mica, E., Todesco, M., Long, J. A., & Weigel, D. (2010). On reconciling the interactions between *APETALA2*, *miR172* and *AGAMOUS* with the ABC model of flower development. *Development, 137*, 3633–3642.

Yamaguchi, T., Lee, D. Y., Miyao, A., Hirochika, H., An, G., & Hirano, H.-Y. (2006). Functional diversification of the two C-class MADS box genes *OSMADS3* and *OSMADS58* in *Oryza sativa*. *Plant Cell*, *18*, 15–28.

Yang, X., Pang, H. B., Liu, B. L., Qiu, Z. J., Gao, Q., Wei, L., et al. (2012). Evolution of double positive autoregulatory feedback loops in *CYCLOIDEA2* clade genes is associated with the origin of floral Zygomorphy. *Plant Cell*, *24*(5), 1834–1847.

Yant, L., Collani, S., Puzey, J. R., Levy, C., & Kramer, E. M. (2015). Molecular basis for three-dimensional elaboration of the *Aquilegia* petal spur. *Proceedings of the Royal Society of London, Series B: Biological Sciences*, *282*(1803), 20142778.

Zhang, R., Guo, C., Zhang, W., Wang, P., Li, L., Duan, X., et al. (2013). Disruption of the petal identity gene *APETALA3-3* is highly correlated with loss of petals within the buttercup family (Ranunculaceae). *Proceedings of the National Academy of Sciences of the United States of America*, *110*(13), 5074–5079.

Zhang, W., Steinmann, V. W., Nikolov, L., Kramer, E. M., & Davis, C. C. (2013). Divergent genetic mechanisms underlie reversals to radial floral symmetry from diverse zygomorphic flowered ancestors. *Frontiers in Plant Science*, *4*.

Zhong, J., Powell, S., & Preston, J. C. (2016). Organ boundary NAC-domain transcription factors are implicated in the evolution of petal fusion. *Plant Biology*, *18*(6), 893–902.

Zhong, J. S., & Preston, J. C. (2015). Bridging the gaps: Evolution and development of perianth fusion. *New Phytologist*, *208*(2), 330–335.

CHAPTER TEN

Pre-meiotic anther development

Karina van der Linde[a], Virginia Walbot[b,*]
[a]Department of Cell Biology and Plant Biochemistry, University of Regensburg, Regensburg, Germany
[b]Department of Biology, Stanford University, Stanford, CA, United States
*Corresponding author: e-mail address: walbot@stanford.edu

Contents

Abstract

Most genetic and molecular analyses of anther development utilize *Arabidopsis thaliana*, *Oryza sativa* (rice), and *Zea mays* (maize). Especially in maize, early stages of anther development are easy to study because: (1) Maize has unisex flowers. (2) Compared to rice or *A. thaliana*, maize anthers are relatively large, making dissection for molecular and biochemical analyses easy. (3) Anther developmental stage is strongly correlated with maize anther length. Besides these technical advantages, understanding anther and pollen development in maize is of significant agricultural importance. Today maize is a worldwide cereal crop: approximately 25% of all consumed food contains maize. Yield stability or even increases depend on maintenance of hybrid vigor, and production of hybrid seed requires manual detasseling or genetic control of pollen development. Knowledge of pollen development can also be used to manage transgene containment.

In the first section of this chapter, we will describe the current model for sequential cell fate specification in maize anther lobes, with reference to rice and *A. thaliana* to point out similarities and differences. In the second section of this chapter, we will review what is known about the individual cell types in anther lobes. The diversity of anther organization is addressed to a limited extent by cytological studies of anthers, often directed to clarify taxonomic relationships. In the third section, we will comment on how new lines of investigation could clarify questions remaining in our current appreciation of anther development.

Current Topics in Developmental Biology, Volume 131
ISSN 0070-2153
https://doi.org/10.1016/bs.ctdb.2018.11.001

1. Orientation to key steps in pre-meiotic anther development

The stamen is the male reproductive organ of the flower (Fig. 1A). It consists of two parts: the anther and a supporting filament. The filaments typically serve two roles; they contain the vasculature from the main plant body that terminates within the anther, and filaments elongate (often dramatically) prior to pollen dispersal. Angiosperm anthers have a typical butterfly shape in the transverse plane, where four identical lobes, two adaxial

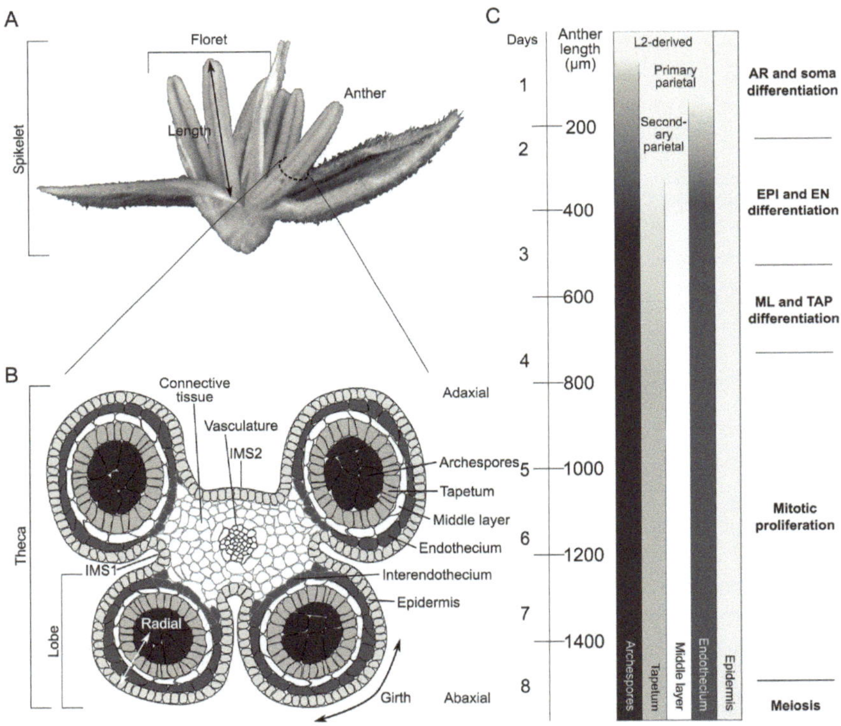

Fig. 1 Orientation to anther terminology. (A) Photo of a maize male spikelet containing two florets. Each floret contains three anthers. (B) Schematic transverse section of an anther before the start of meiosis. (C) Progression of pre-meiotic maize anther development aligned with days since lobe conception and anther length. Color gradients in each column represent advancement in development. *AR*, archesporial cell; *EN*, endothecium; *EPI*, epidermis; *ML*, middle layer; *TAP*, tapetum. *Figure parts modified from Egger, R. L., & Walbot, V. (2016). A framework for evaluating developmental defects at the cellular level: An example from ten maize anther mutants using morphological and molecular data.* Developmental Biology, 419, *26–40.*

and two abaxial, surround the central, sterile connective, and vascular tissue (Fig. 1B). One adaxial (upper lobe in Fig. 1B) and one abaxial lobe together form one theca (Fig. 1B). Before meiosis, within each lobe the archesporial cells (ARs), which later become pollen mother cells (PMCs), are encircled by four layers of somatic tissue: epidermis (EPI), endothecium (EN), middle layer (ML), and tapetum (TAP) (Fig. 1C). In most angiosperms, each individual somatic tissue layer is only one cell thick, and all are required for support of the ARs, PMCs, meiocytes, and successful pollen maturation and dispersal.

All parts of the male flower trace back to the floral meristem. The floral meristem of dicots contains outer L1 and internal L2 and L3 layers, while in monocots there is an L1 and a bilayered L2. As far as it is known, only the L1 and L2 meristem zones contribute cells that ultimately form anther lobes; the dicot L3 is restricted to the vasculature within the filament and the connective tissue at the center of the anther as viewed in transverse section (Fig. 1B). While the EPI develops from L1-derived cells (L1-d), all other anther cells arise from L2-d or L3-d. The stamen primordium first undergoes a series of patterning events to specify the proximal filament, the distal anther, abaxial and adaxial polarity, theca, and lobes (Fig. 1B). The specification of these lobes, which involves initial adaxial/abaxial patterning to establish the two thecae followed by an axis rotation to produce two lobes from each theca, has been established for rice (Toriba et al., 2010) and is reviewed by Walbot and Egger (2016). Then, within each lobe, an inner subset of L2-d acquires AR fate, the first step in anther cell type specification (Fig. 1C) (Kelliher & Walbot, 2011). The surrounding pluripotent L2-d cells then acquire bipotent somatic cell fate as primary parietal cells (PPCs), which then undergo a periclinal division to form the EN and bipotent secondary parietal cells (SPCs) (Kelliher & Walbot, 2011). The SPCs establish the ML and TAP by a second periclinal division (Kelliher & Walbot, 2011). During these cell fate specification events, there is active mitotic proliferation that causes rapid anther growth, and mitotic proliferation remains high in the soma until the PMC stage (Fig. 1C). The sperm produced in the haploid gametophytes after meiosis constitute the germ line, while all previous cells in the lineage are "germinal," that is contribute to reproduction. In concert with germinal cell maturation, meiosis, and pollen development, the anther somatic tissues also differentiate and redifferentiate to play sequential and distinctive supporting roles in both germinal cell differentiation and pollen dispersal.

Most of the signaling components regulating and coordinating anther development remain unclear. It is known that stamen initiation is controlled by phytohormones and a combination of class B and C genes described by the ABCDE model for the determination of floral organ identity (see chapter "Molecular regulation of flower development" by Thomson and Wellmer, this issue). Based on the lack of meristematic tissue, however, once stamen primordia are formed, subsequent signaling does not seem to rely on phytohormones but rather on small signaling peptides and environmental conditions (Kelliher & Walbot, 2012; van der Linde et al., 2018; Zhang et al., 2014). For example, in maize, naturally arising hypoxic conditions around the tassel trigger AR formation in the lobes. Most probably, this environmental signal is integrated by glutaredoxins, namely MALE STERILE CONVERTED ANTHER1 (MSCA1) in maize (Kelliher & Walbot, 2012). The MSCA1 homologs of *A. thaliana*, the two ROXY glutaredoxins, interact with bZIP transcription factors (Li et al., 2009): a logical hypothesis is that reduced glutaredoxin interaction with existing bZIPs activates these transcription factors, resulting in acquisition of AR cell fate through changes in gene expression. By manipulating redox status any maize lobe cell can become an AR, supporting the concept that the signaling system pre-exists in every cell (Kelliher & Walbot, 2012) and that the normal placement of AR cells occurs because these cells are the first to experience a sufficient reducing state to trigger bZIP-mediated transcriptional regulation. Currently, it is unknown if hypoxia is also the trigger for AR specification in other flowering plants.

In maize, subsequent establishment of the somatic niche is controlled by the small protein MULTIPLE ARCHESPORIAL CELLS1 (MAC1) (van der Linde et al., 2018; Wang et al., 2012). MAC1 is produced by newly specified ARs, secreted and recognized by the surrounding L2-d cells, most probably via the leucine-rich repeat receptor kinase (LRR-RK) MULTIPLE SPOROCYTE1 (MSP1) (van der Linde et al., 2018). In response, these cells differentiate as PPCs and then undergo a periclinal division to form EN and SPCs. A key phenotypic observation is that the sterile *mac1* mutant is not only missing the EN and SPCs but also develops excess ARs, indicating that MAC1 either serves as a dual signal or that PPCs and/or SPCs adjacent to the ARs generate a suppressor of cell division acting on the ARs (Wang et al., 2012). Homologs of MAC1 in rice and *A. thaliana* interact with LRR-RKs, and null mutants have somatic layer defects as well as excess ARs just as in maize (Jia, Liu, Owen, & Zhao, 2008; van der Linde et al., 2018; Wang et al., 2012; Yang et al., 2016), suggesting that this signaling module is widely conserved.

Proper differentiation of the tapetum has been investigated by extensive genetic analyses. In *A. thaliana*, rice, and maize, a succession of bHLH-family transcription factors are involved: mutants show excess tapetal precursor proliferation, aberrant cytology, and the absence of functions required to support meiosis or post-meiotic pollen development (Nan et al., 2017). The bHLH transcription factors are moderately well-conserved among the model plant species. They exhibit near parallel genetical and protein interaction hierarchies, such as the ability to form heterodimers, timing of action during initial tapetal development, and subsequent cell redifferentiation for post-meiotic functions (Nan et al., 2017). The large number of tapetal developmental mutants recovered from male sterile mutant collections likely reflects the importance of this tissue in male fertility but may also highlight that there are many tapetal-specific transcription factors. An important consideration in anther genetic analysis is that many interesting types of mutants may never be recovered: anthers are a terminal organ produced at the end of the annual plant lifecycle. Consequently, factors critical for embryo or vegetative organ development will exhibit defects long before flower formation and, if lethal, may be unavailable for analysis of floral phenotypes. Because of the importance of male sterility in crop breeding, there are hundreds of male sterile lines that could be analyzed in the future to better understand other anther cell types and the signaling components underlying their development. Indeed, there are so many cases that many mutants remain unmapped and uncloned, providing a wealth of starting materials for plant biologists interested in working in this area.

2. Stamens lack a meristem

An important feature of stamens is that there is no meristem: once the stamen primordium is established, the terminal anther develops without an organizing center. The lack of a meristem generating cells that sequentially develop into tissues composed of distinct cell types, means that the mass of anther cells must self-organize into the complex lobe architecture. In Section 1, we proposed—and there is growing evidence supporting this hypothesis—that cell-cell signaling is critical in the local regulation of organ patterning, cell division, and cell differentiation. Importantly, anthers develop without bulges or gaps, indicating that local forces, hormones, or other forms of cell-cell communication are tightly regulated to coordinate growth. Furthermore, these local controls are coordinated or propagated throughout the entire anther organ to retain overall morphology. Could the vegetative plant be a source of signals via vascular transport to developing anthers? The vasculature remains a single column within the anther—there is no branching into the lobes. Both water and nutrients from the vasculature must traverse many cell diameters through the connective tissue to reach the lobes, and then across multiple cell layers to reach the germinal cells or the outer arc of the somatic tissues. It is possible that informative gradients are established emanating from the vasculature to pattern lobes. However, without a complex pattern of directed transport, it is difficult to envisage radial gradients (such as EN to EPI and EN to center) being established within lobes. Future mapping of potential auxin or sugar gradients using cellular reporters could be informative in understanding the distribution of materials from the vasculature into the anther, particularly the distinction between diffusion or directed transport into the lobes.

Strikingly, anthers highly defective in cell type specification and differentiation nonetheless retain a normal overall morphology. For example, in the maize *msca1* mutant, defective in the glutaredoxin normally required to trigger AR differentiation, none of the characteristic L2-derived cell types are present (Chaubal et al., 2003). Externally, the organ resembles a normal anther and grows to ~2 mm, the mid-meiotic stage, about 40% of normal length. Maize mutants affecting only one cell type, such as *outer cell layer4* (*ocl4*) (EPI defects) (Vernoud et al., 2009), *male sterile23* (*ms23*) (early TAP defects) (Chaubal et al., 2000), and *ameiotic1* (AR/PMC defects) (Pawlowski et al., 2009) also exhibit a morphology indistinguishable from normal anthers, at least through the 2 mm stage. Some highly defective, male sterile maize anthers continue organ growth to near full size. For example, *callose, somatic,* and *microspore defect1* (*csmd1*), which has both pre-meiotic somatic defects and post-meiotic gametophytic defects, grows to full size and cannot be distinguished from fertile anthers based on external shape (Wang, Skibbe, & Walbot, 2011).

Because overall anther growth can be sustained in the absence of normal cell types, we hypothesize that in addition to local growth control within the anther, there are developmental programs that modulate overall organ growth and morphology. Furthermore, these are postulated to be independent of anther cell composition. The nature of these programs represents a frontier for future research—how can organ shape be sustained if the cellular building blocks are all defective? (Fig. 2).

3. Lobe architecture

After successive steps of cell specification and differentiation, the four lobes of a typical angiosperm anther are identical and contain four somatic tissue layers encircling the central pre-meiotic cells (Fig. 1B). There can be

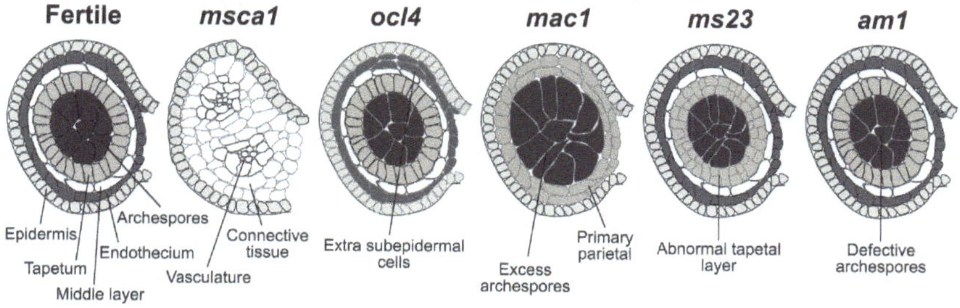

Fig. 2 Schematic overview of lobe transverse sections of five well-characterized male sterile maize mutants and one fertile lobe. All mutants show defects in archesporial cell development or meiosis, additional abnormal cells are indicated in the figure. Overview is based on previous publications (Chaubal et al., 2000; Kelliher & Walbot, 2012; Nan et al., 2017; Pawlowski et al., 2009; Timofejeva et al., 2013; Vernoud et al., 2009; Wang et al., 2012; Zhai et al., 2015).

size dimorphism between the abaxial and adaxial lobes. The somatic layers are only a single cell thick and are readily recognizable by position, by cell shape, and by staining properties. In this section, we will review the properties and proposed functions of each cell type and highlight evolutionary deviations from the typical angiosperm lobe architecture.

3.1 Epidermis

Throughout the plant, the L1 generates the epidermal tissue. In anthers EPI cells typically elongate into a columnar shape (Kelliher & Walbot, 2011). Scant attention has been paid to steps in anther epidermal cell differentiation. In maize, the tissue lacks stomata, but these pores form in the *msca1* mutant (Chaubal et al., 2003), suggesting that the lack of AR cells, which normally pattern the L2-d cells, also results in an epidermal defect, i.e., failure of the EPI to acquire an anther identity. At least in the grasses and model dicots, anther epidermal tissues often lack hairs, trichomes, and other specialized epidermal structures found in leaves and roots. The epidermis is a smooth continuous layer encasing the anther and filament, but there are exceptions such as *Chelone glabra* L. (Arekal, 1963) in which anthers are covered by epidermal hairs. This example highlights an important feature of anther studies: "rules" or generalizations drawn from the three model plants will almost always have exceptions within angiosperms. Initially anther epidermal cells lack a definitive cuticle, and at least in maize, the anthers are embedded in a fluid, perhaps eliminating the requirement for a water-tight seal produced by the epidermis. Later in development a cuticle is produced, and this process has been analyzed by microscopy (Cheng, Greyson, & Walden, 1986).

Interestingly, the rice *wax-deficient anther1* (*wda1*) mutation affects both anther cuticle formation and the exine materials secreted by the tapetum to coat developing pollen grains (Jung et al., 2006), suggesting that there may be anther-specific programs for secretion of complex molecules.

The EPI persists as living tissue throughout anther development, and must ultimately be "opened" to allow pollen dispersal. Epidermal organization is coordinated with the underlying EN, the only other persistent tissue layer in most angiosperms, to first contain the developing pollen and then release it. It has been proposed that the columnar EPI have cell wall reinforcements 90° offset from the wall reinforcements of the underlying EN cells, which are radially elongated and short to add strength to the outer anther wall, much like a steel belted radial tire (Kelliher & Walbot, 2011).

In maize, and likely in the grasses generally, the anther epidermis is key to the production of 21-nt phased small RNAs (phasiRNAs) (Zhai et al., 2015); male reproductive phasiRNAs have recently been detected in some dicots as well (R. Xia and B. C. Meyers, personal communication). In rice, mutations in two of the many 21-nt *PHAS* loci confer day length- and temperature-dependent male sterility (Ding et al., 2012; Fan et al., 2016). This male sterility is the key technology for hybrid rice seed production in China. Mechanistically, the role(s) of the 21-nt phasiRNAs are still obscure; however, given their connection to fertility control, there is growing interest in their analysis. In maize, this class of small RNAs is produced from more than 400 loci and collectively the phasiRNAs reach peak abundance early in anther development. At the 0.4 mm anther length stage, when the PPCs complete periclinal divisions to yield the subepidermal EN cells and SPCs, 21-nt phasiRNAs are about two-thirds of all 21-nt small RNAs in anthers (Zhai et al., 2015). By *in situ* hybridization, three components of 21-nt phasiRNA biogenesis pathway—*microRNA2118*, *Dicer-like 4* (*DCL4*) transcripts, and *PHAS* precursor transcripts—localize to the epidermis (Zhai et al., 2015). The 21-nt products are readily detected in both the EN and SPCs. In male sterile *ocl4* mutants, which lack a bZIP transcription factor expressed in epidermal cells throughout the plant (Vernoud et al., 2009), 21-nt *PHAS* locus transcripts are absent or undetectable and consequently the small RNAs are also missing (Zhai et al., 2015). Collectively, these results suggest that differentiating anther epidermal cells are responsible for 21-phasiRNA biogenesis. One speculation is that the 21-nt phasiRNAs are a signal from the EPI to other somatic tissues. Charting the movement and discovering the targets of 21-nt phasiRNAs might implicate them as coordinating signal molecules.

In terms of anther anatomy, OCL4 appears to repress periclinal division in newly formed endothecial cells: in its absence, the outer hemisphere EN cells, those most distant from the connective tissue, undergo a single ectopic periclinal division (Vernoud et al., 2009; Wang et al., 2012), resulting in male sterility. Consequently, this mutant defines proximal (adjacent to the connective) and distal (far from the connective) hemispheres within anther lobes. Because OCL4 localizes to the epidermis, the anatomical defect is attributed to signaling from the EPI to the underlying EN. Closer inspection of *microRNA2118* localization indicates that the strongest *in situ* hybridization signal occurs in the central cells on the outermost arc of each anther lobe. This is also the distribution of *Ocl4* RNA (Vernoud et al., 2009) suggesting that this pole position is a second landmark in addition to the equatorial hemisphere zonation within lobes. This polar position also marks the zone where *Ms23* transcripts first disappear from the middle layer (Nan et al., 2017). Thus, although the epidermal cells of anther lobes are similar by microscopy, there is a zone of a several cells viewed transversely and as a stripe viewed longitudinally along the outer arc of each lobe that have organizing properties. This zone may serve as a source of cell to cell signals (such as 21-nt phasiRNAs and the unknown OCL4-dependent repressor) that coordinate subepidermal anther development.

3.2 Endothecium

EN cells are generated by the periclinal division of PPCs, programmed by MAC1 peptides secreted by the ARs in maize. The control of this division is not yet established in other plants. EN cells almost immediately begin accumulating starch in amyloplasts, a phenotype readily detected by Lugols′ iodine staining. Flowers are very strong sinks and generally non-photosynthetic. It is likely that the EN serves as an energy storage tissue to sustain anther development when photosynthesis is slow, such as on a cloudy day and at night. In maize, the amyloplasts are faintly green, detected by red chlorophyll fluorescence (Murphy, Egger, & Walbot, 2015). The last step in chlorophyll biosynthesis requires a photon; consequently, despite the many layers of intervening tissue, some light must penetrate through the whorl of immature leaves to reach anthers within the maize tassel. It is also possible that EN and other cell types process photon quality or quantity as a source of information for anther development. Within lobes the several EN cells adjacent to the connective have non-green plastids and less starch. These cells are termed interendothecium (Murphy et al., 2015) and have an ambiguous origin—are they connective or differentiated from PPCs?

Although a partially duplicated EN layer in the outer lobe hemisphere is associated with male sterility in *ocl4* maize mutants, extra EN cells are tolerated in *Panax ginseng* Meyer (Kim et al., 2016). In this dicot (family Apiales), the hemisphere adjacent to the connective can have 1–4 EN layers. This observation is an example of tolerated variation in lobe architecture in some species as detailed in many more cases by Bhandari (1984). In contrast, more modest anatomical variation in the model plants results in male sterility (Egger & Walbot, 2016; Walbot & Egger, 2016) as discussed in more detail in a subsequent section.

3.3 Middle layer

The ML is the least well-understood cell type in anther lobes. Few mutants affect this layer exclusively (Walbot & Egger, 2016), and the major focus has been on the TAP, formed by the sister cells from periclinal division of the SPCs. In the model species, the ML is a short-lived, transient cell type. The layer is crushed mechanically or undergoes programmed cell death (PCD) prior to the completion of meiosis. Furthermore, the ML can be completely absent, as is true in the microsporangia of gymnosperms (Singh, 1979) or multi-layered in diverse angiosperm examples (Bhandari, 1984). In *Magnolia stellata*, there are 3–4 middle layers, and although the ML adjacent to the tapetum degenerates, other layers persist until anthesis (Kapil & Bhandari, 1964).

Despite the lack of evidence for roles, the ML could function to separate the EN from the TAP. Recall that the PPCs divide to directly form an EN cell and a pluripotent SPC. Thus, sister SPCs and EN cells share a common wall and may communicate via primary plasmodesmata. When the SPCs divide periclinally, this common wall is part of the ML, potentially separating the TAP from direct influence of the EN.

3.4 Tapetum

As indicated earlier, TAP cells have many functions: conduit for water and nutrient flow to the ARs and PMCs, remodeling of callose secreted by the PMCs prior to meiosis, release of microspores from the tetrad at the end of meiosis, secretion of fluid to fill the locule to support gametophyte development, and synthesis and secretion of most pollen exine components. In maize, there is a further role. Tapetal cell collapse generates a landing pad for developing gametophytes, which must physically attach to these pads (one gametophyte to each collapsed tapetal cell clone derived from each SPC) to continue development and form viable pollen (Kelliher & Walbot, 2011).

Most male sterile mutants affect the TAP layer. Analysis of diverse tapetal mutants has illuminated the order of steps in cell fate setting, initial acquisition of tapetal-specific traits, and sequential redifferentiation (see Walbot & Egger, 2016 for a review of the tapetal mutants for which the genes are cloned). Although model and most crop plants have only one tapetal layer, once again there are notable exceptions. *Panax ginseng* Meyer has two tapetal layers, as do a number of other non-crop plants (Bhandari, 1984).

In grasses, a second size class of phasiRNAs is produced during meiosis I (Zhai et al., 2015). These 24-nt phasiRNAs appear to be synthesized in the tapetum: by *in situ* hybridization the required *miR2275*, *DCL5*, and the precursor transcripts ultimately processed into phasiRNAs are all enriched in the TAP. The phasiRNAs localize primarily to the TAP, but there is also evidence for their presence in the PMCs and meiocytes (Ono et al., 2018). The precise functions of both the 21-nt and 24-nt phasiRNAs remain a mystery: unlike tasiRNAs, they lack complementarity to cellular mRNAs and may perform roles that depend on their superabundance rather than sequences. Several male sterile maize mutants with defects in TAP cells lack the 24-nt phasiRNAs. Mutants disrupting the bHLH transcription factor MS23, for instance, fail to express DCL5, the normal levels of *miR2275*, and the *PHAS* locus primary transcripts implicating MS23 as a master regulator of tapetal differentiation including 24-nt phasiRNA biogenesis (Nan et al., 2017). 24-nt phasiRNAs have also been proposed to play an important role in the progression of meiosis (Ono et al., 2018). As with the 21-nt phasiRNAs, clarifying possible small RNA movement and targets would illuminate the roles of these molecules.

3.5 Archesporial cells/pollen mother cells/meiocytes

As described earlier, ARs are the first anther-specific differentiated cell type. Through the secreted MAC1 protein in maize, ARs program differentiation of the single layer of L2-d pluripotent subepidermal cells to PPCs, which in turn divide periclinally to generate the EPI and the SPC bipotent stem cells. ARs are individually specified in a column during early anther growth, resulting in multiple germinal lineages within a lobe. This contrasts with the formal hypodermal lineage model of a singular AR in each lobe, resulting in meiocytes that are all derived from the same founding cell. ARs proliferate mitotically, and the length of this period determines the number of PMCs formed (very large in lily, moderate in maize, few in *A. thaliana*). After mitotic divisions of the ARs stop prior to the PMC stage, AR cells initiate pre-meiotic S-phase. DNA synthesis does not guarantee that meiosis will ensue, because mutants such as maize *ameiotic1* (Pawlowski et al., 2009) conduct a mitosis rather than meiosis.

AR and PMC differentiation appears to be cell autonomous, because these cells do not require normal somatic lobe cells to start meiosis. Furthermore, the just specified ARs express a number of genes important for meiosis-specific processes, such as recombination (Kelliher & Walbot, 2014), when only PPCs, EN, and the SPCs are present, excluding any influence from the TAP. In nearly all male sterile mutants with aberrant somatic architecture, however, the meiocytes arrest during prophase I, often at the leptotene/zygotene transition. It can be inferred that at this point there is either a nutritional defect that stalls meiosis or a somatic signal is missing when lobe architecture is defective. After unsynchronized mitotic proliferation in the ARs, PMC maturation is regular but not synchronized, until in maize and other species, there is a nearly synchronous start of meiosis. The start signal could be autonomously generated by the meiocytes, but it is also possible that start and continued synchrony are regulated or aided by somatic signals. Clarifying meiotic start and the requirements for maintaining meiotic progression should be a fruitful area of future study in plants.

As meiosis is covered in detail elsewhere (Mercier, Mézard, Jenczewski, Macaisne, & Grelon, 2015), we will not comment on it further, except to point out that most analysis is focused on chromatin events during pre-meiotic S-phase and meiosis, but little attention has been paid to the rest of the nucleoplasm or to the cytoplasm. Pre-meiotic cells typically have a very prominent and greatly enlarged nucleolus; the cells are very large and the walls are poorly reinforced based on the roundish shape and apparent deformation by the surrounding tapetum. In maize *mac1* mutants, the excess ARs and PMCs assume many shapes in the absence of the normal surrounding somatic layers. In maize, there is a gap between the ARs and neighboring cells (cf. Kelliher & Walbot, 2011); to date no reagent has been found to stain this region for light microscopy (V. Walbot, personal observations). At the PMC stage, this gap region and the spaces between PMCs becomes filled with callose deposited by the PMCs; the callose is subsequently remodeled by callose secreted by the TAP to form a thinner, smoother layer encasing the PMCs.

4. Classes of cell division defects in anther lobes

Many male sterile mutants with defects prior to meiosis exhibit cytological defects viewed by two-dimensional microscopy, such as extra periclinal divisions, too few or too many cells of one or more cell types, and absence of defining characteristics of one or more cell types. With the advent of confocal microscopy and the evaluation of entire anthers rather than transverse sections, it is apparent that extrapolations from transverse sections were often wrong. For example, it was assumed that mutants with extra periclinal cell divisions resulting in extra layers of cells would inevitably have more cells per anther lobe than normal, fertile siblings. Reconstruction of the development of 10 maize anther mutants initially scored as exhibiting extra periclinal divisions and proposed to have more cells than normal, found that only one of these mutants did fit the description originally proposed (Fig. 3) (Egger & Walbot, 2016). The bottom part of Fig. 3A–C shows images from confocal microscopy of small longitudinal optical sections of all 10 maize anther mutants, in which all lobe cell types can be viewed. Extra cells are highlighted by asterisks. The top part of Fig. 3A–C illustrates tracings of the transverse view in confocal imaging. Among the mutants, class A (Fig. 3A) had no extra periclinal divisions but altered anticlinal proliferation: skewed walls, viewed transversely, could fool the observer into thinking that there had been a periclinal division, when there were only misshapen cells within the layer. For class B (Fig. 3B), there are sporadic periclinal division examples as well as examples in which an entire extra layer is formed. Recall that extra layers are characteristic of some species, but in a species that normally has just one layer, periclinal division mutants are associated with male sterility. In the class B examples shown, although mutant anthers were expected to have more cells than wild type, fertile anthers, there were actually fewer cells in all but one of the mutants. In class C (Fig. 3C), there are periclinal division defects and a normal number of total anther lobe cells. This analysis highlights how little we understand about global control of cell division and hence cell number in anthers. When defective cells are produced, what causes compensation to retain anther shape and, in many cases, a normal final number of anther cells?

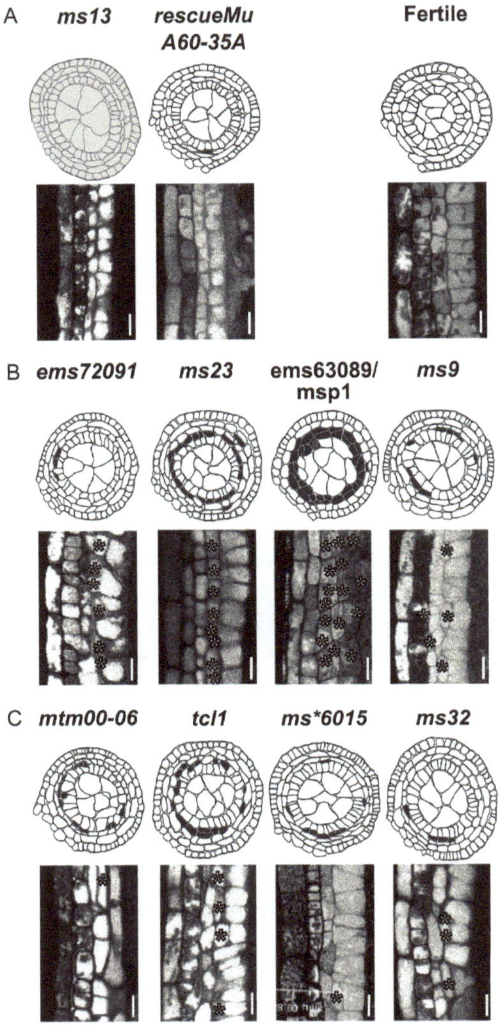

Fig. 3 Example of 10 different maize anther mutants. Transverse diagrams and longitudinal confocal images of fertile or mutant lobes at meiosis. (A) Mutants with anticlinal proliferation defects. (B) Mutants with periclinal division defects and abnormal anther lobe cell numbers. (C) Mutants with periclinal division defects and normal anther lobe cell numbers. Confocal images of sterile individuals of each mutant were traced at anther stages 1300–1500 μm. Cells of undetermined identity are highlighted in gray in transverse sections and indicated by an asterisk in longitudinal images. Scale bar = 20 μm. *Figure modified from Egger, R. L., & Walbot, V. (2016). A framework for evaluating developmental defects at the cellular level: An example from ten maize anther mutants using morphological and molecular data.* Developmental Biology, 419, *26–40.*

5. Future directions

As with many problems in plant biology, analysis of anther development and evolution of the regulatory pathways controlling it would benefit greatly from more researchers participating. Classical light microscopy investigations were immensely useful in categorizing species, in identifying exceptions to the general rules, and in analyzing mutants. Today the emphasis has switched from anatomy to uncovering the mechanisms guiding development. For anthers, these mechanisms do not depend on a meristem, making analysis of anthers a "new branch" of plant development, one that challenges us to discover local cell to cell as well as tissue and organ level signaling and response pathways. New technologies such as *in vivo* sensors for presumptive regulatory molecules (hormones, peptides, ions, nutrients), imaging methods that allow greater precision in cellular analysis, and deployment of new technologies beyond genetics for correcting or perturbing development (cf. van der Linde et al., 2018) can usher in a new era in anther biological studies. We anticipate many new discoveries! Comparative analyses between species and between anthers and other terminal organs in the same plant will clarify mechanisms that may be anther-specific (evolutionarily conserved in anthers but absent in other organs) as well as taxon-specific innovations fundamental in organs lacking meristems (utilized repetitively within a species but absent in distant species). As for the anther-specific mechanisms, the MAC1 protein and its homologs in other plants appear to function exclusively in anthers with an anther-specific receptor; the general mechanism of peptide ligand and cognate receptor is widely deployed, but an apparently anther-specific module has deep roots within angiosperms. Yet, this conserved module seems to be used at slightly different checkpoints of anther development in different flowering plant species, because there are distinctive knockout phenotypes (Jia et al., 2008; van der Linde et al., 2018; Wang et al., 2012; Yang et al., 2016; Zhao et al., 2008). The same holds true for MSCA1 and its homologs (Kelliher & Walbot, 2012; Xing & Zachgo, 2008), proving the necessity to study anther developmental processes in a variety of flowering plants.

As for taxon-specific mechanisms, deployment of phasiRNAs is rapidly evolving within the grasses in terms of the number of *PHAS* loci, and based on limited sampling, phasiRNAs are found sporadically in dicots and non-grass monocots. Because the species definition in angiosperms requires that there be unique floral characteristics to qualify as a species, diversification of anther structures and underlying biochemical mechanisms is to be expected. The flowering plants are not only highly successful but also have evolved exceptionally rapidly into hundreds of thousands (perhaps millions) of species. This almost limitless frontier for future research promises not only to enrich our understanding of development and ecological and evolutionary interplay with development but also to provide insights critical to sustaining the world food supply.

Acknowledgments

Research on anther development and phasiRNAs in the Walbot lab is supported by NSF grant 1649424. K.v.d.L.'s research on anthers is financed by the Leopoldina—Nationale Akademie der Wissenschaften and through the Collaborative Research Centre SFB 924 by the German Research Foundation (http://sfb924.wzw.tum.de). The authors would like to thank R.L. Egger for providing original figure files from Egger and Walbot (2016) as templates for some of the figures presented here.

References

Arekal, G. D. (1963). Contribution to the embryology of *Chelone glabra* L. *Phytomorphology, 13*, 376–388.

Bhandari, N. N. (1984). The microsporangium. In B. M. Johri (Ed.), *Embryology of angiosperms* (pp. 53–121). Berlin/Heidelberg: Springer-Verlag.

Chaubal, R., Anderson, J. R., Trimnell, M. R., Fox, T. W., Albertsen, M. C., & Bedinger, P. (2003). The transformation of anthers in the *msca1* mutant of maize. *Planta, 216*, 778–788.

Chaubal, R., Zanella, C., Trimnell, M. R., Fox, T. W., Albertsen, M. C., & Bedinger, P. (2000). Two male-sterile mutants of *Zea mays* (Poaceae) with an extra cell division in the anther wall. *American Journal of Botany, 87*, 1193–1201.

Cheng, P. C., Greyson, R. I., & Walden, D. B. (1986). The anther cuticle of *Zea mays. Canadian Journal of Botany, 64*, 2088–2097.

Ding, J., Lu, Q., Ouyang, Y., Mao, H., Zhang, P., Yao, J., et al. (2012). A long noncoding RNA regulates photoperiod-sensitive male sterility, an essential component of hybrid rice. *Proceedings of the National Academy of Sciences of the United States of America, 109*, 2654–2659.

Egger, R. L., & Walbot, V. (2016). A framework for evaluating developmental defects at the cellular level: An example from ten maize anther mutants using morphological and molecular data. *Developmental Biology, 419*, 26–40.

Fan, Y., Yang, J., Mathioni, S. M., Yu, J., Shen, J., Yang, X., et al. (2016). *PMS1T*, producing phased small-interfering RNAs, regulates photoperiod-sensitive male sterility in rice. *Proceedings of the National Academy of Sciences of the United States of America, 113*, 15144–15149.

Jia, G., Liu, X., Owen, H. A., & Zhao, D. (2008). Signaling of cell fate determination by the TPD1 small protein and EMS1 receptor kinase. *Proceedings of the National Academy of Sciences of the United States of America, 105*, 2220–2225.

Jung, K. H., Han, M. J., Lee, D. Y., Lee, Y. S., Schreiber, L., Franke, R., et al. (2006). *Wax-deficient anther1* is involved in cuticle and wax production in rice anther walls and is required for pollen development. *The Plant Cell, 18*, 3015–3032.

Kapil, R. N., & Bhandari, N. N. (1964). Morphology and embryology of *Magnolia* Dill. ex Linn. *Proceedings of the National Institute of Sciences of India, Section B: Biological Sciences, 30*, 245–262.

Kelliher, T., & Walbot, V. (2011). Emergence and patterning of the five cell types of the *Zea mays* anther locule. *Developmental Biology, 350*, 32–49.

Kelliher, T., & Walbot, V. (2012). Hypoxia triggers meiotic fate acquisition in maize. *Science, 337*, 345–348.

Kelliher, T., & Walbot, V. (2014). Maize germinal cell initials accommodate hypoxia and precociously express meiotic genes. *The Plant Journal, 77*, 639–652.

Kim, Y. J., Jang, M. G., Zhu, L., Silva, J., Zhu, X., Sukweenadhi, J., et al. (2016). Cytological characterization of anther development in *Panax ginseng* Meyer. *Protoplasma, 253*, 1111–1124.

Li, S., Lauri, A., Ziemann, M., Busch, A., Bhave, M., & Zachgo, S. (2009). Nuclear activity of ROXY1, a glutaredoxin interacting with TGA factors, is required for petal development in *Arabidopsis thaliana*. *The Plant Cell, 21*, 429–441.

Mercier, R., Mézard, C., Jenczewski, E., Macaisne, N., & Grelon, M. (2015). The molecular biology of meiosis in plants. *Annual Review of Plant Biology, 66*, 297–327.

Murphy, K. M., Egger, R. L., & Walbot, V. (2015). Chloroplasts in anther endothecium of *Zea mays* (Poaceae). *American Journal of Botany, 102*, 1931–1937.

Nan, G. L., Zhai, J., Arikit, S., Morrow, D., Fernandes, J., Mai, L., et al. (2017). MS23, a master basic helix-loop-helix factor, regulates the specification and development of the tapetum in maize. *Development, 144*, 163–172.

Ono, S., Liu, H., Tsuda, K., Fukai, E., Tanaka, K., Sasaki, T., et al. (2018). EAT1 transcription factor, a non-cell-autonomous regulator of pollen production, activates meiotic small RNA biogenesis in rice anther tapetum. *PLoS Genetics, 14*, e1007238.

Pawlowski, W. P., Wang, C. J., Golubovskaya, I. N., Szymaniak, J. M., Shi, L., Hamant, O., et al. (2009). Maize AMEIOTIC1 is essential for multiple early meiotic processes and likely required for the initiation of meiosis. *Proceedings of the National Academy of Sciences of the United States of America, 106*, 3603–3608.

Singh, H. (1979). *Embryology of gymnosperms*. Berlin: Gebr. Borntraeger.

Timofejeva, L., Skibbe, D. S., Lee, S., Golubovskaya, I., Wang, R., Harper, L., et al. (2013). Cytological characterization and allelism testing of anther developmental mutants identified in a screen of maize male sterile lines. *G3 (Bethesda, Md.), 3*, 231–249.

Toriba, T., Suzaki, T., Yamaguchi, T., Ohmori, Y., Tsukaya, H., & Hirano, H. Y. (2010). Distinct regulation of adaxial-abaxial polarity in anther patterning in rice. *The Plant Cell, 22*, 1452–1462.

van der Linde, K., Timofejeva, L., Egger, R. L., Ilau, B., Hammond, R., Teng, C., et al. (2018). Pathogen Trojan horse delivers bioactive host protein to alter maize (*Zea mays*) anther cell behavior *in situ*. *The Plant Cell, 30*, 528–542.

Vernoud, V., Laigle, G., Rozier, F., Meeley, R. B., Perez, P., & Rogowsky, P. M. (2009). The HD-ZIP IV transcription factor OCL4 is necessary for trichome patterning and anther development in maize. *The Plant Journal, 59*, 883–894.

Walbot, V., & Egger, R. L. (2016). Pre-meiotic anther development: Cell fate specification and differentiation. *Annual Review of Plant Biology, 67*, 365–395.

Wang, C. J., Nan, G. L., Kelliher, T., Timofejeva, L., Vernoud, V., Golubovskaya, I. N., et al. (2012). Maize *multiple archesporial cells 1* (*mac1*), an ortholog of rice *TDL1A*, modulates cell proliferation and identity in early anther development. *Development, 139*, 2594–2603.

Wang, D., Skibbe, D. S., & Walbot, V. (2011). Maize *csmd1* exhibits pre-meiotic somatic and post-meiotic microspore and somatic defects but sustains anther growth. *Sexual Plant Reproduction, 24*, 297–306.

Xing, S., & Zachgo, S. (2008). *ROXY1* and *ROXY2*, two *Arabidopsis* glutaredoxin genes, are required for anther development. *The Plant Journal, 53*, 790–801.

Yang, L., Qian, X., Chen, M., Fei, Q., Meyers, B. C., Liang, W., et al. (2016). Regulatory role of OsTDL1A-MSP1 signaling in specifying anther cell identity in rice. *Plant Physiology*, *171*, 2085–2100.

Zhai, J., Zhang, H., Arikit, S., Huang, K., Nan, G.-L., Walbot, V., et al. (2015). Spatiotemporally dynamic, cell-type-dependent premeiotic and meiotic phasiRNAs in maize anthers. *Proceedings of the National Academy of Sciences of the United States of America*, *112*, 3146–3151.

Zhang, H., Egger, R. L., Kelliher, T., Morrow, D., Fernandes, J., Nan, G. L., et al. (2014). Transcriptomes and proteomes define gene expression progression in pre-meiotic maize anthers. *G3 (Bethesda, Md.)*, *4*, 993–1010.

Zhao, X., de Palma, J., Oane, R., Gamuyao, R., Luo, M., Chaudhury, A., et al. (2008). OsTDL1A binds to the LRR domain of rice receptor kinase MSP1, and is required to limit sporocyte numbers. *The Plant Journal*, *54*, 375–387.

CHAPTER ELEVEN

The evolution and patterning of male gametophyte development

Dieter Hackenberg*, David Twell*
Department of Genetics and Genome Biology, University of Leicester, Leicester, United Kingdom
*Corresponding authors: e-mail address: dh253@le.ac.uk; twe@le.ac.uk

Contents

Abstract

The reproductive adaptations of land plants have played a key role in their terrestrial colonization and radiation. This encompasses mechanisms used for the production, dispersal and union of gametes to support sexual reproduction. The production of small motile male gametes and larger immotile female gametes (oogamy) in specialized multicellular gametangia evolved in the charophyte algae, the closest extant relatives of land plants. Reliance on water and motile male gametes for sexual reproduction was retained by bryophytes and basal vascular plants, but was overcome in seed plants by the dispersal of pollen and the guided delivery of non-motile sperm to the female gametes. Here we discuss the evolutionary history of male gametogenesis in streptophytes (green plants) and the underlying developmental biology, including

Current Topics in Developmental Biology, Volume 131
ISSN 0070-2153
https://doi.org/10.1016/bs.ctdb.2018.10.008

recent advances in bryophyte and angiosperm models. We conclude with a perspective on research trends that promise to deliver a deeper understanding of the evolutionary and developmental mechanisms of male gametogenesis in plants.

1. The evolutionary history of the male gametophyte

1.1 The origins of spermatogenesis in green plants

The ancestral mode of sexual reproduction in plants is isogamy (gametes of similar size and morphology) and is described in basal chlorophyte and charophyte algae. Oogamy (sperm and egg mating systems) evolved independently in both algal lineages, and the transition from isogamy to oogamy is associated with increasing complexity and colonization (reviewed in Mori, Kawai-Toyooka, Igawa, & Nozaki, 2015; Umen, 2014).

The volvocine algae have been developed as a model for the evolution of reproductive traits among chlorophytes. The unicellular species *Chlamydomonas reinhardtii* is isogamous and one of the most basal species (Nozaki et al., 2000), whereas colonial and multicellular genera can be isogamous, anisogamous or oogamous (Mori et al., 2015). The differentiation of *plus* and *minus* gametes in *C. reinhardtii* is determined by a 200–300 kb multigenic mating type (MT) locus. Gametes are *MT−*, if the *MID* (*minus* dominance) gene is present or *MT+*, if *MID* is absent (Ferris, Armbrust, & Goodenough, 2002; Ferris & Goodenough, 1994). *MID* encodes an RWP-RK domain transcription factor and acts as a master regulator of sexual differentiation (Ferris & Goodenough, 1997). From its ancestral role as a mating-type factor in *C. reinhardtii*, *MID* has evolved to become a determinant of sperm and egg development in oogamous *Volvox carteri* (Geng, De Hoff, & Umen, 2014). Although orthologs of *MID* are present in land plant genomes, their functions are currently unknown (Rövekamp, Bowman, & Grossniklaus, 2016).

In charophyte algae the phylogenetic transition from isogamy to anisogamy (Fig. 1) and the genes responsible for sex determination remain unclear. With the exception of the unicellular isogamous *Mesostigma viride*, sexual reproduction in most charophyte lineages is unknown (Delwiche, 2016; McCourt, Delwiche, & Karol, 2004; Mori et al., 2015). However, late divergent lineages, such as Charales and Coleochaetales, developed oogamous reproduction with bi-flagellated sperm (Renzaglia & Garbary, 2001). The stoneworts of the genus *Chara* produce gametes in morphologically complex gametangia, known as globules (antheridia) (Fig. 2A) or

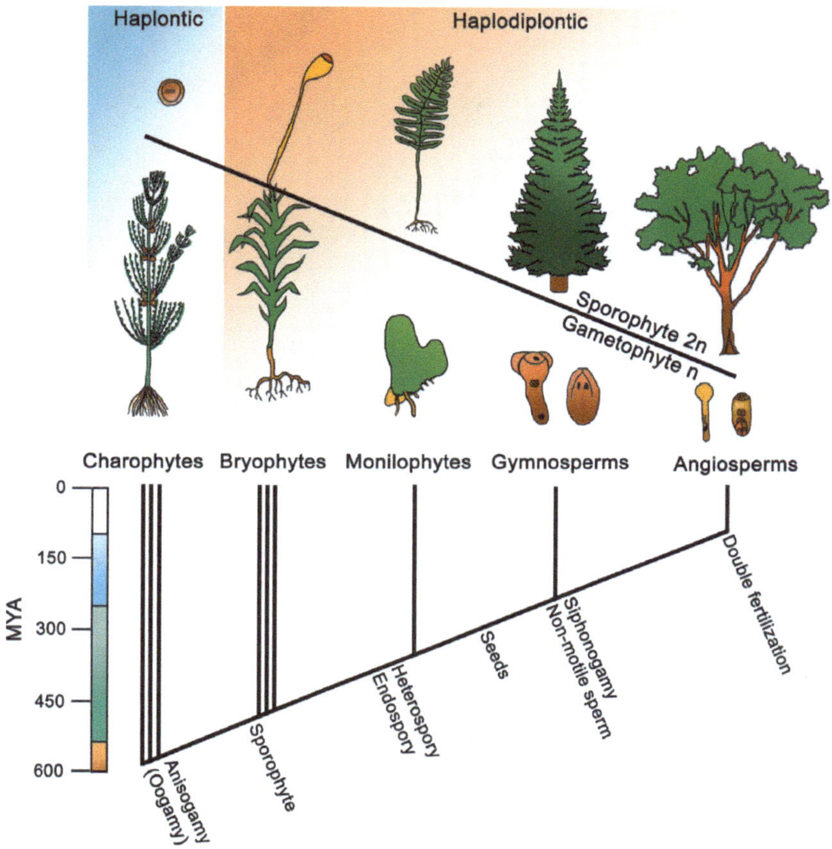

Fig. 1 Evolution of the gametophyte and sporophyte in the land plant lineage. The upper panel illustrates the trend from a dominant haploid gametophyte to that of the diploid sporophyte. Charophyte algae have a haplontic life cycle and possess a multicellular gametophyte, but only the zygote is diploid. The haplodiplontic life cycle and multicellular sporophyte evolved in bryophytes, while complex sporophytes are dominant in vascular plants (e.g., monilophytes and spermatophytes). In spermatophytes (i.e., gymnosperms and angiosperms), the gametophyte is reduced to a few cells and is entirely dependent on the sporophyte. The lower panel shows a cladogram of streptophyte (green plant) taxa indicating their estimated times of divergence and associated reproductive adaptations. Geologic eras are colored: Cenozoic (white), Mesozoic (blue), Paleozoic (green), Neoproterozoic (brown).

nucules (oogonia), that are speculated to be homologous to gametangia of basal land plants (Niklas & Kutschera, 2009). Interestingly, the Zygnematales, an extant charophyte order sister to land plants (Wodniok et al., 2011), secondarily lost flagella and reproduce through isogamous conjugation (McCourt et al., 2004).

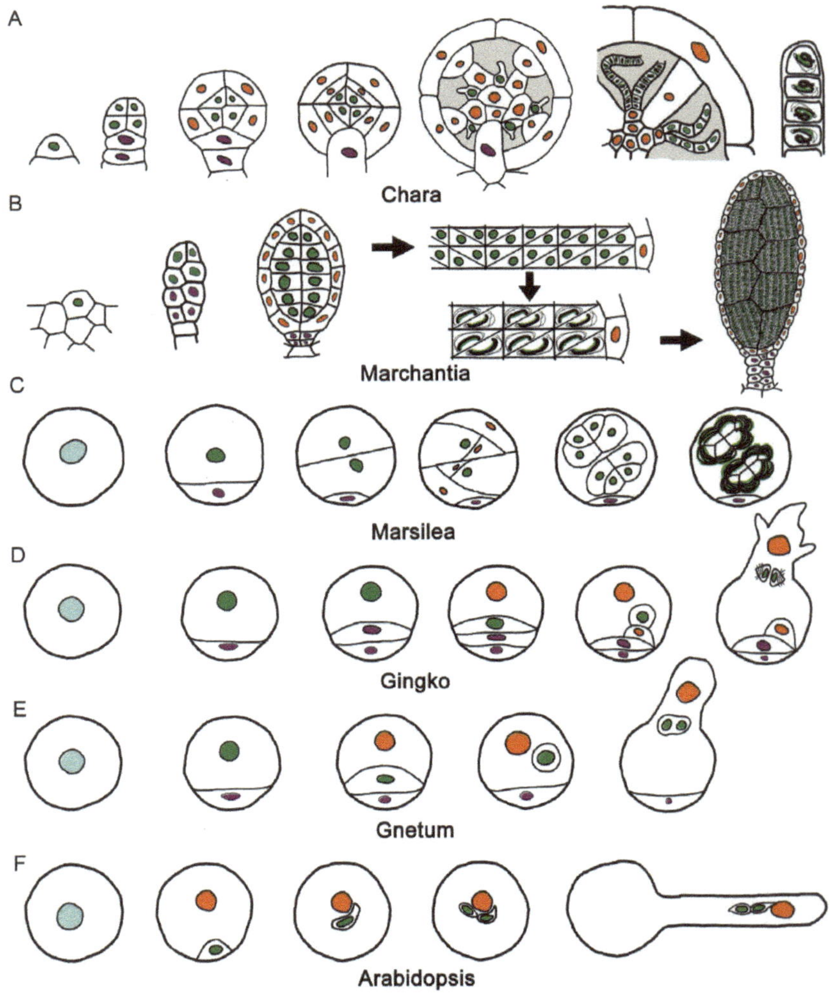

Fig. 2 Overview of cell division patterns during male gametogenesis in representative streptophytes. Sperm cell lineage nuclei are green; primary microspore nucleus (C–F) is blue; basal antheridial/stalk (A and B) and prothallial cell nuclei (C–E) are purple; non-reproductive cell nuclei, derived from the male reproductive lineage, are orange. (A) *Chara* spp. globule (antheridium) development starts with division of an apical initial cell to form apical (green) and basal (purple) antheridial cells. Apical cells form inner primary capitular cells (green) and outer shield cells (orange). Capitular cells give rise to antheridial filament initial cells (green). Antheridial filament cells divide and then differentiate to form biflagellate spermatozoids (green), whereas filament initial cells become non-reproductive (orange). (B) *Marchantia polymorpha* antheridia are formed by division of an antheridial initial to form outer primary antheridial cells (green) and inner stalk cells (purple). Primary antheridial cells divide asymmetrically to generate sterile jacket cells (orange) and spermatogenous cells (green), which divide repeatedly.

1.2 The sporophyte was key for terrestrial plant evolution

Charophyte algae developed a haplontic life cycle, in which the haploid stage is multicellular and the only diploid stage is the unicellular zygote (Shaw, Szovenyi, & Shaw, 2011). The haplodiplontic life cycle, which involves the alternation of multicellular haploid and diploid generations, evolved with the colonization of terrestrial habitats (Fig. 1). This transition requires a delay of meiosis and the intercalation of mitotic cell divisions; the resulting multicellular sporophyte thereby amplifying haploid spore production from each zygote. The limited capacity of charophyte algae for dispersal via motile sperm, therefore gave way to terrestrial colonization via dissemination of durable spores shed by increasingly complex and dominant sporophytes (reviewed in Niklas & Kutschera, 2009).

Each spermatogenous mother cell divides diagonally before differentiating into bi-flagellated spermatozoids. (C) *Marsilea vestita* microspores (blue) divide asymmetrically to form an antheridial initial cell (green) and a prothallial cell (purple), which no longer divides. The antheridal initial divides symmetrically and both daughter cells (green) complete three rounds of asymmetric division to produce three sterile jacket cells (orange) and a primary spermatogenous cell (green). Each spermatogenous cell divides to form four spermatocyte mother cells (green), which divide into eight spermatocytes. These cells divide to form 16 spermatids which differentiate into multiflagellated spermatozoids. (D) *Gingko biloba* microspores (blue) divide twice to form one antheridial (green) cell and two prothallial (purple) cell that do not divide further. The antheridial cell forms the generative cell (green) and tube cell (orange) by asymmetric division. The generative cell divides again to form a sterile stalk cell (orange) and a body cell (green), which divides and differentiates to produce mutliflagellated sperm (see Fig. 3C). (E) *Gnetum gnemon* microspores (blue) divide asymmetrically to form an antheridial cell (green) and a single prothallial cell (purple). The antheridial cell undergoes asymmetric division to produce a generative cell (green) and a tube cell (orange). The generative cell divides to form two non-motile sperm nuclei. (F) *Arabidopsis thaliana* microspores (blue) divide asymmetrically to form the generative cell (green) and the vegetative cell (orange). The generative cell undergoes a second symmetric division to form two non-motile two sperm cells (green). *The figure is adapted and redrawn from (Panel A) Kwiatkowska, M., Kaźmierczak, A., & Popłońska, K. (2002). Ultrastructural, autoradiographic and electrophoretic examinations of Chara tomentosa spermiogenesis.* Acta Societatis Botanicorum Poloniae, 71, *201–209; (Panel B) Durand, E. J. (1908). The development of the sexual organs and sporogonium of Marchantia polymorpha.* Bulletin of the Torrey Botanical Club, 35, *321–335; (Panel C) Tsai, C. W., Van der Weele, C. M., & Wolniak, S. M. (2004). Differential segregation and modification of mRNA during spermiogenesis in Marsilea vestita.* Developmental Biology, 269, *319–330; and (Panels E and F) Rudall, P. J., & Bateman, R. M. (2007). Developmental bases for key innovations in the seed-plant microgametophyte.* Trends in Plant Science, 12, *317–326.*

The production of desiccation-tolerant spores by elevated sporophytes promoted aerial dispersal of gametophytes (Strother, Al-Hajri, & Traverse, 1996; Wellman, 2010). This key innovation was enabled by synthesis of the resistant biopolymer sporopollenin in spore walls (Niklas & Kutschera, 2009). The ability to produce sporopollenin is shared by chlorophyte and charophyte algae (Brooks & Shaw, 1978) and is considered pre-adaptive for terrestrial colonization (Wallace, Fleming, Wellman, & Beerling, 2011). The homology between "lower" plant spores and the pollen of seed plants is bolstered by the presence of homologs of pollen wall associated genes in moss (Wallace et al., 2011). Further, the GAMYB family of transcription factors, which have been shown to regulate sporopollenin biosynthesis genes, are required for the formation of functional spores in moss and for pollen wall formation in angiosperms (Aya et al., 2011, 2009).

The molecular control of the haploid-to-diploid transition seems to share common ancestry in green algae and land plants. In *C. reinhardtii*, *plus* and *minus* gametes express either BEL1-LIKE (BELL) or KNOTTED1-LIKE HOMEOBOX (KNOX) class homeodomain proteins, respectively. After gamete fusion, these proteins heterodimerize to initiate zygote development and meiosis (Lee, Lin, Joo, & Goodenough, 2008). Similarly, KNOX/BELL heterodimers pattern sporophyte development in moss (*Physcomitrella patens*) and in flowering plants, but are neither functional nor expressed during the gametophytic phase (Hay & Tsiantis, 2010; Keiko, Tomoaki, Hironori, & Mitsuyasu, 2008). While antagonistic roles are proposed for the duplicated KNOX1 and KNOX2 gene classes in sporophytic development in angiosperms, KNOX2 genes actively represses the gametophytic body plan in sporophytes of *P. patens* (Furumizu, Alvarez, Sakakibara, & Bowman, 2015; Sakakibara et al., 2013).

In bryophytes, the gametophytic phase is dominant and the mature gametophyte is either a thallus or a leafy shoot with indeterminate growth (Ligrone, Duckett, & Renzaglia, 2012). Male (antheridia) and female (archegonia) gametangia are commonly produced by differentiation of the growing tip of the shoot (gametophore), producing either sperm or eggs cells by mitotic division. In the liverwort *Marchantia polymorpha*, antheridia consist of spermatogenous cells surrounded by a single layer of sterile jacket cells (Fig. 2B; Durand, 1908; Shimamura, 2016). The spermatogenous cells undergo proliferative mitotic divisions, eventually dividing diagonally to form a pair of triangular sperm cells (Durand, 1908). Similar to their charophycean ancestors, the bryophytes (liverworts, hornworts and mosses) release motile sperm into environmental water (Fig. 3A). During

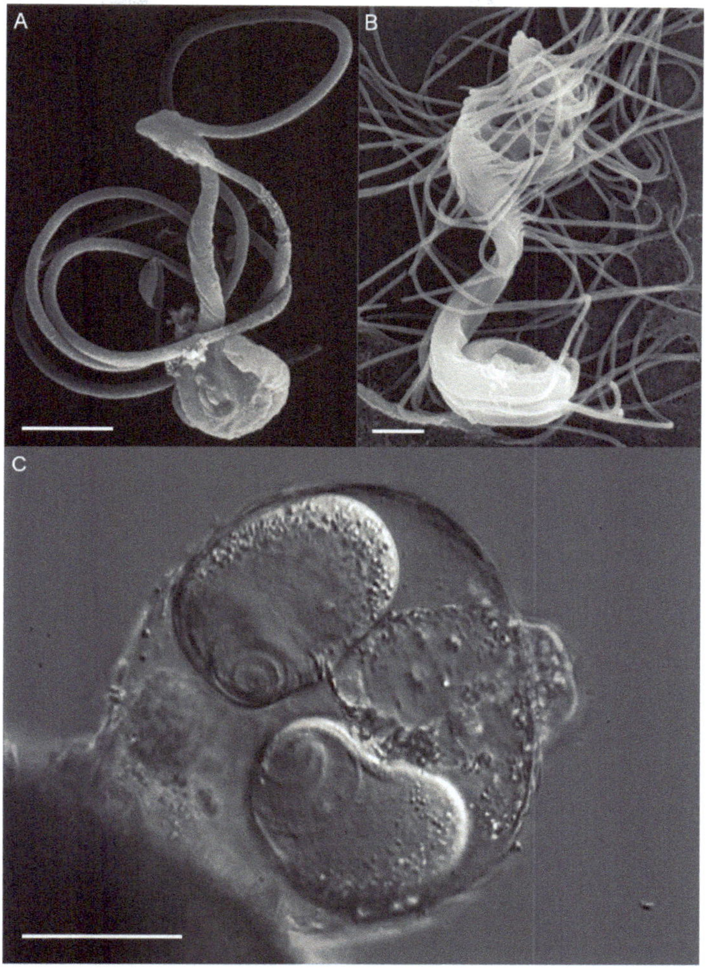

Fig. 3 Micrographs of motile spermatozoids of plants. Scanning electron micrographs of (A) bi-flagellated spermatozoid of the liverwort *Conocephalum conicum*, (B) multi-flagellated spermatozoid of the leptosporangiate fern *Ceratopteris richardii* and (C) light micrograph of motile sperm of the gymnosperm *Ginkgo biloba*. Scale bars: (A and B) 2.0 µm, (C) 80 µm. *Panel A: Reprinted with permission from Renzaglia, K. S., & Garbary, D. J. (2001). Motile gametes of land plants: Diversity, development, and evolution.* Critical Reviews in Plant Sciences, 20, *107–213. Panel B: Reprinted with permission from Lopez-Smith, R., & Renzaglia, K. (2008). Sperm cell architecture, insemination, and fertilization in the model fern, Ceratopteris richardii.* Sexual Plant Reproduction, 21, *153–167. Panel C: Reprinted with permission from Suzuki, Y., Tsuneya, T., Hori, T., Shogo, K., & Okada, K. (2000). The sea in the seed—Sperm of Gingko biloba and reproductive evolution in plants—A scientific video. TokyoCinema, Japan.*

spermatogenesis the cells undergo complex morphological changes, including extreme cytoplasmic reduction, nuclear condensation, and flagella assembly (see Section 3.1; Renzaglia & Garbary, 2001).

1.3 Heterospory and the origin of microgametophytes

The increase in size and complexity of the sporophytes of vascular plants is matched by corresponding reductions in their gametophytes (Fig. 1). Basal extant lycophytes and monilophytes produce free-living exosporic gametophytes (prothallia) with flattened thalloid anatomy, or subterranean tuber-like structures (Ligrone et al., 2012). These taxa retain an oogamous life style and prothallia produce gametes in separate male and female gametangia, similar to those of bryophytes.

The spores of extant bryophytes and basal vascular plants are uniform in size, a condition known as homospory. Heterospory, defined as the production of smaller male microspores and larger female megaspores, arose independently in all major vascular plant lineages (lycophytes, monilophytes and spermatophytes) and is considered a key innovation for terrestrial colonization (Bateman & DiMichele, 1994; Petersen & Burd, 2016). Endospory, the development of gametophytes within the spore wall, is tightly coupled to unisexuality and is proposed as an innovation that allowed the evolution of heterospory (Bateman & DiMichele, 1994). The transition to endospory increased the influence of the diploid parent on gametogenesis, and allowed spore production and release to be synchronized with favorable environmental conditions.

The reduction of the gametophyte and transition to endospory is exemplified by comparison of the male gametophytes of the dioicous liverwort *M. polymorpha* and the heterosporous water fern *Marsilea vestita*. In *M. polymorpha*, antheridial initial cells arise from epidermal cells of the receptacle on a reproductive branch called the antheridiophore (Durand, 1908). Initial cells divide several times prior to a formative asymmetric division, which gives rise to a spermatogenous mother cell and a sterile jacket cell. Each mother cell undergoes multiple rounds of division to produce several thousand spermatozoids within each antheridium (Fig. 2B). In contrast, microspores of *M. vestita* complete one asymmetric and one symmetric division before spermatogenous cells are produced by a formative asymmetric division. The spermatogenous cells undergo four rounds of symmetric division to produce a total of 32 spermatozoids within each microspore (Fig. 2C; Sharp, 1914).

This trend toward fewer antheridial divisions prior to spermatogenesis culminates in the highly reduced male (micro)gametophytes of seed plants (Fig. 2D–F). In gymnosperms, microspores undergo asymmetric cell divisions to produce one or more prothallial cells, prior to segregation of a spermatogenous cell lineage (Fig. 2D and E). Reduction of the microgametophyte is most extreme in angiosperms, which require only two cell divisions to produce two non-motile sperm cells from the microspore (Fig. 2F). This indicates that the asymmetric cell divisions which produce one or more prothallial cells in gymnosperms are absent, and the germline is directly segregated by asymmetric division of the microspore (see Rudall & Bateman, 2007).

1.4 Adaptations of seed plant microgametophytes

The development of male and female gametophytes of seed plants (spermatophytes) takes place in micro- and megasporangia formed on modified leaves (sporophylls) in cones or in flowers. In contrast to most free-sporing plants, immature microgametophytes of seed plants are dispersed as desiccation-tolerant pollen. The evolution of pollen is a milestone in the adaptation of seed plants to terrestrial habitats. By contrast, the endosporic female megagametophyte remains partially or completely enclosed by sporophytic tissues, the nucellus and integument (see chapters "Development and evolution of the unique ovules of flowering plants" by Gasser and Skinner, this issue and "Development and function of the flowering plant female gametophyte" by Erbasol Serbes et al., this issue). Thus, sperm cells must either swim to the megagametophyte enclosed by the integument or be delivered via the pollen tube.

With some important exceptions, extant gymnosperms and angiosperms transitioned from zooidogamy (swimming sperm) to siphonogamy (conduction of non-motile sperm via a pollen tube) (Friedman, 1993; Gifford & Foster, 1989). In cycads and in *Ginkgo biloba*, pollen tubes have a primarily haustorial role to support the development of large flagellated spermatozoids (Fig. 3C; Friedman & Gifford, 1997). Flagellated spermatozoids of these taxa develop slowly after pollination and are released proximally into a pollen chamber. Spermatozoids subsequently migrate to the egg cells at the micropylar end of the ovule (Rudall & Bateman, 2007).

In gymnosperms, pollen reception must be relatively near to the egg as pollen tubes are relatively slow growing. Furthermore, the time between pollination and fertilization can take days to months. In angiosperms, pollen tube growth is significantly faster, shortening the progamic phase to hours or minutes, thereby facilitating the evolution of rapid reproductive cycles (Williams, 2008). The evolution of fast-growing angiosperm pollen tubes is linked with modifications to pollen tube walls, including callose-reinforced secondary tube walls, callose plugs, and extensible pectin tips. Rapid pollen tube growth and longer-distance growth are considered to be key features enabling carpel diversification and the multiplicity of flower and fruit form (Williams, 2008).

Double fertilization *sensu stricto* is a synapomorphy (derived trait shared by two or more taxa) of angiosperms and occurs following the delivery of a pair of sperm cells to the ovule, resulting in the formation of embryo and endosperm. In the Gnetales, however, a modified form of this process produces a supernumerary embryo (Carmichael & Friedman, 1995; Friedman, 1998, 1990). Double fertilization enhances the fitness of the embryo, but also contributes to maternal fitness, as resources are only committed to the ovule after the egg and central cell have been fertilized. The two sperm cells can be functionally equivalent as demonstrated for *Arabidopsis thaliana* (Hamamura et al., 2011; Ingouff et al., 2009; Kong, Lau, & Jurgens, 2015), and the majority of species show an equal distribution of DNA-containing organelles among the two sperm (Saito et al., 2002). However, dimorphic sperm, which show preferential fertilization and differing gene expression profiles, are described for *Plumbago zeylanica* (Gou, Yuan, Wei, & Russell, 2009; Russell, 1985).

An important life history trait for adaptation to particular ecological niches is the length of the life cycle. In time-limited habitats, short lifecycles are commonly associated with an annual habit, shorter bud development time, and reduced flower longevity (i.e., reduced time between flower maturation and fertilization) among other reproductive traits (Snell & Aarssen, 2005). Although gymnosperms disperse sexually immature pollen (Fernando, Quinn, Brenner, & Owens, 2010; Friedman, 1999), angiosperms have evolved mechanisms to minimize the length of the reproductive phase. This involves the advancement of pollen maturity before dispersal. In about 70% of angiosperms, pollen is dispersed in a bicellular state and division of the generative cell takes place during pollen tube growth. The remainder undergo precocious division of the generative cell and disperse tricellular pollen containing two sperm cells (Brewbaker, 1967; Williams, Taylor, & O'Meara, 2014). Tricellular pollen is considered advanced and is associated with shorter reproductive cycles, reduced longevity, rapid germination, and faster pollen tube growth (Brewbaker, 1967;

Rowe, Paul-Victor, & Paul-Victor, 2012; Williams, 2012). Although less common, tricellularity does not appear to be irreversible and multiple instances of secondary shifts from tricellular to bicellular pollen are postulated (Williams et al., 2014). In rare cases, as in *Annona cherimola*, a mixture of bicellular and tricellular pollen is dispersed and the onset of generative cell division is advanced by increases in air temperature (Lora, Herrero, & Hormaza, 2009). This may be adaptive, potentially providing plasticity in response to dispersal conditions.

In summary, the advanced microgametophytes of seed plants evolved from their embryophyte ancestors by the acquisition of several key features. These are, endospory and an extreme reduction of mitotic cell divisions prior to segregation of the male germline *sensu stricto*, increased sporophytic dependency, and adaptation to a siphonogamous gamete delivery system involving the loss of gamete motility. Subsequent diversification in the onset and pace of gametogenesis has enabled the microgametophytes of angiosperms to adapt to a diversity of niches.

2. Male germline development in angiosperms

Molecular and genetic studies of male germline development documented in earlier reviews are dominated by those of angiosperms (Berger & Twell, 2011; Russell & Jones, 2015; Twell, 2011). Other recent reviews focus on omics (Fíla, Záveská Drábková, Gibalová, & Honys, 2017; Rutley & Twell, 2015), chromatin remodeling (Borg & Berger, 2015), post-transcriptional processes (Guan, Guo, Li, & Yang, 2013; Hafidh, Fila, & Honys, 2016; Michard, Simon, Tavares, Wudick, & Feijó, 2017), or wider aspects of reproductive cell lineages (Schmidt, Schmid, & Grossniklaus, 2015). The focus is on regulatory aspects of male gametogenesis including comparative studies and advances in pre-angiosperm models. The roles of asymmetric division and cell cycle control in patterning are discussed, also the roles and regulation of transcription factors in cell specification. Chromatin remodeling and communication between germline and non-germ cells are also highlighted.

2.1 Asymmetric microspore division and cell fate determination

Studies in angiosperms have shown that asymmetric microspore division is formative and essential for the establishment of germ cell fate. The induction of symmetric microspore divisions by environmental or chemical agents, such as the microtubule poison colchicine, results in daughter cells with vegetative cell-like nuclei (Tanaka & Ito, 1980; Terasaka & Niitsu, 1987) and vegetative cell marker expression (Eady, Lindsey, & Twell, 1995). Mutations which disturb microspore polarity or division have also been informative. In *gemini pollen1* (*gem1*) mutants of *A. thaliana*, microspores can undergo equal divisions and both daughter cells adopt vegetative cell fate, similar to the products of chemically-induced symmetric divisions (Park, Howden, & Twell, 1998). The pollen phenotypes of *gem1* result from loss of microspore polarity and cytokinesis defects, arising from mutations affecting the MICROTUBULE-ASSOCIATED PROTEIN215 (MAP215)-family protein MICROTUBULE ORGANIZATION1 (MOR1)/GEM1 (Fig. 4; Park et al., 1998; Twell et al., 2002; Whittington et al., 2001). Vegetative cell fate is also adopted in single-celled pollen resulting from colchicine treatment of tobacco microspores (Eady et al., 1995), or from cytokinesis defects (Park et al., 1998; Twell, Park, & Lalanne, 1998). Recent studies of additional single-celled pollen mutants confirm vegetative cell fate as the default program, and show that pollen tubes are guided and discharge their contents in the absence of gametes (Glöckle et al., 2018; Zhang et al., 2017).

A novel class of cytokinesis mutants, which fail to compartmentalize newly-formed vegetative and generative cell nuclei, provides insight into the mechanisms underlying germ cell fate determination. Microspores polarize in *two-in-one* (*tio*) (Oh et al., 2005) and *Kinesin-12A/B* double mutants (Lee, Li, & Liu, 2007), but failure of cytokinesis results in binucleate pollen unable to differentiate sperm. The TIO *Fused* kinase and KINESIN-12 interact to form a signaling module (Fig. 4), which is thought to act via the canonical NACK-PQR MITOGEN-ACTIVATED PROTEIN KINASE (MAPK) pathway (Soyano, Nishihama, Morikiyo, Ishikawa, & Machida, 2003) to regulate phragmoplast expansion (Oh, Allen et al., 2012; Oh, Bourdon, Das'Pal, Dickinson, & Twell, 2008). Collectively, mutants of this class form a presumptive generative cell nucleus, but this fails to divide and nuclear chromatin is gradually dispersed, consistent with the default vegetative cell fate. Thus, germ cell fate cannot be established without confinement of daughter nuclei to unequal compartments.

The correct developmental timing of mitosis and a properly oriented spindle are essential requirements for asymmetric microspore division and cell fate determination. Mutations in *SIDECAR POLLEN (SCP)* disturb the orientation of microspore division (Fig. 4), resulting in tricellular pollen containing an extra vegetative cell (Chen & McCormick, 1996). Defects

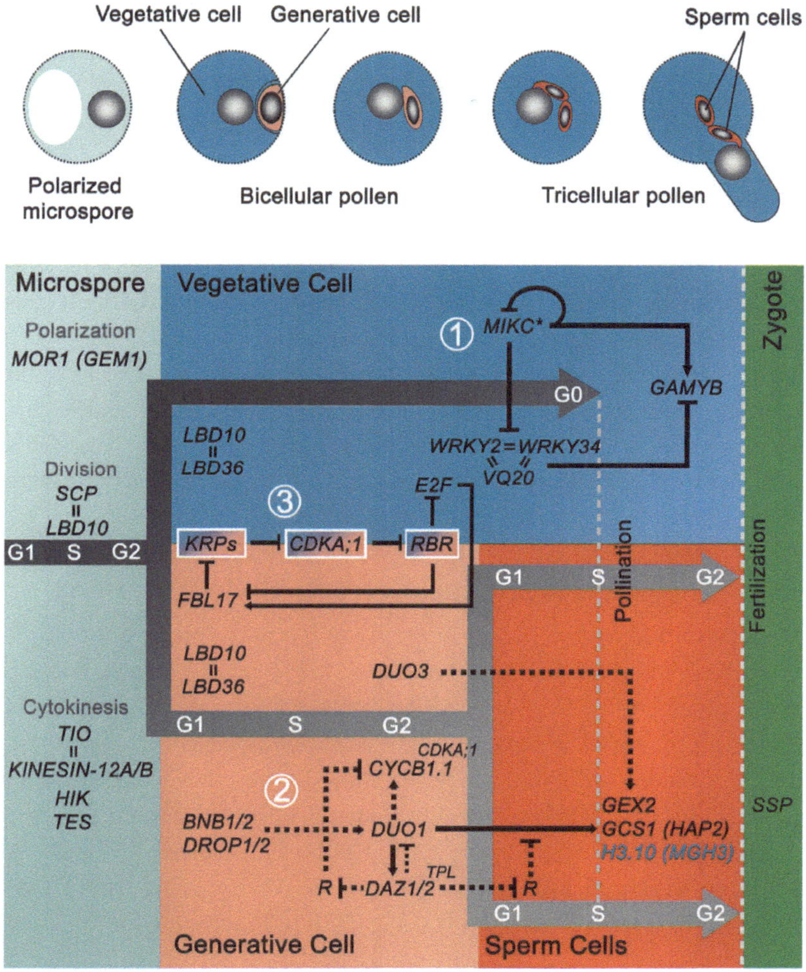

Fig. 4 Overview of the morphological stages and genetic pathways regulating male gametophyte development in *Arabidopsis thaliana*. Polarized microspores (light blue) undergo a highly asymmetric division to produce a bicellular pollen grain with a small generative cell (pink), engulfed within the vegetative cell (dark blue). The generative cell undergoes mitotic division to produce two differentiated sperm cells (red). Genes and regulatory pathways with important functions are shown with direct interactions indicated by solid arrows. The major regulatory pathways indicated are (1) the *WRKY2/34-MIKC** network in the vegetative cell, (2) the *DUO1-DAZ1* differentiation network in germline cells, and (3) the *FBL17-KRPs-CDKA;1-RBR1-E2F* G1/S-phase regulatory module. Physical interactions between proteins are denoted by double dashes. Indirect regulatory pathways are indicated with dotted lines. Cell cycle phases (G1, S, G2) are labeled. Pollination and fertilization events are indicated with gray dotted lines. (1) AtMIKC* MADS-domain protein (AGL65, AGL66 and AGL104) complexes repress early genes and activate late pollen genes in the vegetative cell. Pollen-specific GAMYB genes *(Continued)*

arise following delayed microspore division, indicating close coupling of cell cycle progression and spindle orientation (Oh, Park, Twell, & Park, 2010). SCP is a microspore-specific LATERAL ORGAN BOUNDARIES-DOMAIN/ASYMMETRIC LEAVES2-LIKE (LBD/ASL) domain family protein, whose expression overlaps with that of other family members in developing pollen. SCP (LBD27) and LBD10 form nuclear-localized heterodimers and act synergistically to control pollen development (Kim, Kim, Lee, Park, & Kim, 2015). Thus, combinatorial interactions among LBD proteins are thought to be important for their roles in the male gametophyte (Kim, Kim, Pandey, & Kim, 2016).

In summary, the asymmetric division of angiosperm microspores is critical for the formation and specification of the male germline *sensu stricto*. Studies of a variety of mutants strengthen the hypothesis that germ cell fate is specified by the segregation and confinement of yet unknown determinants by a new cell wall (Twell, 2011; Twell et al., 1998).

2.2 Cell cycle control and male gametophyte patterning

Correct patterning of the male gametophyte requires differential control of the cell cycle in vegetative and generative cells. Failure of generative cell division, resulting in bicellular pollen with a single germ cell, has been described for several *A. thaliana* mutants. Among the genes affected are those encoding the core cell cycle machinery, like the A-type cyclin-dependent kinase (CDK), CDKA;1. In *cdka;1* mutants, S-phase is retarded and generative cells are only able to divide during pollen tube growth (Fig. 4; Aw, Hamamura, Chen, Schnittger, & Berger, 2010; Iwakawa, Shinmyo, & Sekine, 2006; Nowack et al., 2005). CDKA;1 activity and cell cycle

Fig. 4—Cont'd (*MYB97, MYB101* and *MYB120*) are negatively regulated by WRKY2 and WRK34 via interaction with the pollen-specific VQ motif-protein VQ20 (Lei et al., 2017; Lei, Ma, & Yu, 2018). WRKY2/34-VQ20 complex activity is regulated by MPK3/MPK6-mediated phosphorylation (Guan et al., 2014). (2) DUO1 controls gamete specification genes, including *DAZ1* and *DAZ2* (*DAZ1/2*), which counteract repressive mechanisms (R) via EAR-dependent interaction with the corepressor TPL. Through this mechanism, *DAZ1/2* promote the G2/M-phase accumulation of mitotic cyclins, generative cell division, and the *DUO1*-dependent activation of gamete differentiation genes and DUO1 turnover (Borg et al., 2014). (3) *CDKA* activity and generative cell cycle progression are maintained by the proteasome-dependent degradation of CDK inhibitors (KRPs) by the SCF^{FBL17} (Gusti et al., 2009; Kim et al., 2008). The transient expression of *FBL17* in the generative cell promotes G1/S transition, while KRPs persist in the vegetative cell (Kim et al., 2008). The *FBL17-KRPs-CDKA;1* cascade is linked to the *RBR1-E2F* pathway to form a general G1/S regulatory module that also functions in microspores (Zhao et al., 2012).

progression are maintained in the generative cell by the proteasome-dependent degradation of the KIP-RELATED PROTEIN (KRP) CDK inhibitors. KRPs are targeted for degradation by the SCF^{FBL17} (SKP-CULLIN-F-BOX) E3 ubiquitin protein ligase complex, which contains the F-box protein FBL17 (Fig. 4; Kim et al., 2008). In *fbl17* mutants, generative cells fail to divide as CDK inhibitors are not degraded (Gusti et al., 2009; Kim et al., 2008). The transient expression of *FBL17* only in the generative cell soon after microspore division is thought to license proliferation of the germline, while KRPs are maintained in the vegetative cell (Fig. 4; Kim et al., 2008). Importantly, the *FBL17-KRPs-CDKA;1* cascade is linked to the *RBR1-E2F* pathway to form a general G1/S regulatory module that also functions in microspores of *A. thaliana*, but is likely to control entry into S-phase in all cells (Zhao et al., 2012).

Patterning of angiosperm microgametophytes is also determined by the retinoblastoma (*Rb*)-*E2F* pathway, which regulates cell cycle progression in a variety of organisms (Fig. 4; Gutzat, Borghi, & Gruissem, 2012; Harbour & Dean, 2000). *Rb* function is dependent on its phosphorylation by CDK in late G1-phase. This alleviates Rb-mediated repression of the E2F/DP family transcription factors which commit cells to S-phase (Weinberg, 1995). The *A. thaliana Rb* homolog *RETINOBLASTOMA-RELATED1* (*RBR1*) is highly expressed in the vegetative cell along with *KRP* genes encoding CDK inhibitors (Kim et al., 2008). Loss of *RBR1* results in hyperproliferation of the vegetative cell and to a lesser extent the generative cell (Chen, Hafidh, Poh, Twell, & Berger, 2009; Johnston, Matveeva, Kirioukhova, Grossniklaus, & Gruissem, 2008). Hyperproliferation in *rbr1* mutants is caused by *CDKA;1* activity, since introduction of a *cdka;1* mutant allele suppresses this phenotype in *rbr1* pollen (Chen et al., 2009). Interestingly, cell fate is incorrectly specified in a fraction of *rbr1* vegetative cells, but this phenotype is completely reversed in the absence of *CDKA;1*. These findings point to *RBR1* primarily controlling cell cycle progression, while having a secondary impact on cell fate commitment. This places RBR1-mediated repression of E2F/DP downstream of KRP-dependent inhibition of CDKA;1, with both mechanisms enforcing cell-cycle exit, thereby restricting cell fate (Fig. 4). In addition, E2F and RBR1 can both bind to the *FBL17* promoter and are thought act antagonistically to stimulate or repress *FBL17* expression, respectively (Zhao et al., 2012). Thus, increased RBR1 protein in the vegetative cell would result in less FBL17 and reduced degradation of KRPs, thereby reinforcing cell vegetative cell cycle exit.

2.3 Transcriptional control of male gametophyte patterning

2.3.1 Regulators that integrate the control of cell division and germline fate

The male germline-specific R2R3 Myb-domain protein DUO POLLEN1 (DUO1) was the first transcription factor to be identified with a role in generative cell division (Rotman et al., 2005). Mutant generative cells of *duo1* pollen complete S-phase, but fail to enter mitosis due to reduced accumulation of the mitotic cyclin, CYCB1;1 (Brownfield, Hafidh, Borg, et al., 2009; Durbarry, Vizir, & Twell, 2005). *DUO1* also regulates a network of genes required for sperm cell differentiation (Borg et al., 2011; Brownfield, Hafidh, Borg, et al., 2009). Direct targets of DUO1 include genes encoding the sperm-egg adhesion protein, GAMETE EXPRESSED2 (GEX2) (Engel, Holmes-Davis, & McCormick, 2005; Mori, Igawa, Tamiya, Miyagishima, & Berger, 2014), and the sperm cell-surface fusogen, GENERATIVE CELL-SPECIFIC1/HAPLESS2 (GCS1/HAP2) (Takahashi, Honda, Mori, & Igawa, 2017; von Besser, Frank, Johnson, & Preuss, 2006). Thus, *DUO1* has an integrative role, controlling G2 to M-phase transition and shaping the germline transcriptome to promote spermatogenesis (Borg et al., 2011; Brownfield, Hafidh, Borg, et al., 2009). DUO1 forms a regulatory module with two of its direct targets, the C_2H_2 zinc finger repressor protein encoding genes *DUO1-ACTIVATED ZINC FINGER PROTEIN1* (*DAZ1*) and *DAZ2*. Current data support a *DUO1-DAZ1/2* network model in which the *DAZ1/2* node controls G2- to M-phase transition and *DUO1*-dependent gamete differentiation through a negative regulatory circuit involving physical interaction of DAZ1 with the corepressor TOPLESS (TPL) (Fig. 4; Borg et al., 2014).

The expression of a subset of germline genes under *DUO1* control, including *GCS1/HAP2* and *GEX2*, also requires the homeodomain-like protein DUO3 (Brownfield, Hafidh, Durbarry, et al., 2009). DUO1 and DUO3 have overlapping roles in gametogenesis, but unlike in *duo1* mutants, generative cell division cannot be rescued by *CYCB1;1* expression in *duo3* mutants. Moreover, *DUO3* is expressed in sporophytic tissues and has an essential role in embryogenesis. DUO3 is a functionally conserved protein in land plants, as the *A. thaliana duo3* mutant is rescued by the moss *DUO3* ortholog (Brownfield, Hafidh, Durbarry, et al., 2009). The ancestry of DUO3 also extends to protein domains conserved with cell lineage regulators including GONADLESS-4 (GON-4) of *Caenorhabditis elegans* and vertebrate Gon4-like (Gon4l), which are required for gonadogenesis (Liu et al.,

2007; Williams et al., 2018) and hematopoiesis (Friedman, Santa Anna-Arriola, Hodgkin, & Kimble, 2000), respectively. DUO3, in common with Gon4l, is proposed to act as a scaffold, enabling the assembly of distinct transcriptional complexes in restricted cell types (Brownfield, Hafidh, Durbarry, et al., 2009; Lu et al., 2011; Williams et al., 2018).

2.3.2 Regulation of DUO1 expression

The *DUO1-DAZ1* regulatory module is an essential switch in angiosperm spermatogenesis. Strict spatial control of *DUO1* is vital as ectopic expression results in vegetative growth defects or pollen cell death (Brownfield, Hafidh, Borg, et al., 2009; Palatnik et al., 2007). Initial efforts to understand this tight control, identified a 198 bp *DUO1* promoter fragment sufficient for generative and sperm cell-specific transcription (Brownfield, Hafidh, Borg, et al., 2009). Phylogenomic studies have delimited important sequences in this region to an ordered *cis*-regulatory module (CRM), termed *regulatory region of DUO1* (*ROD1*). *ROD1* has a distinct architecture in eudicots; three DNGTGGV motifs are required to initiate germline expression and two YAACYGY motifs, bound by DUO1, enhance expression in a positive feedback loop (Peters et al., 2017). *ROD1* is functionally conserved in *Medicago truncatula*, but absent from pre-angiosperm *DUO1* promoters, indicating a shift in the transcriptional control of *DUO1* in pollen after the divergence of gymnosperms and angiosperms 300–350 Mya (Peters et al., 2017).

The control of *DUO1* expression is fine-tuned by several mechanisms. A negative feedback loop is thought to contribute to the developmental phasing of *DUO1* and *DAZ1/DAZ2* expression, involving the limitation of *DUO1* accumulation in sperm by *DAZ1/DAZ2* (Fig. 4; Borg et al., 2014). A possible mechanism could involve *miR159*-mediated regulation, since the *DUO1* transcript contains a functional *miR159* binding site (Palatnik et al., 2007), and increased expression of *DUO1* has been associated with reduced levels of the Anaphase Promoting Complex APC/C and *miR159* (Zheng, Chen, & McCormick, 2011). There is also evidence for positive regulation of *DUO1* by the pollen-specific AT-RICH INTERACTING DOMAIN1 (ARID1) protein, which binds to a promoter region upstream of *ROD1* (Zheng, He, Zheng, Wu, & McCormick, 2014).

In recent work, mutations in two distinct families of basic helix-loop-helix (bHLH) transcription factors were reported to cause profound defects in *A. thaliana* male gametogenesis. BONOBO1 (BNB1) and BNB2 are pollen-specific members of the bHLH subfamily VIIIa, which are redundantly required for generative cell specification (Yamaoka et al., 2018). DEFECTIVE REGION OF POLLEN1 (DROP1) and DROP2 are members of bHLH subfamily XI (Heim et al., 2003) that were previously shown to function in root hair elongation (Lin et al., 2015). In pollen of *bnb1 bnb2* double mutants, generative cells are not engulfed by vegetative cells (Yamaoka et al., 2018), while generative cell nuclei disintegrate in *drop1 drop2* pollen (Zhang et al., 2017). Generative cells fail to differentiate in both double mutants and sperm cell marker genes like *DUO1* and *GCS1/HAP2* (in *drop1 drop2*) or *HISTONE THREE RELATED10* (*HTR10*) (in *bnb1 bnb2)* are not expressed (Yamaoka et al., 2018; Zhang et al., 2017). In conclusion, *BNB1/2* and *DROP1/2* are required for early generative cell development and for germline differentiation, highlighting their potential as upstream regulators of the *DUO1-DAZ1* module (Fig. 4).

2.3.3 Regulators of pollen maturation that act in the vegetative cell

The MIKC* MADS-domain proteins are an ancient class of transcriptional regulators which play an essential role in pollen maturation and germination in angiosperms (Adamczyk & Fernandez, 2009; Liu et al., 2013; Verelst, Saedler, & Münster, 2007; Verelst, Twell, et al., 2007). Pollen-specific members of the MIKC* gene family in *A. thaliana* and *Oryza sativa* (rice) are restricted to the vegetative cell and form heterodimeric complexes (Liu et al., 2013; Verelst, Saedler, et al., 2007). In *A. thaliana*, AtMIKC* complexes regulate pollen development by repressing early pollen genes and activating mature pollen genes. For example, the male gametophyte-specific transcription factor gene *WRKY34* is upregulated, while *MYB97* is downregulated in the absence of MIKC* complexes (Fig. 4; Verelst, Twell, et al., 2007). Interestingly, MYB97 is a member of a sub-clade of S18 (GAMYB) R2R3 MYB proteins (MYB97, MYB101, MYB120), which act redundantly to promote the discharge of sperm from pollen tubes in the embryo sac (Leydon et al., 2013; Liang et al., 2013). These pollen-specific GAMYB genes are negatively regulated by a ternary complex of WRKY2, WRKY34, and the pollen-specific VQ motif-containing protein VQ20 (Lei et al., 2017, 2018). Moreover, the temporal control of WRKY2/34-VQ20 ternary complex activity by MITOGEN-ACTIVATED PROTEIN KINASE3 (MPK3)/MPK6-mediated phosphorylation, is a compelling example of the post-translational regulation of transcription in developing pollen (Guan et al., 2014).

2.4 Chromatin remodeling in male gametophyte development

Chromatin organization differs dramatically between the two different cell types of angiosperm pollen. The vegetative cell has diffuse nuclear chromatin, whereas sperm cell nuclei possess highly compact chromatin, and these differences can be traced to asymmetric microspore division. After formation of the vegetative cell, centromeric heterochromatin is gradually decondensed, coincident with the loss of the histone H3-like variant CENH3 from centromeres and repressive histone H3 lysine 9 di-methylation (H3K9me2) marks (Schoft et al., 2009). The loss of CENH3 is consistent with terminal differentiation of the vegetative cell, while germ cells maintain centromeric CENH3 and their ability to divide (Ingouff, Jullien, & Berger, 2006; Schoft et al., 2009). CENH3 removal from centromeres was shown to require the CDC48A AAA-ATPase molecular chaperone and a pair of interacting small ubiquitin-like modifier (SUMO) adaptors, NPL4A and NPL4B (Mérai et al., 2014). The $CDC48A^{NPL4}$ complex is proposed to bind and promote the active unloading of sumoylated CENH3. $CDC48A^{NPL4}$-mediated decondensed centromeric heterochromatin was also proposed to "euchromatinize" rDNA loci to facilitate rRNA gene transcription to support the high metabolic needs of pollen tubes (Mérai et al., 2014). In this regard, it is interesting that CENH3 is not excluded from centromeres in vegetative cell nuclei of rye and barley where repressive H3K9me2 marks persist (Houben, Kumke, Nagaki, & Hause, 2011; Pandey, Houben, Kumlehn, Melzer, & Rutten, 2013). In a survey of the immunodetected chromatin marks reported for pollen of different species, there was general agreement on their prevalence, but there were notable differences between eudicots and monocots (Borg & Berger, 2015). Although only monocots were studied, repressive H3K27me3 marks were enriched in subtelomeric chromatin and specific for the vegetative cell (Houben et al., 2011; Pandey et al., 2013; Sano & Tanaka, 2010).

The chromatin of male germ cells shows dynamic changes in histone composition which varies significantly among angiosperms (reviewed in Borg & Berger, 2015). Generative cells of lily incorporate distinct histone H2A, H2B, and H3 variants compared with those of somatic tissues (Ueda et al., 2000; Ueda & Tanaka, 1994; Xu, Swoboda, Bhalla, & Singh, 1999). While chromatin of *A. thaliana* microspores is mainly composed of H3.1 and H3.3 histone variants, sperm cell chromatin is strongly biased toward H3.3 and H3.10 variants (Borg & Berger, 2015; Okada, Endo, Singh, & Bhalla, 2005). Consequently, the dynamics of histone variants are thought to be linked to chromatin reprogramming and the compaction of the sperm cell genome (Borg & Berger, 2015). These roles may arise though functional differences exhibited by histone variants. In particular, the abundant germline-specific histone H3 variant H3.10 is proposed to reshape the chromatin landscape by conferring immunity to the repressive chromatin mark H3K27me3 (Borg & Berger, 2015). The absence of immunodetected H3K27me3 marks observed for germline cells of several monocots is in accord with this notion. Unexpectedly, an insertion mutant in the *HTR10* gene, encoding H3.10, did not show an obvious phenotype, suggesting functional compensation among germline histones (Okada et al., 2005). There is also evidence that rice sperm cells express an unique and diverse combination of histone variants, highlighting the differing chromatin landscapes of sperm among angiosperms (Anderson et al., 2013; Russell et al., 2012). In addition to potential roles in chromatin compaction, germline-specific histones might function to remove somatic chromatin marks to facilitate the specification of germline fate and/or to allow resetting of the chromatin state prior to fertilization (see Borg & Berger, 2015).

2.5 Communication between non-germ and germ cells

Angiosperm pollen is a functional unit in which the vegetative cell nourishes, protects, and transports the germ cells to the embryo sac via the pollen tube. An intrinsic feature of this "cell-within-a-cell" architecture is the reliance of the germline on the vegetative cell for the transfer of nutrients and biomolecules. This unilateral dependence is illustrated by genetic ablation studies, in which the expression of diphtheria toxin A-chain (DTA) was restricted to each cell-type. When vegetative cell development was arrested with DTA, generative cells were not engulfed and most rapidly lost viability (Twell, 1995), but when DTA was used to block germline development, pollen matured normally and discharged single gametes into the embryo sac (Frank & Johnson, 2009). Similarly, pollen tubes of single-celled pollen mutants are guided to the ovule and are competent to discharge their contents in the absence of gametes (Glöckle et al., 2018; Zhang et al., 2017).

The vegetative nucleus and generative cell or sperm cell pair are also physically linked in an assemblage known as the male germ unit (Dumas, Knox, & Gaude, 1985; Jensen & Fisher, 1970; Russell & Cass, 1981). This structure is established in bicellular pollen and is maintained until discharge of the pollen tube contents into the embryo sac (reviewed in McCue, Cresti, Feijo, & Slotkin, 2011; Mogensen, 1992). The potential for molecular communication between pollen cell types, via the physical extension linking one of the sperm cell pair to the vegetative nucleus, was first postulated by Mogensen (1992). Rapid freeze fixation, which can reduce artifacts caused by conventional methods, did not identify plasmodesmata-like channels between the two cells (Cresti, Lancelle, & Hepler, 1987). However, the topic remains controversial (see McCue et al., 2011) and it is evident that mechanisms exist to transport both mRNAs and short interfering RNAs (siRNAs) from the vegetative cell to the developing germline.

While studying the expression of the sperm cell-enriched ABA-related protein phosphatase gene *ABA-HYPERSENSITIVE GERMINATION3* (*AHG3*), Jiang et al. (2015) unexpectedly found that the *AHG3* promoter was transcriptionally active in the vegetative cell, whereas a translational fusion protein was localized to sperm (Jiang et al., 2015). The authors provided evidence that *AHG3* transcripts move from the vegetative cell to sperm cells, and that transport requires sequences in both the 5′ UTR and coding region. The mechanisms involved and the extent of communication via this route are not yet clear, but these results document an additional role for the vegetative cell in shaping the sperm cell transcriptome (Jiang et al., 2015).

Different classes of small RNAs are present in vegetative cells and sperm cells (reviewed in He, Yang, Wu, & Zheng, 2015). These span most conserved microRNA (miRNA) families and include biogenesis genes such as *DICER-LIKE1* (*DCL1*) (Borges, Pereira, Slotkin, Martienssen, & Becker, 2011; Creasey et al., 2014; Grant-Downton, Hafidh, Twell, & Dickinson, 2009; Slotkin et al., 2009). Based on the de-repression of transposable elements (TEs) in the vegetative cell, and the appearance of corresponding 21 nt siRNAs in sperm cells, it was proposed that siRNAs could move between the vegetative cell and sperm cells of *A. thaliana* to maintain TE silencing in sperm (Slotkin et al., 2009). Evidence in support of this proposal was that artificial miRNAs expressed under control of the vegetative cell-expressed tomato *LAT52* promoter could repress gene expression in the sperm cells (Slotkin et al., 2009). However, the *LAT52* promoter is first active in the microspores of *A. thaliana* (Eady et al., 1995), raising the possibility of inheritance of miRNAs by the germline. Truly vegetative cell-specific promoters also failed to silence targets in the germline, indicating that miRNA cannot move between cells (Grant-Downton et al., 2009). However, recent experiments utilizing robustly cell-specific promoters demonstrate that siRNAs produced from vegetative cell transcripts can silence TE reporters in sperm cells (Martínez, Panda, Köhler, & Slotkin, 2016).

Thus, siRNAs, but not miRNAs, are trafficked into the germline, although the mechanisms involved in transfer, or which discriminate between these two classes of small RNAs, remain to be discovered.

3. Molecular evolution of the male germline

3.1 Motile plant and animal sperm share common features

The morphology of sperm cells has changed markedly during the evolution of land plants. Excellent reviews by Renzaglia and Garbary (2001) and Southworth and Cresti (1997) discuss the diversity, development, and evolution of land plant sperm. Only selected aspects and recent molecular and genetic data are highlighted here.

Biflagellate spermatozoids with condensed chromatin, packaged into an elongated and coiled nucleus, are produced by advanced charophyte algae and bryophytes (Fig. 3A). These bi-flagellated spermatozoids are not only superficially reminiscent of mammalian spermatozoa, but recent analysis of antheridia-enriched transcripts in *M. polymorpha,* revealed homologs of a number of motility-related genes that are absent from angiosperms (Higo et al., 2016). These include the mammalian Parkin co-regulated gene (Mp*PACRG*), responsible for axoneme formation (Ikeda, Ikeda, Morikawa, & Kamiya, 2007; Lorenzetti, Bishop, & Justice, 2004; Yanagimachi et al., 2004), light chains of the inner arm dynein of the axoneme (Mp*CL7*), adenylate cyclase, required for sperm motility (Schmid et al., 2007), and human CHD5, which is required for spermatid chromatin remodeling and condensation (Li et al., 2014; Zhuang et al., 2014).

One of the most striking similarities between spermatozoa of most animals and the spermatozoids of charophytes and basal land plants is their nuclear chromatin compaction. This character is thought to improve the hydrodynamic properties of sperm (Erenpreiss, Spano, Erenpreisa, Bungum, & Giwercman, 2006), and to protect the genome against irradiation damage, nuclease degradation and mechanical forces (Rathke et al., 2010; Wykes & Krawetz, 2003). In animal sperm, histone-based nucleosomes are replaced by highly basic proteins called protamines (Johnson et al., 2011; Meistrich, Mohapatra, Shirley, & Zhao, 2003). Protamine-like proteins can be found in spermatozoids of algae (*Chara corallina*), bryophytes (*M. polymorpha*) and ferns (*M. vestita*), but not in seed plants (reviewed in Borg & Berger, 2015). In *M. polymorpha* sperm, histones are replaced by protamines and the exchange mechanism involves transitory proteins of intermediate size and basicity, similar to the situation in animals (Reynolds & Wolfe, 1984, 1978). Variation in the coexistence of intermediate histone/ protamine proteins and somatic histones indicates a remarkable adaptation of nucleoprotein composition exchange mechanism among early divergent species (Borg & Berger, 2015).

The spermatozoids of charophyte algae and basal land plants share the common feature of cytoplasmic reduction via macroautophagy, with animal spermatozoa. Macroautophagy is controlled by autophagy related proteins (ATGs) in mice as well as in the moss *P. patens*, suggesting recurrent adoption of an ancestral mechanism. ATG5 and ATG7 are essential for sperm fertility in *P. patens*, similar to the requirement for ATG7 in mice. Impaired autophagy is thought to impact microtubule and flagella development, as well as nuclear packaging, by reduced clearance of cytoplasmic repressors (Sanchez-Vera et al., 2017). In angiosperms, ATG genes evidently do not influence pollen morphology, but indirectly support pollen maturation and germination. Knock-out of rice *ATG7* resulted in sporophytic pollen defects, arising from reduced autophagic degradation in the anther tapetum (Kurusu et al., 2014). Furthermore, pollen germination defects observed in *A. thaliana ATG6* mutants are gametophytically determined, but attributed to functions other than autophagy (Fujiki, Yoshimoto, & Ohsumi, 2007). Thus, the role of autophagy in male reproduction has diversified with the evolution of non-motile sperm in seed plants. Cytoplasmic reduction is likely to be less important for the larger and non-motile sperm of seed plants than for spermatozoids of basal land plants. However, there is evidence that sperm cells of some angiosperms reduce their cellular volume by shedding cytoplasmic vesicles into the vegetative cell (Yu & Russell, 1992). This may assist in sperm cell movement within the pollen tube and could limit the paternal cytoplasmic contribution to the zygote.

A signal perception apparatus used for chemotactic navigation by land plants may operate in mammalian sperm. The GLUTAMATE-RECEPTOR-LIKE (GLR) ion-channels are essential for moss spermatozoids to navigate to archegonia (Ortiz-Ramírez et al., 2017). In flowering plants, where immotile sperm no longer require chemoattraction, GLRs have been redeployed to regulate pollen tube growth and navigation in response to chemotactic factors released by the pistil (Michard et al., 2011). Although GLRs are expressed in mammalian sperm, their functions are currently unknown (Hu et al., 2004).

In monilophytes, spermatozoids can develop over 100 flagella (Fig. 3B; Myles & Hepler, 1977), while those of the non-siphonogamous gymnosperms possess much larger multiflagellate sperm, with around 1000 flagella for *G. biloba* (Fig. 3C; Gifford & Lin, 1975). The cycads have the largest sperm with a diameter of 200–500 μm and an estimated 40,000 flagella for the genus *Zamia* (Norstog, 1975). The increase in flagella number is associated with increased sperm cell volume, the absence of coiled morphology and reduced chromatin compaction. With the loss of sperm cell motility in siphonogamous gymnosperms (e.g., in *Gnetum gnemon* and the genus *Ephedra*) and angiosperms, only the angiosperms possess highly condensed chromatin (see Southworth & Cresti, 1997), indicating that chromatin compaction may be atavistic in angiosperms. The sperm cells of flowering plants also possess a distinctive cytoskeleton when compared to somatic cells (reviewed in Palevitz & Tiezzi, 1992; Southworth & Cresti, 1997). This consists of microtubule bundles that form a basket or cage around the nucleus. This internal cytoskeleton is thought to be involved in the establishment or maintenance of cell shape, because isolated sperm and generative cells become spheroidal and lose microtubule bundles (Mogensen, 1992; Palevitz & Tiezzi, 1992). It remains unclear whether flowering plant sperm evolved from motile sperm by loss of flagella, but distinct microtubule organizing centers such as blepharoplasts, centrioles, or basal bodies are absent and microtubules do not occur regularly or in a 9 plus 2 arrangement (Palevitz & Tiezzi, 1992).

In summary, the shared features of motile plant and animal sperm, like flagella-based locomotion, protamine-mediated chromatin compaction or cytoplasmic reduction via autophagy, have evolved independently. Chromatin compaction mediated by protamines was apparently lost in vascular plants, while flagellar-based motility was lost in some gymnosperms and in the common ancestor of angiosperms. Furthermore, immotile plant sperm evolved a unique microtubule cytoskeleton with uncertain origins. Autophagosomal pathways and chemoattractant perception were also redeployed to support the complex reproductive systems of angiosperms.

3.2 Gene expression in male germ cells

Transcriptome data for reproductive gametophyte stages of basal land plant species are limited to the water fern *M. vestita* and the liverwort *M. polymorpha*. A remarkable feature of spermatogenesis in *M. vestita* is that the process is largely under post-transcriptional control. This facilitates rapid spermatogenesis, largely independent of the external environment and also shifts the control of gametophyte fitness toward the sporophyte. Messenger

RNAs required for sperm cell development are stored in microspores as partially processed pre-mRNAs. After spores are rehydrated, pre-mRNAs are unmasked, processed, and translated according to a precise spatiotemporal pattern. Stored transcripts include those encoding tubulins and centrins (Tsai, Van der Weele, & Wolniak, 2004), as well as proteins important for mitosis and flagella formation, such as kinesins (Tomei & Wolniak, 2016). More detailed information is available for the transcriptome of developing antheridia of *M. polymorpha* (Higo et al., 2016). Among more than 700 genes specifically or preferentially expressed in antheridia are genes associated with sperm motility (see Section 3.1), cell cycle control, reproduction, regulation of gene expression and epigenetic processes, including homologs of angiosperm genes essential for male gametophyte and gamete formation (see Section 3.3).

The male germ cells of angiosperms possess a distinct transcriptome compared to those of other tissues or organs. Data from cDNA-sequencing are available for isolated generative cells or sperm cells of several species (reviewed in Russell & Jones, 2015; Rutley & Twell, 2015). Larger scale microarray studies of sperm cells of *A. thaliana* or rice detected 5829 or 10,732 different transcripts, respectively (Borges et al., 2008; Russell et al., 2012). A similar proportion of these transcripts (18–22%) were absent from seedlings of both species and were thus considered specific or enriched in sperm. In an independent RNA-sequencing study, 16,985 transcripts were reliably detected in rice sperm, while 18,611 transcripts were present in a vegetative cell-enriched fraction (Anderson et al., 2013). Similar to *A. thaliana*, male gamete-enriched transcripts in rice are associated with ubiquitin-mediated proteolysis, DNA modification, and chromatin remodeling (reviewed in Russell & Jones, 2015; Rutley & Twell, 2015). Ubiquitin mediated protein turnover is critical for male gametogenesis, and disruption of the ubiquitin proteasome system impairs male gametophyte development in *A. thaliana* (Doelling et al., 2007; Gallois et al., 2009). Further compelling evidence for the importance of the ubiquitin proteasome system in gametogenesis includes the role of the SCF^{FBL17} E3 ubiquitin ligase complex, which promotes generative cell cycle progression (see Section 2.2; Fig. 4).

Changes in the relationship between cell cycle progression and sperm development have led to different ontogenetic patterns in seed plants, and this is important for gamete synchrony at fertilization (Friedman, 1999). In essence, the timing of generative cell division and the cell cycle phase of sperm cells can vary in relation to pollination and fertilization between taxa (Friedman, 1999). For example, *A. thaliana* and *P. zeylanica* sperm cells are in S-phase at anthesis and reach the G2-phase prior to fertilization, whereas grass species like rice and maize remain in G1-phase until fertilization (Friedman, 1999; Gou et al., 2009). In accord with predictions based on these ontogenetic patterns, cell cycle gene expression in sperm cells differs between species. Consequently, S-phase associated cell cycle genes are enriched in sperm cells of *A. thaliana* (Borges et al., 2008) and *P. zeylanica* (Gou et al., 2009), but not in rice (Anderson et al., 2013) and transcripts of G1-type cyclins were not detected in isolated sperm of maize (Sauter et al., 1998).

Angiosperm male gametes express a full complement of DNA and chromatin modifying proteins to support gametogenesis and transgenerational epigenetic processes (Anderson et al., 2013; Borges et al., 2008). In common with *A. thaliana*, rice sperm cells preferentially express DNA *METHYLTRANSFERASE1* (*MET1*), consistent with the maintenance and epigenetic inheritance of CG-methylation through the male germline (Calarco et al., 2012; Saze, 2008; Saze, Mittelsten Scheid, & Paszkowski, 2003). Moreover, silencing of the transcription factor ARID1 in *A. thaliana* sperm cells is dependent upon *MET1* and linked to gene body specific CG-methylation (Li, Wu, Zhao, & Zheng, 2017). The epigenetic remodeling of sperm chromatin is also thought to involve the creation and removal of histone marks and the expression of histone lysine methyl transferase and demethylase enzymes (see Anderson et al., 2013). Further aspects of chromatin organization in the male germ cells of angiosperms are discussed in Section 2.4. A key feature is that distinct histone variants and several core histones are specifically or preferentially expressed in the male germline and can be unique even among angiosperms. In contrast, the amino acid sequence of the *M. polymorpha* antheridium-specific H3 variant does not differ from that of somatic H3.3 (Higo et al., 2016). This indicates that the specialized roles proposed for unique histone variants of flowering plant sperm (Section 2.4) may not operate in bryophytes and that the control of histone H3.3 abundance may be more important for gametogenesis.

Adaptation of the transcriptome of male gametes is evident from the recruitment of new genes to a sperm-cell network in *A. thaliana*. The *SHORT SUSPENSOR* (*SSP*) gene of *A. thaliana* encodes a membrane-associated IRAK/Pelle-like kinase, which regulates YODA mitogen-activated protein kinase signaling in early embryogenesis (Bayer et al., 2009). *SSP* is specifically transcribed in sperm, but transcripts are only translated after transmission to the zygote. Phylogenetic analysis has revealed the divergence and neofunctionalization of SSP following a polyploidy event in Brassicaceae

(Liu & Adams, 2010). Moreover, the recruitment of *SSP* to a sperm cell-specific network, leading to the paternal control of zygote development, is a recent evolutionary event (Liu & Adams, 2010).

3.3 The evolution of transcription factors regulating male gametogenesis

Comparative studies are starting to uncover the evolutionary origins of transcriptional networks in male gametophyte development. Functionally conserved modules, as well as the rewiring of gametophyte and germline gene networks, have been described.

Recent studies of MIKC-type MADS-domain transcription factors illustrate how closely related gene families can follow distinct evolutionary trajectories. Three of the six member MIKC^{C}-type MADS-domain family in *P. patens* are redundantly required for the formation of sperm with functional flagella (Koshimizu et al., 2018). Mutations in these genes also disturb cell division and growth in the stems of gametophores, which is important for the uptake of external water and for fertilization. In mutants deleted for all six *P. patens* MIKC^{C}-type genes, the expression of flagellum-related genes is disturbed, leading to defective sperm movement and less efficient entry of sperm into archegonia. In *Chara globularis,* MIKC^{C}-type MADS-box genes are expressed during sperm differentiation, suggesting an ancient and conserved function in the regulation of flagellum genes (Tanabe et al., 2005). In angiosperms, MIKC^{C}-type MADS-box genes seem to have lost their role in fertilization, but through amplification and neofunctionalization, these genes still control cell division and organ growth and identity (Koshimizu et al., 2018). In contrast, the closely related MIKC* MADS-genes are thought to have retained conserved roles in the gametophyte (Kwantes, Liebsch, & Verelst, 2012). MIKC* MADS-genes evolved from an MIKC^{C}-like ancestor (Henschel et al., 2002) and function in heteromeric complexes regulating pollen development (see Section 2.4). Unlike the MIKC^{C} class, the MIKC* MADS-genes did not acquire sporophytic expression in angiosperms (Kwantes et al., 2012). Although the function of MIKC* MADS-genes in bryophytes is yet to be determined, the successful complementation of *A. thaliana* mutants lacking MIKC* complexes with the single *M. polymorpha* ortholog supports a conserved molecular mechanism in land plants (Zobell, Faigl, Saedler, & Münster, 2010).

Comparative analysis is also beginning to uncover the evolutionary origins of germline specification factors. In *M. polymorpha,* the bHLH-family transcription factor BNB is required for the initiation of specialized male and female reproductive shoots (gametangiophores), which bear male and female gametangia, while two *A. thaliana* orthologs of *BNB* are redundantly required for male gametogenesis (Yamaoka et al., 2018). Mp*BNB* is expressed in gametangiophore and gametangia initial cells of both sexes, and in the germ cell lineage of antheridia and archegonia. Although Mp*BNB* is expressed in immature eggs cells and in spermatogenous cells prior to proliferative cell divisions, it is not maintained in maturing gametes. In *A. thaliana, AtBNB1* expression is restricted to the generative cell of developing pollen and is absent from sperm cells. This shift in expression pattern of *BNB* from *M. polymorpha* to *A. thaliana* reflects the absence of gametangiophores and antheridial initial cells from angiosperms, and the maintenance of expression in the immature sperm cell lineage (i.e., generative cells) (Fig. 2B and F). Mutation of both *AtBNB* paralogs abolishes the expression of the sperm cell differentiation marker H3.10, which is under the control of the *DUO1-DAZ1* module (Yamaoka et al., 2018). Moreover, *DUO1* and *DAZ1* are specifically expressed in antheridia of *M. polymorpha* (Higo et al., 2016). The overlapping expression of *BNB* and *DUO1* in generative cells of *A. thaliana* highlights the potential role of *BNB* as a conserved upstream regulator of the *DUO1-DAZ1* module (Fig. 4).

DUO1 and *DAZ1* orthologs are also co-expressed in sperm cells of rice, together with homologs of several DUO1 target genes, indicating conservation of a key germline regulatory network in angiosperms (Li, Zong, Zhou, Twell, & Zhang, 2010; Russell et al., 2012; Rutley & Twell, 2015). The antheridia-specific expression of the single copy *DUO1* and *DAZ1* orthologs in *M. polymorpha* (Higo et al., 2016) and the presence of two *DUO1* orthologues in *P. patens* (Brownfield, Hafidh, Borg, et al., 2009) suggest a conserved role for the *DUO1-DAZ1* module in bryophytes. Moreover, homologs of several genes essential for angiosperm gametogenesis are specifically or preferentially expressed in antheridia of *M. polymorpha* (Higo et al., 2016). These include homologs of the DUO1-target genes, *GEX2* and *GCS1/HAP2,* which highlight the question of when these fertilization genes were recruited to the *DUO1-DAZ1* network.

There is growing evidence for the adaptation and rewiring of germline gene networks in the 450–500 million years of evolution separating liverworts and angiosperms. The transcription factors *RKD* and *BNB* evolved sex-specific roles in angiosperms from broader ancestral roles in bryophytes. In *M. polymorpha,* Mp*RKD* is required for egg and sperm differentiation, while in *A. thaliana,* RKD family genes are associated with embryogenesis and egg cell formation, but not with sperm differentiation (Koi et al., 2016;

Rövekamp et al., 2016). Similarly, the *M. polymorpha* bHLH protein BNB has an essential role in male and female gametangia development, while the function of the *A. thaliana* orthologues (*BNB1/2*) is restricted to the male germline (Yamaoka et al., 2018). Evolutionary rewiring has also changed the regulatory network of GAMYB transcription factors. GAMYB proteins have retained an ancient role in the control of sporopollenin biosynthesis, which is required for spore and pollen grain development in land plants (Aya et al., 2011, 2009). In vascular plants, sporopollenin biosynthesis controlled by the GAMYB transcriptional network was adopted by the newly evolved gibberellin (GA) signal transduction module (Aya et al., 2011). Duplication and neofunctionalization diversified the roles of GAMYBs in angiosperms to include inflorescence and anther development (Gocal et al., 2001; Kaneko et al., 2004), pollen tube differentiation (see Section 2.3; Fig. 4; Leydon et al., 2013; Liang et al., 2013), and non-reproductive pathways, including GA-signaling during seed germination (Tsuji et al., 2006).

4. Perspective

The evolutionary history of male gametogenesis has left a trail of adaptive changes, which enabled plants to safeguard their reproduction on land. Our understanding of these changes has progressed with advances in microscopy, genetics, and omics-approaches, but knowledge at the molecular level has been largely limited to angiosperms. Important evolutionary questions remain unresolved. How and when did anisogamy evolve in streptophyte algae? Which regulators and gene networks are responsible for the distinct development of male and female gametangia? A central question in angiosperms is how the germline *sensu stricto* is specified by asymmetric division of the microspore. The role and diversity of the changing chromatin landscape of germline versus non-germ cells is another fundamental issue for which there is limited knowledge.

Given the reduced costs of DNA sequencing, progress can be made to clarify the origins and the evolution of genes and networks responsible for sexual reproduction. Sexual reproduction for most basal charophytes is unknown, and nuclear genome sequences are only available for one charophyte species, *Klebsormidium nitens* (Hori et al., 2014). Whole genome sequencing and transcriptome studies of gametangia of charophytes, as well as basal land plants, like hornworts, will help to uncover pathways crucial for reproductive adaptation to terrestrial habitats. The availability of a full range of tools for gene manipulation in bryophytes, like *M. polymorpha* and *P. patens*, is already beginning to advance understanding of reproductive development and the adaptations of basal land plants (Koi et al., 2016; Koshimizu et al., 2018; Rövekamp et al., 2016; Yamaoka et al., 2018). These comparative studies span the major reproductive transitions, from exo- to endosporic gametophytes, from motile to immotile sperm, and the innovation of double fertilization. These key transitions are expected to involve the rewiring of gene networks such as those controlled by deeply conserved regulators such as *BNB* (Yamaoka et al., 2018), *DUO1* and *DAZ1* (Brownfield, Hafidh, Borg, et al., 2009; Higo et al., 2016).

Collaborative efforts, such as the multinational EVOREPRO consortium (http://www.evorepro.org/), are providing new tools for comparative transcriptomics. Transcriptome data from various stages of gametogenesis from model and from crop plants, representing major phylogenetic clades of land plants, will accelerate discovery through an open source database (Proost & Mutwil, 2018). The inclusion of data from *M. polymorpha* and *P. patens,* alongside *Amborella trichopda*, maize, rice, *A. thaliana,* and tomato, will provide new opportunities to identify novel and missing components involved in gamete development and fertilization. Moreover, rapidly developing methods, like chromatin immunoprecipitation sequencing (ChIPseq) (Kaufmann et al., 2010) and Transposase-Accessible Chromatin using sequencing (ATACseq) (Bajic, Maher, & Deal, 2018; Buenrostro, Giresi, Zaba, Chang, & Greenleaf, 2013; Lu, Hofmeister, Vollmers, DuBois, & Schmitz, 2017), offer new horizons in the decoding of pollen and gamete regulatory networks. These approaches would allow the binding between transcription factors and their target loci to be resolved, and to decipher how these interactions are integrated with the changing chromatin landscape. The precision of such methods will be further enhanced by the use of fluorescence-activated cell sorting (FACS) and single cell transcriptome analysis (Borges et al., 2012; Santos, Bispo, & Becker, 2017; Schoft et al., 2015). Future studies using established and newly emerging models promise major advances in our understanding of the development and evolution of male gametogenesis in plants.

Acknowledgments

We thank all members of the EVOREPRO consortium for their cooperation, stimulating discussion and for sharing of unpublished results. We also thank Karen Renzaglia (Southern Illinois University) and Kazuo Okada (TokyoCinema Inc.) for granting reprint permission to include motile sperm micrographs. D.H. was supported within the ERA-CAPS project EVOREPRO by the Biotechnology and Biological Research Council (BB/N005090/1 to D.T.).

References

Adamczyk, B. J., & Fernandez, D. E. (2009). MIKC* MADS domain heterodimers are required for pollen maturation and tube growth in Arabidopsis. *Plant Physiology*, *149*, 1713–1723.

Anderson, S. N., Johnson, C. S., Jones, D. S., Conrad, L. J., Gou, X., Russell, S. D., et al. (2013). Transcriptomes of isolated *Oryza sativa* gametes characterized by deep sequencing: Evidence for distinct sex-dependent chromatin and epigenetic states before fertilization. *The Plant Journal*, *76*, 729–741.

Aw, S. J., Hamamura, Y., Chen, Z., Schnittger, A., & Berger, F. (2010). Sperm entry is sufficient to trigger division of the central cell but the paternal genome is required for endosperm development in *Arabidopsis*. *Development*, *137*, 2683–2690.

Aya, K., Hiwatashi, Y., Kojima, M., Sakakibara, H., Ueguchi-Tanaka, M., Hasebe, M., et al. (2011). The Gibberellin perception system evolved to regulate a pre-existing GAMYB-mediated system during land plant evolution. *Nature Communications*, *2*, 544.

Aya, K., Ueguchi-Tanaka, M., Kondo, M., Hamada, K., Yano, K., Nishimura, M., et al. (2009). Gibberellin modulates anther development in rice via the transcriptional regulation of GAMYB. *The Plant Cell*, *21*, 1453–1472.

Bajic, M., Maher, K. A., & Deal, R. B. (2018). Identification of open chromatin regions in plant genomes using ATAC-Seq. *Methods in Molecular Biology*, *1675*, 183–201.

Bateman, R. M., & DiMichele, W. A. (1994). Heterospory: The most iterative key innovation in the evolutionary history of the plant kingdom. *Biological Reviews*, *69*, 345–417.

Bayer, M., Nawy, T., Giglione, C., Galli, M., Meinnel, T., & Lukowitz, W. (2009). Paternal control of embryonic patterning in *Arabidopsis thaliana*. *Science*, *323*, 1485–1488.

Berger, F., & Twell, D. (2011). Germline specification and function in plants. *Annual Review of Plant Biology*, *62*, 461–484.

Borg, M., & Berger, F. (2015). Chromatin remodelling during male gametophyte development. *The Plant Journal*, *83*, 177–188.

Borg, M., Brownfield, L., Khatab, H., Sidorova, A., Lingaya, M., & Twell, D. (2011). The R2R3 MYB transcription factor DUO1 activates a male germline-specific regulon essential for sperm cell differentiation in *Arabidopsis*. *The Plant Cell*, *23*, 534–549.

Borg, M., Rutley, N., Kagale, S., Hamamura, Y., Gherghinoiu, M., Kumar, S., et al. (2014). An EAR-dependent regulatory module promotes male germ cell division and sperm fertility in *Arabidopsis*. *The Plant Cell*, *26*, 2098–2113.

Borges, F., Gardner, R., Lopes, T., Calarco, J. P., Boavida, L. C., Slotkin, R. K., et al. (2012). FACS-based purification of Arabidopsis microspores, sperm cells and vegetative nuclei. *Plant Methods*, *8*, 44.

Borges, F., Gomes, G., Gardner, R., Moreno, N., McCormick, S., Feijó, J. A., et al. (2008). Comparative transcriptomics of *Arabidopsis* sperm cells. *Plant Physiology*, *148*, 1168–1181.

Borges, F., Pereira, P. A., Slotkin, R. K., Martienssen, R. A., & Becker, J. D. (2011). MicroRNA activity in the *Arabidopsis* male germline. *Journal of Experimental Botany*, *62*, 1611–1620.

Brewbaker, J. L. (1967). The distribution and phylogenetic significance of binucleate and trinucleate pollen grains in the angiosperms. *American Journal of Botany*, *54*, 1069–1083.

Brooks, J., & Shaw, G. (1978). Sporopollenin: A review of its chemistry, palaeochemistry and geochemistry. *Grana*, *17*, 91–97.

Brownfield, L., Hafidh, S., Borg, M., Sidorova, A., Mori, T., & Twell, D. (2009). A plant germline-specific integrator of sperm specification and cell cycle progression. *PLoS Genetics*, *5*, e1000430.

Brownfield, L., Hafidh, S., Durbarry, A., Khatab, H., Sidorova, A., Doerner, P., et al. (2009). Arabidopsis *DUO POLLEN3* is a key regulator of male germline development and embryogenesis. *The Plant Cell*, *21*, 1940–1956.

Buenrostro, J. D., Giresi, P. G., Zaba, L. C., Chang, H. Y., & Greenleaf, W. J. (2013). Transposition of native chromatin for fast and sensitive epigenomic profiling of open chromatin, DNA-binding proteins and nucleosome position. *Nature Methods*, *10*, 1213–1218.

Calarco, J. P., Borges, F., Donoghue, M. T. A., Van Ex, F., Jullien, P. E., Lopes, T., et al. (2012). Reprogramming of DNA methylation in pollen guides epigenetic inheritance via small RNA. *Cell*, *151*, 194–205.

Carmichael, J. S., & Friedman, W. E. (1995). Double fertilization in *Gnetum gnemon*: The relationship between the cell cycle and sexual reproduction. *The Plant Cell*, *7*, 1975–1988.

Chen, Z., Hafidh, S., Poh, S. H., Twell, D., & Berger, F. (2009). Proliferation and cell fate establishment during *Arabidopsis* male gametogenesis depends on the retinoblastoma protein. *Proceedings of the National Academy of Sciences of the United States of America*, *106*, 7257–7262.

Chen, Y. C., & McCormick, S. (1996). *Sidecar pollen*, an *Arabidopsis thaliana* male gametophytic mutant with aberrant cell divisions during pollen development. *Development*, *122*, 3243–3253.

Creasey, K. M., Zhai, J., Borges, F., Van Ex, F., Regulski, M., Meyers, B. C., et al. (2014). miRNAs trigger widespread epigenetically-activated siRNAs from transposons in *Arabidopsis*. *Nature*, *508*, 411–415.

Cresti, M., Lancelle, S. A., & Hepler, P. K. (1987). Structure of the generative cell wall complex after freeze substitution in pollen tubes of *Nicotiana* and *Impatiens*. *Journal of Cell Science*, *88*, 373–378.

Delwiche, C. F. (2016). The genomes of charophyte green algae. In S. Rensing (Ed.), *Genomes and evolution of charophytes, bryophytes, lycophytes and ferns* (pp. 255–270). Academic Press.

Doelling, J. H., Phillips, A. R., Soyler-Ogretim, G., Wise, J., Chandler, J., Callis, J., et al. (2007). The ubiquitin-specific protease subfamily UBP3/UBP4 is essential for pollen development and transmission in *Arabidopsis*. *Plant Physiology*, *145*, 801–813.

Dumas, C., Knox, R. B., & Gaude, T. (1985). The spatial association of the sperm cells and vegetative nucleus in the pollen grain of *Brassica*. *Protoplasma*, *124*, 168–174.

Durand, E. J. (1908). The development of the sexual organs and sporogonium of *Marchantia polymorpha*. *Bulletin of the Torrey Botanical Club*, *35*, 321–335.

Durbarry, A., Vizir, I., & Twell, D. (2005). Male germ line development in *Arabidopsis*. *duo pollen* mutants reveal gametophytic regulators of generative cell cycle progression. *Plant Physiology*, *137*, 297–307.

Eady, C., Lindsey, K., & Twell, D. (1995). The significance of microspore division and division symmetry for vegetative cell-specific transcription and generative cell differentiation. *The Plant Cell*, *7*, 65–74.

Engel, M. L., Holmes-Davis, R., & McCormick, S. (2005). Green sperm. Identification of male gamete promoters in *Arabidopsis*. *Plant Physiology*, *138*, 2124–2133.

Erenpreiss, J., Spano, M., Erenpreisa, J., Bungum, M., & Giwercman, A. (2006). Sperm chromatin structure and male fertility: Biological and clinical aspects. *Asian Journal of Andrology*, *8*, 11–29.

Fernando, D., Quinn, C., Brenner, E., & Owens, J. (2010). Male gametophyte development and evolution in extant gymnosperms. In D. D. Fernando (Ed.), *International journal of plant developmental biology: Vol. 4 (Special Issue 1). Plant development and evolution* (pp. 47–63). Global Science Books.

Ferris, P. J., Armbrust, E. V., & Goodenough, U. W. (2002). Genetic structure of the mating-type locus of *Chlamydomonas reinhardtii*. *Genetics*, *160*, 181–200.

Ferris, P. J., & Goodenough, U. W. (1994). The mating-type locus of *Chlamydomonas reinhardtii* contains highly rearranged DNA sequences. *Cell*, *76*, 1135–1145.

Ferris, P. J., & Goodenough, U. W. (1997). Mating type in *Chlamydomonas* is specified by mid, the minus-dominance gene. *Genetics, 146*, 859–869.
Fíla, J., Záveská Drábková, L., Gibalová, A., & Honys, D. (2017). When simple meets complex: Pollen and the -omics BT. In G. Obermeyer & J. Feijó (Eds.), *Pollen tip growth: From biophysical aspects to systems biology* (pp. 247–292). Cham: Springer International Publishing.
Frank, A. C., & Johnson, M. A. (2009). Expressing the diphtheria toxin a subunit from the *HAP2(GCS1)* promoter blocks sperm maturation and produces single sperm-like cells capable of fertilization. *Plant Physiology, 151*, 1390–1400.
Friedman, W. E. (1990). Double fertilization in *ephedra*, a nonflowering seed plant: Its bearing on the origin of angiosperms. *Science, 247*, 951–954.
Friedman, W. E. (1993). The evolutionary history of the seed plant male gametophyte. *Trends in Ecology & Evolution, 8*, 15–21.
Friedman, W. E. (1998). The evolution of double fertilization and endosperm: An "historical" perspective. *Sexual Plant Reproduction, 11*, 6–16.
Friedman, W. E. (1999). Expression of the cell cycle in sperm of *Arabidopsis*: Implications for understanding patterns of gametogenesis and fertilization in plants and other eukaryotes. *Development, 126*, 1065–1075.
Friedman, W. E., Gifford, E. M., & Del Tredici, P. (1997). Development of the male gametophyte of *Ginkgo biloba*: A window into the reproductive biology of early seed plants. In T. Hori, R. W. Ridge, W. Tulecke, J. Trémouillaux-Guiller, & H. Tobe (Eds.), *Ginkgo biloba: A global treasure: From biology to medicine* (pp. 29–49). Tokyo: Springer Japan.
Friedman, L., Santa Anna-Arriola, S., Hodgkin, J., & Kimble, J. (2000). *gon-4*, a cell lineage regulator required for gonadogenesis in *Caenorhabditis elegans*. *Developmental Biology, 228*, 350–362.
Fujiki, Y., Yoshimoto, K., & Ohsumi, Y. (2007). An Arabidopsis homolog of yeast *ATG6/VPS30* is essential for pollen germination. *Plant Physiology, 143*, 1132–1139.
Furumizu, C., Alvarez, J. P., Sakakibara, K., & Bowman, J. L. (2015). Antagonistic roles for *KNOX1* and *KNOX2* genes in patterning the land plant body plan following an ancient gene duplication. *PLoS Genetics, 11*, e1004980.
Gallois, J.-L., Guyon-Debast, A., Lécureuil, A., Vezon, D., Carpentier, V., Bonhomme, S., et al. (2009). The Arabidopsis proteasome RPT5 subunits are essential for gametophyte development and show accession-dependent redundancy. *The Plant Cell, 21*, 442–459.
Geng, S., De Hoff, P., & Umen, J. G. (2014). Evolution of sexes from an ancestral mating-type specification pathway. *PLoS Biology, 12*, e1001904.
Gifford, E. M., & Foster, A. S. (1989). *Morphology and evolution of vascular plants*. New York: W.H. Freeman.
Gifford, M. E., & Lin, J. (1975). Light microscope and ultrastructural studies of the male gametophyte in *Ginkgo biloba*: The spermatogenous cell. *American Journal of Botany, 62*, 974–981.
Glöckle, B., Urban, W. J., Nagahara, S., Andersen, E. D., Higashiyama, T., Grini, P. E., et al. (2018). Pollen differentiation as well as pollen tube guidance and discharge are independent of the presence of gametes. *Development, 145*, dev152645.
Gocal, G. F., Sheldon, C. C., Gubler, F., Moritz, T., Bagnall, D. J., MacMillan, C. P., et al. (2001). *GAMYB-like* genes, flowering, and gibberellin signaling in Arabidopsis. *Plant Physiology, 127*, 1682–1693.
Gou, X., Yuan, T., Wei, X., & Russell, S. D. (2009). Gene expression in the dimorphic sperm cells of *Plumbago zeylanica*: Transcript profiling, diversity, and relationship to cell type. *The Plant Journal, 60*, 33–47.

Grant-Downton, R., Hafidh, S., Twell, D., & Dickinson, H. G. (2009). Small RNA pathways are present and functional in the angiosperm male gametophyte. *Molecular Plant, 2*, 500–512.

Guan, Y., Guo, J., Li, H., & Yang, Z. (2013). Signaling in pollen tube growth: Crosstalk, feedback, and missing links. *Molecular Plant, 6*, 1053–1064.

Guan, Y., Meng, X., Khanna, R., LaMontagne, E., Liu, Y., & Zhang, S. (2014). Phosphorylation of a WRKY transcription factor by MAPKs is required for pollen development and function in *Arabidopsis*. *PLoS Genetics, 10*, e1004384.

Gusti, A., Baumberger, N., Nowack, M., Pusch, S., Eisler, H., Potuschak, T., et al. (2009). The *Arabidopsis thaliana* F-box protein FBL17 is essential for progression through the second mitosis during pollen development. *PLoS One, 4*, e4780.

Gutzat, R., Borghi, L., & Gruissem, W. (2012). Emerging roles of RETINOBLASTOMA-RELATED proteins in evolution and plant development. *Trends in Plant Science, 17*, 139–148.

Hafidh, S., Fila, J., & Honys, D. (2016). Male gametophyte development and function in angiosperms: A general concept. *Plant Reproduction, 29*, 31–51.

Hamamura, Y., Saito, C., Awai, C., Kurihara, D., Miyawaki, A., Nakagawa, T., et al. (2011). Live-cell imaging reveals the dynamics of two sperm cells during double fertilization in *Arabidopsis thaliana*. *Current Biology, 21*, 497–502.

Harbour, J. W., & Dean, D. C. (2000). The Rb/E2F pathway: Expanding roles and emerging paradigms. *Genes & Development, 14*, 2393–2409.

Hay, A., & Tsiantis, M. (2010). *KNOX* genes: Versatile regulators of plant development and diversity. *Development, 137*, 3153–3165.

He, H., Yang, T., Wu, W., & Zheng, B. (2015). Small RNAs in pollen. *Science China. Life Sciences, 58*, 246–252.

Heim, M. A., Jakoby, M., Werber, M., Martin, C., Weisshaar, B., & Bailey, P. C. (2003). The basic helix-loop-helix transcription factor family in plants: A genome-wide study of protein structure and functional diversity. *Molecular Biology and Evolution, 20*, 735–747.

Henschel, K., Kofuji, R., Hasebe, M., Saedler, H., Münster, T., & Theißen, G. (2002). Two ancient classes of MIKC-type MADS-box genes are present in the Moss *Physcomitrella patens*. *Molecular Biology and Evolution, 19*, 801–814.

Higo, A., Niwa, M., Yamato, K. T., Yamada, L., Sawada, H., Sakamoto, T., et al. (2016). Transcriptional framework of male gametogenesis in the liverwort *Marchantia polymorpha L. Plant and Cell Physiology, 57*, 325–338.

Hori, K., Maruyama, F., Fujisawa, T., Togashi, T., Yamamoto, N., Seo, M., et al. (2014). *Klebsormidium flaccidum* genome reveals primary factors for plant terrestrial adaptation. *Nature Communications, 5*, 3978.

Houben, A., Kumke, K., Nagaki, K., & Hause, G. (2011). CENH3 distribution and differential chromatin modifications during pollen development in rye (*Secale cereale L.*). *Chromosome Research, 19*, 471–480.

Hu, J. H., Yang, N., Ma, Y. H., Jiang, J., Zhang, J. F., Fei, J., et al. (2004). Identification of glutamate receptors and transporters in mouse and human sperm. *Journal of Andrology, 25*, 140–146.

Ikeda, K., Ikeda, T., Morikawa, K., & Kamiya, R. (2007). Axonemal localization of *Chlamydomonas* PACRG, a homologue of the human Parkin-coregulated gene product. *Cell Motility, 64*, 814–821.

Ingouff, M., Jullien, P. E., & Berger, F. (2006). The female gametophyte and the endosperm control cell proliferation and differentiation of the seed coat in Arabidopsis. *The Plant Cell, 18*, 3491–3501.

Ingouff, M., Sakata, T., Li, J., Sprunck, S., Dresselhaus, T., & Berger, F. (2009). The two male gametes share equal ability to fertilize the egg cell in *Arabidopsis thaliana*. *Current Biology, 9*, 19–20.

Iwakawa, H., Shinmyo, A., & Sekine, M. (2006). *Arabidopsis CDKA;1*, a cdc2 homologue, controls proliferation of generative cells in male gametogenesis. *The Plant Journal, 45*, 819–831.

Jensen, W. A., & Fisher, D. B. (1970). Cotton embryogenesis: The pollen tube in the stigma and style. *Protoplasma, 69*, 215–235.

Jiang, H., Yi, J., Boavida, L. C., Chen, Y., Becker, J. D., Kohler, C., et al. (2015). Intercellular communication in Arabidopsis thaliana pollen discovered via *AHG3* transcript movement from the vegetative cell to sperm. *Proceedings of the National Academy of Sciences of the United States of America, 112*, 13378–13383.

Johnson, G. D., Lalancette, C., Linnemann, A. K., Leduc, F., Boissonneault, G., & Krawetz, S. A. (2011). The sperm nucleus: Chromatin, RNA and the nuclear matrix. *Reproduction, 141*, 21–36.

Johnston, A. J., Matveeva, E., Kirioukhova, O., Grossniklaus, U., & Gruissem, W. (2008). A dynamic reciprocal RBR-PRC2 regulatory circuit controls arabidopsis gametophyte development. *Current Biology, 18*, 1680–1686.

Kaneko, M., Inukai, Y., Ueguchi-Tanaka, M., Itoh, H., Izawa, T., Kobayashi, Y., et al. (2004). Loss-of-function mutations of the rice *GAMYB* gene impair alpha-amylase expression in aleurone and flower development. *The Plant Cell, 16*, 33–44.

Kaufmann, K., Muiño, J. M., Østerås, M., Farinelli, L., Krajewski, P., & Angenent, G. C. (2010). Chromatin immunoprecipitation (ChIP) of plant transcription factors followed by sequencing (ChIP-SEQ) or hybridization to whole genome arrays (ChIP-CHIP). *Nature Protocols, 5*, 457.

Keiko, S., Tomoaki, N., Hironori, D., & Mitsuyasu, H. (2008). Class 1 *KNOX* genes are not involved in shoot development in the moss *Physcomitrella patens* but do function in sporophyte development. *Evolution & Development, 10*, 555–566.

Kim, M., Kim, M., Lee, M. R., Park, S. K., & Kim, J. (2015). LATERAL ORGAN BOUNDARIES DOMAIN (LBD)10 interacts with SIDECAR POLLEN/LBD27 to control pollen development in Arabidopsis. *The Plant Journal, 81*, 794–809.

Kim, M., Kim, M.-J., Pandey, S., & Kim, J. (2016). Expression and protein interaction analyses reveal combinatorial interactions of LBD transcription factors during *Arabidopsis* pollen development. *Plant and Cell Physiology, 57*, 2291–2299.

Kim, H. J., Oh, S. A., Brownfield, L., Hong, S. H., Ryu, H., Hwang, I., et al. (2008). Control of plant germline proliferation by SCFFBL17 degradation of cell cycle inhibitors. *Nature, 455*, 1134.

Koi, S., Hisanaga, T., Sato, K., Shimamura, M., Yamato, K. T., Ishizaki, K., et al. (2016). An evolutionarily conserved plant *RKD* factor controls germ cell differentiation. *Current Biology, 26*, 1775–1781.

Kong, J., Lau, S., & Jurgens, G. (2015). Twin plants from supernumerary egg cells in *Arabidopsis*. *Current Biology, 25*, 225–230.

Koshimizu, S., Kofuji, R., Sasaki-Sekimoto, Y., Kikkawa, M., Shimojima, M., Ohta, H., et al. (2018). *Physcomitrella* MADS-box genes regulate water supply and sperm movement for fertilization. *Nature Plants, 4*, 36–45.

Kurusu, T., Koyano, T., Hanamata, S., Kubo, T., Noguchi, Y., Yagi, C., et al. (2014). *OsATG7* is required for autophagy-dependent lipid metabolism in rice postmeiotic anther development. *Autophagy, 10*, 878–888.

Kwantes, M., Liebsch, D., & Verelst, W. (2012). How MIKC* MADS-box genes originated and evidence for their conserved function throughout the evolution of vascular plant gametophytes. *Molecular Biology and Evolution, 29*, 293–302.

Lee, Y.-R. J., Li, Y., & Liu, B. (2007). Two Arabidopsis phragmoplast-associated kinesins play a critical role in cytokinesis during male gametogenesis. *The Plant Cell, 19*, 2595–2605.

Lee, J.-H., Lin, H., Joo, S., & Goodenough, U. (2008). Early sexual origins of homeoprotein heterodimerization and evolution of the plant *KNOX/BELL* family. *Cell, 133*, 829–840.

Lei, R., Li, X., Ma, Z., Lv, Y., Hu, Y., & Yu, D. (2017). *Arabidopsis* WRKY2 and WRKY34 transcription factors interact with VQ20 protein to modulate pollen development and function. *The Plant Journal, 91*, 962–976.

Lei, R., Ma, Z., & Yu, D. (2018). WRKY2/34–VQ20 modules in *Arabidopsis thaliana* negatively regulate expression of a trio of related MYB transcription factors during pollen development. *Frontiers in Plant Science, 9*, 331.

Leydon, A. R., Beale, K. M., Woroniecka, K., Castner, E., Chen, J., Horgan, C., et al. (2013). Three MYB transcription factors control pollen tube differentiation required for sperm release. *Current Biology, 23*, 1209–1214.

Li, W., Wu, J., Kim, S.-Y., Zhao, M., Hearn, S. A., Zhang, M. Q., et al. (2014). *Chd5* orchestrates chromatin remodelling during sperm development. *Nature Communications, 5*, 3812.

Li, L., Wu, W., Zhao, Y., & Zheng, B. (2017). A reciprocal inhibition between *ARID1* and *MET1* in male and female gametes in Arabidopsis. *Journal of Integrative Plant Biology, 59*, 657–668.

Li, S.-L., Zong, J., Zhou, Z.-G., Twell, D., & Zhang, D.-B. (2010). Cloning and expression analysis of OsDUO1 encoding a rice MYB transcription factor. *Plant Physiology Journal, 46*, 1033–1039.

Liang, Y., Tan, Z.-M. M., Zhu, L., Niu, Q.-K. K., Zhou, J.-J. J., Li, M., et al. (2013). *MYB97*, *MYB101* and *MYB120* function as male factors that control pollen tube-Synergid interaction in Arabidopsis thaliana fertilization. *PLoS Genetics, 9*, e1003933.

Ligrone, R., Duckett, J. G., & Renzaglia, K. S. (2012). Major transitions in the evolution of early land plants: A bryological perspective. *Annals of Botany, 109*, 851–871.

Lin, Q., Ohashi, Y., Kato, M., Tsuge, T., Gu, H., Qu, L.-J., et al. (2015). GLABRA2 directly suppresses basic helix-loop-helix transcription factor genes with diverse functions in root hair development. *The Plant Cell, 27*, 2894–2906.

Liu, S.-L., & Adams, K. L. (2010). Dramatic change in function and expression pattern of a gene duplicated by polyploidy created a paternal effect gene in the Brassicaceae. *Molecular Biology and Evolution, 27*, 2817–2828.

Liu, Y., Cui, S., Wu, F., Yan, S., Lin, X., Du, X., et al. (2013). Functional conservation of MIKC*-type MADS box genes in *Arabidopsis* and rice pollen maturation. *The Plant Cell, 25*, 1288–1303.

Liu, Y., Du, L., Osato, M., Teo, E. H., Qian, F., Jin, H., et al. (2007). The zebrafish *udu* gene encodes a novel nuclear factor and is essential for primitive erythroid cell development. *Blood, 110*, 99–106.

Lora, J., Herrero, M., & Hormaza, J. I. (2009). The coexistence of bicellular and tricellular pollen in *Annona cherimola* (Annonaceae): Implications for pollen evolution. *American Journal of Botany, 96*, 802–808.

Lorenzetti, D., Bishop, C. E., & Justice, M. J. (2004). Deletion of the Parkin coregulated gene causes male sterility in the quaking(viable) mouse mutant. *Proceedings of the National Academy of Sciences of the United States of America, 101*, 8402–8407.

Lu, P., Hankel, I. L., Hostager, B. S., Swartzendruber, J. A., Friedman, A. D., Brenton, J. L., et al. (2011). The developmental regulator protein Gon4l associates with protein YY1, co-repressor Sin3a, and histone deacetylase 1 and mediates transcriptional repression. *The Journal of Biological Chemistry, 286*, 18311–18319.

Lu, Z., Hofmeister, B. T., Vollmers, C., DuBois, R. M., & Schmitz, R. J. (2017). Combining ATAC-seq with nuclei sorting for discovery of cis-regulatory regions in plant genomes. *Nucleic Acids Research, 45*, e41.

Martínez, G., Panda, K., Köhler, C., & Slotkin, R. K. (2016). Silencing in sperm cells is directed by RNA movement from the surrounding nurse cell. *Nature Plants, 2*, 16030.

McCourt, R. M., Delwiche, C. F., & Karol, K. G. (2004). Charophyte algae and land plant origins. *Trends in Ecology & Evolution, 19*, 661–666.

McCue, A. D., Cresti, M., Feijo, J. A., & Slotkin, R. K. (2011). Cytoplasmic connection of sperm cells to the pollen vegetative cell nucleus: Potential roles of the male germ unit revisited. *Journal of Experimental Botany, 62*, 1621–1631.

Meistrich, M. L., Mohapatra, B., Shirley, C. R., & Zhao, M. (2003). Roles of transition nuclear proteins in spermiogenesis. *Chromosoma, 111*, 483–488.

Mérai, Z., Chumak, N., García-Aguilar, M., Hsieh, T.-F., Nishimura, T., Schoft, V. K., et al. (2014). The AAA-ATPase molecular chaperone Cdc48/p97 disassembles sumoylated centromeres, decondenses heterochromatin, and activates ribosomal RNA genes. *Proceedings of the National Academy of Sciences of the United States of America, 111*, 16166–16171.

Michard, E., Lima, P. T., Borges, F., Silva, A. C., Portes, M. T., Carvalho, J. E., et al. (2011). Glutamate receptor–like genes form Ca2+ channels in pollen tubes and are regulated by pistil d-serine. *Science, 332*, 434–437.

Michard, E., Simon, A. A., Tavares, B., Wudick, M. M., & Feijó, J. A. (2017). Signaling with ions: The keystone for apical cell growth and morphogenesis in pollen tubes. *Plant Physiology, 173*, 91–111.

Mogensen, H. L. (1992). The male germ unit: Concept, composition, and significance. In S. D. Russell & Dumas C. (Eds.), *International review of cytology* (pp. 129–147). Academic Press.

Mori, T., Igawa, T., Tamiya, G., Miyagishima, S., & Berger, F. (2014). Gamete attachment requires *GEX2* for successful fertilization in *Arabidopsis*. *Current Biology, 24*, 170–175.

Mori, T., Kawai-Toyooka, H., Igawa, T., & Nozaki, H. (2015). Gamete dialogs in green lineages. *Molecular Plant, 8*, 1442–1454.

Myles, D. G., & Hepler, P. K. (1977). Spermiogenesis in the fern Marsilea microtubules, nuclear shaping, and cytomorphogenesis. *Journal of Cell Science, 23*, 57–83.

Niklas, K. J., & Kutschera, U. (2009). The evolution of the land plant life cycle. *New Phytologist, 185*, 27–41.

Norstog, K. J. (1975). The motility of cycad spermatozoids in relation to structure and function. In J. G. Duckett & P. A. Racey (Eds.), *The biology of the male gamete* (pp. 135–142). London: Academic Press.

Nowack, M. K., Grini, P. E., Jakoby, M. J., Lafos, M., Koncz, C., & Schnittger, A. (2005). A positive signal from the fertilization of the egg cell sets off endosperm proliferation in angiosperm embryogenesis. *Nature Genetics, 38*, 63.

Nozaki, H., Misawa, K., Kajita, T., Kato, M., Nohara, S., & Watanabe, M. M. (2000). Origin and evolution of the colonial volvocales (Chlorophyceae) as inferred from multiple, chloroplast gene sequences. *Molecular Phylogenetics and Evolution, 17*, 256–268.

Oh, S. A., Allen, T., Kim, G. J., Sidorova, A., Borg, M., Park, S. K., et al. (2012). *Arabidopsis* fused kinase and the kinesin-12 subfamily constitute a signalling module required for phragmoplast expansion. *The Plant Journal, 72*, 308–319.

Oh, S.-A., Bourdon, V., Das'Pal, M., Dickinson, H., & Twell, D. (2008). *Arabidopsis* kinesins HINKEL and TETRASPORE act redundantly to control cell plate expansion during cytokinesis in the male gametophyte. *Molecular Plant, 1*, 794–799.

Oh, S. A., Johnson, A., Smertenko, A., Rahman, D., Park, S. K., Hussey, P. J., et al. (2005). A divergent cellular role for the FUSED kinase family in the plant-specific cytokinetic phragmoplast. *Current Biology, 15*, 2107–2111.

Oh, S. A., Park, K. S., Twell, D., & Park, S. K. (2010). The *SIDECAR POLLEN* gene encodes a microspore-specific LOB/AS2 domain protein required for the correct timing and orientation of asymmetric cell division. *The Plant Journal, 64*, 839–850.

Okada, T., Endo, M., Singh, M. B., & Bhalla, P. L. (2005). Analysis of the histone H3 gene family in Arabidopsis and identification of the male-gamete-specific variant *AtMGH3*. *The Plant Journal, 44*, 557–568.

Ortiz-Ramírez, C., Michard, E., Simon, A. A., Damineli, D. S. C., Hernández-Coronado,-M., Becker, J. D., et al. (2017). GLUTAMATE RECEPTOR-LIKE channels are essential for chemotaxis and reproduction in mosses. *Nature, 549*, 91.

Palatnik, J. F., Wollmann, H., Schommer, C., Schwab, R., Boisbouvier, J., Rodriguez, R., et al. (2007). Sequence and expression differences underlie functional specialization of Arabidopsis MicroRNAs *miR159* and *miR319*. *Developmental Cell, 13*, 115–125.

Palevitz, B. A., & Tiezzi, A. (1992). In S. D. Russell & Dumas C. (Eds.), *Organization, composition, and function of the generative cell and sperm cytoskeleton* (pp. 149–185). Academic Press.

Pandey, P., Houben, A., Kumlehn, J., Melzer, M., & Rutten, T. (2013). Chromatin alterations during pollen development in Hordeum vulgare. *Cytogenetic and Genome Research, 141*, 50–57.

Park, S. K., Howden, R., & Twell, D. (1998). The Arabidopsis thaliana gametophytic mutation gemini pollen1 disrupts microspore polarity, division asymmetry and pollen cell fate. *Development, 125*, 3789–3799.

Peters, B., Casey, J., Aidley, J., Zohrab, S., Borg, M., Twell, D., et al. (2017). A conserved cis-regulatory module determines germline fate through activation of the transcription factor *DUO1* promoter. *Plant Physiology, 173*, 280–293.

Petersen, K. B., & Burd, M. (2016). Why did heterospory evolve? *Biological Reviews, 92*, 1739–1754.

Proost, S., & Mutwil, M. (2018). CoNekT: An open-source framework for comparative genomic and transcriptomic network analyses. *Nucleic Acids Research, 46*, W133–W140.

Rathke, C., Barckmann, B., Burkhard, S., Jayaramaiah-Raja, S., Roote, J., & Renkawitz-Pohl, R. (2010). Distinct functions of Mst77F and protamines in nuclear shaping and chromatin condensation during *Drosophila* spermiogenesis. *European Journal of Cell Biology, 89*, 326–338.

Renzaglia, K. S., & Garbary, D. J. (2001). Motile gametes of land plants: Diversity, development, and evolution. *Critical Reviews in Plant Sciences, 20*, 107–213.

Reynolds, W. F., & Wolfe, S. L. (1978). Changes in basic proteins during sperm maturation in a plant, *Marchantia polymorpha*. *Experimental Cell Research, 116*, 269–273.

Reynolds, W. F., & Wolfe, S. L. (1984). Protamines in plant sperm. *Experimental Cell Research, 152*, 443–448.

Rotman, N., Durbarry, A., Wardle, A., Yang, W. C., Chaboud, A., Faure, J.-E., et al. (2005). A novel class of MYB factors controls sperm-cell formation in plants. *Current Biology, 15*, 244–248.

Rövekamp, M., Bowman, J. L., & Grossniklaus, U. (2016). Marchantia MpRKD regulates the gametophyte-sporophyte transition by keeping egg cells quiescent in the absence of fertilization. *Current Biology, 26*, 1782–1789.

Rowe, N., Paul-Victor, C., & Paul-Victor, C. (2012). Herbs and secondary woodiness—Keeping up the cambial habit. *New Phytologist, 193*, 3–5.

Rudall, P. J., & Bateman, R. M. (2007). Developmental bases for key innovations in the seed-plant microgametophyte. *Trends in Plant Science, 12*, 317–326.

Russell, S. D. (1985). Preferential fertilization in Plumbago: Ultrastructural evidence for gamete-level recognition in an angiosperm. *Proceedings of the National Academy of Sciences of the United States of America, 82*, 6129–6132.

Russell, S. D., & Cass, D. D. (1981). Ultrastructure of the sperms of *Plumbago zeylanica*. *Protoplasma, 107*, 85–107.

Russell, S. D., Gou, X., Wong, C. E., Wang, X., Yuan, T., Wei, X., et al. (2012). Genomic profiling of rice sperm cell transcripts reveals conserved and distinct elements in the flowering plant male germ lineage. *New Phytologist, 195*, 560–573.

Russell, S. D., & Jones, D. S. (2015). The male germline of angiosperms: Repertoire of an inconspicuous but important cell lineage. *Frontiers in Plant Science*, *6*, 173.
Rutley, N., & Twell, D. (2015). A decade of pollen transcriptomics. *Plant Reproduction*, *28*, 73–89.
Saito, C., Nagata, N., Sakai, A., Mori, K., Kuroiwa, H., & Kuroiwa, T. (2002). Angiosperm species that produce sperm cell pairs or generative cells with polarized distribution of DNA-containing organelles. *Sexual Plant Reproduction*, *15*, 167–178.
Sakakibara, K., Ando, S., Yip, H. K., Tamada, Y., Hiwatashi, Y., Murata, T., et al. (2013). *KNOX2* genes regulate the haploid-to-diploid morphological transition in land plants. *Science*, *339*, 1067–1070.
Sanchez-Vera, V., Kenchappa, C. S., Landberg, K., Bressendorff, S., Schwarzbach, S., Martin, T., et al. (2017). Autophagy is required for gamete differentiation in the moss *Physcomitrella patens*. *Autophagy*, *13*, 1939–1951.
Sano, Y., & Tanaka, I. (2010). Distinct localization of histone H3 methylation in the vegetative nucleus of lily pollen. *Cell Biology International*, *34*, 253–259.
Santos, M. R., Bispo, C., & Becker, J. D. (2017). Isolation of Arabidopsis pollen, sperm cells, and vegetative nuclei by fluorescence-activated cell sorting (FACS). In A. Schmidt (Ed.), *Plant germline development: Methods and protocols* (pp. 193–210). New York: Springer New York.
Sauter, M., Wiegen, P. v., Lorz, H., Kranz, E., von Wiegen, P., Lörz, H., et al. (1998). Cell cycle regulatory genes from maize are differentially controlled during fertilization and first embryonic cell division. *Sexual Plant Reproduction*.
Saze, H. (2008). Epigenetic memory transmission through mitosis and meiosis in plants. *Seminars in Cell & Developmental Biology*, *19*, 527–536.
Saze, H., Mittelsten Scheid, O., & Paszkowski, J. (2003). Maintenance of CpG methylation is essential for epigenetic inheritance during plant gametogenesis. *Nature Genetics*, *34*, 65–69.
Schmid, A., Sutto, Z., Nlend, M.-C., Horvath, G., Schmid, N., Buck, J., et al. (2007). Soluble adenylyl cyclase is localized to cilia and contributes to ciliary beat frequency regulation via production of cAMP. *The Journal of General Physiology*, *130*, 99–109.
Schmidt, A., Schmid, M. W., & Grossniklaus, U. (2015). Plant germline formation: Common concepts and developmental flexibility in sexual and asexual reproduction. *Development*, *142*, 229–241.
Schoft, V. K., Chumak, N., Bindics, J., Slusarz, L., Twell, D., Kohler, C., et al. (2015). SYBR green-activated sorting of *Arabidopsis* pollen nuclei based on different DNA/RNA content. *Plant Reproduction*, *28*, 61–72.
Schoft, V. K., Chumak, N., Mosiolek, M., Slusarz, L., Komnenovic, V., Brownfield, L., et al. (2009). Induction of RNA-directed DNA methylation upon decondensation of constitutive heterochromatin. *EMBO Reports*, *10*, 1015–1021.
Sharp, L. W. (1914). Spermatogenesis in *Marsilia*. *Botanical Gazette*, *58*, 419–431.
Shaw, A. J., Szovenyi, P., & Shaw, B. (2011). Bryophyte diversity and evolution: Windows into the early evolution of land plants. *American Journal of Botany*, *98*, 352–369.
Shimamura, M. (2016). *Marchantia polymorpha*: Taxonomy, phylogeny and morphology of a model system. *Plant and Cell Physiology*, *57*, 230–256.
Slotkin, R. K., Vaughn, M., Borges, F., Tanurdžić, M., Becker, J. D., Feijó, J. A., et al. (2009). Epigenetic reprogramming and small RNA silencing of transposable elements in pollen. *Cell*, *136*, 461–472.
Snell, R., & Aarssen, L. W. (2005). Life history traits in selfing versus outcrossing annuals: Exploring the "time-limitation" hypothesis for the fitness benefit of self-pollination. *BMC Ecology*, *5*, 2.
Southworth, D., & Cresti, M. (1997). Comparison of flagellated and nonflagellated sperm in plants. *American Journal of Botany*, *84*, 1301–1311.

Soyano, T., Nishihama, R., Morikiyo, K., Ishikawa, M., & Machida, Y. (2003). NQK1/NtMEK1 is a MAPKK that acts in the NPK1 MAPKKK-mediated MAPK cascade and is required for plant cytokinesis. *Genes & Development*, *17*, 1055–1067.

Strother, P. K., Al-Hajri, S., & Traverse, A. (1996). New evidence for land plants from the lower middle Ordovician of Saudi Arabia. *Geology*, *24*, 55–58.

Takahashi, T., Honda, K., Mori, T., & Igawa, T. (2017). Loss of *GCS1/HAP2* does not affect the ovule-targeting behavior of pollen tubes. *Plant Reproduction*, *30*, 147–152.

Tanabe, Y., Hasebe, M., Sekimoto, H., Nishiyama, T., Kitani, M., Henschel, K., et al. (2005). Characterization of MADS-box genes in charophycean green algae and its implication for the evolution of MADS-box genes. *Proceedings of the National Academy of Sciences of the United States of America*, *102*, 2436–2441.

Tanaka, I., & Ito, M. (1980). Induction of typical cell division in isolated microspores of *Lilium longiflorum* and *Tulipa gesneriana*. *Plant Science Letters*, *17*, 279–285.

Terasaka, O., & Niitsu, T. (1987). Unequal cell division and chromatin differentiation in pollen grain cells—I. Centrifugal, cold and caffeine treatments. *The Botanical Magazine Tokyo*, *100*, 205–216.

Tomei, E. J., & Wolniak, S. M. (2016). Transcriptome analysis reveals a diverse family of kinesins essential for spermatogenesis in the fern *Marsilea*. *Cytoskeleton*, *73*, 145–159.

Tsai, C. W., Van der Weele, C. M., & Wolniak, S. M. (2004). Differential segregation and modification of mRNA during spermiogenesis in *Marsilea vestita*. *Developmental Biology*, *269*, 319–330.

Tsuji, H., Aya, K., Ueguchi-Tanaka, M., Shimada, Y., Nakazono, M., Watanabe, R., et al. (2006). GAMYB controls different sets of genes and is differentially regulated by microRNA in aleurone cells and anthers. *The Plant Journal*, *47*, 427–444.

Twell, D. (1995). Diphtheria toxin-mediated cell ablation in developing pollen: Vegetative cell ablation blocks generative cell migration. *Protoplasma*, *187*, 144–154.

Twell, D. (2011). Male gametogenesis and germline specification in flowering plants. *Sexual Plant Reproduction*, *24*, 149–160.

Twell, D., Park, S. K., Hawkins, T. J., Schubert, D., Schmidt, R., Smertenko, A., et al. (2002). *MOR1/GEM1* has an essential role in the plant-specific cytokinetic phragmoplast. *Nature Cell Biology*, *4*, 711.

Twell, D., Park, S. K., & Lalanne, E. (1998). Asymmetric division and cell-fate determination in developing pollen. *Trends in Plant Science*, *3*, 305–310.

Ueda, K., Kinoshita, Y., Xu, Z.-J., Ide, N., Ono, M., Akahori, Y., et al. (2000). Unusual core histones specifically expressed in male gametic cells of *Lilium longiflorum*. *Chromosoma*, *108*, 491–500.

Ueda, K., & Tanaka, I. (1994). The basic proteins of male gametic nuclei isolated from pollen grains of *Lilium longiflorum*. *Planta*, *192*, 446–452.

Umen, J. G. (2014). Green algae and the origins of multicellularity in the plant kingdom. *Cold Spring Harbor Perspectives in Biology*, *6*, a016170.

Verelst, W., Saedler, H., & Münster, T. (2007). MIKC* MADS-protein complexes bind motifs enriched in the proximal region of late pollen-specific Arabidopsis promoters. *Plant Physiology*, *143*, 447–460.

Verelst, W., Twell, D., de Folter, S., Immink, R., Saedler, H., & Münster, T. (2007). MADS-complexes regulate transcriptome dynamics during pollen maturation. *Genome Biology*, *8*, 249.

von Besser, K., Frank, A. C., Johnson, M. A., & Preuss, D. (2006). Arabidopsis *HAP2 (GCS1)* is a sperm-specific gene required for pollen tube guidance and fertilization. *Development*, *133*, 4761–4769.

Wallace, S., Fleming, A., Wellman, C. H., & Beerling, D. J. (2011). Evolutionary development of the plant spore and pollen wall. *AoB PLANTS*, *2011*, plr027.

Weinberg, R. A. (1995). The retinoblastoma protein and cell cycle control. *Cell, 81*, 323–330.

Wellman, C. H. (2010). The invasion of the land by plants: When and where? *New Phytologist, 188*, 306–309.

Whittington, A. T., Vugrek, O., Wei, K. J., Hasenbein, N. G., Sugimoto, K., Rashbrooke, M. C., et al. (2001). *MOR1* is essential for organizing cortical microtubules in plants. *Nature, 411*, 610.

Williams, J. H. (2008). Novelties of the flowering plant pollen tube underlie diversification of a key life history stage. *Proceedings of the National Academy of Sciences of the United States of America, 105*, 11259–11263.

Williams, J. H. (2012). Pollen tube growth rates and the diversification of flowering plant reproductive cycles. *International Journal of Plant Sciences, 173*, 649–661.

Williams, M. L. K., Sawada, A., Budine, T., Yin, C., Gontarz, P., & Solnica-Krezel, L. (2018). *Gon4l* regulates notochord boundary formation and cell polarity underlying axis extension by repressing adhesion genes. *Nature Communications, 9*, 1319.

Williams, J. H., Taylor, M. L., & O'Meara, B. C. (2014). Repeated evolution of tricellular (and bicellular) pollen. *American Journal of Botany, 101*, 559–571.

Wodniok, S., Brinkmann, H., Glöckner, G., Heidel, A. J., Philippe, H., Melkonian, M., et al. (2011). Origin of land plants: Do conjugating green algae hold the key? *BMC Evolutionary Biology, 11*, 104.

Wykes, S. M., & Krawetz, S. A. (2003). The structural organization of sperm chromatin. *Journal of Biological Chemistry, 278*, 29471–29477.

Xu, H., Swoboda, I., Bhalla, P. L., & Singh, M. B. (1999). Male gametic cell-specific expression of H2A and H3 histone genes. *Plant Molecular Biology, 39*, 607–614.

Yamaoka, S., Nishihama, R., Yoshitake, Y., Ishida, S., Inoue, K., Saito, M., et al. (2018). Generative cell specification requires transcription factors evolutionarily conserved in land plants. *Current Biology, 28*, 479–486.

Yanagimachi, R., Wakayama, T., Kishikawa, H., Fimia, G. M., Monaco, L., & Sassone-Corsi, P. (2004). Production of fertile offspring from genetically infertile male mice. *Proceedings of the National Academy of Sciences of the United States of America, 101*, 1691–1695.

Yu, H.-S., & Russell, S. D. (1992). Male cytoplasmic diminution and male germ unit in young and mature pollen of *Cymbidium goeringii*: A 3-dimensional and quantitative study. *Sexual Plant Reproduction, 5*, 169–181.

Zhang, J., Huang, Q., Zhong, S., Bleckmann, A., Huang, J., Guo, X., et al. (2017). Sperm cells are passive cargo of the pollen tube in plant fertilization. *Nature Plants, 3*, 17079.

Zhao, X., Harashima, H., Dissmeyer, N., Pusch, S., Weimer, A. K., Bramsiepe, J., et al. (2012). A general G1/S-phase cell-cycle control module in the flowering plant *Arabidopsis thaliana. PLoS Genetics, 8(8)*, e1002847.

Zheng, B., Chen, X., & McCormick, S. (2011). The anaphase-promoting complex is a dual integrator that regulates both MicroRNA-mediated transcriptional regulation of cyclin B1 and degradation of cyclin B1 during Arabidopsis male gametophyte development. *The Plant Cell, 23*, 1033–1046.

Zheng, B., He, H., Zheng, Y., Wu, W., & McCormick, S. (2014). An ARID domain-containing protein within nuclear bodies is required for sperm cell formation in *Arabidopsis thaliana. PLoS Genetics, 10*, e1004421.

Zhuang, T., Hess, R. A., Kolla, V., Higashi, M., Raabe, T. D., & Brodeur, G. M. (2014). *CHD5* is required for spermiogenesis and chromatin condensation. *Mechanisms of Development, 131*, 35–46.

Zobell, O., Faigl, W., Saedler, H., & Münster, T. (2010). MIKC* MADS-box proteins: Conserved regulators of the gametophytic generation of land plants. *Molecular Biology and Evolution, 27*, 1201–1211.

Further reading

An, R., Liu, X., Wang, R., Wu, H., Liang, S., Shao, J., et al. (2014). The over-expression of two transcription factors, *ABS5/bHLH30* and *ABS7/MYB101*, leads to upwardly curly leaves. *PLoS One*, *9*, e107637.

Bowman, J. L., Kohchi, T., Yamato, K. T., Jenkins, J., Shu, S., Ishizaki, K., et al. (2017). Insights into land plant evolution garnered from the *Marchantia polymorpha* genome. *Cell*, *171*, 287–304.

Xue, T., Liu, Z., Dai, X., & Xiang, F. (2017). Primary root growth in *Arabidopsis thaliana* is inhibited by the miR159 mediated repression of MYB33, MYB65 and MYB101. *Plant Science*, *262*, 182–189.

CHAPTER TWELVE

Evolution of development of pollen performance

Joseph H. Williams*, John B. Reese
Department of Ecology and Evolutionary Biology, University of Tennessee, Knoxville, TN, United States
*Corresponding author: e-mail address: joewill@utk.edu

Contents

Abstract

With the origin of pollination in ancient seed plants, the male gametophyte ("pollen") began to evolve a new and unique life history stage, the progamic phase, a post-pollination period in which pollen sexual maturation occurs in interaction with sporophyte-derived tissues. Pollen performance traits mediate the timing of the fertilization process, often in competition with other pollen, via the speed of pollen germination, sperm development, and pollen tube growth. Studies of pollen development rarely address the issue of performance or its evolution, which involves linking variation in developmental rates to relative fitness within populations or to adaptations on a macroevolutionary scale. Modifications to the pollen tube pathway and changes in the intensity of pollen competition affect the direction and strength of selection on pollen performance. Hence, pollen developmental evolution is always contextual—it involves both the population biology of pollen reaching stigmas and the co-evolution of sporophytic traits, such as the pollen tube pathway and mating system. For most species, performance evolution generally reflects a wandering history of periods of directional selection and relaxed selection, channeled by developmental limitations, a pattern that favors the accumulation of diversity and redundancy in developmental mechanisms and the genetic machinery. Developmental biologists are focused on finding universal

Current Topics in Developmental Biology, Volume 131
ISSN 0070-2153
https://doi.org/10.1016/bs.ctdb.2018.11.012

mechanisms that underlie pollen function, and these are largely mechanisms that have evolved through their effects on performance. Here, we suggest ways in which studies of pollen performance or function could progress by cross-fertilization between the "evo" and "devo" fields.

1. Introduction

Evolution of development as a field has its roots in comparative life history analyses at the organismal level (Alberch, Gould, Oster, & Wake, 1979; Gould, 1977; McKinney & McNamara, 1991; Stearns, 1992; Wake, 2003). The term "evo-devo" marks a point in the early 1980s when new molecular tools allowed deeper insights into the molecular basis of development, at which point the field became much more reductionist. At the same time, evolutionary studies began to consider developmental constraints or biases (Maynard-Smith et al., 1985), and researchers became cautious about invoking selection and adaptation as the sole process for the spread of such variants (Gould & Lewontin, 1979). Today, evo-devo is a field that broadly seeks a mechanistic understanding of the developmental bases of phenotypic evolution, including phenomena such as drift, developmental biases, plasticity, evolvability, and natural or sexual selection (Abouheif et al., 2014; Hall & Olson, 2003; Hendrikse, Parsons, & Hallgrimsson, 2007; Olson, 2012; West-Eberhard, 2003). In plants, development and life history are inseparable, because viability and fertility are the products of continuous or recurrent developmental processes that take place throughout the life span of an individual. Based on the many books on evo-devo written in the past 30 years, one might think that the major questions in plant evo-devo concern only the evolution of plant sporophytes, despite a rich literature on comparative plant "embryology"—the study of gametophytes and embryos (Johri, 1984; Maheshwari, 1950).

In land plants, gametophytes are multicellular plants that produce gametes by mitosis, whereas sporophytes are multicellular plants that produce spores by meiosis. Plant life history models usually consider the gametophyte "generation" of seed plants to be subsumed within the sporophyte "generation," as a quirk of seed plant sporophyte reproduction—the indirect production of eggs and sperms. It has even been argued that the gametophytes of angiosperms should eventually be lost, such that sporophytes will produce gametes directly as in most animals (Dickinson & Grant-Downton, 2009). Even though gametophyte and sporophyte are best described as phases within a single life cycle (Petersen & Burd, 2017), from an evo-devo and a life history perspective, losing the view of the gametophyte as an organism is a conceptual mistake. Gametophytes are likely to have been the dominant multicellular lifeform in ancestral land plants, and if so, the simple, multicellular body plans and reproductive biology of early land plant sporophytes have likely been co-opted from gametophytic ontogenies (Bowman, Briginshaw, & Florent, 2019; Kenrick, 2017; Niklas & Kutschera, 2010; Pires et al., 2013; Szövényi, Rensing, Lang, Wray, & Shaw, 2010; Szövényi, Waller, & Kirbis, 2019). In all land plants, both sporophytes and gametophytes initiate development from a single cell (zygote and meiospore, respectively) and produce distinct plant bodies that have their own modes of reproduction. Even at the most extreme extent of size reduction and early onset of reproduction known in multicellular land plants, represented by the sexually mature, three-celled male gametophyte dispersed in pollen grains of some 30% of angiosperms (Brewbaker, 1967), the gametophyte retains a birth, growth, free-living stage, reproduction, and death as an independent organism.

The male gametophyte (hereafter "pollen" for ease of use) in all seed plants reproduces only once, and in all but a few species (Friedman, 1993; Russell, 1992) is capable of siring only one embryo. Pollen performance is a developmental process that involves pollen germination, pollen tube growth, and sperm production and release. Sperm transport to the egg occurs by zooidogamy (swimming sperm) in *Ginkgo* and Cycads, and by siphonogamy (sperm delivery by pollen tube) in conifers, Gnetales, and angiosperms. Performance is important because male gametophyte development must occur within the timeframe of egg receptivity relative to the time of pollination, and because male gametophytes often compete for access to eggs.

Pollen performance evolution, especially in light of pollen competition, seems to crop up as an issue every generation. Even before our understanding of fertilization and inheritance, Darwin (1876) observed that outcross pollen tubes reached ovules earlier than self pollen tubes. In the 1900s, a number of studies reported that pollen competition caused biased sex ratios (reviewed in Jones, 1928; Stehlik & Barrett, 2006). Pollen performance mutations also became a topic of study (Buchholz & Blakeslee, 1927, 1930, 1936), and Haldane (1932) considered the theoretical implications of performance mutants under pollen competition. The 1970s saw renewed interest in haploid selection as a driver of sporophyte evolution (Mulcahy, 1971, 1979; Mulcahy & Mulcahy, 1987), whereas sexual selection became a topic that dominated the 1980s (Charnov, 1979; Delph & Ashman, 2006; Skogsmyr & Lankinen, 2002; Snow, 1994; Willson & Burley, 1983).

Interspecific pollen competition was studied as a mechanism for reinforcement of reproductive isolation, and as a potential contributor to sympatric or parapatric speciation (Arnold, 1997; Howard, 1999). Although each era had its favored topics, efforts toward a unified view of the processes shaping the progamic phase—in both animals and plants—were also made (Bernasconi et al., 2004).

From a purely developmental perspective, pollen performance refers to physical traits of an individual male gametophyte that have measurable effects on its ability to complete its life cycle. For an individual gametophyte, performance traits affect either viability up to the time of first (and only) reproduction, or the probability of fertilization. Performance is more commonly equated with competitive ability and measured under competitive conditions. Pollen performance studies rarely measure individuals, but instead cohorts of male gametophytes that can be identified after or during competition, such as pollen from one sporophytic parent or pollen carrying a gametophytic mutation. Evolutionary biologists have been more interested in the relative siring success of parent sporophytes, without reference to individual gametophytes or their specific mechanisms for achieving strong performance. Developmental biologists have been more interested in mechanisms, but without considering how performance evolution might shape these mechanisms or their interpretation. Below we attempt to bring a comparative developmental perspective to the evolution of pollen performance.

2. Pollen competition, a pervasive force

Pollen performance traits evolve in response to changes in the physical and the biophysical environment (which can include maternal and paternal sporophytic traits), but also to changes in the intensity of pollen competition, as shown in Fig. 1. Most pollen performance studies involve comparisons under *in vitro* conditions or using *in vivo* hand-pollinations, in which pollen grains vastly outnumber eggs. It is often assumed that competition has shaped performance, but relative performance is almost never interpreted in light of the likely historical pollination conditions of the study species. Thus, a first step is to assess whether or not performance traits in a study species/group have likely evolved under conditions that promote pollen competition. In an ideal world, plants would produce the optimal number of flowers that could mature seed and export pollen, and stigmas would receive no more or less than the amount of pollen they required for full seed set. In the real

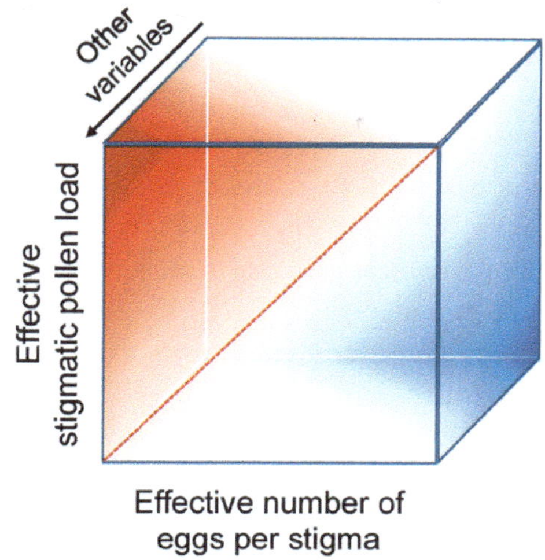

Fig. 1 Factors shaping the intensity of pollen competition. Average stigmatic (or ovular) pollen load and the number of eggs accessible to those pollen grains determine pollen competition intensity (*y*- and *x*-axes, respectively). Other genetically based maternal and paternal sporophytic and male gametophytic traits also affect competition intensity, with higher intensity values in the direction of the arrow on the *z*-axis and listed in Table 1. Pollen competition is more intense in deeper red areas above the red dotted line, and increasingly unimportant in deeper blue areas below the red dotted line.

world, stigmas regularly receive more or less pollen than needed to fertilize all ovules. In theory, the average stigmatic-pollen-load-to-egg (SPL:E) ratio must be >1:1 for pollen performance to evolve via pressure from competition (Fig. 1). Pollen viability is usually <100%, so the effective pollen load on a stigma is usually lower than the pollen load (Thomson, 1989), however, egg (ovule) viability is also usually <100% (Palser, Rouse, & Williams, 1989). The effective pollen load can be measured by assessing *in vivo* pollen germination success or by counting the number of pollen tubes in the style (Alonso, Herrera, & Ashman, 2012; Field, Pickup, & Barrett, 2012; Wagner et al., 2016). Ovule or egg viability is much harder to quantify visually (e.g., Palser et al., 1989), but a minimum can be estimated from the outcrossed-seed-set to ovule ratio after pollination with artificially high pollen loads on healthy plants (Harder & Aizen, 2010; Harder, Hobbhahn, & Richards, 2012).

From a gametophyte perspective, the 1:1 ratio line in Fig. 1 is actually a fuzzy line for two reasons. First, if the average SPL:E ratio is below one but the magnitude of the variance in reproductive success is large, substantial seed siring might still have occurred under competitive conditions (Herrera, 2002). Second, it is commonly found that maximum seed set does not occur unless the SPL:E ratio is substantially above one (Erbar, 2003). Important to note is that the intensity of competition is not the converse of the intensity of "pollen limitation," which is the reduction in seed set that occurs due to insufficient pollen receipt over the lifetime of a sporophyte. From a gametophyte perspective, studies of pollen limitation usually confound the opportunity for competition and the intensity of competition (Snow, 1986a, 1986b). To male gametophytes, siring success always occurs in the context of the average intensity of competition, so only the seed that was actually set is relevant to the evolution of gametophyte performance traits. Hence, from a pollen performance perspective, one should measure the intensity of competition as the average effective SPL:E ratio from flowers that set seed (i.e., after excluding flowers that got no pollen at all; Bernasconi, Lang, & Schmid, 2007). The average seed-to-pollen-load ratio under natural pollination is a measure of the percentage of pollen in natural pollen loads that actually produced seeds (Mitchell, 1997). The lower the percentage, the greater the strength of selection due to pollen competition. Such data is rarely ever reported, but it could be.

In Table 1, we list a number of genetically based traits that can reflect on the historical levels of pollen competition in a particular taxon. Such traits supplement the observed SPL:E ratio, which may or may not reflect the pollination conditions under which the taxon evolved. We consider that both the mean and the variance of pollen-based traits affect competition intensity, since low variances increase the effects of environment and chance in determining outcomes. For example, the relatedness of pollen affects competition intensity because the greater the mean and variance in relatedness in the average pollen load, the greater the opportunity for a response to selection (Mazer et al., 2010). Pollen relatedness on the stigma can be approximated by determining the breeding system, the mating system, or by population genetic measures of historical inbreeding (Table 1).

Pollen competition imposes directional selection on gametophytic traits that increase the speed of pollen germination or pollen tube growth, which can in turn enable sporophytic traits, such as earlier egg maturation and/or a shorter progamic phase to evolve. As Fig. 1 and Table 1 suggest, however, there are many combinations of conditions in which pollen competition does not govern performance evolution. In the absence of pollen competition, performance traits can evolve by neutral processes, by stabilizing selection imposed by a fixed timing of fertilization relative to pollination, or by directional selection imposed solely by sporophytic traits that change the timing of fertilization relative to pollination. Furthermore, species that have

Table 1 Some plant traits hypothesized to affect the intensity of pollen competition.

	Trait or condition	Low/high intensity trait states	Hypothesized cause	Citations
1	Stigmatic pollen load (SPL)	Small/large	For a given ovule number, larger average loads increase SPL:E ratio and variability of competitors on stigma	Snow (1986a), Walsh and Charlesworth (1992), Hormaza and Herrero (1994), Herrera (2002, 2004), Harder, Aizen, and Richards (2016), Burd (2016), refs. in 2
2	Number of eggs (E) accessible to one SPL (=no. ovules in angiosperms)	Many/few	For a given SPL, lower numbers of ovules (eggs) increases the SPL:E ratio	Burd (1995), Burd et al. (2009), Rosenheim, Williams, and Schreiber (2014), Rosenheim, Schreiber, and Williams (2016), and refs. in 1
3	Pollen relatedness on stigma (~Mating system) (~Inbreeding)	Related/unrelated Selfing/outcrossing Inbred/outbred	The less related the pollen, the more variation there is for selection to act on; but see 4	Mazer, Hove, Miller, and Barbet-Massin (2010) and Aizen and Harder (2007)
4	Conflict of interest between pollen and egg parents	Conflict high/low	Greater genetic conflict between pollen and maternal tissue reduces effect of pollen genotype; but see 3	Waser (1993), fig. 9.2
5	Duration of progamic phase (PP) (fertilization delay)	Short/long	Longer PP increases time for small performance differences to take effect (see also 7)	Willson and Burley (1983)
6	Pollen-pistil interactions	Strong/weak	Strong pollen-pistil interactions reduce consistency of male gametophytic effects	Walsh and Charlesworth (1992)
7	Pollen tube pauses/ dormant periods	Non-specific locations/specific locations	Pollen tube pauses in specific locations may indicate maternal strategy to provide "fair start" after staggered germination; dormant phases not in specific locations imply that environment is more a factor than competition	Dahl and Fredrikson (1996), Herrero and Arbeloa (1989), and Marshall and Folsom (1991)

Continued

Table 1 Some plant traits hypothesized to affect the intensity of pollen competition.—cont'd

	Trait or condition	Low/high intensity trait states	Hypothesized cause	Citations
8	Balance of maternal versus paternal interests	Paternal/maternal	Maternal interests favor exposing male genotypes by enhancing competition; paternal interests favor masking male genotypes	Willson and Burley (1983); Otto, Scott, and Immler (2015)
9	Duration of autotrophic relative to heterotrophic phase	Short/long	Competition during the autotrophic phase is less affected by maternal interactions; and when very short, may not allow time for pollen gene expression to be involved	This study, Herrero (2003), Stephenson, Travers, Mena-Ali, and Winsor (2003), and Cruden (2009)
10	Space/resource limitation in style	Low limitation/high limitation	Resource limitation in the pollen tube pathway, whatever its cause, magnifies competitive differences	Heslop-Harrison (1982), Hormaza and Herrero (1994), Erbar (2003), and Harder, Aizen, Richards, Joseph, and Busch (2016)
11	Density-independent (DI) mortality before competition	High/low	DI mortality early in progamic phase reduces number of competing tubes in style	Harder, Aizen, and Richards (2016), Harder, Aizen, Richards, Joseph, et al. (2016)
12	Style length	Short/long	Longer styles allow more time for competitive differences to be expressed	Mulcahy (1979, 1983); but see Lankinen and Skogsmyr (2001), and Cruden (2009)
13	Duration of stigma receptivity	Long/short	Short receptivity evolves when pollination is reliable and sufficient; long receptivity may be a sign of pollen limitation (but see 14)	Willson and Burley (1983)
14	"Delayed" stigma receptivity	Short/long	Stigmatic delay of pollen germination allows greater pollen diversity to accumulate (but see 13)	Willson and Burley (1983), Galen, Shykoff, and Plowright (1986), Skogsmyr and Lankinen (2002), and Goodwillie and Weber (2018)

15	Stigma condition	Wet/dry	Dry stigmas require more specific recognition signals to adhere and germinate, reducing interference from non-competitors	This study
16	Dichogamy	Protogyny/protandry	Similar to 15, protandry reduces interference, so density of outcross pollen is increased	This study, Bertin and Newman (1993)
17	Stigma shape	Elongate/platform	Platform shape gives all pollen grains equal distance to egg; but elongate stigmas might also have evolved to increase pollen loads	Endress (1994), Armbruster (1996) and Friedman and Barrett (2009)
18	Carpel fusion	Apocarpy/syncarpy	For a given *floral* pollen load, syncarpy forces all, rather than subsets, of the pollen load into competition (even though floral PL:E ratio is the same)	Endress (1982) and Armbruster, Debevec, and Willson (2002)
19	Pollination vector	Wind/animal	Animals transfer higher pollen loads at one time, wind transfers pollen sequentially	Mulcahy (1979)
20	Pollen arrival schedule	Staggered/simultaneous	Simultaneous arrival provides fair start	Willson and Burley (1983) and Lankinen and Madjidian (2011)
21	Pollen germination schedule	Staggered/simultaneous	Simultaneous germination ensures all competitors get a fair start	Thomson (1989), Hormaza and Herrero (1994)
22	Floral longevity	Long/short	Longer-lived flowers must present longer-lived pollen, which increases variance in germination times (see 23 and 27)	This study; Wagner, Lechleitner, and Hosp (2016) and Arroyo, Pacheco, and Dudley (2017)
23	Degree of pollen dehydration (pollen water content)	Dehydrated/hydrated	Pollen with high water content hydrates on stigma faster and simultaneously; variance in germination is lower and better reflects genotype differences	Dafni and Firmage (2000), Nepi, Franchi, and Pacini (2001), and Franchi, Nepi, Dafni, and Pacini (2002)

Continued

Table 1 Some plant traits hypothesized to affect the intensity of pollen competition.—cont'd

	Trait or condition	Low/high intensity trait states	Hypothesized cause	Citations
24	Pollen cell number	2/3	Tricellular pollen is more metabolically active at pollination, reducing variance in germination times	Hoekstra (1983), Nepi et al. (2001), and Franchi et al. (2002)
25	Pollen sexual maturity at dispersal	Mature/immature	The more developmental processes shifted to pre-pollination phase, the fewer gametophytically expressed genes exposed to competition or female choice	This study, Williams and Brown (2018)
26	Pollen or pollen tube size	Large/small	Larger pollen takes up more space in anther and on stigma or in style, reducing SPL:E ratio Larger pollen tubes may grow more slowly	This study, Scribailo and Barrett (1991a, 1991b), Sarkissian and Harder (2001) Reese and Williams (2018)
27	Pollen longevity	Long/short	Longer-lived pollen is more dehydrated and metabolically inactive, reducing competition due to staggered germination	Dafni and Firmage (2000)
28	Ploidy of pollen	Polyploid/haploid Polyploid/haploid	Polyploidy can mask expression of deleterious alleles for performance Polyploidy can increase cell dimensions resulting in smaller SPLs and pollen tube densities	Husband (2016) and Otto et al. (2015) This study, see also 26

Intensity is determined first by the stigmatic-pollen-load-to-egg (SPL:E) ratio (*y*- and *x*-axes in Fig. 1) and then by other factors (*z*-axis in Fig. 1), such as pollen quality and the nature of floral organs and tissues with which the male gametophyte interacts (traits and references not comprehensive).

evolved fast germination or pollen tube growth rates (PTGRs) in the past might now be evolving under relaxed selection (Lahti et al., 2009), after a shift in one or more traits that weaken competition, such as transitions from outcrossing to selfing or from lower to higher ploidy (Table 1).

From a cell biological perspective, the angiosperm vegetative (tube) cell, from pollen germination to tube growth and double fertilization, may be one of the most studied of all plant cell types. In evo-devo studies, however, the question shifts from "how does it work?" to "how does functional variation originate, become fixed, and persist in lineages?" Studies of pollen development and function in model systems would do well to consider the degree to which pollen competition has shaped performance traits in nature (Fig. 1; Table 1).

Arabidopsis thaliana has become a model for studying mechanistic aspects of pollen tube germination and growth, with some attention given to performance mutants (e.g., Albert et al., 2018; A. Edlund, unpublished; Prieu et al., 2016; Qin, Ting, Shieh, & McCormick, 2012; Swanson, Hammond, Carlson, Gong, & Donovan, 2016). Performance mutants are defined as having lower fertilization success under competition than without competition, as opposed to mutants that do not affect fertilization success or that cause 100% attrition (cell death). Yet, *A. thaliana* exhibits a number of traits that indicate weak pollen competition, such as being highly selfing, a low effective SPL:E due to sequential self-pollination and many ovules/stigma, and variable germination speed due to pollen dehydration and variable aperture formation (Edlund et al., 2017, 2016). Though *A. thaliana* is often touted as a model for "very rapid [pollen tube] growth rates" (e.g., Reinders, 2016), when compared to other angiosperms, it actually has quite slow PTGR (Williams, 2012a). The highly inbred mating system of *A. thaliana* does offer one advantage: fewer genes can contribute to variation in competitive ability and, hence, genes for competitive ability might be easier to find (Carlson, Gerald, Telligman, Roshanmanesh, & Swanson, 2011). A major point to consider is that shifts to selfing relax directional selection on performance, and reduced performance is likely to evolve via a diversity of causes, making the "model" less generalizable. Comparative developmental biologists have a golden opportunity to compare *A. thaliana* with other diploid and polyploid outcrossing and selfing relatives (Arunkumar, Josephs, Williamson, & Wright, 2013; Fulgione & Hancock, 2018; Gossmann, Saleh, Schmid, Spence, & Schmid, 2016; Koch, Wernisch, & Schmickl, 2008).

On the opposite end of the spectrum, pollen competition has likely been important in the evolution of *Zea mays* (maize) (Mulcahy, 1971;

Valdivia, Wu, Li, Cosgrove, & Stephenson, 2007), a good model taxon for exceptionally rapid pollen germination and with PTGRs up to 100 × faster than in *A. thaliana* (Dresselhaus, Lausser, & Márton, 2011; Mól, Filek, Dumas, & Matthys-Rochon, 2004; Wu, Diggle, & Friedman, 2011). Maize is highly outcrossed (Sleper & Poehlman, 2006), with an SPL:E ratio ≥13 (Ottaviano, Gorla, Frova, & Pe, 1988), a longer duration of pollen tube growth within perhaps the longest known transmitting tract in angiosperms, and with high production of fully hydrated pollen grains (Aylor, Baltazar, & Schoper, 2005) and one ovule per stigma—all indicators of pollen competition. Its pollen performance speeds were likely under selection during the domestication process, due to lengthening of the silks (styles) to accommodate elongation of the inflorescence relative to that of its wild, sub-specific relative. Other cereal grasses also have rapid pollen germination and tube growth (Heslop-Harrison, 1979), but some of these do not exhibit conditions for intense competition, due to inbred mating systems (Dresselhaus et al., 2011) and to very short styles. On the other hand, in *Oryza sativa* (rice), fertilization occurs so rapidly (within 15–30 min) that performance is determined entirely by paternal/gametophytic transcripts and proteins present in mature pollen (Russell et al., 2012; Wei et al., 2010).

These few examples illustrate how little we know about the evolution of pollen developmental mechanisms in the context of performance, despite a wealth of well-developed model systems (Chang, Bowman, & Meyerowitz, 2016; Higashiyama & Inatsugi, 2006). Comparative developmental studies of performance in non-model systems are mostly at the phenotypic level, the level at which fitness via differential performance is realized. We argue below that understanding the nature of (developmental) phenotypic change in performance variation can provide new and functionally relevant phenotypic targets for developmental genetic studies in model systems.

3. Consequences of microgametogenesis for pollen longevity and germination speed

From a life history perspective, pollen ontogeny occurs in three phases: growth and development within the anther (pre-dispersal), presentation and dispersal (free-living stage), and post-pollination growth and development (progamic phase) (Fig. 2). An embryologist or geneticist might divide these differently, into microsporogenesis and microgametogenesis, without respect to variation in life history stages.

As shown in Fig. 2, pollen can be dispersed in a variety of developmental stages and metabolic states, a parallel to the great diversity in seed dispersal states (Baskin & Baskin, 2014; Franchi et al., 2011). A common-sense hypothesis, with some support from comparative studies (Franchi et al., 2002; Williams, Taylor, & O'Meara, 2014, unpublished), is that ancestral seed plant pollen was dispersed in a sexually immature and dehydrated, metabolically inactive state (Fig. 2). Subsequent developmental modifications have enabled extreme reduction of the gametophyte to one vegetative cell plus two sperm cells in a few conifers and in all angiosperms (Friedman, 1993; Rudall & Bateman, 2007). Within angiosperms, there has been repeated evolution of early sexual maturity of pollen, such that the two sperm cells are formed before pollen dispersal (e.g., "tricellular pollen"; Brewbaker, 1967; Williams et al., 2014). There have also been many origins of metabolically active pollen with high water content in phylogenetically derived taxa (Franchi et al., 2002; Hoekstra & Bruinsma, 1975).

Pollen dehydration is an active process in which one or more mechanisms operate to protect cellular structures during gradual desiccation in

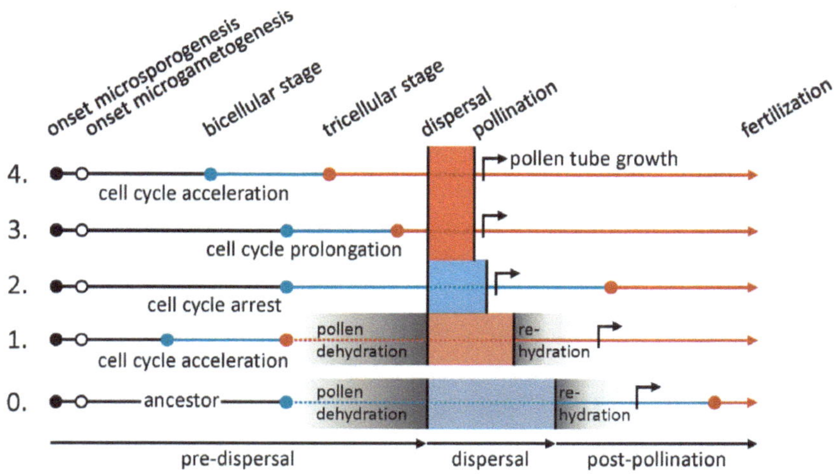

Fig. 2 Potential pathways to the evolution of tricellular pollen. Shown at bottom is the likely ancestral male gametophytic ontogeny (0), in which pollen undergoes dehydration in a bicellular state, enabling a long period of presentation/dispersal, which is followed by a long pollen germination period and division of the generative cell during pollen tube growth. Four potential derived states are indicated. (1) Pollen could achieve the tricellular state before dehydration causes developmental arrest. (2–4) Pollen dehydration could be lost, and either the bicellular stage is retained (2), or the tricellular condition is reached via either prolongation (3) or acceleration (4) of development before dispersal. Black circles, initial microspore stage; blue, bicellular stage; red, tricellular stage.

the anther (Firon, Nepi, & Pacini, 2012; Hoekstra & Golovina, 1999; Hoekstra, Golovina, & Buitink, 2001; Lin & Dickinson, 1984). Dehydration prolongs pollen lifespan (Hoekstra et al., 2001), but also imposes a penalty on pollen germination speed and hence competitive ability (Dafni & Firmage, 2000; Fig. 2). The evolutionary loss of controlled pollen dehydration can occur in favorable dispersal environments, such as in highly humid conditions or when pollen dispersal is reliably rapid (even if inefficient, as in many wind-pollinated species). Dispersal of hydrated pollen is said to confer greater competitive ability via its effects on pollen germination speed (Fig. 2), but this has never been tested in lines with different capacities for desiccation.

Why have there been so many transitions to extreme early sexual maturity in angiosperm pollen (and so few reversions)? If pollen dispersal states evolve along a longevity-to-competitive-ability trade-off axis, as often hypothesized, then early onset of sexual maturity would be selected when the pollination context begins to favor short lifespan and/or high competitive ability. It is well known that pollen with high water content and in a terminally differentiated state (i.e., with the sperm already formed) are difficult to store and have short lifespans (Dafni & Firmage, 2000). Sexual maturity and water content were hypothesized to be developmentally linked traits (Hoekstra & Bruinsma, 1975), because sexually mature, tricellular pollen had high metabolic activity (implying high water content), whereas bicellular pollen was quiescent and likely dehydrated. It was pointed out that the correlation must be weak, since bicellular-hydrated and tricellular-dehydrated pollen are relatively common in nature (Franchi et al., 2002; Nepi et al., 2001). But a recent study found that tricellular pollen had a 30% higher hydration index (an indicator of relative water content) than bicellular pollen, after accounting for the size of pollen (larger pollen has a smaller surface area to volume ratio and had higher inferred water content) (Williams & Brown, 2018). Furthermore, a literature survey of pollen water content in 344 seed plant species found that bicellular-dehydrated and tricellular-hydrated combinations were over-represented in the data, as predicted if the traits were developmentally linked and subject to the longevity-competitive ability trade-off (Williams & Brown, 2018). Given that high pollen water content can evolve rapidly (Franchi et al., 2002), whereas pollen cell number changes have been much more conservative (Williams et al., 2014), it was hypothesized that a likely route to the developmental evolution of tricellular pollen involves first acquiring the hydrated state (Williams & Brown, 2018). The gain of the hydrated state removes a potential cell cycle block (dehydration) and prolongs metabolic activity, using up stored energy reserves, which imposes a shorter lifespan on pollen and leaves it prone to being invaded by mutations that enable cell cycle acceleration or prolongation to the tricellular stage in the anther (the pathway from sequence 0 to 2 to 3/4 in Fig. 2). It would be important to discover the role of pollen dehydration relative to other mechanisms for regulating the cell cycle of the generative cell, in pollen in the anther and on the stigma (Higashiyama & Inatsugi, 2006; Velappan, Signorelli, & Considine, 2017).

Tricellular pollen has long been described as advantageous, based mostly on *in vitro* studies showing it was resistant to premature germination in moist environments but germinated rapidly on very specific stigmatic surfaces (Brewbaker, 1967; Hoekstra, 1983; Mulcahy & Mulcahy, 1983). A survey of *in vivo* germination speeds is consistent with the prediction that tricellular pollen has faster germination than bicellular pollen (Fig. 3). The average (backtransformed) germination time for tricellular pollen was 8.5 min versus 52.4 min for bicellular pollen ($P < 0.0001$). Notably, germination time was also more variable in bicellular pollen, ranging from 1 to 5760 min versus 1 to 120 min for tricellular pollen. The physiological

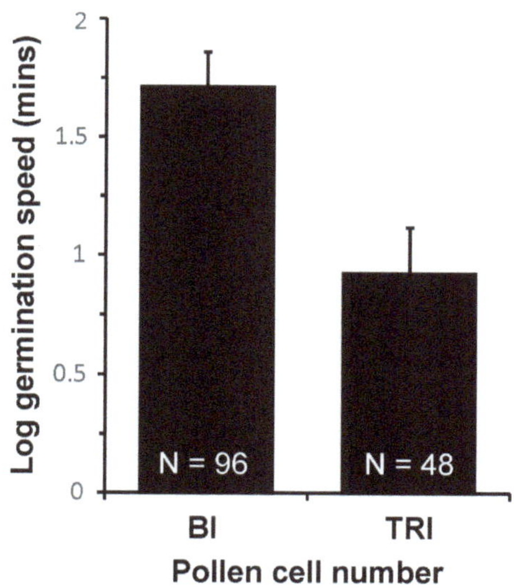

Fig. 3 Tricellular pollen germination is faster than that of bicellular pollen. Log_{10} pollen germination speed (time from pollination until pollen tube emergence) from $N=96$ genera with bicellular (BI) pollen and $N=48$ for genera with tricellular (TRI) pollen. Values of species within the same genus were averaged ($N=127$ and 55 species, respectively). Gymnosperm pollen (not shown), which is also sexually immature at the time of pollination, has strikingly longer germination times (mean $=48 \pm 16$ days; $N=16$ genera [29 species]). Error bars represent 95% confidence intervals.

explanation for the longer lag time in bicellular pollen is thought to be due to later timing of its autotrophic to heterotrophic transition (Mulcahy & Mulcahy, 1982, 1983; but see Russell et al., 2012), though actually very few studies have ever measured the transition timing directly.

In fact, the mechanisms for rapid germination of tricellular pollen are largely due to the correlation with high water content (Franchi et al., 2002, 2011). Tricellular-hydrated pollen (sequence 3 and 4 in Fig. 2) skips the re-hydration process, the re-activation of metabolic machinery, and the need to support the generative cell cycle to form sperm after germination. Species with tricellular-dehydrated pollen (sequence 1 in Fig. 2), such as *A. thaliana*, only have the latter advantage of being fully differentiated at dispersal. Thus, it would be important to compare the relative contributions of water content and metabolic activity versus sexual differentiation to post-pollination performance, starting with germination speed. Does the tricellular condition actually result in greater competitive ability? This is difficult to study, since intra-specific variation in pollen cell number usually reflects variation in the degree of pollen maturation, not genetically based polymorphism (Lora, Herrero, & Hormaza, 2012; Williams et al., 2014).

The evolutionary developmental interpretation of the rapid germination-tricellular state association is not so clear. Is the target of selection competitive ability or is it rapid development, irrespective of competition? It is striking that short-lived, tricellular pollen is almost always associated with sporophytes with the shortest progamic phases (Fig. 4). This is especially true of annuals, ephemerals, and other ruderal species with rapid reproductive cycles. As pointed out in Fig. 2, the duration of the progamic phase is set by the timing of development of stigma, transmitting tract and ovule, sporophytic tissues that must develop rapidly in sporophytes with extremely accelerated life cycles (Aarssen, 2000; Guerrant, 1988). Thus, an alternative, not mutually exclusive, hypothesis is that the tricellular condition has arisen repeatedly due to shifts in sporophyte life cycles that select for rapid reproduction, irrespective of pollen competition intensity.

Species with progamic phases longer than 3 days rarely have tricellular pollen (Fig. 4). Assuming pollen germination competition depends only on the pollination conditions on the stigma (i.e., SPL:E ratio and pollen quality) and selects for rapid germination, then one should expect to see tricellular pollen originate whenever pollen competition is strong, across all types of life histories. That tricellular pollen is almost never found in plants with long reproductive cycles or even long flower longevities suggests its advantages are only present when the reproductive cycle is short, or that competitive ability is never important in species with long progamic phases.

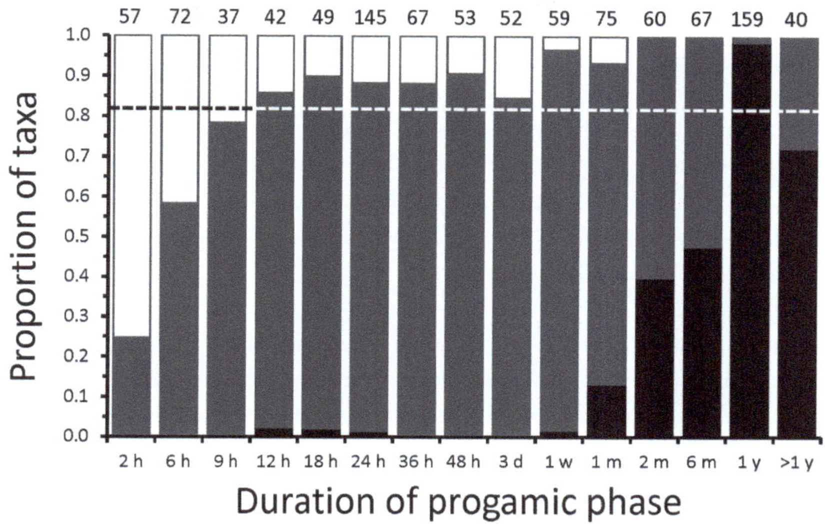

Fig. 4 Tricellular pollen is more frequent in species with short progamic phases. Dark bars indicate the proportion of taxa with sexually immature pollen in angiosperms (gray) and gymnosperms (black), whereas the remaining white bar indicates the proportion of taxa with sexually mature, tricellular pollen. The dashed line indicates the proportion of the total dataset (0.82) with sexually immature pollen (i.e., without sperm at anthesis). Taxa were sorted into relatively even bins according to the duration of their progamic phase, which ranges from <2 h to >12 months (sample sizes for each bin at top).

The latter seems unlikely at first glance, but major groups with long progamic phases, such as the Fagales, Orchidaceae, and gymnosperms, are known for having extended stigmatic (or ovular) receptivity, an indicator of chronic pollen limitation, which implies low SPL:E ratios and weak pollen competition (Table 1).

In summary, angiosperm pollen displays an intriguing pattern of repeated shifts to earlier sexual maturity (Williams et al., 2014) and higher water content (Franchi et al., 2002), and both are associated with exceptionally rapid germination speeds. Evo-devo studies of these convergent or parallel origins of faster performance provide a clear window into the origin and maintenance of developmental mechanisms and their linkage to function and fitness.

4. Pollen tube growth rates and the evolution of "accelerated" and "delayed" fertilization

Pollen germination typically takes up <8% of the duration of the progamic phase, whereas the period of pollen tube growth determines >92% of its duration (Williams, 2012b). Pollen germination is a function of both hydration speed and the speed at which the pollen tube tip can be organized to form a tube that emerges through the pollen wall (Heslop-Harrison & Heslop-Harrison, 1992; Hoedemaekers et al., 2015). The latter process involves cell wall modification and extension, which indicates germination speed and PTGR have some physiological correlation (Addicott, 1943). For example, among 183 angiosperms there are no examples of slow germinators (≥3 h) with fast PTGRs (≥median of 625 μm/h) (Williams, 2012a, 2012b, unpublished).

Angiosperm *in vivo* PTGRs are orders of magnitude faster than elongation rates of any other type of green plant tip-growing cell, including rhizoids, protonema, root hairs, and gymnosperm pollen tubes (Williams, Edwards, & Ramsey, 2016). Among these, only angiosperm pollen tubes function in a competitive context. It has been argued that pollen tube competition is the primary causal force driving the pattern of exceptionally rapid PTGRs in angiosperms (Mulcahy, 1979).

Gymnosperm pollen tubes are quite different in cell wall structure and cytology from those of angiosperms (Derksen, Li, Knuiman, & Geurts, 1999; Fernando, Lazzaro, & Owens, 2005; Mollet, Leroux, Dardelle, & Lehner, 2013), which might have constrained their ability to evolve similarly fast rates (Williams, 2008). Alternatively, gymnosperms might have the developmental potential for fast PTGR but no selection for fast growth. It has long been argued that this is so—gymnosperm ovules can receive very few pollen grains and these compete for more than one egg, indicating a very low pollen-load-to-egg ratio (Haig, 1992).

If the evolution of PTGRs beyond the slow values reported for gymnosperms was driven by pollen competition prior to or early in extant angiosperm history (Mulcahy, 1979), then one prediction is that species from ancient angiosperm lineages, or early-angiosperm fossils, would display reproductive traits that intensify pollen competition (Table 1). Hereafter, we use the arbitrary term "ancient angiosperms" to refer to the grade of five extant lineages (Amborellales, Nymphaeales, Austrobaileyales, and Chloranthales plus Magnoliidae) that diverge from the deepest nodes of the angiosperm phylogenetic tree, prior to the monocot-eudicot split (APG IV, 2016; Magallón, Gómez-Acevedo, Sánchez-Reyes, & Hernández-Hernández, 2015). Two traits that intensify competition are the carpel-derived stigma as a pollen reception surface and the reduction of egg number per ovule to one. On the other hand, ancient angiosperms carpels have very short styles (Williams, 2008), self-compatibility is the rule (Allen & Hiscock, 2008; Endress, 2001), wind or wind/insect pollination is common

(Thien et al., 2009), and flowers have many small stigmas, rather than one large stigma, due to apocarpy (Endress & Doyle, 2009). Importantly, almost all gymnosperms mature their eggs days to months after pollination, whereas ancient angiosperms have evolved eggs that mature just before or very near the time of pollination (Williams, 2009; Willson & Burley, 1983). Shifts to earlier timing of egg receptivity also reduce the intensity of competition (Table 1). These facts suggest that changing sporophyte life histories played a large role in driving faster PTGRs in the lineage leading to extant angiosperms.

Ancient angiosperm pollen tube transmitting tissues are structurally quite diverse (Endress, 2005; Endress & Igersheim, 2000; Frame, 2003; Prychid et al., 2011; Sage et al., 2009; Taylor, MacFarlane, & Williams, 2010; Taylor & Williams, 2009; Williams, McNeilage, Lettre, & Taylor, 2010), yet their PTGRs always exceed those of any gymnosperm (Williams, 2012a), which implies that early angiosperm PTGRs were strongly affected by modifications of the pollen tube. A good case can be made that the critical pollen tube growth rate innovations trace to a combination of (1) the callose wall, synthesized by an ancient *CalS-5* paralog (Abercrombie, O'Meara, Moffatt, & Williams, 2011), which is cheaper and faster to sythesize (Schlüpmann, Bacic, & Read, 1994), and which enables callose plugs to form (gymnosperm pollen tubes do not have callose walls or plugs); and (2) the highly localized action of one or more self-regulating (type II) pectin methylesterase (*PME*) genes, which also encodes one or more co-transcribed PME inhibitors (*PMEI*) that prevent premature rigidification of the pollen tube wall at the tip apex (Chebli, Kaneda, Zerzour, & Geitmann, 2012; Wallace & Williams, 2017). The combination of targeted PME activity and callose synthesis just behind the tip give the pollen tube its hard shell-soft tip morphology (Chebli et al., 2012; Vogler et al., 2013), enabling invasive growth and the maintenance of turgor by a very thin circumferential wall. These and other structural apomorphies (unique, derived features), such as reverse-fountain cytoplasmic streaming and the subapical actin fringe (Stephan, 2017), are present in most angiosperm pollen tubes. Thus, an outstanding question is whether universally shared pollen tube structural innovations were a cause or a consequence of the acceleration of early angiosperm PTGRs.

It is striking that ancient angiosperm pollen tubes always have strongly aniline blue staining walls, whereas species with far faster PTGRs sometimes stain weakly for callose (e.g., *Hibiscus moscheutos*; Snow & Spira, 1991) or not at all, such as in some lamiids, also known as euasterids I (Prósperi &

Cocucci, 1979). Knock-out experiments have shown that loss of either callose or PME activity in pollen tubes causes reduced function, such as disoriented and/or slower growth, but not complete loss of the ability to fertilize (Bosch & Hepler, 2006; Jiang et al., 2005; Nishikawa, Zinkl, Swanson, Maruyama, & Preuss, 2005; Röckel, Wolf, Kost, Rausch, & Greiner, 2008). Reduced function rather than loss of function is predicted if innovation came about, not by replacement, but by modifications or additions to a previously evolved pollen tube growth pattern. Reduced function is also consistent with the observation of great redundancy in both functional energetic and wall construction traits, and in potential pollen tube growth- and PTGR-determining genes (Drábková & Honys, 2017; Geitmann, 2011; Hoedemaekers et al., 2015; Honys, Renak, & Twell, 2006; Mollet et al., 2013; Selinski & Scheibe, 2014). For these reasons, it is likely that the callose cell wall and/or PME/PMEI activity arose and initially functioned within a very different pollen tube pathway context, compared to modern transmitting tracts that now support exceptionally rapid PTGRs. Now that joint pollen tube and pistil gene expression studies are possible (e.g., Boavida, Borges, Becker, & Feijo, 2011; Leydon et al., 2017; Qin et al., 2009), an ancient angiosperm model system would be especially informative (Chen et al., 2017; Povilus, Losada, & Friedman, 2014; Vialette-Guiraud et al., 2011), keeping in mind the ancient, independent histories of these diverse lineages.

A reconstruction of PTGR evolution in seed plants is shown in Fig. 5A (left pane). Though uncertainty in ancestral reconstructions of continuous traits can be large (Cunningham, Omland, & Oakley, 1998; Losos, 2011), newer model-based approaches can test hypotheses without relying on specific ancestral reconstructions (O'Meara, 2012). Such approaches offer several clear take-aways on the pattern of PTGR evolution (Reese & Williams, 2018). First, models in which PTGRs evolved randomly or by Brownian Motion (BM) processes were strongly rejected in favor of Ornstein-Uhlenbeck (OU) models. An OU process indicates PTGRs have evolved under a stabilizing-selection-like effect (in that less among-lineage variation has accumulated than one would expect if PTGRs evolved from an ancestral value following a neutral BM process). Second, closely related species tend to have similar PTGRs (i.e., there is significant phylogenetic signal). Third, PTGRs of angiosperms have evolved around a significantly higher selective optimum relative to those of gymnosperms ($\log_{10}$ PTGR = $2.69 \pm 0.048\,\mu m/h$ versus $0.187 \pm 0.123\,\mu m/h$, respectively). Finally, within angiosperms, a separate analysis found both slow-downs and speed-ups of

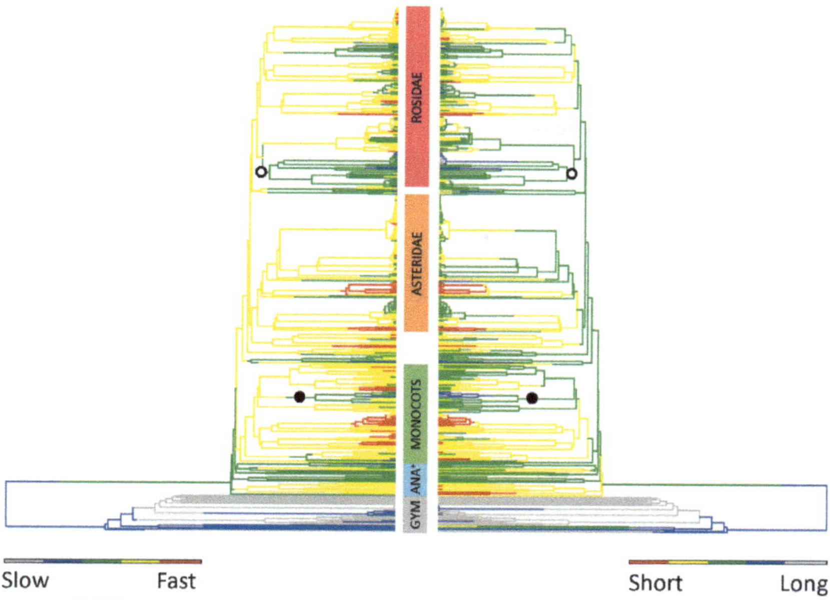

Fig. 5 Phylogenetic distribution of pollen tube growth rate and progamic phase duration. Maximum likelihood reconstructions using continuous character mapping. Back-transformed values of PTGR (μm/h) quintiles are: gray ≤2, blue ≤24, green ≤274, yellow ≤3094, red >3094; and progamic phase (h) quintiles are: red ≤2, yellow ≤17, green ≤136, blue ≤1110, gray >1110. Black dots, Orchidaceae; white dots, Fagales. Abbreviations: GYM = Acrogymnospermae; ANA^{+} = Amborellales, Nymphaeales, Austrobaileyales, Chloranthales, eumagnoliids. Phylogenetic tree for PTGR extracted from Reese and Williams (2018).

PTGR (29 and 22, respectively; not all are apparent in Fig. 5A). These results give a picture of gradual evolution of angiosperm PTGRs by the expansion of variance away from a left wall of very slow growth, inherited from an ancient gymnosperm-like ancestor. An important point for developmental biologists to consider is the issue of context: the pattern of PTGR evolution is not one of monotonically increasing PTGR in angiosperms, but instead one of highly variable histories of slowdowns and speedups among different lineages.

5. The evolution of "delayed fertilization"

Species such as pines and oaks have long been described as having "delayed fertilization," a long period between pollination and the initiation and completion of female gametogenesis (Benson, 1894; Endress, 1977;

Schnarf, 1929; Sogo & Tobe, 2005, 2006). Willson and Burley (1983) called attention to the fact that fertilization delay can be a tactic for accumulation of more pollen on the stigma or in the style, thereby increasing the opportunity for pollen competition or female choice. "Fertilization delay" is a term borrowed from the animal literature (Austin & Short, 1978) that also applies to seed plants, because it refers to the time internal fertilizers take between copulation and fertilization, relative to external fertilizers where copulation is equivalent to gamete fusion. Delay in a comparative context and interpreted as a reproductive strategy (Willson & Burley, 1983) implies the evolution of greater time between pollination and egg receptivity, which means the ancestors of such taxa had shorter progamic phases. Is that true?

The progamic phase ranges from 15 min to well over 1 year in duration, and a comparative analysis indicates many transitions to both shorter and longer cycles (Fig. 5B). Gymnosperms have long progamic phases, with some transitions to shorter cycles, especially with in Gnetales. Although the long reproductive cycles of gymnosperms have almost universally been seen as "delayed" (Willson & Burley, 1983), the phylogenetic pattern suggests that they were ancestrally on the order of months long, followed by transitions to both shorter and longer cycles (Fig. 5B). Fernando (2014) noted that the exceptional year-long progamic phases of conifers likely have a strong genetic basis, because their long developmental sequences are maintained in several derived lineages that have transitioned from temperate to tropical environments (where long winter dormant periods for pollen tubes would seem to be unwarranted). The reconstruction of ancestrally long progamic phase duration is consistent with the fact that gametophytes in extant seed plant outgroups are perennial or take many months between spore dispersal and the formation of sperm in antheridia (Haig & Westoby, 1988; Lloyd & Klekowski, 1970). Thus, we argue caution in interpreting gymnosperm progamic phases as being "delayed," when in fact the opposite may be true. The transition to endospory and pollination in an ancient seed plant ancestor likely occurred from an exosporic, free-living predecessor that required a long time to reach reproductive maturity after spore dispersal.

Irrespective of the most ancient transitions, the pattern in Fig. 5B suggests that within extant seed plants, there has been interplay between the interests of gametophytes and sporophytes, another reason to consider that gametophytes have their own life histories. Plant life history theory indicates that age at reproductive maturity (roughly equivalent to progamic phase duration) evolves according to a trade-off: rapid sexual maturation confers the benefits of less exposure to premature death and a shorter generation time, whereas delayed maturation generally confers larger size and hence of more gametes and more seeds (Lewontin, 1965; Stearns, 1992). Yet, angiosperm male gametophytes reproduce only once with a maximum of one embryo sired, so delaying fertilization cannot result in more offspring, whereas the benefits of early reproduction are always potentially present. Under pollen competition, rapid sexual maturation is even more beneficial. Hence, transitions to long delays must be sporophyte-driven shifts, driven by sexual selection (Willson & Burley, 1983), or by the advantages of partitioning reproductive processes among different seasons (Endress, 1977).

Within angiosperms there have been many "reversions" from ancestrally short to longer progamic phases, i.e., true fertilization delays. These are hardly true reversions since each has a unique developmental sequence (as so clearly demonstrated in Endress, 1977). The best-known examples of large groups with long delays are the oak order, Fagales (Dahl & Fredrikson, 1996; Sogo & Tobe, 2005, 2006), and the orchids (Arditti, 1992; Zhang & O'Neill, 1993). Why have these groups evolved and retained such long periods between pollination and fertilization? One explanation might be that both have long-lived flowers with long periods of stigmatic receptivity and strong maternal control over pollen tube pauses and re-starts, all indicators of extreme efforts to deal with chronic pollen limitation. There are other indicators as well, such as the over-production of female flowers in most wind-pollinated Fagalean taxa (Stephenson, 1981), or the over-production of ovules coupled with pollination by massive pollen aggregates from a single father in many orchids (Arditti, 1992; Harder & Johnson, 2008).

If groups that have evolved long, maternally controlled fertilization delays produce seed under low levels of pollen competition, then relaxed directional selection on PTGR should also cause PTGR slowdowns. A "long" progamic phase could be conservatively defined as being at least 2 weeks long, because in such taxa female gametogenesis has not yet begun at the time of pollination (in angiosperms, female gametogenesis is remarkably consistent in lasting about 6–14 days; Endress, 1977; Palser et al., 1989). For 17 angiosperms with long progamic phases, the median PTGR was 24.1 μm/h and the fastest was 381 μm/h, values that are well below the angiosperm median of 625.0 μm/h ($N=535$ species). These results are consistent with PTGR slowdowns occurring as a consequence of relaxed directional selection on performance speed, but if so, then there must be a cost to fast performance. Alternatively, displacement of egg receptivity to later times might incur greater maternal control, which could directly cause slower PTGR irrespective of pollen competition intensity.

6. The evolution of "accelerated fertilization"

Based on the discussion above, germination speed and PTGR are correlated performance traits, and rapid performance is necessary for very rapid reproductive cycles, whereas very long reproductive cycles seem to be associated with shifts to slower PTGR. In fact, as shown in Fig. 6, there is a strong negative correlation between PTGR and progamic phase duration across all seed plants, and in gymnosperms and angiosperms separately, and with or without correcting for phylogeny. The empty spaces in bottom left and upper right of Fig. 6 might indicate developmental limitations to evolving faster (or slower) growth rate. Does the duration of pollen tube growth impose metabolic limits on the rate of growth? Tricellular pollen is known to have a shorter lifespan than bicellular pollen, and species with tricellular pollen have short durations of growth (Fig. 4). Yet, among the fastest 10% of PTGRs ($N=359$ angiosperms), 44% are from species with bicellular pollen, which presumably is longer-lived. Given heterotrophic

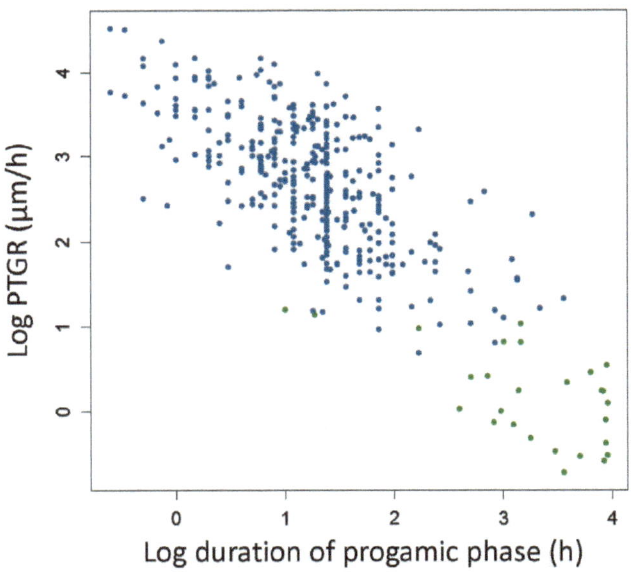

Fig. 6 Relationship between duration of progamic phase and pollen tube growth rate. Angiosperm datapoints are blue; gymnosperms are green. The best model in a Phylogenetic Least Squares analysis for all seed plants showed a significant negative relationship: $\log_{10}$ PTGR $= -0.629x + 2.62$ (Lambda model weight = 74.2%; $P \ll 0.0001$), where x indicates $\log_{10}$ progamic phase duration. For angiosperms-only: $\log_{10}$ PTGR $= -0.628x + 3.28$ (Lambda model weight = 98.5%; $P \ll 0.0001$).

pollen tube growth in both bi- and tricellular pollen, as commonly assumed, maternal resources are more likely to limit PTGR than to limit pollen tube lifespan. A probable scenario is that style longevity limits the duration of growth (e.g., Valdivia, Cosgrove, & Stephenson, 2006), which forces longer styles to evolve faster PTGRs. No species have evolved long styles with slow PTGRs (Williams, 2012a).

The major developmental questions being addressed today are focused on deeply conserved mechanisms of tip growth (Bascom, Hepler, & Bezanilla, 2018; Michard, Simon, Tavares, Wudick, & Feijó, 2017), usually in species with slow to moderate PTGRs, say from ~265 μm/h in *A. thaliana* to ~1200 μm/h in *Lilium longiflorum* (lily). The mechanics of tube growth are still the subject of experimental and modeling studies (Hill, Shachar-Hill, Skepper, Powell, & Shachar-Hill, 2012; Hu et al., 2017; Michard et al., 2017; Van Hemelryck, Bernal, Rojas, Dumais, & Kroeger, 2017; Winship, Obermeyer, Geitmann, & Hepler, 2010, 2011; Zonia & Munnik, 2011), and the energetics underlying PTGR are also not yet fully understood (Colaço, Moreno, & Feijó, 2012; Obermeyer, Fragner, Lang, & Weckwerth, 2013; Rounds, Hepler, Fuller, & Winship, 2010; Rounds, Winship, & Hepler, 2011; Selinski & Scheibe, 2014). One could argue that understanding the basic process of tip-growth must come before we can truly discover what aspects of development and energy use cause changes in growth rates. Still, a major problem faced by experimental biologists is the great redundancy of mechanisms at both the phenotypic and genotypic levels of reproduction. Comparative studies of evolved variation can shed light in areas where mutational studies are difficult to interpret, especially with respect to rate-limiting aspects of growth rate.

Comparative approaches have suggested pollen tube architecture can limit PTGR. For example, growth rates were correlated with the frequency of callose plug deposition among competing pollen tubes from unrelated sporophytes (Snow & Spira, 1991), and with the position of the first callose plug among different *A. thaliana* lines (Qin et al., 2012). Among three water lilies, *Nymphaea odorata* had the smallest pollen tube but the highest wall production rate (WPR) per unit of wall circumference—the combination of fast wall production and small tube size resulted in its faster tip growth rate (Williams et al., 2016). The volume of wall material needed for growth can be modified by changes in wall thickness or tube diameter. In lily, Hu et al. (2017) found that tube diameter was negatively correlated with turgor and wall stiffness, and hence with wall thickness. However, a literature survey found that wall thicknesses in many species were already near their minimum of ~0.15–0.23 μm (as measured on TEM micrographs), whereas tube diameters varied much more, from about ~3 to 23 μm (Williams et al., 2016). If tube circumferential wall thickness is already near the minimal threshold needed to resist tension and compression stress (Hu et al., 2017), then growth efficiency—the relative amount of wall material used to achieve a particular PTGR—must evolve largely through changes in pollen tube diameter. Growth efficiency may not be an issue for the large (~23 μm diameter) pollen tubes of *H. moscheutos*, which produce exorbitant amounts of wall material to achieve a PTGR 10 times faster than pollen tubes of *N. odorata* (~10 μm diameter). In contrast, maize pollen tubes (~7–8 μm diameter) are much more growth efficient and have an even faster PTGR than *H. moscheutos* (Williams et al., 2016).

What kinds of intra-cellular limitations are there to evolving exceptionally rapid PTGR? In lily, Benkert, Obermeyer, and Bentrup (1997) famously showed no correlation between turgor pressure and PTGR, and similarly, Vidali, McKenna, and Hepler (2001) found no correlation between the rate of cytoplasmic streaming and PTGR. Yet, clearly turgor and cytoplasmic streaming rates can evolve, since there is interspecific variation across cells and cell types in both (Benkert et al., 1997; Rounds et al., 2011; Tadege & Kuhlemeier, 1997; Turner, 2018; Vogler et al., 2013). That turgor pressure can evolve is also suggested by the great variation in pollen tube wall design and osmotic and physical conditions of *in vivo* growth. In general, it is thought that rates of vesicle transport must far exceed the rate of tip growth to produce reverse-fountain streaming and the pattern of exocytosis and endocytosis seen in angiosperm pollen tube tips (Geitmann, Wojciechowicz, & Cresti, 1996; Van Hemelryck et al., 2017). At some point, secretion rates of pectins and wall enzymes carried in vesicles must be a bottleneck for rapid growth. Vesicles accumulate at the pollen tube tip due to the action of myosin motors, which travel in a polar direction along rapidly synthesized longitudinally oriented actin cables (Madison & Nebenführ, 2013). Myosin motors have a range of speeds (Ryan & Nebenführ, 2018) and are associated with a broad range of intracellular transport rates of vesicles in pollen tubes (Geitmann & Nebenführ, 2015). Vesicle velocities of between 3600 and 14,400 μm/h have been measured in pollen tubes, but given that vesicle velocity must greatly exceed tip extension velocity (Rounds, Hepler, & Winship, 2014; Vidali et al., 2009), cytoskeleton assembly rates must be much faster in many grasses and asterids that have PTGRs of >20,000 μm/h. The subapical actin fringe is assembled and dissembled to match tip extension rate and comparative studies show that it is a unique aspect of rapid tip-growth in angiosperms (Stephan, 2017). Actin assembly might be rate-limiting for a number of reasons, given its many proposed functions (Bascom et al., 2018; Stephan, 2017; Vidali & Hepler, 2001).

In summary, tremendous variation in PTGR has evolved within angiosperms. Yet, there are few cases of closely related species evolving dramatically different PTGRs, consistent with the fact that a great number of genetically based, internal and external traits contribute to the maintenance of PTGR. Comparative studies can offer insights into the developmental avenues for the evolution of faster or slower PTGRs. As the extremes of fast PTGR are approached, unrelated taxa with long histories of PTGR increases are more likely to reveal universal developmental determinants of growth rate. For all other taxa with slower PTGRs, there are always two possibilities to consider: a particular PTGR evolved from even slower ancestors or from even faster ancestors. Examples of unrelated taxa with equivalent PTGRs that come from lineages with a history of accelerations include many ancient angiosperms and probably many species of annual ephemerals with rapid life cycles. On the other hand, many monocots and eudicots have evolved in lineages with decelerating PTGRs, due to relaxed directional selection on competitive ability and/or increased maternal control. Because there are many ways to reduce performance of a polygenic trait, taxa that have a mixed history of accelerations and decelerations may be less likely to display universal features of PTGR. But comparative studies of taxa that control for historical effects could reveal critical features of PTGR that are not accessible to studies of mutant lines in model species.

Acknowledgments

We thank Ueli Grossnikaus for the invitation to participate in this volume. J.H.W. is grateful to Leonardo Versieux and the Department of Botany and Zoology at Universidade Federal do Rio Grande do Norte, Brazil, for academic support during the writing of this manuscript.

References

Aarssen, L. W. (2000). Why are most selfers annuals? A new hypothesis for the fitness benefit of selfing. *Oikos*, *89*(3), 606–612.

Abercrombie, J. M., O'Meara, B. C., Moffatt, A. R., & Williams, J. H. (2011). Developmental evolution of flowering plant pollen tube cell walls: Callose synthase (*CalS*) gene expression patterns. *EvoDevo*, *2*(1), 14.

Abouheif, E., Fave, M. J., Ibarraran-Viniegra, A. S., Lesoway, M. P., Rafiqi, A. M., & Rajakumar, R. (2014). Eco-Evo-Devo: The time has come. In C. R. Landry & N. AubinHorth (Eds.), *Vol. 781. Ecological genomics: Ecology and the evolution of genes and genomes*. (pp. 107–125).

Addicott, F. T. (1943). Pollen germination and pollen tube growth, as influenced by pure growth substances. *Plant Physiology*, *18*(2), 270–279.

Aizen, M. A., & Harder, L. D. (2007). Expanding the limits of the pollen-limitation concept: Effects of pollen quantity and quality. *Ecology*, *88*(2), 271–281.

Alberch, P., Gould, S. J., Oster, G. F., & Wake, D. B. (1979). Size and shape in ontogeny and phylogeny. *Paleobiology*, *5*, 296–317.

Albert, B., Ressayre, A., Dillmann, C., Carlson, A. L., Swanson, R. J., Gouyon, P.-H., et al. (2018). Effect of aperture number on pollen germination, survival and reproductive success in *Arabidopsis thaliana*. *Annals of Botany*, *121*(4), 733–740.

Allen, A. M., & Hiscock, S. J. (2008). Evolution and phylogeny of self-incompatibility systems in angiosperms. In V. E. Franklin-Tong (Ed.), *Self-incompatibility in flowering plants—Evolution, diversity, and mechanisms* (pp. 73–101). Berlin: Springer-Verlag.

Alonso, C., Herrera, C. M., & Ashman, T. L. (2012). A piece of the puzzle: A method for comparing pollination quality and quantity across multiple species and reproductive events. *New Phytologist*, *193*(2), 532–542.

Angiosperm Phylogeny Group IV. (2016). An update of the Angiosperm Phylogeny Group classification for the orders and families of flowering plants: APG IV. *Botanical Journal of the Linnean Society*, *181*(1), 1–20.

Arditti, J. (1992). *Fundamentals of orchid biology*. New York: Wiley.

Armbruster, W. S. (1996). Evolution of floral morphology and function: An integrative approach to adaptation, constraint, and compromise in *Dalechampia* (Euphorbiaceae). In D. G. Lloyd & S. C. H. Barrett (Eds.), *Floral biology* (pp. 241–272). Boston, MA: Springer.

Armbruster, W. S., Debevec, E. M., & Willson, M. F. (2002). Evolution of syncarpy in angiosperms: Theoretical and phylogenetic analyses of the effects of carpel fusion on offspring quantity and quality. *Journal of Evolutionary Biology*, *15*(4), 657–672.

Arnold, M. L. (1997). *Natural hybridization and evolution*. Oxford, UK: Oxford University Press.

Arroyo, M. T., Pacheco, D. A., & Dudley, L. S. (2017). Functional role of long-lived flowers in preventing pollen limitation in a high elevation outcrossing species. *AoB Plants*, *9*(6), plx050.

Arunkumar, R., Josephs, E. B., Williamson, R. J., & Wright, S. I. (2013). Pollen-specific, but not sperm-specific, genes show stronger purifying selection and higher rates of positive selection than sporophytic genes in *Capsella grandiflora*. *Molecular Biology and Evolution*, *30*(11), 2475–2486.

Austin, C. R., & Short, R. V. (1978). *Reproduction in mammals: Germ cells and fertilization*. London: Cambridge University Press.

Aylor, D. E., Baltazar, B. M., & Schoper, J. B. (2005). Some physical properties of teosinte (*Zea mays* subsp. *parviglumis*) pollen. *Journal of Experimental Botany*, *56*(419), 2401–2407.

Bascom, C. S., Hepler, P. K., & Bezanilla, M. (2018). Interplay between ions, the cytoskeleton, and cell wall properties during tip growth. *Plant Physiology*, *176*(1), 28–40.

Baskin, C. C., & Baskin, J. M. (2014). *Seeds: Ecology, biogeography, and evolution of dormancy and germination* (2nd ed.). . San Diego, CA: Academic Press.

Benkert, R., Obermeyer, G., & Bentrup, F. W. (1997). The turgor pressure of growing lily pollen tubes. *Protoplasma*, *198*(1–2), 1–8.

Benson, M. (1894). Contributions to the embryology of the Amentiferae 1. *Transactions of the Linnaean Society London*, *3*, 409–432.

Bernasconi, G., Ashman, T. L., Birkhead, T. R., Bishop, J. D. D., Grossniklaus, U., Kubli, E., et al. (2004). Evolutionary ecology of the prezygotic stage. *Science*, *303*, 971–975.

Bernasconi, G., Lang, D. J., & Schmid, B. (2007). Microgametophyte population sizes and plant reproductive output in the insect-pollinated *Prunella grandiflora* (Lamiaceae). *New Phytologist*, *173*(2), 393–400.

Bertin, R. I., & Newman, C. M. (1993). Dichogamy in angiosperms. *Botanical Review, 59*, 112–152.
Boavida, L. C., Borges, F., Becker, J. D., & Feijo, J. A. (2011). Whole genome analysis of gene expression reveals coordinated activation of signaling and metabolic pathways during pollen-pistil interactions in *Arabidopsis*. *Plant Physiology, 155*(4), 2066–2080.
Bosch, M., & Hepler, P. K. (2006). Silencing of the tobacco pollen pectin methylesterase *NtPPME1* results in retarded in vivo pollen tube growth. *Planta, 223*, 736–745.
Bowman, J. L., Briginshaw, L. N., & Florent, S. N. (2019). Evolution and co-option of developmental regulatory networks in early land plants. In U. Grossniklaus (Ed.), *Current topics in developmental biology*: *Vol. 131*. Plant development and evolution (pp. 35–53). Elsevier.
Brewbaker, J. L. (1967). Distribution and phylogenetic significance of binucleate and trinucleate pollen grains in angiosperms. *American Journal of Botany, 54*(9), 1069–1083.
Buchholz, J. T., & Blakeslee, A. F. (1927). Abnormalities in pollen-tube growth in *Datura* due to the gene "*tricarpel*" *Proceedings of the National Academy of Sciences of the United States of America, 13*, 242–249.
Buchholz, J. T., & Blakeslee, A. F. (1930). Pollen tube growth and control of gametophytic selection in cocklebur, a 25-chromosome *Datura*. *Botanical Gazette, 90*, 0366–0383.
Buchholz, J. T., & Blakeslee, A. F. (1936). Genes from radium treatment affecting pollen tube growth in *Datura*. *Genetics, 21*, 731–751.
Burd, M. (1995). Ovule packaging in stochastic pollination and fertilization environments. *Evolution, 49*(1), 100–109.
Burd, M. (2016). Pollen limitation is common—Should it be? (A comment on Rosenheim et al., "parental optimism versus parental pessimism in plants: How common should we expect pollen limitation to be?"). *American Naturalist, 187*(3), 388–396.
Burd, M., Ashman, T. L., Campbell, D. R., Dudash, M. R., Johnston, M. O., Knight, T. M., et al. (2009). Ovule number per flower in a world of unpredictable pollination. *American Journal of Botany, 96*(6), 1159–1167.
Carlson, A. L., Gerald, J. N. F., Telligman, M., Roshanmanesh, J., & Swanson, R. J. (2011). Defining the genetic architecture underlying female and male mediated nonrandom mating and seed yield traits in *Arabidopsis thaliana*. *Plant Physiology, 157*, 1956–1964.
Chang, C., Bowman, J. L., & Meyerowitz, E. M. (2016). Field guide to plant model systems. *Cell, 167*(2), 325–339.
Charnov, E. L. (1979). Simultaneous hermaphroditism and sexual selection. *Proceedings of the National Academy of Sciences of the United States of America, 76*, 2480–2484.
Chebli, Y., Kaneda, M., Zerzour, R., & Geitmann, A. (2012). The cell wall of the *Arabidopsis* pollen tube—Spatial distribution, recycling, and network formation of polysaccharides. *Plant Physiology, 160*(4), 1940–1955.
Chen, F., Liu, X., Yu, C., Chen, Y., Tang, H., & Zhang, L. (2017). Water lilies as emerging models for Darwin's abominable mystery. *Horticulture Research, 4*, 17051.
Colaço, R., Moreno, N., & Feijó, J. A. (2012). On the fast lane: Mitochondria structure, dynamics and function in growing pollen tubes. *Journal of Microscopy, 247*(1), 106–118.
Cruden, R. W. (2009). Pollen grain size, stigma depth, and style length: The relationships revisited. *Plant Systematics and Evolution, 278*(3–4), 223–238.
Cunningham, C. W., Omland, K. E., & Oakley, T. H. (1998). Reconstructing ancestral character states: A critical reappraisal. *Trends in Ecology & Evolution, 13*(9), 361–366.
Dafni, A., & Firmage, D. (2000). Pollen viability and longevity: Practical, ecological and evolutionary implications. *Plant Systematics and Evolution, 222*, 113–132.
Dahl, A. E., & Fredrikson, M. (1996). The timetable for development of maternal tissues sets the stage for male genomic selection in *Betula pendula* (Betulaceae). *American Journal of Botany, 83*, 895–902.

Darwin, C. (1876). *The effects of cross and self fertilisation in the vegetable kingdom*. London: J. Murray.
Delph, L. F., & Ashman, T.-L. (2006). Trait selection in flowering plants: How does sexual selection contribute? *Integrative and Comparative Biology*, *46*(4), 465–472.
Derksen, J., Li, Y. Q., Knuiman, B., & Geurts, H. (1999). The wall of *Pinus sylvestris* L. pollen tubes. *Protoplasma*, *208*(1–4), 26–36.
Dickinson, H. G., & Grant-Downton, R. (2009). Bridging the generation gap: Flowering plant gametophytes and animal germlines reveal unexpected similarities. *Biological Reviews*, *84*(4), 589–615.
Drábková, L. Z., & Honys, D. (2017). Evolutionary history of callose synthases in terrestrial plants with emphasis on proteins involved in male gametophyte development. *PLoS One*, *12*(11), e0187331.
Dresselhaus, T., Lausser, A., & Márton, M. L. (2011). Using maize as a model to study pollen tube growth and guidance, cross-incompatibility and sperm delivery in grasses. *Annals of Botany*, *108*(4), 727–737.
Edlund, A. F., Olsen, K., Mendoza, C., Wang, J., Buckley, T., Nguyen, M., et al. (2017). Pollen wall degradation in the Brassicaceae permits cell emergence after pollination. *American Journal of Botany*, *104*(8), 1266–1273.
Edlund, A. F., Zheng, Q., Lowe, N., Kuseryk, S., Ainsworth, K. L., Lyles, R. H., et al. (2016). Pollen from *Arabidopsis thaliana* and other Brassicaceae are functionally omniaperturate. *American Journal of Botany*, *103*(6), 1006–1019.
Endress, P. K. (1977). Evolutionary trends in the Hamamelidales-Fagales-group. *Plant Systematics and Evolution*, (1), 321–347. Suppl. 1.
Endress, P. K. (1982). Syncarpy and alternative modes of escaping disadvantages of apocarpy in primitive angiosperms. *Taxon*, *31*(1), 48–52.
Endress, P. K. (1994). *Diversity and evolutionary biology of tropical flowers*. Cambridge, UK: Cambridge University Press.
Endress, P. K. (2001). The flowers in extant basal angiosperms and inferences on ancestral flowers. *International Journal of Plant Sciences*, *162*(5), 1111–1140.
Endress, P. K. (2005). Carpels in *Brasenia* (Cabombaceae) are completely ascidiate despite a long stigmatic crest. *Annals of Botany*, *96*(2), 209–215.
Endress, P. K., & Doyle, J. A. (2009). Reconstructing the ancestral angiosperm flower and its initial specializations. *American Journal of Botany*, *96*(1), 22–66.
Endress, P. K., & Igersheim, A. (2000). Gynoecium structure and evolution in basal angiosperms. *International Journal of Plant Sciences*, *161S*, S211–S223.
Erbar, C. (2003). Pollen tube transmitting tissue: Place of competition of male gametophytes. *International Journal of Plant Sciences*, *164*, S265–S277.
Fernando, D. D. (2014). The pine reproductive process in temperate and tropical regions. *New Forests*, *45*(3), 333–352.
Fernando, D. D., Lazzaro, M. D., & Owens, J. N. (2005). Growth and development of conifer pollen tubes. *Sexual Plant Reproduction*, *18*(4), 149–162.
Field, D. L., Pickup, M., & Barrett, S. C. (2012). The influence of pollination intensity on fertilization success, progeny sex ratio, and fitness in a wind-pollinated, dioecious plant. *International Journal of Plant Sciences*, *173*(2), 184–191.
Firon, N., Nepi, M., & Pacini, E. (2012). Water status and associated processes mark critical stages in pollen development and functioning. *Annals of Botany*, *109*(7), 1201–1213.
Frame, D. (2003). The pollen tube pathway in *Tasmannia insipida* (Winteraceae): Homology of the male gametophyte conduction tissue in angiosperms. *Plant Biology*, *5*(3), 290–296.
Franchi, G. G., Nepi, M., Dafni, A., & Pacini, E. (2002). Partially hydrated pollen: Taxonomic distribution, ecological and evolutionary significance. *Plant Systematics and Evolution*, *234*(1–4), 211–227.

Franchi, G. G., Piotto, B., Nepi, M., Baskin, C. C., Baskin, J. M., & Pacini, E. (2011). Pollen and seed desiccation tolerance in relation to degree of developmental arrest, dispersal, and survival. *Journal of Experimental Botany*, *62*(15), 5267–5281.

Friedman, W. E. (1993). The evolutionary history of the seed plant male gametophyte. *Trends in Ecology & Evolution*, *8*, 15–20.

Friedman, J., & Barrett, S. C. H. (2009). Wind of change: New insights on the ecology and evolution of pollination and mating in wind-pollinated plants. *Annals of Botany*, *103*(9), 1515–1527.

Fulgione, A., & Hancock, A. M. (2018). Archaic lineages broaden our view on the history of *Arabidopsis thaliana*. *New Phytologist*, *219*, 1194–1198.

Galen, C., Shykoff, J., & Plowright, R. (1986). Consequences of stigma receptivity schedules for sexual selection in flowering plants. *American Naturalist*, *127*(4), 462–476.

Geitmann, A. (2011). Generating a cellular protuberance: Mechanics of tip growth. In P. Wojtaszek (Ed.), *Mechanical integration of plant cells and plants* (pp. 117–132). Berlin: Springer-Verlag.

Geitmann, A., & Nebenführ, A. (2015). Navigating the plant cell: Intracellular transport logistics in the green kingdom. *Molecular Biology of the Cell*, *26*, 3373–3378.

Geitmann, A., Wojciechowicz, K., & Cresti, M. (1996). Inhibition of intracellular pectin transport in pollen tubes by monensin, brefeldin A and cytochalasin D. *Botanica Acta: Journal of the German Botanical Society*, *109*(5), 373–381.

Goodwillie, C., & Weber, J. J. (2018). The best of both worlds? A review of delayed selfing in flowering plants. *American Journal of Botany*, *105*, 641–655.

Gossmann, T. I., Saleh, D., Schmid, M. W., Spence, M. A., & Schmid, K. J. (2016). Transcriptomes of plant gametophytes have a higher proportion of rapidly evolving and young genes than sporophytes. *Molecular Biology and Evolution*, *33*(7), 1669–1678.

Gould, S. J. (1977). *Ontogeny and phylogeny*. Cambridge, MA: Harvard University Press.

Gould, S. J., & Lewontin, R. C. (1979). The spandrels of San Marco and the Panglossian paradigm: A critique of the adaptationist programme. *Proceedings of the Royal Society of London B*, *205*, 581–598.

Guerrant, E. O. (1988). Heterochrony in plants: The intersection of evolution, ecology and ontogeny. In M. L. McKinney (Ed.), *Heterochrony in evolution: A multidisciplinary approach* (pp. 111–133). New York: Plenum Press.

Haig, D. (1992). Brood reduction in gymnosperms. In M. A. Elgar & B. J. Crespi (Eds.), *Cannibalism: Ecology and evolution among diverse taxa* (pp. 63–84). Oxford, UK: Oxford University Press.

Haig, D., & Westoby, M. (1988). A model for the origin of heterospory. *Journal of Theoretical Biology*, *134*(2), 257–272.

Haldane, J. B. S. (1932). *The causes of evolution*. London, UK: Longmans, Green.

Hall, B. K., & Olson, W. M. (2003). *Keywords and concepts in evolutionary developmental biology*. Cambridge, MA: Harvard University Press.

Harder, L. D., & Aizen, M. A. (2010). Floral adaptation and diversification under pollen limitation. *Philosophical Transactions of the Royal Society B*, *365*(1539), 529–543.

Harder, L. D., Aizen, M. A., & Richards, S. A. (2016). The population ecology of male gametophytes: The link between pollination and seed production. *Ecology Letters*, *19*(5), 497–509.

Harder, L. D., Aizen, M. A., Richards, S. A., Joseph, M. A., & Busch, J. W. (2016). Diverse ecological relations of male gametophyte populations in stylar environments. *American Journal of Botany*, *103*, 484–497.

Harder, L. D., Hobbhahn, N., & Richards, S. A. (2012). How depressed? Estimates of inbreeding effects during seed development depend on reproductive conditions. *Evolution*, *66*(5), 1375–1386.

Harder, L. D., & Johnson, S. D. (2008). Function and evolution of aggregated pollen in angiosperms. *International Journal of Plant Sciences*, *169*(1), 59–78.

Hendrikse, J. L., Parsons, T. E., & Hallgrimsson, B. (2007). Evolvability as the proper focus of evolutionary developmental biology. *Evolution & Development*, *9*(4), 393–401.

Herrera, C. M. (2002). Censusing natural microgametophyte populations: Variable spatial mosaics and extreme fine-graininess in winter-flowering *Helleborus foetidus* (Ranunculaceae). *American Journal of Botany*, *89*(10), 1570–1578.

Herrera, C. M. (2004). Distribution ecology of pollen tubes: Fine-grained, labile spatial mosaics in southern Spanish Lamiaceae. *New Phytologist*, *161*(2), 473–484.

Herrero, M. (2003). Male and female synchrony and the regulation of mating in flowering plants. *Philosophical Transactions of the Royal Society of London B*, *358*, 1019–1024.

Herrero, M., & Arbeloa, A. (1989). Influence of the pistil on pollen tube kinetics in peach (*Prunus persica*). *American Journal of Botany*, *76*(10), 1441–1447.

Heslop-Harrison, J. (1979). Aspects of the structure, cytochemistry and germination of the pollen of rye (*Secale cereale* L.). *Annals of Botany*, *44*(S1), 1–47.

Heslop-Harrison, J. (1982). Pollen-stigma interaction and cross-incompatibility in the grasses. *Science*, *215*, 1358–1364.

Heslop-Harrison, Y., & Heslop-Harrison, J. (1992). Germination of monocolpate angiosperm pollen - evolution of the actin cytoskeleton and wall during hydration, activation and tube emergence. *Annals of Botany*, *69*(5), 385–394.

Higashiyama, T., & Inatsugi, R. (2006). Comparative analysis of biological models used in pollen tube growth. *Plant Cell Monographs*, *3*, 265–286.

Hill, A. E., Shachar-Hill, B., Skepper, J. N., Powell, J., & Shachar-Hill, Y. (2012). An osmotic model of the growing pollen tube. *PLoS One*, 7(5), e36585.

Hoedemaekers, K., Derksen, J., Hoogstrate, S. W., Wolters-Arts, M., Oh, S. A., Twell, D., et al. (2015). BURSTING POLLEN is required to organize the pollen germination plaque and pollen tube tip in *Arabidopsis thaliana*. *New Phytologist*, *206*(1), 255–267.

Hoekstra, F. A. (1983). Physiological evolution in angiosperm pollen: Possible role of pollen vigor. In D. L. Mulcahy & E. Ottaviano (Eds.), *Pollen: Biology and implications for plant breeding* (pp. 35–41). Amsterdam, Netherlands: Elsevier.

Hoekstra, F. A., & Bruinsma, J. (1975). Respiration and vitality of binucleate and trinucleate pollen. *Physiologia Plantarum*, *34*(3), 221–225.

Hoekstra, F., & Golovina, E. (1999). Membrane behavior during dehydration: Implications for desiccation tolerance. *Russian Journal of Plant Physiology*, *46*, 295–306.

Hoekstra, F. A., Golovina, E. A., & Buitink, J. (2001). Mechanisms of plant desiccation tolerance. *Trends in Plant Science*, *6*(9), 431–438.

Honys, D., Renak, D., & Twell, D. (2006). Male gametophyte development and function. In *Vol. 1. Floriculture, ornamental and plant biotechnology: Advances and topical issues* (pp. 76–87).

Hormaza, J. I., & Herrero, M. (1994). Gametophytic competition and selection. In E. G. Williams (Ed.), *Genetic control of self-incompatibility and reproductive development in flowering plants* (pp. 372–400). Netherlands: Kluwer.

Howard, D. L. (1999). Conspecific sperm and pollen precedence. *Annual Review of Ecology and Systematics*, *30*, 109–132.

Hu, C., Munglani, G., Vogler, H., Fabrice, T. N., Shamsudhin, N., Wittel, F. K., et al. (2017). Characterization of size-dependent mechanical properties of tip-growing cells using a lab-on-chip device. *Lab on a Chip*, *17*, 82–90.

Husband, B. C. (2016). Effect of inbreeding on pollen tube growth in diploid and tetraploid *Chamerion angustifolium*: Do polyploids mask mutational load in pollen? *American Journal of Botany*, *103*(3), 532–540.

Jiang, L. X., Yang, S. L., Xie, L. F., Puah, C. S., Zhang, X. Q., Yang, W. C., et al. (2005). *VANGUARD1* encodes a pectin methylesterase that enhances pollen tube growth in the *Arabidopsis* style and transmitting tract. *Plant Cell*, *17*(2), 584–596.

Johri, B. M. (1984). *Embryology of angiosperms*. Berlin: Springer-Verlag.

Jones, D. F. (1928). *Selective fertilization*. Chicago: University of Chicago Press.
Kenrick, P. (2017). How land plant life cycles first evolved. *Science*, *358*(6370), 1538–1539.
Koch, M. A., Wernisch, M., & Schmickl, R. (2008). *Arabidopsis thaliana*'s wild relatives: An updated overview on systematics, taxonomy and evolution. *Taxon*, *57*(3), 933.
Lahti, D. C., Johnson, N. A., Ajie, B. C., Otto, S. P., Hendry, A. P., Blumstein, D. T., et al. (2009). Relaxed selection in the wild. *Trends in Ecology & Evolution*, *24*(9), 487–496.
Lankinen, Å., & Madjidian, J. A. (2011). Enhancing pollen competition by delaying stigma receptivity: Pollen deposition schedules affect siring ability, paternal diversity, and seed production in *Collinsia heterophylla* (Plantaginaceae). *American Journal of Botany*, *98*(7), 1191–1200.
Lankinen, A., & Skogsmyr, I. (2001). Evolution of pistil length as a choice mechanism for pollen quality. *Oikos*, *92*(1), 81–90.
Lewontin, R. C. (1965). Selection for colonizing ability. In H. G. Baker & G. L. Stebbins (Eds.), *The genetics of colonizing species* (pp. 77–91). New York, NY: Academic Press.
Leydon, A. R., Weinreb, C., Venable, E., Reinders, A., Ward, J. M., & Johnson, M. A. (2017). The molecular dialog between flowering plant reproductive partners defined by SNP-informed RNA-sequencing. *Plant Cell*, *29*, 984–1006.
Lin, J.-J., & Dickinson, D. B. (1984). Ability of pollen to germinate prior to anthesis and effect of desiccation on germination. *Plant Physiology*, *74*(3), 746–748.
Lloyd, R. M., & Klekowski, E. J., Jr. (1970). Spore germination and viability in Pteridophyta: Evolutionary significance of chlorophyllous spores. *Biotropica*, *2*, 129–137.
Lora, J., Herrero, M., & Hormaza, J. I. (2012). Pollen performance, cell number, and physiological state in the early-divergent angiosperm *Annona cherimola* Mill. (Annonaceae) are related to environmental conditions during the final stages of pollen development. *Sexual Plant Reproduction*, *25*(3), 157–167.
Losos, J. B. (2011). Seeing the forest for the trees: The limitations of phylogenies in comparative biology. *American Naturalist*, *177*(6), 709–727.
Madison, S. L., & Nebenführ, A. (2013). Understanding myosin functions in plants: Are we there yet? *Current Opinion in Plant Biology*, *16*, 710–717.
Magallón, S., Gómez-Acevedo, S., Sánchez-Reyes, L. L., & Hernández-Hernández, T. (2015). A metacalibrated time-tree documents the early rise of flowering plant phylogenetic diversity. *New Phytologist*, *207*(2), 437–453.
Maheshwari, P. (1950). *An introduction to the embryology of angiosperms*. New York: McGraw-Hill.
Marshall, D. L., & Folsom, M. W. (1991). Mate choice in plants—An anatomical to population perspective. *Annual Review of Ecology and Systematics*, *22*, 37–63.
Maynard-Smith, J., Burian, R., Kauffman, S., Alberch, P., Campbell, J., Goodwin, B., et al. (1985). Developmental constraints and evolution. *Quarterly Review of Biology*, *60*(3), 265–287.
Mazer, S. J., Hove, A. A., Miller, B. S., & Barbet-Massin, M. (2010). The joint evolution of mating system and pollen performance: Predictions regarding male gametophytic evolution in selfers vs. outcrossers. *Perspectives in Plant Ecology, Evolution and Systematics*, *12*(1), 31–41.
McKinney, M. L., & McNamara, K. J. (1991). *Heterochrony: The evolution of ontogeny*. New York, NY: Plenum Press.
Michard, E., Simon, A. A., Tavares, B., Wudick, M. M., & Feijó, J. A. (2017). Signaling with ions: The keystone for apical cell growth and morphogenesis in pollen tubes. *Plant Physiology*, *173*(1), 91–111.
Mitchell, R. J. (1997). Effects of pollen quantity on progeny vigor: Evidence from the desert mustard *Lesquerella fendleri*. *Evolution*, *51*, 1679–1684.

Mól, R., Filek, M., Dumas, C., & Matthys-Rochon, E. (2004). Cytoplasmic calcium in silk trichomes after pollen grain reception and post-pollination changes of the electric potential in pistil tissues of maize. *Plant Science, 166*(6), 1461–1469.

Mollet, J.-C., Leroux, C., Dardelle, F., & Lehner, A. (2013). Cell wall composition, biosynthesis and remodeling during pollen tube growth. *Plants, 2*(1), 107–147.

Mulcahy, D. L. (1971). A correlation between gametophytic and sporophytic characteristics in *Zea mays* L. *Science, 171*(3976), 1155–1156.

Mulcahy, D. L. (1979). The rise of the angiosperms—A genecological factor. *Science, 206*(4414), 20–23.

Mulcahy, D. L. (1983). Models of pollen tube competition in *Geranium maculatum*. In L. Real (Ed.), *Pollination biology* (pp. 151–161). Orlando, FL: Academic Press.

Mulcahy, G. B., & Mulcahy, D. L. (1982). Two phases of growth of *Petunia hybrida* (Hort. Vilm-Andz.) pollen tubes through compatible styles. *Journal of Palynology, 18*, 61–64.

Mulcahy, G. B., & Mulcahy, D. L. (1983). A comparison of pollen tube growth in bi- and trinucleate pollen. In D. L. Mulcahy & E. Ottaviano (Eds.), *Pollen: Biology and implications for plant breeding* (pp. 29–33). Amsterdam, Netherlands: Elsevier.

Mulcahy, D. L., & Mulcahy, G. B. (1987). The effects of pollen competition. *American Scientist, 75*, 44–50.

Nepi, M., Franchi, G. G., & Pacini, E. (2001). Pollen hydration status at dispersal: Cytophysiological features and strategies. *Protoplasma, 216*(3–4), 171–180.

Niklas, K. J., & Kutschera, U. (2010). The evolution of the land plant life cycle. *New Phytologist, 185*, 27–41.

Nishikawa, S. I., Zinkl, G. M., Swanson, R. J., Maruyama, D., & Preuss, D. (2005). Callose (beta-1,3 glucan) is essential for *Arabidopsis* pollen wall patterning, but not tube growth. *BMC Plant Biology, 5*, 22.

Obermeyer, G., Fragner, L., Lang, V., & Weckwerth, W. (2013). Dynamic adaption of metabolic pathways during germination and growth of lily pollen tubes after inhibition of the electron transport chain. *Plant Physiology, 162*(4), 1822–1833.

Olson, M. E. (2012). The developmental renaissance in adaptationism. *Trends in Ecology & Evolution, 27*(5), 278–287.

O'Meara, B. C. (2012). Evolutionary inferences from phylogenies: A review of methods. In D. J. Futuyma (Ed.), *Vol. 43. Annual review of ecology, evolution, and systematics* (pp. 267–285).

Ottaviano, E., Gorla, M. S., Frova, C., & Pe, E. (1988). Male gametophytic selection in higher plants. In M. Cresti, P. Gori, & E. Pacini (Eds.), *Sexual reproduction in higher plants* (pp. 35–42). Berlin: Springer.

Otto, S. P., Scott, M. F., & Immler, S. (2015). Evolution of haploid selection in predominantly diploid organisms. *Proceedings of the National Academy of Sciences of the United States of America, 112*(52), 15952–15957.

Palser, B. F., Rouse, J. L., & Williams, E. G. (1989). Coordinated timetables for megagametophyte development and pollen tube growth in *Rhododendron nuttallii* from anthesis to early postfertilization. *American Journal of Botany, 76*(8), 1167–1202.

Petersen, K. B., & Burd, M. (2017). Why did heterospory evolve? *Biological Reviews, 92*(3), 1739–1754.

Pires, N. D., Yi, K., Breuninger, H., Catarino, B., Menand, B., & Dolan, L. (2013). Recruitment and remodeling of an ancient gene regulatory network during land plant evolution. *Proceedings of the National Academy of Sciences of the United States of America, 110*(23), 9571–9576.

Povilus, R. A., Losada, J. M., & Friedman, W. E. (2014). Floral biology and ovule and seed ontogeny of *Nymphaea thermarum*, a water lily at the brink of extinction with potential as a model system for basal angiosperms. *Annals of Botany, 115*(2), 211–226.

Prieu, C., Matamoro-Vidal, A., Raquin, C., Dobritsa, A., Mercier, R., Gouyon, P. H., et al. (2016). Aperture number influences pollen survival in *Arabidopsis* mutants. *American Journal of Botany*, *103*(3), 452–459.

Prósperi, C. H., & Cocucci, A. E. (1979). Importancia taxonomica de la calosa de los tubos polinicos en Tubiflorae. *Kurtziana*, *12–13*, 75–81.

Prychid, C. J., Sokoloff, D. D., Remizowa, M. V., Tuckett, R. E., Yadav, S. R., & Rudall, P. J. (2011). Unique stigmatic hairs and pollen tube growth within the stigmatic cell wall in the early-divergent angiosperm family Hydatellaceae. *Annals of Botany*, *108*(4), 599–608.

Qin, Y., Leydon, A. R., Manziello, A., Pandey, R., Mount, D., Denic, S., et al. (2009). Penetration of the stigma and style elicits a novel transcriptome in pollen tubes, pointing to genes critical for growth in a pistil. *PLoS Genetics*, *5*(8), e1000621.

Qin, P., Ting, D., Shieh, A., & McCormick, S. (2012). Callose plug deposition patterns vary in pollen tubes of *Arabidopsis thaliana* ecotypes and tomato species. *BMC Plant Biology*, *12*, 178.

Reese, J. B., & Williams, J. H. (2018). How does genome size affect the evolution of pollen tube growth rate, a haploid performance trait? Preprint available at *BioRxiv, 462663*. https://doi.org/10.1101/462663.

Reinders, A. (2016). Fuel for the road—Sugar transport and pollen tube growth. *Journal of Experimental Botany*, *67*(8), 2121.

Röckel, N., Wolf, S., Kost, B., Rausch, T., & Greiner, S. (2008). Elaborate spatial patterning of cell-wall PME and PMEI at the pollen tube tip involves PMEI endocytosis, and reflects the distribution of esterified and de-esterified pectins. *Plant Journal*, *53*(1), 133–143.

Rosenheim, J. A., Schreiber, S. J., & Williams, N. M. (2016). Does an 'oversupply' of ovules cause pollen limitation? *New Phytologist*, *210*(1), 324–332.

Rosenheim, J. A., Williams, N. M., & Schreiber, S. J. (2014). Parental optimism versus parental pessimism in plants: How common should we expect pollen limitation to be? *American Naturalist*, *184*(1), 75–90.

Rounds, C. M., Hepler, P. K., Fuller, S. J., & Winship, L. J. (2010). Oscillatory growth in lily pollen tubes does not require aerobic energy metabolism. *Plant Physiology*, *152*(2), 736–746.

Rounds, C. M., Hepler, P. K., & Winship, L. J. (2014). The apical actin fringe contributes to localized cell wall deposition and polarized growth in the lily pollen tube. *Plant Physiology*, *166*, 139–151.

Rounds, C. M., Winship, L. J., & Hepler, P. K. (2011). Pollen tube energetics: Respiration, fermentation and the race to the ovule. *AoB Plants*, *2011*, Plr019.

Rudall, P. J., & Bateman, R. M. (2007). Developmental bases for key innovations in the seed-plant microgametophyte. *Trends in Plant Science*, *12*(7), 317–326.

Russell, S. D. (1992). Double fertilization. *International Review of Cytology*, *140*, 357–388.

Russell, S. D., Gou, X., Wong, C. E., Wang, X., Yuan, T., Wei, X., et al. (2012). Genomic profiling of rice sperm cell transcripts reveals conserved and distinct elements in the flowering plant male germ lineage. *New Phytologist*, *195*(3), 560–573.

Ryan, J. M., & Nebenführ, A. (2018). Update on myosin motors: Molecular mechanisms and physiological functions. *Plant Physiology*, *176*, 119–127.

Sage, T. L., Hristova-Sarkovski, K., Koehl, V., Lyew, J., Pontieri, V., Bernhardt, P., et al. (2009). Transmitting tissue architecture in basal relictual angiosperms: Implications for transmitting tissue origins. *American Journal of Botany*, *96*(1), 183–206.

Sarkissian, T., & Harder, L. (2001). Direct and indirect responses to selection on pollen size in *Brassica rapa* L. *Journal of Evolutionary Biology*, *14*(3), 456–468.

Schlüpmann, H., Bacic, A., & Read, S. M. (1994). Uridine-diphosphate glucose metabolism and callose synthesis in cultured pollen tubes of *Nicotiana alata* Link Et Otto. *Plant Physiology*, *105*(2), 659–670.

Schnarf, K. (1929). *Embryologie der angiospermen*. Berlin: Gebrüder Borntraeger.

Scribailo, R. W., & Barrett, S. C. (1991a). Pollen-pistil interactions in tristylous *Pontederia sagittata* (Pontederiaceae). I. Floral heteromorphism and structural features of the pollen tube pathway. *American Journal of Botany*, *78*(12), 1643–1661.

Scribailo, R. W., & Barrett, S. C. H. (1991b). Pollen-pistil interactions in tristylous *Pontederia sagittata* (Pontederiaceae). 2. Patterns of pollen tube growth. *American Journal of Botany*, *78*(12), 1662–1682.

Selinski, J., & Scheibe, R. (2014). Pollen tube growth: Where does the energy come from? *Plant Signaling & Behavior*, *9*(12), e977200.

Skogsmyr, I., & Lankinen, A. (2002). Sexual selection: An evolutionary force in plants. *Biological Reviews*, 77(4), 537–562.

Sleper, D. A., & Poehlman, J. M. (2006). Breeding corn (maize). In D. A. Sleper & J. M. Poehlman (Eds.), *Breeding field crops* (pp. 277–296). Oxford, UK: Blackwell.

Snow, A. A. (1986a). Evidence for and against pollen tube competition in natural populations. In D. L. Mulcahy, G. B. Mulcahy, & E. M. Ottaviano (Eds.), *Biotechnology and ecology of pollen* (pp. 405–410). New York: Springer-Verlag.

Snow, A. A. (1986b). Pollination dynamics in *Epilobium canum* (Onagraceae): Consequences for gametophytic selection. *American Journal of Botany*, *73*, 139–151.

Snow, A. A. (1994). Postpollination selection and male fitness in plants. *American Naturalist*, *144*(Suppl), S69–S83.

Snow, A. A., & Spira, T. P. (1991). Differential pollen tube growth rates and nonrandom fertilization in *Hibiscus moscheutos* (Malvaceae). *American Journal of Botany*, *78*(10), 1419–1426.

Sogo, A., & Tobe, H. (2005). Intermittent pollen-tube growth in pistils of alders (*Alnus*). *Proceedings of the National Academy of Sciences of the United States of America*, *102*(24), 8770–8775.

Sogo, A., & Tobe, H. (2006). Delayed fertilization and pollen-tube growth in pistils of *Fagus japonica* (Fagaceae). *American Journal of Botany*, *93*(12), 1748–1756.

Stearns, S. C. (1992). *The evolution of life histories*. Oxford, UK: Oxford University Press.

Stehlik, I., & Barrett, S. C. (2006). Pollination intensity influences sex ratios in dioecious *Rumex nivalis*, a wind-pollinated plant. *Evolution*, *60*(6), 1207–1214.

Stephan, O. O. (2017). Actin fringes of polar cell growth. *Journal of Experimental Botany*, *68*(13), 3303–3320.

Stephenson, A. G. (1981). Flower and fruit abortion: Proximate causes and ultimate functions. *Annual Review of Ecology and Systematics*, *12*, 252–279.

Stephenson, A. G., Travers, S. E., Mena-Ali, J. I., & Winsor, J. A. (2003). Pollen performance before and during the autotrophic-heterotrophic transition of pollen tube growth. *Philosophical Transactions of the Royal Society of London B*, *358*, 1009–1018.

Swanson, R. J., Hammond, A. T., Carlson, A. L., Gong, H., & Donovan, T. K. (2016). Pollen performance traits reveal pre-zygotic nonrandom mating and interference competition in *Arabidopsis thaliana*. *American Journal of Botany*, *103*, 498–513.

Szövényi, P., Waller, M., & Kirbis, A. (2019). Evolution of the plant body plan. In U. Grossniklaus (Ed.), *Current topics in developmental biology*: *Vol. 131*. Plant development and evolution (pp. 1–34). Elsevier.

Szövényi, P., Rensing, S. A., Lang, D., Wray, G. A., & Shaw, A. J. (2010). Generation-biased gene expression in a bryophyte model system. *Molecular Biology and Evolution*, *28*(1), 803–812.

Tadege, M., & Kuhlemeier, C. (1997). Aerobic fermentation during tobacco pollen development. *Plant Molecular Biology*, *35*(3), 343–354.

Taylor, M. L., MacFarlane, T. D., & Williams, J. H. (2010). Reproductive ecology of the basal angiosperm *Trithuria submersa* (Hydatellaceae). *Annals of Botany*, *106*(6), 909–920.

Taylor, M. L., & Williams, J. H. (2009). Consequences of pollination syndrome evolution for post-pollination biology in an ancient angiosperm family. *International Journal of Plant Sciences, 170*(5), 584–598.
Thien, L. B., Bernhardt, P., Devall, M. S., Chen, Z. D., Luo, Y. B., Fan, J. H., et al. (2009). Pollination biology of basal angiosperms (ANITA grade). *American Journal of Botany, 96*(1), 166–182.
Thomson, J. D. (1989). Germination schedules of pollen grains—Implications for pollen selection. *Evolution, 43*(1), 220–223.
Turner, N. C. (2018). Turgor maintenance by osmotic adjustment: 40 years of progress. *Journal of Experimental Botany, 69*, 3223–3323.
Valdivia, E. R., Cosgrove, D. J., & Stephenson, A. G. (2006). Role of accelerated style senescence in pathogen defense. *American Journal of Botany, 93*(11), 1725–1729.
Valdivia, E. R., Wu, Y., Li, L. C., Cosgrove, D. J., & Stephenson, A. G. (2007). A group-1 grass pollen allergen influences the outcome of pollen competition in maize. *PLoS One, 2*, e154.
Van Hemelryck, M., Bernal, R., Rojas, E., Dumais, J., & Kroeger, J. H. (2017). A fresh look at growth oscillations in pollen tubes: Kinematic and mechanistic descriptions. In G. Obermeyer & J. Feijo (Eds.), *Pollen tip growth* (pp. 369–389). Cham: Springer.
Velappan, Y., Signorelli, S., & Considine, M. J. (2017). Cell cycle arrest in plants: What distinguishes quiescence, dormancy and differentiated G1? *Annals of Botany, 120*(4), 495–509.
Vialette-Guiraud, A. C., Alaux, M., Legeai, F., Finet, C., Chambrier, P., Brown, S. C., et al. (2011). *Cabomba* as a model for studies of early angiosperm evolution. *Annals of Botany, 108*(4), 589–598.
Vidali, L., & Hepler, P. K. (2001). Actin and pollen tube growth. *Protoplasma, 215*, 64–76.
Vidali, L., McKenna, S. T., & Hepler, P. K. (2001). Actin polymerization is essential for pollen tube growth. *Molecular Biology of the Cell, 12*, 2534–2545.
Vidali, L., van Gisbergen, P. A. C., Gue, C., Franco, P., Li, M., Burkart, G. M., et al. (2009). Rapid formin-mediated actin-filament elongation is essential for polarized plant cell growth. *Proceedings of the National Academy of Sciences of the United States of America, 106*, 13341–13346.
Vogler, H., Draeger, C., Weber, A., Felekis, D., Eichenberger, C., Routier-Kierzkowska, A. L., et al. (2013). The pollen tube: A soft shell with a hard core. *Plant Journal, 73*(4), 617–627.
Wagner, J., Lechleitner, M., & Hosp, D. (2016). Pollen limitation is not the rule in nival plants: A study from the European Central Alps. *American Journal of Botany, 103*(3), 375–387.
Wake, M. (2003). Life history evolution. In B. K. Hall & W. M. Olson (Eds.), *Keywords and concepts in evolutionary developmental biology* (pp. 234–242). Cambridge, MA: Harvard University Press.
Wallace, S., & Williams, J. H. (2017). Evolutionary origins of pectin methylesterase genes associated with novel aspects of angiosperm pollen tube walls. *Biochemical and Biophysical Research Communications, 487*(3), 509–516.
Walsh, N. E., & Charlesworth, D. (1992). Evolutionary interpretations of differences in pollen tube growth rates. *Quarterly Review of Biology, 67*, 19–37.
Waser, N. (1993). Population structure, optimal outbreeding, and assortative mating in angiosperms. In N. W. Thornhill (Ed.), *The natural history of inbreeding and outbreeding* (pp. 173–199). Chicago, IL: University of Chicago Press.
Wei, L. Q., Xu, W. Y., Deng, Z. Y., Su, Z., Xue, Y., & Wang, T. (2010). Genome-scale analysis and comparison of gene expression profiles in developing and germinated pollen in *Oryza sativa*. *BMC Genomics, 11*(1), 338.
West-Eberhard, M. J. (2003). *Developmental plasticity and evolution*. New York, NY: Oxford University Press.

Williams, J. H. (2008). Novelties of the flowering plant pollen tube underlie diversification of a key life history stage. *Proceedings of the National Academy of Sciences of the United States of America*, *105*(32), 11259–11263.

Williams, J. H. (2009). *Amborella trichopoda* (Amborellaceae) and the evolutionary developmental origins of the angiosperm progamic phase. *American Journal of Botany*, *96*(1), 144–165.

Williams, J. H. (2012a). Pollen tube growth rates and the diversification of flowering plant reproductive cycles. *International Journal of Plant Sciences*, *173*(6), 649–661.

Williams, J. H. (2012b). The evolution of pollen germination timing in flowering plants: *Austrobaileya scandens* (Austrobaileyaceae). *AoB Plants*, *2012*, pls010. 1–12.

Williams, J. H., & Brown, C. D. (2018). Pollen has higher water content when dispersed in a tricellular state than in a bicellular state. *Acta Botânica Brasílica*, *32*(3), 454–461.

Williams, J. H., Edwards, J. A., & Ramsey, A. J. (2016). Economy, efficiency, and the evolution of pollen tube growth rates. *American Journal of Botany*, *103*(3), 471–483.

Williams, J. H., McNeilage, R. T., Lettre, M. T., & Taylor, M. L. (2010). Pollen tube growth and the pollen-tube pathway of *Nymphaea odorata* (Nymphaeaceae). *Botanical Journal of the Linnean Society*, *162*(4), 581–593.

Williams, J. H., Taylor, M. L., & O'Meara, B. C. (2014). Repeated evolution of tricellular (and bicellular) pollen. *American Journal of Botany*, *101*(4), 559–571.

Willson, M. F., & Burley, N. (1983). Mate choice in plants: Tactics, mechanisms, and consequences. *Vol. 19*. Princeton, NJ: Princeton University Press.

Winship, L. J., Obermeyer, G., Geitmann, A., & Hepler, P. K. (2010). Under pressure, cell walls set the pace. *Trends in Plant Science*, *15*(7), 363–369.

Winship, L. J., Obermeyer, G., Geitmann, A., & Hepler, P. K. (2011). Pollen tubes and the physical world. *Trends in Plant Science*, *16*(7), 353–355.

Wu, C.-C., Diggle, P. K., & Friedman, W. E. (2011). Female gametophyte development and double fertilization in *Balsas teosinte*, *Zea mays* subsp. *parviglumis* (Poaceae). *Sexual Plant Reproduction*, *24*, 219–229.

Zhang, X. S., & O'Neill, S. D. (1993). Ovary and gametophyte development are coordinately regulated by auxin and ethylene following pollination. *Plant Cell*, *5*, 403–418.

Zonia, L., & Munnik, T. (2011). Understanding pollen tube growth: The hydrodynamic model versus the cell wall model. *Trends in Plant Science*, *16*(7), 347–352.

CHAPTER THIRTEEN

Female reproductive organ formation: A multitasking endeavor

Sara Simonini[†], Lars Østergaard*
Department of Crop Genetics, John Innes Centre, Norwich, United Kingdom
*Corresponding author: e-mail address: lars.ostergaard@jic.ac.uk

Contents

Abstract

Multicellular organisms, such as plants, fungi, and animals, develop organs with specialized functions. Major challenges in developing such structures include establishment of polarity along three axes (apical-basal, medio-lateral, and dorso-ventral/abaxial-adaxial), specification of tissue types and their coordinated growth, and maintenance of communication between the organ and the entire organism. The gynoecium of the model plant *Arabidopsis thaliana* embodies the female reproductive organ and has proven an excellent model system for studying organ establishment and development, given its division into different regions with distinct symmetries and highly diverse tissue types.

[†] Present address: Department of Plant and Microbial Biology & Zurich-Basel Plant Science Center, University of Zurich, Zollikerstrasse 107, CH-8008, Zurich, Switzerland.

Current Topics in Developmental Biology, Volume 131
ISSN 0070-2153
https://doi.org/10.1016/bs.ctdb.2018.10.004

Upon pollination, the gynoecium undergoes dramatic changes in morphology and developmental programming to form the seed-containing fruit. In this review, we wish to provide a detailed overview of the molecular and genetic mechanisms that are known to guide gynoecium and fruit development in *A. thaliana*. We describe networks of key genetic regulators and their interactions with hormonal dynamics in driving these developmental processes. The discoveries made to date clearly demonstrate that conclusions drawn from studying gynoecium and fruit development in flowering plants can be used to further our general understanding of organ formation across the plant kingdom. Importantly, this acquired knowledge is increasingly being used to improve fruit and seed crops, facilitated by the recent profound advances in genomics, cloning, and gene-editing technologies.

1. Introduction

The female reproductive organ in plants is called the gynoecium and is one of the most complex plant structures. It arises at the center of the flower and is fundamental for ensuring ovule formation and fertilization, and eventually forms the major tissues of the fruit, sustaining the development, and dispersal of the seeds. Spatiotemporal regulation of gynoecium and fruit development involves intricate connections and crosstalk among plant hormones, transcription factors, and mechanisms of epigenetically regulated gene expression. In this review, we describe the events of gynoecium and fruit formation from a morphological and a molecular point of view, highlighting recent progress toward a mechanistic understanding of networks of transcription factors, hormones, and epigenetic regulation in these processes.

2. Gynoecium/carpel and fruit

Flowering plants belonging to the angiosperms develop a gynoecium—the female reproductive organ—in the center of the flower (see Thomson & Wellmer, 2019). The gynoecium harbors the developing ovules that become fertilized upon pollination and is formed from one or more fused carpels, whereby the carpel is a single structural unit bearing ovules. The evolution of carpel structures confers several advantages to the plant: (i) carpels provide structural and molecular protection for the developing ovules, serving as a physical barrier to predators, and producing defense compounds against insects and microorganisms (Scutt et al., 2003); (ii) carpels ensure efficient pollination and fertilization, developing tissues specifically dedicated to pollen capture, germination, and growth; and (iii) ultimately, following fertilization, the carpels develop into the fruit, a protective structure that nurtures the developing seeds and provides for their efficient and timely dispersal to the environment.

Although an amazing variety of shapes can be observed in gynoecia belonging to different angiosperm families (some examples are illustrated in Fig. 1), the tissue composition of gynoecia generally follows a few simple, common rules. For instance, based on the way through which multiple carpels fuse together to form the gynoecium, gynoecia can be considered apocarpic, when their carpels remain free at anthesis, or syncarpic, when their carpels are fused to form a single female structure (Armbruster et al., 2002). Apocarpy is mainly observed in monocotyledonous plants and is considered an ancestral character (Bessey, 1915; Endress & Doyle, 2009), whereas syncarpic gynoecia represent the vast majority of gynoecia observed in angiosperms. In addition, syncarpic carpels can be either congenital, when their fusion is apparent from the earliest emergence of the gynoecium primordia, or post-genital, when the fusion takes place during development. Among these two types, congenital carpel fusion is the more common (Armbruster et al., 2002; Lolle & Pruitt, 1999).

The gynoecium is a highly complex organ composed of a diverse range of tissues that change dynamically throughout development. Although impressive advances have been made toward uncovering the evolution of angiosperm structures, knowledge of the cellular and molecular pathways and the components that facilitate their coordinated development is now beginning to emerge (Table 1).

3. Gynoecium morphology and development

The mature gynoecium of *A. thaliana* is composed of distinct tissues with specific functions (Fig. 2). It is topped by stigmatic tissue (or a stigma), which is comprised of a single layer of elongated papillary cells. The stigma facilitates pollen adhesion, recognition, and germination and prevents germination of pollen coming from a different plant species (inter-specific incompatibility barrier). The stigma also marks the beginning of the transmitting tract, a tissue enriched in an extracellular polysaccharide matrix that provides guidance and a physical path for the pollen tubes growing toward the ovules (Lord & Russell, 2002; Palanivelu & Preuss, 2000; Sessions & Zambryski, 1995).

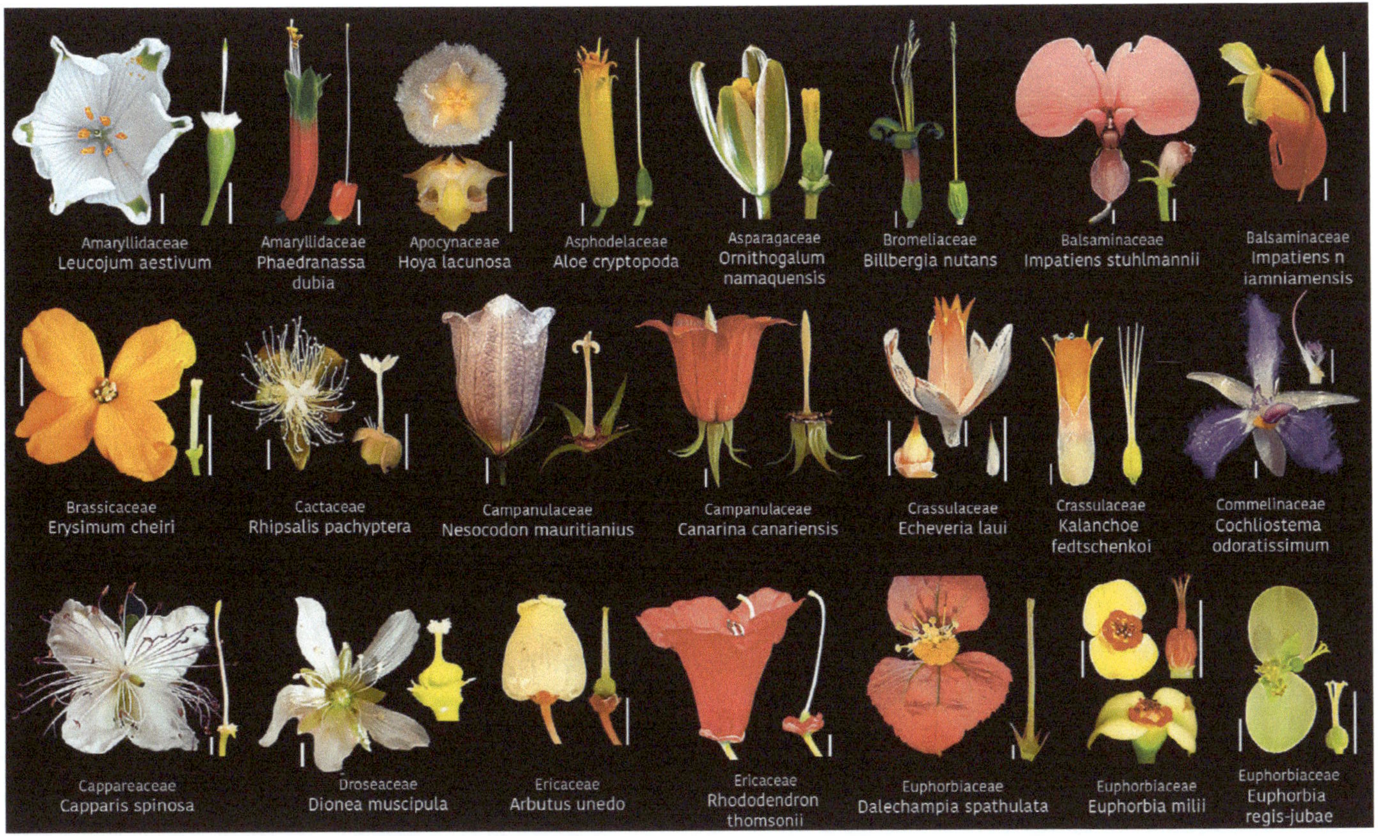

Fig. 1 Diversity among gynoecium shapes among angiosperms. Collection of images of flowers and dissected pistils of plants belonging to the angiosperms. Images were taken with a Leica 13MP/f/1.9 lens camera. Scale bar: 5 mm.

Euphorbiaceae
Pedilanthus tithymaloides
Eupomatiaceae
Eupomatia bennettii
Fabaceae
Bauhinia purpurea
Geraniaceae
Pelargonium desertorum
Fabaceae
Clianthus maximus
Gesneriaceae
Gesneria ventricosa
Gesneriaceae
Kohleria digitiflora
Gesneriaceae
Nematanthus brasiliensis
Gesneriaceae
Sinningia reitzii
Gesneriaceae
Nematanthus fissus
Iridaceae
Gladiolus palustrid
Liliaceae
Fritillaria imperialis
Liliaceae
Tulipa clusiana
Linaceae
Reinwardtia indica
Magnoliaceae
Magnolia stellata
Malvaceae
Pavonia intermedia
Papaveraceae
Dicentra spectabilis
Papaveraceae
Stylophorum diphyllum
Passifloraceae
Passiflora vitifolia
Philesiaceae
Lapageria rosea
Rutaceae
Citrus medica digitata
Rutaceae
Erythrochiton brasiliensis
Saxifragaceae
Bergenia stracheyi

Fig. 1—Cont'd

Table 1 Genes and family nomenclature.

Gene name	Family	AG code	Abbreviation
AUXIN RESPONSE FACTOR3/ ETTIN	ARF	AT2G33860	ARF3/ETT
AUXIN RESPONSE FACTOR4	ARF	AT5G60450	ARF4
AUXIN RESPONSE FACTOR5/ MONOPTEROS	ARF	AT1G19850	ARF5/MP
INDOLE-3-ACETIC ACID INDUCIBLE12	AUX/IAA	AT1G04550	IAA12
REPLUMLESS	HOMEOBOX	AT5G02030	RPL
BREVIPEDICELLUS	KNOX	AT4G08150	BP
HOMEOBOX-LEUCINE ZIPPER PROTEIN3	HOMEOBOX	AT3G60390	HAT3
HOMEOBOX-LEUCINE ZIPPER PROTEIN4	HOMEOBOX	AT2G44910	ATHB4
SHOOTMERISTEMLESS	KNOX	AT1G62360	STM
ALCATRAZ	bHLH	AT5G67110	ALC
INDEHISCENT	bHLH	AT4G00120	IND
HECATE1-2-3	bHLH	AT5G67060 AT3G50330 AT5G09750	HEC1, HEC2, HEC3
SPATULA	bHLH	AT4G36930	SPT
CRABS CLAW	YABBY	AT1G69180	CRC
FILAMENTOUS FLOWER	YABBY	AT2G45190	FIL
YABBY3	YABBY	AT4G00180	YAB3
ASYMMETRIC LEAVES1-2	MYB	AT2G37630 AT1G65620	AS1, AS2
NGATHA1-4	AP2/B3-like	AT2G46870 AT3G61970 AT1G01030 AT4G01500	NGA
STYLISH/SHORT INTERNODE	STY/SHI	AT5G66350	STY/SHI
CUPSHAPED COTYLEDON1-2	NAC	AT3G15170 AT5G53950	CUC1, CUC2
FRUITFULL	MADS	AT5G60910	FUL
SHATTERPROOF1-2	MADS	AT3G58780 AT2G42830	SHP1, SHP2

Table 1 Genes and family nomenclature.—cont'd

Gene name	Family	AG code	Abbreviation
AGAMOUS	MADS	AT4G18960	AG
KANADI	KANADI	AT5G16560	KAN
JAGGED	Zinc finger	AT1G68480	JAG
NO TRANSMITTING TRACT	C2H2/C2HC zinc finger	AT3G57670	NTT
PIN-FORMED1/3	PIN	AT1G73590 AT1G70940	PIN1, PIN3
TRYPTOPHAN AMINOTRANSFERASE OF ARABIDOPSIS1		AT1G70560	TAA1
YUCCA4	YUC	AT5G11320	YUC4
PINOID	Serine/threonine kinase	AT2G34650	PID
AINTEGUMENTA	AP2-domain	AT4G37750	ANT
LEUNIG	Transcriptional co-repressor	AT4G32551	LUG
SEUSS	Transcriptional co-regulator	AT1G43850	SEU
TEOSINTE BRANCHED1/CYCLOIDEA/PCF15	TCP	AT1G69690	TCP15
BRAHMA	SNF2-type chromatin remodeler	AT2G46020	BRM
SPLAYED	SWI2/SNF2-like protein	AT2G28290	SPY
PICKLE		AT2G25170	PKL
SQUAMOSA PROMOTER BINDING PROTEIN-LIKE8	SPL	AT1G02065	SPL8
ISOPENTENYLTRANSFERASE7		AT3G23630	IPT7
BRASSINOSTEROID-6-OXIDASE2		AT3G30180	CYP85A2
TOPLESS	WD-40 protein	AT1G15750	TPL
HISTONE DEACETYLASE19		AT4G38130	HDA19
LEAFY		AT5G61850	LFY
ARABIDOPSIS TRITHORAX1		AT2G31650	ATHX1

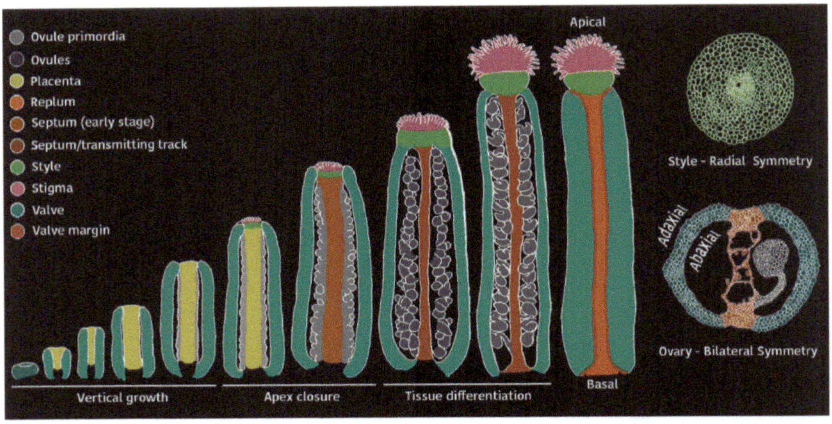

Fig. 2 Gynoecium development: developmental stages, tissues definitions, and polarity. Schematic representation of the developmental stages during gynoecium growth. The different tissues are colored as follows: light gray, ovule primordial; dark gray, ovules; yellow, placenta; orange, replum; light and dark brown, septum/transmitting track; bright green, style; pink, stigma; blue, valves; red, valve margins.

Below the stigma, the style develops as a solid cylinder providing structural support to the transmitting tract, which runs through its center. The style of *A. thaliana* is short compared to the majority of angiosperms where the style is more elongated, making up a significant proportion of the entire gynoecium length.

Below the style is the elongated tubular ovary, which encloses, protects, and nurtures the developing ovules. Based on anatomical and morphological features, the ovary of *A. thaliana* can be further divided into different tissues: valves, replum, septum, and valve margins. The valves make up the lateral walls of the ovary and are separated externally by the replum and internally by the septum. The former is a strip of vertically elongated cells that offer mechanical support and enclose the medial vascular bundle. In contrast, the septum stretches from the inner side of one replum to the other replum and supports the connection between transmitting tract and style. On each side of the replum, a strip of narrow cells forms the border between valve and replum. These cells comprise the valve margins that can be observed as a vertical indentation on the ovary surface. The valve margins allow the valves to separate from the replum when the fruit is mature, allowing for the release of the seeds. Ultimately, at the base of the ovary, the gynophore develops, consisting of a short stalk that connects the gynoecium to the rest of the plant.

Gynoecium development in *A. thaliana* has been divided into distinct stages (Smyth, Bowman, & Meyerowitz, 1990; schematically represented in Fig. 2). The earliest stages involve lateral expansion of the cells located in the center of the floral meristem. Cell division in this region allows the formation of the first visible gynoecium primordium, which consists of a ridge-shaped protrusion that surrounds an internal cleft where the rate of cell division is very low (Sessions, 1997; Smyth et al., 1990). Once organ identity has been specified, the gynoecium primordium develops along different polarity axes that are tightly coordinated by interactions between genetic and hormonal pathways. These axes are apical-basal, medio-lateral, and abaxial-adaxial (outer-inner) (Fig. 2).

A critical event during gynoecium development is the closure of the apex. This event relies on a bilateral-to-radial symmetry change facilitated by genetic and hormonal interactions (Moubayidin & Ostergaard, 2014). During subsequent developmental stages, the gynoecium continues to grow, the style becomes morphologically distinct with stigmatic papillae arising on its surface, and the ovules reach maturity. Eventually, the gynoecium becomes ready for pollination.

4. Transcription factor networks in gynoecium development

Over the last 3 decades, it has become clear that an integral network of transcription factors controls carpel development in *A. thaliana*. Figs. 3 and 4 summarize these networks within and among the different gynoecium tissues and the expression pattern of the involved factors during gynoecium development, respectively.

Among the first factors identified with a prominent role in gynoecium patterning and identity is AUXIN RESPONSE FACTOR3/ETTIN (ARF3/ETT; henceforth referred to as ETT) (Sessions et al., 1997). In agreement with the broad expression pattern of *ETT* in the entire developing gynoecium, the *ett* mutant shows dramatic polarity defects with overproliferation of stigmatic tissue, a long gynophore, and a reduced ovary (Nemhauser, Feldman, & Zambryski, 2000; Sessions et al., 1997; Simonini et al., 2016). In the ovary, *ETT*, in synergy with *ARF4*, regulates the adaxial/abaxial polarity of the valves (Pekker, Alvarez, & Eshed, 2005). This leads to activation of members of the KANADI (KAN) polarity gene family on the carpel abaxial side (Kerstetter, Bollman, Taylor, Bomblies, & Poethig, 2001; Pekker et al., 2005). Loss of expression of the KAN genes in

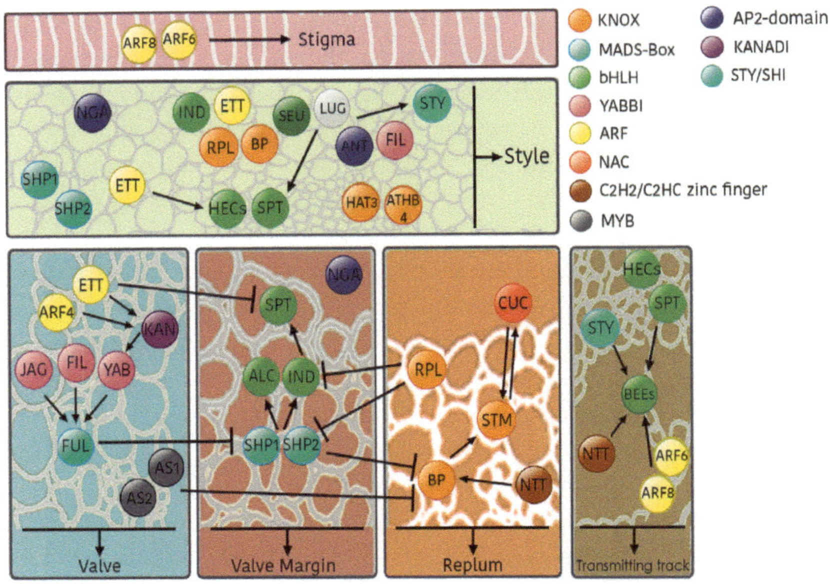

Fig. 3 Transcription factors network in gynoecium development. Connections between several transcription factors involved in the specification and identity of gynoecium tissues. The gynoecium areas are depicted in different colors: blue, valves; red, valve margin; orange, replum; brown, transmitting track; green, style; pink, stigma. Each factor is represented with a full colored circle. The color of the circle indicates the family to which they belong: orange, KNOX; light blue, MADS-domain; green, bHLH; pink, YABBI; yellow, ARF; red, NAC; brown, C2H2/C2H2 zinc finger; gray, MYB; dark blue, AP2-domain; purple, KANADI; cyan, STY/SHI. Lines ending in arrowheads indicate positive regulation, whereas vertical straight lines indicate negative regulation.

the *ett arf4* double mutant triggers the formation of adaxialized gynoecia, with ovules emerging on their exterior parts (Pekker et al., 2005). From the abaxial side of the ovary, KAN factors in turn antagonize the activity of the class III HD-ZIP (C3HDZ) genes, restricting their activity to the adaxial side of the growing gynoecium (Pekker et al., 2005). Other factors contributing to the establishment of valve polarity are the YABBY (YAB) family transcription factors JAGGED (JAG), FILAMENTOUS FLOWER (FIL), and YAB3, which, together with the KAN genes, promote abaxialization of the valves. *JAG*, *FIL*, and *YAB3* are also involved in valve identity through activation of the MADS-box gene *FRUITFULL* (*FUL*), which establishes valve identity by restricting the expression of the MADS-box valve margin-identity genes *SHATTERPROOF1* (*SHP1*) and *SHP2* (Dinneny & Yanofsky, 2004; Ferrándiz, Liljegren, & Yanofsky, 2000; Liljegren et al., 2000).

Fig. 4 Expression patterns of factors involved in gynoecium tissues identity. Graphic representation of expression patterns of transcription factors involved in gynoecium development and identity. Expression is depicted in bright green color. Expression outside the gynoecium domain is not indicated. *IND* expression in the middle cross section indicates expression at the top of the gynoecium.

The MYB-transcription factors ASYMMETRIC LEAVES1 (AS1) and AS2 are involved in the patterning of lateral organs, including the gynoecium (Alonso-Cantabrana et al., 2007). AS1/AS2 restrict the expression of replum identity KNOX genes, such as *BREVIPEDICELLUS* (*BP*) from the valves. *BP* expression is activated by NO TRANSMITTING TRACT (NTT) to specify replum identity together with the homeodomain-containing protein REPLUMLESS (RPL) (Ferrándiz, Liljegren, & Yanofsky, 2000; Roeder, Ferrándiz, & Yanofsky, 2003). Mutations in these replum identity factors lead to a reduced replum width, caused by a reduction in the number of replum cell files (Alonso-Cantabrana et al., 2007; Marsch-Martínez et al., 2014; Roeder et al., 2003).

The *SHP1/SHP2* genes are specifically expressed in the valve margin, where they direct the formation of the dehiscence zone, at least partially by activating expression of the *INDEHISCENT* (*IND*) and *ALCATRAZ* (*ALC*) genes, both encoding members of the basic helix-loop-helix (bHLH) transcription factor family (Groszmann, Paicu, Alvarez, Swain, & Smyth, 2011; Liljegren et al., 2000, 2004; Rajani & Sundaresan, 2001). Indeed, the fruits of the *shp1 shp2* double mutant, as well as *ind* and *alc* single mutants, do not open due to failure in the differentiation of the dehiscence zone (Liljegren et al., 2000; Rajani & Sundaresan, 2001). Additionally, SHP1 and SHP2 are also involved in style, stigma, and medial carpel tissue development (Colombo et al., 2010; Liljegren et al., 2000).

Besides their role in the specification of the valve, valve margin, and replum domain, many of the aforementioned genes, e.g., the bHLH factors SPATULA (SPT), IND, HECATE1 (HEC1), HEC2, HEC3, and ETT, also function in medial tissue development (septum, transmitting tract) and the establishment of style symmetry. Mutations in genes encoding these factors affect marginal tissue development: *spt* mutants are deficient in septum and transmitting tract formation and develop styles that fail to undergo radialization. They become "split" at the apex with reduced stigmatic tissue (Alvarez & Smyth, 1999; Girin et al., 2011; Heisler, Atkinson, Bylstra, Walsh, & Smyth, 2001), a phenotype greatly enhanced by mutations in the *IND* and *HEC* genes (Gremski, Ditta, & Yanofsky, 2007; Girin et al., 2011; Moubayidin & Ostergaard, 2014; Schuster, Gaillochet, & Lohmann, 2015). Similarly, *ett* mutant also develop styles that are partially bilateral (Sessions et al., 1997), and this phenotype is enhanced by mutations in the *IND*, *BP*, and *RPL* genes (Simonini et al., 2016; Simonini, Stephenson, & Østergaard, 2018). Therefore, *ETT* fulfills multiple roles in gynoecium development: it controls style polarity together with *IND* and in parallel to the *HEC-SPT* pathway, whereas in the valves, it promotes tissue identity and polarity, and in the medial region, it represses the activity of *SPT* and the HEC gene family to prevent activation of the apical development program (Alvarez & Smyth, 1999; Gremski et al., 2007; Heisler et al., 2001).

A number of additional genes encoding transcription factors control medial region formation, including the C2H2 zinc finger protein encoding gene *NTT*, the YABBY gene *CRC*, the boundary genes *CUPSHAPED COTYLEDON1* (*CUC1*) and *CUC2*, the homeobox-leucine zipper genes *HAT3* and *ATHB4*, and members of the STYLISH/SHORT INTERNODE (STY/SHI), NGATHA (NGA), and SPLAYED (SPL) gene families. Transmitting tract tissue is completely absent in *ntt* mutants (Crawford, Ditta, & Yanofsky, 2007; Marsch-Martínez et al., 2014), rendering pollen tube growth very inefficient. Intriguingly, *NTT* has recently been shown to also be involved in root meristem cell identity (Crawford et al., 2015), making a clear example on how key regulators of gynoecium development can be involved in other developmental processes.

Last but not least, FIL together with the AP2-domain transcription factors ANT and the transcriptional repressors LEUNIG (LUG) and SEUSS (SEU) are major players in the development of the carpel margins (Elliott et al., 1996; Nole-Wilson & Krizek, 2006; Sridhar, Surendrarao, Gonzalez, Conlan, & Liu, 2004; Wynn, Rueschhoff, & Franks, 2011). The *LUG* and *SEU* genes encode two interacting transcriptional co-repressors (Sridhar et al., 2004) that recruit histone deacetylases, leading to transcriptional inactivation (Liu & Karmarkar, 2008). Genetic interaction studies suggest that *LUG* has a positive effect on *STY* expression (Kuusk, Sohlberg, Magnus Eklund, & Sundberg, 2006), possibly by repressing a repressor of *STY*. In *lug* mutants, the gynoecium is partially unfused at the apex and presents defects in septum and ovule development (Krizek, Prost, & Macias, 2000; Liu, Franks, & Klink, 2000; Liu & Meyerowitz, 1995), whereas *seu* mutants develop apical clefts at the gynoecium top (Franks, Wang, Levin, & Liu, 2002; Pfluger & Zambryski, 2004; Simonini et al., 2018) and enhance the split-style defects of the *ett ind* double mutant (Simonini et al., 2018).

Also, stigmatic tissue development is controlled at transcriptional level among the main players are the two auxin response factors ARF6 and ARF8. Indeed, gynoecia of the *arf6 arf8* double mutant fail to properly form stigmatic papillae and consequently exhibit severe defects in pollen reception (Nagpal et al., 2005).

5. Hormonal networks in gynoecium development

Auxin and cytokinin (CK) are two plant hormones known to play major roles in gynoecium patterning. Auxin is involved in the establishment of apical-basal, medio-lateral, and abaxial-adaxial polarity in the gynoecium. Visualization of auxin signaling is possible using the DR5 synthetic reporter: a tandem repetition of the auxin response element recognized by ARF proteins (Guilfoyle, 2015; Ulmasov, Murfett, Hagen, & Guilfoyle, 1997). Therefore, the DR5 reporter does not reflect auxin levels, but rather auxin signaling; however, for simplicity we will henceforth refer to the DR5 signal as reflecting the presence auxin.

At early stages of development, auxin accumulates in two distinct lateral foci at the apex of the gynoecium primordium (Fig. 5), generated by local auxin biosynthesis and active acropetal (from the base to the apex) transport. At these stages, *CRC* controls auxin homeostasis in the medial region of the developing gynoecium to generate proper auxin maxima (Yamaguchi,

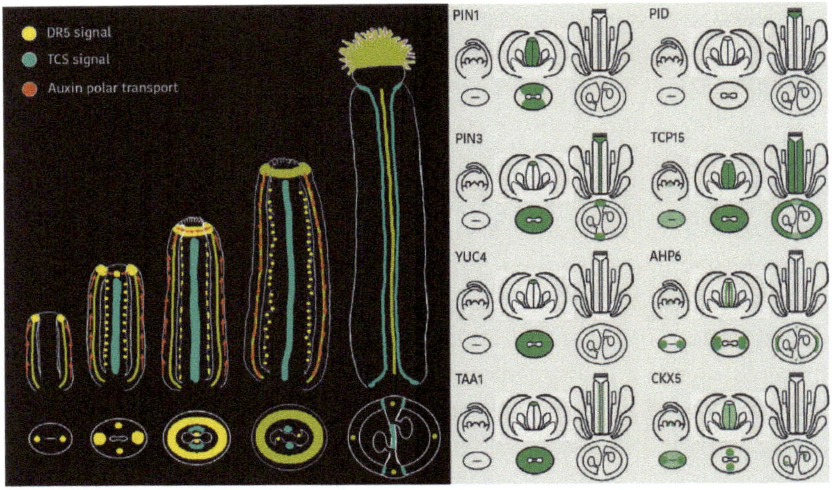

Fig. 5 Hormonal networks in gynoecium development. Left: graphic representation of DR5 (yellow) and TCS (cyan) markers during gynoecium development; direction of polar auxin transport is represented by red arrows. Right: schematic representation of the expression patterns of genes involved in auxin and CK transport, synthesis, and signaling. Expression is depicted in bright green color. Expression outside the gynoecium domain is not indicated.

Huang, Xu, Tanoi, & Ito, 2017), whereas local auxin biosynthesis is mediated mainly by the flavonoid monooxygenase YUCCA4 (YUC4) and the TRYPTOPHAN AMINOTRANSFERASE OF ARABIDOPSIS1 (TAA1), whose expression patterns clearly localize to the top of the growing gynoecium (Cheng, Dai, & Zhao, 2006; Stepanova et al., 2008, 2011). Among the positive regulators of auxin biosynthesis are the STY, NGA, and SPL factors (Eklund et al., 2010; Sohlberg et al., 2006; Ståldal et al. 2008; Trigueros et al., 2009; Xing et al., 2013). At these early developmental stages, auxin transport is facilitated by the PIN1 efflux carrier, which exhibits polar localization at the apical side of epidermal cells (Larsson, Roberts, Claes, Franks, & Sundberg, 2014; Moubayidin & Ostergaard, 2014). While auxin is accumulated in this region, it is also transported downward internally in the gynoecium in a drainage movement that is necessary for the formation of the central vasculature and prevents the lateral domains from developing medial domain identity (Larsson et al., 2014). Later, two additional auxin foci appear in the medial domain of the gynoecium apex, which eventually connect to the first two foci forming a continuous "auxin ring" at the gynoecium top (Larsson et al., 2014; Moubayidin & Ostergaard, 2014) (Fig. 5). This event is accompanied by a predominantly apolar cellular localization of PIN1 and PIN3, thereby facilitating and sustaining auxin accumulation in this region (Moubayidin & Ostergaard, 2014). Fundamental for the acquisition of the apolar distribution of PIN proteins at the gynoecium top, is the SPT-mediated repression of the *PINOID* (*PID*) gene, which encodes a class 3 AGC protein kinase that promotes the apical localization of proteins (Friml et al., 2003; Moubayidin & Ostergaard, 2014). In this mechanism, both SPT-IND and SPT-HEC1 dimers are required for the proper transcriptional activation of *PIN1* and *PIN3* to direct style and stigma formation (Moubayidin & Ostergaard, 2014; Schuster et al., 2015). Premature expression of *PID* in *spt*, *spt ind*, and *spt hec* mutants is accompanied by maintenance of PINs apical distribution, thus impairing auxin ring formation and radial symmetry establishment (Moubayidin & Ostergaard, 2014; Schuster et al., 2015). Formation of the auxin ring precedes and guides a symmetry transition from bilateral-to-radial symmetry, leading to gynoecium closure at the apex (Moubayidin & Ostergaard, 2014). Subsequently, auxin flows back down from the gynoecium top (Fig. 5). After fertilization of the mature gynoecium, auxin is absent from the gynoecium top and the valves, while accumulating in the replum (Sorefan et al., 2009) (Fig. 5).

Inhibition of auxin transport during gynoecium development causes a severe reduction of valve formation and the expansion of style and gynophore tissue. In severe cases, such as in strong *pid* mutants or treatment with the auxin transport inhibitor *N*-1-naphthylphthalamic acid (NPA), a complete transformation of the gynoecium into a radial structure is observed (Bennett, Alvarez, Bossinger, & Smyth, 1995; Nemhauser et al., 2000; Okada, Ueda, Komaki, Bell, & Shimura, 1991).

ETT plays a key role in gynoecium patterning and functions as a transcriptional regulator as well as a mediator of auxin signaling in the gynoecium (Nemhauser et al., 2000; Sessions et al., 1997; Simonini, Bencivenga, Trick, & Østergaard, 2017; Simonini et al., 2016). ETT is an atypical ARF that lacks the Phox/Bem1p (PB1) domain required for interaction with AUXIN/INDOLE ACETIC ACID (Aux/IAA) transcriptional repressor proteins during canonical auxin signaling (Guilfoyle, 2015). It has recently been demonstrated that ETT mediates organ patterning through an alternative mechanism, which appears not to require the canonical inhibition of ARFs by the Aux/IAA repressors (Simonini et al., 2016). ETT dimerizes with protein partners in an auxin-sensitive manner, and the fine-tuning of ETT activities at different auxin levels contributes to the diversification of transcriptional and developmental responses (Simonini et al., 2017). For example, in the style, ETT dimerizes with IND to precisely regulate when and where auxin maxima are formed via *PID* transcriptional control (Simonini et al., 2016).

CK controls aspects of gynoecium development in a manner that, in many ways, appears complementary to that of auxin. Visualization of the CK signaling is possible through the synthetic transcriptional CK response marker TCS (Müller & Sheen, 2007; Zürcher et al., 2013), and during gynoecium development, TCS::GFP signal is specifically observed at the carpel margin meristem (CMM), showing a localization pattern contrasting that of auxin (Marsch-Martínez et al., 2012) (Fig. 5). Known regulators of CK biosynthesis at these stages include KNOX transcription factors, such as BP and SHOOTMERISTEMLESS (STM), whose expression in the CMM ensures correct spatial and temporal CK signaling (Yanai et al., 2005). As gynoecium development progresses, TCS::GFP expression is detectable in the septum, transmitting tract, and in the valve margins of mature gynoecia (Marsch-Martínez et al., 2012). In the style and stigmatic region, CK signaling is antagonized by the activity of SPT and HEC factors that activate the negative regulators of CK signaling belonging to the type A ARABIDOPSIS RESPONSE REGULATORs (ARR-As) (Schuster et al., 2015). Indeed, *spt* and *hec* mutants exhibit hypersensitivity in the style/stigma region to CK applications (Schuster et al., 2015).

Consistent with the role of CK in stem cell maintenance and inhibition of cell differentiation, increasing levels of CK can have drastic consequences on gynoecium development. Studies using either ectopic expression of CK biosynthesis genes or exogenous application of synthetic CK induce replum width enlargement, outgrowth of proliferating tissue from the replum domain, and affect overall polarity (Bartrina, Otto, Strnad, Werner, & Schmülling, 2011; Marsch-Martínez et al., 2012; Schuster et al., 2015). Apical-basal gynoecium patterning is also affected by CK applications, with a phenotype that is reminiscent of mutant gynoecia in which polar auxin transport is affected (Zúñiga-Mayo, Reyes-Olalde, Marsch-Martinez, & de Folter, 2014). Indeed, exogenous treatment with CK can shift PIN1 polarity in the gynoecium and other tissues (Bencivenga, Simonini, Benková, & Colombo, 2012; Dello Ioio et al., 2008; Laplaze et al., 2007; Ruzicka et al., 2009). This is not the only example of auxin-CK crosstalk in the gynoecium. Indeed, the transcription factor TCP15 of the TEOSINTE BRANCHED1/CYCLOIDEA/PCF (TCP) family, which is expressed in the valve and weakly in the replum, is induced by CK and, in turn, represses auxin biosynthesis (Lucero et al., 2015). In addition, auxin promotes the expression of *ARABIDOPSIS HISTIDINE*

PHOSPHOTRANSFER PROTEIN6 (*AHP6*) that, in turn, represses CK responses in the lateral domain, thereby confining CK signaling to the medial domain (Reyes-Olalde et al., 2017).

Auxin and CK may be the most prominent hormones involved in gynoecium development. However, other hormones may also contribute. For example, mutations in the *CYP85A2* gene, encoding an enzyme of the brassinosteroids synthesis pathway, have been shown to cause defects at the gynoecium apex and develop horn-like protrusions on the valves, suggesting a role for brassinosteroids in gynoecium development (Kim et al., 2005; Nole-Wilson, Rueschhoff, Bhatti, & Franks, 2010).

6. Epigenetic control of gynoecium development

Epigenetic regulation refers to the control of gene activity and expression that is not dependent on DNA sequence and can include covalent modifications of DNA and histone proteins. Histones can be decorated with a variety of post-translational modifications, such as acetylation, methylation, ubiquitylation, and phosphorylation, all with a specific effect on chromatin state and DNA accessibility. For instance, acetylation and the tri-methylation on lysine 4 of histone H3 is associated with an open chromatin state and increased DNA accessibility. In contrast, tri-methylation on lysine 27 of histone H3 is a repressive mark associated with a closed chromatin state. Histone modifications are detected and interpreted by a series of chromatin remodelers, which can alter nucleosome occupancy and positioning. In *A. thaliana*, BRAHMA (BRM) and SYD, belonging to the SWITCH/SUCROSE NONFERMENTING (SWI/SNF) family of ATP-dependent chromatin remodelers, and PICKLE (PKL), a member of the SNF2-like family of ATPases, are among the best studied. Very little is known about the epigenetic control of gynoecium development, and the majority of the work describing connections between epigenetic regulation and tissue differentiation focus on flower and seed development. During flower primordium initiation, a complex containing the transcriptional co-repressor TOPLESS (TPL) and HISTONE DEACETYLASE19 (HDA19) interacts with the Aux/IAA repressors of auxin signaling (Wu et al., 2015). Because Aux/IAA proteins bind to ARFs, TPL/HDA19 is brought into close proximity of ARF target loci, thus establishing a repressive chromatin context through the deacetylation of histone tails. This mechanism has been exemplified for the MONOPTEROS (MP)/IAA12 module (Wu et al., 2015). An increase in auxin concentration initiates differentiation of floral primordia through degradation of the Aux/IAA repressors and, consequently, the removal of TPL/HDA19 from MP target regions (Wu et al., 2015). This allows for recruitment of SWI/SNF complexes at the MP target loci *LEAFY* and *FIL*, thus promoting the opening of chromatin, active transcription, and activation of floral homeotic genes of the B and C class (Wu et al., 2012, 2015).

Another level of regulation to precisely activate floral organ differentiation is through H3K4 tri-methylation mediated by *ARABIDOPSIS TRITHORAX1* (ATHX1) at homeotic gene loci (Alvarez-Venegas et al., 2003). In this context, also the chromatin remodeler PKL promotes floral organ differentiation via activation of floral homeotic genes (Aichinger, Villar, Di Mambro, Sabatini, & Köhler, 2011).

7. Developmental depiction of fruit growth

Initiation of fruit development is defined as the time of ovule fertilization and, in *A. thaliana*, marks the morphological transformation of the ovary to form a silique with two separate locules. Fertilization occurs at anthesis, when the ovules are mature and the stigma is receptive for pollen germination. Pollen grains germinate on the stigma and the pollen tubes grow into the ovary through the transmitting tract (Kandasamy, Nasrallah, & Nasrallah, 1994). As fertilization of the ovules occurs, signals from the developing embryos and seeds are exchanged with the plant to initiate fruit elongation and differentiation of specific tissues along the carpel margins (Dorcey, Urbez, Blázquez, Carbonell, & Perez-Amador, 2009; Goetz, Vivian-Smith, Johnson, & Koltunow, 2006; Vivian-Smith, Luo, Chaudhury, & Koltunow, 2001) (Fig. 6). The cells of the valves and the replum divide and expand primarily along the longitudinal axis, contributing to the lengthening of the fruit (Eldridge et al., 2016). In the first phase after fertilization, the fruit grows mainly due to cell division, and subsequently by cell expansion (Azzi et al., 2015; Gillaspy, Ben-David, & Gruissem, 1993; Vivian-Smith & Koltunow, 1999).

Concomitant with valve elongation, the valve margin starts to differentiate and will form narrow files of lignified cells as well as a layer of separation cells (Fig. 6). Both cell layers will be fundamental for seed dispersal when the fruit reaches maturity and dries out. At that point in development, the valves detach from the replum, releasing the brown and mature seeds to fall from the replum in a process called fruit dehiscence (Fig. 6).

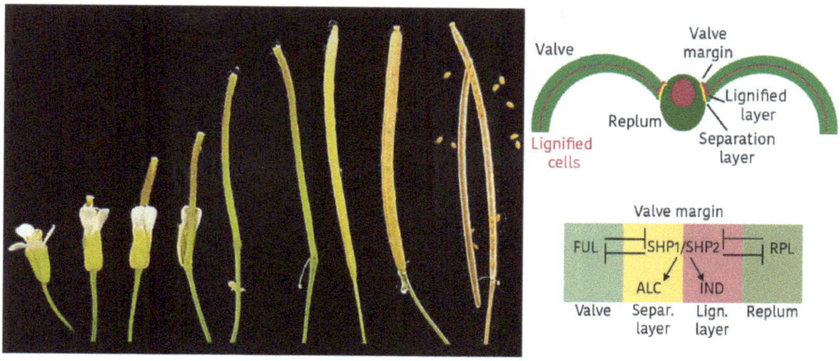

Fig. 6 Fruit development: developmental stages and tissue definitions. Left: *A. thaliana* fruit at different developmental stages, from fertilization to dehiscence. Right: schematic representation of tissue organization in the *A. thaliana* fruit and the genes controlling the identity of the different tissue domains.

Simple fruits can be classified as dry or fleshy, dehiscent or indehiscent, although this classification can vary considerably, taking different shapes and sizes into account. Development and final fruit size depend on (i) genotype, (ii) number of cell within the ovary prior to fertilization, (iii) number of successful fertilizations that occur in the ovary, and (iv) the number of cells composing the fruit (Srivastava & Handa, 2005). Moreover, seeds themselves are known to impact the development of fruits. Indeed, size and shape of many fruits are being determined by seed number and distribution within the ovary (Srivastava & Handa, 2005).

Both dry and fleshy fruits undergo the developmental phases of fruit set, fruit growth, maturation, and ripening. Fleshy fruits are believed to have evolved from dry fruits, and a high level of conservation exists between the genetic and molecular circuits that guide the development of fruits in both classes (Knapp, 2002; Seymour et al., 2013). Fleshy fruits are classified as climacteric if ripening happens concomitantly with an increase in respiration and ethylene biosynthesis, and non-climacteric if these events do not occur.

8. Transcriptional network in fruit development

Over the last decades, a number of genes have been identified that control fruit patterning and tissue identity. One major regulator of fruit elongation and shape is the MADS-box transcription factor FUL. During gynoecium development, *FUL* specifies valve identity, whereas during fruit development, *FUL* plays a role in controlling fruit shape and patterning through the regulation of stomata formation and cell expansion (Gu, Ferrándiz, Yanofsky, & Martienssen, 1998). Indeed, *ful* mutants develop fruit that fails to elongate upon fertilization, due to ectopic expression of valve margin-identity genes in ovary tissues (Gu et al., 1998). Defects in fruit elongation in the *ful* mutant are established prior to fertilization, with a deformed gynoecium and an increased number of cells in the inner epidermis of the valves, thus excluding the involvement of a stage-specific post-fertilization role of *FUL* in the control of fruit development (Gu et al., 1998). A clonal analysis of *ful* mutant gynoecia in *Capsella rubella* identified *FUL* as master regulator of directional cell growth and expansion in the valves, possibly being an important factor for generating diversification of gynoecium and fruit shape in Brassicaceae (Eldridge et al., 2016). Interestingly, *FUL* has been recently identified as a key player in the regulation of global meristem arrest, an age-based mechanism that controls the life span of *A. thaliana* plants (Balanzà et al., 2018), further supporting its pleiotropic role in plant development. An additional intriguing role of FUL that has recently emerged, is its interaction with ARF proteins (such as ARF8) to directly activate the transcription of *miR172C*, which targets *AP2* in order to promote valve growth (Ripoll et al., 2015). In maize miR172, encoded by the *tasselseed4* (*ts4*) locus, also targets the AP2 homolog *indeterminate spikelet1* and thus plays an important role in regulating maize inflorescence development (Chuck, Meeley, Irish, Sakai, & Hake, 2007). Indeed, *ts4* mutants develop male inflorescences lacking stamens and female inflorescences with unfused carpels (Chuck et al., 2007).

In *Solanum lycopersicum* (tomato), several MADS-domain transcription factors are involved in fruit development: SlFUL1, SlFUL2, TOMATO AGAMOUS-LIKE1, ENHANCER OF JOINTLESS2, and RIPENING INHIBITOR are master regulators of fruit shape and inflorescence branching, and of tomato fruit ripening through the regulation of ethylene biosynthetic genes (Jaakola et al., 2010; McAtee, Karim, Schaffer, & David, 2013; Schaffer, Ireland, Ross, Ling, & David, 2013; Seymour et al., 2011; Vrebalov et al., 2002). Similar to *A. thaliana*, *SlAP2a* and *COLORLESS NON-RIPENING* transcript abundance is regulated by *miR172* and *miR156*, respectively (Karlova et al., 2013).

Genetic studies in *A. thaliana* have highlighted that transcription factors belonging not only to the MADS-domain class but also to the bHLH family are involved in the regulation of fruit development. Particularly, valve margin development is under strict regulation by bHLH transcription factors, such as IND (Liljegren et al., 2000, 2004). Indeed, loss-of-function *ind* mutants lack the entire valve margin, leading to the development of fruits that fail to disperse their seeds. *IND* controls valve margin development at least partially by regulating hormone homeostasis. First, *IND* mediates the formation of an auxin minimum specifically in the valve margin through the control of PIN-mediated transport (Girin et al., 2011; Sorefan et al., 2009; van Gelderen, van Rongen, Liu, Otten, & Offringa, 2016). Second, IND induces GA biosynthesis, which leads to the de-repression of another bHLH transcription factor, ALC, required for the development of the separation layer (Arnaud et al., 2010; Rajani & Sundaresan, 2001). Moreover, temporal regulation of *IND* expression has recently been associated with increasing temperature: when grown at high temperature conditions, fruits of Brassicaceae dehisce prematurely, correlating with changes in the chromatin state at the *IND* locus and precocious expression of *IND* at the valve margins (Li, Deb, Kumar, & Østergaard, 2018).

Also, ARF and Aux/IAA proteins have prominent roles during the different phases of fruit development. In *A. thaliana*, ARF8 is a repressor of fruit initiation before fertilization. In tomato, down-regulation of *SlARF7* inhibits cell division and promotes cell expansion during the fruit set phase (de Jong, Mariani, & Vriezen, 2009; de Jong, Wolters-Arts, et al., 2009; de Jong et al., 2011), whereas *SlARF4* regulates fruit firmness during the ripening phase (Guillon et al., 2008).

Interestingly, loss-of-function mutants of *ARF8* in *A. thaliana* and of *Aux/IAA9* or *ARF7* in tomato result in parthenocarpy, where fruits develop in the absence of fertilization (de Jong, Mariani, et al., 2009; de Jong, Wolters-Arts, et al., 2009; de Jong et al., 2011; Wang et al. 2005). Indeed, parthenocarpy is a process tightly controlled by plant hormones, and parthenocarpic fruits can be obtained by the combined exogenous application of auxin, GA, and CK (Crane, 1964; Gillaspy et al., 1993; Mariotti et al., 2011; Nitsch, 1952; Vivian-Smith & Koltunow 1999). Over-accumulation of GA is enough to drive seedless fruit development in both *A. thaliana* and tomato (Fuentes et al. 2012; Martí et al. 2007).

9. Hormonal network in fruit development

The fruits of *A. thaliana* and tomato have been extensively studied as examples of dry and fleshy fruits, respectively. A finely tuned balance of hormones governs each step of fruit development: fruit set, fruit growth, fruit maturation, and finally fruit ripening and dispersal. Fig. 7 summarizes

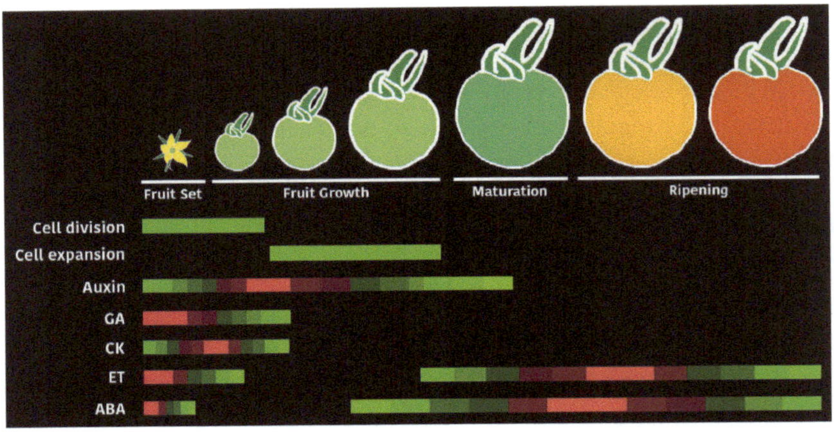

Fig. 7 Hormonal networks in fruit development. Schematic representation of the stages of fruit development in the fleshy fruit of tomato and the involvement of the different hormones at each stage. Green to magenta color in the bar signifies an increase in the corresponding hormone level.

the events during tomato fruit growth and the concomitant hormone increase/decrease during the different developmental phases.

The hormones auxin, GA, and CK, as well as crosstalk between them, play a major role in the regulation of fruit set and in the determination of fruit size (Carrera et al., 2012; Crane, 1964; de Jong, Mariani, et al., 2009; de Jong, Wolters-Arts, et al., 2009; Ozga et al., 2003; Ruan et al., 2012; Srivastava & Handa, 2005; Vivian-Smith & Koltunow, 1999). After fertilization, auxin produced in the developing seeds positively regulates GA biosynthesis, which is transported to the valves and promotes silique growth (Dorcey et al., 2009; Gallego-Giraldo et al., 2014). In this developmental context, ethylene acts negatively on GA perception and signaling, thus restricting the window of fruit responsiveness to GA (Serrani et al., 2008; Wang et al., 2009).

At the end of the cell division phase, seeds start to produce auxin and CK, which stimulate fruit growth and placental expansion inside the locular cavity (Blumenfeld & Gazit, 1970; Devoghalaere et al., 2012; Pattison & Catalá, 2012), thus governing the cell expansion phase together with GA and, to some extent, abscisic acid (ABA) (Gillaspy et al., 1993; Nitsch et al., 2012). In *A. thaliana*, CK signaling as visualized by the TCS reporter (Müller & Sheen, 2007) is detected in the valve margins of developing fruits, suggesting a role for CK in fruit dehiscence (Marsch-Martínez et al., 2012). Interestingly, although proper fruit dehiscence requires auxin depletion from the valve margins (Sorefan et al., 2009; van Gelderen et al., 2016), the auxin biosynthetic enzyme TAA1 shows expression in the valve margins (Martínez-Fernández et al., 2014).

Auxin together with CK and ethylene together with ABA are the primary hormones governing the fruit maturation and ripening phases, respectively (Davey & van Staden, 1977; Giovannoni, 2004; McAtee et al., 2013). Fruit ripening is a process that promotes efficient seed dispersal with fruits developing distinct characteristics, such as bright colors, aroma, flavor, and succulence, to become palatable for animals. Ripening includes conversion of complex carbohydrates into sugars, thereby counterbalancing fruit acidity, conversion of greenish fruits into fruits of orange-red color, and cell-wall modification that facilitates fruit dehiscence or softening (Klee & Giovannoni, 2011; Seymour et al., 2013). Based on the involvement of ethylene in the ripening process, fruits can be classified as climacteric or non-climacteric. In climacteric fruits, ripening is controlled by ethylene in crosstalk with auxin (Gillaspy et al., 1993; Trainotti et al., 2007), whereas ABA seems to have a stronger role during ripening of non-climacteric fruits by up-regulation of ethylene biosynthesis genes (McAtee et al., 2013; Sun et al., 2012). Similarly, in *A. thaliana*, ABA promotes fruit dehiscence through regulation of ethylene production in the siliques (Kanno et al., 2010). In addition to promoting ripening, ABA and ethylene also regulate fruit softening and shelf life (Lohani et al., 2004; López-Gómez et al., 2009; Nishiyama et al., 2007; Xiong et al., 2005).

10. Epigenetic regulation of fruit development

The fruit epigenome exhibits significant variation over the course of development and ripening (Manning et al., 2006), suggesting the existence of a fine equilibrium between DNA methylation and active DNA demethylation as a key regulator of fruit development (Liu et al., 2015). Indeed, extensive DNA demethylation has been shown to trigger a vast reprogramming of gene expression during the initiation of the fruit growth and ripening phases (Liu et al., 2015; Zhong et al., 2013). DNA methylation, which is at least partly associated with transcriptional repression, is hypothesized to be necessary to prevent premature fruit ripening before seed formation (Gallusci et al., 2016). Several genes involved in methylation maintenance, including the *DEMETER-LIKE2*, *METHYLTRANSFERASE1*, *CHROMO-METHYLTRANSFERASE*, and *DOMAINS-REARRANGED METHYLTRANSFERASE* genes of tomato, are expressed during early fruit development and ripening, supporting a role of DNA methylation/demethylation in fruit development (Teyssier et al., 2008). DNA methylation is also involved in controlling post-harvest events, such as cuticle biosynthesis and its deposition during senescence (Cherian et al., 2014; Osorio et al., 2013), as well as abiotic/biotic stress responses (Dowen et al., 2012; Jones-Rhoades & Bartel, 2004).

A clear connection between transcription factor activity and epigenetic control of fruit development has recently emerged. Master regulators of fruit development in tomato, such as RIN, NON-RIPENING (NOR), and CNR, are under epigenetic control (Giovannoni, 2007; Manning et al., 2006; Vrebalov et al., 2002), and changes in DNA methylation at the promoters of *NOR* and *CNR* have been observed during ripening (Manning et al., 2006; Zhong et al., 2013). Moreover, recent data suggest that auxin, GA, and brassinosteroids might influence the epigenome via the promotion of deacetylation/acetylation or demethylation/methylation of histones (Yamamuro et al., 2016), thus contributing to regulating gene expression through chromatin modifications. Dynamic post-translational modifications of histone tails have been implicated in the regulation of fruit development, and genes encoding both histone methyltransferases and acetylases/deacetylases are prevalently expressed during cell division and early developmental phases of fruit ripening in tomato, apple, and grape (Aquea et al., 2010, 2011; Cigliano et al., 2013; Janssen et al., 2008; Teyssier et al., 2008).

In summary, there is a tight control of the chromatin state in a wide range of species at the level of DNA methylation and histone modifications at target genes controlling fruit development and ripening. Given that fruits undergo a number of developmental switches, including fertilization, ripening, and dispersal, we speculate that chromatin remodeling provides a particularly responsive mechanism for changing gene expression patterns and gene networks, allowing rapid response to developmental and possibly environmental cues.

11. Technology advances and molecular knowledge facilitate crop improvement

One of the major global challenges today is the implementation of strategies that will allow the improvement and acceleration of crop production to meet the constantly increasing human population. In parallel to a general improvement of crop productivity and quality, an additional challenge is to create crops that perform better under adverse environmental conditions. Developing fruit crops adapted to the increasing demands for high yield with low input, is unrealistic if purely based on traditional plant breeding methods using existing genotypes. Another obstacle in the production of new fruit crop varieties, is that many fruit crops are hybrids and rely on in vitro regeneration, which can be difficult to achieve. Gene-editing technologies, such as CRISPR/Cas9 (Cong et al., 2013), provide an enormous potential for the highly targeted modification of specified DNA target sequences and are increasingly used in agricultural crops. CRISPR/Cas9 is part of an adaptive immune system in prokaryotes, working as a defense barrier against foreign DNA through site-specific cleavage of the latter. A single guide RNA directs the Cas9 endonuclease to specific genomic sites, resulting in DNA cleavage and the formation of double-strand breaks (Jinek et al., 2012). Therefore CRISPR/Cas9 has the potential to introduce mutations and deletions at specific sites in the genome. The first CRISPR/Cas9-mediated genome editing reported in a crop was in 2014, where the *ARGONAUTE7* gene of tomato (*SlAGO7*) was disrupted (Brooks et al., 2014). A similar approach has been used to target genes of the tomato *BLADE-ON-PETIOLE* family, involved in progressive meristem maturation to promote inflorescence complexity (Xu et al., 2016), and *SELF PRUNING 5G* (Soyk, Lemmon, et al., 2017; Soyk, Müller, et al., 2017), a flowering repressor involved in day-length perception. Another trait that has been improved in tomato cultivars by targeting a MADS-box transcription factor gene similar to *SEPALLATA4* in *A. thaliana* (Soyk, Lemmon, et al., 2017; Soyk, Müller, et al., 2017), is the length of the joint, i.e., the swollen part on the tomato plant where a branch meets a stem. This trait was first described in the 1950s, when researchers observed the development of a wild tomato variety native to the Galápagos Islands, which was "jointless." These plants lacked the joint, which somehow impaired the dropping of the fruit, a trait that breeders incorporated in tomato cultivars. Although the resulting plants display the *jointless* phenotype, they also produce many more flowers, which negatively affect fruit size and yield. By uncovering the gene responsible for the *jointless* trait, it is now possible to specifically achieve *jointless* plants that maintain a high yield (Soyk, Lemmon, et al., 2017; Soyk, Müller, et al., 2017). CRISPR/Cas9 has also been adopted to target developmental genes in *Brassica oleracea* (Lawrenson et al., 2015), biosynthetic pathways in tomato (Čermák et al., 2015; Hayut et al., 2017; Pan et al., 2016), and the development of parthenocarpic tomato fruits (Klap et al., 2017; Ueta et al., 2017). In conclusion, the gene-editing technology is developing at an incredible speed and appears to be only limited by the transformability of the target crop. This technology has an enormous potential to benefit the development of crops for many years to come, depending on the policies adopted for its regulation.

12. Concluding remarks

Advances in understanding the genetics underlying gynoecium and fruit development have been impressive, in particular over the last 2–3 decades when many of the key regulatory genes were identified. Moreover, it is becoming increasingly clear that discoveries made while studying gynoecium/fruit development provide generalizable information on organ initiation, tissue specification, and growth. For instance, recent observations suggest that genetic and hormonal interaction networks observed during gynoecium formation are likely relevant to a wide range of biological processes during plant growth and development (Simonini et al., 2016). The advent of powerful new sequencing technologies combined with gene-editing and mathematical modeling will further accelerate the speed at which important discoveries are being made, thus generating knowledge toward a better understanding of development in multicellular organisms and for applications in crop improvement.

Acknowledgments

We thank Peter Enz, Manfred Knabe, and René Stalder from the Botanischer Garten der Universität Zürich for assistance. This work has been supported by the UK Biological and Biotechnology Research Council (BBSRC) via grant BB/P013511/1 to the John Innes Centre.

References

Aichinger, E., Villar, C. B., Di Mambro, R., Sabatini, S., & Köhler, C. (2011). The CHD3 chromatin remodeler PICKLE and polycomb group proteins antagonistically regulate meristem activity in the *Arabidopsis* root. *Plant Cell*, *23*, 1047–1060.

Alonso-Cantabrana, H., Ripoll, J. J., Ochando, I., Vera, A., Ferrándiz, C., & Martínez-Laborda, A. (2007). Common regulatory networks in leaf and fruit patterning revealed by mutations in the *Arabidopsis* ASYMMETRIC LEAVES1 gene. *Development*, *134*, 2663–2671.

Alvarez, J., & Smyth, D. R. (1999). CRABS CLAW and SPATULA, two *Arabidopsis* genes that control carpel development in parallel with AGAMOUS. *Development*, *126*, 2377–2386.

Alvarez-Venegas, R., Pien, S., Sadder, M., Witmer, X., Grossniklaus, U., & Avramova, Z. (2003). ATX-1, an *Arabidopsis* homolog of trithorax, activates flower homeotic genes. *Current Biology*, *13*, 627–637.

Aquea, F., Timmermann, T., & Arce-Johnson, P. (2010). Analysis of histone acetyltransferase and deacetylase families of *Vitis vinifera*. *Plant Physiology and Biochemistry*, *48*, 194–199.

Aquea, F., Vega, A., Timmermann, T., Poupin, M. J., & Arce-Johnson, P. (2011). Genome-wide analysis of the SET DOMAIN GROUP family in grapevine. *Plant Cell Reports*, *30*, 1087–1097.

Armbruster, W. S., Mulder, C. P., Baldwin, B. G., Kalisz, S., Wessa, B., & Nute, H. (2002). Comparative analysis of late floral development and mating-system evolution in tribe Collinsieae (Scrophulariaceae s.l.). *American Journal of Botany*, *89*, 37–49.

Arnaud, N., Girin, T., Sorefan, K., Fuentes, S., Wood, T. A., Lawrenson, T., et al. (2010). Gibberellins control fruit patterning in *Arabidopsis thaliana*. *Genes & Development*, *24*, 2127–2132.

Azzi, L., Deluche, C., Gévaudant, F., Frangne, N., Delmas, F., Hernould, M., et al. (2015). Fruit growth-related genes in tomato. *Journal of Experimental Botany*, *66*, 1075–1086.

Balanzà, V., Martínez-Fernández, I., Sato, S., Yanofsky, M. F., Kaufmann, K., Angenent, G. C., et al. (2018). Genetic control of meristem arrest and life span in *Arabidopsis* by a FRUITFULL-APETALA2 pathway. *Nature Communications*, *9*, 565.

Bartrina, I., Otto, E., Strnad, M., Werner, T., & Schmülling, T. (2011). Cytokinin regulates the activity of reproductive meristems, flower organ size, ovule formation, and thus seed yield in *Arabidopsis thaliana*. *Plant Cell*, *23*, 69–80.

Bencivenga, S., Simonini, S., Benková, E., & Colombo, L. (2012). The transcription factors BEL1 and SPL are required for cytokinin and auxin signaling during ovule development in *Arabidopsis*. *Plant Cell*, *24*, 2886–2897.

Bennett, S. R. M., Alvarez, J., Bossinger, G., & Smyth, D. R. (1995). Morphogenesis in pinoid mutants of *Arabidopsis thaliana*. *Plant Journal*, *8*, 505–520.

Bessey, C. E. (1915). Another applied botany book. *Science*, *41*, 364–365.

Blumenfeld, A., & Gazit, S. (1970). Cytokinin activity in avocado seeds during fruit development. *Plant Physiology*, *46*, 331–333.

Brooks, C., Nekrasov, V., Lippman, Z. B., & Van Eck, J. (2014). Efficient gene editing in tomato in the first generation using the clustered regularly interspaced short palindromic repeats/CRISPR-associated9 system. *Plant Physiology*, *166*, 1292–1297.

Carrera, E., Ruiz-Rivero, O., Peres, L. E., Atares, A., & Garcia-Martinez, J. L. (2012). Characterization of the procera tomato mutant shows novel functions of the SlDELLA protein in the control of flower morphology, cell division and expansion, and the auxin-signaling pathway during fruit-set and development. *Plant Physiology*, *160*, 1581–1596.

Čermák, T., Baltes, N. J., Čegan, R., Zhang, Y., & Voytas, D. F. (2015). High-frequency, precise modification of the tomato genome. *Genome Biology*, *16*, 232.

Cheng, Y., Dai, X., & Zhao, Y. (2006). Auxin biosynthesis by the YUCCA flavin monooxygenases controls the formation of floral organs and vascular tissues in *Arabidopsis*. *Genes & Development*, *20*, 1790–1799.

Cherian, S., Figueroa, C. R., & Nair, H. (2014). Movers and shakers' in the regulation of fruit ripening: A cross-dissection of climacteric versus non-climacteric fruit. *Journal of Experimental Botany*, *65*, 4705–4722.

Chuck, G., Meeley, R., Irish, E., Sakai, H., & Hake, S. (2007). The maize tasselseed4 microRNA controls sex determination and meristem cell fate by targeting tasselseed6/indeterminate spikelet1. *Nature Genetics*, *39*, 1517–1521.

Cigliano, R. A., Cremona, G., Paparo, R., Termolino, P., Perrella, G., Gutzat, R., et al. (2013). Histone deacetylase AtHDA7 is required for female gametophyte and embryo development in *Arabidopsis*. *Plant Physiology*, *163*, 431–440.

Colombo, M., Brambilla, V., Marcheselli, R., Caporali, E., Kater, M. M., & Colombo, L. (2010). A new role for the SHATTERPROOF genes during *Arabidopsis* gynoecium development. *Developmental Biology*, *337*, 294–302.

Cong, L., Ran, F. A., Cox, D., Lin, S., Barretto, R., Habib, N., et al. (2013). Multiplex genome engineering using CRISPR/Cas systems. *Science*, *339*, 819–823.

Crane, J. C. (1964). Growth substances in fruit setting and development. *Annual Review of Plant Physiology*, *15*, 303–326.

Crawford, B. C., Ditta, G., & Yanofsky, M. F. (2007). The NTT gene is required for transmitting-tract development in carpels of *Arabidopsis thaliana*. *Current Biology*, *17*, 1101–1108.

Crawford, B. C., Sewell, J., Golembeski, G., Roshan, C., Long, J. A., & Yanofsky, M. (2015). Plant development. Genetic control of distal stem cell fate within root and embryonic meristems. *Science*, *347*, 655–659.

Davey, J. E., & van Staden, J. (1977). A cytokinin complex in the developing fruits of *Lupinus albus*. *Physiologia Plantarum*, *39*, 221–224.

de Jong, M., Mariani, C., & Vriezen, W. H. (2009). The role of auxin and gibberellin in tomato fruit set. *Journal of Experimental Botany*, *60*, 1523–1532.

de Jong, M., Wolters-Arts, M., Feron, R., Mariani, C., & Vriezen, W. H. (2009). The *Solanum lycopersicum* auxin response factor 7 (SlARF7) regulates auxin signaling during tomato fruit set and development. *Plant Journal*, *57*, 160–170.

de Jong, M., Wolters-Arts, M., García-Martínez, J. L., Mariani, C., & Vriezen, W. H. (2011). The *Solanum lycopersicum* AUXIN RESPONSE FACTOR 7 (SlARF7) mediates cross-talk between auxin and gibberellin signalling during tomato fruit set and development. *Journal of Experimental Botany*, *62*, 617–626.

Dello Ioio, R., Nakamura, K., Moubayidin, L., Perilli, S., Taniguchi, M., Morita, M. T., et al. (2008). A genetic framework for the control of cell division and differentiation in the root meristem. *Science*, *322*, 1380–1384.

Devoghalaere, F., Doucen, T., Guitton, B., Keeling, J., Payne, W., Ling, T. J., et al. (2012). A genomics approach to understanding the role of auxin in apple (Malus x domestica) fruit size control. *BMC Plant Biology*, *12*, 7.

Dinneny, J. R., & Yanofsky, M. F. (2004). Floral development: An ABC gene chips in downstream. *Current Biology*, *14*, R840–R841.

Dorcey, E., Urbez, C., Blázquez, M. A., Carbonell, J., & Perez-Amador, M. A. (2009). Fertilization-dependent auxin response in ovules triggers fruit development through the modulation of gibberellin metabolism in *Arabidopsis*. *Plant Journal*, *58*, 318–332.

Dowen, R. H., Pelizzola, M., Schmitz, R. J., Lister, R., Dowen, J. M., Nery, J. R., et al. (2012). Widespread dynamic DNA methylation in response to biotic stress. *Proceedings of the National Academy of Sciences of the United States of America*, *109*, 2183–2191.

Eklund, D. M., Ståldal, V., Valsecchi, I., Cierlik, I., Eriksson, C., Hiratsu, K., et al. (2010). The *Arabidopsis thaliana* STYLISH1 protein acts as a transcriptional activator regulating auxin biosynthesis. *Plant Cell*, *22*, 349–363.

Eldridge, T., Łangowski, Ł., Stacey, N., Jantzen, F., Moubayidin, L., Sicard, A., et al. (2016). Fruit shape diversity in the *Brassicaceae* is generated by varying patterns of anisotropy. *Development*, *143*, 3394–3406.

Elliott, R. C., Betzner, A. S., Huttner, E., Oakes, M. P., Tucker, W. Q. J., Gerente, D., et al. (1996). AINTEGUMENTA and APETALA2-like gene of *Arabidopsis* with pleiotropic roles in ovule development and floral organ growth. *Plant Cell*, *8*, 155–168.

Endress, P. K., & Doyle, J. A. (2009). Reconstructing the ancestral angiosperm flower and its initial specializations. *American Journal of Botany*, *96*, 22–66.

Ferrándiz, C., Liljegren, S. J., & Yanofsky, M. F. (2000). Negative regulation of the SHATTERPROOF genes by FRUITFULL during *Arabidopsis* fruit development. *Science*, *289*, 436–438.

Franks, R. G., Wang, C., Levin, J. Z., & Liu, Z. (2002). SEUSS, a member of a novel family of plant regulatory proteins, represses floral homeotic gene expression with LEUNIG. *Development*, *129*, 253–263.

Friml, J., Vieten, A., Sauer, M., Weijers, D., Schwarz, H., Hamann, T., et al. (2003). Efflux-dependent auxin gradients establish the apical-basal axis of *Arabidopsis*. *Nature*, *426*, 147–153.

Fuentes, S., Ljung, K., Sorefan, K., Alvey, E., Harberd, N. P., & Østergaard, L. (2012). Fruit growth in *Arabidopsis* occurs via DELLA-dependent and DELLA-independent gibberellin responses. *Plant Cell*, *24*, 3982–3996.

Gallego-Giraldo, C., Hu, J., Urbez, C., Gomez, M. D., Sun, T. P., & Perez-Amador, M. A. (2014). Role of the gibberellin receptors GID1 during fruit-set in *Arabidopsis*. *Plant Journal*, *79*, 1020–1032.

Gallusci, P., Hodgman, C., Teyssier, E., & Seymour, G. B. (2016). DNA methylation and chromatin regulation during fleshy fruit development and ripening. *Frontiers in Plant Science*, *7*, 807.

Gillaspy, G., Ben-David, H., & Gruissem, W. (1993). Fruits: A developmental perspective. *Plant Cell*, *5*, 1439–1451.

Giovannoni, J. J. (2004). Genetic regulation of fruit development and ripening. *Plant Cell*, *16*, S170–S180.

Giovannoni, J. J. (2007). Fruit ripening mutants yield insights into ripening control. *Current Opinion in Plant Biology*, *10*, 283–289.

Girin, T., Paicu, T., Stephenson, P., Fuentes, S., Körner, E., O'Brien, M., et al. (2011). INDEHISCENT and SPATULA interact to specify carpel and valve margin tissue and thus promote seed dispersal in *Arabidopsis*. *Plant Cell*, *23*, 3641–3653.

Goetz, M., Vivian-Smith, A., Johnson, S. D., & Koltunow, A. M. (2006). AUXIN RESPONSE FACTOR8 is a negative regulator of fruit initiation in *Arabidopsis*. *Plant Cell*, *18*, 1873–1886.

Gremski, K., Ditta, G., & Yanofsky, M. F. (2007). The HECATE genes regulate female reproductive tract development in *Arabidopsis thaliana*. *Development*, *134*, 3593–3601.

Groszmann, M., Paicu, T., Alvarez, J. P., Swain, S. M., & Smyth, D. R. (2011). SPATULA and ALCATRAZ, are partially redundant, functionally diverging bHLH genes required for *Arabidopsis* gynoecium and fruit development. *Plant Journal*, *68*, 816–829.

Gu, Q., Ferrándiz, C., Yanofsky, M. F., & Martienssen, R. (1998). The FRUITFULL MADS-box gene mediates cell differentiation during *Arabidopsis* fruit development. *Development*, *125*, 1509–1517.

Guilfoyle, T. J. (2015). The PB1 domain in auxin response factor and Aux/IAA proteins: A versatile protein interaction module in the auxin response. *Plant Cell*, *27*, 33–43.

Guillon, F., Philippe, S., Bouchet, B., Devaux, M. F., Frasse, P., Jones, B., et al. (2008). Down-regulation of an auxin response factor in the tomato induces modification of fine pectin structure and tissue architecture. *Journal of Experimental Botany*, *59*, 273–288.

Hayut, S., Melamed Bessudo, C., & Levy, A. A. (2017). Targeted recombination between homologous chromosomes for precise breeding in tomato. *Nature Communications*, *8*, 15605.

Heisler, M. G., Atkinson, A., Bylstra, Y. H., Walsh, R., & Smyth, D. R. (2001). SPATULA, a gene that controls development of carpel margin tissues in *Arabidopsis*, encodes a bHLH protein. *Development*, *128*, 1089–1098.

Jaakola, L., Poole, M., Jones, M. O., Kämäräinen-Karppinen, T., Koskimäki, J. J., Hohtola, A., et al. (2010). A SQUAMOSA MADS box gene involved in the regulation of anthocyanin accumulation in bilberry fruits. *Plant Physiology*, *153*, 1619–1629.

Janssen, B. J., Thodey, K., Schaffer, R. J., Alba, R., Balakrishnan, L., Bishop, R., et al. (2008). Global gene expression analysis of apple fruit development from the floral bud to ripe fruit. *BMC Plant Biology*, *8*, 16.

Jinek, M., Chylinski, K., Fonfara, I., Hauer, M., Doudna, J. A., & Charpentier, E. (2012). A programmable dual-RNA-guided DNA endonuclease in adaptive bacterial immunity. *Science*, *337*, 816–821.

Jones-Rhoades, M. W., & Bartel, D. P. (2004). Computational identification of plant microRNAs and their targets, including a stress-induced miRNA. *Molecular Cell*, *14*, 787–799.

Kandasamy, M. K., Nasrallah, J. B., & Nasrallah, M. E. (1994). Pollen-pistil interactions and developmental regulation of pollen tube growth in *Arabidopsis*. *Development*, *120*, 3405–3418.

Kanno, Y., Jikumaru, Y., Hanada, A., Nambara, E., Abrams, S. R., Kamiya, Y., et al. (2010). Comprehensive hormone profiling in developing *Arabidopsis* seeds: Examination of the site of ABA biosynthesis, ABA transport and hormone interactions. *Plant & Cell Physiology*, *51*, 1988–2001.

Karlova, R., van Haarst, J. C., Maliepaard, C., van de Geest, H., Bovy, A. G., Lammers, M., et al. (2013). Identification of microRNA targets in tomato fruit development using high-throughput sequencing and degradome analysis. *Journal of Experimental Botany, 64*, 1863–1878.

Kerstetter, R. A., Bollman, K., Taylor, R. A., Bomblies, K., & Poethig, R. S. (2001). KANADI regulates organ polarity in *Arabidopsis*. *Nature, 411*, 706–709.

Kim, T.-W., Hwang, J.-Y., Kim, Y.-S., Joo, S.-H., Chang, S. C., Lee, J. S., et al. (2005). *Arabidopsis* CYP85A2, a cytochrome P450, mediates the Baeyer-Villiger oxidation of castasterone to brassinolide in brassinosteroid biosynthesis. *Plant Cell, 17*, 2397–2412.

Klap, C., Yeshayahou, E., Bolger, A. M., Arazi, T., Gupta, S. K., Shabtai, S., et al. (2017). Tomato facultative parthenocarpy results from SlAGAMOUS-LIKE 6 loss of function. *Plant Biotechnology Journal, 15*, 634–647.

Klee, H. J., & Giovannoni, J. J. (2011). Genetics and control of tomato fruit ripening and quality attributes. *Annual Review of Genetics, 45*, 41–59.

Knapp, S. (2002). Tobacco to tomatoes: A phylogenetic perspective on fruit diversity in the *Solanaceae*. *Journal of Experimental Botany, 53*, 2001–2022.

Krizek, B. A., Prost, V., & Macias, A. (2000). AINTEGUMENTA promotes petal identity and acts as a negative regulator of AGAMOUS. *Plant Cell, 12*, 1357–1366.

Kuusk, S., Sohlberg, J. J., Magnus Eklund, D., & Sundberg, E. (2006). Functionally redundant SHI family genes regulate *Arabidopsis* gynoecium development in a dose-dependent manner. *Plant Journal, 47*, 99–111.

Laplaze, L., Benkova, E., Casimiro, I., Maes, L., Vanneste, S., Swarup, R., et al. (2007). Cytokinins act directly on lateral root founder cells to inhibit root initiation. *Plant Cell, 19*, 3889–3900.

Larsson, E., Roberts, C. J., Claes, A. R., Franks, R. G., & Sundberg, E. (2014). Polar auxin transport is essential for medial versus lateral tissue specification and vascular-mediated valve outgrowth in *Arabidopsis* gynoecia. *Plant Physiology, 166*, 1998–2012.

Lawrenson, T., Shorinola, O., Stacey, N., Li, C., Østergaard, L., Patron, N., et al. (2015). Induction of targeted, heritable mutations in barley and *Brassica oleracea* using RNA-guided Cas9 nuclease. *Genome Biology, 16*, 258.

Li, X. R., Deb, J., Kumar, S. V., & Østergaard, L. (2018). Temperature modulates tissue-specification program to control fruit dehiscence in *Brassicaceae*. *Molecular Plant, 11*, 598–606.

Liljegren, S. J., Ditta, G. S., Eshed, Y., Savidge, B., Bowman, J. L., & Yanofsky, M. F. (2000). SHATTERPROOF MADS-box genes control seed dispersal in *Arabidopsis*. *Nature, 404*, 766–770.

Liljegren, S. J., Roeder, A. H., Kempin, S. A., Gremski, K., Østergaard, L., Guimil, S., et al. (2004). Control of fruit patterning in *Arabidopsis* by INDEHISCENT. *Cell, 116*, 843–853.

Liu, Z., Franks, R. G., & Klink, V. P. (2000). Regulation of gynoecium marginal tissue formation by LEUNIG and AINTEGUMENTA. *Plant Cell, 12*, 1879–1892.

Liu, R., How-Kit, A., Stammitti, L., Teyssier, E., Rolin, D., Mortain-Bertrand, A., et al. (2015). A DEMETER-like DNA demethylase governs tomato fruit ripening. *Proceedings of the National Academy of Sciences of the United States of America, 112*, 10804–10809.

Liu, Z., & Karmarkar, V. (2008). Groucho/Tup1 family co-repressors in plant development. *Trends in Plant Science, 13*, 137–144.

Liu, Z., & Meyerowitz, E. M. (1995). LEUNIG regulates AGAMOUS expression in *Arabidopsis* flowers. *Development, 121*, 975–991.

Lohani, S., Trivedi, P. K., & Nath, P. (2004). Changes in activities of cell wall hydrolases during ethylene-induced ripening in banana: Effect of 1-MCP, ABA and IAA. *Postharvest Biology and Technology, 31*, 119–126.

Lolle, S. J., & Pruitt, R. E. (1999). Epidermal cell interactions: A case for local talk. *Trends in Plant Science*, *4*, 14–20.

López-Gómez, R., Cabrera-Ponce, J. L., Saucedo-Arias, L. J., Carreto-Montoya, L., Villanueva-Arce, R., Díaz-Perez, J. C., et al. (2009). Ripening in papaya fruit is altered by ACC oxidase cosuppression. *Transgenic Research*, *18*, 89–97.

Lord, E. M., & Russell, S. D. (2002). The mechanisms of pollination and fertilization in plants. *Annual Review of Cell and Developmental Biology*, *18*, 81–105.

Lucero, L. E., Uberti-Manassero, N. G., Arce, A. L., Colombatti, F., Alemano, S. G., & Gonzalez, D. H. (2015). *TCP15* modulates cytokinin and auxin responses during gynoecium development in *Arabidopsis*. *Plant Journal*, *84*, 267–282.

Manning, K., Tör, M., Poole, M., Hong, Y., Thompson, A. J., King, G. J., et al. (2006). A naturally occurring epigenetic mutation in a gene encoding an SBP-box transcription factor inhibits tomato fruit ripening. *Nature Genetics*, *38*, 948–952.

Mariotti, L., Picciarelli, P., Lombardi, L., & Ceccarelli, N. (2011). Fruit-set and early fruit growth in tomato are associated with increases in indoleacetic acid, cytokinin, and bioactive gibberellin contents. *Journal of Plant Growth Regulation*, *30*, 405–415.

Marsch-Martínez, N., Ramos-Cruz, D., Irepan Reyes-Olalde, J., Lozano-Sotomayor, P., Zúñiga-Mayo, V. M., & de Folter, S. (2012). The role of cytokinin during *Arabidopsis* gynoecia and fruit morphogenesis and patterning. *Plant Journal*, *72*, 222–234.

Marsch-Martínez, N., Zúñiga-Mayo, V. M., Herrera-Ubaldo, H., Ouwerkerk, P. B., Pablo-Villa, J., Lozano-Sotomayor, P., et al. (2014). The NTT transcription factor promotes replum development in *Arabidopsis* fruits. *Plant Journal*, *80*, 69–81.

Martí, C., Orzáez, D., Ellul, P., Moreno, V., Carbonell, J., & Granell, A. (2007). Silencing of DELLA induces facultative parthenocarpy in tomato fruits. *Plant Journal*, *52*, 865–876.

Martínez-Fernández, I., Sanchís, S., Marini, N., Balanzá, V., Ballester, P., Navarrete-Gómez, M., et al. (2014). The effect of NGATHA altered activity on auxin signaling pathways within the *Arabidopsis* gynoecium. *Frontiers in Plant Science*, *5*, 210.

McAtee, P., Karim, S., Schaffer, R., & David, K. (2013). A dynamic interplay between phytohormones is required for fruit development, maturation, and ripening. *Frontiers in Plant Science*, *17*, 79.

Moubayidin, L., & Ostergaard, L. (2014). Dynamic control of auxin distribution imposes a bilateral-to-radial symmetry switch during gynoecium development. *Current Biology*, *24*, 2743–2748.

Müller, B., & Sheen, J. (2007). Advances in cytokinin signaling. *Science*, *318*, 68–69.

Nagpal, P., Ellis, C. M., Weber, H., Ploense, S. E., Barkawi, L. S., Guilfoyle, T. J., et al. (2005). Auxin response factors *ARF6* and *ARF8* promote jasmonic acid production and flower maturation. *Development*, *132*, 4107–4118.

Nemhauser, J. L., Feldman, L. J., & Zambryski, P. C. (2000). Auxin and ETTIN in *Arabidopsis* gynoecium morphogenesis. *Development*, *127*, 3877–3888.

Nishiyama, K., Guis, M., Rose, J. K., Kubo, Y., Bennett, K. A., Wangjin, L., et al. (2007). Ethylene regulation of fruit softening and cell wall disassembly in *Charentais melon*. *Journal of Experimental Botany*, *58*, 1281–1290.

Nitsch, J. P. (1952). Plant hormones in the development of fruits. *The Quarterly Review of Biology*, *27*, 33–57.

Nitsch, L., Kohlen, W., Oplaat, C., Charnikhova, T., Cristescu, S., Michieli, P., et al. (2012). ABA-deficiency results in reduced plant and fruit size in tomato. *Journal of Plant Physiology*, *169*, 878–883.

Nole-Wilson, S., & Krizek, B. A. (2006). AINTEGUMENTA contributes to organ polarity and regulates growth of lateral organs in combination with YABBY genes. *Plant Physiology*, *141*, 977–987.

Nole-Wilson, S., Rueschhoff, E. E., Bhatti, H., & Franks, R. G. (2010). Synergistic disruptions in seuss cyp85A2 double mutants reveal a role for brassinolide synthesis during gynoecium and ovule development. *BMC Plant Biology*, *10*, 198.

Okada, K., Ueda, J., Komaki, M. K., Bell, C. J., & Shimura, Y. (1991). Requirement of the auxin polar transport system in early stages of *Arabidopsis* floral bud formation. *Plant Cell, 3*, 677–684.

Osorio, S., Scossa, F., & Fernie, A. R. (2013). Molecular regulation of fruit ripening. *Frontiers in Plant Science, 4*, 198.

Ozga, J. A., Yu, J., & Reinecke, D. M. (2003). Pollination-, development-, and auxin-specific regulation of gibberellin 3beta-hydroxylase gene expression in pea fruit and seeds. *Plant Physiology, 131*, 1137–1146.

Palanivelu, R., & Preuss, D. (2000). Pollen tube targeting and axon guidance: Parallels in tip growth mechanisms. *Trends in Cell Biology, 10*, 517–524.

Pan, C., Ye, L., Qin, L., Liu, X., He, Y., Wang, J., et al. (2016). CRISPR/Cas9-mediated efficient and heritable targeted mutagenesis in tomato plants in the first and later generations. *Scientific Reports, 6*, 24765.

Pattison, R. J., & Catalá, C. (2012). Evaluating auxin distribution in tomato (*Solanum lycopersicum*) through an analysis of the PIN and AUX/LAX gene families. *Plant Journal, 70*, 585–598.

Pekker, I., Alvarez, J. P., & Eshed, Y. (2005). Auxin response factors mediate *Arabidopsis* organ asymmetry via modulation of KANADI activity. *Plant Cell, 17*, 2899–2910.

Pfluger, J., & Zambryski, P. (2004). The role of SEUSS in auxin response and floral organ patterning. *Development, 131*, 4697–4707.

Rajani, S., & Sundaresan, V. (2001). The *Arabidopsis* myc/bHLH gene ALCATRAZ enables cell separation in fruit dehiscence. *Current Biology, 11*, 1914–1922.

Reyes-Olalde, J. I., Zúñiga-Mayo, V. M., Serwatowska, J., Chavez Montes, R. A., Lozano-Sotomayor, P., Herrera-Ubaldo, H., et al. (2017). The bHLH transcription factor SPATULA enables cytokinin signaling, and both activate auxin biosynthesis and transport genes at the medial domain of the gynoecium. *PLoS Genetics, 13*, e1006726.

Ripoll, J., Bailey, L. J., Mai, Q. A., Wu, S. L., Hon, C. T., Chapman, E. J., et al. (2015). microRNA regulation of fruit growth. *Nature Plants, 1*, 15036.

Roeder, A. H., Ferrándiz, C., & Yanofsky, M. F. (2003). The role of the REPLUMLESS homeodomain protein in patterning the *Arabidopsis* fruit. *Current Biology, 13*, 1630–1635.

Ruan, Y. L., Patrick, J. W., Bouzayen, M., Osorio, S., & Fernie, A. R. (2012). Molecular regulation of seed and fruit set. *Trends in Plant Science, 17*, 656–665.

Ruzicka, K., Simásková, M., Duclercq, J., Petrásek, J., Zazímalová, E., Simon, S., et al. (2009). Cytokinin regulates root meristem activity via modulation of the polar auxin transport. *Proceedings of the National Academy of Sciences of the United States of America, 106*, 4284–4289.

Schaffer, R. J., Ireland, H. S., Ross, J. J., Ling, T. J., & David, K. M. (2013). *SEPALLATA1/2* suppressed mature apples have low ethylene, high auxin and reduced transcription of ripening-related genes. *AoB Plants, 5*, pls047.

Schuster, C., Gaillochet, C., & Lohmann, J. U. (2015). *Arabidopsis* HECATE genes function in phytohormone control during gynoecium development. *Development, 142*, 3343–3350.

Scutt, C. P., Vinauger-Douard, M., Fourquin, C., Ailhas, J., Kuno, N., Uchida, K., et al. (2003). The identification of candidate genes for a reverse genetic analysis of development and function in the *Arabidopsis* gynoecium. *Plant Physiology, 132*, 653–665.

Serrani, J. C., Ruiz-Rivero, O., Fos, M., & García-Martínez, J. L. (2008). Auxin-induced fruit-set in tomato is mediated in part by gibberellins. *Plant Journal, 56*, 922–934.

Sessions, R. A. (1997). *Arabidopsis* (*Brassicaceae*) flower development and gynoecium patterning in wild type and Ettin mutants. *American Journal of Botany, 84*, 1179.

Sessions, A., Nemhauser, J. L., McColl, A., Roe, J. L., Feldmann, K. A., & Zambryski, P. C. (1997). ETTIN patterns the *Arabidopsis* floral meristem and reproductive organs. *Development, 124*, 4481–4491.

Sessions, R. A., & Zambryski, P. C. (1995). *Arabidopsis* gynoecium structure in the wild and in ettin mutants. *Development, 121*, 1519–1532.

Seymour, G. B., Østergaard, L., Chapman, N. H., Knapp, S., & Martin, C. (2013). Fruit development and ripening. *Annual Review of Plant Biology, 64*, 219–241.

Seymour, G. B., Ryder, C. D., Cevik, V., Hammond, J. P., Popovich, A., King, G. J., et al. (2011). A SEPALLATA gene is involved in the development and ripening of strawberry (Fragaria x ananassa Duch.) fruit, a non-climacteric tissue. *Journal of Experimental Botany, 62*, 1179–1188.

Simonini, S., Bencivenga, S., Trick, M., & Østergaard, L. (2017). Auxin-induced modulation of ETTIN activity orchestrates gene expression in *Arabidopsis. Plant Cell, 29*, 1864–1882.

Simonini, S., Deb, J., Moubayidin, L., Stephenson, P., Valluru, M., Freire-Rios, A., et al. (2016). A noncanonical auxin-sensing mechanism is required for organ morphogenesis in *Arabidopsis. Genes & Development, 30*, 2286–2296.

Simonini, S., Stephenson, P., & Østergaard, L. (2018). A molecular framework controlling style morphology in *Brassicaceae. Development, 145*, dev158105.

Smyth, D. R., Bowman, J. L., & Meyerowitz, E. M. (1990). Early flower development in *Arabidopsis. Plant Cell, 2*, 755–767.

Sohlberg, J. J., Myrenås, M., Kuusk, S., Lagercrantz, U., Kowalczyk, M., Sandberg, G., et al. (2006). STY1 regulates auxin homeostasis and affects apical-basal patterning of the *Arabidopsis* gynoecium. *Plant Journal, 47*, 112–123.

Sorefan, K., Girin, T., Liljegren, S. J., Ljung, K., Robles, P., Galván-Ampudia, C. S., et al. (2009). A regulated auxin minimum is required for seed dispersal in *Arabidopsis. Nature, 459*, 583–586.

Soyk, S., Lemmon, Z. H., Oved, M., Fisher, J., Liberatore, K. L., Park, S. J., et al. (2017). Bypassing negative epistasis on yield in tomato imposed by a domestication gene. *Cell, 169*, 1142–1155.

Soyk, S., Müller, N. A., Park, S. J., Schmalenbach, I., Jiang, K., Hayama, R., et al. (2017). Variation in the flowering gene SELF PRUNING 5G promotes day-neutrality and early yield in tomato. *Nature Genetics, 49*, 162–168.

Sridhar, V. V., Surendrarao, A., Gonzalez, D., Conlan, R. S., & Liu, Z. (2004). Transcriptional repression of target genes by LEUNIG and SEUSS, two interacting regulatory proteins for *Arabidopsis* flower development. *Proceedings of the National Academy of Sciences of the United States of America, 101*, 11494–11499.

Srivastava, A., & Handa, A. K. (2005). Hormonal regulation of tomato fruit development: A molecular perspective. *Journal of Plant Growth Regulation, 24*, 67–82.

Ståldal, V., Sohlberg, J. J., Eklund, D. M., Ljung, K., & Sundberg, E. (2008). Auxin can act independently of CRC, LUG, SEU, SPT and STY1 in style development but not apical-basal patterning of the *Arabidopsis* gynoecium. *The New Phytologist, 180*, 798–808.

Stepanova, A. N., Robertson-Hoyt, J., Yun, J., Benavente, L. M., Xie, D. Y., Dolezal, K., et al. (2008). TAA1-mediated auxin biosynthesis is essential for hormone crosstalk and plant development. *Cell, 133*, 177–191.

Stepanova, A. N., Yun, J., Robles, L. M., Novak, O., He, W., Guo, H., et al. (2011). The *Arabidopsis* YUCCA1 flavin monooxygenase functions in the indole-3-pyruvic acid branch of auxin biosynthesis. *Plant Cell, 23*, 3961–3973.

Sun, L., Sun, Y., Zhang, M., et al. (2012). Suppression of 9-cis-epoxycarotenoid dioxygenase, which encodes a key enzyme in abscisic acid biosynthesis, alters fruit texture in transgenic tomato. *Plant Physiology, 158*, 283–298.

Teyssier, E., Bernacchia, G., Maury, S., How Kit, A., Stammitti-Bert, L., Rolin, D., et al. (2008). Tissue dependent variations of DNA methylation and endoreduplication levels during tomato fruit development and ripening. *Planta, 228*, 391–399.

Thomson, B., & Wellmer, F. (2019). Molecular regulation of flower development. In U. Grossniklaus (Ed.), *Current topics in developmental biology: Vol. 131*. Plant development and evolution (pp. 185–210). Elsevier.

Trainotti, L., Tadiello, A., & Casadoro, G. (2007). The involvement of auxin in the ripening of climacteric fruits comes of age: The hormone plays a role of its own and has an intense interplay with ethylene in ripening peaches. *Journal of Experimental Botany*, *58*, 3299–3308.

Trigueros, M., Navarrete-Gómez, M., Sato, S., Christensen, S. K., Pelaz, S., Weigel, D., et al. (2009). The NGATHA genes direct style development in the *Arabidopsis* gynoecium. *Plant Cell*, *21*, 1394–1409.

Ueta, R., Abe, C., Watanabe, T., Sugano, S. S., Ishihara, R., Ezura, H., et al. (2017). Rapid breeding of parthenocarpic tomato plants using CRISPR/Cas9. *Scientific Reports*, *7*, 507.

Ulmasov, T., Murfett, J., Hagen, G., & Guilfoyle, T. J. (1997). Aux/IAA proteins repress expression of reporter genes containing natural and highly active synthetic auxin response elements. *Plant Cell*, *9*, 1963–1971.

van Gelderen, K., van Rongen, M., Liu, A., Otten, A., & Offringa, R. (2016). An INDEHISCENT-controlled auxin response specifies the separation layer in early *Arabidopsis* fruit. *Molecular Plant*, *9*, 857–869.

Vivian-Smith, A., & Koltunow, A. M. (1999). Genetic analysis of growth-regulator-induced parthenocarpy in *Arabidopsis*. *Plant Physiology*, *121*, 437–451.

Vivian-Smith, A., Luo, M., Chaudhury, A., & Koltunow, A. (2001). Fruit development is actively restricted in the absence of fertilization in *Arabidopsis*. *Development*, *128*, 2321–2331.

Vrebalov, J., Ruezinsky, D., Padmanabhan, V., White, R., Medrano, D., Drake, R., et al. (2002). A MADS-box gene necessary for fruit ripening at the tomato ripening-inhibitor (rin) locus. *Science*, *296*, 343–346.

Wang, H., Jones, B., Li, Z., Frasse, P., Delalande, C., Regad, F., et al. (2005). The tomato Aux/IAA transcription factor *IAA9* is involved in fruit development and leaf morphogenesis. *Plant Cell*, *17*, 2676–2692.

Wang, H., Schauer, N., Usadel, B., Frasse, P., Zouine, M., Hernould, M., et al. (2009). Regulatory features underlying pollination-dependent and independent tomato fruit set revealed by transcript and primary metabolite profiling. *Plant Cell*, *21*, 1428–1452.

Wu, M. F., Sang, Y., Bezhani, S., Yamaguchi, N., Han, S. K., Li, Z., et al. (2012). SWI2/SNF2 chromatin remodeling ATPases overcome polycomb repression and control floral organ identity with the LEAFY and SEPALLATA3 transcription factors. *Proceedings of the National Academy of Sciences of the United States of America*, *109*, 3576–3581.

Wu, M. F., Yamaguchi, N., Xiao, J., Bargmann, B., Estelle, M., Sang, Y., et al. (2015). Auxin-regulated chromatin switch directs acquisition of flower primordium founder fate. *eLife*, *4*, e09269.

Wynn, A. N., Rueschhoff, E. E., & Franks, R. G. (2011). Transcriptomic characterization of a synergistic genetic interaction during carpel margin meristem development in *Arabidopsis thaliana*. *PLoS One*, *6*, e26231.

Xing, S., Salinas, M., Garcia-Molina, A., Höhmann, S., Berndtgen, R., & Huijser, P. (2013). SPL8 and miR156-targeted SPL genes redundantly regulate *Arabidopsis* gynoecium differential patterning. *Plant Journal*, *75*, 566–577.

Xiong, A. S., Yao, Q. H., Peng, R. H., Li, X., Han, P. L., & Fan, H. Q. (2005). Different effects on ACC oxidase gene silencing triggered by RNA interference in transgenic tomato. *Plant Cell Reports*, *23*, 639–646.

Xu, C., Park, S. J., Van Eck, J., & Lippman, Z. B. (2016). Control of inflorescence architecture in tomato by BTB/POZ transcriptional regulators. *Genes & Development*, *30*, 2048–2061.

Yamaguchi, N., Huang, J., Xu, Y., Tanoi, K., & Ito, T. (2017). Fine-tuning of auxin homeostasis governs the transition from floral stem cell maintenance to gynoecium formation. *Nature Communications*, *8*, 1125.

Yamamuro, C., Zhu, J. K., & Yang, Z. (2016). Epigenetic modifications and plant hormone action. *Molecular Plant*, *9*, 57–70.

Yanai, O., Shani, E., Dolezal, K., Tarkowski, P., Sablowski, R., Sandberg, G., et al. (2005). *Arabidopsis* KNOXI proteins activate cytokinin biosynthesis. *Current Biology*, *15*, 1566–1571.

Zhong, S., Fei, Z., Chen, Y.-R., Zheng, Y., Huang, M., Vrebalov, J., et al. (2013). Single-base resolution methylomes of tomato fruit development reveal epigenome modifications associated with ripening. *Nature Biotechnology*, *31*, 154–159.

Zúñiga-Mayo, V. M., Reyes-Olalde, J. I., Marsch-Martinez, N., & de Folter, S. (2014). Cytokinin treatments affect the apical-basal patterning of the *Arabidopsis* gynoecium and resemble the effects of polar auxin transport inhibition. *Frontiers in Plant Science*, *5*, 191.

Zürcher, E., Tavor-Deslex, D., Lituiev, D., Enkerli, K., Tarr, P. T., & Müller, B. (2013). A robust and sensitive synthetic sensor to monitor the transcriptional output of the cytokinin signaling network in planta. *Plant Physiology*, *161*, 1066–1075.

Further reading

Sagar, M., Chervin, C., Mila, I., Hao, Y., Roustan, J. P., Benichou, M., et al. (2013). SlARF4, an auxin response factor involved in the control of sugar metabolism during tomato fruit development. *Plant Physiology*, *161*, 1362–1374.

Yanofsky, M. F., Ma, H., Bowman, J. L., Drews, G. N., Feldmann, K. A., & Meyerowitz, E. M. (1990). The protein encoded by the *Arabidopsis* homeotic gene agamous resembles transcription factors. *Nature*, *346*, 35–39.

CHAPTER FOURTEEN

Development and evolution of the unique ovules of flowering plants

Charles S. Gasser*, Debra J. Skinner
Department of Molecular and Cellular Biology, University of California, Davis, Davis, CA, United States
*Corresponding author: e-mail address: csgasser@ucdavis.edu

Contents

Abstract

Ovules are the precursors to seeds and as such are critical to plant propagation and food production. Mutant studies have led to the identification of numerous genes regulating ovule development. Genes encoding transcription factors have been shown to direct ovule spacing, ovule identity and integument formation. Particular co-regulators have now been associated with activities of some of these transcription factors, and other protein families including cell surface receptors have been shown to regulate ovule development. Hormone levels and transport, especially of auxin, have also been shown to play critical roles in ovule emergence and morphogenesis and to interact with the transcriptional regulators. Ovule diversification has been studied using orthologs of regulatory genes in divergent angiosperm groups. Combining modern genetic evidence with expanding knowledge of the fossil record illuminates the possible origin of the unique bitegmic ovules of angiosperms.

Current Topics in Developmental Biology, Volume 131
ISSN 0070-2153
https://doi.org/10.1016/bs.ctdb.2018.10.007

1. Introduction

Seeds represent a stage of the angiosperm lifecycle with special significance for humans. Direct consumption of seeds accounts for at least 60% of human caloric intake and 50% of human protein consumption (Bruinsma, 2003). The contribution of seeds to the human diet is greater if seeds fed to animals for meat and dairy production are also included. Seeds additionally are the primary means of propagation of crop plants and many fruit crops. Seeds form from their developmental precursors, the ovules, which develop within the carpels in angiosperms (see chapter "Molecular regulation of flower development" by Thomson and Wellmer, this issue).

The ovules of most angiosperms consist of three functional regions: the terminal nucellus, in which the embryo sac and its included egg cell form; the chalaza, the region subtending the nucellus from which most commonly two integuments emerge and extend to cover the nucellus; and the funiculus, a stalk connecting the ovule to the placental region (defined as the site of emergence of the ovules) within a carpel. While this common ovule form appears to be ancestral within the angiosperms (Endress, 2011), there has been significant divergence in ovule form, especially with respect to the number and shape of the integuments. Herein we will discuss progress in understanding regulation of ovule development, ovule diversification and the origin of ovules. A majority of the developmental work has been performed in *Arabidopsis thaliana* and studies in this system will inform a major part of our discussion. Researchers have also used the sequences of *A. thaliana* genes controlling ovule development to extend the work to other species with divergent ovule morphology, illuminating ovule evolution.

2. Ovule development

2.1 Ovule initiation

Ovules arise from the placenta as initially featureless, finger-like primordia, most commonly in a regularly spaced pattern characteristic of each species. Mutations affecting ovule initiation and spacing provide clues to the regulation of these processes. One gene closely associated with ovule initiation is *AINTEGUMENTA* (*ANT*). While the primary effect on ovules of *ant* mutations is loss of the integuments, these mutants also have fewer ovules than the wild type (Elliott et al., 1996; Klucher, Chow, Reiser, & Fischer, 1996). The precise role of *ANT* in ovule initiation may be obscured by the activity of paralogous genes with overlapping expression patterns, such as *AINTEGUMENTA-LIKE6* (*AIL6*). The ANT and closely related AIL proteins are putative transcription factors that appear to bind to the same consensus sequence (Nole-Wilson & Krizek, 2000). Unfortunately, it has not been possible to evaluate the effects on ovule development of combining mutations in these two genes because the profound effects of this combination on floral organ development led to highly aberrant carpels that do not produce ovules (Krizek, 2009).

The spacing of the ovules is further controlled by boundary determinants, most notably the products of the *CUP SHAPED COTYLEDON2* (*CUC2*) and *CUC3* genes. Expression domains of these two NAC transcription factor genes mark the borders of the regions from which individual ovules arise. Mutations in these two genes lead to an aberrant spacing of ovules and frequently to fused ovule primordia that eventually form fused seeds (Goncalves et al., 2015). Proper spacing of ovules ensures optimum ability of seeds to develop and expand free of spatial interference from other developing seeds.

2.2 Ovule patterning

Following initiation, a pattern is established to differentiate the three main regions of the ovule that is then followed by elaboration of the integuments and formation of differentiated cells (Fig. 1) (Robinson-Beers, Pruitt, & Gasser, 1992; Schneitz, Hulskamp, & Pruitt, 1995).

While a complete loss of ovule patterning has not been observed in mutants (all appear to form a funicular region), several major patterning genes have been identified. Expression of *ANT* and *WUSCHEL* (*WUS*) was found to be restricted to the chalaza and nucellus, respectively, and appears to be important for establishing the functions of these two domains (Elliott et al., 1996; Gross-Hardt, Lenhard, & Laux, 2002). In mutants of either gene, integuments fail to emerge from the chalaza, indicating disruption of the function of this region (Fig. 2A and B). The mutants also fail to form functional embryo sacs, but whether this is a defect in the nucellar domain, or a secondary effect of the absence of integuments has not been determined. *WUS* and *ANT* do not appear to participate in patterning the expression of each other since mutation of either gene does not alter the expression zone of the other one (Gross-Hardt et al., 2002). Confinement of *WUS* expression to the nucellus is at least partially mediated by polarity determinants, the Class III homeodomain leucine zipper (C3HDZ) genes, and the *BELL1* (*BEL1*)

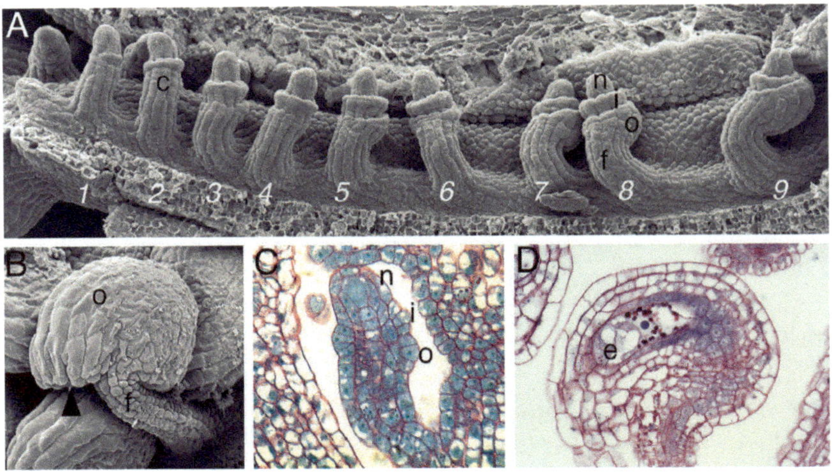

Fig. 1 *A. thaliana* ovule development. (A, B) Scanning election micrographs of *lug* and wild-type ovules. (C, D) Stained plastic sections of wild-type ovules. (A) A developmental series of *A. thaliana* ovules can be observed in a *lug* mutant carpel in which multiple stages are present simultaneously. The base of the carpel is to the right. Ovule primordia emerge from the placenta, along the length of the carpel. After elongation of the fingerlike ovule primordium (1), the inner integument initiates as a ring around the upper half of the chalazal (central) region (2). Just below the inner integument primordium, the outer integument initiates on the side of the ovule oriented toward the base of the carpel (gynobasal) (3, 5). Above the integuments, the distal region is defined as the nucellus, and below the chalaza the funiculus connects the developing ovule with the carpel. Both integuments grow through anticlinal divisions toward the apex of the ovule (8). The inner integument forms a cylinder around the nucellus while the outer integument continues its asymmetric growth, finally covering the inner integument and contributing to the curvature of the ovule (9). (B) In the final form of the ovule, the outer integument has completed its growth to cover the inner integument and to position the micropyle (arrowhead) near the funiculus to facilitate pollen tube entry. (C) Longitudinal section early in ovule development showing the division of the two to three surface cells that initiate formation of the inner and outer integument primordia, and the enlargement of the megasporocyte in the center of the nucellus. (D) Mature ovule showing the multiple cell layers of the integuments enclosing the expanded mature embryo sac. c, chalaza; e, embryo sac; f, funiculus; i, inner integument (primordium); o, outer integument (primordium); n, nucellus. *Panels (C) and (D): Reprinted from Skinner, D. J., Hill, T. A., & Gasser, C. S. (2004). Regulation of ovule development.* The Plant Cell, 16, *S32–S45 (www.plantcell.org), © American Society of Plant Biologists.*

gene involved in integument identity. Combining a *bel1* mutant with mutations in two C3HDZ genes results in expansion of the *WUS* expression domain into and below the chalaza, resulting in the subsequent formation of ectopic growths from the ovule (Yamada, Sasaki, Hashimoto, Nakajima, & Gasser, 2016).

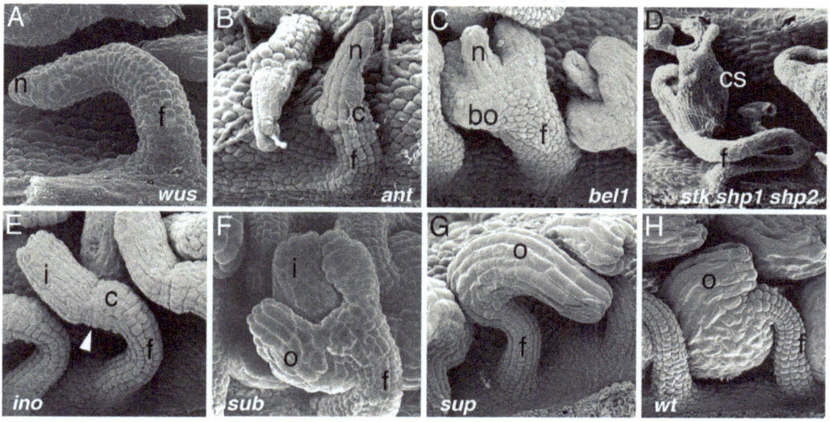

Fig. 2 Some *A. thaliana* ovule mutants. Scanning electron micrographs of: (A) *wus*, (B) *ant*, (C) *bel1*, (D) *stk shp1 shp2*, (E) *ino*, (F) *sub*, (G) *sup*, (H) wild-type ovules. *wus* (A) and *ant* (B) mutants lack integuments, but in the *wus* mutant, the ovule is elongated and the chalazal region is not well differentiated, while in *ant*, the chalaza is visible as enlarged cells, with nucellus above and funiculus below. In *bel1* mutants (C) the integuments are replaced by an irregular outgrowth from the chalazal region and the funiculus expands and loses cell file organization. In *stk shp1 shp2* triple mutants (D), integuments are replaced by a carpelloid structure and the funiculus elongates. Severe *ino* mutants (E) fail to initiate an outer integument from the gynobasal side (arrow), leaving the inner integument exposed. Outer integument growth is aberrant in *sub* mutants (F), and in *sup* mutants (G), the outer integument grows on both sides of the ovule. bo, *bel1* outgrowth; c, chalaza; cs, carpelloid structure; f, funiculus; i, inner integument; o, outer integument; n, nucellus. *Panel (A): Reprinted from Gross-Hardt, R., Lenhard, M., & Laux, T. (2002). WUSCHEL signaling functions in interregional communication during Arabidopsis ovule development.* Genes and Development, 16, *1129–1138. Panel (D): Courtesy Martin Yanofsky, reprinted from Skinner, D. J., Hill, T. A., & Gasser, C. S. (2004). Regulation of ovule development.* The Plant Cell, 16, *S32–S45 (www.plantcell.org), © American Society of Plant Biologists. Panel (F): Courtesy Kay Schneitz.*

In the *sporocyteless/nozzle* (*spl/nzz*) mutant, the terminal nucellar domain is reduced in size and fails to form a megasporocyte (Schiefthaler et al., 1999; Yang, Ye, Xu, & Sundaresan, 1999). That this mutant also fails to form microsporocytes may indicate that the reduced nucellus results from a failure to form sporocytes in general (Schiefthaler et al., 1999; Yang et al., 1999). However, observations indicating that *SPL/NZZ* also participates in the regulation of integument development genes (Balasubramanian & Schneitz, 2000) and *WUS* (Sieber, Gheyselinck, et al., 2004), and influences auxin movement in the nucellus (Bencivenga, Simonini, Benkova, & Colombo, 2012), imply a larger developmental regulatory role in ovule patterning for this putative transcription factor gene (Balasubramanian & Schneitz, 2002).

2.3 Ovule identity

Ovule primordia have formed in the context of floral and gynoecium identity and must be directed to follow the ovule development program. The carpel homeotic regulatory gene *AGAMOUS*, a MADS-box gene, is expressed in ovule primordia (Bowman, Drews, & Meyerowitz, 1991). Thus, other regulators must act with *AG* to distinguish ovule from carpel fate (Ray et al., 1994; Western & Haughn, 1999). Individual genes and gene combinations promoting the identity of ovule parts have been identified through mutant studies. Mutations in *BEL1* lead to formation of an initially amorphous structure in place of the integuments (Robinson-Beers et al., 1992) (Fig. 2C). In further development, this structure can form a carpelloid organ (Modrusan, Reiser, Feldmann, Fischer, & Haughn, 1994; Ray et al., 1994)—even to the point of producing a secondary set of also aberrant ovules (Ray et al., 1994). This conversion appears to be stochastic as some ovules instead produce more linear outgrowths from the aberrant structure, and these have sometimes been interpreted as abortive nucelli (Herr, 1995). One interpretation of these variable outcomes is that in the absence of *BEL1* activity, the chalaza is still directed to produce growth, but this growth does not have ovule identity. In the absence of such identity, the structure initiates a pathway for development of the prior (carpel) or subsequent (nucellus) structure and these pathways are self-reinforcing once initiated.

An even more complete loss of integument identity is produced by a combination of mutations in the three closely related MADS-box genes *SHATTERPROOF1* (*SHP1*), *SHP2*, and *SEEDSTICK* (*STK*). In the triple mutant, the funiculus is still usually visible, but the integuments are converted to a carpel-like organ (Brambilla et al., 2007; Pinyopich et al., 2003) (Fig. 2D). This work shows that the combined function of these three genes is a primary determinant of ovule identity. Interestingly, partial loss of function of the combined activity of the *SEPALATA* (*SEP*) genes can produce a similar phenotype, indicating that *SEP* activity may be necessary for the combined *SHP1 SHP2 STK* activity (Favaro et al., 2003). Mutations in these genes interact synergistically with *bel1* and there is evidence for a physical interaction between the protein products of these genes, indicating that they may work together to establish integument identity (Brambilla et al., 2007).

2.4 Integument development

2.4.1 Transcription factors

The formation of the integuments in *A. thaliana* has been extensively studied. As noted above, the *WUS* and *ANT* activities are essential to integument formation as neither integument forms when either gene is inactivated, and these mutations are epistatic to other mutations affecting integument development (such as *bel1*) (Baker, Robinson-Beers, Villanueva, Gaiser, & Gasser, 1997; Elliott et al., 1996; Klucher et al., 1996). While these activities appear to ensure that growth occurs from the chalazal region, they do not specify the identity and form of the resulting structures. Thus, the structure that grows from the chalaza in the *bel1* mutant is amorphous and can later adopt non-integument identity as noted above. With the growth and identity factors present, the structures that form have integument identity, but still depend on additional factors for proper formation.

Generation of integument shape from a ring or partial ring primordium requires differential growth of cell layers closest to the nucellus (adaxial layers) and those closest to the base of the ovule (abaxial). Integuments therefore have ab-adaxial polarity, which determines the direction of cell growth and which can also be observed later in development by differentiation of specific cell layers in the integuments. Ab-adaxial polarity in lateral organs such as leaves, sepals, petals and carpels has been shown to be affected by the function and genetic interaction of a few key gene families. Members of these families also affect growth and shape of the integuments.

The most important known genes for formation of the outer integuments are *INO* (Villanueva et al., 1999), and *KANADI1* (*KAN1*) and *KAN2* (Eshed, Baum, Perea, & Bowman, 2001; McAbee et al., 2006). These genes are members of the YABBY and KANADI families, respectively, that were initially identified by their functions in lateral organ polarity and expansion (Bowman & Smyth, 1999; Eshed et al., 2001; Sawa et al., 1999). In leaves, members of these families are expressed in abaxial tissues where they promote abaxial identity and blade expansion (Eshed, Izhaki, Baum, Floyd, & Bowman, 2004; Sarojam et al., 2010). *INO* is expressed abaxially in the outer integument (Villanueva et al., 1999), and mutant studies indicate the same for *KAN1* and *KAN2* (McAbee et al., 2006). *INO*, the only YABBY gene expressed in ovules, appears to be the more essential of the two classes of genes as severe *ino* mutations lead to a complete absence of the outer integument (Baker et al., 1997; Villanueva et al., 1999) (Fig. 2E). This phenotype is more severe than the incomplete loss of laminar expansion observed in leaves with no YABBY function (Sarojam et al., 2010). The *kan1 kan2* double mutant grows an amorphous structure in place of the outer integument (Eshed et al., 2001; McAbee et al., 2006). This indicates that *INO* is essential for cell division in the outer integument, while *KAN* function is dispensable for cell division, but is essential for directed laminar growth. None of these three mutations affect the formation of the inner integument, which still grows to cover the nucellus.

In contrast to the above mutations which interrupt/fail to establish outer integument growth, the shape of the outer integument in the *superman* (*sup*) mutant is altered by excessive growth on the gynoapical side of the ovule (Gaiser, Robinson-Beers, & Gasser, 1995) (Fig. 2G). The activity of SUP, a transcriptional repressor (Hiratsu, Ohta, Matsui, & Ohme-Takagi, 2002), was shown to confine *INO* expression to the gynobasal side of the ovule, with expression expanding to both sides of the ovule in the *sup* mutant (Meister, Kotow, & Gasser, 2002). This interaction confirms the role of *INO* as sufficient as well as essential for outer integument outgrowth in the ovule context.

Another KANADI gene, *ABERRANT TESTA SHAPE* (*ATS* aka *KAN4*), plays a role in inner integument development and integument separation. *ats* mutants have a single integument that results from congenital fusion of the inner and outer integuments into a single structure (Léon-Kloosterziel, Keijzer, & Koornneef, 1994; McAbee et al., 2006). ATS was shown to form a protein complex with AUXIN RESPONSE FACTOR3/ETTIN (ARF3/ETT), and mutation of the *ETT* gene produces the same integument fusion phenotype seen in *ats* (Kelley, Arreola, Gallagher, & Gasser, 2012). Similarly, KAN function in leaves depends on physical interaction with the ARF proteins ETT and ARF4 (Pekker, Alvarez, & Eshed, 2005). Both *ETT* and *ATS* are expressed at the border between the inner and outer integuments, and in the outer (abaxial) layer of the inner integument (Kelley et al., 2012; McAbee et al., 2006). If the outer integument is eliminated (by the *ino* mutation), the absence of *ATS* activity leads to an amorphously growing inner integument, similar to the effects of *kan1 kan2* mutants on the outer integument (McAbee et al., 2006). Thus, the KANADI genes appear to have similar roles in appropriately directing growth into planar structures for both the outer and inner integuments.

In leaves, C3HDZ family genes define the adaxial zone of primordia and influence the continued life of the shoot apical meristem (reviewed in Bowman & Floyd, 2008). *KAN* and C3HDZ family genes have an antagonistic spatial relationship in leaves that ensures the presence and juxtaposition of ab- and adaxial zones in the primordia. The border between these zones is important for laminar outgrowth of lateral organs (Waites & Hudson, 1995), which also depends on YABBY activity (Eshed et al., 2004; Sarojam et al., 2010). As seen for leaf development, integument shape and outgrowth are promoted by the abaxial functions provided by YABBY and KAN family genes, and adaxial functions of the C3HDZ genes *PHABULOSA* (*PHB*), *PHAVOLUTA* (*PHV*), *CORONA* (*CNA*), and *REVOLUTA* (*REV*). However, there are differences in how these genes participate in this interaction.

PHB, *PHV* and *CNA* are expressed adaxially in the inner integument but cause reduced growth in both integuments when multiple family members are mutated (Kelley, Skinner, & Gasser, 2009). Similarly, ectopic expression via dominant alleles *phb1-d* and *phv1-d* reduces integument growth (Sieber, Gheyselinck, et al., 2004). Surprisingly, the *ATS* zone of expression is not affected in the dominant mutants, and C3HDZ expression domains do not expand in *ats* mutants. This suggests another adaxial factor in the inner integument. The role of *REV*, expressed across both integuments, remains unclear and may be to promote adaxial activity in the outer as well as inner integument, and to help establish the gynoapical zone of the chalaza, thereby keeping *INO* from being expressed there (Kelley et al., 2009). One structural difference as compared with leaf primordia is the close association between the two integument primordia, resulting in multiple borders between ab- and adaxial zones. In this case, a balance between the activities of abaxial and adaxial regulators may be needed to define the zones of growth as well as the area of no growth between the integuments. Expression domains of C3HDZ genes are regulated by the *miR165/6* genes and, specifically in the ovule, a subset of *miR166* genes are active in intricate patterns that combine to confine *PHB* expression to the inner integument (Hashimoto, Miyashima, Sato-Nara, Yamada, & Nakajima, 2018).

C3HDZ gene mutants show an interesting interaction with *bel1* in acting to maintain the chalazal area free from *WUS* expression (Yamada et al., 2016). *WUS* expands into the chalaza in mutants affecting *BEL1* and the C3HDZ genes, and this expansion is at least partly responsible for the reduction in integument growth in C3HDZ loss-of-function mutants. In *cna phb phv bel1* mutants, ectopic *WUS* expression in the chalaza leads to a striking branched ovule phenotype, where new zones of *WUS* expression precede formation of additional ovule primordia below the integuments (Yamada et al., 2016). This intriguing phenotype suggests that C3HDZ genes not only define adaxial regions of the planar integuments, but also act to separate the chalazal zone from the nucellus.

2.4.2 Coactivators, corepressors and receptors

Transcription factors rarely act by themselves in controlling gene expression, rather acting by recruiting coregulators to their target genes. The SPL/NZZ protein has been shown to interact with the TOPLESS (TPL) and TOPLESS-RELATED (TPR) corepressors that are necessary for its action in repression of target genes (Bonaccorso, Lee, Puah, Scutt, & Golz, 2012; Wei et al., 2015). These results further support the role of SPL/NZZ as a transcription factor and provide a mechanism for its function. Evidence has been presented indicating that YABBY proteins have both positive and negative effects on the expression of downstream target genes (Bonaccorso et al., 2012). YABBY proteins, including orthologs of INO, have been shown to interact with the LEUNIG (LUG) corepressor and orthologs in both *A. thaliana* (Stahle, Kuehlich, Staron, von Arnim, & Golz, 2009) and *Antirrhinum majus* (Navarro et al., 2004; Sridhar, Surendrarao, & Liu, 2006). The interaction of LUG with INO was confirmed in *A. thaliana*, where INO was also shown to interact with SEUSS (SEU), a known co-repressive partner of LUG (Simon, Skinner, Gallagher, & Gasser, 2017). Both LUG and SEU are necessary for normal outer integument growth, indicating a participation in part of the activity of INO. Furthermore, INO was shown to interact with ADA2b, a partner of the coactivator GCN5, and loss of ADA2b function also leads to decreased outer integument growth (Simon et al., 2017). The more extreme phenotype of the *ada2b lug* double mutant indicates that the two proteins participate in parallel processes of positive and negative gene regulation by *INO*. That both classes of coregulators also interact with other YABBY proteins provides a possible explanation for the positive and negative regulatory activities of other members of this family (Simon et al., 2017).

In addition to transcription factors, other types of regulatory proteins have been associated with regulation of integument development. Two different putative cell surface receptor classes, containing STRUBBELIG (SUB) and ERECTA (ER) and related proteins, have been shown to be important for ovule development. The ER class proteins are leucine-rich repeat receptor kinases that have multiple roles in plant development, including the well-known role of ER in plant stature and cell elongation (Shpak, 2013). Mutation of single genes of the family does not noticeably affect ovule development, but combinations such as an *er erl1 erl2/ERL2* mutant line have reduced inner and outer integuments (Pillitteri, Bemis, Shpak, & Torii, 2007). While members of the SUB class of genes also have multiple roles in plant development, *sub* mutants have severe effects on outer integument growth, leading to formation of a blade-like structure on only one side of the ovule (Chevalier et al., 2005) (Fig. 2F). SUB and related proteins are unusual in that they include a region with clear similarity to the kinase domains of proteins like ER but they do not exhibit kinase activity (Chevalier et al., 2005). Further results show that the protein sorting component HAPLESS13 (HAP13) is necessary for proper membrane localization of the SUB protein to support ovule development (Wang et al., 2016). The participation of two classes of putative membrane-bound receptors suggests roles for the sensing of diffusible receptor ligands in integument growth, but the identity of such ligands remains undetermined.

2.4.3 Hormones

Plant hormones affect almost every aspect of plant development including ovules. Local auxin maxima have been shown to be associated with the emergence of plant organs (Reinhardt, Mandel, & Kuhlemeier, 2000). Accumulations that result in emergence of leaf primordia are driven by directional localization of the auxin transporter PIN-FORMED1 (PIN1) in surface cells at the shoot apex (Benkova et al., 2003). Similar accumulations of auxin were found to be present at the growing tips of ovule primordia, and at the sites of initiation of the integuments (Benkova et al., 2003). Accumulations of PIN1 were present in adjacent surface cells oriented such that auxin would be directed to the growing tips (Bencivenga et al., 2012; Benkova et al., 2003). That these gradients drive growth of ovules and integuments was demonstrated by the severe effects on ovule development produced by treatment with auxin transport inhibitors (Benkova et al., 2003). Similarly, while strong *pin1* mutants do not produce ovules, the weak *pin1–5* mutant does produce ovules, with about 10% completely lacking integuments (Bencivenga et al., 2012).

PIN1 also appears to serve as a link to other regulators of ovule development. Notably, plants mutant in three cytokinin receptors, *ARABIDOPSIS HISTIDINE KINASE2* (*AHK2*), *AHK3* and *AHK4*, produce aberrant ovules resembling those of the *pin1–5* mutant (Bencivenga et al., 2012). Furthermore, these plants fail to accumulate PIN1 protein in ovules. Similarly, treatment of plants with benzylaminopurine (BAP, a synthetic cytokinin) led to ectopic accumulation of PIN1 in the chalazal region, and formation of a single amorphous structure in place of the integuments (Bencivenga et al., 2012). This shows that cytokinin signaling promotes PIN1 accumulation and localization in ovules during integument development, such that cytokinin signaling is an indirect regulator of auxin gradient formation during ovule development (Bencivenga et al., 2012). The misdistribution of PIN1 and resulting altered auxin accumulation could explain the aberrant growth from the chalaza in BAP treated plants. The *spl/nzz* mutant also exhibits dramatically reduced accumulation of PIN1 in ovules, and BAP did not induce PIN1 expression in the *spl/nzz* mutant (Bencivenga et al., 2012). Thus, SPL/NZZ may be a critical regulator of PIN1 accumulation and an intermediate player in cytokinin induction of PIN1 accumulation. The pattern of PIN1 accumulation is also mediated by *CUC1* and *CUC2* as mutation of these two genes leads to a failure to focus PIN1 into spots predicting ovule primordium emergence in the placenta (Galbiati et al., 2013). The spacing of the resulting ovules is aberrant. *CUC1* and *CUC2* were also able to activate transcription from the *PIN1* promoter in a heterologous system and thus may directly regulate *PIN1* expression to define locations of ovule initiation (Galbiati et al., 2013). BAP is shown to restore ovule primordium formation in *cuc1 cuc2* but not *pin1–5* mutants, respectively, indicating that cytokinin signaling is likely downstream of *CUC* activity, but upstream of *PIN1* activity (Galbiati et al., 2013).

The participation of auxin gradients in ovule and integument development is clear, but its mode of regulating development remains largely undetermined. One mode of action of auxin is through modulation of the activity of ARF proteins (Liscum & Reed, 2002; Weijers & Wagner, 2016). Two ARFs, MONOPTEROS (MP/ARF5) and ETT/ARF3, have been associated with ovule development. MP has been shown to directly bind the promoters of *ANT* (Yamaguchi et al., 2013), *CUC1* and *CUC2* (Galbiati et al., 2013), and expression of all three genes is reduced in *mp* mutants (Galbiati et al., 2013). Thus, MP could be involved in a feedback loop between these three proteins and auxin in promotion of ovule primordium formation (Galbiati et al., 2013). *ETT* is required for normal extension of the inner integument and for separation of the two integuments (Kelley et al., 2012). ETT has also been shown to be able to directly respond to auxin, such that its interaction with transcription factor partners, including ATS, is disrupted (Simonini et al., 2016). KAN proteins like ATS appear to be repressors of growth (Eshed et al., 2001) and have been shown to down-regulate growth-promoting and auxin pathway genes (Merelo et al., 2013; Reinhart et al., 2013). The sensing ability of ETT could cause the growth-repressive activity of the ATS-ETT complex to be confined to the point between the two auxin maxima at the growing points of the two integuments (since auxin would disrupt this activity), promoting the separation of the two integuments and laminar growth of the inner integument. The auxin maximum could also be affected by direct action of ETT (likely in a complex with ATS) on the transcription of *PIN1* (Galbiati et al., 2013; Simonini, Bencivenga, Trick, & Ostergaard, 2017). This could provide a mechanistic explanation of the interplay of these transcription factors and auxin action in integument growth.

Plants defective in the function of multiple DELLA gibberellin-sensing proteins have abnormal ovules, indicating a role for gibberellin sensing in ovule development (Gomez, Ventimilla, Sacristan, & Perez-Amador, 2016). DELLA proteins were shown to interact with ATS, but the DELLA loss-of-function phenotype differs from that of *ats* (Galbiati et al., 2013). This indicates that the DELLA proteins must have other activities in the ovule besides a possible role in modulating ATS activity.

3. Ovule evolution

3.1 Angiosperm ovule diversification

The ovules of the earliest diverging groups of angiosperms uniformly have two integuments (bitegmic) and are mostly recurved so that the micropyle is proximal to the funiculus (anatropous). However, notable exceptions of erect ovules with the micropyle directed away from the funiculus (orthotropous) are also observed (Endress & Igersheim, 2000). The plesiomorphic (ancestral) state in the angiosperms is thus clearly bitegmic, and most likely anatropous (Endress, 2011). While ovule form is relatively conserved within numerous clades among the angiosperms, significant divergences in morphology, anatomy, size, and integument number are also observed. Reduction in integument number is one aspect that has engendered considerable study and can serve to illustrate such divergence and its analysis. Multiple separate lineages include basal groups with the plesiomorphic bitegmic state and more derived species with ovules with reduced integument number, indicating multiple independent derivations of a reduced integument number (Fig. 3) (Bouman, 1984; Philipson, 1974; Stebbins, 1974). One such example is seen in the rosids where a small number of unitegmic species have arisen within this otherwise uniformly bitegmic group (Lora, Hormaza, & Herrero, 2015). The large asterid clade is primarily unitegmic with bitegmy being observed only in the most basal groups (McAbee, Kuzoff, & Gasser, 2005), indicating an independent derivation of unitegmy. The most extreme examples of derived alteration in integument number are observed in the Santalales, where basal groups have two integuments but there is progressive reduction among the crown groups to one or no integuments (Bouman, 1984; Brown, Nickrent, & Gasser, 2010) (Fig. 3).

Ovule reductions from two to one integument appear to result from either the loss of an integument or a failure of integuments to form as separate entities (Bouman, 1984; McAbee et al., 2005). Notably, these two mechanisms for a shift from bitegmy to unitegmy have been observed in

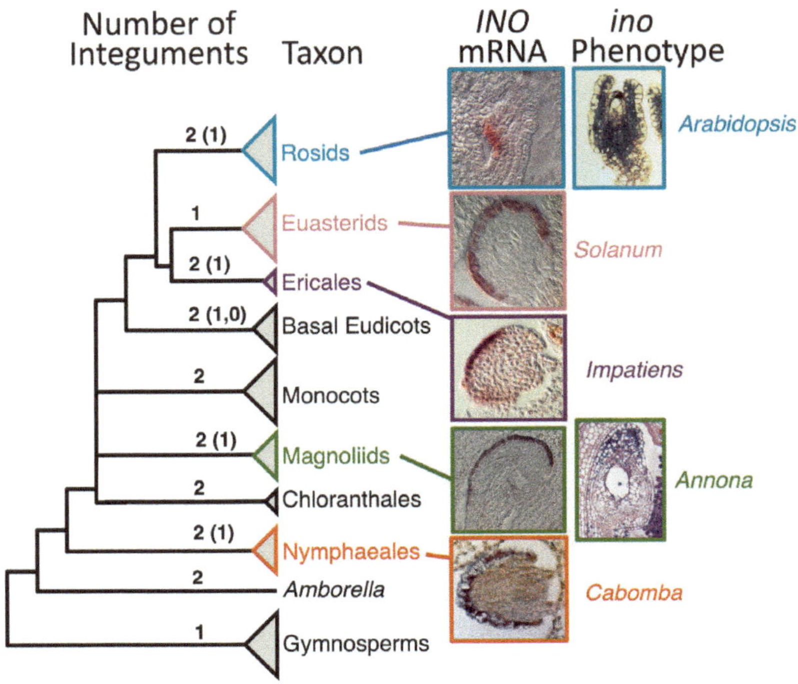

Fig. 3 Simplified phylogeny of angiosperms showing integument number variation and conservation of *INO* expression and function. The simplified phylogeny was based on multiple recent analyses including Kuzoff and Gasser (2000) and Moore, Bell, Soltis, and Soltis (2007). Numbers on clades indicate the number of integuments in ovules of that group with plesiomorphic and also the predominant number first, with derived states in parentheses. "*INO* mRNA" panels are *INO* expression patterns in developing ovules of the indicated genera as revealed by *in situ* hybridization. Expression is uniformly in a single outer layer of the developing outer or single integument in each examined species. "*ino* Phenotype" indicates the unitegmic, orthotropous (erect) ovule form that results from loss of *INO* function in the normally bitegmic, anatropous ovules of both *A. thaliana* and *A. squamosa*. Cabomba *image courtesy of Toshihiro Yamada and* Annona *images courtesy of Jorge Lora.*

A. thaliana mutants. In *ino* mutants, the outer integument fails to form, resulting in a unitegmic erect ovule (Baker et al., 1997; Villanueva et al., 1999) (Fig. 2E). In the *ats* and *ett* mutants, a single integument forms as a result of a failure to form the separation between the inner and outer integument primordia (Kelley et al., 2012; Léon-Kloosterziel et al., 1994; McAbee et al., 2006). All three mutants are able to produce functional seeds and hence recapitulate the processes expected for evolutionary derivation of unitegmy. Complete absence of integuments requires a failure of either integument to emerge (Bouman, 1984; Brown et al., 2010). A complete loss of integuments in observed in the *ant* (Baker et al., 1997; Elliott et al., 1996; Klucher et al., 1996) (Fig. 2B) and *wus* (Gross-Hardt et al., 2002) (Fig. 2A) mutants of *A. thaliana*, but these mutants are infertile (in contrast to the fertile ovules of naturally ategmic plants).

If the functions of orthologs of the genes in model systems described above were conserved among angiosperms, then evolutionary variation in ovule form could result from changes in their expression patterns or functions. Conservation of expression pattern could also enable the utilization of the genes as markers to study structures and events in ovule development. The highly refined pattern, and easy phylogenetic identification of *INO* orthologs, has enabled an analysis of the pattern of *INO* expression across widely divergent angiosperms. Fig. 3 shows expression of *INO* orthologs in the outer layer of the outer or single integument of species spanning most of the phylogenetic range of angiosperms, including Nymphaeales, magnoliids, and crown eudicots in both the rosid and asterid clades. This pattern is observed in every species of angiosperm evaluated to date, providing a first indication of conservation of *INO* function across the angiosperms. Studies of *ino* mutants in *A. thaliana* showed that *INO* is necessary for growth of the outer integument (Meister et al., 2002; Villanueva et al., 1999). More recently, analysis of a spontaneous mutant in the magnoliid *Annona squamosa* showed that this function of *INO* genes is conserved to near the base of the angiosperms (Lora, Hormaza, Herrero, & Gasser, 2011) as illustrated in Fig. 3. A similar function for *INO* was demonstrated for the single integument of *Nicotiana benthamiana*, a representative euasterid (Skinner, Brown, Kuzoff, & Gasser, 2016). In combination with the extensive demonstration of a conserved expression pattern, these results indicate that the role of *INO* genes is conserved across the angiosperms. Some evidence of conservation of function in ovule development has also been shown for several other genes. For example, a mutation in an ortholog of *SUP* was examined in petunia and found to make a change in ovule morphology highly similar to that observed in the sup mutant of *A. thaliana* (Nakagawa, Ferrario, Angenent, Kobayashi, & Takatsuji, 2004). This indicates conservation of this gene's function in ovule development across the crown eudicots. The pattern of *CUC* expression during ovule initiation was conserved between *A. thaliana* (rosid) and pea and tomato (euasterids) (Goncalves et al., 2015). Although functional conservation was not demonstrated, the conservation of expression pattern indicates the possibility of conservation of function of this gene class in ovule initiation among crown eudicots.

Evolutionary reduction in integument number has been studied in several systems. Reduction from two to one integument was examined in the genus *Impatiens*, a member of the Ericales, an early branching group of asterids that shows variation in integument number from two to one (McAbee et al., 2005). This contrasts with the larger sister euasterid clade that uniformly has unitegmic ovules. McAbee et al. (2005) used anatomical studies and molecular and histochemical markers to show that the single integument of unitegmic species included tissues derived from the outer integument (identified by *INO* ortholog expression) and the inner integument (identified by the darkly staining endothelium). The derived "congenital fusion" of the integuments resulted from a failure to form a separation between the two primordia during development due to a shift in patterns of cell division and expansion (McAbee et al., 2005). These results also explained *Impatiens* species with intermediate morphology where the two integuments were largely fused and only separated at the tips. Skinner et al. (2016) studied ovule development in the solanaceous euasterids *N. benthamiana* and *Solanum lycopersicum* and concluded that a similar congenital fusion of two ancestrally separate integuments was responsible for the single integument in these species. Because unitegmy is common to all euasterids, this process is likely to have occurred in the last common ancestor (LCA) of the euasterids. Thus, a similar process occurred in the examined *Impatiens* species and the euasterid LCA, but these two events must be independent as they do not lie along the same lineage. Further examination of the distribution of integument number among *Impatiens* species shows that unitegmy evolved independently several times within this lineage, and even that apparent reversals from partial unitegmy to full bitegmy occurred (McAbee et al., 2005). These relatively rapid changes are observed despite a prior prolonged conservation of the bitegmic state since the origin of the angiosperms, implying a shift to a more plastic state in the common ancestor of the asterids. This plastic state persisted in the Ericales, and particularly *Impatiens* (see below concerning a similar plastic state in the Santalales), but a new stably fixed unitegmic state appears to have been established in the euasterid lineage.

Notably, the process responsible for unitegmy in asterids is remarkably similar to that observed in the *ats* and *ett* mutants of *A. thaliana* (Kelley et al., 2012; McAbee et al., 2006), although no work to link these mutations to the phenotypes in these naturally unitegmic species was performed in the above studies on asterids. However, evidence of such a link was found in a different case in the elegant experiments of Lora et al. (2015) on the derived unitegmy in *Prunus*, a rosid genus. They examined *Prunus* species with bitegmic, unitegmic, and intermediate bifid integumented (integuments separated only at the tips) ovules. They further identified and evaluated the expression of orthologs of *INO*, *ATS*, and *ETT* in all three ovule types. They found in the bitegmic and intermediate ovules that the patterns of expression of all three genes were conserved relative to *A. thaliana*, with *INO* expressed in the outer layer of the outer integument, and *ATS* and *ETT* expressed in the outer layer of the inner integument and inner layer of the outer integument. This indicated likely conservation of function of these genes in the bitegmic *Prunus* species. *INO* was expressed in the outer layer of the single integument of the unitegmic species, and developmental studies indicated that unitegmy resulted from congenital fusion of the two integuments (Lora et al., 2015), similar to what was seen for unitegmic asterids (McAbee et al., 2006; Skinner et al., 2016). Expression of *ATS* was only at the tip of the single integument in the unitegmic species, while *ETT* expression was absent from the developing integument. Since loss of *ETT* function in ovules of *A. thaliana* results in the fused integument phenotype, this correlation of loss of *ETT* expression with the fusion of the integuments could provide a mechanistic explanation for integument fusion. The actual genetic change could be in the *ETT* gene itself, or in a regulator of *ETT*, but the absence of expression could be the direct cause of fusion.

More extreme cases of integument reduction are seen in the Santalales, a large group of mostly parasitic plants (Nickrent, Malecot, Vidal-Russell, & Der, 2010). Bitegmic ovules are the plesiomorphic state in the Santalales, but multiple derivations of unitegmy and complete loss of integuments (ategmy), as well as more extreme ovule reductions are observed (Bouman, 1984; Brown et al., 2010; Eames, 1977). Ategmic ovules of this group superficially resemble *ant* and *wus* mutant ovules in *A. thaliana*, except that the former are fertile. To evaluate the presence of different ovule tissues in reduced ovules and to test hypotheses of possible genes involved in derived ategmy, Brown et al. (2010) isolated orthologs of the *A. thaliana* genes *BEL1* and *ANT* from representative Santalalean species. They found that in the bitegmic species, the patterns of expression were similar to those observed in *A. thaliana*, consistent with a conserved role for the genes in ovule development. These genes also showed similar expression in unitegmic ovules. Furthermore, both genes were expressed in the surface cells of ategmic ovules in regions where integuments would have emerged in integumented ovules. This indicates that loss of expression of *ANT* or *BEL1* was not responsible for the failure in integument formation. It also may indicate that integument properties are maintained in these tissues of the ovule, performing their functions in the absence of morphologically distinct integuments. Further ovule reductions are observed in this group. Some species not only lack integuments, but morphologically distinct ovules never emerge and archesporial cells and embryo sacs form directly from subepidermal cells at multiple locations in a placenta (Bhandari & Nanda, 1968). The most extreme known case of ovule and gynoecium reduction occurs in species of the *Balanophora* genus, where an entire female flower consists simply of a gynoecium comprising a single carpel that is only two to four cell layers wide. The embryo sac differentiates directly within this structure without ovule formation (Eberwein, Nickrent, & Weber, 2009; Fagerlind, 1945).

The multiple independent changes in integument number in the Santalales indicate an especially plastic state in this group, mirroring the plastic state observed in the basal asterids. The emergence of a plastic state after millennia of conservation of integument number implies some genetic change in each of these groups that allows for easier alteration to reduced integument number. This can be followed by an additional change, leading to a new stable state, as in the persistence of derived unitegmy in the euasterids (Fig. 3).

3.2 Origin of the angiosperm ovule

The extant seed plants are monophyletic in recent plant phylogenies with ovules, and the resulting seeds, as a common and defining feature (synapomorphy) of this group (Doyle, 2013). The origins of the parts of the angiosperm ovule remain under discussion as evolutionary intermediates have not survived and would only be represented among fossils. Models consistently view the nucellus as a megasporangium. On the basis of fossil forms, the single integument of gymnosperms enclosing the nucellus is most commonly hypothesized to derive from the fusion of surrounding sterile or sterilized terminal appendages, or telomes, of a dichotomous ultimate branch (Andrews, 1963; Doyle, 2013; Herr, 1995; Kenrick & Crane, 1997). Notably the evolution of leaves and ovules was roughly contemporaneous, so that the derivation of gymnosperm integuments and leaves was independent, and while they both are hypothesized to derive from similar structures (dichotomous branch systems terminating in telomes), one is not believed to have been derived from the other.

Among extant seed plants, the outer integument is a synapomorphy of angiosperms as the unitegmic state is derived within this group (Fig. 3). Because the outer integument is a laminar structure surrounding a unitegmic ovule, ovule-bearing enclosures, referred to as cupules, in Permian and Mesozoic plants such as peltasperms, corystosperms, glossopterids, *Caytonia*, and *Petriellaea* have been examined as possible bitegmic ovule precursors (Andrews, 1963; Doyle, 2008, 2013). Notably, cupules of different lineages may not be homologous and are interpreted variously as having ovules borne on the abaxial or adaxial surfaces. The recurved cupules of *Caytonia nathorsti* (Harris, 1933) (Fig. 4A), a Middle Jurassic seed plant, have been examined as progenitors of outer integuments. The cupule shape resembles the presumably plesiomorphic anatropous shape of the ovules of early angiosperms, and *C. nathorsti* also has other angiosperm-like features (Doyle, 2008). While *Caytonia* cupules include multiple unitegmic erect ovules, a hypothesized reduction to a single ovule in each cupule would result in an angiosperm-like structure with the cupule becoming the outer integument (Doyle, 2008). Additional fossils with recurved cupules have been uncovered that support the hypothesis of reduction in enclosed ovule number in some lineages. For example, *Umkomasia* cupules contain fewer ovules than observed in *Caytonia*,

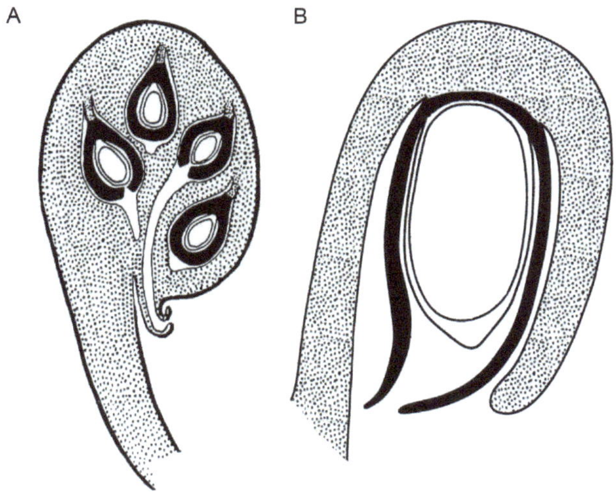

Fig. 4 Reconstructions of cupules of Mesozoic seed ferns. (A) The cupule (stippled) of *Caytonia nathorsti* is recurved and includes several orthotropous ovules with single integuments (black) housing the nucelli (unfilled). (B) A cupule (stippled) of *Umkomasia resinosa* is recurved and includes one ovule with a single integument (black) housing the nucellus (unfilled). *Panel (A): Reproduced from Harris, T. M. (1933). A new member of the Caytoniales.* The New Phytologist, 32, *97–113.* © John Wiley and Sons. *Panel (B): Redrawn from Klavins, S. D., Taylor, T. N., & Taylor, E. L. (2002). Anatomy of Umkomasia (Corystospermales) from the Triassic of Antarctica.* American Journal of Botany, 89, *664–676. © John Wiley and Sons.*

with some cupules containing only a single ovule (Klavins, Taylor, & Taylor, 2002; Zan, Axsmith, Fraser, Liu, & Xing, 2008). Fig. 4B shows a reconstruction of a recurved cupule of *Umkomasia resinosa* containing a single ovule (Klavins et al., 2002), with an overall appearance much like an angiosperm ovule. While this shows that reduction to a single ovule in a cupule was a feature of seed ferns, its affinity to an angiosperm ovule is unclear because, in contrast to the cupules of *Caytonia*, *Umkomasia* cupules are interpreted as having ovules on the abaxial surface (Klavins et al., 2002; Taylor, Delfueyo, & Taylor, 1994).

The cupules of the Permian and Mesozoic fossil plants are hypothesized to be modified foliar organs (Crane, 1985; Doyle, 2008), and some molecular evidence is consistent with such an origin for the outer integument. As noted in Section 2.4.1, the development of the outer integument in *A. thaliana* is dependent on the abaxial expression of YABBY (*INO*) and KANADI (*KAN1/2*) genes (Eshed et al., 2004; McAbee et al., 2006). This is similar to abaxial requirements of leaves for blade expansion (Eshed et al., 2004; Sarojam et al., 2010), and the origin of YABBY genes appears to be associated with the origin of the megaphylls of seed plants (Floyd & Bowman, 2007). As previously noted, the function of *INO* appears to be conserved across the angiosperms, so abaxial YABBY gene expression is a conserved feature of outer integument development. Notably, there is no evidence of any *YABBY* gene expressed in inner integuments. Rather, of these polarity genes, only abaxial KANADI (*ATS*) expression is required for inner integument development (Kelley et al., 2012; McAbee et al., 2006). Evolution of abaxial KANADI gene expression has been shown to precede the evolution of leaves and outer integuments (Floyd & Bowman, 2007) and would therefore likely have been a feature of telomes hypothesized to fuse to form the first integument of all seed plants. Adaxial expression of C3HDZ genes is also required for normal inner integument formation, and this pattern of expression also appears to precede the evolution of leaves (Floyd & Bowman, 2007). Thus, the molecular evidence is consistent with the inner integument of angiosperm ovules being homologous with the single integument of gymnosperm ovules as a structure directly evolving from fusion or planation of precursor telomes. Similarly, the cupule/leaf derivation of the outer integument is supported by the common expression pattern of a YABBY gene in these structures, with the ovules being borne on the adaxial side. While C3HDZ gene expression is required for normal outer integument growth (Kelley et al., 2009), it has not been found confined to the adaxial side (Sieber, Petrascheck, Barberis, & Schneitz, 2004).

Because adaxial C3HDZ gene expression is a feature of leaves (McConnell et al., 2001; Prigge et al., 2005), this is one inconsistency in the model of derivation of the outer integument from a leaf-like organ, but this could be a derived feature of cupules/outer integuments in general, or simply a derived feature in *A. thaliana* and related plants in particular. Overall, the presence of ovule-like features in the cupules of plants that precede angiosperms in the fossil record, and the molecular markers differentiating the inner and outer integuments make a sound case for the model of a type of cupule being the precursor of the angiosperm outer integument.

It is useful to evaluate the implications of the "cupule to outer integument" model for angiosperm ovule evolution. The outer integument is a synapomorphy of angiosperms, and we note that the same is true for the presence of an *INO* gene. Phylogenetic analyses of YABBY sequences show that the individual clades of such genes that are readily recognized among angiosperms are not differentiated in extant gymnosperms (Bartholmes, Hidalgo, & Gleissberg, 2012; Finet et al., 2016; Yamada et al., 2011). The *INO* clade is present only in angiosperms, linking the origin of this YABBY gene clade with the origin of the outer integument, and perhaps with origin of precursor cupules. Notably, the YABBY gene *CRABS CLAW* is closely associated with carpels and is also present only among angiosperms, linking derivation of this clade to a second major synapomorphy of angiosperms (Fourquin, Vinauger-Douard, Fogliani, Dumas, & Scutt, 2005; Yamada et al., 2011). Thus, elaboration of the YABBY family may have had a pivotal role in the evolution from leaves of the novel angiosperm reproductive structures of outer integuments and carpels.

The "cupule to outer integument" model also bears on the homologies between angiosperm and gymnosperm ovule structures. First, it indicates that the gymnosperm ovule is not homologous with the angiosperm ovule, but rather only with the nucellus, inner integument, and subtending tissue adjacent to the outer integument. This angiosperm homolog of the gymnosperm ovule is uniformly orthotropous as the outer integument is the primary determinant of anatropy in angiosperms (as seen in *A. thaliana* (Villanueva et al., 1999) and *A. squamosa* (Lora et al., 2011)) (Fig. 3). We note that loss of YABBY function in the outer integument leads to a failure of this structure to form (Lora et al., 2011; Villanueva et al., 1999), whereas loss of vegetative YABBY function does not prevent leaf formation, but rather prevents elaboration of the leaf blade (Sarojam et al., 2010). This could indicate that the outer integument (and hence the cupule) is a structure deriving from a leaf blade, rather than an entire leaf. This is consistent with a previously hypothesized homology of a cupule to a pinna or pinnule (a terminal structure in a compound leaf) (Crane, 1985; Doyle, 1978). The funiculus of the angiosperm ovule would not be homologous to the funiculus of a gymnosperm ovule, but rather with the axis, petiole, or petiolule supporting a single cupule. As noted above, a cupule can bear multiple ovules. Some *A. thaliana* mutants could potentially be atavistic representations of such a structure relative to the angiosperm ovule. The combination of the *tousled* and *ett* mutants eliminates the external carpel structures (carpel wall, stigma, and style) in *A. thaliana*, leaving only a small stipitate structure with multiple ovule primordia emerging from the edges (Roe, Nemhauser, & Zambryski, 1997). This structure has been interpreted as a "naked placenta" which would represent an axis harboring multiple ovules (Skinner, Hill, & Gasser, 2004). Yamada et al. (2016) found that combination of the *bel1* mutation with loss of three different C3HDZ genes led to ectopic ovule primordia emerging from the funiculus. These branched structures with multiple terminal ovules could also represent an earlier state where a branching axis produced multiple ovules. Interpretation of these mutant phenotypes as atavistic is, of course, speculative, but it does provide some additional support for the idea of branching ovulate axes in plants like *Caytonia*, *Umkomasia*, and other Mesozoic seed ferns as being representative of reproductive structures in angiosperm precursors.

Acknowledgments

We thank Martin Yanofsky, Toshihiro Yamada, Kay Schneitz, Rita Gross-Hardt and Jorge Lora for permission to use images, and James Doyle for helpful comments. This work was supported by U.S. National Science Foundation grant IOS-1354014 to C.S.G.

References

Andrews, H. N. J. (1963). Early seed plants. *Science*, *142*, 925–931.

Baker, S. C., Robinson-Beers, K., Villanueva, J. M., Gaiser, J. C., & Gasser, C. S. (1997). Interactions among genes regulating ovule development in *Arabidopsis thaliana*. *Genetics*, *145*, 1109–1124.

Balasubramanian, S., & Schneitz, K. (2000). *NOZZLE* regulates proximal-distal pattern formation, cell proliferation and early sporogenesis during ovule development in *Arabidopsis thaliana*. *Development*, *127*, 4227–4238.

Balasubramanian, S., & Schneitz, K. (2002). *NOZZLE* links proximal-distal and adaxial-abaxial pattern formation during ovule development in *Arabidopsis thaliana*. *Development*, *129*, 4291–4300.

Bartholmes, C., Hidalgo, O., & Gleissberg, S. (2012). Evolution of the *YABBY* gene family with emphasis on the basal eudicot *Eschscholzia californica* (Papaveraceae). *Plant Biology (Stuttgart)*, *14*, 11–23.

Bencivenga, S., Simonini, S., Benkova, E., & Colombo, L. (2012). The transcription factors BEL1 and SPL are required for cytokinin and auxin signaling during ovule development in *Arabidopsis*. *The Plant Cell, 24*, 2886–2897.

Benkova, E., Michniewicz, M., Sauer, M., Teichmann, T., Seifertova, D., Jurgens, G., et al. (2003). Local, efflux-dependent auxin gradients as a common module for plant organ formation. *Cell, 115*, 591–602.

Bhandari, N. N., & Nanda, K. (1968). Studies on the Viscaceae. II. A reinvestigation of the female gametophyte of *Arceuthobium douglasii*. *American Journal of Botany, 55*, 1028–1030.

Bonaccorso, O., Lee, J. E., Puah, L., Scutt, C. P., & Golz, J. F. (2012). *FILAMENTOUS FLOWER* controls lateral organ development by acting as both an activator and a repressor. *BMC Plant Biology, 12*, 176.

Bouman, F. (1984). The ovule. In B. M. Johri (Ed.), *Embryology of the angiosperms* (pp. 123–157). New York: Springer-Verlag.

Bowman, J. L., Drews, G. N., & Meyerowitz, E. M. (1991). Expression of the *Arabidopsis* floral homeotic gene *AGAMOUS* is restricted to specific cell types late in flower development. *The Plant Cell, 3*, 749–758.

Bowman, J. L., & Floyd, S. K. (2008). Patterning and polarity in seed plant shoots. *Annual Review of Plant Biology, 59*, 67–88.

Bowman, J. L., & Smyth, D. R. (1999). *CRABS CLAW*, a gene that regulates carpel and nectary development in *Arabidopsis*, encodes a novel protein with zinc finger and helix-loop-helix domains. *Development, 126*, 2387–2396.

Brambilla, V., Battaglia, R., Colombo, M., Masiero, S., Bencivenga, S., Kater, M. M., et al. (2007). Genetic and molecular interactions between BELL1 and MADS box factors support ovule development in *Arabidopsis*. *The Plant Cell, 19*, 2544–2556.

Brown, R. H., Nickrent, D. L., & Gasser, C. S. (2010). Expression of ovule and integument-associated genes in reduced ovules of Santalales. *Evolution and Development, 12*, 231–240.

Bruinsma, J. (2003). *World agriculture: Towards 2015/2030, Food and Agricultural Organization*. London, UK: Earthscan Publications Ltd.

Chevalier, D., Batoux, M., Fulton, L., Pfister, K., Yadav, R. K., Schellenberg, M., et al. (2005). *STRUBBELIG* defines a receptor kinase-mediated signaling pathway regulating organ development in *Arabidopsis*. *Proceedings of the National Academy of Sciences of the United States of America, 102*, 9074–9079.

Crane, P. R. (1985). Phylogenetic analysis of seed plants and the origin of angiosperms. *Annals of the Missouri Botanical Garden, 72*, 716–793.

Doyle, J. A. (1978). Origin of angiosperms. *Annual Review of Ecology and Systematics, 9*, 365–392.

Doyle, J. A. (2008). Integrating molecular phylogenetic and paleobotanical evidence on the origin of the flower. *International Journal of Plant Sciences, 169*, 816–843.

Doyle, J. A. (2013). Phylogenetic analyses and morphological innovations in land plants. *Annual Plant Reviews, 45*, 1–50.

Eames, A. J. (1977). *The ovule morphology of the angiosperms* (2nd ed.), (pp. 256–289). Huntington, NY: Robert E. Krieger Publishing Company, Inc.

Eberwein, R., Nickrent, D. L., & Weber, A. (2009). Development and morphology of flowers and inflorescences in *Balanophora papuana* and *B. elongata* (Balanophoraceae). *American Journal of Botany, 96*, 1055–1067.

Elliott, R. C., Betzner, A. S., Huttner, E., Oakes, M. P., Tucker, W. Q. J., Gerentes, D., et al. (1996). *AINTEGUMENTA*, an *APETALA2*-like gene of Arabidopsis with pleiotropic roles in ovule development and floral organ growth. *The Plant Cell, 8*, 155–168.

Endress, P. K. (2011). Angiosperm ovules: Diversity, development, evolution. *Annals of Botany, 107*, 1465–1489.

Endress, P. K., & Igersheim, A. (2000). Gynoecium structure and evolution in basal angiosperms. *International Journal of Plant Sciences, 161*, S211–S223.

Eshed, Y., Baum, S. F., Perea, J. V., & Bowman, J. L. (2001). Establishment of polarity in lateral organs of plants. *Current Biology*, *11*, 1251–1260.

Eshed, Y., Izhaki, A., Baum, S. F., Floyd, S. K., & Bowman, J. L. (2004). Asymmetric leaf development and blade expansion in *Arabidopsis* are mediated by KANADI and YABBY activities. *Development*, *131*, 2997–3006.

Fagerlind, F. (1945). Blüte und Blütenstand der Gattung *Balanaphora*. *Botaniska Notiser*, 330–350.

Favaro, R., Pinyopich, A., Battaglia, R., Kooiker, M., Borghi, L., Ditta, G., et al. (2003). MADS-box protein complexes control carpel and ovule development in Arabidopsis. *The Plant Cell*, *15*, 2603–2611.

Finet, C., Floyd, S. K., Conway, S. J., Zhong, B., Scutt, C. P., & Bowman, J. L. (2016). Evolution of the YABBY gene family in seed plants. *Evolution and Development*, *18*, 116–126.

Floyd, S. K., & Bowman, J. L. (2007). The ancestral developmental tool kit of land plants. *International Journal of Plant Sciences*, *168*, 1–35.

Fourquin, C., Vinauger-Douard, M., Fogliani, B., Dumas, C., & Scutt, C. P. (2005). Evidence that *CRABS CLAW* and *TOUSLED* have conserved their roles in carpel development since the ancestor of the extant angiosperms. *Proceedings of the National Academy of Sciences of the United States of America*, *102*, 4649–4654.

Gaiser, J. C., Robinson-Beers, K., & Gasser, C. S. (1995). The Arabidopsis *SUPERMAN* gene mediates asymmetric growth of the outer integument of ovules. *The Plant Cell*, *7*, 333–345.

Galbiati, F., Sinha Roy, D., Simonini, S., Cucinotta, M., Ceccato, L., Cuesta, C., et al. (2013). An integrative model of the control of ovule primordia formation. *Plant Journal*, *76*, 446–455.

Gomez, M. D., Ventimilla, D., Sacristan, R., & Perez-Amador, M. A. (2016). Gibberellins regulate ovule integument development by interfering with the transcription factor ATS. *Plant Physiology*, *172*, 2403–2415.

Goncalves, B., Hasson, A., Belcram, K., Cortizo, M., Morin, H., Nikovics, K., et al. (2015). A conserved role for *CUP-SHAPED COTYLEDON* genes during ovule development. *Plant Journal*, *83*, 732–742.

Gross-Hardt, R., Lenhard, M., & Laux, T. (2002). *WUSCHEL* signaling functions in interregional communication during *Arabidopsis* ovule development. *Genes and Development*, *16*, 1129–1138.

Harris, T. M. (1933). A new member of the Caytoniales. *The New Phytologist*, *32*, 97–113.

Hashimoto, K., Miyashima, S., Sato-Nara, K., Yamada, T., & Nakajima, K. (2018). Functionally diversified members of the *MIR165/6* gene family regulate ovule morphogenesis in *Arabidopsis thaliana*. *Plant and Cell Physiology*, *59*, 1017–1026.

Herr, J. M. (1995). The origin of the ovule. *American Journal of Botany*, *82*, 547–564.

Hiratsu, K., Ohta, M., Matsui, K., & Ohme-Takagi, M. (2002). The SUPERMAN protein is an active repressor whose carboxy-terminal repression domain is required for the development of normal flowers. *FEBS Letters*, *514*, 351–354.

Kelley, D. R., Arreola, A., Gallagher, T. L., & Gasser, C. S. (2012). ETTIN (ARF3) physically interacts with KANADI proteins to form a functional complex essential for integument development and polarity determination in Arabidopsis. *Development*, *139*, 1105–1109.

Kelley, D. R., Skinner, D. J., & Gasser, C. S. (2009). Roles of polarity determinants in ovule development. *Plant Journal*, *57*, 1054–1064.

Kenrick, P., & Crane, P. R. (1997). *The origin and early diversification of land plants: A cladistic study*. Washington, DC: Smithsonian Institution Press.

Klavins, S. D., Taylor, T. N., & Taylor, E. L. (2002). Anatomy of *Umkomasia* (Corystospermales) from the Triassic of Antarctica. *American Journal of Botany*, *89*, 664–676.

Klucher, K. M., Chow, H., Reiser, L., & Fischer, R. L. (1996). The *AINTEGUMENTA* gene of Arabidopsis required for ovule and female gametophyte development is related to the floral homeotic gene *APETALA2*. *The Plant Cell*, *8*, 137–153.

Krizek, B. (2009). *AINTEGUMENTA* and *AINTEGUMENTA-LIKE6* act redundantly to regulate Arabidopsis floral growth and patterning. *Plant Physiology*, *150*, 1916–1929.

Kuzoff, R. K., & Gasser, C. S. (2000). Recent progress in reconstructing angiosperm phylogeny. *Trends in Plant Science*, *5*, 330–336.

Léon-Kloosterziel, K. M., Keijzer, C. J., & Koornneef, M. (1994). A seed shape mutant of Arabidopsis that is affected in integument development. *The Plant Cell*, *6*, 385–392.

Liscum, E., & Reed, J. W. (2002). Genetics of Aux/IAA and ARF action in plant growth and development. *Plant Molecular Biology*, *49*, 387–400.

Lora, J., Hormaza, J. I., & Herrero, M. (2015). Transition from two to one integument in *Prunus* species: Expression pattern of *INNER NO OUTER* (*INO*), *ABERRANT TESTA SHAPE* (*ATS*) and *ETTIN* (*ETT*). *The New Phytologist*, *208*, 584–595.

Lora, J., Hormaza, J. I., Herrero, M., & Gasser, C. S. (2011). Seedless fruits and the disruption of a conserved genetic pathway in angiosperm ovule development. *Proceedings of the National Academy of Sciences of the United States of America*, *108*, 5461–5465.

McAbee, J. M., Hill, T. A., Skinner, D. J., Izhaki, A., Hauser, B. A., Meister, R. J., et al. (2006). *ABERRANT TESTA SHAPE* encodes a KANADI family member, linking polarity determination to separation and growth of Arabidopsis ovule integuments. *Plant Journal*, *46*, 522–531.

McAbee, J. M., Kuzoff, R. K., & Gasser, C. S. (2005). Mechanisms of derived unitegmy among *Impatiens* species. *The Plant Cell*, *17*, 1674–1684.

McConnell, J. R., Emery, J., Eshed, Y., Bao, N., Bowman, J., & Barton, M. K. (2001). Role of *PHABULOSA* and *PHAVOLUTA* in determining radial patterning in shoots. *Nature*, *411*, 709–713.

Meister, R. J., Kotow, L. M., & Gasser, C. S. (2002). SUPERMAN attenuates positive *INNER NO OUTER* autoregulation to maintain polar development of *Arabidopsis* ovule outer integuments. *Development*, *129*, 4281–4289.

Merelo, P., Xie, Y., Brand, L., Ott, F., Weigel, D., Bowman, J. L., et al. (2013). Genome-wide identification of KANADI1 target genes. *PLoS One*, *8*, e77341.

Modrusan, Z., Reiser, L., Feldmann, K. A., Fischer, R. L., & Haughn, G. W. (1994). Homeotic transformation of ovules into carpel-like structures in *Arabidopsis*. *The Plant Cell*, *6*, 333–349.

Moore, M. J., Bell, C. D., Soltis, P. S., & Soltis, D. E. (2007). Using plastid genome-scale data to resolve enigmatic relationships among basal angiosperms. *Proceedings of the National Academy of Sciences of the United States of America*, *104*, 19363–19368.

Nakagawa, H., Ferrario, S., Angenent, G. C., Kobayashi, A., & Takatsuji, H. (2004). The petunia ortholog of Arabidopsis *SUPERMAN* plays a distinct role in floral organ morphogenesis. *The Plant Cell*, *16*, 920–932.

Navarro, C., Efremova, N., Golz, J. F., Rubiera, R., Kuckenberg, M., Castillo, R., et al. (2004). Molecular and genetic interactions between *STYLOSA* and *GRAMINIFOLIA* in the control of *Antirrhinum* vegetative and reproductive development. *Development*, *131*, 3649–3659.

Nickrent, D. L., Malecot, V., Vidal-Russell, R., & Der, J. P. (2010). A revised classification of Santalales. *Taxon*, *59*, 538–558.

Nole-Wilson, S., & Krizek, B. A. (2000). DNA binding properties of the *Arabidopsis* floral development protein AINTEGUMENTA. *Nucleic Acids Research*, *28*, 4076–4082.

Pekker, I., Alvarez, J. P., & Eshed, Y. (2005). Auxin response factors mediate *Arabidopsis* organ asymmetry via modulation of KANADI activity. *The Plant Cell*, *17*, 2899–2910.

Philipson, W. R. (1974). Ovular morphology and the major classification of the dicotyledons. *Botanical Journal of the Linnean Society*, *68*, 89–108.

Pillitteri, L. J., Bemis, S. M., Shpak, E. D., & Torii, K. U. (2007). Haploinsufficiency after successive loss of signaling reveals a role for ERECTA-family genes in *Arabidopsis* ovule development. *Development, 134*, 3099–3109.

Pinyopich, A., Ditta, D. S., Savidge, B., Liljegren, S. J., Baumann, E., Wisman, E., et al. (2003). Assessing the redundancy of MADS-box genes during carpel and ovule development. *Nature, 424*, 85–88.

Prigge, M. J., Otsuga, D., Alonso, J. M., Ecker, J. R., Drews, G. N., & Clark, S. E. (2005). Class III homeodomain-leucine zipper gene family members have overlapping, antagonistic, and distinct roles in Arabidopsis development. *The Plant Cell, 17*, 61–76.

Ray, A., Robinson-Beers, K., Ray, S., Baker, S. C., Lang, J. D., Preuss, D., et al. (1994). The *Arabidopsis* floral homeotic gene *BELL* (*BEL1*) controls ovule development through negative regulation of *AGAMOUS* gene (*AG*). *Proceedings of the National Academy of Sciences of the United States of America, 91*, 5761–5765.

Reinhardt, D., Mandel, T., & Kuhlemeier, C. (2000). Auxin regulates the initiation and radial position of plant lateral organs. *The Plant Cell, 12*, 507–518.

Reinhart, B. J., Liu, T., Newell, N. R., Magnani, E., Huang, T., Kerstetter, R., et al. (2013). Establishing a framework for the ad/abaxial regulatory network of Arabidopsis: Ascertaining targets of class III homeodomain leucine zipper and KANADI regulation. *The Plant Cell, 25*, 3228–3249.

Robinson-Beers, K., Pruitt, R. E., & Gasser, C. S. (1992). Ovule development in wild-type *Arabidopsis* and two female-sterile mutants. *The Plant Cell, 4*, 1237–1249.

Roe, J. L., Nemhauser, J. L., & Zambryski, P. C. (1997). *TOUSLED* participates in apical tissue formation during gynoecium development in *Arabidopsis*. *The Plant Cell, 9*, 335–353.

Sarojam, R., Sappl, P. G., Goldshmidt, A., Efroni, I., Floyd, S. K., Eshed, Y., et al. (2010). Differentiating *Arabidopsis* shoots from leaves by combined YABBY activities. *The Plant Cell, 22*, 2113–2130.

Sawa, S., Watanabe, K., Goto, K., Kanaya, E., Morita, E. H., & Okada, K. (1999). *FILAMENTOUS FLOWER*, a meristem and organ identity gene of *Arabidopsis*, encodes a protein with a zinc finger and HMG-related domains. *Genes and Development, 13*, 1079–1088.

Schiefthaler, U., Balasubramanian, S., Sieber, P., Chevalier, D., Wisman, E., & Schneitz, K. (1999). Molecular analysis of *NOZZLE*, a gene involved in pattern formation and early sporogenesis during sex organ development in *Arabidopsis thaliana*. *Proceedings of the National Academy of Sciences of the United States of America, 96*, 11664–11669.

Schneitz, K., Hulskamp, M., & Pruitt, R. E. (1995). Wild-type ovule development in *Arabidopsis thaliana*: A light microscope study of cleared whole-mount tissue. *Plant Journal, 7*, 731–749.

Shpak, E. D. (2013). Diverse roles of *ERECTA* family genes in plant development. *Journal of Integrative Plant Biology, 55*, 1238–1250.

Sieber, P., Gheyselinck, J., Gross-Hardt, R., Laux, T., Grossniklaus, U., & Schneitz, K. (2004). Pattern formation during early ovule development in *Arabidopsis thaliana*. *Developmental Biology, 273*, 321–334.

Sieber, P., Petrascheck, M., Barberis, A., & Schneitz, K. (2004). Organ polarity in *Arabidopsis*. NOZZLE physically interacts with members of the YABBY family. *Plant Physiology, 135*, 2172–2185.

Simon, M. K., Skinner, D. J., Gallagher, T. L., & Gasser, C. S. (2017). Integument development in *Arabidopsis* depends on interaction of YABBY protein INNER NO OUTER with coactivators and corepressors. *Genetics, 207*, 1489–1500.

Simonini, S., Bencivenga, S., Trick, M., & Ostergaard, L. (2017). Auxin-induced modulation of ETTIN activity orchestrates gene expression in Arabidopsis. *The Plant Cell, 29*, 1864–1882.

Simonini, S., Deb, J., Moubayidin, L., Stephenson, P., Valluru, M., Freire-Rios, A., et al. (2016). A noncanonical auxin-sensing mechanism is required for organ morphogenesis in Arabidopsis. *Genes and Development, 30*, 2286–2296.

Skinner, D. J., Brown, R. H., Kuzoff, R. K., & Gasser, C. S. (2016). Conservation of the role of *INNER NO OUTER* in development of unitegmic ovules of the Solanaceae despite a divergence in protein function. *BMC Plant Biology, 16*, 143.

Skinner, D. J., Hill, T. A., & Gasser, C. S. (2004). Regulation of ovule development. *The Plant Cell, 16*, S32–S45.

Sridhar, V. V., Surendrarao, A., & Liu, Z. C. (2006). *APETALA1* and *SEPALLATA3* interact with *SEUSS* to mediate transcription repression during flower development. *Development, 133*, 3159–3166.

Stahle, M. I., Kuehlich, J., Staron, L., von Arnim, A. G., & Golz, J. F. (2009). YABBYs and the transcriptional corepressors LEUNIG and LEUNIG_HOMOLOG maintain leaf polarity and meristem activity in Arabidopsis. *The Plant Cell, 21*, 3105–3118.

Stebbins, G. L. (1974). *Flowering plants: Evolution above the species level.* Cambridge, MA: Belknap Press of Harvard University Press.

Taylor, T. N., Delfueyo, G. M., & Taylor, E. L. (1994). Permineralized seed fern cupules from the Triassic of Antarctica - implications for cupule and carpel evolution. *American Journal of Botany, 81*, 666–677.

Villanueva, J. M., Broadhvest, J., Hauser, B. A., Meister, R. J., Schneitz, K., & Gasser, C. S. (1999). *INNER NO OUTER* regulates abaxial-adaxial patterning in *Arabidopsis* ovules. *Genes and Development, 13*, 3160–3169.

Waites, R., & Hudson, A. (1995). *phantastica*: A gene required for dorsoventrality of leaves in *Antirrhinum majus*. *Development, 121*, 2143–2154.

Wang, J. G., Feng, C., Liu, H. H., Ge, F. R., Li, S., Li, H. J., et al. (2016). HAPLESS13-mediated trafficking of STRUBBELIG is critical for ovule development in *Arabidopsis*. *PLoS Genetics, 12*, e1006269.

Wei, B., Zhang, J., Pang, C., Yu, H., Guo, D., Jiang, H., et al. (2015). The molecular mechanism of SPOROCYTELESS/NOZZLE in controlling *Arabidopsis* ovule development. *Cell Research, 25*, 121–134.

Weijers, D., & Wagner, D. (2016). Transcriptional responses to the auxin hormone. *Annual Review of Plant Biology, 67*, 539–574.

Western, T. L., & Haughn, G. W. (1999). *BELL1* and *AGAMOUS* genes promote ovule identity in *Arabidopsis thaliana*. *Plant Journal, 18*, 329–336.

Yamada, T., Sasaki, Y., Hashimoto, K., Nakajima, K., & Gasser, C. S. (2016). *CORONA*, *PHABULOSA* and *PHAVOLUTA* collaborate with *BELL1* to confine *WUSCHEL* expression to the nucellus in *Arabidopsis* ovules. *Development, 143*, 422–426.

Yamada, T., Yokota, S., Hirayama, Y., Imaichi, R., Kato, M., & Gasser, C. S. (2011). Ancestral expression patterns and evolutionary diversification of YABBY genes in angiosperms. *Plant Journal, 67*, 26–36.

Yamaguchi, N., Wu, M. F., Winter, C. M., Berns, M. C., Nole-Wilson, S., Yamaguchi, A., et al. (2013). A molecular framework for auxin-mediated initiation of flower primordia. *Developmental Cell, 24*, 271–282.

Yang, W.-C., Ye, D., Xu, J., & Sundaresan, V. (1999). The *SPOROCYTELESS* gene of *Arabidopsis* is required for initiation of sporogenesis and encodes a novel nuclear protein. *Genes and Development, 13*, 2108–2117.

Zan, S. Q., Axsmith, B. J., Fraser, N. C., Liu, F. X., & Xing, D. H. (2008). New evidence for laurasian corystosperms: *Umkomasia* from the Upper Triassic of northern China. *Review of Palaeobotany and Palynology, 149*, 202–207.

CHAPTER FIFTEEN

Development and function of the flowering plant female gametophyte

Isil Erbasol Serbes[†], Joakim Palovaara[†], Rita Groß-Hardt*
Centre for Biomolecular Interactions, University of Bremen, Bremen, Germany
*Corresponding author: e-mail address: gross-hardt@uni-bremen.de

Contents

Abstract

Flowering plants constitute an indispensable basis for the existence of most organisms, including humans. In a world characterized by rapid population growth and climate changes, understanding plant reproduction becomes increasingly important in order to respond to the resource shortage associated with this development. New technologies enabling powerful forward genetic approaches, comprehensive genome and transcriptome analyses, and sophisticated cell isolation and imaging have advanced our understanding of the molecular mechanisms underlying gamete formation and fertilization. In addition, these techniques have allowed us to explore the fascinating cellular crosstalk, which coordinates the intra- and interorganismic interactions that secure reproductive success. Here we review the basic principles underlying development of

[†] These authors contributed equally to this work.

Current Topics in Developmental Biology, Volume 131
ISSN 0070-2153
https://doi.org/10.1016/bs.ctdb.2018.11.016

the germ cell-harboring female gametophyte in flowering plants. We start with the selection of the founder cells and end with the formation of a few-celled, highly specialized structure that operates on the basis of division of labor in order to generate the next generation.

1. Introduction

Land plants exhibit a complex life cycle that alternates between a diploid sporophytic phase and a reduced, haploid gametophytic phase. In angiosperms (flowering plants), the most dominant form of land plants, the sporophyte is a result of asexual or sexual reproduction. Sexual reproduction relies on male and female gametophytes, minute structures which are also referred to as pollen and embryo sac, respectively (Johri, Ambegaokoar, & Srivastava, 1992; Maheshwari, 1950). The female gametophyte, also referred to as the embryo sac, develops in the ovule located in the gynoecium. In most angiosperms, including the model plant *Arabidopsis thaliana*, it contains a so-called *Polygonum* type arrangement of seven cells (Fig. 1). These cells exhibit four distinct profiles: centrally positioned egg and central cell flanked by ephemeral antipodal (three) and synergid (two) cells. In contrast, pollen grains develop in the anthers and contain two sperm cells encased in a vegetative cell (see chapter "Pre-meiotic anther development" by van der Linde and Walbot, this issue). As a characteristic feature of flowering plants, not only the egg cell but also the adjacent central cell gets fertilized. The respective sperm cells for this double fertilization are

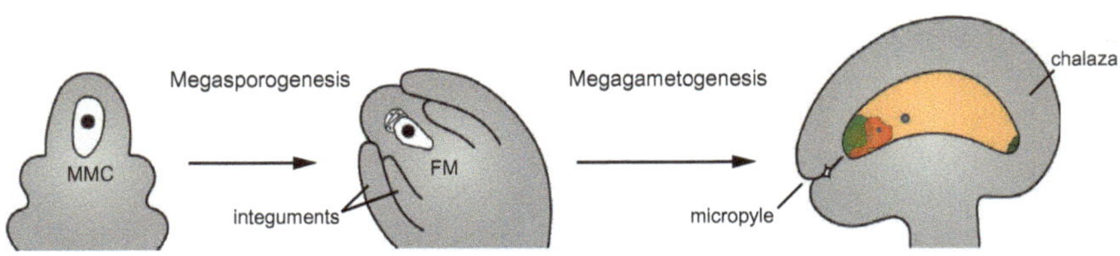

Fig. 1 Schematic illustration of female gametophyte development in *A. thaliana*. During megasporogenesis, the megaspore mother cell (MMC) undergoes meiosis to form the functional megaspore (FM). The mature female gametophyte, containing the gametic cells (egg cell [red], central cell [yellow]) and accessory cells (synergids [green], antipodals [dark green]), is established following megagametogenesis when the FM nucleus mitotically divide, nuclei migrate and cell fates are determined. Sporophytic cell layers called integuments are initiated from the chalazal part of the ovule and enclose the embryo sac, except for a small opening called the micropyle to allow pollen tube entry during fertilization.

delivered by a single pollen tube. During fertilization pollen grains are deposited onto the stigma, germinate and form a pollen tube that grows through the transmitting tract and toward the ovule. This allows for the delivery of the two sperm cells into the embryo sac and subsequent double fertilization of both the egg cell and the central cell (see chapter "Friend or foe: Signaling mechanisms during double fertilization in flowering seed plants" by Zhou and Dresselhaus, this issue). The fusion products then develop into embryo and endosperm, respectively. Fertilization marks the start of a new sporophyte generation in which the embryo, with mechanical and nutritional support from the endosperm (reviewed in Yan, Duermeyer, Leoveanu, & Nambara, 2014), will mature into the seed, from which later a new plant is formed, thereby completing the life cycle.

The gametophytes are central to the plant life cycle and, as such, a significant effort has been made over the last decades to elucidate their development and function, particularly in angiosperms. This has been especially true for the female gametophyte since it is the site of fertilization and embryogenesis. Furthermore, its development exhibits tightly coordinated cell division, specification, and differentiation events that involve unique genetic and epigenetic regulatory pathways, programmed cell death (PCD), and cell-to-cell communication (reviewed in, e.g., Nakajima, 2018; Nonomura, 2018; Schmidt, Schmid, & Grossniklaus, 2015; Tekleyohans, Nakel, & Groß-Hardt, 2017). In this review, we will cover current knowledge and recent advances regarding female gametophyte development in angiosperms, with focus on the molecular mechanism underlying these events.

2. Megasporogenesis

Female gametophyte development is preceded and influenced by megasporogenesis, the completion of which marks the beginning of the gametophytic phase. While there are a multitude of megasporogenesis patterns in land plants (Gifford & Foster, 1989), angiosperms exhibit three main patterns: monosporic (*Polygonum* type), bisporic (*Alisma* type) and tetrasporic (*Drusa* type) (Johri et al., 1992; Maheshwari, 1950). The most common pattern appears to be the *Polygonum* type, which can be found in *A. thaliana* as well as in *Zea mays* (maize), *Oryza sativa* (rice), and other important crop species (Christensen, King, Jordan, & Drews, 1997). It is initiated when, during early ovule development in the sporophyte gynoecium, a single subepidermal nucellus cell, the archesporial cell, differentiates into

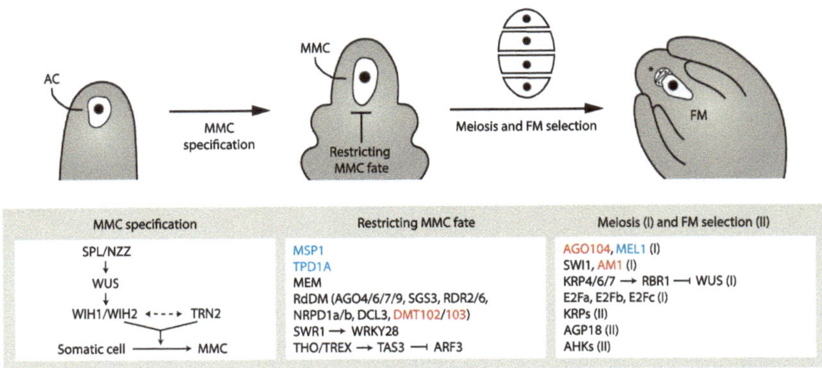

Fig. 2 Schematic illustration of *Polygonum* type megasporogenesis. In the nucellus, the megaspore mother cell (MMC) differentiates from a subepidermal so-called archesporial cell. The MMC undergoes meiosis to produce four one-nucleate cells. Subsequent degradation of three of these cells (asterisk) leaves one functional megaspore (FM) that will give rise to the mature female gametophyte. Selected genes and pathways are indicated. Text color denotes species (*A. thaliana* = black, *O. sativa* = blue, *Z. mays* = red).

the megaspore mother cell (Fig. 2). The megaspore mother cell undergoes meiosis to produce four one-nucleate, haploid megaspore cells. Subsequently, three of the megaspores, usually the ones positioned closest to the micropylar end, degenerate, leaving one functional megaspore to give rise to the embryo sac (see Section 3). Henceforth, we will refer to the megaspore mother cell as MMC and to the functional megaspore as FM.

2.1 Initiating the female germline

The differentiation of the MMC marks the onset of the female germline; however, there are limited data on how it is first specified. Still, one important regulatory pathway has been identified where *WUSCHEL* (*WUS*), a gene required for stem cell fate in shoot and floral meristems (Laux, Mayer, Berger, & Jurgens, 1996; Mayer et al., 1998), acts downstream of the transcription factor *SPOROCYTELESS/NOZZLE* (*SPL/NZZ*) to promote the expression of *WINDHOSE1* (*WIH1*) and *WIH2* in the nucellus of the ovule primordium (Chen, Sun, Liu, Liu, & Yang, 2014; Lieber, Lora, Schrempp, Lenhard, & Laux, 2011; Yang, Ye, Xu, & Sundaresan, 1999). Here, the *WIH1/2* genes, which encode small peptides, and the tetraspanin-encoding gene *TORNADO2* (*TRN2*) appear to direct MMC formation (Lieber et al., 2011). Indeed, in all loss-of-function mutants of the pathway, the female germline is not established, with nucellar cells of somatic appearance replacing the MMC, and ovule development ceases

(Lieber et al., 2011; Yang et al., 1999). The mutants do not show complete penetrance and, thus, additional factors are likely involved.

During differentiation, the MMC enlarges, elongates and its nuclear size doubles (She et al., 2013). When cell enlargement is first observed, the chromatin organization of the MMC has been shown to be distinct from its progenitor cells, with a reduced heterochromatin content and depleted linker histones. This suggests that epigenetic chromatin reprogramming contributes to the specification of the MMC. However, how and to what extent this is accomplished and biologically relevant is currently unknown.

2.2 Restricting germline fate

A substantial amount of data exist regarding how female germline fate is restricted to a single cell. In rice, loss-of-function of the *MULTIPLE SPOROPHYTE1* (*MSP1*) gene results in supernumerary MMC-like cells in the ovule primordium that can undergo meiosis to generate embryo sacs, although they are abnormal and partially sterile (Nonomura et al., 2003). An increased number of archesporial cells are also found in the anthers. The *MSP1* gene, expressed in the nucellar cells surrounding the MMC, encodes a leucine-rich repeat (LRR) receptor-like kinase that binds a ligand encoded by the *TAPETUM DETERMINANT1* gene (*OsTPD1A*) (Nonomura et al., 2003; Yang, Qian, et al., 2016; Zhao et al., 2008). Silencing of this gene in rice and loss-of-function of the equivalent gene in maize phenocopy the *msp1* mutant (Sheridan, Avalkina, Shamrov, Batyea, & Golubovskaya, 1996; Wang, Nan, et al., 2012; Zhao et al., 2008). Given that the maize protein localizes to archesporial cells and can be secreted (Wang, Nan, et al., 2012), this supports the existence of a non-cell-autonomous lateral inhibition mechanism that, in the ovule primordium, allows the MMC to repress the formation of additional MMCs in surrounding tissue and thus restricts germline fate.

The *A. thaliana* gene *MNEME* (*MEM*) has been shown to restrict MMC-adjacent nucellar cells from acquiring MMC identity, likely in a non-cell-autonomous manner (Schmidt et al., 2011). *MEM* is predominantly expressed in the MMC and encodes a putative RNA helicase of the DEAD-box family. Heterozygous mutants exhibit multiple MMC-like cells, which can develop into unreduced FMs and embryo sacs by bypassing meiosis. DEAD-box RNA helicases have been reported to act in RNA-directed DNA methylation (RdDM) (Khan et al., 2014), in which members of the ARGONAUTE4 (AGO4) clade, AGO4, AGO6 and

AGO9, associate with 24-nucleotide small interfering RNAs (siRNAs) to silence transposable elements (TEs) (reviewed in Matzke & Mosher, 2014). In fact, loss-of-function mutants of all members of this clade including AGO8 exhibit defects similar to those observed in *mem* mutants (Hernández-Lagana, Rodríguez-Leal, Lúa, & Vielle-Calzada, 2016; Olmedo-Monfil et al., 2010). However, variable phenotypic frequencies and expression patterns could be observed in different single and double mutants (Hernández-Lagana et al., 2016), suggesting hierarchical and complex genetic interactions, and/or non-autonomous protein dosage effects among AGO4 clade members.

Similar to mutants affecting the AGO4 clade, loss-of-function of other RdDM pathway components, such as *SUPPRESSOR OF GENE SILENCING3* (*SGS3*) and *RNA-DEPENDENT RNA POLYMERASE6* (*RDR6*) (Olmedo-Monfil et al., 2010), results in multiple MMC-like cells and/or unreduced embryo sacs (Table 1). In addition, the MMC-like cells in the *ago9*, *sgs3* and *rdr6* mutants show a distinct chromatin state similar to that of the MMC when compared to surrounding nucellar cells (She et al., 2013). Taken together, these results suggest that epigenetic regulation is necessary to maintain a chromatin and histone state in nucellar cells adjacent to the

Table 1 RNA-directed DNA methylation (RdDM) pathway components implicated in restricting megaspore mother cell (MMC) fate to a single cell in the ovule primordium.

Gene(s)	Species	Protein function	Reference(s)
NRPD1a, *NRPD1b*	*A. thaliana*	Alternative largest subunit of RNA Pol IV	Olmedo-Monfil et al. (2010)
RDR2, *RDR6*	*A. thaliana*	Produces dsRNA precursors	Olmedo-Monfil et al. (2010)
SGS3	*A. thaliana*	Stabilizes dsRNA precursors	Olmedo-Monfil et al. (2010)
DICER-LIKE 3 (*DCL3*)	*A. thaliana*	Cleaves dsRNA into 24 nt siRNAs	Olmedo-Monfil et al. (2010)
AGO4, *AGO6*, *AGO8*, *AGO9*	*A. thaliana*	Binds 24 nt siRNAs	Hernández-Lagana et al. (2016); Olmedo-Monfil et al. (2010)
DMT102, *DMT103*	*Zea mays*	Maintains DNA methylation	Garcia-Aguilar, Michaud, Leblanc, and Grimanelli (2010)

MMC to prevent the ectopic acquisition of germline fate and, thereby, restrict it to a single cell. Supporting this conclusion, it was recently shown that the chromatin remodeling complex SWR1 mediates the incorporation of the linker histone H2A.Z at the locus encoding the WRKY28 transcription factor (Zhao et al., 2018). This incorporation promotes *WRKY28* expression and, thereby, suppresses MMC identity in adjacent nucellar cells. The recruitment of SWR1 to the *WRKY28* locus is presumably dependent on a mobile signal derived from the cytochrome P450 encoding *KLUH* (*KLU*) gene, which is expressed in the inner integument of the ovule primordium. This implies that the regulation of *WRKY28* by *KLU* is non-cell-autonomous (Zhao et al., 2018).

The incomplete penetrance of epigenetic mutants indicates the existence of additional regulatory pathways. Recently, a second-site enhancer mutant of *rdr6* was identified where the casual mutation was mapped to *TEX1*, which encodes a component of the THO/TREX complex. This complex is, together with SGS3 and RDR6, involved in generating *trans*-acting siRNAs (ta-siRNAs; TAS) from longer *TAS* gene transcripts (Su et al., 2017). Ta-siRNAs regulate target genes by mRNA cleavage. In the *tex1* mutant, a *TAS3*-derived ta-siRNA species that targets transcripts encoding AUXIN RESPONSE FACTORs (ARFs), i.e., transcription factors that respond to the plant hormone auxin, is depleted (Su et al., 2017). Simultaneously, *ARF3* expression expands to overlap with wild-type *TEX1* expression in the distal epidermis of the ovule primordium. This suggests that TEX1, as a component of the THO/TREX complex, acts redundantly with RDR6 to restrict MMC fate through a *TAS3-ARF3* repression module.

2.3 Meiosis and the functional megaspore

Many genes associated with epigenetic regulation not only restrict MMC fate but are also involved in the processes governing the meiotic divisions of the MMC. In recent years, several additional genes of the AGO family have been identified to regulate meiosis in the female germline. In maize, loss-of-function of *AGO104*, a homologue of *A. thaliana AGO9*, results in unreduced gametes but not multiple MMC-like cells (Singh et al., 2011). *AGO104* is exclusively expressed in the nucellar cells and, thus, meiotic defects are likely the result of a non-cell-autonomous signal. In addition, the MMC-specific rice gene *MEIOSIS ARRESTED LEPTOTENE1* (*MEL1*), related to the AGO5 clade in *A. thaliana*, is essential for meiosis since loss-of-function leads to meiotic arrest (Nonomura et al., 2007).

MEL1 associates with 21-nucleotide phased siRNAs (pha-siRNA) proposed to regulate meiotic chromosome structure by means of RNA silencing, possibly through DNA methylation (Komiya et al., 2014; Nonomura et al., 2007). This, together with observed changes in chromatin organization in *A. thaliana* (She et al., 2013), indicates the importance of epigenetic regulation for both restricting germline fate and commitment to meiosis.

Other molecular mechanisms relating to cell cycle progression have been shown to be required for meiosis in the female germline. Loss-of-function of the *A. thaliana* gene *DYAD/SWITCH1* (*SWI1*), required for meiotic chromatin organization (Agashe, Prasad, & Siddiqi, 2002; Mercier et al., 2001; Siddiqi, Ganesh, Grossniklaus, & Subbiah, 2000), and its maize homologue *AMEIOTIC1* (*AM1*) results in mitosis-like divisions of the MMC and the establishment of unreduced gametes (Pawlowski et al., 2009; Ravi, Marimuthu, & Siddiqi, 2008). Similar meiotic defects can be observed in genotypes that combine mutants of core meiotic genes (d'Erfurth et al., 2010, 2009), and in mutants of genes associated with meiotic progression (Liu et al., 2014; Makkena, Lee, Sack, & Lamb, 2012; Nonomura et al., 2011; Ren et al., 2018; Stevens et al., 2004). More recently, research in *A. thaliana* has implicated a *WUS*-dependent regulatory cascade in meiotic entry (Zhao et al., 2017). In this cascade, the KIP-RELATED PROTEIN (KRP) genes *KRP4*, *KRP6* and *KRP7*, which encode inhibitors of cyclin-dependent kinases (CDKs), redundantly act in the MMC to restrict the CDKA;1-dependent inactivation of the *RETINOBLASTOMA-RELATED1* (*RBR1*) gene (Ebel, Mariconti, & Gruissern, 2004; Nowack et al., 2012; Zhao et al., 2017). RBR1, a transcriptional repressor known to control cell differentiation (Harashima & Sugimoto, 2016), in turn, directly suppresses *WUS* activity, which allows the MMC to enter meiosis (Zhao et al., 2017). In *rbr1* and *krp* triple mutants, multiple MMCs and, subsequently, several unreduced embryo sacs are established in the ovule when the MMC fails to switch from the mitotic to the meiotic mode of division. This phenotype is partially reversed by depleting *WUS* in *rbr1* mutants. Thus, there appears to be a delicate balance between the activation of *WUS* to promote MMC formation and its inactivation by RBR1 to allow the MMC to enter meiosis. Interestingly, *WUS* misexpression in the MMC is not by itself sufficient to induce mitotic divisions (Zhao et al., 2017), indicating that additional factors downstream of *RBR1* are required. This is substantiated by the discovery that loss-of-function of the E2F transcriptional repressors (E2Fa, E2Fb, E3Fc) in *A. thaliana*, which are cell cycle regulators that form complexes with RBR1, phenocopies the MMC defect seen in the *rbr1* mutant (Yao et al., 2018).

Once the MMC undergoes meiosis, four haploid megaspore cells are generated that are marked by the accumulation of callose in their cell walls (Rodkiewicz, 1970). Three of them degenerate through PCD, while one survives and gives rise to the FM. The chromatin status of the FM suggests that its selection involves both specific and common chromatin characteristics as compared to MMC differentiation in *A. thaliana* (She et al., 2013). In addition to controlling entry into meiosis, the KRPs seem to be important for FM selection. In septuple mutants of the highly redundant KRP family, the multiple MMC phenotype is confirmed and more than one megaspore frequently survive (Cao et al., 2018). The surviving megaspores can form multiple embryo sacs that, in rare cases, results in twin seedlings following fertilization. Notably, in the wild type, KRP4 was specifically detected in the degenerating megaspores but not in the FM. Taken together, this suggest that KRPs are involved in the degeneration of non-functional megaspores. Several other *A. thaliana* mutants affecting FM selection have been identified. In the mutant *antikevorkian*, whose corresponding locus has yet to be determined, degradation of the four megaspore cells is prevented and all meiotic products appear capable of generating an embryo sac (Yang & Sundaresan, 2000). Overexpression of *ARABINOGALACTAN PROTEIN18* (*AGP18*), which encodes a plasma membrane-attached glycosylated protein that localizes to the integumentary cells of the meiotic ovule, results in the survival of multiple megaspore cells that can all acquire FM identity (Demesa-Arévalo & Vielle-Calzada, 2013). Finally, in triple loss-of-function mutants of ARABIDOPSIS HISTONE KINASE (AHK) receptor genes the FM is often not specified, based on morphology and lack of marker expression, resulting in no or abnormal embryo sac development (Cheng, Mathews Dennis, Eric Schaller, & Kieber Joseph, 2012). The expression of all *AHK*s is detected in the chalazal sporophytic tissue and acts in the signaling pathway of the phytohormone cytokinin.

3. Megagametogenesis

Following its establishment, the FM undergoes a series of developmental processes, summarized under the term megagametogenesis, to generate the mature female gametophyte (Fig. 3). This process includes tightly regulated mitotic divisions, repositioning of nuclei along a polar axis, as well as the acquisition of distinct identities by individual cells in order to ensure female gametophyte fertilization and, thereby, plant reproduction.

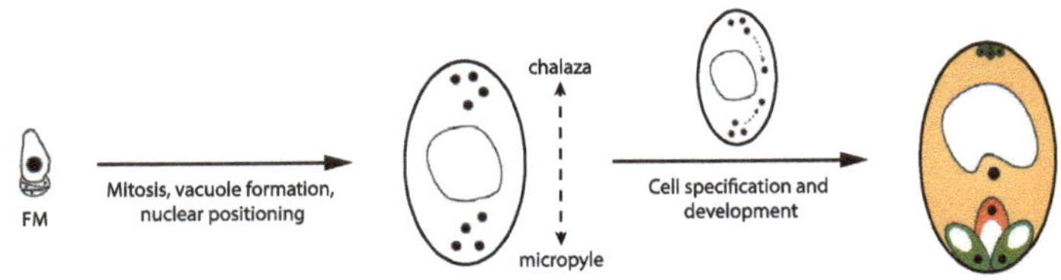

Mitosis (I) and vacuole formation (II)	Nuclear positioning (I) and polar axis establishment (II)	Cell specification and development	
PIN1, AUX1, LAX1, YUC8, TAA1 (I)	IG1 (I)	RKD1, RKD2 (ec)	AMP1
CYP851 (I)	TUBG1/2, GCP2, AtNACK1/2 (I)	BIP1/2, AtERdj3A/B,	BLH1, OFP5
PPR2, AGL23 (I)	YUC1, YUC2 (II)	AtP58^{IPK} (cc)	WYRD
RBR1 - FBL17 - KRPs - CDKA;1 module (I)	CKI1 (II)	SEC22 (cc)	LIS, CLO, ATO
Chromatin-remodeling factors (I)	MYB64, MYB119 (II)	GFA2, NFD1, SYCO, GCD1 (cc)	EAL1
AAA-ATPases (I)		AGL80, AGL61, DME (cc)	
VGL (II)		High ROS levels (cc)	

Fig. 3 Schematic illustration of *Polygonum* type megagametogenesis. The functional megaspore (FM) undergoes three rounds of nuclear divisions, accompanied by nuclear positioning along the micropylar-chalazal axis, to establish an eight-nucleate syncytium with four nuclei at opposite poles separated by a developing vacuole. Following polar nuclei migration and fusion, the final position of the nuclei anticipates the cellularization pattern, with gametic cells (egg cell [ec; red]), central cell [cc; yellow]) flanked by accessory cells (synergid [green], antipodals [dark green]). Selected genes, pathways and processes are indicated (see also Table 2). Text color denotes species (*A. thaliana* = black, *Z. mays* = red).

3.1 Mitotic divisions of the functional megaspore

Polygonum type megagametogenesis is initiated when the FM undergoes three rounds of nuclear division giving rise to an eight-nucleate syncytium (Christensen et al., 1997). Numerous genes have been shown to be important for the entry and progression of FM mitosis, as well as for vacuole biogenesis during this process (Table 2). For example, in *A. thaliana* defect in genes that mediate transport of auxin from the sporophytic tissue [efflux: *PIN-FORMED1* (*PIN1*)] into the embryo sac [influx: *AUX1*, *LIKE AUX1* (*LAX1*)], as well as genes necessary for auxin synthesis in the embryo sac [*YUCCA8* (*YUC8*), *TRYPTOPHAN AMINOTRANSFERASE OF ARABIDOPSIS1* (*TAA1*)] results in mitotic arrest at one-, two-, and four-nucleate stage (Ceccato et al., 2013; Panoli et al., 2015). Mitotic arrest at one-nucleate stage can also be seen in the loss-of-function mutant of *CYP851*, which encodes a key enzyme for brassinosteroid synthesis (Pérez-España, Sánchez-León, & Vielle-Calzada, 2011). Together, this shows that phytohormones are not only crucial for megasporogenesis, as described above for MMC fate and meiosis, but also for early megagametogenesis. In addition, plastids seem to play a role in mitotic entry as evidenced by mutants defective for PENTATRICOPEPTIDE REPEAT2 (PPR2) protein or the Type

Table 2 List of mutant plant lines not discussed in the main text that affect functional megaspore (FM) mitosis progression at one-, two-, and four-nucleate embryo sac stage, as well as nuclei number, proliferation and positioning.

Mutant	Species	Protein function	Phenotype	Reference(s)
acp1, acp2, acp4, acp6, cdc27a hbt	*A. thaliana*	Anaphase-promoting complex/cyclosome subunits	Mitotic arrest/abnormal nuclei number, proliferation, and positioning	Capron et al. (2003); Kwee and Sundaresan (2003); Pérez-Pérez José et al. (2007); Wang, Hou, et al. (2012); Wang et al. (2013)
aog1	*A. thaliana*	Unknown	Mitotic arrest	Cui et al. (2015)
bip 1 bip 2 bip 3	*A. thaliana*	Hsp70 chaperones in the ER	Mitotic arrest	Maruyama, Endo, and Nishikawa (2015)
cdt1a	*A. thaliana*	DNA replication factor	Mitotic arrest	Domenichini et al. (2012)
CTF7 OX[a]	*A. thaliana*	Acetyltransferase involved in sister chromatin cohesion	Mitotic arrest	Jiang, Yuan, Xia, and Makaroff (2010)
fsv1	*Oryza sativa*	Unknown	Mitotic arrest	Yang, Wu, et al. (2016)
gaf1	*A. thaliana*	RNases P/MRP subunit	Mitotic arrest	Wang, Shi, et al. (2012)
lo2	*Zea mays*	Unknown	Mitotic arrest	Sheridan and Huang (1997)
mos7	*A. thaliana*	Nucleoporin associated with the mitotic spindle	Mitotic arrest	Park et al. (2014)
nop10	*A. thaliana*	Assembles H/ACA snoRNPs	Mitotic arrest/unfused polar nuclei	Li et al. (2018)
plc2	*A. thaliana*	Phosphoinositide-specific phospholipase C	Mitotic arrest	Di Fino et al. (2017)
rhf1a rhf2a	*A. thaliana*	RING-type E3 ligases that target KRP6	Mitotic arrest	Liu et al. (2008)

Continued

Table 2 List of mutant plant lines not discussed in the main text that affect functional megaspore (FM) mitosis progression at one-, two-, and four-nucleate embryo sac stage, as well as nuclei number, proliferation and positioning.—cont'd

Mutant	Species	Protein function	Phenotype	Reference(s)
rg1 rg2	*A. thaliana*	GTPase-activating proteins	Mitotic arrest	Rodrigo-Peiris, Xu, Zhao, Wang, and Meier (2011)
rrp42	*A. thaliana*	RNase PH domain-type subunit of the exosome complex	Mitotic arrest	Yan, Yan, and Han (2017)
swa1, swa2, swa3	*A. thaliana*	Ribosome biogenesis	Mitotic arrest/ abnormal nuclei positioning	Huang et al. (2010); Li et al. (2009); Liu, Shi, Yuan, Liu, and Wang (2010); Shi et al. (2005)
yao	*A. thaliana*	Nucleolar WD4-repeat protein	Mitotic arrest	Li, Liu, Shi, Liu, and Yang (2010)
ZmDSUL RNAi	*Zea mays*	diSUMO-like protein	Abnormal nuclei positioning	Srilunchang, Krohn, and Dresselhaus (2010)

[a]OX = overexpression.

I MADS-domain transcription factor AGAMOUS-LIKE23 (AGL23), which exhibit mitotic arrest. PPR2 localizes to plastids in order to support the translational process, and AGL23 has been linked to chloroplast biogenesis (Colombo et al., 2008; Lu et al., 2011).

Recently, a regulatory cascade controlling entry into mitotic S phase, and thus cell cycle progression, was identified, consisting of RBR1, the F-box protein F-BOX-LIKE17 (FBL17), KRPs, and CDKA;1 (Zhao et al., 2012). Loss-of-function of *FBL17* and *CDKA;1* result in mitotic arrest at the one- or two-nucleate embryo sac stage. This supports the existence of a general G1/S-phase module where RBR1 represses *FBL17*, FBL17 acts to degrade KRPs, KRPs inhibits CDKA;1, and CDKA;1 inactivates RBR1 (Nakai, Kato, Shinmyo, & Sekine, 2006; Nowack et al., 2012; Zhao et al., 2012). Since CDKA;1 is essential for G1/S phase transition, the decision to enter division thus depends on the concentration of KRPs (Zhao et al., 2012). How this module relates to meiotic entry of the MMC remains to be investigated. Other genes that are necessary for normal FM mitosis are various chromatin remodeling factors (Cigliano et al., 2013; Huanca-Mamani, Garcia-Aguilar, León-Martínez, Grossniklaus, & Vielle-Calzada, 2005; Latrasse et al., 2008), as well as genes encoding AAA-ATPase subunits of the proteasome regulatory particle (Gallois et al., 2009; Ueda et al., 2011).

The gene *VACUOLELESS GAMETOPHYTES* (*VGL*) was recently identified as being required for the biogenesis of the large central cell vacuole in *A. thaliana* (D'Ippólito, Arias, Casalongué, Pagnussat, & Fiol, 2017). *VGL* encodes a DC1-domain containing protein specifically localized to pre-vacuolar compartments or multivesicular bodies that traffics proteins to vacuoles. Vacuole formation is disrupted in the loss-of-function mutant, likely due to alterations in vesicle fusion. This often results in mitotic arrest at the one- or two-nucleate embryo sac stage. Thus, it appears that the formation of the central vacuole is necessary for mitotic progression in the embryo sac.

3.2 Regulation of nuclear positioning

Megagametogenesis is accompanied by phases of substantial relocation of nuclei along a distinct polar axis (reviewed in Sprunck & Groß-Hardt, 2011). The first mitotic division of the FM generates two nuclei, which become positioned at the micropylar and chalazal end of the female gametophyte, separated by the developing central vacuole (Christensen et al., 1997). Two further incomplete mitotic divisions then generate groups of four nuclei at either pole. This stage is followed by another phase of nuclei migration, which positions one nucleus from either pole in the center. In reference to their spatial origin, these nuclei are referred to as polar nuclei. This final position of the nuclei anticipates the cellularization pattern compartmentalizing the female gametophyte into seven cells. Moreover, it is likely the position of the nuclei and respective cells along the micropylar-chalazal axis that determines cellular identity. This was first suggested by maize mutants defective for the LOB-domain protein INDETERMINATE GAMETOPHYTE 1 (IG1), which generate supernumerary cells that differentiate according to their position into extra synergids, egg cells, or a central cell (Guo, Huang, Han, & Zee, 2004; Lin, 1978).

Already in 1994, Webb and Gunning suggested a critical role for cytoskeleton elements in organizing the cytoplasmic arrangements and nuclei positioning during megagametogenesis (Webb & Gunning, 1994). In fact, nuclear division and cellularization are critically affected upon perturbations of the cytoskeleton: mutations in *A. thaliana* genes encoding γ-tubulin isoforms (*TUBG1, TUBG2*) or a core subunit of the γ-tubulin containing complex (*GCP2*) exhibit female gametophytic defects in mitosis, nuclear positioning, and cellularization (Nakamura & Hashimoto, 2009; Pastuglia et al., 2006). Furthermore, kinesin encoding genes like *AtNACK1/HINKEL* and *STUD/TETRASPORE/AtNACK2* have been shown to be necessary for proper cellularization and nuclear positioning during female gametophyte development (Tanaka et al., 2004).

The 3D position of nuclei within female gametophytic cells has also been investigated. The micropylar polarity of the central cell nucleus is established and/or maintained by balancing F-actin dynamics. This is demonstrated by targeted perturbation of F-acting integrity in *A. thaliana* using a semi-dominant negative form of ACTIN8, which shifts the central cell nucleus toward the center (Kawashima & Berger, 2015). Notably, this approach does not affect the fertilization process, indicating that nuclear fusion does not rely on polar position of the central cell nucleus. Nuclear position does, however, seem to correlate with cell identity, an aspect discussed in Section 3.4.3.

3.3 Specification of the micropylar-chalazal axis

The differentiation of female gametophytic cells according to their micropylar-chalazal axis position suggests that important patterning clues are encoded in, and provided by, the maternal ovule tissue. Supporting this, several studies show that sporophytic auxin production critically affects the development of the female gametophyte. Early stages of female gametophyte development in *A. thaliana* are preceded and accompanied by local auxin response maxima in the nucellus (Benková et al., 2003; Ceccato et al., 2013; Lituiev et al., 2013; Pagnussat, Alandete-Saez, Bowman, & Sundaresan, 2009), and, as previously mentioned, perturbations of auxin transport result in an arrest of female gametophyte development already at an early stage (Ceccato et al., 2013; Panoli et al., 2015). Moreover, a reduction of auxin synthesis in *yuc1 yuc2* double mutants shifts egg cell identity toward the micropylar pole of the female gametophyte while overexpression of *YUC1* has the opposite effect and expands micropylar identity toward the chalazal pole (Panoli et al., 2015). While these observations are consistent with the idea that auxin serves as a morphogen (Larsson, Vivian-Smith, Offringa, & Sundberg, 2017; Pagnussat et al., 2009), there is uncertainty as to whether biologically relevant amounts of auxin are generated or present in the female gametophyte (Ceccato et al., 2013;

Lituiev et al., 2013). The defects caused by modulating auxin biosynthesis or response might hence be indirect (Lituiev et al., 2013).

In addition to auxin, the establishment of the polar axis has been shown to rely on the histidine kinase CYTOKININ INDEPENDENT1 (CKI1), which is active during female gametophyte development in *A. thaliana* (Hejátko, Pernisová, Eneva, Palme, & Brzobohatý, 2003; Liu, Yuan, Song, Yu, & Sundaresan, 2017; Pischke et al., 2002). CKI1 can constitutively activate cytokinin signaling (Kakimoto, 1996; Yuan et al., 2016) even though it lacks a cytokinin binding domain (Yamada et al., 2001). Notably, CKI1 localizes to the perinuclear ER of chalazal nuclei already at the four-nucleate stage (Yuan et al., 2016). Here, loss-of-function studies indicate that CKI1 is necessary to restrict synergid and egg cell identities to the micropylar region and that this is mediated via histidine phosphotransferases. Furthermore, ectopic expression of *CKI1* is sufficient to confer central cell identity to the cells in the position of synergids and egg cell. Following fertilization, these female gametophytes can give rise to seeds that contain endosperm but lack an embryo. These results thus suggest that neither pollen tube attraction and sperm cell release, nor fertilization of the central cell and endosperm initiation require the presence of an egg cell. Conversely, the transcription factors MYB64 and MYB119 restrict chalazal identity (Rabiger & Drews, 2013). While *MYB64* is expressed in all female gametophyte cells, *MYB119* is specifically detected in the chalazal cell types. The observation that *MYB119* is down-regulated in *cki1* mutants suggests that the activation of this potent polarity determinant relies on chalazal signatures in the female gametophyte rather than sporophytic signals.

Besides micropylar-chalaza patterning clues, there might exist a common regulatory mechanism for a gametic module since several genes have been described to be expressed in both the egg and central cell (Groß-Hardt et al., 2007; Haecker et al., 2004; Lorbiecke et al., 2005; Pischke et al., 2002; Yang, Kaur, Kiriakopolos, & McCormick, 2006). Also, mutants affecting the putative transcription factor VERDANDI exhibit defects restricted only to accessory cells, with synergids expressing genes characteristic for antipodal cells and vice versa (Matias-Hernandez et al., 2010).

3.4 Gametophytic cell specification

As in other complex tissues, differentiation of the individual female gametophyte cell types in 3D space requires the integration of information generated within the cell and patterning cues provided by surrounding cells.

It is this communication between cells that provides an impressive potential for female gametophytic cells to respond and adjust to their cellular environment. At the same time, this interdependence makes it difficult to dissect primary and secondary effects in patterning mutants.

3.4.1 The egg cell

The haploid egg cell ensures the transfer of genetic information to the next generation by fusing with one of the two sperm cells released by the pollen tube to form a diploid embryo. The egg cell is polarly organized with a large vacuole located at the micropylar region of the cell while the nucleus is positioned chalazally (Mansfield, Briarty, & Erni, 1991). Interestingly, the central cell-adjoining egg cell wall can in some plants be thicker and slightly indented when compared to the synergid-adjoining egg cell wall (Mogensen & Suthar, 1979; Tekleyohans, Mao, Kägi, Stierhof, & Groß-Hardt, 2017).

Egg cell specification is not well understood, which likely reflects the complex and redundant molecular mechanisms underlying this process. However, recent work have identified RWP-RK domain-containing proteins (RKDs), which represent a subclass of the plant specific RWP-RK transcription factor family, as being involved in egg cell differentiation (Chardin, Girin, Roudier, Meyer, & Krapp, 2014; Koi et al., 2016; Köszegi et al., 2011; Rövekamp, Bowman, & Grossniklaus, 2016; Tedeschi, Rizzo, Rutten, Altschmied, & Bäumlein, 2017). There are five *RKD* members in the *A. thaliana* genome and all are transcriptionally active during ovule development (Köszegi et al., 2011; Tedeschi et al., 2017). Two of them, *AtRKD1* and *AtRKD2*, are preferentially expressed in the *A. thaliana* egg cell (Köszegi et al., 2011). Various mutant combinations suggest that this gene family is critically important for correct megagametogenesis, including specification of the egg cell (Köszegi et al., 2011; Tedeschi et al., 2017). Furthermore, overexpression of *AtRKD1* and *AtRKD2* causes ectopic tissue proliferation, and reporter analysis as well as comprehensive transcriptome studies of *AtRKD2* overexpressing cells suggests that *AtRKD2* is sufficient to confer an egg cell-like expression profile to cells (Kőszegi et al., 2011).

To bypass experimental limitations associated with genetic redundancy, two groups shifted the scene to the liverwort *Marchantia polymorpha*, where Mp*RKD* is a single-copy gene (Koi et al., 2016; Rövekamp et al., 2016). Using homologous recombination-mediated gene targeting, Koi and colleagues generated Mp*rkd* null mutants that exhibited defects in the differentiation of both egg and sperm cells (Koi et al., 2016). In contrast, based on morphological criteria, the RNAi-induced silencing of *RKD* performed by Rövekamp and colleagues did not affect egg cell specification but yielded egg cells that acquired an embryo-like developmental program in the absence of fertilization (Rövekamp et al., 2016). Together, these studies suggest a dual role of *RKD* in the specification of egg cells and the repression of an asexual apogamic program.

3.4.2 The central cell

The central cell is the largest cell in the female gametophyte and is located between the egg and synergid cells on one side and the antipodal cells on the other (Christensen et al., 1997). Like the egg cell, the central cell fuses with one sperm cell, with the fusion product developing into an embryo-nurturing tissue called the endosperm. Notably, however, the molecular program underlying egg and central cell fusion differs. This is evidenced by loss-of-function mutants of the acyl-transferase gene *GLAUCE*, which affects fertilization of the central cell only (Leshem et al., 2012; Ngo, Moore, Baskar, Grossniklaus, & Sundaresan, 2007). The chalazal pole of the central cell is occupied by a large central vacuole while the nuclei are positioned at the micropylar pole close to the egg cell nucleus (Christensen et al., 1997). The prominent central cell appears highly metabolically active, as reflected by a mitochondria-dense cytoplasm, large starch-containing plastids around the nucleus (Mansfield et al., 1991), and high amounts of reactive oxygen species (ROS) (Martin, Fiol, Sundaresan, Zabaleta, & Pagnussat, 2013).

In many angiosperms, including *A. thaliana*, the two polar nuclei fuse before fertilization giving rise to a large central cell nucleus (reviewed in Huang & Russell, 1992). In contrast, polar nuclei fusion is only completed after fertilization in many grasses including rice and maize (Diboll, 1968). The fusion process was described in detail already 50 years ago (Jensen, 1964; Schulz & Jensen, 1973). First, the two polar nuclei adopt a position where the endoplasmic reticulum (ER) of both nuclei can get into physical contact through their outer nuclear membranes. At these contact points, first the outer membranes fuse, followed by the merging of the inner nuclear membranes. How is this sequential process regulated? Polar nuclei fusion in plants shares substantial molecular signatures with nuclear fusion during yeast mating, including the involvement of the molecular chaperone IMMUNOGLOBULIN BINDING PROTEIN (BiP), which is a member of the heat shock protein 70 (Hsp70) family (Maruyama, Endo, & Nishikawa, 2010; Maruyama, Yamamoto, Endo, & Nishikawa, 2014). The *A. thaliana* genome encodes three BIP family members and plants defective for both *BIP1* and *BIP2* exhibit unfused polar nuclei (Maruyama et al., 2010). In yeast and plants, BIP proteins cooperate with ER-resident J-domain containing proteins in a sequential and combinatorial fashion in order to mediate the individual steps of membrane fusion (Maruyama et al., 2014; Yamamoto, Maruyama, Endo, & Nishikawa, 2008). Ultrastructural analysis in plants has revealed that the J proteins AtERdj3A and AtP58^{IPK} mediate outer nuclear membrane fusion, while inner nuclear membrane fusion is regulated by AtERdj3B and AtP58^{IPK} (Maruyama et al., 2014). SNARE proteins involved in ER-derived secretory vesicle fusion also play a role in secondary nucleus formation as evidenced by *sec22* mutants, which fail to undergo polar nuclei fusion (El-Kasmi et al., 2011).

Early evidence for a critical role of mitochondria during polar nuclei fusion came from the analysis of the *A. thaliana GAMETOPHYTIC FACTOR2* (*GFA2*) gene, which encodes a mitochondrial chaperone, similar to the yeast Mdj1p protein (Christensen et al., 2002). In *gfa2* mutants the initial alignment of both polar nuclei is unaffected; however, the fusion of the outer nuclear membrane is impaired (Christensen et al., 2002; Christensen, Subramanian, & Drews, 1998). Similarly, *nuclear fusion defective1* (*nfd1*) ovules show unfused polar nuclei due to a defect in outer membrane fusion (Portereiko, Sandaklie-Nikolova, et al., 2006). *NFD1* codes for the mitochondrial 50S ribosomal subunit L21. In the central cell of *syco-1* and *gamete cell defective1* (gcd*1*) mutants, the failure of polar nuclei fusion is associated with mitochondrial cristae disintegration and this defect correlates with an extended lifespan of adjacent antipodal cells (see Section 3.4.4) (Kägi, Baumann, Nielsen, Stierhof, & Groß-Hardt, 2010; Wu et al., 2012). Moreover, targeted disruption of the electron transport chain in central cell mitochondria revealed that the lifespan of antipodal cells is coupled to the metabolic activity of the central cell (Kägi et al., 2010). *SYCO* encodes a mitochondrial cysteinyl t-RNA synthetase (Kägi et al., 2010), while the function of the mitochondrially localized GCD1 protein is currently unclear (Wu et al., 2012).

In *A. thaliana*, the specification of the central cell critically relies on the MADS-domain transcription factor AGL80. Loss-of-function *agl80* mutants exhibit aberrant nucleus and vacuole size and fail to express two out of three central cell reporter genes tested (Portereiko, Lloyd, et al., 2006). Notably, central cell expressed *AGL80* is required for the expression of *DEMETER* (*DME*) (Portereiko, Lloyd, et al., 2006), which in turn regulates the activation of *FERTILIZATION INDEPENDENT SEED2*

(*FIS2*), *FLOWERING WAGENINGEN* (*FWA*) and *MEDEA* (*MEA*) in the central cell and endosperm (Choi et al., 2002; Jullien, Kinoshita, Ohad, & Berger, 2006). DME mediates DNA demethylation of these and other genes in the central cell (Frost et al., 2018; Ibarra et al., 2012; Park et al., 2016), which is vital for reproduction (Park et al., 2017). Additionally, AGL80 was shown to interact with AGL61 in order to regulate downstream genes essential for central cell and endosperm function (Bemer, Wolters-Arts, Grossniklaus, & Angenent, 2008; Steffen, Kang, Portereiko, Lloyd, & Drews, 2008). The activation of central cell characteristic molecular signatures also seems to be promoted by high amounts of ROS. High ROS levels are typically present in the central cell but can also be found in egg and synergid cells of *oiwa* mutants, where ROS accumulation correlates with the activation of central cell reporter genes (Martin et al., 2013).

3.4.3 The synergid cells

The two micropylar-most cells of the embryo sac differentiate after cellularization into synergid cells (Christensen et al., 1997). Synergids play an important role in pollen tube attraction, pollen tube reception, the termination of pollen tube growth, and sperm cell release (see, e.g., Boisson-Dernier, Frietsch, Kim, Dizon, & Schroeder, 2008; Higashiyama et al., 2001; Kessler & Grossniklaus, 2011; Liu et al., 2016). Their essential role during reproduction and their interaction with the male gametophyte are reviewed in detail in chapter "Friend or foe: Signaling mechanisms during double fertilization in flowering seed plants" by Zhou and Dresselhaus (this issue).

Synergids are highly metabolically active and typically contain a large vacuole at the chalazal region of the cell, while the nucleus lies at the opposite micropylar pole (Mansfield et al., 1991). With respect to nuclear positioning, synergids thus exhibit just the opposite polarity as compared to the egg cell. Notably, reprogramming of synergids into egg cells, which has been observed in several mutants (see Section 3.5), is strictly associated with a relocalization of the synergid nucleus into a position characteristic of the egg cell. It is currently unclear whether nuclear relocalization is cause or consequence of such a cell identity switch. The micropylar end of the synergids, in addition, exhibits a pronounced thick cell wall structure, referred to as filiform apparatus (Mansfield et al., 1991). The filiform apparatus plays an important role during double fertilization (reviewed in Higashiyama & Yang, 2017). This was first demonstrated in *A. thaliana* by mutants defective in the R2R3-type MYB transcription factor MYB98, which show an abnormal filiform apparatus and fail to attract the pollen tube (Kasahara, Portereiko, Sandaklie-Nikolova, Rabiger, & Drews, 2005). Also, *Zea mays* EGG APPARATUS1 (ZmEA1) and the cysteine-rich LURE proteins, which are essential for pollen tube attraction in maize and *A. thaliana*, respectively (Márton, Cordts, Broadhvest, & Dresselhaus, 2005; Okuda et al., 2009), are secreted from the filiform apparatus (Márton et al., 2005; Takeuchi & Higashiyama, 2012).

The life span of synergids is typically short: during double fertilization, the first synergid degenerates upon physical contact with the pollen tube (Sandaklie-Nikolova, Palanivelu, King, Copenhaver, & Drews, 2007), a process which involves vacuolar acidification mediated by V-ATPases and ADAPTOR PROTEIN 1γ (Wang et al., 2017). The disintegration of the second synergid is initiated after at least one of female gametes has successfully been fertilized (Beale, Leydon, & Johnson, 2012; Kasahara et al., 2012). This degeneration of the second synergid occurs in two main steps: in the first step, the synergid fuses to the central cell thereby diluting its cytoplasmic content including potent pollen tube attracting LURE proteins (Maruyama, Völz, et al., 2015). In the second step, the synergid nucleus disintegrates. This biphasic process requires the FIS *Polycomb* Repressive Complex 2 (Maruyama et al., 2013) and ethylene response regulators (Völz, Heydlauff, Ripper, von Lyncker, & Groß-Hardt, 2013), and is necessary for the establishment of a pollen tube block that terminates pollen tube attraction.

3.4.4 The antipodal cells

After cellularization, the three most chalazal cells differentiate into so-called antipodals. In *A. thaliana*, antipodal cells are symplastically connected to each other and to adjacent sporophytic cells through plasmodesmata (Mansfield et al., 1991). These physical connections are in line with their predicted function in aiding nutrient transfer from the nucellus to the female gametophyte. However, functional studies have been hampered by the ephemeral nature of this cell type and its position deep in the maternal tissue. Recently, a protocol for the isolation of antipodals was published opening new avenues for the functional analysis of this cell type (Yu, Luo, & Sun, 2016).

In many grasses, the fate of antipodals differs fundamentally from that of *A. thaliana*. The three initial antipodals of barley enlarge and divide before fertilization to produce up to 100 cells (Cass & Jensen, 1970; Engell, 1994). After fertilization, the antipodals stop dividing but continue to enlarge

(Engell, 1994). In addition, the cell walls disintegrate resulting in a release of antipodal cytoplasm into the developing endosperm (Cass & Jensen, 1970). Also in maize, the three antipodal cells undergo division during female gametophyte development (Diboll, 1968; Diboll & Larson, 1966; Weatherwax, 1926). Due to incomplete cell wall formation, the mature maize embryo sacs consist of 20 or more multinucleate antipodal cells. After fertilization, maize antipodals typically persist and continue to divide (Weatherwax, 1926), and a recent study of the dominant leaf polarity mutant *Lax midrib1-O* (*Lxm1-O*) has suggested a potential role of auxin signaling during this process (Chettoor & Evans, 2015).

In contrast to maize and barley, there is no evidence for antipodal proliferation in *A. thaliana* (Kägi et al., 2010; Song, Yuan, & Sundaresan, 2014). Antipodal nuclei become smaller in maturing unfertilized female gametophytes. When female gametophytes are artificially kept unfertilized, nuclei cannot be detected by light microscopy or ultrastructural analysis 3 days after emasculation (Kägi et al., 2010), suggesting that fertilization is not required for antipodal cell disintegration. With respect to ovule age, 3 days after emasculation corresponds to fertilized ovules having undergone four or more nuclear divisions in the endosperm (Heydlauff & Groß-Hardt, unpublished). Indeed, Song and colleagues detected fluorescence signals associated with antipodal nuclei until the 16-nucleate endosperm stage (Song et al., 2014). As previously mentioned, antipodal lifespan is dependent on the metabolic activity of the central cell. Reduced mitochondrial function resulting from mutations affecting GCD1 or the cysteinyl-tRNA synthetase SYCO, as well as targeted mitochondrial membrane uncoupling, extends antipodal lifespan (Kägi et al., 2010; Wu et al., 2012).

Antipodal cells exhibit a remarkable developmental plasticity. Defects in the egg or central cell can result in a reprogramming of these inconspicuous cells into central cells (Groß-Hardt et al., 2007; Krohn, Lausser, Juranić, & Dresselhaus, 2012; Moll et al., 2008). This reprogramming involves antipodal nuclei fusion and the assimilation of a molecular profile characteristic for the central cell, suggesting that antipodals can act as a backup for the central cell should it fail to differentiate (Groß-Hardt et al., 2007).

3.5 Restricting germ cell fate

There are several lines of evidence showing that all cells in the female gametophyte exhibit a gametic potential and that this potential is repressed in accessory cells of wild-type embryo sacs. In *A. thaliana* plants defective for *ALTERED MERISTEM PROGRAM1* (*AMP1*), female gametophytes can exhibit an extra egg cell that forms at the expense of a synergid cell (Kong, Lau, & Jürgens, 2015). If supernumerary sperm cells are provided, twin plants can be recovered from *amp1* mutants, suggesting that both egg cells are fully functional. Interestingly, sporophytic *AMP1* expression is sufficient to repress the defect, highlighting the importance of sporophytic tissue in delimiting the area where egg cells form. Supernumerary egg cells that form at the expense of synergids have also been observed in *eostre* and *wyrd* mutants (Kirioukhova et al., 2011; Pagnussat, Yu, & Sundaresan, 2007). The defect is preceded by a mislocalization of nuclei at the micropylar end, with two instead of one nucleus occupying an egg cell position. The *eostre* phenotype is caused by ectopic expression of the *BEL1-LIKE HOMEODOMAIN1* (*BLH1*) gene (Pagnussat et al., 2007). While *BLH1* is absent from wild-type embryo sacs, it appears likely that other BLH family members act in the female gametophytes since mutations in the *OVATE FAMILY PROTEIN5* (*OFP5*), predicted to be a negative regulator of *BLH1*, phenocopy the *eostre* mutant (Pagnussat et al., 2007). The *wyrd* mutant is defective in a putative *A. thaliana* ortholog of the inner centrosome protein (INCENP) that is expressed in the female gametes (Kirioukhova et al., 2011).

Very broad gametic reprogramming is observed in female gametophytes defective for the *A. thaliana* genes *LACHESIS* (*LIS*), *CLOTHO* (*CLO*) and *ATROPOS* (*ATO*). Here, synergids and central cells adopt a molecular and morphological profile characteristic to egg cells, while antipodal cells resemble central cells (Groß-Hardt et al., 2007; Moll et al., 2008). *LIS*, *CLO/GFA1* and *ATO* all encode proteins homologous to yeast or human splicing factors (Coury et al., 2007; Groß-Hardt et al., 2007; Moll et al., 2008). While all three genes are initially expressed throughout the female gametophyte, *LIS* becomes specifically enriched in maturing female gametes (Groß-Hardt et al., 2007). In addition, egg cell-specific depletion of *LIS* in wild-type affects the development of all female gametophytic cells in a pattern strongly resembling *lis* mutants (Völz et al., 2012). The specific requirement of *LIS* in the egg cell indicates that the gametic potential of egg-adjoining cells is repressed by signals generated by the egg cell (Groß-Hardt et al., 2007; Völz et al., 2012). A key role for the egg cell during cell specification is further supported by the *Zea mays EA1-like 1* (*ZmEAL1*) gene, which encodes a secreted peptide specifically generated in maize egg cells (Krohn et al., 2012). In female gametophytes defective for *ZmEAL1*, antipodal cells exhibit morphological aspects characteristic of central cells, suggesting that the ZmEAL1 peptide is involved in communication with the antipodal cells to represses their gametic competence. This lateral inhibition from the egg cell is further substantiated by elegant egg ablation studies, which have been shown to activate egg cell signatures in synergid cells (Lawit, Chamberlin, Agee, Caswell, & Albertsen, 2013).

Taken together, these results highlight a substantial crosstalk between cells in the female gametophyte before fertilization, which appears to be at least in part facilitated by symplastic connectivity. Microinjection-based work in *Torenia fournieri* and *A. thaliana* suggests that the young egg cell is initially rather permissive for molecules up to 10 kDa for a short period, but then becomes increasingly symplastically isolated against its neighbors during maturation (Erdmann et al., 2017; Han, Huang, Zee, & Yuan, 2000).

4. Concluding remarks

The female gametophyte not only forms the female gametes but provides the physical and molecular basis for fertilization and the initiation of seed development. This minute structure thus comprises the central hub of plant reproduction, relying on robust molecular programs for its development. In fact, female gametophyte development is not only sustained by redundant gene functions but also by redundant cellular functions, which are encoded by accessory cells that can become gametically active in case of gametic failure.

The last two decades have seen tremendous advances in our understanding of the molecular mechanisms underlying the distinct steps of female gametophyte development. These steps include the selection of the female gametophyte founder cell, nuclear division and migration, as well as patterning and differentiation processes. Still, our understanding only appears to scratch the surface, which is partially due to the fact that research has primarily focused on a very restricted number of model organisms and invariable plant growth conditions that neglect the environmental challenges plants are exposed to in nature. With the advent of high throughput sequencing tools and the 1001 genomes project (http://1001genomes.org/) these limitations can now, to some extent, be overcome. The progress made in the field of plant reproduction has in the past and will in the future dictate the extent to which plant reproduction can be used and/or engineered in order to meet the challenges of a changing world.

Acknowledgments

We thank members of the Groß-Hardt lab for providing critical comments on the manuscript, and gratefully acknowledge the financial support from the European Research Council to R.G. (ERC Consolidator Grant "bi-BLOCK" ID. 64664).

References

Agashe, B., Prasad, C. K., & Siddiqi, I. (2002). Identification and analysis of *DYAD*: A gene required for meiotic chromosome organisation and female meiotic progression in *Arabidopsis*. *Development*, *129*, 3935–3943.

Beale, K. M., Leydon, A. R., & Johnson, M. A. (2012). Gamete fusion is required to block multiple pollen tubes from entering an *Arabidopsis* ovule. *Current Biology*, *22*, 1090–1094.

Bemer, M., Wolters-Arts, M., Grossniklaus, U., & Angenent, G. C. (2008). The MADS domain protein DIANA acts together with AGAMOUS-LIKE80 to specify the central cell in *Arabidopsis* ovules. *The Plant Cell*, *20*, 2088–2101.

Benková, E., Michniewicz, M., Sauer, M., Teichmann, T., Seifertová, D., Jürgens, G., et al. (2003). Local, efflux-dependent auxin gradients as a common module for plant organ formation. *Cell*, *115*, 591–602.

Boisson-Dernier, A., Frietsch, S., Kim, T.-H., Dizon, M. B., & Schroeder, J. I. (2008). The peroxin loss-of-function mutation abstinence by mutual consent disrupts male-female gametophyte recognition. *Current Biology*, *18*, 63–68.

Cao, L., Wang, S., Venglat, P., Zhao, L., Cheng, Y., Ye, S., et al. (2018). *Arabidopsis* ICK/KRP cyclin-dependent kinase inhibitors function to ensure the formation of one megaspore mother cell and one functional megaspore per ovule. *PLoS Genetics*, *14*, e1007230.

Capron, A., Serralbo, O., Fülöp, K., Frugier, F., Parmentier, Y., Dong, A., et al. (2003). The *Arabidopsis* anaphase-promoting complex or cyclosome: Molecular and genetic characterization of the APC2 subunit. *The Plant Cell*, *15*, 2370–2382.

Cass, D. D., & Jensen, W. A. (1970). Fertilization in barley. *American Journal of Botany*, *57*, 62–70.

Ceccato, L., Masiero, S., Sinha Roy, D., Bencivenga, S., Roig-Villanova, I., Ditengou, F. A., et al. (2013). Maternal control of PIN1 is required for female gametophyte development in *Arabidopsis*. *PLoS One*, *8*, e66148.

Chardin, C., Girin, T., Roudier, F., Meyer, C., & Krapp, A. (2014). The plant RWP-RK transcription factors: Key regulators of nitrogen responses and of gametophyte development. *Journal of Experimental Botany*, *65*, 5577–5587.

Chen, G.-H., Sun, J.-Y., Liu, M., Liu, J., & Yang, W.-C. (2014). SPOROCYTELESS is a novel embryophyte-specific transcription repressor that interacts with TPL and TCP proteins in *Arabidopsis*. *Journal of Genetics and Genomics*, *41*, 617–625.

Cheng, C. Y., Mathews Dennis, E., Eric Schaller, G., & Kieber Joseph, J. (2012). Cytokinin-dependent specification of the functional megaspore in the *Arabidopsis* female gametophyte. *The Plant Journal*, *73*, 929–940.

Chettoor, A. M., & Evans, M. M. S. (2015). Correlation between a loss of auxin signaling and a loss of proliferation in maize antipodal cells. *Frontiers in Plant Science*, *6*, 187.

Choi, Y., Gehring, M., Johnson, L., Hannon, M., Harada, J. J., Goldberg, R. B., et al. (2002). DEMETER, a DNA glycosylase domain protein, is required for endosperm gene imprinting and seed viability in *Arabidopsis*. *Cell*, *110*, 33–42.

Christensen, C. A., Gorsich, S. W., Brown, R. H., Jones, L. G., Brown, J., Shaw, J. M., et al. (2002). Mitochondrial GFA2 is required for synergid cell death in *Arabidopsis*. *The Plant Cell*, *14*, 2215–2232.

Christensen, C. A., King, E. J., Jordan, J. R., & Drews, G. N. (1997). Megagametogenesis in *Arabidopsis* wild type and the *Gf* mutant. *Sexual Plant Reproduction, 10,* 49–64.
Christensen, C. A., Subramanian, S., & Drews, G. N. (1998). Identification of gametophytic mutations affecting female gametophyte development in *Arabidopsis. Developmental Biology, 202,* 136–151.
Cigliano, R. A., Cremona, G., Paparo, R., Termolino, P., Perrella, G., Gutzat, R., et al. (2013). Histone deacetylase AtHDA7 is required for female gametophyte and embryo development in *Arabidopsis. Plant Physiology, 163,* 431–440.
Colombo, M., Masiero, S., Vanzulli, S., Lardelli, P., Kater, M. M., & Colombo, L. (2008). *AGL23*, a type I MADS-box gene that controls female gametophyte and embryo development in *Arabidopsis. The Plant Journal, 54,* 1037–1048.
Coury, D. A., Zhang, C., Ko, A., Skaggs, M. I., Christensen, C. A., Drews, G. N., et al. (2007). Segregation distortion in *Arabidopsis gametophytic factor 1* (*gfa1*) mutants is caused by a deficiency of an essential RNA splicing factor. *Sexual Plant Reproduction, 20,* 87–97.
Cui, H. H., Liao, H. Z., Tang, Y., Du, X. Y., Chen, L. Q., Ye, D., et al. (2015). *ABORTED GAMETOPHYTE 1* is required for gametogenesis in *Arabidopsis. Journal of Integrative Plant Biology, 57,* 1003–1016.
Demesa-Arévalo, A., & Vielle-Calzada, J.-P. (2013). The classical arabinogalactan protein AGP18 mediates megaspore selection in *Arabidopsis. The Plant Cell, 25,* 1274–1287.
d'Erfurth, I., Cromer, L., Jolivet, S., Girard, C., Horlow, C., Sun, Y., et al. (2010). The CYCLIN-A CYCA1;2/TAM is required for the meiosis I to meiosis II transition and cooperates with OSD1 for the prophase to first meiotic division transition. *PLoS Genetics, 6,* e1000989.
d'Erfurth, I., Jolivet, S., Froger, N., Catrice, O., Novatchkova, M., & Mercier, R. (2009). Turning meiosis into mitosis. *PLoS Biology, 7,* e1000124.
Di Fino, L. M., D'Ambrosio, J. M., Tejos, R., van Wijk, R., Lamattina, L., Munnik, T., et al. (2017). *Arabidopsis* phosphatidylinositol-phospholipase C2 (PLC2) is required for female gametogenesis and embryo development. *Planta, 245,* 717–728.
Diboll, A. G. (1968). Fine structural development of the megagametophyte of *Zea mays* following fertilization. *American Journal of Botany, 55,* 797–806.
Diboll, A. G., & Larson, D. A. (1966). An electron microscopic study of the mature megagametophyte in *Zea mays. American Journal of Botany, 53,* 391–402.
D'Ippólito, S., Arias, L. A., Casalongué, C. A., Pagnussat, G. C., & Fiol, D. F. (2017). The DC1-domain protein VACUOLELESS GAMETOPHYTES is essential for development of female and male gametophytes in *Arabidopsis. The Plant Journal, 90,* 261–275.
Domenichini, S., Benhamed, M., De Jaeger, G., Van De Slijke, E., Blanchet, S., Bourge, M., et al. (2012). Evidence for a role of *Arabidopsis* CDT1 proteins in gametophyte development and maintenance of genome integrity. *The Plant Cell, 24,* 2779–2791.
Ebel, C., Mariconti, L., & Gruissern, W. (2004). Plant retinoblastoma homologues control nuclear proliferation in the female gametophyte. *Nature, 429,* 776–780.
El-Kasmi, F., Pacher, T., Strompen, G., Stierhof, Y.-D., Müller, L. M., Koncz, C., et al. (2011). *Arabidopsis* SNARE protein SEC22 is essential for gametophyte development and maintenance of Golgi-stack integrity. *The Plant Journal, 66,* 268–279.
Engell, K. (1994). Embryology of barley. IV. Ultrastructure of the antipodal cells of *Hordeum vulgare* L. cv. Bomi before and after fertilization of the egg cell. *Sexual Plant Reproduction, 7,* 333–346.
Erdmann, R. M., Hoffmann, A., Walter, H.-K., Wagenknecht, H.-A., Groß-Hardt, R., & Gehring, M. (2017). Molecular movement in the *Arabidopsis thaliana* female gametophyte. *Plant Reproduction, 30,* 141–146.
Frost, J. M., Kim, M. Y., Park, G. T., Hsieh, P.-H., Nakamura, M., Lin, S. J. H., et al. (2018). FACT complex is required for DNA demethylation at heterochromatin during

reproduction in *Arabidopsis*. *Proceedings of the National Academy of Sciences of the United States of America, 115*, E4720–E4729.

Gallois, J.-L., Guyon-Debast, A., Lécureuil, A., Vezon, D., Carpentier, V., Bonhomme, S., et al. (2009). The *Arabidopsis* proteasome RPT5 subunits are essential for gametophyte development and show accession-dependent redundancy. *The Plant Cell, 21*, 442–459.

Garcia-Aguilar, M., Michaud, C., Leblanc, O., & Grimanelli, D. (2010). Inactivation of a DNA methylation pathway in maize reproductive organs results in apomixis-like phenotypes. *The Plant Cell, 22*, 3249–3267.

Gifford, E. M., & Foster, A. S. (1989). *Morphology and evolution of vascular plants* (1st ed.). New York: W.H. Freeman and Company.

Groß-Hardt, R., Kägi, C., Baumann, N., Moore, J. M., Baskar, R., Gagliano, W. B., et al. (2007). *LACHESIS* restricts gametic cell fate in the female gametophyte of *Arabidopsis*. *PLoS Biology, 5*, e47.

Guo, F., Huang, B.-Q., Han, Y., & Zee, S.-Y. (2004). Fertilization in maize *indeterminate gametophyte1* mutant. *Protoplasma, 223*, 111–120.

Haecker, A., Groß-Hardt, R., Geiges, B., Sarkar, A., Breuninger, H., Herrmann, M., et al. (2004). Expression dynamics of *WOX* genes mark cell fate decisions during early embryonic patterning in *Arabidopsis thaliana*. *Development, 131*, 657–668.

Han, Y.-Z., Huang, B.-Q., Zee, S.-Y., & Yuan, M. (2000). Symplastic communication between the central cell and the egg apparatus cells in the embryo sac of *Torenia fournieri* Lind. before and during fertilization. *Planta, 211*, 158–162.

Harashima, H., & Sugimoto, K. (2016). Integration of developmental and environmental signals into cell proliferation and differentiation through RETINOBLASTOMA-RELATED 1. *Current Opinion in Plant Biology, 29*, 95–103.

Hejátko, J., Pernisová, M., Eneva, T., Palme, K., & Brzobohatý, B. (2003). The putative sensor histidine kinase CKI1 is involved in female gametophyte development in *Arabidopsis*. *Molecular Genetics and Genomics, 269*, 443–453.

Hernández-Lagana, E., Rodríguez-Leal, D., Lúa, J., & Vielle-Calzada, J.-P. (2016). A multigenic network of ARGONAUTE4 clade members controls early megaspore formation in *Arabidopsis*. *Genetics, 204*, 1045–1056.

Higashiyama, T., Yabe, S., Sasaki, N., Nishimura, Y., Miyagishima, S.-y., Kuroiwa, H., et al. (2001). Pollen tube attraction by the synergid cell. *Science, 293*, 1480–1483.

Higashiyama, T., & Yang, W.-c. (2017). Gametophytic pollen tube guidance: Attractant peptides, gametic controls, and receptors. *Plant Physiology, 173*, 112–121.

Huanca-Mamani, W., Garcia-Aguilar, M., León-Martínez, G., Grossniklaus, U., & Vielle-Calzada, J.-P. (2005). CHR11, a chromatin-remodeling factor essential for nuclear proliferation during female gametogenesis in *Arabidopsis thaliana*. *Proceedings of the National Academy of Sciences of the United States of America, 102*, 17231–17236.

Huang, C.-K., Huang, L.-F., Huang, J.-J., Wu, S.-J., Yeh, C.-H., & Lu, C.-A. (2010). A DEAD-box protein, AtRH36, is essential for female gametophyte development and is involved in rRNA biogenesis in *Arabidopsis*. *Plant and Cell Physiology, 51*, 694–706.

Huang, B.-Q., & Russell, S. D. (1992). Female germ unit: Organization, isolation, and function. In S. D. Russel & C. Dumas (Eds.), *International review of cytology* (pp. 233–293): Academic Press.

Ibarra, C. A., Feng, X., Schoft, V. K., Hsieh, T.-F., Uzawa, R., Rodrigues, J. A., et al. (2012). Active DNA demethylation in plant companion cells reinforces transposon methylation in gametes. *Science, 337*, 1360–1364.

Jensen, W. A. (1964). Observations on the fusion of nuclei in plants. *The Journal of Cell Biology, 23*, 669–672.

Jiang, L., Yuan, L., Xia, M., & Makaroff, C. A. (2010). Proper levels of the *Arabidopsis* cohesion establishment factor CTF7 are essential for embryo and megagametophyte, but not endosperm, development. *Plant Physiology, 154*, 820–832.

Johri, B. M., Ambegaokoar, K. B., & Srivastava, P. S. (1992). *Comparative embryology of angiosperms* (1st ed.). Berlin: Springer.

Jullien, P. E., Kinoshita, T., Ohad, N., & Berger, F. (2006). Maintenance of DNA methylation during the *Arabidopsis* life cycle is essential for parental imprinting. *The Plant Cell, 18*, 1360–1372.

Kägi, C., Baumann, N., Nielsen, N., Stierhof, Y.-D., & Groß-Hardt, R. (2010). The gametic central cell of *Arabidopsis* determines the lifespan of adjacent accessory cells. *Proceedings of the National Academy of Sciences of the United States of America, 107*, 22350–22355.

Kakimoto, T. (1996). CKI1, a histidine kinase homolog implicated in cytokinin signal transduction. *Science, 274*, 982–985.

Kasahara, R. D., Maruyama, D., Hamamura, Y., Sakakibara, T., Twell, D., & Higashiyama, T. (2012). Fertilization recovery after defective sperm cell release in *Arabidopsis*. *Current Biology, 22*, 1084–1089.

Kasahara, R. D., Portereiko, M. F., Sandaklie-Nikolova, L., Rabiger, D. S., & Drews, G. N. (2005). *MYB98* is required for pollen tube guidance and synergid cell differentiation in *Arabidopsis*. *The Plant Cell, 17*, 2981–2992.

Kawashima, T., & Berger, F. (2015). The central cell nuclear position at the micropylar end is maintained by the balance of F-actin dynamics, but dispensable for karyogamy in *Arabidopsis*. *Plant Reproduction, 28*, 103–110.

Kessler, S. A., & Grossniklaus, U. (2011). She's the boss: Signaling in pollen tube reception. *Current Opinion in Plant Biology, 14*, 622–627.

Khan, A., Garbelli, A., Grossi, S., Florentin, A., Batelli, G., Acuna, T., et al. (2014). The *Arabidopsis* STRESS RESPONSE SUPPRESSOR DEAD-box RNA helicases are nucleolar- and chromocenter-localized proteins that undergo stress-mediated relocalization and are involved in epigenetic gene silencing. *The Plant Journal, 79*, 28–43.

Kirioukhova, O., Johnston, A. J., Kleen, D., Kägi, C., Baskar, R., Moore, J. M., et al. (2011). Female gametophytic cell specification and seed development require the function of the putative *Arabidopsis* INCENP ortholog *WYRD*. *Development, 138*, 3409–3420.

Koi, S., Hisanaga, T., Sato, K., Shimamura, M., Yamato, K. T., Ishizaki, K., et al. (2016). An evolutionarily conserved plant RKD factor controls germ cell differentiation. *Current Biology, 26*, 1775–1781.

Komiya, R., Ohyanagi, H., Niihama, M., Watanabe, T., Nakano, M., Kurata, N., et al. (2014). Rice germline-specific Argonaute MEL1 protein binds to phasiRNAs generated from more than 700 lincRNAs. *The Plant Journal, 78*, 385–397.

Kong, J., Lau, S., & Jürgens, G. (2015). Twin plants from supernumerary egg cells in *Arabidopsis*. *Current Biology, 25*, 225–230.

Köszegi, D., Johnston, A. J., Rutten, T., Czihal, A., Altschmied, L., Kumlehn, J., et al. (2011). Members of the RKD transcription factor family induce an egg cell-like gene expression program. *The Plant Journal, 67*, 280–291.

Krohn, N. G., Lausser, A., Juranić, M., & Dresselhaus, T. (2012). Egg cell signaling by the secreted peptide ZmEAL1 controls antipodal cell fate. *Developmental Cell, 23*, 219–225.

Kwee, H. S., & Sundaresan, V. (2003). The *NOMEGA* gene required for female gametophyte development encodes the putative APC6/CDC16 component of the anaphase promoting complex in *Arabidopsis*. *The Plant Journal, 36*, 853–866.

Larsson, E., Vivian-Smith, A., Offringa, R., & Sundberg, E. (2017). Auxin homeostasis in *Arabidopsis* ovules in anther-dependent at maturation and changes dynamically upon fertilization. *Frontiers in Plant Science, 8*, 1735.

Latrasse, D., Benhamed, M., Henry, Y., Domenichini, S., Kim, W., Zhou, D.-X., et al. (2008). The MYST histone acetyltransferases are essential for gametophyte development in *Arabidopsis*. *BMC Plant Biology, 8*, 121.

Laux, T., Mayer, K. F., Berger, J., & Jurgens, G. (1996). The *WUSCHEL* gene is required for shoot and floral meristem integrity in *Arabidopsis*. *Development*, *122*, 87–96.

Lawit, S. J., Chamberlin, M. A., Agee, A., Caswell, E. S., & Albertsen, M. C. (2013). Transgenic manipulation of plant embryo sacs tracked through cell-type-specific fluorescent markers: Cell labeling, cell ablation, and adventitious embryos. *Plant Reproduction*, *26*, 125–137.

Leshem, Y., Johnson, C., Wuest, S. E., Song, X., Ngo, Q. A., Grossniklaus, U., et al. (2012). Molecular characterization of the *glauce* mutant: A central cell–specific function is required for double fertilization in *Arabidopsis*. *The Plant Cell*, *24*, 3264–3277.

Li, L.-X., Liao, H.-Z., Jiang, L.-X., Tan, Q., Ye, D., & Zhang, X.-Q. (2018). *Arabidopsis thaliana NOP10* is required for gametophyte formation. *Journal of Integrative Plant Biology*, *60*, 723–736.

Li, H.-J., Liu, N.-Y., Shi, D.-Q., Liu, J., & Yang, W.-C. (2010). YAO is a nucleolar WD40-repeat protein critical for embryogenesis and gametogenesis in *Arabidopsis*. *BMC Plant Biology*, *10*, 169.

Li, N., Yuan, L., Liu, N., Shi, D., Li, X., Tang, Z., et al. (2009). *SLOW WALKER2*, a NOC1/MAK21 homologue, is essential for coordinated cell cycle progression during female gametophyte development in *Arabidopsis*. *Plant Physiology*, *151*, 1486–1497.

Lieber, D., Lora, J., Schrempp, S., Lenhard, M., & Laux, T. (2011). *Arabidopsis WIH1* and *WIH2* genes act in the transition from somatic to reproductive cell fate. *Current Biology*, *21*, 1009–1017.

Lin, B.-Y. (1978). Structural modifications of the female gametophyte associated with the *indeterminate gametophyte* (*ig*) mutant in maize. *Canadian Journal of Genetics and Cytology*, *20*, 249–257.

Lituiev, D. S., Krohn, N. G., Müller, B., Jackson, D., Hellriegel, B., Dresselhaus, T., et al. (2013). Theoretical and experimental evidence indicates that there is no detectable auxin gradient in the angiosperm female gametophyte. *Development*, *140*, 4544–4553.

Liu, X., Castro, C., Wang, Y., Noble, J., Ponvert, N., Bundy, M., et al. (2016). The role of LORELEI in pollen tube reception at the interface of the synergid cell and pollen tube requires the modified eight-cysteine motif and the receptor-like kinase FERONIA. *The Plant Cell*, *28*, 1035–1052.

Liu, M., Shi, D.-Q., Yuan, L., Liu, J., & Yang, W.-C. (2010). *SLOW WALKER3*, encoding a putative DEAD-box RNA helicase, is essential for female gametogenesis in *Arabidopsis*. *Journal of Integrative Plant Biology*, *52*, 817–828.

Liu, M., Shi, S., Zhang, S., Xu, P., Lai, J., Liu, Y., et al. (2014). SUMO E3 ligase AtMMS21 is required for normal meiosis and gametophyte development in *Arabidopsis*. *BMC Plant Biology*, *14*, 153.

Liu, Z., Yuan, L., Song, X., Yu, X., & Sundaresan, V. (2017). *AHP2*, *AHP3*, and *AHP5* act downstream of *CKI1* in *Arabidopsis* female gametophyte development. *Journal of Experimental Botany*, *68*, 3365–3373.

Liu, J., Zhang, Y., Qin, G., Tsuge, T., Sakaguchi, N., Luo, G., et al. (2008). Targeted degradation of the cyclin-dependent kinase inhibitor ICK4/KRP6 by RING-type E3 ligases is essential for mitotic cell cycle progression during *Arabidopsis* gametogenesis. *The Plant Cell*, *20*, 1538–1554.

Lorbiecke, R., Steffens, M., Tomm, J. M., Scholten, S., von Wiegen, P., Kranz, E., et al. (2005). Phytosulphokine gene regulation during maize (*Zea mays* L.) reproduction. *Journal of Experimental Botany*, *56*, 1805–1819.

Lu, Y., Li, C., Wang, H., Chen, H., Berg, H., & Xia, Y. (2011). AtPPR2, an *Arabidopsis* pentatricopeptide repeat protein, binds to plastid 23S rRNA and plays an important role in the first mitotic division during gametogenesis and in cell proliferation during embryogenesis. *The Plant Journal*, *67*, 13–25.

Maheshwari, P. (1950). *An introduction to the embryology of angiosperms* (1st ed.). New York: McGraw-Hill.

Makkena, S., Lee, E., Sack, F. D., & Lamb, R. S. (2012). The R2R3 MYB transcription factors FOUR LIPS and MYB88 regulate female reproductive development. *Journal of Experimental Botany, 63*, 5545–5558.

Mansfield, S. G., Briarty, L. G., & Erni, S. (1991). Early embryogenesis in *Arabidopsis thaliana*. I. The mature embryo sac. *Canadian Journal of Botany, 69*, 447–460.

Martin, M. V., Fiol, D. F., Sundaresan, V., Zabaleta, E. J., & Pagnussat, G. C. (2013). *oiwa*, a female gametophytic mutant impaired in a mitochondrial manganese-superoxide dismutase, reveals crucial roles for reactive oxygen species during embryo sac development and fertilization in *Arabidopsis*. *The Plant Cell, 25*, 1573–1591.

Márton, M. L., Cordts, S., Broadhvest, J., & Dresselhaus, T. (2005). Micropylar pollen tube guidance by egg apparatus 1 of maize. *Science, 307*, 573–576.

Maruyama, D., Endo, T., & Nishikawa, S.-i. (2010). BiP-mediated polar nuclei fusion is essential for the regulation of endosperm nuclei proliferation in *Arabidopsis thaliana*. *Proceedings of the National Academy of Sciences of the United States of America, 107*, 1684–1689.

Maruyama, D., Endo, T., & Nishikawa, S.-i. (2015). BiP3 supports the early stages of female gametogenesis in the absence of BiP1 and BiP2 in *Arabidopsis thaliana*. *Plant Signaling & Behavior, 10*, e1035853.

Maruyama, D., Hamamura, Y., Takeuchi, H., Susaki, D., Nishimaki, M., Kurihara, D., et al. (2013). Independent control by each female gamete prevents the attraction of multiple pollen tubes. *Developmental Cell, 25*, 317–323.

Maruyama, D., Völz, R., Takeuchi, H., Mori, T., Igawa, T., Kurihara, D., et al. (2015). Rapid elimination of the persistent synergid through a cell fusion mechanism. *Cell, 161*, 907–918.

Maruyama, D., Yamamoto, M., Endo, T., & Nishikawa, S.-i. (2014). Different sets of ER-resident J-proteins regulate distinct polar nuclear-membrane fusion events in *Arabidopsis thaliana*. *Plant and Cell Physiology, 55*, 1937–1944.

Matias-Hernandez, L., Battaglia, R., Galbiati, F., Rubes, M., Eichenberger, C., Grossniklaus, U., et al. (2010). *VERDANDI* is a direct target of the MADS domain ovule identity complex and affects embryo sac differentiation in *Arabidopsis*. *The Plant Cell, 22*, 1702–1715.

Matzke, M. A., & Mosher, R. A. (2014). RNA-directed DNA methylation: An epigenetic pathway of increasing complexity. *Nature Reviews Genetics, 15*, 394.

Mayer, K. F. X., Schoof, H., Haecker, A., Lenhard, M., Jürgens, G., & Laux, T. (1998). Role of *WUSCHEL* in regulating stem cell fate in the *Arabidopsis* shoot meristem. *Cell, 95*, 805–815.

Mercier, R., Vezon, D., Bullier, E., Motamayor, J. C., Sellier, A., Lefèvre, F., et al. (2001). SWITCH1 (SWI1): A novel protein required for the establishment of sister chromatid cohesion and for bivalent formation at meiosis. *Genes & Development, 15*, 1859–1871.

Mogensen, H. L., & Suthar, H. K. (1979). Ultrastructure of the egg apparatus of *Nicotiana tabacum* (*Solanaceae*) before and after fertilization. *Botanical Gazette, 140*, 168–179.

Moll, C., von Lyncker, L., Zimmermann, S., Kägi, C., Baumann, N., Twell, D., et al. (2008). *CLO/GFA1* and *ATO* are novel regulators of gametic cell fate in plants. *The Plant Journal, 56*, 913–921.

Nakai, T., Kato, K., Shinmyo, A., & Sekine, M. (2006). *Arabidopsis* KRPs have distinct inhibitory activity toward cyclin D2-associated kinases, including plant-specific B-type cyclin-dependent kinase. *FEBS Letters, 580*, 336–340.

Nakajima, K. (2018). Be my baby: Patterning toward plant germ cells. *Current Opinion in Plant Biology, 41*, 110–115.

Nakamura, M., & Hashimoto, T. (2009). A mutation in the *Arabidopsis* γ-tubulin-containing complex causes helical growth and abnormal microtubule branching. *Journal of Cell Science*, *122*, 2208–2217.
Ngo, Q. A., Moore, J. M., Baskar, R., Grossniklaus, U., & Sundaresan, V. (2007). *Arabidopsis GLAUCE* promotes fertilization-independent endosperm development and expression of paternally inherited alleles. *Development*, *134*, 4107–4117.
Nonomura, K.-I. (2018). Small RNA pathways responsible for non-cell-autonomous regulation of plant reproduction. *Plant Reproduction*, *31*, 21–29.
Nonomura, K.-I., Eiguchi, M., Nakano, M., Takashima, K., Komeda, N., Fukuchi, S., et al. (2011). A novel RNA-recognition-motif protein is required for premeiotic G1/S-phase transition in rice (*Oryza sativa* L.). *PLoS Genetics*, *7*, e1001265.
Nonomura, K.-I., Miyoshi, K., Eiguchi, M., Suzuki, T., Miyao, A., Hirochika, H., et al. (2003). The *MSP1* gene is necessary to restrict the number of cells entering into male and female sporogenesis and to initiate anther wall formation in rice. *The Plant Cell*, *15*, 1728–1739.
Nonomura, K.-I., Morohoshi, A., Nakano, M., Eiguchi, M., Miyao, A., Hirochika, H., et al. (2007). A germ cell–specific gene of the *ARGONAUTE* family is essential for the progression of premeiotic mitosis and meiosis during sporogenesis in rice. *The Plant Cell*, *19*, 2583–2594.
Nowack, M. K., Harashima, H., Dissmeyer, N., Zhao, X. A., Bouyer, D., Weimer, A. K., et al. (2012). Genetic framework of cyclin-dependent kinase function in *Arabidopsis*. *Developmental Cell*, *22*, 1030–1040.
Okuda, S., Tsutsui, H., Shiina, K., Sprunck, S., Takeuchi, H., Yui, R., et al. (2009). Defensin-like polypeptide LUREs are pollen tube attractants secreted from synergid cells. *Nature*, *458*, 357.
Olmedo-Monfil, V., Durán-Figueroa, N., Arteaga-Vázquez, M., Demesa-Arévalo, E., Autran, D., Grimanelli, D., et al. (2010). Control of female gamete formation by a small RNA pathway in *Arabidopsis*. *Nature*, *464*, 628.
Pagnussat, G. C., Alandete-Saez, M., Bowman, J. L., & Sundaresan, V. (2009). Auxin-dependent patterning and gamete specification in the *Arabidopsis* female gametophyte. *Science*, *324*, 1684–1689.
Pagnussat, G. C., Yu, H.-J., & Sundaresan, V. (2007). Cell-fate switch of synergid to egg cell in *Arabidopsis eostre* mutant embryo sacs arises from misexpression of the BEL1-like homeodomain gene *BLH1*. *The Plant Cell*, *19*, 3578–3592.
Panoli, A., Martin, M. V., Alandete-Saez, M., Simon, M., Neff, C., Swarup, R., et al. (2015). Auxin import and local auxin biosynthesis are required for mitotic divisions, cell expansion and cell specification during female gametophyte development in *Arabidopsis thaliana*. *PLoS One*, *10*, e0126164.
Park, G. T., Frost, J. M., Park, J.-S., Kim, T. H., Lee, J. S., Oh, S. A., et al. (2014). Nucleoporin MOS7/Nup88 is required for mitosis in gametogenesis and seed development in *Arabidopsis*. *Proceedings of the National Academy of Sciences of the United States of America*, *111*, 18393–18398.
Park, J.-S., Frost, J. M., Park, K., Ohr, H., Park, G. T., Kim, S., et al. (2017). Control of DEMETER DNA demethylase gene transcription in male and female gamete companion cells in *Arabidopsis thaliana*. *Proceedings of the National Academy of Sciences of the United States of America*, *114*, 2078–2083.
Park, K., Kim, M. Y., Vickers, M., Park, J.-S., Hyun, Y., Okamoto, T., et al. (2016). DNA demethylation is initiated in the central cells of *Arabidopsis* and rice. *Proceedings of the National Academy of Sciences of the United States of America*, *113*, 15138–15143.
Pastuglia, M., Azimzadeh, J., Goussot, M., Camilleri, C., Belcram, K., Evrard, J.-L., et al. (2006). γ-Tubulin is essential for microtubule organization and development in *Arabidopsis*. *The Plant Cell*, *18*, 1412–1425.

Pawlowski, W. P., Wang, C.-J. R., Golubovskaya, I. N., Szymaniak, J. M., Shi, L., Hamant, O., et al. (2009). Maize AMEIOTIC1 is essential for multiple early meiotic processes and likely required for the initiation of meiosis. *Proceedings of the National Academy of Sciences of the United States of America, 106*, 3603–3608.

Pérez-España, V. H., Sánchez-León, N., & Vielle-Calzada, J.-P. (2011). *CYP85A1* is required for the initiation of female gametogenesis in *Arabidopsis thaliana*. *Plant Signaling & Behavior, 6*, 321–326.

Pérez-Pérez José, M., Serralbo, O., Vanstraelen, M., González, C., Criqui, M. C., Genschik, P., et al. (2007). Specialization of CDC27 function in the *Arabidopsis thaliana* anaphase-promoting complex (APC/C). *The Plant Journal, 53*, 78–89.

Pischke, M. S., Jones, L. G., Otsuga, D., Fernandez, D. E., Drews, G. N., & Sussman, M. R. (2002). An *Arabidopsis* histidine kinase is essential for megagametogenesis. *Proceedings of the National Academy of Sciences of the United States of America, 99*, 15800–15805.

Portereiko, M. F., Lloyd, A., Steffen, J. G., Punwani, J. A., Otsuga, D., & Drews, G. N. (2006). *AGL80* is required for central cell and endosperm development in *Arabidopsis*. *The Plant Cell, 18*, 1862–1872.

Portereiko, M. F., Sandaklie-Nikolova, L., Lloyd, A., Dever, C. A., Otsuga, D., & Drews, G. N. (2006). *NUCLEAR FUSION DEFECTIVE1* encodes the *Arabidopsis* RPL21M protein and is required for karyogamy during female gametophyte development and fertilization. *Plant Physiology, 141*, 957–965.

Rabiger, D. S., & Drews, G. N. (2013). *MYB64* and *MYB119* are required for cellularization and differentiation during female gametogenesis in *Arabidopsis thaliana*. *PLoS Genetics, 9*, e1003783.

Ravi, M., Marimuthu, M. P. A., & Siddiqi, I. (2008). Gamete formation without meiosis in *Arabidopsis*. *Nature, 451*, 1121.

Ren, L., Tang, D., Zhao, T., Zhang, F., Liu, C., Xue, Z., et al. (2018). *OsSPL* regulates meiotic fate acquisition in rice. *New Phytologist, 218*, 789–803.

Rodkiewicz, B. (1970). Callose in cell walls during megasporogenesis in angiosperms. *Planta, 93*, 39–47.

Rodrigo-Peiris, T., Xu, X. M., Zhao, Q., Wang, H.-J., & Meier, I. (2011). RanGAP is required for post-meiotic mitosis in female gametophyte development in *Arabidopsis thaliana*. *Journal of Experimental Botany, 62*, 2705–2714.

Rövekamp, M., Bowman, J. L., & Grossniklaus, U. (2016). *Marchantia* Mp*RKD* regulates the gametophyte-sporophyte transition by keeping egg cells quiescent in the absence of fertilization. *Current Biology, 26*, 1782–1789.

Sandaklie-Nikolova, L., Palanivelu, R., King, E. J., Copenhaver, G. P., & Drews, G. N. (2007). Synergid cell death in *Arabidopsis* is triggered following direct interaction with the pollen tube. *Plant Physiology, 144*, 1753–1762.

Schmidt, A., Schmid, M. W., & Grossniklaus, U. (2015). Plant germline formation: Common concepts and developmental flexibility in sexual and asexual reproduction. *Development, 142*, 229–241.

Schmidt, A., Wuest, S. E., Vijverberg, K., Baroux, C., Kleen, D., & Grossniklaus, U. (2011). Transcriptome analysis of the *Arabidopsis* megaspore mother cell uncovers the importance of RNA helicases for plant germline development. *PLoS Biology, 9*, e1001155.

Schulz, P., & Jensen, W. A. (1973). *Capsella* embryogenesis: The central cell. *Journal of Cell Science, 12*, 741–763.

She, W., Grimanelli, D., Rutowicz, K., Whitehead, M. W. J., Puzio, M., Kotliński, M., et al. (2013). Chromatin reprogramming during the somatic-to-reproductive cell fate transition in plants. *Development, 140*, 4008–4019.

Sheridan, W. F., Avalkina, N. A., Shamrov, I. I., Batyea, T. B., & Golubovskaya, I. N. (1996). The *mac1* gene: Controlling the commitment to the meiotic pathway in maize. *Genetics, 142*, 1009–1020.

Sheridan, W. F., & Huang, B.-Q. (1997). Nuclear behavior is defective in the maize (*Zea mays* L.) *lethal ovule2* female gametophyte. *The Plant Journal*, *11*, 1029–1041.

Shi, D.-Q., Liu, J., Xiang, Y.-H., Ye, D., Sundaresan, V., & Yang, W.-C. (2005). *SLOW WALKER1*, essential for gametogenesis in *Arabidopsis*, encodes a WD40 protein involved in 18S ribosomal RNA biogenesis. *The Plant Cell*, *17*, 2340–2354.

Siddiqi, I., Ganesh, G., Grossniklaus, U., & Subbiah, V. (2000). The *dyad* gene is required for progression through female meiosis in *Arabidopsis*. *Development*, *127*, 197–207.

Singh, M., Goel, S., Meeley, R. B., Dantec, C., Parrinello, H., Michaud, C., et al. (2011). Production of viable gametes without meiosis in maize deficient for an ARGONAUTE protein. *The Plant Cell*, *23*, 443–458.

Song, X., Yuan, L., & Sundaresan, V. (2014). Antipodal cells persist through fertilization in the female gametophyte of *Arabidopsis*. *Plant Reproduction*, *27*, 197–203.

Sprunck, S., & Groß-Hardt, R. (2011). Nuclear behavior, cell polarity, and cell specification in the female gametophyte. *Sexual Plant Reproduction*, *24*, 123–136.

Srilunchang, K. O., Krohn, N. G., & Dresselhaus, T. (2010). DiSUMO-like DSUL is required for nuclei positioning, cell specification and viability during female gametophyte maturation in maize. *Development*, *137*, 333–345.

Steffen, J. G., Kang, I.-H., Portereiko, M. F., Lloyd, A., & Drews, G. N. (2008). AGL61 interacts with AGL80 and is required for central cell development in *Arabidopsis*. *Plant Physiology*, *148*, 259–268.

Stevens, R., Grelon, M., Vezon, D., Oh, J., Meyer, P., Perennes, C., et al. (2004). A *CDC45* homolog in *Arabidopsis* is essential for meiosis, as shown by RNA interference–induced gene silencing. *The Plant Cell*, *16*, 99–113.

Su, Z., Zhao, L., Zhao, Y., Li, S., Won, S., Cai, H., et al. (2017). The THO complex non-cell-autonomously represses female germline specification through the *TAS3-ARF3* module. *Current Biology*, *27*, 1597–1609.e2.

Takeuchi, H., & Higashiyama, T. (2012). A species-specific cluster of defensin-like genes encodes diffusible pollen tube attractants in *Arabidopsis*. *PLoS Biology*, *10*, e1001449.

Tanaka, H., Ishikawa, M., Kitamura, S., Takahashi, Y., Soyano, T., Machida, C., et al. (2004). The *AtNACK1/HINKEL* and *STUD/TETRASPORE/AtNACK2* genes, which encode functionally redundant kinesins, are essential for cytokinesis in *Arabidopsis*. *Genes to Cells*, *9*, 1199–1211.

Tedeschi, F., Rizzo, P., Rutten, T., Altschmied, L., & Bäumlein, H. (2017). RWP-RK domain-containing transcription factors control cell differentiation during female gametophyte development in *Arabidopsis*. *New Phytologist*, *213*, 1909–1924.

Tekleyohans, D. G., Mao, Y., Kägi, C., Stierhof, Y.-D., & Groß-Hardt, R. (2017). Polyspermy barriers: A plant perspective. *Current Opinion in Plant Biology*, *35*, 131–137.

Tekleyohans, D. G., Nakel, T., & Groß-Hardt, R. (2017). Patterning the female gametophyte of flowering plants. *Plant Physiology*, *173*, 122–129.

Ueda, M., Matsui, K., Ishiguro, S., Kato, T., Tabata, S., Kobayashi, M., et al. (2011). *Arabidopsis RPT2a* encoding the 26S proteasome subunit is required for various aspects of root meristem maintenance, and regulates gametogenesis redundantly with its homolog, *RPT2b*. *Plant and Cell Physiology*, *52*, 1628–1640.

Völz, R., Heydlauff, J., Ripper, D., von Lyncker, L., & Groß-Hardt, R. (2013). Ethylene signaling is required for synergid degeneration and the establishment of a pollen tube block. *Developmental Cell*, *25*, 310–316.

Völz, R., von Lyncker, L., Baumann, N., Dresselhaus, T., Sprunck, S., & Groß-Hardt, R. (2012). *LACHESIS*-dependent egg-cell signaling regulates the development of female gametophytic cells. *Development*, *139*, 498–502.

Wang, J.-G., Feng, C., Liu, H.-H., Feng, Q.-N., Li, S., & Zhang, Y. (2017). AP1G mediates vacuolar acidification during synergid-controlled pollen tube reception. *Proceedings of the National Academy of Sciences of the United States of America*, *114*, E4877–E4883.

Wang, Y., Hou, Y., Gu, H., Kang, D., Chen, Z., Liu, J., et al. (2012). The *Arabidopsis* APC4 subunit of the anaphase-promoting complex/cyclosome (APC/C) is critical for both female gametogenesis and embryogenesis. *The Plant Journal, 69*, 227–240.
Wang, Y., Hou, Y., Gu, H., Kang, D., Chen, Z. L., Liu, J., et al. (2013). The *Arabidopsis* anaphase-promoting complex/cyclosome subunit 1 is critical for both female gametogenesis and embryogenesis. *Journal of Integrative Plant Biology, 55*, 64–74.
Wang, C.-J. R., Nan, G.-L., Kelliher, T., Timofejeva, L., Vernoud, V., Golubovskaya, I. N., et al. (2012). Maize *multiple archesporial cells1* (*mac1*), an ortholog of rice *TDL1A*, modulates cell proliferation and identity in early anther development. *Development, 139*, 2594–2603.
Wang, S.-Q., Shi, D.-Q., Long, Y.-P., Liu, J., & Yang, W.-C. (2012). GAMETOPHYTE DEFECTIVE1, a putative subunit of RNases P/MRP, is essential for female gametogenesis and male competence in *Arabidopsis*. *PLoS One*, 7, e33595.
Weatherwax, P. (1926). Persistence of the antipodal tissue in the development of the seed of maize. *Bulletin of the Torrey Botanical Club, 53*, 381–384.
Webb, M. C., & Gunning, B. E. S. (1994). Embryo sac development in *Arabidopsis thaliana* II. The cytoskeleton during megagametogenesis. *Sexual Plant Reproduction*, 7, 153–163.
Wu, J.-J., Peng, X.-B., Li, W.-W., He, R., Xin, H.-P., & Sun, M.-X. (2012). Mitochondrial GCD1 dysfunction reveals reciprocal cell-to-cell signaling during the maturation of *Arabidopsis* female gametes. *Developmental Cell, 23*, 1043–1058.
Yamada, H., Suzuki, T., Terada, K., Takei, K., Ishikawa, K., Miwa, K., et al. (2001). The *Arabidopsis* AHK4 histidine kinase is a cytokinin-binding receptor that transduces cytokinin signals across the membrane. *Plant and Cell Physiology, 42*, 1017–1023.
Yamamoto, M., Maruyama, D., Endo, T., & Nishikawa, S.-i. (2008). *Arabidopsis thaliana* has a set of J proteins in the endoplasmic reticulum that are conserved from yeast to animals and plants. *Plant and Cell Physiology, 49*, 1547–1562.
Yan, D., Duermeyer, L., Leoveanu, C., & Nambara, E. (2014). The functions of the endosperm during seed germination. *Plant and Cell Physiology, 55*, 1521–1533.
Yan, X., Yan, Z., & Han, Y. (2017). RRP42, a subunit of exosome, plays an important role in female gametophytes development and mesophyll cell morphogenesis in *Arabidopsis*. *Frontiers in Plant Science, 8*, 981.
Yang, H., Kaur, N., Kiriakopolos, S., & McCormick, S. (2006). EST generation and analyses towards identifying female gametophyte-specific genes in *Zea mays* L. *Planta, 224*, 1004–1014.
Yang, L., Qian, X., Chen, M., Fei, Q., Meyers, B. C., Liang, W., et al. (2016). Regulatory role of a receptor-like kinase in specifying anther cell identity. *Plant Physiology, 171*, 2085–2100.
Yang, W.-C., & Sundaresan, V. (2000). Genetics of gametophyte biogenesis in *Arabidopsis*. *Current Opinion in Plant Biology, 3*, 53–57.
Yang, L., Wu, Y., Yu, M., Mao, B., Zhao, B., & Wang, J. (2016). Genome-wide transcriptome analysis of female-sterile rice ovule shed light on its abortive mechanism. *Planta, 244*, 1011–1028.
Yang, W.-C., Ye, D., Xu, J., & Sundaresan, V. (1999). The *SPOROCYTELESS* gene of *Arabidopsis* is required for initiation of sporogenesis and encodes a novel nuclear protein. *Genes & Development, 13*, 2108–2117.
Yao, X., Yang, H., Zhu, Y., Xue, J., Wang, T., Song, T., et al. (2018). The canonical E2Fs are required for germline development in *Arabidopsis*. *Frontiers in Plant Science, 9*, 638.
Yu, X., Luo, A., & Sun, M.-x. (2016). An efficient antipodal cell isolation method for screening of cell type-specific genes in *Arabidopsis thaliana*. *PLoS One, 11*, e0166390.
Yuan, L., Liu, Z., Song, X., Johnson, C., Yu, X., & Sundaresan, V. (2016). The CKI1 histidine kinase specifies the female gametic precursor of the endosperm. *Developmental Cell, 37*, 34–46.

Zhao, X. A., Bramsiepe, J., Van Durme, M., Komaki, S., Prusicki, M. A., Maruyama, D., et al. (2017). RETINOBLASTOMA RELATED1 mediates germline entry in *Arabidopsis*. *Science*, *356*, eaaf6532.

Zhao, L., Cai, H., Su, Z., Wang, L., Huang, X., Zhang, M., et al. (2018). *KLU* suppresses megasporocyte cell fate through SWR1-mediated activation of *WRKY28* expression in *Arabidopsis*. *Proceedings of the National Academy of Sciences of the United States of America*, *115*, E526–E535.

Zhao, X., de Palma, J., Oane, R., Gamuyao, R., Luo, M., Chaudhury, A., et al. (2008). OsTDL1A binds to the LRR domain of rice receptor kinase MSP1, and is required to limit sporocyte numbers. *The Plant Journal*, *54*, 375–387.

Zhao, X. A., Harashima, H., Dissmeyer, N., Pusch, S., Weimer, A. K., Bramsiepe, J., et al. (2012). A general G1/S-phase cell-cycle control module in the flowering plant *Arabidopsis thaliana*. *PLoS Genetics*, *8*, e1002847.

CHAPTER SIXTEEN

Self-incompatibility in the Brassicaceae: Regulation and mechanism of self-recognition

June B. Nasrallah*
Section of Plant Biology, School of Integrative Plant Science, Cornell University, Ithaca, NY, United States
*Corresponding author: e-mail address: jbn2@cornell.edu

Contents

Abstract

Self-incompatibility is one of the most common mechanisms used by plants to prevent self-fertilization. In the Brassicaceae, the inhibition of self-pollen is triggered right at the stigma surface by interaction of two highly polymorphic self-recognition proteins that are encoded by tightly linked genes of the *S*-locus haplotype: a receptor protein kinase displayed at the surface of stigma epidermal cells and its small diffusible ligand that is localized in the outer coat of pollen grains. It is the specific interaction between receptor and ligand encoded in the same *S* haplotype that determines specificity in the rejection of self-pollen. The chapter reviews recent results that have shed light on the genetic control, cell biology, and regulation of the self-recognition molecules, as well as the structural basis of ligand recognition. Models that aim to explain how diversification of the self-recognition repertoire can occur in this two-gene self-recognition system are discussed.

Current Topics in Developmental Biology, Volume 131
ISSN 0070-2153
https://doi.org/10.1016/bs.ctdb.2018.10.002

1. Introduction

Evolutionary pressures that favor the evolution and maintenance of outcrossing in flowering plants have resulted in a bewildering array of adaptations, both morphological and physiological, that enforce out-crossing and reduce or prevent selfing. Morphological adaptations include the rewiring of developmental programs for physical separation of male and female flowers on the same plant or in different plants, temporal separation by differential maturation of male and female flowers or organs, or remolding of flower structure (Barrett, 2002). Among the best-known physiological outbreeding devices are the genetic self-incompatibility (SI) systems of hermaphroditic plants, in which pollen grains develop in close proximity to the pistil. These widespread systems, which occur in approximately half of flowering plant taxa, are based on pre-fertilization self/nonself-discrimination events that allow cells of the pistil to specifically prevent "self" pollen, but not cross (nonself) pollen, from effecting fertilization. Despite sharing a similar outcome, however, recognition specificity in the SI systems that operate in diverged taxa exhibit a number of differences (reviewed in Rea & Nasrallah, 2008). They may differ in genetic control by one or more genetic loci, in the site and mode of self-pollen inhibition, and in exhibiting uniform floral traits (homomorphic SI) or morphologically differentiated flowers (heteromorphic SI). Consistent with these differences, the SI systems of diverged plant families are now known to be controlled by evolutionarily unrelated molecular mechanisms. Thus, SI emerged independently multiple times during the evolution of angiosperms.

This chapter focuses on the recognition phase of the homomorphic SI response of the Brassicaceae (crucifer) family, in which approximately half of the species exhibit SI. The family includes several economically important crop species. For example, oilseed *Brassica napus* and *Brassica rapa* cultivars are a source of oil products that are ranked among the most valuable commodities in agriculture, and the edible cultivars of *B. rapa* and *Brassica oleracea* are a source of vegetables and leafy greens for billions of people around the world. In many of these species, breeding programs rely extensively on SI as a pollination control system. A mechanistic understanding of SI is therefore important for improving schemes used for the generation of hybrids and the production of pure hybrid seed on a commercial scale.

After presenting an overview of the molecular features of the receptor/ligand-based crucifer SI system, the chapter discusses the genetic control, cell biology, and regulation of the self-recognition molecules, the structural basis of ligand recognition, and models for diversification of the recognition repertoire.

2. SI in the Brassicaceae: Selective activation of stigma receptors by pollen ligands

Self-incompatible members of the Brassicaceae share an SI system having the following characteristics: pollen recognition is controlled by variants of a single locus (the *S* locus), pollen SI phenotype is determined sporophytically, and the recognition and inhibition of self-pollen occur at the stigma surface within a few minutes after pollen-stigma contact due to failure of pollen hydration, pollen germination, and pollen tube growth into the pistil (Fig. 1A). The *S* locus contains two highly polymorphic and tightly linked genes that together form the "*S* haplotype" and are responsible for the ability of the stigma to discriminate between self-pollen (i.e., from a pollen donor that expresses the same *S* haplotype as that expressed in the stigma) and nonself-pollen (i.e., from a pollen donor that expresses an *S* haplotype different from that expressed in the stigma). One gene, the *S*-locus receptor kinase (*SRK*) gene (Stein, Howlett, Boyes, Nasrallah, & Nasrallah, 1991), is expressed predominantly in the stigma epidermis, it encodes a single-pass transmembrane serine/threonine protein kinase, and it determines SI specificity in the stigma. The second gene, the *S*-locus cysteine-rich protein (*SCR*, also designated *SP-11*) gene (Schopfer, Nasrallah, & Nasrallah, 1999), is expressed in the anther tapetum, it produces a diffusible pollen coat-localized protein that serves as the ligand for SRK, and it determines SI specificity in the pollen. In any one species, functionally distinct *S* haplotypes typically occur in large numbers and their SRK and SCR proteins can diverge by as much as 35% and 80%, respectively, and *trans*-specific polymorphisms are frequently observed. These *S* haplotypes have been maintained over long evolutionary timescales due to strong negative frequency-dependent selection, whereby plants harboring a new and therefore rare *S* haplotype are able to cross with many more mates than those harboring a common *S* haplotype (Charlesworth, 2006).

SRK and *SCR* are both necessary and sufficient for the recognition of self-pollen. This conclusion is supported by loss- and gain-of-function studies in *Brassica*, and by the observation that evolutionary switches from a self-incompatible to a self-fertile mode of mating is accompanied by inactivation of one or both of these genes. A case in point is *Arabidopsis thaliana*, which

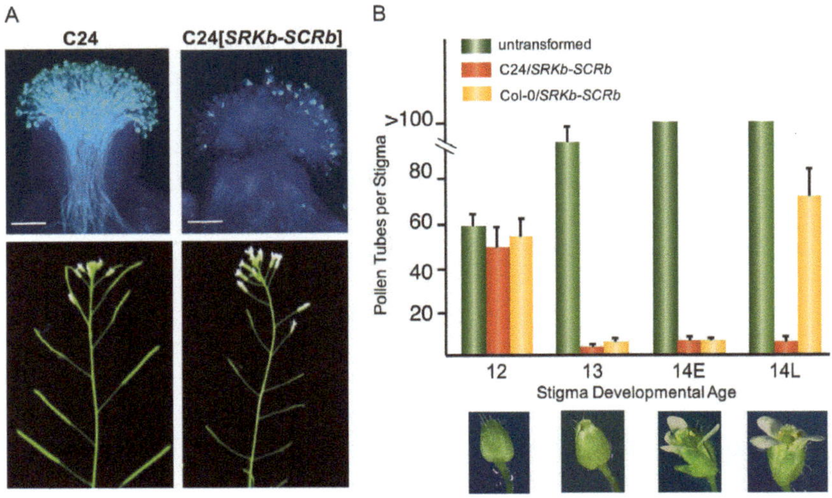

Fig. 1 The SI response in the Brassicaceae, as exemplified by transgenic self-incompatible *A. thaliana* plants transformed with the *A. lyrata SRKb-SCRb* gene pair. (A) The SI phenotype of transgenic C24[*SRKb-SCRb*] plants, which express an intense and developmentally stable SI. Untransformed C24 pistils exhibit profuse growth of self-pollen tubes, resulting in expanded seed-filled siliques. By contrast, no self-pollen tubes are observed at the surface of transgenic stigmas, resulting in unexpanded siliques lacking seed. Bar = 10 μ. (B) Developmental regulation of SI as a function of stigma development in *SRKb-SCRb* transformants of different *A. thaliana* accessions. The graph shows the number of pollen tubes observed upon self-pollination. Note that C24[*SRKb-SCRb*] plants exhibit intense SI starting in the stigmas of stage-13 flower buds and persisting into older flowers. By contrast, Col-0[*SRKb-SCRb*] plants express intense SI only in the stigmas of stage-13 and early stage-14 (14E) flower buds. Representative images of Col-0 of developing buds are shown below the graph. *The microscopic images in (A) are reprinted from Nasrallah, J. B. (2017). Plant mating systems: Self-incompatibility and evolutionary transitions to self-fertility in the mustard family.* Current Opinion in Genetics and Development, 47, *54–60. The plant images in (A) are reprinted from Tantikanjana, T., & Nasrallah, J. B. (2015). Ligand-mediated* cis*-inhibition of receptor signaling in the self-incompatibility response of the Brassicaceae.* Plant Physiology, 169, *1141–1154 and the floral-bud images in (B) from Nasrallah, J. B., & Nasrallah, M. E. (2014). Robust self-incompatibility in the absence of a functional ARC1 in* Arabidopsis thaliana. The Plant Cell, 26, *3838–3841 (www.plantphysiol.org and www.plantcell.org; Copyright American Society of Plant Biologists).*

diverged from its closest self-incompatible relative *Arabidopsis lyrata* approximately 5 million years ago and is thought to have transitioned to self-fertility sometime between 150,000 and 1 million years ago (Payne & Alvarez-Ponce, 2018). *A. thaliana* harbors non-functional *S* haplotypes that contain loss-of-function mutations in the *SRK* and/or *SCR* genes, or partial or complete deletions of these genes (Boggs, Nasrallah, & Nasrallah, 2009; Kusaba et al., 2001; Tsuchimatsu et al., 2017). Importantly, this self-fertile species can be reverted to be self-incompatible by transformation with functional *SRK-SCR* gene pairs from self-incompatible *A. lyrata* and *Capsella grandiflora* (Boggs, Dwyer, Nasrallah, & Nasrallah, 2009; Nasrallah, Liu, & Nasrallah, 2002; Nasrallah, Liu, Sherman-Broyles, Boggs, & Nasrallah, 2004; Nasrallah & Nasrallah, 2014) (Fig. 1). Although many *A. thaliana* accessions have accumulated additional mutations at SI modifier loci unlinked to the *S* locus, it is evident that the *S* locus was the primary target of selection for self-fertility in *A. thaliana* (Nasrallah, 2017). Thus, *SRK* and *SCR* are not only the sole determinants of SI specificity, but they are also major determinants of the out-crossing mode of mating in the Brassicaceae.

The interaction between SCR and the extracellular domain of SRK (hereafter eSRK) is *S* haplotype-specific, and it is this specific interaction that determines the specific recognition of self-pollen in the SI response. Contact between a pollen grain and a stigma epidermal cell leads to the flow of pollen coat material onto the surface of the stigma cell and the formation of an adhesion zone or "foot" through which SCR proteins reach the stigma surface. From there, SCRs translocate through the cell wall toward the plasma membrane where they can access the eSRK. While these events occur in both self- and cross-pollinations, an SCR will only bind and activate an SRK when the two proteins are encoded by the same *S* haplotype. Of course, this only occurs when stigma and pollen donor express the same *S* haplotype, whether they are derived from the same flower, the same plant, or different plants that express the same *S* haplotype.

The nature of the signaling cascade that is triggered by the SRK-SCR interaction within the stigma epidermal cells and culminates in the inhibition of self-pollen has been the subject of debate. So far, a few proteins have been proposed to function as SRK signaling components based on studies in *Brassica* species: the membrane-associated kinase MLPK (Kakita et al., 2007) and the ARM-repeat and U-box protein ARC1 (Samuel et al., 2009). ARC1 is an E3 ligase that is phosphorylated by SRK and ubiquitinates the Exo70A1 subunit of the exocyst complex, which is involved in vesicle trafficking. These activities have been proposed to cause rejection of self-pollen by inhibiting the secretion of factors required for compatible pollen tube growth. However, none of these proteins were found to function in the SI response of self-incompatible *A. thaliana SRK-SCR* transformants: the *ARC1* gene is deleted in *A. thaliana*, and neither loss-of-function mutations in the *A. thaliana* ortholog or closest paralog of *MLPK*, nor over-expression of *A. thaliana* Exo70A1 disrupted or weakened the SI response of *SRK-SCR* plants (Kitashiba, Liu, Nishio, Nasrallah, & Nasrallah, 2011; Nasrallah & Nasrallah, 2014). A resolution of these conflicting results will require assessing the involvement of ARC1 and Exo70A1 in SI signaling in additional crucifer genera.

3. Genetic regulation of SI recognition genes

In addition to sporophytic control of pollen SI phenotype, crucifers exhibit intricate genetic interactions between *S* haplotypes in the stigmas and pollen of *S*-locus heterozygotes, ranging from codominance, mutual weakening, or dominance-recessiveness. Sporophytic control is explained by expression of the *SCR* gene and synthesis of its protein product in the sporophytic anther tapetum (Schopfer & Nasrallah, 2000; Schopfer et al., 1999). In plants heterozygous for codominant *S* haplotypes, e.g., *S1* and *S2*, two SCR variants, SCR1 and SCR2, are produced in the tapetum and transferred to the outer coating of pollen grains, resulting in genotypically *S1* or *S2* haploid pollen grains having an *S1S2* phenotype. This predicted outcome of sporophytic control is disturbed when *S* haplotypes exhibit dominant-recessive interactions. Extensive diallel crosses in various crucifer species have shown that *S* haplotypes typically form dominance hierarchies, especially in pollen. Importantly, these dominance relationships have a major influence on the distribution of *S* haplotypes in nature. Indeed, recessive *S* haplotypes attain high frequencies in populations because their recessiveness allows pollen carrying them to evade inhibition by the cognate stigmatic SRKs. In both *Arabidopsis* (Kusaba, Tung, Nasrallah, & Nasrallah, 2002) and *Brassica* species (Fujii & Takayama, 2018), dominant-recessive interactions in anthers have been shown to result from monoallelic expression of the dominant *SCR* allele. This phenomenon stems from the activity of multiple *S*-locus-linked small-RNA genes that cause epigenetic silencing of the recessive *SCR* allele (Fujii & Takayama, 2018).

In contrast to this detailed understanding of the allelic interactions responsible for the pollen SI phenotype, the basis of dominant-recessive interactions or weakening of *S* haplotypes in the stigma remains elusive, as no reduction in the expression of the recessive *SRK* alleles is observed in heterozygous stigmas beyond that expected for reduced gene dosage (Kusaba et al., 2002). It has been proposed that codominance might be due to a preference for homodimerization of each of the two SRKs expressed in heterozygotes. By contrast, dominance-recessiveness might result from both an increased propensity for SRK heterodimerization and a reduced affinity of heterodimers for the recessive SCR variant (Naithani, Chookajorn, Ripoll, & Nasrallah, 2007).

4. Spatiotemporal regulation of the SRK receptor

4.1 Developmental regulation of SRK and integration of SI signaling with pistil developmental pathways

A well-known feature of crucifer SI is its developmental regulation. Stigmas are self-compatible early in their development, and their ability to reject self-pollen is acquired just prior to flower opening and persists in older flowers. This feature is routinely exploited for the generation and maintenance of *S*-locus homozygotes by manual pollination of the self-compatible stigmas of young floral buds. At the molecular level, the developmental regulation of SI is determined by relatively low levels of SRK in young self-compatible stigmas, with maximal levels attained at the onset of SI. The importance of adequate levels of *SRK* expression for a robust SI response that is sustained over the course of flower development is well illustrated by the SI response conferred by *SRK-SCR* transgene pairs in different geographical accessions of *A. thaliana* (Fig. 1) (Nasrallah & Nasrallah, 2014; Rea & Nasrallah, 2015). Relatively high levels of *SRK* transcripts are observed in *SRK-SCR* transformants of several accessions, such as C24, which express an SI response that is identical in intensity and developmental regulation to that of naturally self-incompatible species (Fig. 1A and B). By contrast, lower levels of *SRK* transcripts, especially at later stages of flower development, are typically observed in *SRK-SCR* transformants of accessions, such as Col-0, which exhibit a transient SI phenotype that is characterized by intense SI during a narrow window of stigma maturation and a subsequent breakdown of SI at later stages of flower development (Fig. 1B). This phenotype, which also characterizes *SRK-SCR* transformants of one-third of tested accessions, is caused by a hypomorphic allele of *PUB8*, an *S*-locus-linked gene encoding an ARM-repeat and U-box protein that regulates *SRK* transcript levels (Liu, Sherman-Broyles, Nasrallah, & Nasrallah, 2007).

The developmental regulation of SI and *SRK* expression as a function of stigma development suggests some level of genetic integration of SI signaling with pistil developmental pathways. This notion is supported by the fact that in nature, SI is often associated with adaptations in floral architecture, such as stigmas protruding above the anthers (known as herkogamy or stigma exsertion), which reduce the chance of self-pollination and increase the likelihood that the stigma will receive cross pollen (Barrett, 2002). An extreme example of these associations is observed in taxa that exhibit heteromorphic SI, such as *Primula*, in which morphologically differentiated types (or morphs) exhibit reciprocal style and stamen lengths (Darwin, 1877). This reciprocal herkogamy has been ascribed to an *S*-locus gene that encodes a cytochrome P450 enzyme whose putative activity is to degrade brassinosteroids, a class of phytohormone that promotes cell elongation (Huu et al., 2016). Interestingly, this gene is found in the short-style morph but is deleted in the long-style morph (Li et al., 2016), which explains the different lengths of the two morphs.

A similar connection was uncovered in the Brassicaceae between SI and another phytohormone, auxin. Mutagenesis of *A. thaliana* Col-0 plants transformed with the *A. lyrata SRKb* and *SCRb* genes, which express transient SI, identified a recessive mutation that caused the simultaneous enhancement of SI and stigma exsertion (Tantikanjana, Rivzi, Nasrallah, & Nasrallah, 2009). Enhanced stigma exsertion resulted from increased pistil elongation, was observed in the absence of the *SCRb* transgene, and required the expression of a catalytically active SRKb protein, all of which point to a role for SCR-independent SRK signaling in pistil elongation. Although such a role for SRK was unexpected, it is consistent with the known fact that *SRK* is expressed at low levels in the style. In any case, analysis of the mutant phenotype uncovered a role for AUXIN RESPONSE FACTOR3 (ARF3) in the enhancement of both stigma exsertion and SI. ARF3 was found to act non-cell-autonomously from its location in the stylar vasculature to enhance SI and simultaneously down-regulate auxin responses in stigma epidermal cells, probably by controlling the production of at least one mobile signal that negatively regulates SI in these cells (Tantikanjana & Nasrallah, 2012). The results suggest that auxin functions in pollen-stigma interactions and in the integration of these interactions with the pistil developmental program. In this role, auxin would enhance pollen tube growth at the stigma surface, while suppression of auxin signaling in stigma epidermal cells would promote inhibition of self-pollen.

4.2 Subcellular distribution of the SRK and its disruption in strategies for pollination control

The subcellular distribution of SRK in stigmas was described by live-cell imaging of self-incompatible *A. thaliana* C24 stigmas that express a functional SRK protein carrying a C-terminal YFP tag (Rea & Nasrallah, 2015). Imaging of unpollinated stigmas showed SRK-YFP to be primarily and uniformly distributed at the plasma membrane (Fig. 2A–C), implying

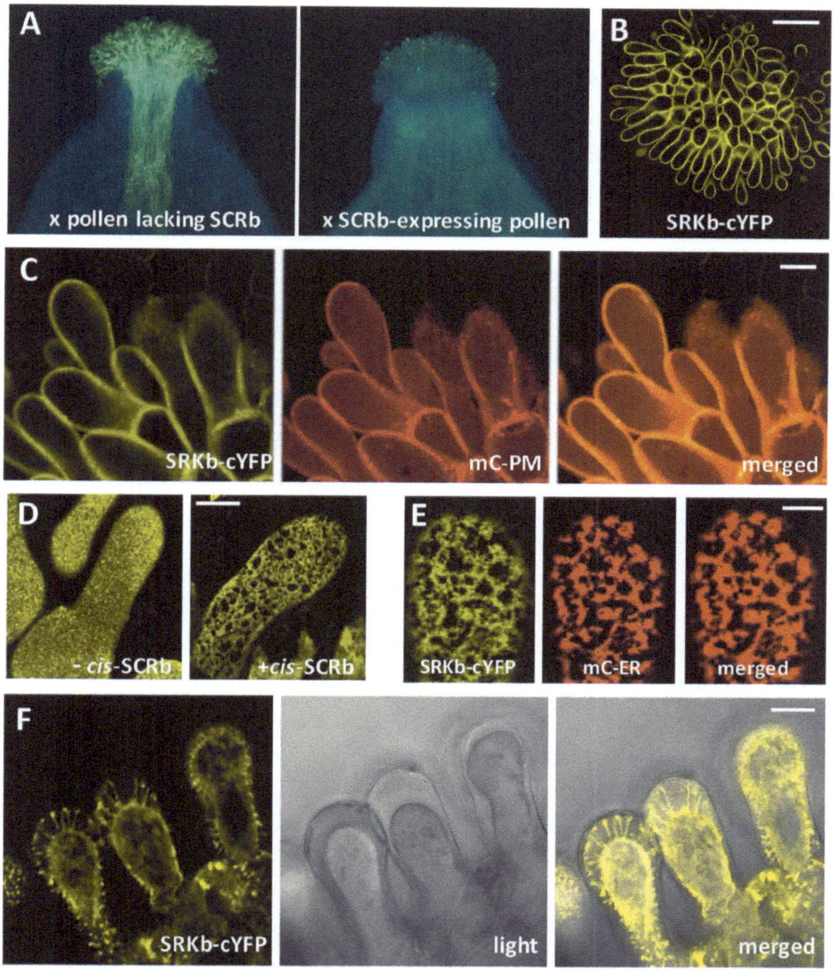

Fig. 2 Subcellular localization of an SRKb-YFP fusion in *A. thaliana* C24 plants. (A) Pollination assays show severe inhibition of self-pollen on C24 stigmas expressing SRKb-YFP, demonstrating that this protein fusion is functional. (B) Expression of SRKb-YFP at the periphery of unpollinated stigma epidermal cells. Bar = 50 μ. (C) Co-localization of SRKb-YFP with an mCherry-PM (mC-PM) marker. Bar = 10 μ. (D–E) Entrapment of SRKb-YFP in the ER by co-expression of SCR (*cis*-SCR). (D) Surface views of stigma epidermal cells showing SRKb-YFP to be uniformly localized at the cell surface in the absence of *cis*-SCR and to be trapped in the ER in the presence of *cis*-SCR. (E) Co-localization of SRKb-YFP with an mCherry-ER (mC-ER) marker in the presence of *cis*-SCR. (F) Plasmolysis of stigma epidermal cells by treatment with 5 M NaCl. Note that SRKb-YFP is observed in Hechtian strands, which are stretched portions of the PM that form upon retraction of the protoplast from the cell wall. Bars = 50 μ (B), 10 μ (C, D, F), 5 μ (E). *Panels (A–C) are reprinted from Rea, A. C., & Nasrallah, J. B. (2015).* In vivo *imaging of the S-locus receptor kinase, the female specificity determinant of self-incompatibility, in transgenic self-incompatible* Arabidopsis thaliana. Annals of Botany, 115, *789–805 with permission. Panels (D–F) are reprinted from Tantikanjana, T., & Nasrallah, J. B. (2015). Ligand-mediated* cis-*inhibition of receptor signaling in the self-incompatibility response of the Brassicaceae.* Plant Physiology, 169, *1141–1154 (www.plantphysiol.org; Copyright American Society of Plant Biologists).*

signaling from the plasma membrane. Interestingly, this localization was not influenced by self-pollination or by saturating the stigma surface with recombinant SCR protein. Signaling by cell surface receptors often involves internalization of ligand-bound receptor molecules as a means of signal attenuation for desensitizing the cell in readiness for a further response to the ligand. In the case of SRK, internalization for signal attenuation may not play a role in the regulation of signaling due to unique features of the pollen-stigma interaction. Indeed, SRK activation occurs in the region of the plasma membrane that subtends the site of contact with a self-pollen grain, and any site that is already occupied will not be exposed to another pollen grain because adhered pollen grains are not typically dislodged from the stigma surface by incoming pollen grains.

It should be noted that the localization of SRK at the plasma membrane and in close juxtaposition to the cell wall is critical for SRK function and for the perception of self-pollen because it makes the eSRK accessible to its pollen coat-derived SCR ligand. Indeed, breakdown of SI has been observed when SRK fails to be targeted to the plasma membrane and is instead trapped in the endoplasmic reticulum, as occurs when N-glycosylation sites in the eSRK are eliminated (Yamamoto, Tantikanjana, Nishio, Nasrallah, & Nasrallah, 2014) or when SRK is co-expressed with its cognate SCR in stigma epidermal cells, resulting in ligand-mediated allele-specific *cis*-inhibition of SI signaling (Tantikanjana & Nasrallah, 2015) (Fig. 2D and E). Breakdown of SI is also observed upon treatment of stigmas with 5 M NaCl, which is explained by NaCl-induced plasmolysis. This causes retraction of the plasma membrane from the cell wall of stigma epidermal cells (Rea & Nasrallah, 2015), thus precluding SCR from accessing the eSRK (Fig. 2F).

Breakdown of SI caused by mis-localization of SRK has applications for pollination control in breeding programs and for mechanistic studies of SI. For example, treatment of stigmas with NaCl has been used by breeders for decades as an effective means of overcoming SI in crucifers, even though the underlying cause of NaCl-induced breakdown of SI was revealed only recently. Additionally, the phenomenon of allele-specific ligand-mediated *cis*-inhibition of SRK may be exploited in novel schemes for hybrid breeding and hybrid seed production in crucifer crops (Tantikanjana & Nasrallah, 2015), and as a conditional SI system in mutagenesis strategies designed to overcome the challenge to standard mutagenesis posed by lack of seed in self-incompatible plants.

5. The structural basis of SCR perception by SRK

5.1 Structure of the SRK-SCR complex

The mechanism of SCR binding to the eSRK has recently been clarified by a crystallographic structural analysis of a complex formed by the BreSRK9 and BrSCR9 variants derived from the *B. rapa S9* haplotype (Ma et al., 2016). Consistent with predictions from previous structure-based sequence alignments, homology modeling, and threading studies of various eSRKs and SCRs, the crystallographic analysis demonstrates that the eSRK9 consists of distinct structural modules (Fig. 3A and B) and SCR9 has a defensin-like compact configuration held together by four intramolecular disulfide bonds (Fig. 3B). The crystal structure further reveals that the eSRK9-SCR9 complex is a 2:2 heterotetrameric structure consisting of an eSRK9 dimer between which are sandwiched two SCR9 molecules that make contacts with the eSRK9 monomers but not with each other (Fig. 3C). The two eSRK9 monomers interact with each other through direct contacts, but the major contribution to eSRK9 dimerization is provided by the simultaneous interaction of SCR9 with the two eSRK9 monomers. Interestingly, this mode of receptor-ligand interaction differs from that typically observed in mammals, where receptor monomers are often cross-linked by a dimeric ligand.

The structure of the eSRK9-SCR9 complex, together with the observation that SCR9 induces eSRK9 homodimerization in solution (Ma et al., 2016), demonstrates that eSRK homodimerization is triggered by its interaction with SCR and suggests that this homodimerization is responsible for activation of SRK. It should be noted, however, that SRK dimers form spontaneously in stigmas in the absence of the SCR ligand, suggesting that domains other than the eSRK, such as the transmembrane or cytoplasmic domains, might contribute to homodimerization of the SRK receptor *in vivo*. Moreover, the *in vitro* binding of SCR to eSRK does not always translate into *in vivo* activation of SRK (Chookajorn, Kachroo, Ripoll, Clark, & Nasrallah, 2004).

5.2 Structural basis of selective ligand recognition

Attempts to identify the amino acid residues that determine *S*-haplotype specificity in the SRK-SCR interaction were initiated soon after the

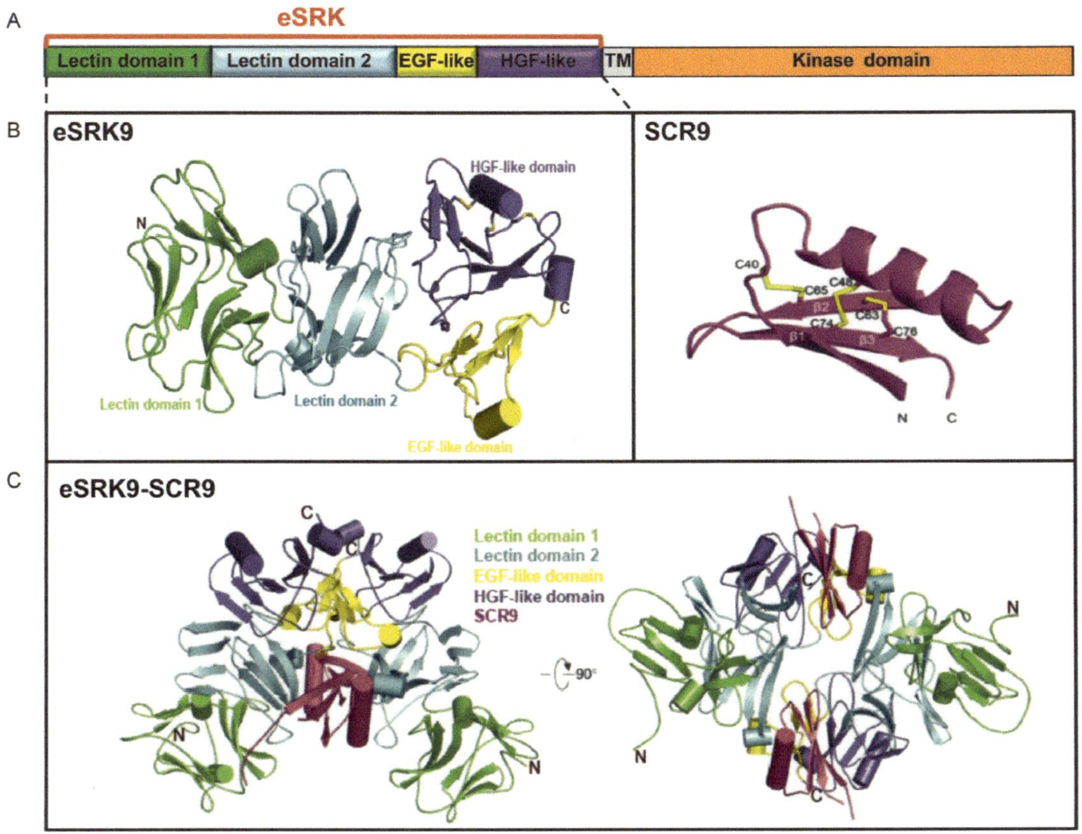

Fig. 3 Structures of SRK and SCR. (A) Domain architecture of SRK, showing the modular organization of the eSRK. The vertical lines above the diagram show 12 conserved cysteine residues in eSRK. (B) Three-dimensional structures of *B. rapa* eSRK9 and SCR9. (C) Three-dimensional structure of the eSRK9-SCR9 tetrameric complex. Note that the two SCR molecules interact with the two eSRK9 monomers but not with each other. *Panels (B and C) are reprinted from Ma, R., Han, Z., Hu, Z., Lin, G., Gong, L., Zhang, H., et al. (2016). Structural basis for specific self-incompatibility response in* Brassica. Cell Research, 26, *1320–1329 with permission.*

isolation of these molecules, but were complicated by the extreme polymorphism of these proteins. In the case of SCR, sequence alignments are fruitless because only its cysteine residues are conserved in most variants (Fig. 4A). In the case of eSRK, computational analyses identified hypervariable (hv) regions that are enriched for residues predicted to be under positive selection for change (Fig. 4B), but failed to pinpoint specificity-determining residues. More progress was made by assaying the *in planta* activity of receptor and ligand mutants generated by inter-allelic domain and single-residue swaps.

Using a pollination bioassay for assessing the activity of *Brassica* SCR variants, several amino acid residues were identified along the length of protein which are essential for SCR function and pointed to the importance of the region located between the fifth and sixth cysteine residues for SI specificity (Chookajorn et al., 2004; Sato, Okamoto, & Nishio, 2004) (Fig. 4A). In the case of SRK, *A. thaliana* C24[*SRK-SCR*] transformants that exhibit complete reversion to SI were used for structure-function assays of *A. lyrata* and *C. grandiflora* variants (Boggs, Dwyer, et al., 2009). These experiments showed that the SI specificity-determining region of the eSRK spanned three hv regions, and that a small number of amino acid residues located in two physically separated segments of the hvI and hvII regions are essential for the ability of SRK to activate the SI response in stigmas (Fig. 4B).

The prediction that these residues are brought together in the three-dimensional eSRK structure to form an SCR-binding pocket was confirmed by the recently solved crystal structure of the *B. rapa* eSRK9-SCR9 complex. Most of the eSRK9 residues that form contact points with SCR9 are located within the hv regions and are particularly concentrated in hvI and hvII (Fig. 4B). Moreover, the highest density of interacting residues occurs at the interface between hvII in eSRK9 and the region between the fifth and sixth cysteine residues in SCR9 (Fig. 4A). This extensive overlap between the results of *in planta* assays and crystallography of diverged SRKs and SCRs suggests that the inferred mechanism of SCR recognition by SRK is applicable to other SRK-SCR pairs. However, differences in the location of specificity-determining residues in different SRK and SCR variants have been observed (Boggs, Dwyer, et al., 2009; Chookajorn et al., 2004). Thus, given the extreme polymorphisms of these proteins, it is likely that some variation on the basic structure of the eSRK-SCR complex revealed by analysis of eSRK9-SCR9 will be observed as more SRK-SCR pairs are analyzed.

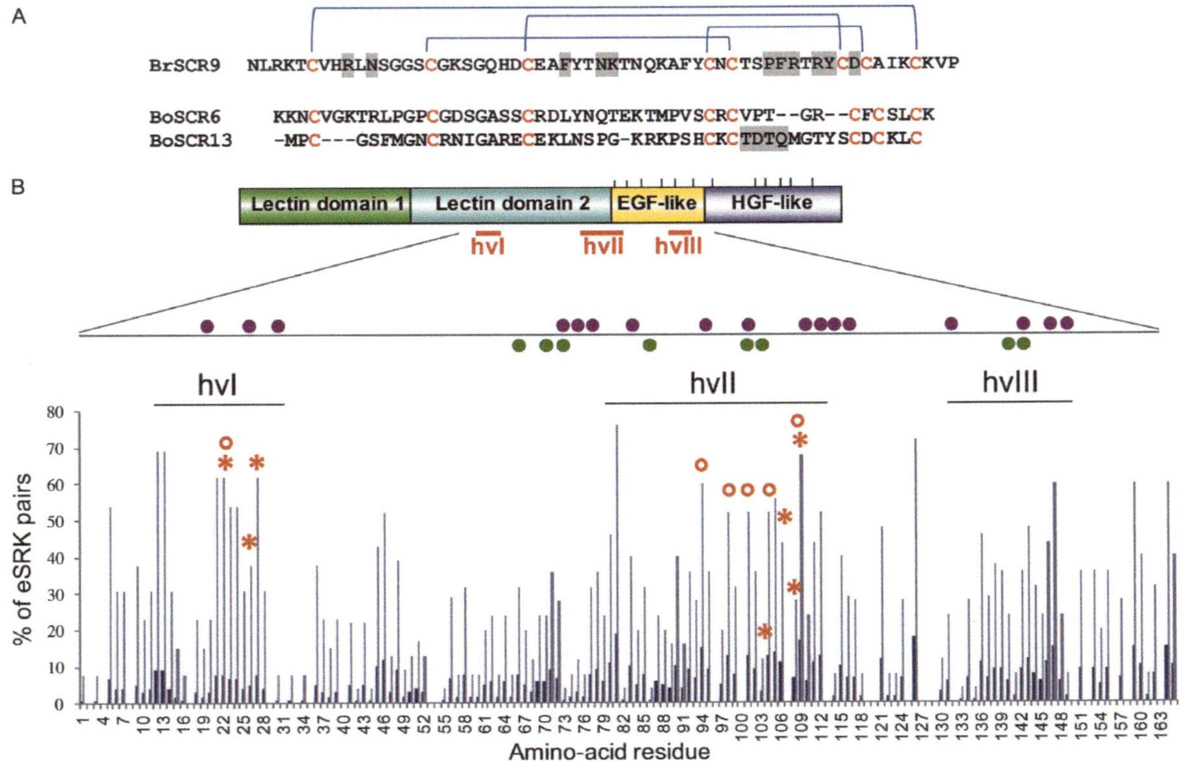

Fig. 4 SI specificity determinants in SCR (A) and eSRK (B). (A) In the *B. rapa* SCR9 sequence, shaded letters indicate amino acids that form contact points with eSRK9, and the brackets above the sequence show the disulfide bridges linking the eight conserved cysteines. The alignment of the *B. oleracea* SCR6 and SCR13 variants shows the extreme intra-specific sequence divergence of SCRs. The four shaded amino acids in SCR13 were shown to determine *SCR13* specificity *in planta*. (B) The diagram of the eSRK shows the location of three hypervariable (hv) regions. The graph shows variability at individual sites within the eSRK hvI-hvIII region based on alignments of pairs of eSRK sequences, with 0 and 100 indicating that 0% and 100% of the pairs differ at a particular site, respectively. The circles and asterisks on the graph indicate the residues that were shown to be essential for *in planta* activation of SI by two different SRK variants. Above the graph, amino acids that form contact points with SCR9 in the eSRK9-SCR9 complex and those involved in eSRK9 homodimerization are shown by circles above and below the line, respectively. Residues in hvIII were not tested for activity *in planta*. *Modified from Nasrallah, J. B. (2017). Plant mating systems: Self-incompatibility and evolutionary transitions to self-fertility in the mustard family.* Current Opinion in Genetics and Development, 47, *54–60.*

6. Co-evolution and diversification of self-recognition molecules

The extensive sequence polymorphism of SRK and SCR variants, the congruence of their phylogenetic trees, and the *S* haplotype-specific nature of the SRK-SCR interaction have long been viewed as reflecting the co-evolution of this receptor-ligand pair. The fact that robust interaction interfaces are observed in the eSRK9-SCR9 crystal structure, while severe steric clashes are detected when modeling the eSRK9 in complex with non-self SCRs (Ma et al., 2016), now leaves no doubt about the co-evolution of these proteins. Clearly, a co-adapted *SRK-SCR* gene pair is essential for the recognition of self-pollen. However, it is the tight genetic linkage of the *SRK* and *SCR* genes that is critical for the persistence of SI over time. This tight linkage appears to be maintained by the structural heteromorphism of *S* haplotypes, which likely serves to suppress recombination in the *S*-locus region and thus prevents disruption of the co-adapted *SRK-SCR* gene complex. Notably, *S* haplotypes can differ in overall physical size and in the arrangement of their genes, they contain haplotype-specific sequences, and they are enriched for repetitive and transposon-like sequences, all of which are features often observed in other self/nonself recognition loci that exhibit suppressed recombination (Nasrallah, 2005).

The co-evolution of *SRK* and *SCR* raises the question of how new SI specificities are generated. This is an intriguing question that is difficult to answer, not only for crucifer SI, but also for other two-component SI systems (Charlesworth, 2000). One hypothesis is that a mutation in one of the two genes disrupts the SRK-SCR interaction and is followed by a compensatory mutation in the second gene that restores the interaction. This hypothesis is problematic because the non-functional *S* haplotypes generated by the first mutation are unlikely to persist long enough in natural populations to allow for the second mutation to occur. An alternative hypothesis that does not invoke passage through self-compatible intermediates was suggested by the results of structure-function studies of SCR (Chookajorn et al., 2004). First, only a small number of residues are essential for the *in planta* function of SRK and SCR, consistent with the similarly small number of residues that form contact points between receptor and ligand in the eSRK9-SCR9 complex. Second, mutations can disrupt *in planta* function with no or only minor effects on the affinity of SCR for SRK. Third, the SCR protein has a striking capacity to tolerate even drastic changes to its sequence as long as its overall structure is maintained: its ability to activate the SI response can persist even when ~30% of its sequence is replaced and its specificity determinants can retain their recognition function in the context of highly diverged SCR backbones. Thus, new SI specificities might be generated through self-incompatible intermediates by small gradual readjustments in the eSRK-SCR interface without detrimental effects on allelic recognition or signaling outcome.

7. Future perspectives

Parallels have often been noted between plant SI systems and other eukaryotic self/nonself-discrimination systems (Larsen, 1986), such as plant-pathogen interactions, fungal and algal mating systems, and histocompatibility reactions in colonial marine invertebrates and vertebrates, and they have even been discussed in the context of the evolutionary origins of adaptive immunity (Burnet, 1971). Indeed, the SI system of the Brassicaceae and these other recognition systems share common features stemming from similar selective pressures for diversification, co-evolution of recognition molecules, and suppressed recombination for maintaining the linkage of co-adapted gene complexes (Nasrallah, 2005). Recent years have witnessed significant progress toward understanding the genetic control and spatio-temporal regulation of the *SRK/SCR* recognition genes and the structural basis of specificity in the SRK-SCR interaction. However, several questions remain. It will be important to determine how far the crystal structure of the eSRK9-SCR9 complex and the SI-determining residues identified in structure-function studies of a limited number of *S*-locus variants can be extrapolated to other variants. Additionally, the conflicting results obtained for *Brassica* and *A. thaliana* regarding candidate downstream components of SRK-mediated signaling must be resolved. Most rewarding of all will be gaining a mechanistic understanding of how new SI specificities are generated. The identification of residues critical for the SRK-SCR interaction and for *in planta* activation of the SI response represents a major advance that sets the stage for future elucidation of the co-evolution of SRK and SCR and the diversification of the SI recognition repertoire. Undoubtedly, the required analysis of many additional naturally occurring and engineered SRK-SCR pairs is a daunting task. Nevertheless, an answer to this most intriguing of evolutionary puzzles will have far-reaching implications for our understanding of the co-evolution of interacting proteins in a variety of self/nonself-discrimination systems.

Acknowledgments

Research in the author's laboratory is supported by grant IOS-1146725 from the US National Science Foundation (http://www.nsf.gov). Any opinions, findings, and conclusions or recommendations expressed in this material are those of the author(s) and do not necessarily reflect the views of the National Science Foundation.

References

Barrett, S. C. H. (2002). Evolution of sex: The evolution of plant sexual diversity. *Nature Reviews Genetics*, *3*, 274–284.

Boggs, N. A., Dwyer, K. G., Nasrallah, J. B., & Nasrallah, M. E. (2009). *In vivo* detection of residues required for ligand-selective activation of the S-locus receptor in *Arabidopsis*. *Current Biology*, *19*, 786–791.

Boggs, N. A., Nasrallah, J. B., & Nasrallah, M. E. (2009). Independent *S*-locus mutations caused self-fertility in *Arabidopsis thaliana*. *PLoS Genetics*, *5*(3), e1000426.

Burnet, F. M. (1971). "Self-recognition" in colonial marine forms and flowering plants in relation to the evolution of immunity. *Nature*, *232*, 230–235.

Charlesworth, D. (2000). How can two-gene models of self-incompatibility generate new specificities? *The Plant Cell*, *12*, 309–310.

Charlesworth, D. (2006). Balancing selection and its effects on sequences in nearby genome regions. *PLoS Genetics*, *2*, 379–384.

Chookajorn, T., Kachroo, A., Ripoll, D. R., Clark, A. G., & Nasrallah, J. B. (2004). Specificity determinants and diversification of the *Brassica* self-incompatibility pollen ligand. *Proceedings of the National Academy of Sciences of the United States of America*, *101*, 911–917.

Darwin, C. (1877). *On the different forms of flowers on the same species*. London: John Murray.

Fujii, S., & Takayama, S. (2018). Multilayered dominance hierarchy in plant self-incompatibility. *Plant Reproduction*, *31*, 15–19.

Huu, C. N., Kappel, C., Keller, B., Sicard, A., Takebayashi, Y., Breuninger, H., et al. (2016). Presence versus absence of CYP734A50 underlies the style-length dimorphism in primroses. *eLife*, *5*. https://doi.org/10.7554/eLife.17956. e17956.

Kakita, M., Murase, K., Iwano, M., Matsumoto, T., Watanabe, M., Shiba, H., et al. (2007). Two distinct forms of M-locus protein kinase localize to the plasma membrane and interact directly with S-locus receptor kinase to transduce self-incompatibility signaling in *Brassica rapa*. *The Plant Cell*, *19*, 3961–3973.

Kitashiba, H., Liu, P., Nishio, T., Nasrallah, J. B., & Nasrallah, M. E. (2011). Functional test of *Brassica* self-incompatibility modifiers in *Arabidopsis thaliana*. *Proceedings of the National Academy of Sciences of the United States of America*, *108*, 18173–18178.

Kusaba, M., Dwyer, K. D., Hendershot, J., Vrebalov, J., Nasrallah, J. B., & Nasrallah, M. E. (2001). Self-incompatibility in the genus Arabidopsis: Characterization of the *S* locus in the outcrossing *A. lyrata* and its autogamous relative *A. thaliana*. *The Plant Cell*, *13*, 627–643.

Kusaba, M., Tung, C.-W., Nasrallah, M. E., & Nasrallah, J. B. (2002). Monoallelic expression and dominance interactions in anthers of self-incompatible *Arabidopsis lyrata*. *Plant Physiology*, *128*, 17–20.

Larsen, K. (1986). Cell-cell recognition and compatibility between heterogenic and homogenic incompatibility. *Hereditas*, *105*, 115–133.

Li, J., Cocker, J. M., Wright, J., Webster, M. A., McMullan, M., Dyer, S., et al. (2016). Genetic architecture and evolution of the *S* locus supergene in *Primula vulgaris*. *Nature Plants*, *2*, 16188. https://doi.org/10.1038/nplants.2016.188.

Liu, P., Sherman-Broyles, S., Nasrallah, M. E., & Nasrallah, J. B. (2007). A cryptic modifier causing transient self-incompatibility in *Arabidopsis thaliana*. *Current Biology*, *17*, 734–740.

Ma, R., Han, Z., Hu, Z., Lin, G., Gong, L., Zhang, H., et al. (2016). Structural basis for specific self-incompatibility response in *Brassica*. *Cell Research*, *26*, 1320–1329.

Naithani, S., Chookajorn, T., Ripoll, D. R., & Nasrallah, J. B. (2007). Structural modules for receptor dimerization in the S-locus receptor kinase extracellular domain. *Proceedings of the National Academy of Sciences of the United States of America, 104*, 12211–12216.

Nasrallah, J. B. (2005). Recognition and rejection of self in plant self-incompatibility: Comparisons to animal histocompatibility. *Trends in Immunology, 26*, 412–418.

Nasrallah, J. B. (2017). Plant mating systems: Self-incompatibility and evolutionary transitions to self-fertility in the mustard family. *Current Opinion in Genetics and Development, 47*, 54–60.

Nasrallah, M. E., Liu, P., & Nasrallah, J. B. (2002). Generation of self-incompatible *Arabidopsis thaliana* by transfer of two *S* locus genes from *A. lyrata*. *Science, 297*, 247–249.

Nasrallah, M. E., Liu, P., Sherman-Broyles, S., Boggs, N. A., & Nasrallah, J. B. (2004). Natural variation in expression of self-incompatibility in *Arabidopsis thaliana:* Implications for the evolution of selfing. *Proceedings of the National Academy of Sciences of the United States of America, 101*, 16070–16074.

Nasrallah, J. B., & Nasrallah, M. E. (2014). Robust self-incompatibility in the absence of a functional ARC1 in *Arabidopsis thaliana*. *The Plant Cell, 26*, 3838–3841.

Payne, B. L., & Alvarez-Ponce, D. (2018). Higher rates of protein evolution in the self-fertilizing plant *Arabidopsis thaliana* than in the out-crossers *Arabidopsis lyrata* and *Arabidopsis halleri*. *Genome Biology and Evolution, 10*, 895–900.

Rea, A., & Nasrallah, J. B. (2008). Self-incompatibility systems: Barriers to self-fertilization in flowering plants. *International Journal of Developmental Biology, 52*, 627–636.

Rea, A. C., & Nasrallah, J. B. (2015). *In vivo* imaging of the S-locus receptor kinase, the female specificity determinant of self-incompatibility, in transgenic self-incompatible *Arabidopsis thaliana*. *Annals of Botany, 115*, 789–805.

Samuel, M. A., Chong, Y. T., Haasen, K. E., Aldea-Brydges, M. G., Stone, S. L., & Goring, D. R. (2009). Cellular pathways regulating responses to compatible and self-incompatible pollen in *Brassica* and *Arabidopsis* stigmas intersect at Exo70A1, a putative component of the exocyst complex. *The Plant Cell, 21*, 2655–2671.

Sato, Y., Okamoto, S., & Nishio, T. (2004). Diversification and alteration of recognition specificity of the pollen ligand SP11/SCR in self-incompatibility of *Brassica* and *Raphanus*. *The Plant Cell, 16*, 3230–3241.

Schopfer, C. R., & Nasrallah, J. B. (2000). Self-incompatibility: Prospects for a novel peptide signaling molecule. *Plant Physiology, 124*, 935–939.

Schopfer, C. R., Nasrallah, M. E., & Nasrallah, J. B. (1999). The male determinant of self-incompatibility in *Brassica*. *Science, 286*, 1697–1700.

Stein, J. C., Howlett, B., Boyes, D. C., Nasrallah, M. E., & Nasrallah, J. B. (1991). Molecular cloning of a putative receptor protein kinase gene encoded at the self-incompatibility locus of *Brassica oleracea*. *Proceedings of the National Academy of Sciences of the United States of America, 88*, 8816–8820.

Tantikanjana, T., & Nasrallah, J. B. (2012). Non-cell-autonomous regulation of crucifer self-incompatibility by Auxin response factor ARF3. *Proceedings of the National Academy of Sciences of the United States of America, 109*, 19468–19473.

Tantikanjana, T., & Nasrallah, J. B. (2015). Ligand-mediated *cis*-inhibition of receptor signaling in the self-incompatibility response of the Brassicaceae. *Plant Physiology, 169*, 1141–1154.

Tantikanjana, T., Rivzi, N., Nasrallah, M. E., & Nasrallah, J. (2009). A dual role for the S-locus receptor kinase in self-incompatibility and pistil development revealed by an *Arabidopsis rdr6* mutation. *The Plant Cell, 21*, 2642–2654.

Tsuchimatsu, T., Goubet, P. M., Gallina, S., Holl, A. C., Fobis-Loisy, I., Berges, H., et al. (2017). Patterns of polymorphism at the self-incompatibility locus in 1,083 *Arabidopsis thaliana* genomes. *Molecular Biology and Evolution, 34*, 1878–1889. https://doi.org/10.1093/molbev/msx122.

Yamamoto, M., Tantikanjana, T., Nishio, T., Nasrallah, M. E., & Nasrallah, J. B. (2014). Site-specific N-glycosylation of the S-locus receptor kinase and its role in the self-incompatibility response of the Brassicaceae. *The Plant Cell, 26*, 4749–4762.

CHAPTER SEVENTEEN

Friend or foe: Signaling mechanisms during double fertilization in flowering seed plants

Liang-zi Zhou, Thomas Dresselhaus*
Cell Biology and Plant Biochemistry, University of Regensburg, Regensburg, Germany
*Corresponding author: e-mail address: thomas.dresselhaus@ur.de

Contents

Abstract

Since the first description of double fertilization 120 years ago, the processes of pollen tube growth and guidance, sperm cell release inside the receptive synergid cell, as well as fusion of two sperm cells to the female gametes (egg and central cell) have been well documented in many flowering plants. Especially microscopic techniques, including live cell imaging, were used to visualize these processes. Molecular as well as genetic methods were applied to identify key players involved. However, compared to the first 11 decades since its discovery, the past decade has seen a tremendous advancement in our understanding of the molecular mechanisms regulating angiosperm fertilization. Whole signaling networks were elucidated including secreted ligands, corresponding receptors, intracellular interaction partners, and further downstream signaling events

Current Topics in Developmental Biology, Volume 131
ISSN 0070-2153
https://doi.org/10.1016/bs.ctdb.2018.11.013

involved in the cross-talk between pollen tubes and their cargo with female reproductive cells. Biochemical and structural biological approaches are now increasingly contributing to our understanding of the different signaling processes required to distinguish between compatible and incompatible interaction partners. Here, we review the current knowledge about signaling mechanisms during above processes with a focus on the model plants *Arabidopsis thaliana* and *Zea mays* (maize). The analogy that many of the identified "reproductive signaling mechanisms" also act partly or fully in defense responses and/or cell death is also discussed.

1. Introduction

Double fertilization is a major characteristic of flowering plants (angiosperms). This complex process involves intensive cross-talk between the male gametophyte (pollen and pollen tube, respectively) with the female tissues of the pistil (e.g., stigma, style, and ovule) and the female gametophyte (embryo sac), respectively (Dresselhaus & Franklin-Tong, 2013). Development and function of male and female gametophytes are described in chapters "The evolution and patterning of male gametophyte development" by Hackenberg and Twell (this issue) and "Development and function of the flowering plant female gametophyte" by Serbes et al. (this issue). Fertilization begins with the adhesion of pollen grains at the stigma of the pistil, continues with pollen tube germination and growth toward and inside ovules, culminating in sperm cell release. This is also called the progamic phase of fertilization. Fertilization is completed by the syngamic phase after successful fusion of one sperm cell with the egg cell and the second sperm cell with the central cell, respectively.

The pollen tube is an invention of seed plants and it plays a major role in signaling during fertilization. Its main task is to deliver two sperm cells to accomplish double fertilization. Whether sperm cells themselves contribute to pollen tube growth and guidance has been a debate since many years. Mutant pollen tubes lacking sperm cells that were able to grow toward ovules were recently used to show that sperm cells are indeed dispensable for fertilization (Glöckle et al., 2018; Zhang, Huang, et al., 2017). Thus, it is the vegetative pollen tube cell that interacts and exchanges signals with the female tissues of the pistil and the embryo sac, respectively. Signaling mechanisms during fertilization therefore caused most attention during the past years to understand the various components of fertilization in angiosperms. 10 years ago, only one signaling ligand and two cell surface receptors with known functions in the final steps of double fertilization of angiosperms were known (Márton & Dresselhaus, 2008). Since then, the research community has made tremendous progress to identify molecules secreted by the pollen tube as well as the surrounding tissues, and identified a number of receptors/co-receptors as well as downstream signaling components involved in both, the progamic and syngamic phases. This research progress was regularly accompanied by reviews highlighting the progress made (e.g., Bleckmann, Alter, & Dresselhaus, 2014; Dresselhaus & Franklin-Tong, 2013; Dresselhaus, Sprunck, & Wessel, 2016; Hamamura, Nagahara, & Higashiyama, 2012; Li, Meng, & Yang, 2018; Mizuta & Higashiyama, 2018). A key finding was that some of the signaling mechanisms are specific to plant species/families, while other components are highly conserved and even occur in unicellular organisms with sexually differentiated gametes. Another important finding was that many of the components identified in the context of fertilization were previously known to play roles in defense responses and cell death, confirming previous hypothesis that reproductive signaling mechanisms may have partially evolved from ancient defense signaling mechanisms (see details below).

Most progress has been made using the model plant *Arabidopsis thaliana*. Due to its morphology, size, and synchronous development of reproductive organs and gametes, as well as the increasing number of molecular, cellular, and genetic tools available, maize was suggested as a model to study fertilization in grass species (Begcy & Dresselhaus, 2017; Dresselhaus, Lausser, & Márton, 2011; Zhou, Juranic, & Dresselhaus, 2017). Indeed, a number of molecular players have been identified in maize, showing that especially the progamic phase of fertilization involves plant family-specific, and partly even species-specific, proteins. Fig. 1 shows a comparison of the pollen tube pathway between a typical grass and an *A. thaliana* pistil, respectively. This review aims to update on our current understanding of the signaling mechanisms involved in pollen tube growth and guidance, pollen tube reception, as well as gamete interaction.

2. Signaling during pollen tube growth and guidance

The journey of the pollen tube was divided into five distinct phases (Lausser, Kliwer, Srilunchang, & Dresselhaus, 2010) and starts from the first interaction between pollen grains and papilla cells at the surface of the stigma (Fig. 1). After capture and adhesion in *Phase I*, compatible pollen grains hydrate, germinate, and form a tubular structure named pollen tube,

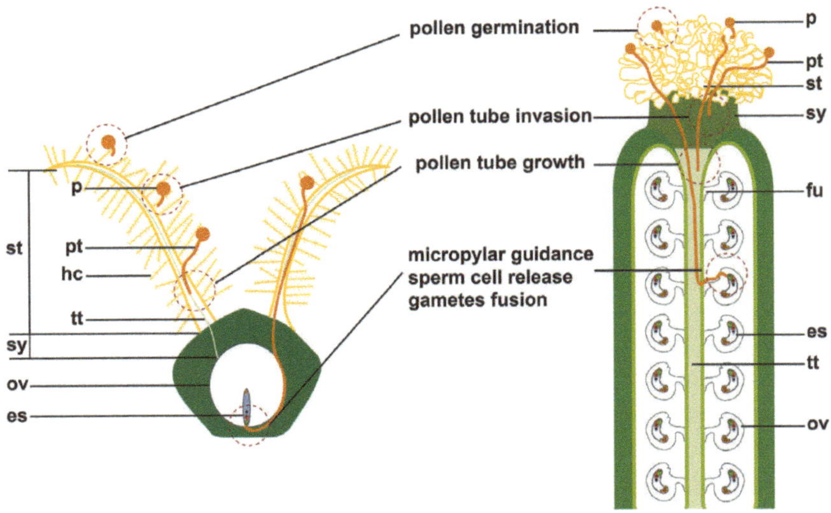

Fig. 1 Comparative pollen tube pathways in grasses and *Arabidopsis thaliana*. Left: a typical grass pistil consisting of two stigmatic branches or stigmata, an ovary (in green) containing a single ovule that harbors the female gametophyte (embryo sac). Each stigma contains a transmitting tract. Note that the stigmatic branches in maize are fused and elongated. Right: *A. thaliana* pistil containing around 50 ovules. The path of pollen tubes is indicated from germination, invasion, growth, guidance, and burst. Abbreviations: es, embryo sac; fu, funiculus; hc, hair cells; ov, ovule; p, pollen grain; pt, pollen tube; st, stigma; sy, style; tt, transmitting tract.

which represents a transport vehicle for non-motile male gametes (sperm cells). In *Phase II* pollen tubes penetrate floral tissues and grow, in *Phase III*, through the transmitting tract. Exit from the transmitting tract occurs in *Phase IV*, which includes guidance toward the ovule. In *Phase V* pollen tubes grow around synergid cells, culminating in sperm cells release. The signaling processes required to distinguish between compatible and incompatible pollen grains, to support or inhibit their germination and growth in *Phases I* and *II* are thoroughly discussed in chapter "Self-incompatibility in the Brassicaceae: Regulation and mechanism of self-recognition" by Nasrallah (this issue) about self-incompatibility and will not be included here. The focus of this review is on pollen tube growth in *Phases III–V*, after successful initial recognition and after pollen tubes reached the transmitting tract(s), which represent highways that lead pollen tubes deeper into the maternal tissues of the pistil and ultimately toward the ovule and female gametophyte, respectively. A scheme showing the pollen tube containing the so-called male germ unit (two sperm cells attached to the nucleus of the vegetative tube cell) as well as an ovule containing the various embryo sac cells during *Phases IV* and *V* is shown in Fig. 2A and B.

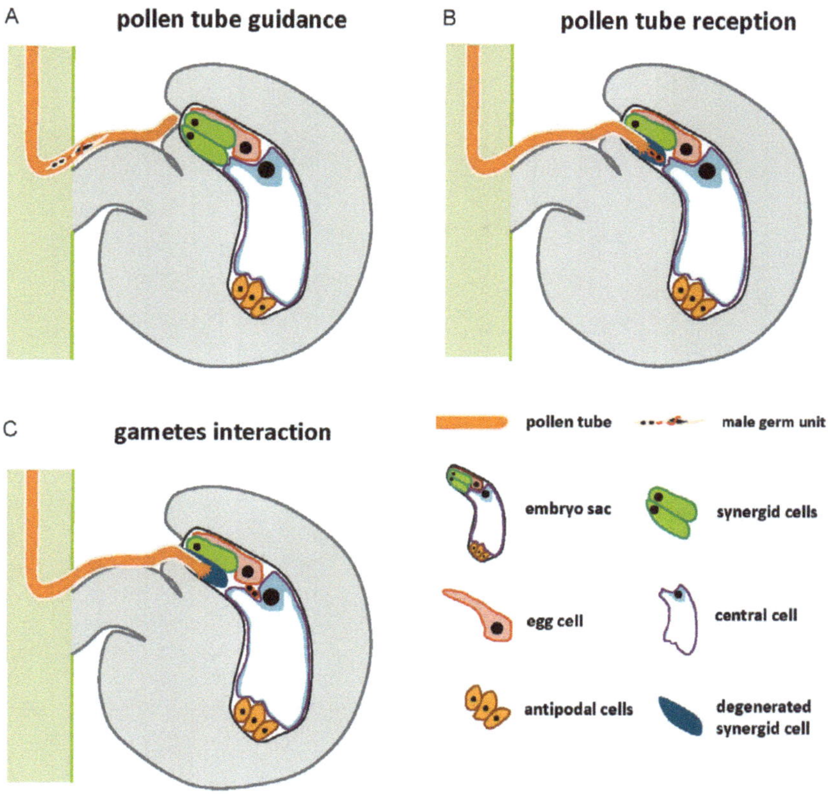

Fig. 2 Short distance communication between male and female gametophytes using *Arabidopsis thaliana* as an example. (A) Micropylar pollen tube guidance, (B) pollen tube reception, and (C) gamete interaction. The various cell types and gametophyte structures are indicated.

2.1 Preovular growth and guidance

During their journey, pollen tubes continuously sense their environment, perceive and secrete signaling molecules themselves. Table 1 provides an overview of proteins involved in pollen tube growth and guidance. The pollen-specific membrane localized receptor-like kinases LePRK1 and LePRK2, for example, were first identified in *Solanum lycopersicum* (tomato). Together with a Rho of plant guanine nucleotide exchange factor (RopGEF, also known as Kinase Partner Protein, KPP) and the actin binding protein PLIM2a, they were shown to be essential for pollen tube growth (Gui et al., 2014; Muschietti, Eyal, & McCormick, 1998). Ligands of LePRK1 and LePRK2 include the pollen-specific, secreted protein LAT52, the

Table 1 Genes involved in signaling during pollen tube growth and guidance, pollen tube reception as well as gamete interaction.

Gene	Expression	Protein family	Species	References
Pollen tube growth				
AGPs	Stigmatic cells	AGPs	*M. virginiana*	Losada et al. (2014)
LAT52	Pollen grain	Ole e I CRPs	*S. lycopersicum*	Muschietti et al. (1994)
LeSTIG1	Pistil	STIG1 CRP	*S. lycopersicum*	Huang et al. (2014) and Tang et al. (2004)
LePRK1/2	Pollen tube	LRR RLKs	*S. lycopersicum*	Muschietti et al. (1998)
KPP	Pollen tube	RopGEF	*S. lycopersicum*	Gui et al. (2014)
PLIM2a	Pollen tube	LIM protein	*S. lycopersicum*	Gui et al. (2014)
LeSHY	Pollen tube	LRR protein	*S. lycopersicum*	Tang et al. (2004)
ANXUR1/2	Pollen tube	*Cr*RLK1Ls	*A. thaliana*	Boisson-Dernier et al. (2009) and Miyazaki et al. (2009)
ERU	Pollen tube	*Cr*RLK1Ls	*A. thaliana*	Schoenaers et al. (2017)
MARIS	Pollen tube	RLCKs	*A. thaliana*	Boisson-Dernier, Franck, Lituiev, and Grossniklaus (2015) and Liao et al. (2016)
BUPS1/2	Pollen tube	*Cr*RLK1Ls	*A. thaliana*	Ge et al. (2017)
RALF4/19	Pollen tube	RALF CRPs	*A. thaliana*	Ge et al. (2017) and Mecchia et al. (2017)
LRX8/9/10/11	Pollen tube	LRR extensin	*A. thaliana*	Mecchia et al. (2017)
SR1	Ovule	Serine-racemase	*A. thaliana*	Michard et al. (2011)
RUPO	Pollen tube	*Cr*RLK1Ls	*O. sativa*	Liu, Zheng, et al. (2016)
PELPIII	Transmitting tract	AGPs	*N. tabacum*	Noyszewski et al. (2017)
TT	Transmitting tract	AGPs	*N. tabacum*	Cheung et al. (1995) and Wu et al. (2000)

120 K	Transmitting tract	AGPs	*N. tabacum*	Noyszewski et al. (2017)
NtPRP	Transmitting tract	AGPs	*N. tabacum*	Noyszewski et al. (2017)
PMEI1	Pollen tube	PME inhibitor	*Z. mays*	Woriedh et al. (2013)
Pollen tube guidance				
LUREs1.1-5	Synergid cells	DEFL (CRPs)	*A. thaliana*	Takeuchi and Higashiyama (2012)
MYB98	Synergid cells	MYB TF	*A. thaliana*	Kasahara, Portereiko, Sandaklie-Nikolova, Rabiger, and Drews (2005)
MAA3	Ovule	RNA helicase	*A. thaliana*	Shimizu, Ito, Ishiguro, and Okada (2008)
CCG	Central cell	TFIIB-like TF	*A. thaliana*	Chen et al. (2007)
CBP1	Central cell	CCG/Pol II interactor	*A. thaliana*	Li, Zhu, et al. (2015)
PRK3/6	Pollen tube	LRR RLKs	*A. thaliana*	Takeuchi and Higashiyama (2016)
LIP1/2	Pollen tube	RLCKs	*A. thaliana*	Liu et al. (2013)
MDIS1	Pollen tube	LRR RLKs	*A. thaliana*	Wang et al. (2016)
MIK1/2	Pollen tube	LRR RLKs	*A. thaliana*	Wang et al. (2016)
CNGC18	Pollen tube	CNGC channel	*A. thaliana*	Gao et al. (2016)
PSKR1/2	Pollen tube	LRR RLKs	*A. thaliana*	Stührwohldt et al. (2015)
LURE1–2	Synergid cells	DEFL (CRPs)	*T. fournieri*	Okuda et al. (2009)
AMOR	Ovule	AGP derivate	*T. fournieri*	Mizukami et al. (2016)
EA1	Egg apparatus	EA1-like	*Z. mays*	Márton, Cordts, Broadhvest, and Dresselhaus (2005) and Uebler, Dresselhaus, and Márton (2013)

Continued

Table 1 Genes involved in signaling during pollen tube growth and guidance, pollen tube reception as well as gamete interaction.—cont'd

Gene	Expression	Protein family	Species	References
Pollen tube growth arrest				
FER	Synergid cells	*Cr*RLK1Ls	*A. thaliana*	Escobar-Restrepo et al. (2007)
LRE	Synergid cells	GAP	*A. thaliana*	Capron et al. (2008)
ENODL14/15	Synergid cells	GAP	*A. thaliana*	Hou et al. (2016)
TUN	Synergid cells	UDP-glycosyltransferase	*A. thaliana*	Lindner et al. (2015)
EVN	Synergid cells	Dolichol kinase	*A. thaliana*	Lindner et al. (2015)
ARU	Synergid cells	OST3/6 subunit	*A. thaliana*	Müller et al. (2016)
NTA/AtMLO7	Synergid cells	MLO	*A. thaliana*	Kessler et al. (2010) and Davis et al. (2017)
EA1	Egg apparatus	EA1-like	*Z. mays*	Márton, Fastner, Uebler, and Dresselhaus (2012)
Pollen tube burst				
RALF34	Ovule	RALF (CRPs)	*A. thaliana*	Ge et al. (2017)
MYB97/101/120	Pollen tube	MYB TF	*A. thaliana*	Leydon et al. (2013)
ES1–4	Embryo sac	DEFL (CRPs)	*Z. mays*	Amien et al. (2010)
PMEI1	Egg apparatus	PME inhibitor	*Z. mays*	Woriedh et al. (2013)

Synergid degeneration				
AP1G1/2	Synergid cells	Gamma-adaptin	*A. thaliana*	Wang et al. (2017)
VHA-A	Pollen tube	V-ATPase subunit	*A. thaliana*	Wang et al. (2017)
EIN2	Synergid cells	Nramp transporter	*A. thaliana*	Völz et al. (2013)
EIN3	Synergid cells	EIL TF	*A. thaliana*	Völz et al. (2013)
VAL	Synergid cells	REM TF	*A. thaliana*	Mendes et al. (2016)
VDD	Synergid cells	REM TF	*A. thaliana*	Mendes et al. (2016)
GFA2	Ovule	DnaJ protein	*A. thaliana*	Christensen et al. (2002)
AGP4	Pistil	AGPs	*A. thaliana*	Pereira, Lopes, & Coimbra (2016)
Gamete activation				
EC1	Egg cell	ECA (CRPs)	*A. thaliana*	Sprunck et al. (2012)
Gamete recognition, adhesion and fusion				
GEX2	Sperm cell	Filamin-repeat IG protein	*A. thaliana*	Mori, Igawa, Tamiya, Miyagishima, and Berger (2014)
HAP2	Sperm cell	HAP2-GCS1	*A. thaliana*	von Besser, Frank, Johnson, and Preuss (2006)
GCS1	Sperm cell	HAP2-GCS1	*L. longiflorum*	Mori, Kuroiwa, Higashiyama, and Kuroiwa (2006)

See text for abbreviations of genes and protein families.

leucine-rich repeat (LRR) protein LeSHY, as well as LeSTIG1, a small cysteine-rich protein (CRP) from the pistil (Huang, Liu, McCormick, & Tang, 2014; Muschietti, Dircks, Vancanneyt, & McCormick, 1994; Tang, Kelley, Ezcurra, Cotter, & McCormick, 2004) (Fig. 3A). Experimental evidence revealed that the receptor-like kinases may interact with different ligands at different steps, in order to regulate pollen germination and pollen tube growth very precisely, given that LAT52 only binds LePRK2 before pollen germination, and its binding can be displaced by LeSTIG1, which was shown to promote pollen tube growth *in vitro* (Tang, Ezcurra, Muschietti, & McCormick, 2002; Tang et al., 2004). Besides, external phosphatidylinositol 3-phosphate [PI(3)P] was also shown to be required to promote the growth of pollen tubes together with LeSTIG1. *In vitro* incubation analysis showed that LeSTIG1 colocalizes with PI(3)P at the subapical region of pollen tubes. Protein-lipid overlay assays revealed direct binding between LeSTIG1 and PI(3)P, which leads to an increase of the intracellular redox potential and thus supports pollen tube growth (Heilmann & Ischebeck, 2016; Huang et al., 2014). However, it is unclear how PI(3)P is presented on the outer surface of the pollen tube as it usually serves as a cytoplasmic second messenger.

The rapid growth of pollen tubes requires a highly dynamic balance between cell wall integrity maintenance and tip elongation mechanisms (Hepler, Rounds, & Winship, 2013). Pollen tubes have to be flexible, they require a very stable cell wall due to the high turgor pressure of its cytoplasm, and also need to burst at the right time point to release the sperm cells for fertilization. Recent studies have shown that plant-specific receptor-like kinases (RLKs) of the *Catharanthus roseus* RLK1-like (*Cr*RLK1L) subfamily named as ANXUR1 (ANX1) and ANX2 as well as ERULUS/$[Ca^{2+}]_{cyt}$-ASSOCIATED PROTEIN KINASE1 (ERU/CAP1) are involved in regulating this process by coordinating the production of reactive oxygen species (ROS), Ca^{2+}-homeostasis, and exocytosis. Over-expression of ANXs causes over-activated exocytosis, leading to over-accumulation of secreted cell wall material in the extracellular matrix, which eventually arrests pollen tube growth. On the contrary, in *anx1/2* double mutants pollen tube integrity is impaired, causing premature bursting (Boisson-Dernier et al., 2009; Miyazaki et al., 2009). During pollen tube growth ANX1/2 function upstream of NADPH oxidases, which generate tip-localized ROS, probably through an unknown Ca^{2+} channel (Boisson-Dernier et al., 2013). ROS production in pollen tubes depends on Ca^{2+} induction for polarized growth (Kaya et al., 2014), suggesting that ANX1/2 and ERU signaling regulates Ca^{2}

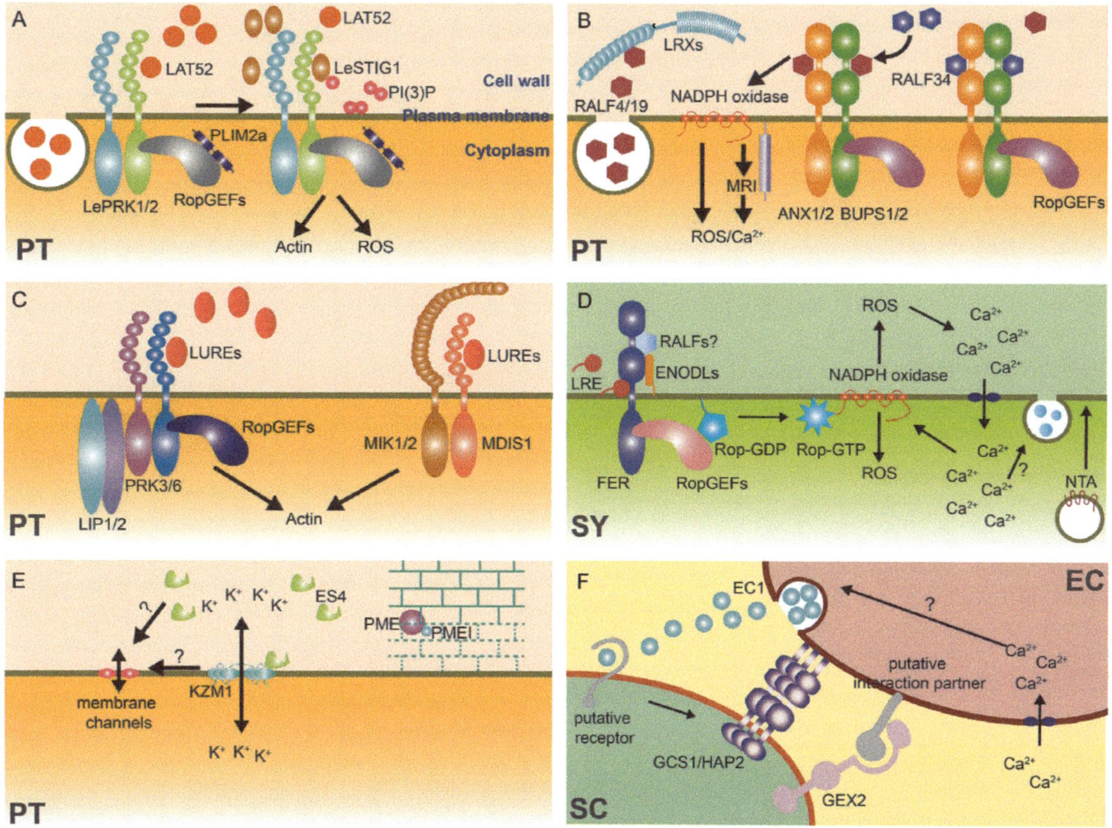

Fig. 3 Signaling mechanisms during angiosperm fertilization. (A) Signaling during pollen tube-stigma interactions in tomato. (B) Signaling required to maintain pollen tube cell wall integrity in *A. thaliana*. (C) Signaling during pollen tube guidance in *A. thaliana*. (D) The *FER* signaling during pollen tube reception in *A. thaliana*. (E) Signaling during pollen tube burst in maize. (F) Signaling during gamete activation, adhesion, and fusion. See text for names of signaling ligands, receptors, and downstream signaling processes. Abbreviations: EC, egg cell; PT, pollen tube; SC, sperm cell; SY, synergid cell.

$^{+}$ fluxes. Further analysis of the ANX1/2 downstream signaling pathway revealed a novel cytoplasmic receptor-like kinase named MARIS (MRI), which appears to represent a *Cr*RLK1L-dependent downstream signaling component required to coordinate cell wall integrity and tip growth (Boisson-Dernier et al., 2015). The function of *Cr*RLK1L family members for cell wall integrity appears to be conserved, as a *Cr*RLK1L receptor kinase named Ruptured Pollen Tube (RUPO) from rice also contributes to pollen tube growth and integrity via regulating potassium transporters. However, RUPO is a member of a different phylogenetic branch compared to ANX1/2 (Liu, Zheng, et al., 2016). It will now be important to study the close homologs of ANX1/2 in other plant species. ERU mentioned above appears to be involved in regulating Ca^{2+}-dependent pollen tube growth, but its precise role remains unclear (Schoenaers, Balcerowicz, Costa, & Vissenberg, 2017).

In addition to above mentioned small CRPs from tomato, secreted CRPs from pollen tubes were also identified in *A. thaliana* and shown to be required for the balanced regulation of cell wall integrity, and thus also the control of pollen tube burst and sperm cells release via their interaction with pollen tube-specific receptor-like kinases (Muschietti & Wengier, 2018; Qu, Li, Lan, & Dresselhaus, 2015). Like ANX1 and ANX2, BUDDHA'S PAPER SEAL1 (BUPS1) and BUPS2 are receptors of the *Cr*RLK1L family, containing two typical extracellular malectin-like domains (Boisson-Dernier, Kessler, & Grossniklaus, 2011). Their closest family member is ERU mentioned above. The *bups1/2* double mutant showed pollen tube integrity defects and the ligands of BUPS1/2 belong to the Rapid Alkalinization Factor (RALF) family of CRPs (Ge et al., 2017), namely RALF4 and RALF19, and the corresponding double mutant showed the same phenotype as the *bups1/2* and *anx1/2* double mutants (Ge et al., 2017; Mecchia et al., 2017). BUPS1/2 are capable to bind to ANX1/2 via their ectodomains (ECDs) and were shown to form a heteromer. Both, BUPSs and ANXs are able to interact with the RALF4/19 ligands (Fig. 3B) (Ge et al., 2017). It was further shown that BUPS1 is capable to form homomers and heteromers with BUPS2, while BUPS2 only forms heteromers with BUPS1. Moreover, BUPS1/2 interact with pollen-expressed RopGEFs, indicating that they may also be involved in the polarized growth of pollen tubes (Zhu et al., 2018). Pollen tube-expressed LEUCINE-RICH REPEAT EXTENSIN (LRX) proteins are also involved in cell wall integrity. Recently, it was reported that LRX proteins interact with RALF4/19 to monitor cell wall changes and to coordinate pollen tube growth and integrity via a *Cr*RLK1L signaling pathway

(Mecchia et al., 2017). Finally, structure analysis revealed that the malectin domains of the ANXs and likely other RLKs of the *Cr*RLK1L family form a cleft that likely serves as the binding site for extracellular ligands (Du, Qu, & Xiao, 2018; Moussu, Augustin, Roman, Broyart, & Santiago, 2018). This finding not only indicates that ANX1/2 interact with extracellular ligands, but moreover suggest that the ECDs of ERU and BUPS1/2 possess similar structures. Altogether, these findings indicate that multiple and fine-tuned ligand-receptor interactions, as well as interactions with cell wall related proteins and/or cell wall polysaccharides, are required to maintain pollen tube growth and integrity during the long journey of the pollen tubes (Higashiyama, 2018). Considering that CRPs of various classes belong to the most abundant proteins in both male and female gametophytes (Huang, Dresselhaus, Gu, & Qu, 2015), it can be expected that they possess many more roles during gametophyte signaling, which will hopefully be elucidated in the near future.

Arabinogalactan proteins (AGPs) are sugar chain coated extracellular hydroxyproline-rich glycoproteins, which are often anchored to the cell surface. These features make them ideal components of receptor complexes for interacting with the cell wall and thus contributing to signal sensing (Pereira, Lopes, et al., 2016). In *Nicotiana tabacum* (tobacco), the transmitting tract controls pollen tube growth via secreting various AGPs into the extracellular matrix of the style. These AGPs include pistil extensin-like protein III (PELPIII), transmitting-tract-specific (TTS) protein, 120 kDa glycoprotein (120K), and also recently discovered *N. tabacum* Proline-Rich Protein (NtPRP) (Noyszewski, Liu, Tamura, & Smith, 2017; Pereira, Lopes, et al., 2016). It was previously shown that TTS could attract pollen tubes and promote their growth rate in tobacco (Cheung, Wang, & Wu, 1995; Wu, Wong, Ogdahl, & Cheung, 2000). Impaired production of AGPs in *A. thaliana* caused defects in pollen germination *in vitro* and male gamete transmission *in vivo* (Stonebloom et al., 2016). Expression of AGPs could also be detected in mature styles of *Magnolia virginiana*, indicating the functional conservation of this gene family during evolution (Losada, Herrero, Hormaza, & Friedman, 2014). Recent structure analysis of these AGPs revealed that their N-terminal domain (NTD) with predicted O-glycosylation sites is likely responsible for protein-protein interactions in the style (Noyszewski et al., 2017).

In the model plant *A. thaliana*, the style possesses multiple cell layers and acts as a connecting bridge between papilla cells and the transmitting tract (Fig. 1). Growing through the style is necessary for pollen tube activation in *A. thaliana*, otherwise pollen tubes are not able to perceive attraction signals at the final stages of their journey (Higashiyama, Kuroiwa, Kawano, & Kuroiwa, 1998; Okuda et al., 2013) (see also below). In summary, at present we understand little about the communication mechanisms between growing pollen tubes and the maternal cell layers of the style. To which extent is growth of compatible pollen tubes supported and that of incompatible tubes prevented? Is there competition among pollen tubes and do compatible tubes trigger maternal tissues to secrete nutrients, such as sugars and cell wall degrading enzymes, to support their growth? General molecules generated by maternal tissues of pistil and ovule, such as brassinosteroids and gamma-amino butyric acid (GABA), were shown to significantly promote pollen tube growth and guidance *in vitro* (Palanivelu, Brass, Edlund, & Preuss, 2003; Vogler, Schmalzl, Englhart, Bircheneder, & Sprunck, 2014; Yu et al., 2014). In the case of GABA, modulation of pollen tube expressed putative Ca^{2+}-permeable membrane channels was demonstrated, which provides a first hint to how pollen tubes communicate with maternal tissues (Palanivelu et al., 2003; Vogler et al., 2014; Yu et al., 2014). However, such general molecules do not explain the different growth behaviors of self and foreign pollen tubes, and suggest the existence of species-specific communication and/or defense systems.

Transcriptome analysis of pistils from various *Arabidopsis* spp. pollinated with self or foreign pollen or infected with fungal spores showed that 79% of down-regulated genes are shared between different conditions. Especially the CRP classes of thionines and (Defensin-like) DEFLs were significantly up-regulated, both after interspecific pollination and infection (Mondragon-Palomino, John-Arputharaj, Pallmann, & Dresselhaus, 2017). These findings indicate that similar to fungal hyphae, foreign pollen tubes are considered as intruders and their growth is supported less than that of compatible pollen tubes. Functional studies of such candidate genes should be performed to test this hypothesis. Transcriptome studies in *A. thaliana* also pointed out that a number of critical genes are expressed in pollen tubes after pistil penetration, which are neither expressed in any other pollen developmental stages nor in *in vitro* growing pollen tubes. Especially genes predicted to be involved in signal transduction were overrepresented (Qin et al., 2009), indicating both the importance of signaling during pollen tube growth and the observation that pistil signals trigger gene expression changes. Functional studies of identified genes will help to elucidate cell-cell communication systems between self/foreign pollen tubes and the style. Moreover, our current understanding is almost exclusively derived from studies in *A. thaliana,* and it is urgently necessary to increase efforts in other plant species to understand to which extent the above described mechanisms are conserved and to which extent they might play a role in preventing growth of closely related pollen tubes and thus to represent speciation mechanisms.

2.2 Ovular guidance

2.2.1 Attraction signals from the female gametophyte

Usually, the journey of pollen tubes in transmitting tracts is quite straightforward and ovules are reached the latest within a few hours. Once pollen tubes sense attraction signals from the ovule, they make a sharp turn, exit the transmitting tract and enter the micropyle of the ovule (Higashiyama & Takeuchi, 2015). Using *Torenia fournieri* as a model, which generates a half-naked embryo sac protruding from the ovule, it was demonstrated that ovules can attract pollen tubes *in vitro* (Higashiyama et al., 1998). The attraction signal is generated by the synergid cells, as laser ablation experiments revealed that a single synergid cell is sufficient to attract pollen tubes toward the egg apparatus (Higashiyama et al., 2001). Since the discovery of the attraction molecules in *T. fournieri* named as LUREs (Okuda et al., 2009), a lot of progress has been made to understand the underlying signaling mechanisms. LURE1 and LURE2 encode DEFLs of 62 and 70 amino acids, respectively, whose transcripts belong to the most abundant ones in *T. fournieri* synergid cells. Recombinant LURE protein synthesized in *Escherichia coli* was shown to be able to attract pollen tubes *in vitro* in a concentration-dependent and species-specific manner. While pollen tubes of other genera are not being attracted *in vitro*, tubes of related species such as *T. concolor* are attracted, although at a significantly reduced efficiency and *vice versa* (Kanaoka et al., 2011). LUREs were later also reported in *A. thaliana* and *Arabidopsis lyrata* (Takeuchi & Higashiyama, 2012). Although they both belong to the DEFL group of CRPs, *T. fournieri* and *A. thaliana* LUREs show little sequence homology, indicating fast evolution and specificity.

Molecules generated by the ovule also contribute to pollen tube attraction. *In vitro* germinated pollen tubes from *T. fournieri* and *A. thaliana* are not able to respond to LUREs and first have to grow through a cut style to obtain competence. In *T. fournieri*, Activation Molecule for Response Capability (AMOR), a methyl-glucuronosyl arabinogalactan derived from ovular AGPs, was found to be capable to induce pollen tube competency to LURE peptides. The terminal disaccharide 4-*O*-methylglucuronosyl residue was necessary for its activity. Chemically synthesized 4-Me-GlcA-b(1,6)-Gal also showed AMOR activity (Mizukami et al., 2016). Structure-activity studies later demonstrated that, with the exception of some hydroxy groups, all components of the disaccharide are essential for its function (Jiao et al., 2017). As previously discussed, AGPs are also abundant in the transmitting tract. It is thus likely that AMOR derived from transmitting tract localized AGPs also contributes to pollen tube growth and guidance. Although the role of polymorphic LUREs in pollen tube attraction appears conserved in eudicots species, the function of AMOR triggering competency in pollen tubes to sense attractants was so far only reported in *T. fournieri*. Further studies need to be done to investigate AMOR function in other species, including monocots to understand its general role in the transmitting tract and ovule. It will be exciting to elucidate the AMOR receptor(s) and associated downstream signaling pathway(s), which likely result in gene expression changes.

LURE CRPs now represent the best studied signaling peptides during pollen tube guidance in eudicots and it was questioned whether they exist in other monocot plant species (Bircheneder & Dresselhaus, 2016; Qu et al., 2015). Notably, the DEFL subgroup of CRPs was first discovered in the embryo sac of maize and named as EMBRYO SAC 1–4 (ES1–4) (Cordts et al., 2001). Initially, their role was discussed as defense peptides required to protect the embryo sac from microorganisms, which might be carried by the pollen tube. Later, it was shown that ES1–4 induce pollen tube burst and do not possess guidance activity (Amien et al., 2010) (see also below). Notably, ES1–4 still possess antifungal activity at high concentrations (Woriedh, Merkl, & Dresselhaus, 2015). In maize, a small non-CRP peptide specifically expressed in the egg apparatus named EGG APPARATUS1 (EA1) was reported as a pollen tube attractant acting in a species-specific manner (Márton et al., 2005). The predicted mature peptide consists of 49 amino acids and was shown to bind at the pollen tube apex to an unknown receptor (Márton et al., 2012; Uebler et al., 2013). Notably, secreted EA1 peptides do not exist outside of the grass family (Uebler, Márton, & Dresselhaus, 2015). In conclusion, the identification of the first ovular pollen tube attraction and guidance molecule in plants indicates that genera use similar secreted peptides, but self-pollen tubes are better supported and guided, while distinct families appear to use different peptides that are not capable to bind to pollen tube surface receptors and to activate downstream signaling pathways.

According to their morphology, synergid cells appear to be secretory cells designed for pollen tube attraction and reception. A thickened cell wall formed by numerous finger-like projections—known as filiform apparatus—is present at their micropylar pole to facilitate cellular communication. The chalazal pole contains little cell wall material and is flexible during sperm release and transport (Punwani & Drews, 2007). In *A. thaliana*, synergid-expressed *MYB98*, which encodes a R2R3 MYB transcription factor, is necessary for pollen tube attraction. The *myb98* mutant has an abnormal filiform apparatus structure and is defective in pollen tube attraction (Kasahara et al., 2005). It is thus very likely that MYB98 acts upstream of the LUREs and other synergid-expressed CRPs (Huang, Dresselhaus, et al., 2015). Mutation of MAGATAMA3 (MAA3) also affects pollen tube attraction in *A. thaliana*. *MAA3* encodes a Sen1p-like RNA helicase and was speculated to regulate target mRNA molecules, whose products are required for pollen tube guidance (Shimizu et al., 2008). Expression analysis indeed revealed that LURE transcript levels were significantly reduced in both *myb98* and *maa3* mutants. These findings indicate the existence of a complicated regulatory pathway generating secreted attraction signals from synergid cells (Takeuchi & Higashiyama, 2012).

Finally, synergid cells are not the only embryo sac cells contributing to pollen tube attraction. While *EA1* in maize is expressed in all embryo sac cells except antipodal cells, central cell expressed *CENTRAL CELL GUIDANCE* (*CCG*) in *A. thaliana*, which encodes a nuclear protein, also contributes to pollen tube re-orientation (Chen et al., 2007). Later, CCG BINDING PROTEIN1 (CBP1), a novel protein without known function, was identified as an interaction partner of CCG. CCG and CBP1 seem to co-regulate the expression level of a subset of CRPs including LUREs (Li, Zhu, et al., 2015), thus representing key transcription factors at the beginning of the ovular guidance machinery.

2.2.2 Receptor complexes at the pollen tube surface

To achieve pollen tube re-orientation, pollen tubes must first perceive attraction signals from the ovule and/or female gametophyte and respond accordingly. As reported earlier in stomata patterning, CRPs bind as ligands directly to RLKs in order to trigger downstream signaling pathways (Lee et al., 2012). It was therefore speculated that such mechanisms might also exist in pollen tube guidance.

Over the last few years, several pollen tube-expressed RLKs were identified to respond to LUREs in *A. thaliana* (Fig. 3C). The tip-localized POLLEN RECEPTOR-LIKE KINASE6 (PRK6) is a receptor essential for sensing LURE1. PRK6 encodes a LRR-domain containing RLK located only at the tip of pollen tubes (Takeuchi & Higashiyama, 2016).

Structure analysis revealed that LURE1 is binding to the extracellular LRR domain of PRK6 (Zhang, Liu, et al., 2017). Pollen tubes of *prk6* mutants showed compromised ovule targeting *in vivo* and a lack of response to LURE1 in *semi-in vivo* analysis. Applying LURE1 *in vitro* induced asymmetric accumulation of tip-localized PRK6 before reorientation of the pollen tubes. PRK6 interacts directly with its homolog PRK3, with RopGEFs, and with the intracellular receptor-like kinase LOST IN POLLEN TUBE GUIDANCE1 (LIP1) and LIP2 (Liu et al., 2013; Takeuchi & Higashiyama, 2016). These findings suggest that PRK6 acts as a critical central receptor at the pollen tube tip sensing secreted guidance molecule and recruits the tip-growth machinery via Rop signaling proteins to achieve re-orientation of pollen tubes. However, PRK6 is not the only LURE1 binding receptor at the pollen tube surface. A MALE-DISCOVERER1 (MDIS1)—MDIS1-INTERACTING RECEPTOR-LIKE KINASE (MIK) heteromer was also found as an interaction partner of LURE1. MDIS1, MIK1 and MIK2 all belong to the same LRR RLK family containing PRK6 and PRK3. Pollen tube sensitivity to LURE1 was reduced in the corresponding RLK mutants (Wang et al., 2016). These findings showed the complexity of the pollen tube attraction pathway in *A. thaliana* and may also represent an explanation for the evolution of species-specific attraction mechanisms (Higashiyama & Yang, 2017).

In addition to LURE receptors, further components involved in pollen tube attraction were identified. It was reported that a steep calcium gradient is essential for pollen tube growth and guidance (Steinhorst & Kudla, 2013a). But until recently, there was no genetic evidence endorsing that calcium channels are directly involved in guidance. The first calcium channel described to be involved in pollen tube guidance was CYCLIC NUCLEOTIDE-GATED CHANNEL18 (CNGC18). Mutation of CNGC18 caused an abnormal calcium gradient, leading to severe pollen tube guidance defects (Gao et al., 2016). General peptides generated by the pollen tube are also involved in ovular guidance. Phytosulfokine (PSK), a 5 aa sulfated peptide (YIYTQ) plant growth factor that is being processed from a 90–100 amino acid precursor which requires tyrosine sulfation for its activity, is perceived by two pollen tube expressed LRR-RLKs, PHYTOSULFOKIN RECEPTOR1 (PSKR1) and PSKR2. The *pskr1/2* double mutant has reduced fertility and leads to defects in seed production, indicating the requirement for PSK peptide signaling during fertilization. Detailed analysis showed that exogenously applied PSK could enhance pollen tube growth *in vitro*, while pollen tubes from knockout mutants were shorter compared to those of the wild type.

However, although their growth rate was reduced in the double mutant, the pollen tubes eventually reached the end of the transmitting tract, suggesting that pollen tube growth rate may not be the key factor influencing fertility. On the other hand, unfertilized ovules in the mutants displayed compromised funicular guidance, thus indicating that PSK signaling is required to guide the pollen tube from the transmitting tract toward the embryo sac (Stührwohldt et al., 2015).

Nitric oxide (NO) and D-serine are other general components discussed to be involved in pollen tube guidance. NO was reported to possess diverse effects on plant development as a growth regulator, but it was also demonstrated to act as a negative chemotropic cue during *in vitro* pollen tube growth and re-orientation, which was mediated by a cGMP signaling pathway (Prado, Porterfield, & Feijo, 2004). Mutations in *A. thaliana NITRIC OXIDE SYNTHASE1* (*AtNOS1*) cause reduced fertility as NO production was affected (Guo, Okamoto, & Crawford, 2003). These findings suggest that NO may have certain functions during pollen tube guidance. *Semi-in vivo* assays with isolated ovules and pollen tubes further confirmed that NO is necessary for ovular pollen tube guidance in *Lilium longiflorum* (lily), and this signaling pathway is dependent on Ca^{2+} signaling (Prado, Colaco, Moreno, Silva, & Feijo, 2008). It remains unclear how NO is perceived at the pollen tube surface. D-serine accumulating in the micropylar region of the ovule also appears to control *in vitro* pollen tube behavior in *A. thaliana* by increasing the Ca^{2+} concentration in pollen tube as an agonist of glutamate receptor-like channel (GLR) Ca^{2+} activity (Michard et al., 2011). The corresponding biosynthesis enzyme, SERINE RACEMASE1 (SR1), displays a similar localization pattern as its product D-serine. Pollination of *sr1* mutant pistils with wild-type pollen showed deformation and branching of pollen tubes similar to those observed in various mutants affecting GLRs. These findings suggest that ovule/pistil derived D-serine is involved in pollen tube growth and/or guidance by modulating GLR activity.

2.3 Novel techniques for analyzing guidance *in vitro*

In the last decade, conventional approaches including forward and reverse genetics, transcriptomic studies, and biochemistry were used to identify and study pollen tube attraction signals. Such approaches were very successful in the discovery of EA1 and LURE peptides, receptors, and downstream signaling components. However, these approaches are technically challenging and it is difficult to generate high-throughput data that could result in the identification and understanding of pollen tube guidance mechanisms in many plant species.

The development of lab-on-chip techniques now offers a great opportunity for searching and testing different guidance cues in a reproducible and high-throughput manner in many plant species (Sato, Sugimoto, Higashiyama, & Arata, 2015). The specificity of above described secreted signals can now be quickly tested on various pollen tubes, including inbred lines and/or accessions of the same species. Moreover, functions of general molecules like carbohydrates, ROS, NO, GABA, D-serine, PSK, and protons (H^+) can be tested easily. Protons, for example, were long thought to play an important role in regulating pollen tube growth as an external stimulus. However, it was challenging to investigate and observe the direct effect of proton gradients and flux on pollen tube behavior. With the help of a newly invented microelectrode device, different growth behaviors of pollen tubes can be examined under changing conditions. Pollen tubes at different positions of the microelectrode device have been exposed to a proton gradient or pH change, which leads to various growth behaviors including tip bursting, pollen tube re-orientation, or growth arrest. Bursting and arresting of pollen tubes indicate that extracellular H^+ concentrations likely affect cell wall integrity and actin polymerization, and the direction change suggests that pollen tubes are capable to sense proton gradients as a signal and to respond accordingly (Hu et al., 2017). To observe the influence of electrical signals on pollen tube growth, even electrical lab-on-chip devices can now be used to test how different current flows affect pollen tube growth behavior. Taking pollen germination ratio, pollen tube length and pollen burst ratio as assessment criterions, experiments applying various Alternating Current (AC) or Direct Current (DC) field treatments suggested that pollen tubes respond more strongly to DC fields compared with AC fields at the same electric strength. It was shown, for example, that AC electric fields restore or even promote the growth of pollen tubes at a certain frequency range, and different AC frequencies triggered pollen tube responses correlated with the conductivity of the growth medium, which might be due to the effect on motion of ions (Agudelo, Packirisamy, & Geitmann, 2016). In order to test how mechanical signals affect pollen tube growth, lab-on-chip techniques were also combined either with a cellular force microscope (CFM) or an atomic force microscope (AFC), which allows the rapid measurement of cell wall stiffness from growing pollen tubes (Shamsudhin et al., 2016; Vogler et al., 2013).

Lab-on-chip systems are nowadays often used to test substances at the micrometer/nanometer scale. The flexibility in shape and size of such devices is also advantages, as they can be modified according to a certain experimental design and allow precise quantification (Kanaoka, 2018). The recent release of secretome data from tobacco pollen tubes (Hafidh et al., 2016) and probably from further species in the near future, as well as of ovular secretomes, will provide lists of many candidate guidance molecules that can be tested. It is thus very likely that pollen tube guidance molecules will soon also be identified in genera/species other than *Torenia* spp., *Arabidopsis* spp., and maize.

3. Signaling during pollen tube reception

Ultimately, pollen tubes arrive at the egg apparatus composed of one egg cell and two synergid cells, representing secretory cells that attract and guide the pollen tube (Fig. 2B). Intensive communication between the pollen tube and female tissues is extended after arrival, culminating in pollen tube growth arrest, degeneration of the receptive synergid cell, pollen tube rupture, and sperm cell release. This process named as pollen tube reception initiates after physical contact between the tip of the arriving pollen tube and the receptive synergid cell, whose degeneration process is initiated (Sandaklie-Nikolova, Palanivelu, King, Copenhaver, & Drews, 2007). Synergid cell death is also triggered by pollen tubes without cargo discharge, although pollen tube burst strongly accelerates degeneration (Leydon et al., 2015), supporting the idea that signaling between pollen tube and receptive synergid occurs via direct contact and/or the secretion of signaling factors from both interaction partners. In this context, the cell wall of the synergid cells plays a central role. As described above, synergid cells are strongly invaginated at their micropylar pole, leading to a large increase of their surface area. Cell wall material at this pole known as the filiform apparatus represents a hub for gametophytic cross-talk, accumulating large amounts of synergid-secreted substances, including guidance factors and cell surface receptors. Additionally, it represents the first direct contact area between both male and female gametophytes.

3.1 Synergid activation by the FERONIA signaling pathway

With the idea to identify components involved in gametophytic signaling pathways, genetic screens were established in the model plant *A. thaliana* to identify mutants defective in pollen tube growth arrest and burst. An overview of proteins involved in pollen tube reception is shown in Table 1. *FERONIA* (*FER*) was the first gene shown to be required for pollen tube growth arrest (Huck, Moore, Federer, & Grossniklaus, 2003). The FER protein accumulates at the micropylar region of the synergid cells associated with the filiform apparatus. *fer* mutant ovules showed normal pollen tube attraction, but a pollen tube overgrowth phenotype was detected inside of the female gametophyte (Escobar-Restrepo et al., 2007). The same phenotype was observed during interspecific crosses using pollen from other Brassicaceae, such as *A. lyrata* and *Cardamine flexuosa*. FER is one of 17 RLKs of the *Cr*RLK1L family in *A. thaliana*. Typical *Cr*RLK1Ls contain two malectin-like extracellular domains predicted to bind carbohydrate moieties (Boisson-Dernier et al., 2011). Notably, ECDs of three members studied (FER, ANX1, and HERKULES1) are not interchangeable for pollen tube reception, while their intracellular kinase domains can be functionally swapped (Kessler, Lindner, Jones, & Grossniklaus, 2015). This finding indicated distinct extracellular ligands and/or co-factors including co-receptors, but activation of similar intracellular signaling pathways. In roots, it was shown that the secreted peptide RALF1 directly interacts with FER (Haruta, Sabat, Stecker, Minkoff, & Sussman, 2014). Whether RALF peptides activate the FER signaling pathway also in the context of pollen tube reception remained unclear. However, from 34 known RALF peptide genes in *A. thaliana*, at least 14, but not *RALF1*, are expressed in pollen tubes (Ge et al., 2017). It is thus not unlikely that either one or several RALF family members may interact with the FER receptor complex in synergid cells.

FER itself is a component of a larger receptor complex in *A. thaliana* (Fig. 3D). Its first co-receptor was identified in a similar genetic screen as described above, displaying a pollen tube overgrowth phenotype in the mutant female gametophyte. The responsible synergid-expressed gene encodes a small plant-specific glycosylphosphatidylinositol (GPI)-anchored protein (GAP) named LORELEI (LRE) (Capron et al., 2008). Via a C-terminal GPI anchor, GAPs like LRE are thought to be exposed on the extracellular surface of the plasma membrane. LRE localizes to the synergid cell plasma membrane-rich filiform apparatus (Liu, Castro, et al., 2016), where it probably physically interacts with the extracellular juxtamembrane region of FER. This interaction was shown to play a pivotal role for FER function (Li, Yeh, et al., 2015; Liu, Castro, et al., 2016). Early nodulin-like proteins (ENODLs) are another class of GAPs required for pollen tube reception. In addition to their C-terminal GPI anchor,

ENODLs contain a plastocyanin-like domain and an arabinogalactan glycomodule (Hou et al., 2016). ENODL14 and ENODL15 are asymmetrically distributed in the plasma membrane of synergid cells, accumulating at the filiform apparatus. Similar to *fer* and *lre* mutants, ovules of RNAi lines downregulating ENODLs display a pollen tube overgrowth phenotype. ENODL14 strongly interacts with the ECD of FER, but not with other tested members of the *Cr*RLK1L family. The interaction between ENODL14 and LRE is significantly weaker (Hou et al., 2016). Like *Cr*RLK1Ls, AGP ENODLs likely contain a number of polysaccharide side chains (Su & Higashiyama, 2018). Notably, mutants of TURAN (TUN) and EVAN (EVN), two synergid-expressed genes encoding enzymes of the N-glycosylation machinery in the endoplasmic reticulum, also display the pollen tube overgrowth phenotype. Although it was not shown whether they attach sugar moieties to ENODLs, FER, LRE or other proteins enriched in synergid cells, they were shown to be necessary for pollen tube integrity maintenance by affecting the stability of the pollen-specific FER homologs ANX1 and ANX2 (Lindner et al., 2015). It remains unclear whether stabilization is associated with glycosylation, but this observation suggests that FER and likely other components of the receptor complex contain sugar side chains, which might contribute to ligand perception, receptor/co-receptor interaction, and/or cell wall sensing.

A recent report in roots has indicated that the ECD of FER directly interacts with pectin in the cell wall and senses the occurrence of cell wall damage, for example, by sodium ion stress (salinity) leading to cell wall softening (Feng et al., 2018). Calcium and boron-mediated bridges between pectin chains are softened by sodium, which might also occur in loss of FER-pectin interaction leading to receptor activation. A receptor complex containing pectin-like side chains could sense local changes in cell wall stiffness, which also occurs during pollen tube penetration of the filiform apparatus. Further support for the existence and necessity of a glycosylated receptor-complex is derived from the analysis of the *ARTUMES* (*ARU*) gene. *ARU*, which was identified in a genome-wide association study (GWAS) investigating the extent of pollen tube overgrowth in crosses between various *A. thaliana* accessions and related species such as *A. lyrata*, *Arabidopsis halleri*, and *A. arenosa*, appears to be involved in the regulation of interspecific, but not intraspecific pollen tube reception (Müller, Lindner, Pires, Gagliardini, & Grossniklaus, 2016). *ARU* encodes the ER-localized OST3/6 subunit of the oligosaccharyltransferase complex playing a key role for N-glycosylation. Although it has not been shown whether components of the FER receptor-complex are N-glycosylated, this study further supports the idea that glycosylation patterns play an important role in sensing pollen tube arrival. However, the observation that arriving pollen tubes from non-compatible species do not burst indicates that sensing of cell wall damage is not sufficient and point to additional more specific interactions involving, for example, peptide ligands such as RALFs and/or direct receptor-receptor interactions.

NORTIA (NTA or AtMLO7), a member of the MILDEW RESISTANCE LOCUS O (MLO) protein family, is another player in *A. thaliana* pollen tube perception. NTA is also expressed in the synergid cells and the corresponding mutant shows a *fer*-like phenotype as described above (Davis et al., 2017; Kessler et al., 2010). It contains seven predicted transmembrane (TM) domains and a cytoplasmic calmodulin-binding domain close to its C-terminus. Whether it is part of the FER signaling complex is not known. However, this is rather unlikely as a direct interaction has not been reported and its localization pattern is different. In contrast to FER, LRE and ENODLs, NTA is localized to a Golgi-associated compartment in the synergid cell and becomes redistributed to the micropylar region of the synergid cells during pollen tube arrival (Jones et al., 2017; Kessler et al., 2010). This process is dependent on *FER* activity, indicating a process downstream of *FER* signaling (Kessler et al., 2010).

The first component of the intracellular *FER* signaling pathway has also been identified. The kinase domain of FER interacts with the GUANINE NUCLEOTIDE EXCHANGE FACTOR1 (GEF1), also referred to as RopGEF1. GEF1 activates membrane localized plant RHO-like GTPases (RAC/ROPs) by exchanging GDP with GTP. This leads to the activation of further downstream responses. Notably, active Rho GTPase level and NADPH oxidase-dependent ROS accumulation were reduced in *fer* mutant roots and root hairs (Duan, Kita, Li, Cheung, & Wu, 2010). In seedlings, it was shown that the LRE sister protein LRE-LIKE GAP1 (LLG1) is also a component of the FER-regulated Rho GTPase signaling complex interacting with ROP2 (Li, Yeh, et al., 2015). Moreover, the authors demonstrated that GTP-saturated ROP2 interacts with the *RbohD*-encoded NADPH oxidase, indicating *FER* signaling mediated ROS production. In roots, it was further shown that interaction of RALF1 and FER activates phosphorylation of AUTOINHIBITED PLASMA MEMBRANE H^+-ATPase2 (AHA2) and CALCIUM-DEPENDENT PROTEIN KINASE9 (CPK9) (Haruta et al., 2014). Whether these downstream signaling processes described in other plant tissues also exist in synergid cells remains to be demonstrated. However, the above described observation that kinase domains of *Cr*RLK1Ls are interchangeable for synergid cell functions suggests functional similarity and a similar downstream signaling machinery.

Cytoplasmic Ca^{2+} levels $\left(Ca^{2+}{}_{cyt}\right)$ are strongly altered as a consequence of *FER* signaling. Oscillation of $Ca^{2+}{}_{cyt}$ in synergid cells of *A. thaliana* was first described in 2012 occurring at their micropylar pole and spreading toward the chalazal pole. A maximum was detected at pollen tube rupture (Iwano et al., 2012). Later reports showed that oscillations were initiated in synergid cells after physical contact with the pollen tube apex. While the pollen tube grows around one or both synergid cells, $Ca^{2+}{}_{cyt}$ transients continue to occur with a maximum at the pollen tube tip and synergid cell interaction zone over a period of 30–50 min until burst of both interaction partners takes place (Denninger et al., 2014). Notably, $Ca^{2+}{}_{cyt}$ transients of a low amplitude did not lead to burst. This observation further supports the hypothesis that $Ca^{2+}{}_{cyt}$ is a downstream messenger of FER receptor activation. A more detailed study using the $Ca^{2+}{}_{cyt}$ chelator BAPTA-AM indicated that depletion of $Ca^{2+}{}_{cyt}$ abolished synergid cell death (Ngo, Vogler, Lituiev, Nestorova, & Grossniklaus, 2014). Moreover, $Ca^{2+}{}_{cyt}$ transients are also absent or reduced in magnitude in both *fer* and *lre* mutants, suggesting that the effected receptor proteins are both requited for initiating and modulating $Ca^{2+}{}_{cyt}$ dynamics. These dynamics are not changed in *nta* mutant ovules, supporting the hypothesis that *NTA* acts downstream of the signaling module and may even represent a direct target of *FER*-mediated calcium regulation via its calmodulin-binding domain. In conclusion, various signal transmitters containing calcium-binding domains such as CPKs and *RbohD*-encoded NADPH oxidase are activated by FER receptor complex signaling, which ultimately induces death of the receptive synergid cell.

3.2 Death of pollen tube and receptive synergid cell

But which signaling mechanisms ultimately lead to death of both, pollen tube and receptive synergid cell, to release the sperm cells for fertilization? In *A. thaliana*, a recent report showed that in parallel to the *FER* signaling pathway, vacuolar acidification in synergid cells is impaired if expression levels of *AP1G*, the γ-subunit of the tetrameric ADAPTOR PROTEIN1 (AP1), are reduced. This leads to a delay in synergid cell degeneration upon pollen tube arrival. AP1 regulates protein sorting at the *trans*-Golgi network/early endosome. Functional loss of V-ATPases in synergid cells phenocopies the *ap1g* mutant with a defect in pollen tube reception, indicating that acidification of the vacuole represents an important mechanism for its degeneration (Wang et al., 2017).

In maize, it was shown that DEFL subclass genes of CRPs ES1–4 mediate pollen tube burst via opening the K^+ channel *Zea mays* 1 (KZM1) potassium channel (Amien et al., 2010) (Fig. 3E). ES1–4 are localized in vesicles at the secretory zone of the synergid cells and are secreted upon pollen tube arrival. Predicted mature peptides induce pollen tube plasma membrane depolarization and pollen tube burst *in vitro* at the very tip in a species-preferential manner. Pollen tube burst is abolished in RNAi knock-down lines. It was further shown that ES4 is capable to open the potassium channel KZM1, which is likely the cause of membrane depolarization and ultimately osmotic burst. It remains to be determined whether ES1–4 peptides are simultaneously capable to burst synergid cells once released to the extracellular space thereby committing cell suicide. It is also unclear whether functional orthologs of these toxin-like peptides exist in other plant species. ES1–4 peptides are highly polymorphic and the domain between cysteine 5 and 6, which was most effective to induce pollen tube burst (Woriedh et al., 2015), does not exist in other plant species. However, species such as *A. thaliana* express large numbers of DEFL and other CRP subclass genes in the embryo sac (Huang, Dresselhaus, et al., 2015). It is thus not unlikely that, similar to ES1–4, other CRPs possess toxin-like functions to induce cell death after secretion. pH-dependence, further modifications, and absence of targets in vesicles may explain why toxin-like activity only occurs after secretion. A novel concept that contributes to synergid cell death was recently postulated in *A. thaliana*. As described above, autocrine signaling via the pollen tube expressed CRPs RALF4/19 that interact with the pollen tube receptors ANX1/2 and BUPS1/2, respectively, is necessary for pollen tube growth and cell wall integrity (Ge et al., 2017; Mecchia et al., 2017). Interestingly, ovule expressed RALF34 is capable to induce pollen tube burst *in vitro* as it also interacts and likely competes with RALF4/19 for the ANX-BUPS receptor complex (Fig. 3B). It was thus suggested that RALF34 could interfere the *RALF4/19* signaling pathway resulting in pollen tube burst and sperm cell release (Ge et al., 2017). It will now be interesting to investigate whether similar mechanisms also occur in other species.

In *A. thaliana*, it was further shown that three closely related transcription factors, *MYB97*, *MYB101* and *MYB120*, which are expressed in the pollen tube nucleus, are required for pollen tube reception. *myb* triple mutant pollen tubes fail to stop growing in synergid cells and to release their sperm cell cargo (Leydon et al., 2013; Liang et al., 2013). A comparison between wild-type and *myb* triple mutant pollen tubes identified a number of differentially expressed candidate genes including those encoding proteins involved in sugar transport and metabolism as well as CRP peptides known to be involved in defense responses (a thionine and a PR-1a CRP) and self-incompatibility (a S1-family member). Whether these candidates are indeed involved in male-female signaling and/or pollen tube burst remains to be determined. In pollen tubes, small proteins with few conserved cysteine residues encode pectin methylesterase inhibitors (PMEIs) play an essential role at the pollen tube tip to prevent pectin methylesterase activity, which itself plays a critical role for cell wall stability and integrity. In maize, it was shown that recombinant PMEI1, which is strongly expressed in the egg apparatus as well as in the pollen tube, did not arrest pollen tube growth, but is capable to induce tube burst at the subapical region in a concentration-dependent manner. A model was proposed in which vesicles containing CRPs including ES1–4 and PMEIs are released from synergid cells upon pollen tube arrival to act in concert on pollen tube burst (Woriedh et al., 2013). To which extent vesicles secreted from the pollen tube tip lead to synergid cell suicide and whether vesicle secretion of both interaction partners depends on receptor complexes such as the FER complex described above and/or downstream Ca^{2+} signaling pathway remain exciting, but challenging experiments for the future. Secretory markers were recently developed/improved for *A. thaliana* synergid cells (Jones et al., 2018), which might help to monitor the detailed time-course of synergid cell activation, vesicle secretion, and cell disintegration.

Recently, it was also shown that an ETHYLENE INSENSITIVE2 (EIN2)/EIN3-dependent ethylene-response pathway is involved in synergid cell death (Völz, Heydlauff, Ripper, von Lyncker, & Gross-Hardt, 2013). Endoplasmic reticulum-located EIN2, which is similar to members of the disease-related Nramp metal transporter family, and EIN3, a transcriptional regulator, have previously been shown to be involved in cross-talk of several methyl-jasmonate-induced senescence and defense processes via activation of several genes including PLANT DEFENSIN1.2 (PDF1.2) (He, Jiang, Wang, & Dehesh, 2017; Zhang, Liu, Chai, & Xing, 2016). Whether they activate defensins or other CRPs in synergid cells or whether they induce programmed cell death (PCD) is unclear. However, ethylene precursors injected into the embryo sac resulted in premature synergid cell death, and likely also triggered fusion of the persisting synergid cell with the large central cell. The latter was indicated by the division of the persisting synergid nucleus in a developing endosperm-specific manner and coincides with the observation that additional pollen tubes were no longer attracted (Völz et al., 2013). A genetic "synergid cell-death module" was recently postulated consisting of two REM transcription factors, VALKYRIE (VAL) and VERDANDI (VDD). Both transcription factors interact physically to control death of the receptive synergid cell. In heterozygous *vdd-1/VDD* mutant and *VAL-RNAi* lines, *GAMETOPHYTIC FACTOR 2* (*GFA2*), a mitochondrial matrix chaperone previously shown to be required for synergid cell death (Christensen et al., 2002), is downregulated (Mendes et al., 2016). Whether the "synergid cell-death module" is also activated upon $Ca^{2+}{}_{cyt}$ signaling and/or an ethylene response pathway is not known and will be a task for future studies. In summary, a number of players have been identified, which are involved in cell death of both interaction partners. To which extent the described mechanisms are conserved in different plant species is unclear. Studies in the two model plants *A. thaliana* and maize indicate that signaling mechanisms involving, for example, RALFs, *Cr*RLK1Ls, ethylene, and $Ca^{2+}{}_{cyt}$ are conserved and expressed in gametophytic cells of both species, while specific mechanisms have evolved involving species-specific toxin-like molecules.

3.3 Death of the persistent synergid cell and prevention of polytubey

After successful discharge, sperm cells are delivered toward a cell wall-free cleft between egg and central cell where they can fuse with the female gametes. The attraction of additional pollen tubes (polytubey) has to be prevented in order to avoid simultaneous release of too many sperm cells and the occurrence of fertilization by multiple sperms (polyspermy), which usually results in sterile progeny. Degeneration of the persistent synergid cell is therefore initiated shortly after sperm cell release and successful gamete fusion to prevent the continuous secretion of pollen tube attraction signals. These mechanisms are tightly controlled as polytubey occurs only at a rate of about 1% in *A. thaliana* (Beale, Leydon, & Johnson, 2012).

3.3.1 Prevention of polytubey

Usually, successful fertilization terminates pollen tube attraction. However, if defective sperm cells are delivered or fertilization failures occur, additional or supernumerary pollen tubes are attracted to a single ovule by the persisting synergid cell, e.g., in *fer* and the allelic *sirene* mutant (Huck et al., 2003; Rotman et al., 2003). This phenomenon was named as polytubey and was also observed in crosses with *duo1, duo3, germ cell specific1/hapless2* (*gcs1/hap2*) defective sperm cells (see also below). Defective sperm cells were not capable to fuse with female gametes, resulting in a significant delay of persistent synergid cell degeneration (Beale et al., 2012; Kasahara et al., 2012). Within a few hours delay, additional pollen tubes arrive and burst in the second synergid cell to release their sperm cell cargo. Thus, fertilization is recovered, maximizing reproductive success and providing an explanation for the occurrence of a second synergid cell in most angiosperms. Notably, this fertilization recovery system requires an excess (saturated number) of pollen tubes as it only occurs in *A. thaliana* pistils pollinated with an excess of pollen grains compared with the number of ovules (Kasahara, Maruyama, & Higashiyama, 2013).

These above reports also showed a delay of polytubey, pointing toward the existence of repellent molecules initially preventing the exit and growth of additional pollen tubes toward an ovule. Such repellent molecules have to be degraded before additional tubes can grow toward and inside of ovules. NO and ROS have previously been discussed as potential repellent molecules (Dresselhaus et al., 2016), but till now a clear role of a molecule as a repellent has not been shown. In *A. thaliana*, AGPs appear to also be involved in the prevention of polytubey. Reduced expression of *AGP4*, which is expressed in the pistil, especially in the transmitting tract and micropylar region of the ovule, leads to an increase of polytubey from about 3% in the wild type to about 16% in *agp4* mutants in one study (Pereira, Nobre, et al., 2016). Note that the rate in wild-type plants is 2% higher compared to another study reported above (Beale et al., 2012). Degeneration of the persisting synergid cell is delayed and was suggested as the cause of increased polytubey. Whether derivatives of AGPs, such as AMOR described above, are generated after successful fertilization and thus represent such repellents or possess different function(s) is unclear, but adds to the candidate list of putative repellents.

It was also debated whether fertilization of both gametes is required to terminate pollen tube attraction. Again, by using defective sperm cells and successive pollination, it could be shown in *A. thaliana* that both female gametes control the occurrence of polytubey. It was also demonstrated that recovery of a half-successful fertilization can result in hetero-fertilization (fusion of female gametes with sperm cells from different pollen tubes) (Maruyama et al., 2013). But to which extent are findings observed in

A. thaliana also conserved in other plants including monocots? In grasses, for example, representing one of the largest and economically most important monocot families, ovaries contain only a single ovule and thus a single female gametophyte. A number of competing pollen tubes are guided via two transmitting tracts toward the egg apparatus (Fig. 1) (Zhou et al., 2017). In maize, it was shown that many pollen tubes arrive almost simultaneously at the micropyle, but only one pollen tube penetrates the nucellus cell files covering the egg apparatus at the micropylar region. Further pollen tubes appear to continue their growth in random directions (Lausser et al., 2010). This observation suggests that either pollen tubes may secrete repellents themselves or nucellus cells are triggered to secrete such molecules. More research efforts are needed in the grasses to elucidate the detailed mechanisms of polytubey prevention.

3.3.2 Elimination of persistent synergid cell function(s)

As described above, the processes of persistent synergid cell degeneration and prevention of polytubey are closely linked in *A. thaliana*. A recent breakthrough report nicely explained how this is achieved (Maruyama et al., 2015). The authors could show that in addition to the double fertilization event, a third cell fusion event takes place between the persisting synergid cell and the fertilized central cell. Fusion of the relatively small synergid cell with a very large cell quickly dilutes the pre-secreted pollen tube attractant and thus is thought to prevent polytubey. Synergid-endosperm fusion was shown to be triggered after successful central cell fertilization, while egg cell fusion was shown to activate ethylene signaling involved in synergid cell disintegration (see above). Thus, both female gametes activate different signaling pathways to eliminate persistent synergid cell function(s) after successful fertilization.

4. Signaling during gamete interaction

Before male and female gametophytes are capable to fuse and merge their cellular contents, gamete recognition, attachment, and activation takes place (Fig. 2C). But which molecular players are involved, and which mechanisms lead to activation of gametes after sperm cell release? The following section will try to provide a couple of answers and summarize our current understanding, which was mainly obtained during the past few years.

4.1 Gamete activation

In animals, Ca^{2+}_{cyt} waves/transients have been reported as key signaling events triggering egg cell activation and initiating mechanism leading to a polyspermy block (Tosti & Menezo, 2016). In *A. thaliana*, a short Ca^{2+}_{cyt} transient occurring in both female gametes, egg and central cell, was shown to be associated with pollen tube burst and sperm cell release (Denninger et al., 2014; Hamamura et al., 2014). A second extended Ca^{2+}_{cyt} transient that occurs solely in the egg cell was correlated with the initiation of plasmogamy (fusion of the cytoplasm of two gametes). Ca^{2+}_{cyt} transients can be interpreted in numerous ways by reproductive cells (Chen, Gutjahr, Bleckmann, & Dresselhaus, 2015). Thus, what happens after sperm cell arrival at the boundary between both female gametes and before fusion occurs? Tracking analysis of the two sperm cells showed that they remain in this position for 5–10 min (average about 8.4 min for the egg cell and 8.5 min for the central cell) before plasmogamy takes place and movement of sperm nuclei toward female nuclei can be observed (Hamamura et al., 2011). Notably, during this period vesicles containing EC1 proteins are released from the egg cell (Sprunck et al., 2012). Whether vesicle secretion is activated by the first Ca^{2+}_{cyt} transient is unclear (Fig. 3F), but EC1 appears in the degenerated synergid cell quickly after sperm delivery. The five *EC1* genes in *A. thaliana* encode CRPs required to activate sperm cells. This is indicated, for example, by an EC1-triggered change in the localization of the fusogen GCS1/HAP2 (see below) from the endomembrane system to the sperm cell surface (Sprunck et al., 2012). It is unlikely that the central cell secretes similar proteins after sperm cell delivery, as EC1-related proteins named ECRs are especially abundant in the synergid cells but not in the central cell (Sprunck, Hackenberg, Englhart, & Vogler, 2014). The central cell also generates many CRPs, but these are especially from the DEFL subgroup of CRPs (Schmid et al., 2012) with unknown functions, and it is not known whether they are secreted during fertilization.

In conclusion, similar to animal oocytes and sperm, which are initially in a quiescent state before motility and chemotaxis sensitivity are activated in the sperm and the second meiotic division is completed in the oocyte, respectively (Tosti & Menezo, 2016), plant gametes also require activation before fusion can be achieved.

4.2 Gamete recognition, adhesion, and fusion

After discharge, the initial positioning of the sperm cells in the embryo sac of *A. thaliana* (Fig. 2C) appears inopportune for gamete fusion and was suggested to require adjustment after initial recognition of the interaction partners. It was further suggested that release and re-adhesion of one sperm cell are necessary, while the other one remains in a waiting position until proper adhesion of each sperm cell to a female gamete is completed (Huang, Ju, Wang, Zhang, & Sodmergen, 2015). This hypothesis has to be validated by experimental data. Until now, the first and so far only molecule identified in plants required for gamete recognition and attachment is GAMETE EXPRESSED2 (GEX2) (Fig. 3F), a conserved plant-specific membrane protein containing extracellular immunoglobulin-like domains. In *A. thaliana*, it is located specifically at the sperm surface and was shown to be necessary for female gamete attachment (Mori et al., 2014). GEX2 is a rapidly evolving protein and it will now be interesting to find out whether the *gex2* mutant of *A. thaliana* can be complemented by corresponding genes from other plant species.

The first gene described in plants required for gamete fusion is *GCS1/HAP2*, which was identified independently by two research teams in pollen of lily and *A. thaliana*, respectively (Mori et al., 2006; von Besser et al., 2006). GCS1/HAP2 contains a carboxy-terminal transmembrane domain and was shown to be localized in the plasma membrane of sperm cells in both plant species. More detailed studies revealed that GCS1/HAP2 is initially mainly located in the endomembrane system of sperm cells and becomes exposed to the cell surface only after sperm cell activation (Sprunck et al., 2012; see also above). It was further reported that, in *A. thaliana*, the *gcs1/hap2* mutant sperm cells fail to fuse with female gametes, resulting in male sterility (Mori et al., 2006; von Besser et al., 2006).

GCS1/HAP2 is a highly conserved protein and homologs were identified in all major eukaryotic taxa except fungi. Gametes of protozoa and algae containing mutations in homologous proteins also fail to fuse with gametes of the opposite sex and the protein is therefore widely considered as an ancient fusogen (Wong & Johnson, 2010). By using GCS1/HAP2 variants containing modifications to N- and/or C-terminal sequences, as well as the single transmembrane domain, it was further reported that the highly conserved N-terminal sequence is essential for fusion. It was further proposed that this interaction likely occurs via egg/central cell-expressed proteins and that the positively charged C-terminus likely functions via electrostatic interactions with the sperm cell plasma membrane. However, this region is less critical for GCS1/HAP2 function (Mori, Hirai, Kuroiwa, & Miyagishima, 2010; Wong, Leydon, & Johnson, 2010). More detailed bioinformatic and structural studies using HAP2 of the unicellular alga *Chlamydomonas reinhardtii* showed that GCS1/HAP2 are class II viral membrane fusion proteins. The fusion mechanism was also elucidated and it could be shown that GCS1/HAP2 inserts into the target gamete membrane via a hydrophobic fusion loop of domain DII, composed of N-terminal and central protein regions (Fedry et al., 2017) (Fig. 3F). Notably, this finding also indicates that gamete membrane fusion mechanisms are highly conserved as they may have evolved (at least partly) from ancient viral cell entry systems.

In conclusion, only two gamete adhesion and fusion proteins have been identified to date. Approaches to isolate the membrane proteome of male and female gametes are currently ongoing and will soon help to increase the list of candidate players involved. The corresponding functional studies will show whether this component of the fertilization process is more conserved compared to other processes, as it might not be necessary to distinguish between "friend and foe" once sperm cells have been released.

4.3 Prevention of polyspermy

Plants not only evolved signaling mechanism to prevent polytubey, they also possess strategies to avoid polyspermy (supernumerary gamete fusion or fusion of a female gamete with multiple sperm cells). After successful fertilization of the egg cell, a block to polyspermy has to be established to avoid sterility effects associated with lethal genome imbalance and chromosome segregation defects during zygote and embryo development. Embryos generated from multiply fertilized egg cells are indeed extremely rare in nature and polyspermy was calculated to occur only at a frequency of significantly below 0.1% in the model plants maize and *A. thaliana*, respectively (Grossniklaus, 2017; Nakel et al., 2017), indicating a very precise signaling-recognition-fusion machinery. In animals, a block to polyspermy is triggered by $Ca^{2+}{}_{cyt}$ transients inducing a cortical secretion reaction that generates a fertilization membrane/envelope preventing the entry of additional sperms. In maize, cell wall material has been observed at the egg cell surface within seconds after *in vitro* gamete fusion, which has been discussed as a quick block to polyspermy in flowering plants (Dresselhaus & Johnson, 2018; Tekleyohans, Mao, Kagi, Stierhof, & Gross-Hardt, 2017). It is unclear whether $Ca^{2+}{}_{cyt}$ transients, which are associated with the beginning of plasmogamy, are also involved in plants in the regulation of vesicle secretion containing cell wall material. Notably, a $Ca^{2+}{}_{cyt}$ transient consisting of a few tight oscillations occurs during egg-sperm cell plasmogamy, but not during sperm-central cell fusion in *A. thaliana* (Denninger et al., 2014;

Hamamura et al., 2014). This observation correlates with the occurrence of a polyspermy block on the egg cell that does not exist on the central cell: by using pollen grains of the polyspermic *tetraspore* (*tes*) mutant of *A. thaliana*, which contains up to four pairs of sperm cells, it was shown that developing seeds contained endosperm derived from multiple fertilizations, while the embryo was exclusively generated after a single fertilization event (Scott, Armstrong, Doughty, & Spielman, 2008). This finding is remarkable as the two sperm cells in *A. thaliana* are isomorphic and functionally equivalent for fertilizing the egg cell (Hamamura et al., 2011; Ingouff et al., 2009). In summary, these findings support the idea that in addition to pollen tube growth and perception, $Ca^{2+}{}_{cyt}$ also may play a key role as a second messenger to establish a block to polyspermy on the egg cell. It is obvious that $Ca^{2+}{}_{cyt}$ regulates the release of vesicles, but it is completely unclear which signaling mechanisms induce its influx into the egg cell and how $Ca^{2+}{}_{cyt}$ transients are transduced into cellular responses, including regulated vesicle secretion. Moreover, it also remains unclear whether the generation of cell wall material is the only reaction or one of a couple of mechanisms to prevent polyspermy in plants. The comparison between egg and central cell membrane proteins might provide candidate cell surface receptors for future studies that will help to understand the underlying signaling and recognition events. Tools are now available to visualize polytubey, gamete fusion, and polyspermy at least in the model plant *A. thaliana*, which also allows the distinction between gamete adhesion, and between single and supernumerary gamete fusion.

5. Concluding remarks

Here, we summarized our current knowledge about signaling events occurring during pollen tube growth and guidance, its reception, and during gamete activation and fusion. Signaling mechanisms during pollen adhesion, hydration and growth toward the transmitting tract have not been considered as they are comprehensively described in chapter "Self-incompatibility in the Brassicaceae: Regulation and mechanism of self-recognition" by Nasrallah (this issue) about self-incompatibility. We also renounced to discuss the activation of seed development, which has been outlined in chapter "Genetic, molecular and parent-of-origin regulation of early embryogenesis in flowering plants" by Armenta-Medina and Gillmor (this issue) about the molecular control of embryogenesis.

Many of the signaling mechanisms discussed above involve general second messengers such as ROS and Ca^{2+}_{cyt}, as well as associated transporters, biosynthesis enzymes and signal transducers, which play a pivotal role both, during fertilization and defense-related processes (Chen et al., 2015; Steinhorst & Kudla, 2013b). Moreover, members of various CRP families play important roles as extracellular ligands both, in most reproductive communication processes (except gamete fusion) and as major components of the plant innate immune system. It is therefore discussed already since a number of years whether reproductive signaling mechanisms were adapted during evolution of land plants from ancient defense signaling mechanisms (Dresselhaus & Márton, 2009; Kessler et al., 2010). CRPs represent a pool of highly polymorphic signaling peptides with a rapid evolution rate and thus likely also represent key players in reproductive isolation and speciation mechanisms (Bircheneder & Dresselhaus, 2016). Some "reproductive CRPs" such as the DEFLs are required for pollen tube burst but still possess antifungal activity (Amien et al., 2010; Woriedh et al., 2015), while RALFs are required for pollen tube cell wall integrity, pollen tube burst, and infection (Ge et al., 2017; Masachis et al., 2016; Mecchia et al., 2017). Similarly, the receptors of CRPs including FER and ANX1/2, as well as further membrane proteins such as NTA, possess dual roles both during reproductive signaling as outlined above and during plant immunity (Kessler et al., 2010; Mang et al., 2017; Stegmann et al., 2017). During infection of stigmas, styles and ovules, it seems that invading fungi block and mimic fertilization processes in rice (Song et al., 2016), indicating that pathogens successfully use the reproductive signaling machinery. Similar results have been obtained after comparing transcriptomes of pollinated and infected pistils in different *Arabidopsis* spp. (Mondragon-Palomino et al., 2017), emphasizing the importance to distinguish between "friend or foe" during reproductive signaling.

Unfortunately, most of our knowledge about signaling mechanisms during double fertilization in flowering plants is derived from investigations of a single model plant, *A. thaliana*. The few studies in maize discussed above have shown already that both the morphology of reproductive organs and many of the involved molecular players are different. It is thus urgently needed to extend the number of plant species being used to study reproductive signaling. The *A. thaliana* mutants described above will help to study the conservation of mechanisms and signaling components by using homologous genes from other species. Novel high-throughput approaches such as the lab-on-chip techniques described above, the purification of mRNAs undergoing translation (the translatome), and the generation of membrane protein maps have a strong potential to identify novel players (Lin et al., 2014; Yang & Wang, 2017) in the near future. The *A. thaliana* research community has developed an excellent toolkit to investigate and understand fertilization mechanisms in plants. Although it will be difficult to adapt all tools to more "difficult" species including, cereals and other important crops, for example from the economically important Fabaceae and Solanaceae families, research will remain exciting as it can be expected that more species/family-specific signaling mechanisms will be uncovered.

Acknowledgments

Work in the Dresselhaus lab on signaling during fertilization with a focus on CRPs is funded by the German Research Council DFG via Collaborative Research Center SFB924, the University of Regensburg, and the Alexander von Humboldt-Foundation to L.Z.

References

Agudelo, C., Packirisamy, M., & Geitmann, A. (2016). Influence of electric fields and conductivity on pollen tube growth assessed via electrical lab-on-chip. *Scientific Reports*, *6*, 19812.

Amien, S., Kliwer, I., Márton, M. L., Debener, T., Geiger, D., Becker, D., et al. (2010). Defensin-like ZmES4 mediates pollen tube burst in maize via opening of the potassium channel KZM1. *PLoS Biology*, *8*, e1000388.

Beale, K. M., Leydon, A. R., & Johnson, M. A. (2012). Gamete fusion is required to block multiple pollen tubes from entering an Arabidopsis ovule. *Current Biology*, *22*, 1090–1094.

Begcy, K., & Dresselhaus, T. (2017). Tracking maize pollen development by the Leaf Collar Method. *Plant Reproduction*, *30*, 171–178.

Bircheneder, S., & Dresselhaus, T. (2016). Why cellular communication during plant reproduction is particularly mediated by CRP signalling. *Journal of Experimental Botany*, *67*, 4849–4861.

Bleckmann, A., Alter, S., & Dresselhaus, T. (2014). The beginning of a seed: Regulatory mechanisms of double fertilization. *Frontiers in Plant Science*, *5*, 452.

Boisson-Dernier, A., Franck, C. M., Lituiev, D. S., & Grossniklaus, U. (2015). Receptor-like cytoplasmic kinase MARIS functions downstream of *Cr*RLK1L-dependent signaling during tip growth. *Proceeding of the National Academy of Sciences of the United States of America*, *112*, 12211–12216.

Boisson-Dernier, A., Kessler, S. A., & Grossniklaus, U. (2011). The walls have ears: The role of plant CrRLK1Ls in sensing and transducing extracellular signals. *Journal of Experimental Botany*, *62*, 1581–1591.

Boisson-Dernier, A., Lituiev, D. S., Nestorova, A., Franck, C. M., Thirugnanarajah, S., & Grossniklaus, U. (2013). ANXUR receptor-like kinases coordinate cell wall integrity with growth at the pollen tube tip via NADPH oxidases. *PLoS Biology*, *11*, e1001719.

Boisson-Dernier, A., Roy, S., Kritsas, K., Grobei, M. A., Jaciubek, M., Schroeder, J. I., et al. (2009). Disruption of the pollen-expressed FERONIA homologs ANXUR1 and ANXUR2 triggers pollen tube discharge. *Development*, *13*, 3279–3288.

Capron, A., Gourgues, M., Neiva, L. S., Faure, J. E., Berger, F., Pagnussat, G., et al. (2008). Maternal control of male-gamete delivery in Arabidopsis involves a putative GPI-anchored protein encoded by the *LORELEI* gene. *Plant Cell*, *20*, 3038–3049.

Chen, J., Gutjahr, C., Bleckmann, A., & Dresselhaus, T. (2015). Calcium signaling during reproduction and biotrophic fungal interactions in plants. *Molecular Plant*, *8*, 595–611.

Chen, Y. H., Li, H. J., Shi, D. Q., Yuan, L., Liu, J., Sreenivasan, R., et al. (2007). The central cell plays a critical role in pollen tube guidance in Arabidopsis. *Plant Cell*, *19*, 3563–3577.

Cheung, A. Y., Wang, H., & Wu, H. M. (1995). A floral transmitting tissue-specific glycoprotein attracts pollen tubes and stimulates their growth. *Cell*, *82*, 383–393.

Christensen, C. A., Gorsich, S. W., Brown, R. H., Jones, L. G., Brown, J., Shaw, J. M., et al. (2002). Mitochondrial GFA2 is required for synergid cell death in Arabidopsis. *Plant Cell*, *14*, 2215–2232.

Cordts, S., Bantin, J., Wittich, P. E., Kranz, E., Lörz, H., & Dresselhaus, T. (2001). *ZmES* genes encode peptides with structural homology to defensins and are specifically expressed in the female gametophyte of maize. *Plant Journal*, *25*, 103–114.

Davis, T. C., Jones, D. S., Dino, A. J., Cejda, N. I., Yuan, J., Willoughby, A. C., et al. (2017). *Arabidopsis thaliana MLO* genes are expressed in discrete domains during reproductive development. *Plant Reproduction*, *30*, 185–195.

Denninger, P., Bleckmann, A., Lausser, A., Vogler, F., Ott, T., Ehrhardt, D. W., et al. (2014). Male-female communication triggers calcium signatures during fertilization in Arabidopsis. *Nature Communications*, *5*, 4645.

Dresselhaus, T., & Franklin-Tong, N. (2013). Male-female crosstalk during pollen germination, tube growth and guidance, and double fertilization. *Molecular Plant*, *6*, 1018–1036.

Dresselhaus, T., & Johnson, M. A. (2018). Reproduction: Plant parentage a trois. *Current Biology*, *28*, R28–R30.

Dresselhaus, T., Lausser, A., & Márton, M. L. (2011). Using maize as a model to study pollen tube growth and guidance, cross-incompatibility and sperm delivery in grasses. *Annals of Botany*, *108*, 727–737.

Dresselhaus, T., & Márton, M. L. (2009). Micropylar pollen tube guidance and burst: Adapted from defense mechanisms? *Current Opinion in Plant Biology*, *12*, 773–780.

Dresselhaus, T., Sprunck, S., & Wessel, G. M. (2016). Fertilization mechanisms in flowering plants. *Current Biology*, *26*, R125–R139.

Du, S., Qu, L. J., & Xiao, J. (2018). Crystal structures of the extracellular domains of the *Cr*RLK1L receptor-like kinases ANXUR1 and ANXUR2. *Protein Science*, *27*, 886–892.

Duan, Q., Kita, D., Li, C., Cheung, A. Y., & Wu, H. M. (2010). FERONIA receptor-like kinase regulates RHO GTPase signaling of root hair development. *Proceeding of the National Academy of Sciences of the United States of America*, *107*, 17821–17826.

Escobar-Restrepo, J. M., Huck, N., Kessler, S., Gagliardini, V., Gheyselinck, J., Yang, W. C., et al. (2007). The FERONIA receptor-like kinase mediates male-female interactions during pollen tube reception. *Science*, *317*, 656–660.

Fedry, J., Liu, Y., Pehau-Arnaudet, G., Pei, J., Li, W., Tortorici, M. A., et al. (2017). The ancient gamete fusogen HAP2 is a eukaryotic class II fusion protein. *Cell*, *168*, 904–915.

Feng, W., Kita, D., Peaucelle, A., Cartwright, H. N., Doan, V., Duan, Q., et al. (2018). The FERONIA receptor kinase maintains cell-wall integrity during salt stress through Ca($^{2+}$) signaling. *Current Biology*, *28*, 666–675.

Gao, Q. F., Gu, L. L., Wang, H. Q., Fei, C. F., Fang, X., Hussain, J., et al. (2016). Cyclic nucleotide-gated channel 18 is an essential Ca^{2+} channel in pollen tube tips for pollen tube guidance to ovules in Arabidopsis. *Proceeding of the National Academy of Sciences of the United States of America*, *113*, 3096–3101.

Ge, Z., Bergonci, T., Zhao, Y., Zou, Y., Du, S., Liu, M. C., et al. (2017). Arabidopsis pollen tube integrity and sperm release are regulated by RALF-mediated signaling. *Science*, *358*, 1596–1600.

Glöckle, B., Urban, W. J., Nagahara, S., Andersen, E. D., Higashiyama, T., Grini, P. E., et al. (2018). Pollen differentiation as well as pollen tube guidance and discharge are independent of the presence of gametes. *Development*, *145*. https://doi.org/10.1242/dev.152645.

Grossniklaus, U. (2017). Polyspermy produces tri-parental seeds in maize. *Current Biology*, *27*, R1300–R1302.

Gui, C. P., Dong, X., Liu, H. K., Huang, W. J., Zhang, D., Wang, S. J., et al. (2014). Overexpression of the tomato pollen receptor kinase LePRK1 rewires pollen tube growth to a blebbing mode. *Plant Cell*, *26*, 3538–3555.

Guo, F. Q., Okamoto, M., & Crawford, N. M. (2003). Identification of a plant nitric oxide synthase gene involved in hormonal signaling. *Science*, *302*, 100–103.

Hafidh, S., Potesil, D., Fila, J., Capkova, V., Zdrahal, Z., & Honys, D. (2016). Quantitative proteomics of the tobacco pollen tube secretome identifies novel pollen tube guidance proteins important for fertilization. *Genome Biology*, *17*, 81.

Hamamura, Y., Nagahara, S., & Higashiyama, T. (2012). Double fertilization on the move. *Current Opinion in Plant Biology*, *15*, 70–77.

Hamamura, Y., Nishimaki, M., Takeuchi, H., Geitmann, A., Kurihara, D., & Higashiyama, T. (2014). Live imaging of calcium spikes during double fertilization in Arabidopsis. *Nature Communications*, *5*, 4722.

Hamamura, Y., Saito, C., Awai, C., Kurihara, D., Miyawaki, A., Nakagawa, T., et al. (2011). Live-cell imaging reveals the dynamics of two sperm cells during double fertilization in *Arabidopsis thaliana*. *Current Biology*, *21*, 497–502.

Haruta, M., Sabat, G., Stecker, K., Minkoff, B. B., & Sussman, M. R. (2014). A peptide hormone and its receptor protein kinase regulate plant cell expansion. *Science*, *343*, 408–411.

He, X., Jiang, J., Wang, C. Q., & Dehesh, K. (2017). ORA59 and EIN3 interaction couples jasmonate-ethylene synergistic action to antagonistic salicylic acid regulation of PDF expression. *Journal of Integrative Plant Biology*, *59*, 275–287.

Heilmann, I., & Ischebeck, T. (2016). Male functions and malfunctions: The impact of phosphoinositides on pollen development and pollen tube growth. *Plant Reproduction*, *29*, 3–20.

Hepler, P. K., Rounds, C. M., & Winship, L. J. (2013). Control of cell wall extensibility during pollen tube growth. *Molecular Plant*, *6*, 998–1017.

Higashiyama, T. (2018). Plant reproduction: Autocrine machinery for the long journey of the pollen tube. *Current Biology*, *28*, R266–R269.

Higashiyama, T., Kuroiwa, H., Kawano, S., & Kuroiwa, T. (1998). Guidance *in vitro* of the pollen tube to the naked embryo sac of *Torenia fournieri*. *Plant Cell*, *10*, 2019–2031.

Higashiyama, T., & Takeuchi, H. (2015). The mechanism and key molecules involved in pollen tube guidance. *Annual Review of Plant Biology*, *66*, 393–413.

Higashiyama, T., Yabe, S., Sasaki, N., Nishimura, Y., Miyagishima, S., Kuroiwa, H., et al. (2001). Pollen tube attraction by the synergid cells. *Science*, *293*, 1480–1483.

Higashiyama, T., & Yang, W. C. (2017). Gametophytic pollen tube guidance: Attractant peptides, gametic controls, and receptors. *Plant Physiology*, *173*, 112–121.

Hou, Y., Guo, X., Cyprys, P., Zhang, Y., Bleckmann, A., Cai, L., et al. (2016). Maternal ENODLs are required for pollen tube reception in Arabidopsis. *Current Biology*, *26*, 2343–2350.

Hu, C., Vogler, H., Aellen, M., Shamsudhin, N., Jang, B., Burri, J. T., et al. (2017). High precision, localized proton gradients and fluxes generated by a microelectrode device induce differential growth behaviors of pollen tubes. *Lab on a Chip*, *17*, 671–680.

Huang, J., Ju, Y., Wang, X., Zhang, Q., & Sodmergen. (2015). A one-step rectification of sperm cell targeting ensures the success of double fertilization. *Journal of Integrative Plant Biology*, *57*, 496–503.

Huang, Q., Dresselhaus, T., Gu, H., & Qu, L. J. (2015). Active role of small peptides in Arabidopsis reproduction: Expression evidence. *Journal of Integrative Plant Biology*, *57*, 518–521.

Huang, W. J., Liu, H. K., McCormick, S., & Tang, W. H. (2014). Tomato pistil factor STIG1 promotes in vivo pollen tube growth by binding to phosphatidylinositol 3-phosphate and the extracellular domain of the pollen receptor kinase LePRK2. *Plant Cell*, *26*, 2505–2523.

Huck, N., Moore, M. J., Federer, M., & Grossniklaus, U. (2003). The Arabidopsis mutant *feronia* disrupts the female gametophytic control of pollen tube reception. *Development*, *130*, 2149–2159.

Ingouff, M., Sakata, T., Li, J., Sprunck, S., Dresselhaus, T., & Berger, F. (2009). The two male gametes share equal ability to fertilize the egg cell in *Arabidopsis thaliana*. *Current Biology*, *19*, R19–R20.

Iwano, M., Ngo, Q. A., Entani, T., Shiba, H., Nagai, T., Miyawaki, A., et al. (2012). Cytoplasmic Ca^{2+} changes dynamically during the interaction of the pollen tube with synergid cells. *Development*, *139*, 4202–4209.

Jiao, J., Mizukami, A. G., Sankaranarayanan, S., Yamguchi, J., Itami, K., & Higashiyawma, T. (2017). Structure-activity relation of AMOR sugar molecule that activates pollen-tubes for ovular guidance. *Plant Physiology*, *173*, 354–363.

Jones, D. S., Liu, X., Willoughby, A. C., Smith, B. E., Palanivelu, R., & Kessler, S. A. (2018). Cellular distribution of secretory pathway markers in the haploid synergid cells of *Arabidopsis thaliana*. *Plant Journal*, *94*, 192–202.

Jones, D. S., Yuan, J., Smith, B. E., Willoughby, A. C., Kumimoto, E. L., & Kessler, S. A. (2017). MILDEW RESISTANCE LOCUS O function in pollen tube reception is linked to its oligomerization and subcellular distribution. *Plant Physiology*, *175*, 172–185.

Kanaoka, M. M. (2018). Cell-cell communications and molecular mechanisms in plant sexual reproduction. *Journal of Plant Research*, *131*, 37–47.

Kanaoka, M. M., Kawano, N., Matsubara, Y., Susaki, D., Okuda, S., Sasaki, N., et al. (2011). Identification and characterization of TcCRP1, a pollen tube attractant from *Torenia concolor*. *Annals of Botany*, *108*, 739–747.

Kasahara, R. D., Maruyama, D., Hamamura, Y., Sakakibara, T., Twell, D., & Higashiyama, T. (2012). Fertilization recovery after defective sperm cell release in Arabidopsis. *Current Biology*, *22*, 1084–1089.

Kasahara, R. D., Maruyama, D., & Higashiyama, T. (2013). Fertilization recovery system is dependent on the number of pollen grains for efficient reproduction in plants. *Plant Signaling & Behavior*, *8*, e23690.

Kasahara, R. D., Portereiko, M. F., Sandaklie-Nikolova, L., Rabiger, D. S., & Drews, G. N. (2005). MYB98 is required for pollen tube guidance and synergid cell differentiation in Arabidopsis. *Plant Cell*, *17*, 2981–2992.

Kaya, H., Nakajima, R., Iwano, M., Kanaoka, M. M., Kimura, S., Takeda, S., et al. (2014). Ca^{2+}-activated reactive oxygen species production by Arabidopsis RbohH and RbohJ is essential for proper pollen tube tip growth. *Plant Cell*, *26*, 1069–1080.

Kessler, S. A., Lindner, H., Jones, D. S., & Grossniklaus, U. (2015). Functional analysis of related *Cr*RLK1L receptor-like kinases in pollen tube reception. *EMBO Reports*, *16*, 107–115.

Kessler, S. A., Shimosato-Asano, H., Keinath, N. F., Wuest, S. E., Ingram, G., Panstruga, R., et al. (2010). Conserved molecular components for pollen tube reception and fungal invasion. *Science*, *330*, 968–971.

Lausser, A., Kliwer, I., Srilunchang, K. O., & Dresselhaus, T. (2010). Sporophytic control of pollen tube growth and guidance in maize. *Journal of Experimental Botany*, *61*, 673–682.

Lee, J. S., Kuroha, T., Hnilova, M., Khatayevich, D., Kanaoka, M. M., McAbee, J. M., et al. (2012). Direct interaction of ligand-receptor pairs specifying stomatal patterning. *Genes and Development*, *26*, 126–136.

Leydon, A. R., Beale, K. M., Woroniecka, K., Castner, E., Chen, J., Horgan, C., et al. (2013). Three MYB transcription factors control pollen tube differentiation required for sperm release. *Current Biology*, *23*, 1209–1214.

Leydon, A. R., Tsukamoto, T., Dunatunga, D., Qin, Y., Johnson, M. A., & Palanivelu, R. (2015). Pollen tube discharge completes the process of synergid degeneration that is initiated by pollen tube-synergid interaction in Arabidopsis. *Plant Physiology*, *169*, 485–496.

Li, C., Yeh, F. L., Cheung, A. Y., Duan, Q., Kita, D., Liu, M. C., et al. (2015). Glycosylphosphatidylinositol-anchored proteins as chaperones and co-receptors for FERONIA receptor kinase signaling in Arabidopsis. *eLife*, *4*, e06587.

Li, H. J., Meng, J. G., & Yang, W. C. (2018). Multilayered signaling pathways for pollen tube growth and guidance. *Plant Reproduction*, *31*, 31–41.

Li, H. J., Zhu, S. S., Zhang, M. X., Wang, T., Liang, L., Xue, Y., et al. (2015). Arabidopsis CBP1 is a novel regulator of transcription initiation in central cell-mediated pollen tube guidance. *Plant Cell*, *27*, 2880–2893.

Liang, Y., Tan, Z. M., Zhu, L., Niu, Q. K., Zhou, J. J., Li, M., et al. (2013). MYB97, MYB101 and MYB120 function as male factors that control pollen tube-synergid interaction in *Arabidopsis thaliana* fertilization. *PLoS Genetics*, *9*, e1003933.

Liao, H. Z., Zhu, M. M., Cui, H. H., Du, X. Y., Tang, Y., Chen, L. Q., et al. (2016). MARIS plays important roles in Arabidopsis pollen tube and root hair growth. *Journal of Integrative Plant Biology*, *58*, 14.

Lin, S. Y., Chen, P. W., Chuang, M. H., Juntawong, P., Bailey-Serres, J., & Jauh, G. Y. (2014). Profiling of translatomes of *in vivo*-grown pollen tubes reveals genes with roles in micropylar guidance during pollination in Arabidopsis. *Plant Cell*, *26*, 602–618.

Lindner, H., Kessler, S. A., Müller, L. M., Shimosato-Asano, H., Boisson-Dernier, A., & Grossniklaus, U. (2015). TURAN and EVAN mediate pollen tube reception in Arabidopsis synergids through protein glycosylation. *PLoS Biology*, *13*, e1002139.

Liu, J., Zhong, S., Guo, X., Hao, L., Wei, X., Huang, Q., et al. (2013). Membrane-bound RLCKs LIP1 and LIP2 are essential male factors controlling male-female attraction in Arabidopsis. *Current Biology*, *23*, 993–998.

Liu, L., Zheng, C., Kuang, B., Wei, L., Yan, L., & Wang, T. (2016). Receptor-like kinase RUPO interacts with potassium transporters to regulate pollen tube growth and integrity in rice. *PLoS Genetics*, *12*, e1006085.

Liu, X., Castro, C., Wang, Y., Noble, J., Ponvert, N., Bundy, M., et al. (2016). The role of LORELEI in pollen tube reception at the interface of the synergid cell and pollen tube requires the modified eight-cysteine motif and the receptor-like kinase FERONIA. *Plant Cell*, *28*, 1035–1052.

Losada, J. M., Herrero, M., Hormaza, J. I., & Friedman, W. E. (2014). Arabinogalactan proteins mark stigmatic receptivity in the protogynous flowers of *Magnolia virginiana* (Magnoliaceae). *American Journal of Botany*, *101*, 1963–1975.

Mang, H., Feng, B., Hu, Z., Boisson-Dernier, A., Franck, C. M., Meng, X., et al. (2017). Differential regulation of two-tiered plant immunity and sexual reproduction by ANXUR receptor-like kinases. *Plant Cell*, *29*, 3140–3156.

Márton, M. L., Cordts, S., Broadhvest, J., & Dresselhaus, T. (2005). Micropylar pollen tube guidance by egg apparatus 1 of maize. *Science*, *307*, 573–576.

Márton, M. L., & Dresselhaus, T. (2008). A comparison of early molecular fertilization mechanisms in animals and flowering plants. *Sexual Plant Reproduction*, *21*, 37–52.

Márton, M. L., Fastner, A., Uebler, S., & Dresselhaus, T. (2012). Overcoming hybridization barriers by the secretion of the maize pollen tube attractant ZmEA1 from Arabidopsis ovules. *Current Biology*, *22*, 1194–1198.

Maruyama, D., Hamamura, Y., Takeuchi, H., Susaki, D., Nishimaki, M., Kurihara, D., et al. (2013). Independent control by each female gamete prevents the attraction of multiple pollen tubes. *Developmental Cell, 25*(3), 317–323.

Maruyama, D., Völz, R., Takeuchi, H., Mori, T., Igawa, T., Kurihara, D., et al. (2015). Rapid elimination of the persistent synergid through a cell fusion mechanism. *Cell, 161*, 907–918.

Masachis, S., Segorbe, D., Turra, D., Leon-Ruiz, M., Furst, U., El Ghalid, M., et al. (2016). A fungal pathogen secretes plant alkalinizing peptides to increase infection. *Nature Microbiology, 1*, 16043.

Mecchia, M. A., Santos-Fernandez, G., Duss, N. N., Somoza, S. C., Boisson-Dernier, A., Gagliardini, V., et al. (2017). RALF4/19 peptides interact with LRX proteins to control pollen tube growth in Arabidopsis. *Science, 358*, 1600–1603.

Mendes, M. A., Guerra, R. F., Castelnovo, B., Silva-Velazquez, Y., Morandini, P., Manrique, S., et al. (2016). Live and let die: A REM complex promotes fertilization through synergid cell death in Arabidopsis. *Development, 143*, 2780–2790.

Michard, E., Lima, P. T., Borges, F., Silva, A. C., Portes, M. T., Carvalho, J. E., et al. (2011). Glutamate receptor-like genes form Ca^{2+} channels in pollen tubes and are regulated by pistil D-serine. *Science, 332*, 434–437.

Miyazaki, S., Murata, T., Sakurai-Ozato, N., Kubo, M., Demura, T., Fukuda, H., et al. (2009). ANXUR1 and 2, sister genes to FERONIA/SIRENE, are male factors for coordinated fertilization. *Current Biology, 19*, 1327–1331.

Mizukami, A. G., Inatsugi, R., Jiao, J., Kotake, T., Kuwata, K., Ootani, K., et al. (2016). The AMOR arabinogalactan sugar chain induces pollen-tube competency to respond to ovular guidance. *Current Biology, 26*, 1091–1097.

Mizuta, Y., & Higashiyama, T. (2018). Chemical signaling for pollen tube guidance at a glance. *Journal of Cell Science, 131*. https://doi.org/10.1242/jcs.208447.

Mondragon-Palomino, M., John-Arputharaj, A., Pallmann, M., & Dresselhaus, T. (2017). Similarities between reproductive and immune pistil transcriptomes of Arabidopsis species. *Plant Physiology, 174*, 1559–1575.

Mori, T., Hirai, M., Kuroiwa, T., & Miyagishima, S. Y. (2010). The functional domain of GCS1-based gamete fusion resides in the amino terminus in plant and parasite species. *PLoS One, 5*, e15957.

Mori, T., Igawa, T., Tamiya, G., Miyagishima, S. Y., & Berger, F. (2014). Gamete attachment requires GEX2 for successful fertilization in Arabidopsis. *Current Biology, 24*, 170–175.

Mori, T., Kuroiwa, H., Higashiyama, T., & Kuroiwa, T. (2006). GENERATIVE CELL SPECIFIC 1 is essential for angiosperm fertilization. *Nature Cell Biology, 8*, 64–71.

Moussu, S., Augustin, S., Roman, A. O., Broyart, C., & Santiago, J. (2018). Crystal structures of two tandem malectin-like receptor kinases involved in plant reproduction. *Acta Crystallographica Section D Structural Biology, 74*, 671–680.

Müller, L. M., Lindner, H., Pires, N. D., Gagliardini, V., & Grossniklaus, U. (2016). A subunit of the oligosaccharyltransferase complex is required for interspecific gametophyte recognition in Arabidopsis. *Nature Communications*, 7, 10826.

Muschietti, J., Dircks, L., Vancanneyt, G., & McCormick, S. (1994). LAT52 protein is essential for tomato pollen development: Pollen expressing antisense *LAT52* RNA hydrates and germinates abnormally and cannot achieve fertilization. *Plant Journal, 6*, 321–338.

Muschietti, J., Eyal, Y., & McCormick, S. (1998). Pollen tube localization implies a role in pollen–pistil interactions for the tomato receptor-like protein kinases LePRK1 and LePRK2. *Plant Cell, 10*, 319–330.

Muschietti, J. P., & Wengier, D. L. (2018). How many receptor-like kinases are required to operate a pollen tube. *Current Opinion in Plant Biology, 41*, 73–82.

Nakel, T., Tekleyohans, D. G., Mao, Y., Fuchert, G., Vo, D., & Gross-Hardt, R. (2017). Triparental plants provide direct evidence for polyspermy induced polyploidy. *Nature Communications*, *8*, 1033.

Ngo, Q. A., Vogler, H., Lituiev, D. S., Nestorova, A., & Grossniklaus, U. (2014). A calcium dialog mediated by the FERONIA signal transduction pathway controls plant sperm delivery. *Developmental Cell*, *29*, 491–500.

Noyszewski, A. K., Liu, Y. C., Tamura, K., & Smith, A. G. (2017). Polymorphism and structure of style-specific arabinogalactan proteins as determinants of pollen tube growth in *Nicotiana*. *BMC Evolutionary Biology*, *17*, 186.

Okuda, S., Suzuki, T., Kanaoka, M. M., Mori, H., Sasaki, N., & Higashiyama, T. (2013). Acquisition of LURE-binding activity at the pollen tube tip of *Torenia fournieri*. *Molecular Plant*, *6*, 1074–1090.

Okuda, S., Tsutsui, H., Shiina, K., Sprunck, S., Takeuchi, H., Yui, R., et al. (2009). Defensin-like polypeptide LUREs are pollen tube attractants secreted from synergid cells. *Nature*, *458*, 357–361.

Palanivelu, R., Brass, L., Edlund, A. F., & Preuss, D. (2003). Pollen tube growth and guidance is regulated by POP2, an Arabidopsis gene that controls GABA levels. *Cell*, *114*, 47–59.

Pereira, A. M., Lopes, A. L., & Coimbra, S. (2016). Arabinogalactan proteins as interactors along the crosstalk between the pollen tube and the female tissues. *Frontiers in Plant Science*, 7, 1895.

Pereira, A. M., Nobre, M. S., Pinto, S. C., Lopes, A. L., Costa, M. L., Masiero, S., et al. (2016). "Love is strong, and you're so sweet": JAGGER is essential for persistent synergid degeneration and polytubey block in *Arabidopsis thaliana*. *Molecular Plant*, *9*, 601–614.

Prado, A. M., Colaco, R., Moreno, N., Silva, A. C., & Feijo, J. A. (2008). Targeting of pollen tubes to ovules is dependent on nitric oxide (NO) signaling. *Molecular Plant*, *1*, 703–714.

Prado, A. M., Porterfield, D. M., & Feijo, J. A. (2004). Nitric oxide is involved in growth regulation and re-orientation of pollen tubes. *Development*, *131*, 2707–2714.

Punwani, J. A., & Drews, G. N. (2007). Development and function of the synergid cell. *Sexual Plant Reproduction*, *21*, 7–15.

Qin, Y., Leydon, A. R., Manziello, A., Pandey, R., Mount, D., Denic, S., et al. (2009). Penetration of the stigma and style elicits a novel transcriptome in pollen tubes, pointing to genes critical for growth in a pistil. *PLoS Genetics*, *5*, e1000621.

Qu, L. J., Li, L., Lan, Z., & Dresselhaus, T. (2015). Peptide signalling during the pollen tube journey and double fertilization. *Journal of Experimental Botany*, *66*, 5139–5150.

Rotman, N., Rozier, F., Boavida, L., Dumas, C., Berger, F., & Faure, J. (2003). Female control of male gamete delivery during fertilization in *Arabidopsis thaliana*. *Current Biology*, *13*, 432–436.

Sandaklie-Nikolova, L., Palanivelu, R., King, E. J., Copenhaver, G. P., & Drews, G. N. (2007). Synergid cell death in Arabidopsis is triggered following direct interaction with the pollen tube. *Plant Physiology*, *144*, 1753–1762.

Sato, Y., Sugimoto, N., Higashiyama, T., & Arata, H. (2015). Quantification of pollen tube attraction in response to guidance by female gametophyte tissue using artificial microscale pathway. *Journal of Bioscience and Bioengineering*, *120*, 697–700.

Schmid, M. W., Schmidt, A., Klostermeier, U. C., Barann, M., Rosenstiel, P., & Grossniklaus, U. (2012). A powerful method for transcriptional profiling of specific cell types in eukaryotes: Laser-assisted microdissection and RNA sequencing. *PLoS One*, 7, e29685.

Schoenaers, S., Balcerowicz, D., Costa, A., & Vissenberg, K. (2017). The kinase ERULUS controls pollen tube targeting and growth in *Arabidopsis thaliana*. *Frontiers in Plant Science*, *8*, 1942.

Scott, R. J., Armstrong, S. J., Doughty, J., & Spielman, M. (2008). Double fertilization in *Arabidopsis thaliana* involves a polyspermy block on the egg but not the central cell. *Molecular Plant, 1*, 611–619.

Shamsudhin, N., Laeubli, N., Atakan, H. B., Vogler, H., Hu, C., Haeberle, W., et al. (2016). Massively parallelized pollen tube guidance and mechanical measurements on a lab-on-a-chip platform. *PLoS One, 11*, e0168138.

Shimizu, K. K., Ito, T., Ishiguro, S., & Okada, K. (2008). MAA3 (MAGATAMA3) helicase gene is required for female gametophyte development and pollen tube guidance in *Arabidopsis thaliana. Plant and Cell Physiology, 49*, 1478–1483.

Song, J. H., Wei, W., Lv, B., Lin, Y., Yin, W. X., Peng, Y. L., et al. (2016). Rice false smut fungus hijacks the rice nutrients supply by blocking and mimicking the fertilization of rice ovary. *Environmental Microbiology, 18*, 3840–3849.

Sprunck, S., Hackenberg, T., Englhart, M., & Vogler, F. (2014). Same same but different: Sperm-activating EC1 and ECA1 gametogenesis-related family proteins. *Biochemical Society Transactions, 42*, 401–407.

Sprunck, S., Rademacher, S., Vogler, F., Gheyselinck, J., Grossniklaus, U., & Dresselhaus, T. (2012). Egg cell-secreted EC1 triggers sperm cell activation during double fertilization. *Science, 338*, 1093–1097.

Stegmann, M., Monaghan, J., Smakowska-Luzan, E., Rovenich, H., Lehner, A., Holton, N., et al. (2017). The receptor kinase FER is a RALF-regulated scaffold controlling plant immune signaling. *Science, 355*, 287–289.

Steinhorst, L., & Kudla, J. (2013a). Calcium—A central regulator of pollen germination and tube growth. *Biochimica et Biophysica Acta, 1833*, 1573–1581.

Steinhorst, L., & Kudla, J. (2013b). Calcium and reactive oxygen species rule the waves of signaling. *Plant Physiology, 163*, 471–485.

Stonebloom, S., Ebert, B., Xiong, G., Pattathil, S., Birdseye, D., Lao, J., et al. (2016). A DUF-246 family glycosyltransferase-like gene affects male fertility and the biosynthesis of pectic arabinogalactans. *BMC Plant Biology, 16*, 90.

Stührwohldt, N., Dahlke, R. I., Kutschmar, A., Peng, X., Sun, M. X., & Sauter, M. (2015). Phytosulfokine peptide signaling controls pollen tube growth and funicular pollen tube guidance in *Arabidopsis thaliana. Physiologia Plantarum, 153*, 643–653.

Su, S., & Higashiyama, T. (2018). Arabinogalactan proteins and their sugar chains: Functions in plant reproduction, research methods, and biosynthesis. *Plant Reproduction, 31*, 67–75.

Takeuchi, H., & Higashiyama, T. (2012). A species-specific cluster of defensin-like genes encodes diffusible pollen tube attractants in Arabidopsis. *PLoS Biology, 10*, e1001449.

Takeuchi, H., & Higashiyama, T. (2016). Tip-localized receptors control pollen tube growth and LURE sensing in Arabidopsis. *Nature, 531*, 245–248.

Tang, W., Ezcurra, I., Muschietti, J., & McCormick, S. (2002). A cysteine-rich extracellular protein, LAT52, interacts with the extracellular domain of the pollen receptor kinase LePRK2. *Plant Cell, 14*, 2277–2287.

Tang, W., Kelley, D., Ezcurra, I., Cotter, R., & McCormick, S. (2004). LeSTIG1, an extracellular binding partner for the pollen receptor kinases LePRK1 and LePRK2, promotes pollen tube growth *in vitro. Plant Journal, 39*, 343–353.

Tekleyohans, D. G., Mao, Y., Kagi, C., Stierhof, Y. D., & Gross-Hardt, R. (2017). Polyspermy barriers: A plant perspective. *Current Opinion in Plant Biology, 35*, 131–137.

Tosti, E., & Menezo, Y. (2016). Gamete activation: Basic knowledge and clinical applications. *Human Reproduction Update, 22*, 420–439.

Uebler, S., Dresselhaus, T., & Márton, M. L. (2013). Species-specific interaction of EA1 with the maize pollen tube apex. *Plant Signaling & Behavior, 8*, e25682.

Uebler, S., Márton, M. L., & Dresselhaus, T. (2015). Classification of EA1-box proteins and new insights into their role during reproduction in grasses. *Plant Reproduction, 28*, 183–197.

Vogler, F., Schmalzl, C., Englhart, M., Bircheneder, M., & Sprunck, S. (2014). Brassinosteroids promote Arabidopsis pollen germination and growth. *Plant Reproduction*, *27*, 153–167.

Vogler, H., Draeger, C., Weber, A., Felekis, D., Eichenberger, C., Routier-Kierzkowska, A. L., et al. (2013). The pollen tube: A soft shell with a hard core. *Plant Journal*, *73*, 617–627.

Völz, R., Heydlauff, J., Ripper, D., von Lyncker, L., & Gross-Hardt, R. (2013). Ethylene signaling is required for synergid degeneration and the establishment of a pollen tube block. *Developmental Cell*, *25*, 310–316.

von Besser, K., Frank, A. C., Johnson, M. A., & Preuss, D. (2006). Arabidopsis *HAP2* (*GCS1*) is a sperm-specific gene required for pollen tube guidance and fertilization. *Development*, *133*, 4761–4769.

Wang, J. G., Feng, C., Liu, H. H., Feng, Q. N., Li, S., & Zhang, Y. (2017). AP1G mediates vacuolar acidification during synergid-controlled pollen tube reception. *Proceeding of the National Academy of Sciences of the United States of America*, *114*, E4877–E4883.

Wang, T., Liang, L., Xue, Y., Jia, P. F., Chen, W., Zhang, M. X., et al. (2016). A receptor heteromer mediates the male perception of female attractants in plants. *Nature*, *531*, 241–244.

Wong, J. L., & Johnson, M. A. (2010). Is HAP2-GCS1 an ancestral gamete fusogen? *Trends in Cell Biology*, *20*, 134–141.

Wong, J. L., Leydon, A. R., & Johnson, M. A. (2010). HAP2(GCS1)-dependent gamete fusion requires a positively charged carboxy-terminal domain. *PLoS Genetics*, *6*, e1000882.

Woriedh, M., Merkl, R., & Dresselhaus, T. (2015). Maize EMBRYO SAC family peptides interact differentially with pollen tubes and fungal cells. *Journal of Experimental Botany*, *66*, 5205–5216.

Woriedh, M., Wolf, S., Márton, M. L., Hinze, A., Gahrtz, M., Becker, D., et al. (2013). External application of gametophyte-specific ZmPMEI1 induces pollen tube burst in maize. *Plant Reproduction*, *26*, 255–266.

Wu, H. M., Wong, E., Ogdahl, J., & Cheung, A. Y. (2000). A pollen tube growth-promoting arabinogalactan protein from *Nicotiana alata* is similar to the tobacco TTS protein. *Plant Journal*, *22*, 165–176.

Yang, N., & Wang, T. (2017). Comparative proteomic analysis reveals a dynamic pollen plasma membrane protein map and the membrane landscape of receptor-like kinases and transporters important for pollen tube growth and interaction with pistils in rice. *BMC Plant Biology*, *17*, 2.

Yu, G. H., Zou, J., Feng, J., Peng, X. B., Wu, J. Y., Wu, Y. L., et al. (2014). Exogenous gamma-aminobutyric acid (GABA) affects pollen tube growth via modulating putative Ca^{2+}-permeable membrane channels and is coupled to negative regulation on glutamate decarboxylase. *Journal of Experimental Botany*, *65*, 3235–3248.

Zhang, J., Huang, Q., Zhong, S., Bleckmann, A., Huang, J., Guo, X., et al. (2017). Sperm cells are passive cargo of the pollen tube in plant fertilization. *Nature Plants*, *3*, 17079.

Zhang, X., Liu, W., Nagae, T. T., Takeuchi, H., Zhang, H., Han, Z., et al. (2017). Structural basis for receptor recognition of pollen tube attraction peptides. *Nature Communications*, *8*, 1331.

Zhang, Y., Liu, J., Chai, J., & Xing, D. (2016). Mitogen-activated protein kinase 6 mediates nuclear translocation of ORE3 to promote *ORE9* gene expression in methyl jasmonate-induced leaf senescence. *Journal of Experimental Botany*, *67*, 83–94.

Zhou, L. Z., Juranic, M., & Dresselhaus, T. (2017). Germline development and fertilization mechanisms in maize. *Molecular Plant*, *10*, 389–401.

Zhu, L., Chu, L. C., Liang, Y., Zhang, X. Q., Chen, L. Q., & Ye, D. (2018). The Arabidopsis *Cr*RLK1L protein kinases BUPS1 and BUPS2 are required for normal growth of pollen tubes in the pistil. *Plant Journal*, *95*, 474–486.

CHAPTER EIGHTEEN

Genetic, molecular and parent-of-origin regulation of early embryogenesis in flowering plants

Alma Armenta-Medina, C. Stewart Gillmor*

Laboratorio Nacional de Genómica para la Biodiversidad (Langebio), Unidad de Genómica Avanzada, Centro de Investigación y Estudios Avanzados del Instituto Politécnico Nacional (CINVESTAV-IPN), Irapuato, Guanajuato, México

*Corresponding author: e-mail address: stewart.gillmor@cinvestav.mx

Contents

Abstract

Embryogenesis in flowering plants has fascinated biologists since at least the 19th century. Embryos of almost all flowering plants share common characteristics, including an asymmetric first division of the zygote, and multiple rounds of cell divisions that generate the major tissue types of the adult plant, usually within a few days of fertilization. This review focuses on early embryogenesis, including fertilization, the contributions of maternal and paternal genomes to the zygote and early embryo, cell fate decisions that create the apical and basal lineages, establishment of the shoot and root meristems, and formation of the other major tissue types in the adult plant. Because most genetic and molecular research on embryogenesis in plants has been conducted on the model species *Arabidopsis thaliana,* we highlight work on this species as well as research with *Zea mays* (maize) and *Oryza sativa* (rice).

Current Topics in Developmental Biology, Volume 131
ISSN 0070-2153
https://doi.org/10.1016/bs.ctdb.2018.11.008

1. Zygote formation: Fusion of the egg and sperm

In flowering plants, seed development begins with a double fertilization event, where the pollen tube releases two sperm cells into the female gametophyte. Fertilization of the egg by one sperm produces the embryo, while fertilization of the homo-diploid central cell by the other sperm produces the triploid endosperm, an extraembryonic nutritive tissue analogous to the placenta in mammals (see chapter "Friend or foe: Signaling mechanisms during double fertilization in flowering seed plants" by Zhou and Dresselhaus, this volume, for a comprehensive review on fertilization).

Fertilization occurs deep within maternal ovule tissues, making observation of gamete fusion challenging. For this reason, initial research on fertilization relied on *in vitro* studies (reviewed in Dresselhaus, Sprunck, & Wessel, 2016; Lord & Russell, 2002). Subsequent *in vivo* experiments determined that in *Zea mays* (maize), fusion of egg and sperm cells (plasmogamy) occurs about 1 h (h) after release of the sperm cells from the pollen tube, and fusion of egg and sperm nucleoli (karyogamy) is complete after about 5 h (Mòl, Dumas, & Matthys-Rochon, 1994). The kinetics of gamete cell and nuclear fusion are roughly similar in *Torenia fournieri*, a plant where the embryo sac protrudes from the ovule, allowing easier visualization of fertilization (Higashiyama, Kuroiwa, Kawano, & Kuroiwa, 1997). In *Arabidopsis thaliana*, confocal microscopy showed that fusion of egg and sperm cells occurs at about 5 h after pollination (hap), while karyogamy initiates at 6–8 hap, and is completed at about 9 hap (Faure, Rotman, Fortuné, & Dumas, 2002; Ingouff, Hamamura, Gourgues, Higashiyama, & Berger, 2007). Division of the zygote is delayed for about 24 h after fertilization, while division of the central cell (producing the endosperm) begins almost immediately after fertilization (Aw, Hamamura, Chen, Schnittger, & Berger, 2010). A study in *A. thaliana* found that at the time of fertilization sperm nuclei have 2C DNA content, corresponding to the G2 phase of the cell cycle, suggesting that at fertilization the egg and central cells of this species are also in G2 (Friedman, 1999). In rice, transcriptome evidence suggests that the egg cell is in the S phase of the cell cycle, while the sperm at anthesis is in G1 (Anderson et al., 2013). By contrast, a recent study of maize found the egg cell transcriptome to be characteristic of G0 (Chen et al., 2017). To prevent problems with nuclear DNA content in the zygote, the gametes must be at the same stage of the cell cycle during nuclear fusion, and thus determining the exact stage of the cell cycle during gamete fusion in different species is an important goal for future research.

Recently, advanced microscopy techniques have enabled *in vivo* visualization of gamete cell and nuclear fusion in intact siliques of *A. thaliana* at high resolution (reviewed in Berger, 2011). After growth of the pollen tube through the gynoecium, the pollen tube enters the ovule through the micropyle, and bursts to release the two sperm cells into the degenerating synergid cell (reviewed in Dresselhaus et al., 2016). <1 min later, the sperm move to a position between the egg and central cell, where they remain for about 7 min, until fusion with the egg and central cell occurs (plasmogamy) (Hamamura et al., 2011). At the molecular level, gamete interactions in *A. thaliana* depend on small cysteine-rich EGG CELL 1 (EC1) proteins, which accumulate in storage vesicles of the egg cell (Sprunck et al., 2012). When sperm cells are delivered to the female gametophyte, EC1 protein is detected extracellularly in the apical region of the degenerating synergid cells, and is required for fusion of sperm with both the egg and central cell. The sperm endomembrane system responds to exogenously applied EC1 peptides by redistributing the potential fusogen HAPLESS2/GENERATIVE CELL SPECIFIC1 to the cell surface (Mori, Kuroiwa, Higashiyama, & Kuroiwa, 2006; Sprunck et al., 2012). Unlike in animals, microtubules are dispensable for the nuclear migration that leads to karyogamy (Kawashima et al., 2014). Instead, filamentous actin and myosin are required for movement of the sperm nucleus toward the egg nucleus (Kawashima et al., 2014). In *A. thaliana*, the two sperm cells have equal capacity to fertilize the egg cell (Hamamura et al., 2011; Ingouff et al., 2009; Kong, Lau, & Jürgens, 2015), yet polyspermy events resulting in a zygote are very rare (Grossniklaus, 2017; Nakel et al., 2017), demonstrating that plants have several mechanisms to block multiple fertilization events (reviewed in Tekleyohans, Mao, Kägi, Stierhof, & Gross-Hardt, 2017).

2. Zygotic genome activation

In animals, zygotes are transcriptionally quiescent immediately after fertilization, with RNAs and proteins inherited from the egg driving early development (reviewed in Lee, Bonneau, & Giraldez, 2014). Regulation of embryogenesis is transferred from the maternal to the zygotic genome during the maternal-to-zygotic transition (MZT), a process that consists of degradation of maternal gene products and zygotic genome activation (ZGA) (reviewed in Baroux, Autran, Gillmor, Grimanelli, & Grossniklaus, 2008). The timing of the MZT varies between species, from a few hours after fertilization to up to a day, and from the first cell division to after multiple cell divisions (Tadros & Lipshitz, 2009).

In flowering plants, there is less consensus on the timing of ZGA, and whether it varies between species. Some experiments have suggested that there is a delay between fertilization and ZGA (as in animals), while other studies have concluded that transcription begins almost immediately after fertilization (reviewed in Baroux & Grossniklaus, 2015; Del Toro-De León, Lepe-Soltero, & Gillmor, 2016). Transcriptomes from early embryos have been studied in wheat, tobacco, maize, rice and *A. thaliana*. In wheat, cDNA libraries of 2-celled proembryos at 1 day after pollination (dap) had transcripts for 325 genes absent in egg cells (Sprunck, Baumann, Edwards, Langridge, & Dresselhaus, 2005), and at 4 dap in tobacco, zygotes had transcripts for 1234 genes not detected in eggs (Zhao et al., 2011). While these studies relied on sequencing of cDNA libraries, and are thus not highly quantitative, they do suggest a difference in transcript populations between eggs and early embryos, with timing of the onset of zygotic transcription that differs between species. In maize, a recent RNA sequencing experiment found transcripts for 3489 genes upregulated at least twofold in the zygote compared to the egg cell at 12 hap, and 3247 genes upregulated at least twofold at 24 hap, demonstrating that about 10% of the genome is transcriptionally active in the first day after fertilization (Chen et al., 2017). In rice, an RNA sequencing comparison of eggs and zygotes found 499 genes with transcripts upregulated in zygotes at 2.5 hap (during karyogamy), and 2485 genes upregulated by 9 hap (G2 phase). This same study used hybrid zygotes to investigate parental contributions to zygotic transcriptomes, finding >95% of genes with a maternal bias at 2.5 and 9 hap ($n > 14{,}000$ genes) (Anderson et al., 2017). Thus, evidence in wheat, maize and rice supports a model where zygotic transcription begins early after fertilization, and recent results in rice show a strong maternal bias in zygotic transcriptomes. The longer pause between fertilization and the first division of the zygote in tobacco may be correlated with a relative delay in ZGA in this species compared to others.

Research on ZGA in *A. thaliana* has used molecular, genetic and transcriptomic approaches. Gene reporter experiments found >20 reporters with higher expression from the maternal allele than the paternal allele in early embryos (Autran et al., 2011; Baroux, Blanvillain, & Gallois, 2001; Golden et al., 2002; Ngo et al., 2012; Vielle-Calzada, Baskar, & Grossniklaus, 2000), as well as a few reporters showing equal expression from maternal and paternal alleles (Aw et al., 2010; Weijers, Geldner, Offringa, & Jürgens, 2001; Xiang et al., 2011). To understand the parental contributions to ZGA at the genome level, two parent-of-origin transcriptome studies in *A. thaliana* have been conducted. Autran et al. (2011) sequenced RNA from *Ler* × Col hybrid embryos at the 2–4-cell stage (~2 dap) and globular stage (~3 dap), finding a maternal transcript bias for 84.9% of 3973 genes at the 2–4 cell stage, decreasing to about 48.3% of 3078 genes at the globular stage. Combined with reporter gene studies in isogenic backgrounds (Autran et al., 2011), these results argued for a maternal transcript bias in early embryogenesis which decreases beginning at the globular stage. Nodine and Bartel (2012) sequenced RNA from CVI × Col and Col × CVI hybrid embryos at the 1–2 cell stage (~1.5 dap), 8 cell stage (~2.5 dap), and 32 cell stage (~3 dap). >7000 genes were profiled at each stage, and averaging of maternal and paternal transcripts from both crosses produced essentially equal amounts of maternal and paternal reads for all stages, arguing for equal contributions from maternal and paternal genomes to transcript populations in early embryogenesis.

An alternative assay for parental contributions to early embryogenesis is to test the function of paternal alleles in early embryogenesis, by determining if wild-type paternal alleles can complement maternally mutant alleles of *embryo defective (emb)* genes. Using this assay, 7 genes were found to have early paternal activity, and 2 were found to have delayed paternal activity (Baroux et al., 2008; Lukowitz, Roeder, Parmenter, & Somerville, 2004; Ronceret, Gadea-Vacas, Guilleminot, & Devic, 2008; Ronceret, Gadea-Vacas, Guilleminot, Lincker, et al., 2008; Ronceret, Guilleminot, Lincker, et al., 2005; Ueda, Zhang, & Laux, 2011; Vielle-Calzada et al., 2000; Weijers et al., 2001; Xu et al., 2005). Subsequently, Del Toro-De León, Garcia-Aguilar, and Gillmor (2014) used this functional assay in a systematic study of 49 *emb* genes, to test the hypothesis that the maternal and paternal genomes make equivalent contributions to early embryogenesis, as proposed by Nodine and Bartel (2012). 40 genes showed delayed paternal activation, and 9 genes showed paternal activity as soon as their function was required (Del Toro-De León et al., 2014). Thus, of the 58 total genes tested for early paternal allele activity in the above studies, 16 showed paternal activity as soon as their function was required (typically by 2 dap), and the remaining 42 genes showed delayed paternal allele activity (typically beginning at 3 or 5 dap). These functional results are consistent with a bias toward maternal transcripts for many genes in early embryogenesis, as proposed by Autran et al. (2011). This maternal bias is likely to be due to both maternal transcripts inherited from the egg, as well as a maternal bias in zygotic transcription, as found in rice (Anderson et al., 2017). Because of the much smaller size of the sperm cell compared to the egg cell, carryover of paternal transcripts from the sperm is likely to make a minor contribution to the zygotic transcriptome compared to the egg cell, though paternal transcripts have been detected in zygotes (Bayer et al., 2009; Xin et al., 2011). Fig. 1 summarizes published transcriptome data in rice, maize and *Arabidopsis*, and functional data in *Arabidopsis*.

To generate a consensus on the timing of ZGA in *A. thaliana*, and whether zygotic transcription initiates with a maternal bias, additional transcriptome studies are required. The lack of agreement between the two published transcriptome studies has variously been attributed to seed coat contamination of embryo samples causing a maternal transcript bias in the L*er* × Col transcriptome (Nodine & Bartel, 2012; Schon & Nodine, 2017), and speculation that the unusual CG DNA methylation and chromatin state of CVI (Pignatta et al., 2014; Schmitz et al., 2013; Tessadori et al., 2009) leads to early activation of paternal transcription in CVI × Col embryos (Baroux, Autran, Raissig, Grimanelli, & Grossniklaus, 2013; Del Toro-De León, Lepe-Soltero, & Gillmor, 2016). In the functional assay performed by Del Toro-De León, Garcia-Aguilar, and Gillmor (2014), Col × CVI embryos were found to have significantly earlier paternal allele activation than isogenic Col embryos, and also to have paternal allele activation patterns different than L*er* × Col embryos, suggesting that the differences in the L*er* × Col and CVI × Col transcriptomes reflect authentic differences in transcript populations, pointing to a fascinating effect of hybridization on gene activity in early embryogenesis (Baroux et al., 2013; Del Toro-De León, Lepe-Soltero, & Gillmor, 2016). For this reason, future transcriptome studies will need to consider possible effects of hybridization when using hybrid embryos as a proxy for parent-of-origin effects on transcription in isogenic embryos.

3. Zygotic, maternal, and paternal regulation of early embryogenesis

Maternal and paternal effect genes determine the phenotype of the embryo regardless of the zygotic genotype, while the phenotype of zygotic effect genes reflects the genotype of the zygote. In animals such as *Drosophila melanogaster*, maternal or zygotic effect genes are common (St Johnston, 2002),

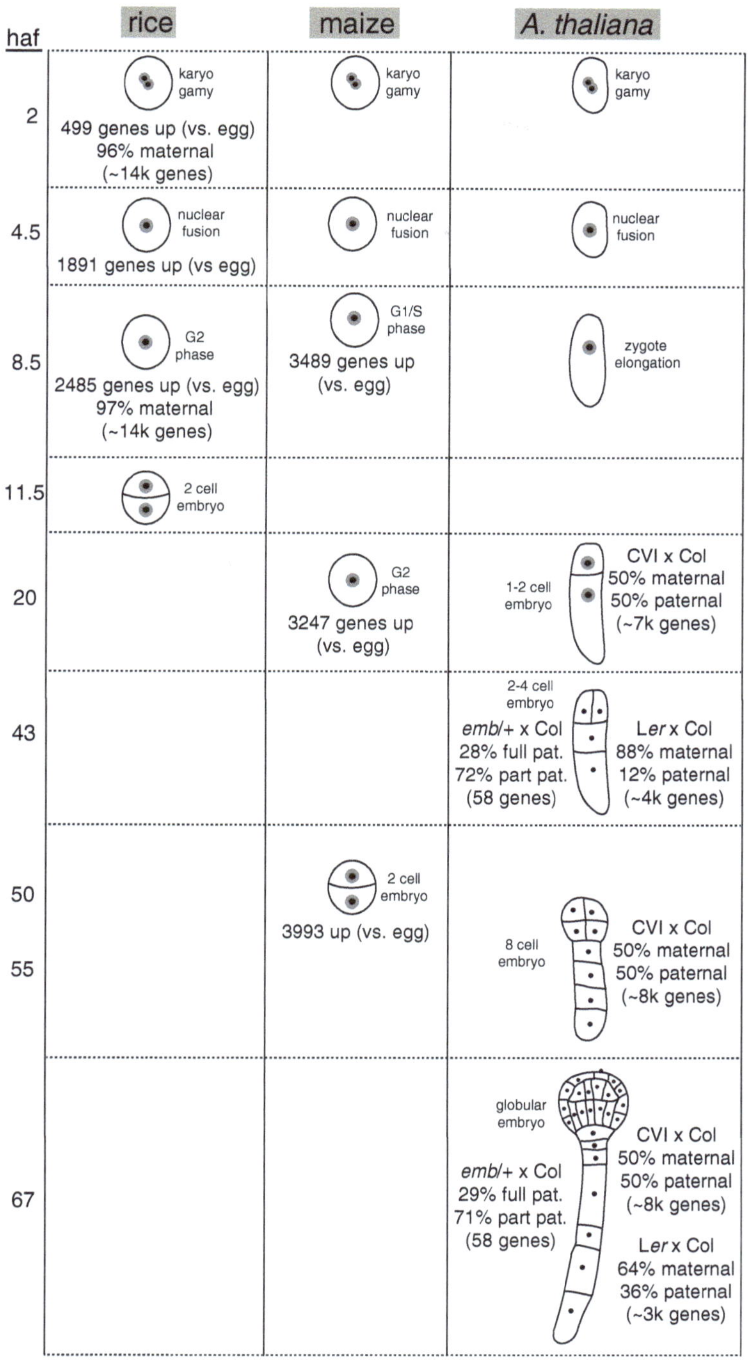

Fig. 1 See legend on next page.

while paternal effect mutants exist, but are uncommon (Fitch, Yasuda, Owens, & Wakimoto, 1997; Loppin, Lepetit, Dorus, Couble, & Karr, 2005). This is likely due to the huge cytoplasm of the egg, which stores large amounts of transcripts and proteins and can regulate development of the zygote until the MZT. In plants, zygotic, maternal and paternal effect genes also exist. Maternal effect genes can be classified as sporophytic, when the genotype of the mother plant (sporophyte) determines the phenotype of the embryo, and gametophytic, when the genotype of the gametophyte determines the phenotype of the embryo (Grossniklaus & Schneitz, 1998). In most cases, maternal gametophytic effects will be due to the egg cell, but it is also possible that other cells of the gametophyte, such as the central cell, influence

Fig. 1 Summary of transcriptional and functional data on genome activation and parent-of-origin gene expression in rice, maize and *A. thaliana*. Data for rice are taken from Anderson et al. (2017), and show the cell cycle stages for the zygote, the number of genes that were significantly upregulated in the zygote compared to an egg cell transcriptome, and the number of genes with maternal transcript bias observed in hybrid zygote transcriptomes. Data for maize are taken from Chen et al. (2017), and show the number of transcripts upregulated in zygotes compared to an egg cell transcriptome. Parent-of-origin expression data are not available for maize. Timing of maize karyogamy and nuclear fusion is estimated from Mòl et al. (1994), and cell cycle stages are taken from Chen et al. (2017). *A. thaliana* transcriptome data for *Ler* × Col are taken from Autran et al. (2011), and for CVI × Col from Nodine and Bartel (2012). These data list the total number of maternal and paternal reads obtained, and the number of genes that were profiled, but do not show the number of genes with maternal or paternal bias *per se* (rice and maize data do show bias per gene). *A. thaliana* functional data taken from Del Toro-De León et al. (2014), and from single gene functional studies (Baroux et al., 2008; Lukowitz et al., 2004; Ronceret, Gadea-Vacas, Guilleminot, & Devic, 2008; Ronceret, Gadea-Vacas, Guilleminot, Lincker, et al., 2008; Ronceret, Guilleminot, Lincker, et al., 2005; Ueda, Zhang, & Laux, 2011; Vielle-Calzada et al., 2000; Weijers et al., 2001; Xu et al., 2005). These data are listed as genes with full paternal activity ("full pat.," i.e., full complementation of maternally conditioned *emb* phenotype by wild type paternal allele), and genes with partial paternal activity ("part pat.," i.e., complementation of some but not all maternally mutant *emb* embryos by the wild type paternal allele). The column on the left of the figure lists the estimated hours after fertilization (haf) for each data point or zygote illustration. The haf for rice were taken from Anderson et al. (2017). Haf for karyogamy and nuclear fusion in maize were taken from Mòl et al. (1994), and for subsequent stages from Chen et al. (2017). Haf for karyogamy and nuclear fusion in *A. thaliana* are from Faure et al. (2002) and Ingouff et al. (2007); for zygote elongation and the first division of the zygote from Kimata et al. (2016); and estimations for haf at 1–2 cell; 2–4 cell, 8 cell, and early globular stages are from work in our laboratory. The spacing of haf stages in the figure is not linear, allowing the figure to fit on one page.

the development of the embryo. Maternal and paternal effects also exist in the form of gene imprinting, when one parental allele is expressed more than the other. Many imprinted genes are known in the endosperm, and a few exist in embryos (Jahnke & Scholten, 2009; Raissig, Bemer, Baroux, & Grossniklaus, 2013; reviewed in Garcia-Aguilar & Gillmor, 2015). Here we focus on functional effects of zygotic, maternal, and paternal regulation of embryogenesis.

Mutants of maize which affect embryo development have been isolated (Chettoor, Phillips, Coker, Dilkes, & Evans, 2016; Clark & Sheridan, 1988; Neuffer & Sheridan, 1980), but the opaque seed coat and persistent endosperm of maize make visualizing embryonic phenotypes challenging. By contrast, the transparent seed coat and transitory endosperm of *A. thaliana* allow facile imaging of embryos (Meinke & Sussex, 1979a), and thus most genetics studies of embryogenesis have been carried out in this species (reviewed in Palovaara, de Zeeuw, & Weijers, 2016). The first studies of *embryo defective (emb)* mutants in *A. thaliana* found 25% mutant seeds in selfed heterozygous plants, characteristic of a zygotic effect on embryo development (Meinke & Sussex, 1979a, 1979b). A saturating genetic screen to identify pattern formation genes in *A. thaliana* also concluded that most genes that regulate embryogenesis act zygotically (Jürgens, Mayer, Torres Ruíz, Berleth, & Miséra, 1991; Mayer & Jürgens, 1993; Mayer, Torrez Ruíz, Berleth, Miséra, & Jürgens, 1991), and genes with well-studied roles in pattern formation, such as *YODA* and *WRKY2*, have been shown to act zygotically (Lukowitz et al., 2004; Ueda et al., 2011). Of the collection of hundreds of *emb* mutants assembled in *A. thaliana* (McElver et al., 2001), almost all segregate close to 25% mutant seeds at late stages of embryogenesis, consistent with a zygotic effect (Muralla, Lloyd, & Meinke, 2011). However, a subsequent large-scale analysis of maternal and paternal contributions to early *A. thaliana* embryos demonstrated that many previously analyzed *emb* mutants show a transient maternal effect, due to delayed activation of the paternal allele (Del Toro-De León et al., 2014). Thus, for many genes that are active in early embryos, the maternal allele initially plays a more important role than the paternal allele. This maternal contribution is likely to depend on both maternal transcripts inherited from the egg cell, as well as zygotic maternal transcripts (Autran et al., 2011). Recent results in rice and maize have demonstrated that the egg cell makes a large contribution to the zygote transcriptome (Anderson et al., 2017; Chen et al., 2017) and in rice, zygotic transcription was demonstrated to begin with a maternal bias, consistent with greater importance for the maternal genome in early embryogenesis (Anderson et al., 2017).

Most genes with sporophytic maternal effects on seed development regulate the growth of the seed coat, the maternal tissue that surrounds the embryo and endosperm, and plays a crucial role in determining seed size during late embryogenesis (reviewed in Li & Li, 2015). Genes expressed in the seed coat also regulate the transport of nutrients from the mother plant to the seed (Chen et al., 2015; Olsen et al., 2016; Vogiatzaki, Baroux, Jung, & Poirier, 2017). In early embryogenesis, the Mitogen Activated Protein (MAP) kinase kinase genes *MKK4* and *MKK5,* and the MAP kinase *MPK6* all have maternal sporophytic effects on the elongation of the zygote (Zhang, Wu, et al., 2017). The maternal sporophytic nature of *mpk6* mutants is curious, because MPK6 is thought to function downstream of the MAP kinase kinase kinase (MAPKKK) YODA (YDA) (Jeong, Eilbert, Bolbol, & Lukowitz, 2016), which is a zygotic effect gene (Lukowitz et al., 2004). Loss of miRNAs has also been shown to have a sporophytic maternal effect on embryogenesis: when crossed with wild-type pollen, the homozygous mutants *dicer-like1* and *serrate* have defects in cotyledon formation during late embryogenesis, suggesting that at least some miRNAs move from the sporophyte to the embryo to regulate development (Prigge & Wagner, 2001; Ray, Golden, & Ray, 1996). 24 nt small interfering RNAs (siRNAs) have also been implicated in the regulation of embryogenesis. Mutations affecting the histone 3 lysine 9 methyltransferase KRYPTONITE and the CHG DNA methyltransferase CHROMOMETHYLASE 3, both of which are associated with the RNA-dependent DNA methylation (RdDM) pathway, cause transient effects in cell divisions in preglobular embryos (Autran et al., 2011; Pillot et al., 2010). In the *A. thaliana* relative *Brassica rapa*, maternal sporophytic RdDM mutants show abnormal embryogenesis and seed abortion, demonstrating that sporophytic 24 nt siRNAs in this species play an important role in promoting embryogenesis, perhaps by repressing activity of transposons, or regulating expression of transcription factors located near transposons (Grover et al., 2018).

Gametophytic maternal effect genes include *FERTILIZATION INDEPENDENT ENDOSPERM (FIE), FERTILIZATION INDEPENDENT SEED 2 (FIS2),* and *MEDEA (MEA). fie* and *fis2* were isolated in screens for fertilization-independent seed development, while *mea* mutants were isolated based on their maternal gametophytic effect on development of the endosperm and embryo (Chaudhury et al., 1997; Grossniklaus, Vielle-Calzada, Hoeppner, & Gagliano, 1998; Ohad et al., 1996). All three mutants have a maternal gametophytic lethal phenotype in the embryo: when fertilized with wild-type sperm, *fie*, *fis2* and *mea* eggs produce defective embryos that arrest at the heart stage of development. Identification of the *FIE*, *FIS2* and *MEA* genes revealed that they encode components of the *Polycomb* Repressive Complex 2 (PRC2) complex in *A. thaliana*, demonstrating that one of the functions of *Polycomb* group genes in plants is to repress development of the endosperm in the absence of fertilization (Grossniklaus et al., 1998; Luo et al., 1999; Ohad et al., 1999). One of the most important PRC2 targets is the *PHERES1* MADS-box transcription factor, whose overexpression in *mea* mutants was shown to be responsible for much of the embryo abortion in this mutant (Köhler et al., 2003).

A large genetic screen for gametophyte development in *A. thaliana* uncovered 70 *maternal effect embryo arrest* mutants, where the phenotype was solely dependent on the maternal allele, and not rescuable by wild-type pollen (Pagnussat et al., 2005). *glauce*, a gametophytic maternal effect mutant isolated in this screen, allows embryo development in the absence of endosperm development, represses the autonomous endosperm development seen in *fis* mutants, and prevents expression of several paternal markers in early embryogenesis (Ngo, Moore, Baskar, Grossniklaus, & Sundaresan, 2007). Mutants in the Mcm7 protein *PROLIFERA* also have a maternal gametophytic effect, where a small percentage of ovules are fertilized and produce embryos that arrest early in development (Springer, Holding, Groover, Yordan, & Martienssen, 2000). Maternal gametophytic effect mutants which alter embryo development have also been identified in maize, though the gene products affected by these mutants have not yet been identified (Bai et al., 2016; Chettoor et al., 2016; Evans & Kermicle, 2001). These mutations, as well as *fie*, *fis2* and *medea*, affect endosperm development as well as embryo development, suggesting that at least in some cases the primary defect may be a lack of nourishment of the embryo by the defective endosperm.

Maternally imprinted genes of *A. thaliana* also have maternal gametophytic effects on early embryo development. The *ZAK IXIK* gene, which encodes a conserved Armadillo repeat protein, is a maternally expressed, imprinted gene required for early embryo and endosperm development (Ngo et al., 2012). *LORELEI (LRE)*, which encodes a glycosylphosphatidylinositol-anchored membrane protein required for pollen tube reception, is maternally expressed in the zygote, and loss of maternal *LRE* in the gametophyte causes a delay in embryo development (Wang et al., 2017). The *NUWA* gene, which is expressed in the gametophyte and maternally *de novo* in early embryos, encodes a mitochondrial protein. *nuwa* mutant seed shows defects in division patterns in the apical cell lineage of the embryo, as well as in endosperm development (He et al., 2017). As discussed above, *MEA* is a maternally expressed imprinted gene that represses embryo and endosperm growth in the absence of fertilization (Grossniklaus et al., 1998). When pollinated with wild-type pollen of the same accession, gametophytic maternally mutant *mea* seed arrests embryo development at the heart stage. However, when plants which are maternally mutant for the *mea* gene (in the L*er* background) are pollinated with the CVI and C24 accessions, many of these hybrid seed develop almost normally. QTL mapping of paternal modifiers of *mea* using L*er*/CVI recombinant inbred lines identified 6 different loci which contribute small effects to paternally rescue *mea* maternal effect seed abortion (Pires et al., 2016).

The *SHORT SUSPENSOR (SSP)* gene, which encodes a receptor associated kinase, is the only strict paternal effect gene in plants whole molecular identity is known (Bayer et al., 2009). *ssp* mutants cause a reduction in suspensor cell fate, which is entirely dependent on the paternal genotype of the zygote. The *SSP* gene is transcribed in sperm cells, but is only translated in the zygote after fertilization, providing an explanation for its paternal effect (see Section 4 for more on the role of *SSP* in suspensor development) (Bayer et al., 2009). The *NIMNA (NMA)* gene was also shown to have a partial paternal effect on suspensor development: $+/nma^{p}$ (maternally wild-type, paternally mutant) embryos show shortened suspensors, while homozygous *nma* embryos have even stronger phenotypes, demonstrating that the maternal *NMA* allele is also functional. *NMA* encodes a polygalacturonase presumably involved in cell wall remodeling, but the mechanism for the partial paternal effect of *nma* is unknown (Babu, Musielak, Henschen, & Bayer, 2013).

4. Polarity of the zygote and first asymmetric division

The *A. thaliana* egg cell is polarized, with the nucleus and cytoplasm at the apex, and a large vacuole at the base (Mansfield & Briarty, 1991a). After fertilization, the zygote shrinks and loses the polarity inherent in the egg: the nucleus moves to center of the cell, microtubules and actin filaments are disorganized, and vacuoles are evenly distributed. A few hours later, polarity is re-established, the zygote elongates about threefold, and about 24 h after fertilization the zygote divides asymmetrically to form a small, cytoplasmic apical cell and an elongated vacuolate basal cell (Kimata et al., 2016; Mansfield & Briarty, 1991b; Ueda et al., 2011). Polar growth of the zygote and cell division is dependent on microtubules, while migration of the nucleus to the apex of the zygote and positioning of the cell division plane requires actin filaments (Kimata et al., 2016; Webb & Gunning, 1991). These early cell fates are marked by expression of *WUSCHEL HOMEOBOX (WOX)* transcription factors. *WOX2* and *WOX8* are expressed in the egg cell and the zygote. After the first division, *WOX2* is expressed in the apical cell, and *WOX8* and *WOX9* are expressed in the basal cell (Haecker et al., 2004) (Fig. 2A). The apical and basal cell lineages make different contributions to the mature embryo. The apical cell lineage produces most of the tissue types of the embryo, while the basal cell lineage produces the suspensor and part of the root meristem (Fig. 3).

The most important component of the signaling cascade that promotes zygote elongation and asymmetric cell fates is the MAPKKK gene *YDA*. *yda* zygotes do not to elongate, and the first division produces a normal sized apical cell, and a short basal cell that has decreased suspensor cell fate. Dominant active versions of YDA have the opposite effect, producing embryos with long suspensors, or in the most extreme cases, elongated suspensors lacking apical cell fate (Lukowitz et al., 2004). The paternal effect gene *SSP* acts upstream of *YDA*. *SSP* encodes a membrane associated receptor-like kinase whose membrane localization, but not kinase activity, is required for function, suggesting that the principal function of SSP may be as a transient paternal signal to promote zygote elongation and suspensor identity (Bayer et al., 2009). Maternal effect *EMBRYO SURROUNDING FACTOR1* (*ESF1*) genes also act upstream of *YDA* to promote suspensor identity. The three tandem *ESF1* genes encode small cysteine-rich peptides that are expressed in the central cell, the maternal gamete located next to the egg cell. Loss of *ESF1* function causes a decrease in suspensor cell identity, and overexpression of ESF1 peptides produces more suspensor cells (Costa et al., 2014). Genetic analysis indicates that *ESF1* peptides and *SSP* act independently to promote *YDA* signaling, demonstrating that both maternal and paternal factors have an important role in pattern formation in early embryogenesis. The receptor for ESF1 peptides is currently unknown, while SSP interacting proteins have recently been discovered (Yu et al., 2016) (Fig. 2B). The signal for zygote elongation downstream of *YDA* is likely to be transmitted by *MKK4*, *MKK5*, *MPK3* and *MPK6* genes (Wang, Ngwenyama, Liu, Walker, & Zhang, 2007). *mkk4/mkk5* double mutants and *mpk3/mpk6* double mutants display phenotypes like *yda*: zygotes which do not elongate and then divide symmetrically (Wang et al., 2007; Zhang, Wu, et al., 2017). Interestingly, *mkk4*, *mkk5* and *mpk6* all have maternal

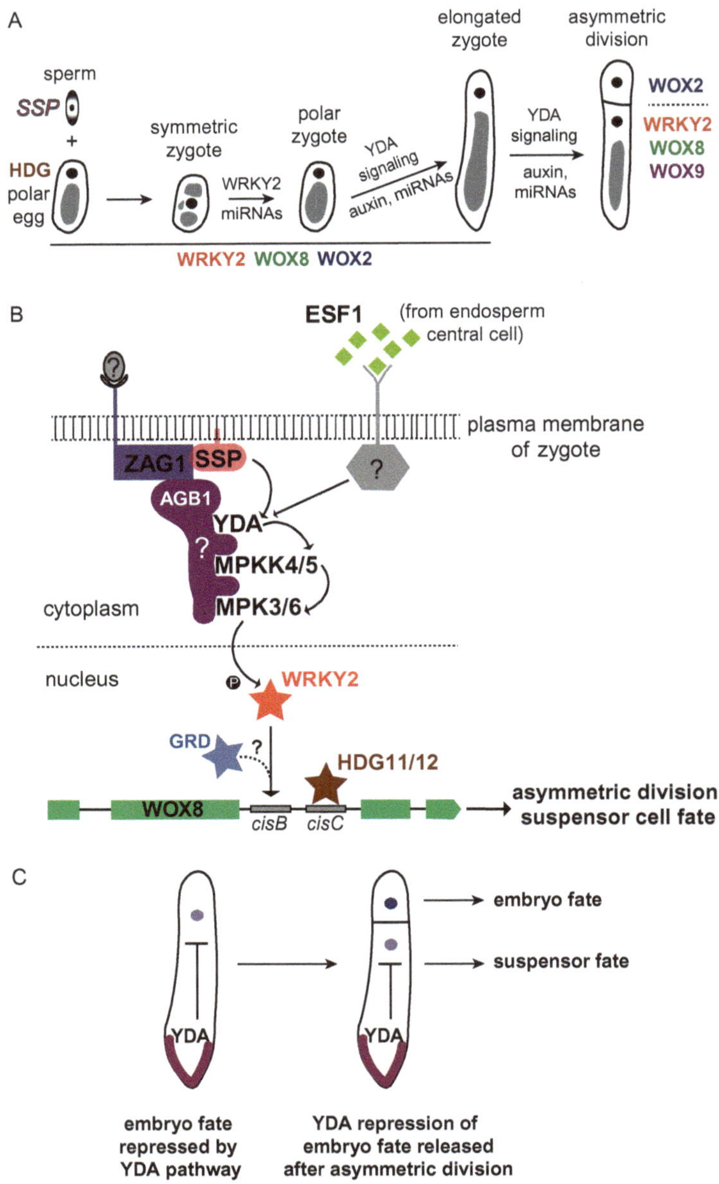

Fig. 2 Polarity and asymmetric division of the *A. thaliana* zygote. (A) The sperm cell expresses *SSP* mRNA, which is translated after fertilization, while the egg cell expresses *HDG11* and *HDG12*. Before fertilization, the egg is polarized, with the nucleus (black dot) at the apex and the vacuole (gray) at the base. Upon fertilization, the zygote depolarizes, and then *WRKY2* and miRNAs promote reestablishment of the cellular asymmetry present in the egg. The zygote then elongates, and undergoes an asymmetric division to produce a small, cytoplasmic apical cell and a larger, vacuolated basal cell. These

sporophytic effects on zygote development, demonstrating a role for the mother plant in zygote polarity.

Jeong et al. (2016) proposed that localization of the *YDA* signaling cascade at the base of the zygote through interaction with a scaffold protein might be required for YDA to promote elongation and asymmetric division of the zygote. This model is based on the mechanism for asymmetric division in the epidermis of *A. thaliana*, where the polarly localized transcription factor BREAKING OF ASYMMETRY IN THE STOMATAL LINEAGE (BASL) serves as a scaffold for YDA and MPK3/6 (Dong, MacAlister, & Bergmann, 2009; Zhang, Wang, Shao, Zhu, & Dong, 2015). The polar localization of this scaffold means that it is inherited by only one daughter cell, allowing the second daughter cell to adopt a new cell fate (Fig. 2C). One candidate for this scaffold function would be a Rho-like small G protein such as ROP3, whose loss of function results in symmetric divisions of the zygote (Huang et al., 2014; Jeong et al., 2016). Alternatively, recent evidence exists for scaffolding of the *YDA* cascade by the heterotrimeric Gβ protein AGB1, which has been shown to interact *in vivo* with

subsequent steps depend on the YDA signaling pathway; *WRKY2* and its downstream targets *WOX8* and *WOX9*; and auxin and miRNAs. The egg cell and zygote express *WRKY2*, *WOX8* and *WOX2*. After the asymmetric division of the zygote, the basal cell expresses *WRKY2*, *WOX8* and *WOX9*, while the apical cell expresses *WOX2*. (B) In the cytoplasm of the zygote, plasma membrane localized SSP acts upstream of the YDA-MPKK4/5-MPK6 kinase cascade, transducing a signal that results in the phosphorylation of WRKY2, which then activates *WOX8* transcription in conjunction with *HDG11* and *HDG12*. The endosperm central cell secretes ESF1 peptides, which promote YDA activity through an unknown receptor. ZAG1, a plasma membrane localized receptor kinase with extracellular leucine rich repeats, interacts with SSP and may convey extracellular signals to the YDA pathway through SSP. The heterotrimeric Gβ protein AGB1 interacts with ZAG1, as well as the kinases of the YDA pathway, and may serve as a scaffold for this pathway. Phosphorylation of WRKY2 by the YDA cascade is required for WRKY2 to activate transcription of *WOX8* through a *cisB* element located in intron 2. HDG11 and HDG12 bind the *cisC* element located in intron 2 of *WOX8*, and are also required for *WOX8* transcription. GRD promotes YDA signaling, but is not itself a target of the YDA pathway, and thus may promote the activity of another downstream component, such as WRKY2. (C) A hypothesis for initiation of the apical cell lineage after asymmetric division of the zygote, based on the mechanism for YDA signaling in leaf epidermal cells (Jeong et al., 2016). In this model, a scaffold protein such as AGB1 or ROP3 (purple) anchors the YDA signaling cascade to the basal end of the zygote. After the asymmetric division of the zygote, YDA promotes suspensor cell fate in the basal cell, while the apical cell is released from YDA repression of embryo cell fate, allowing formation of the embryo lineage.

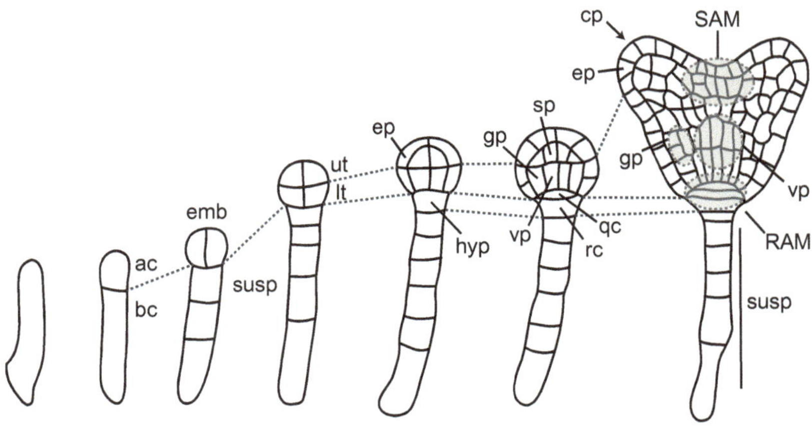

Fig. 3 Oriented cell division, cell expansion, and tissue specification in early *A. thaliana* embryogenesis. Tracings of medial optical sections of *A. thaliana* embryos from the zygote to early heart stage. The first division of the zygote produces an apical cell (ac) and a basal cell (bc). The apical cell lineage produces the entire embryo, except for the quiescent center (qc) and root cap (rc), which are produced by the basal cell lineage. The basal cell lineage also produces the suspensor (susp), a single file of cells which divides laterally until the heart stage of embryogenesis. At the octant stage, the embryo consists of four upper tier (ut) cells and 4 lower tier (lt) cells. The upper tier lineage produces the shoot apical meristem (SAM) and the cotyledon primordia (cp), while the lower tier produces the vascular tissue primordia (vp) and ground tissue primordia (gp). By the dermatogen stage, each cell of the octant embryo has divided periclinally to produce epidermal primordia (ep). The uppermost cell of the suspensor of the dermatogen stage embryo is called the hypophysis (hyp). In the globular stage embryo, the hypophysis divides once to produce an upper cell lens-shaped cell, which will form the quiescent center (qc) of the root meristem, and a lower cell which will produce the root cap (rc). The innermost cells of the upper tier constitute the shoot meristem primordia (sp), while the cells between the shoot primordia and the epidermis will form the cotyledons. The innermost cells of the lower tier form the vascular primordia (vp), while the next outer cell layer will form ground primordia (gp). At the heart stage, the shoot apical meristem (SAM) and root apical meristem (RAM) have been established, and cell proliferation on the flanks of the SAM has produced cotyledon primordia (cp). The SAM consists of epidermal (L1) cells located between the cotyledon primordia, as well as two cell layers (L2 and L3) below the epidermis. The RAM consists of the upper quiescent center cells, as well as the root cap initials (below the qc), and ground and vascular initials cells which surround the qc. Elongated vascular primordia (vp) are in the center of the embryo, with ground tissue primordia (gp) surrounding the vascular primordia. Epidermal cells (ep) surround the embryo. Lineages established by the apical cell, basal cell, upper tier, lower tier and hypophysis are marked by dotted gray lines. The approximate number of hours after fertilization and the cell numbers for each stage are shown.

the regulatory domain of YDA, as well as with MKK4, MKK5, MPK3, and MPK6 (Yuan, Li, & Yang, 2017). *agb1* mutants result in short zygotes with symmetric divisions, and *yda* and *mpk6* were both shown to have phenotypes epistatic to *agb1*, consistent with *AGB1* acting upstream of both *YDA* and *MPK6* (Yu et al., 2016; Yuan et al., 2017). AGB1 also interacts with ZYGOTIC ARREST1 (ZAG1), a plasma membrane localized receptor kinase with extracellular leucine-rich repeats (Yu et al., 2016). *zag1* mutations affect the asymmetry of the first division, genetic interactions suggest that *ZAG1* acts upstream of *SSP*, and the SSP and ZAG1 proteins were shown to interact *in vitro*, raising the tantalizing possibility that ZAG1 links SSP to extracellular signals (Yu et al., 2016) (Fig. 2B). ZAG1 also interacts with calmodulin, an intriguing result because calcium is an important regulator of fertilization in animals, and a calcium spike in the egg cell is correlated with successful fertilization in *A. thaliana* (Denninger et al., 2014; Hamamura et al., 2014). Though they do affect expression of *WOX2* and *WOX8* genes, *agb1* and *zag1* mutants do not have a fully penetrant effect on the asymmetry of the first division or the elongation of the suspensor, pointing to the difficulty in pinpointing the effect of all these new signaling components on zygote asymmetry.

The SSP-YDA-MKK4/5-MPK3/6 signaling pathway is connected to the asymmetric division of the zygote and activation of *WOX* genes through the transcription factor *WRKY2*. Unlike wild-type zygotes, where after fertilization the vacuole coalesces at the base and the nucleus moves to the apex for the asymmetric first division, *wrky2* zygotes do not repolarize after fertilization and divide symmetrically. WRKY2 activates the *WUSCHEL HOMEOBOX 8 (WOX8)* and *WOX9* transcription factors through a *cisB* element located in their second intron (Ueda et al., 2011). WRKY2 binds MPK3 and MPK6, and is activated by Serine phosphorylation by these kinases, promoting transcription of *WOX8*. A phospho-mimic version of WRKY2 can rescue the elongation and symmetric division phenotype of *ssp* zygotes, and *wrky2/ssp* double mutants are indistinguishable from *wrky2* and *ssp* single mutants, indicating that *WRKY2* acts downstream of *SSP* (Ueda et al., 2017). *WOX8* is also a target of the HOMEODOMAIN GLABROUS11 (HDG11) and HDG12 transcription factors, which promote *WOX8* expression by binding a *cisC* element in its second intron. *HDG11* and *HDG12* are expressed in the egg cell, zygote and suspensor, and *hdg11/12* double mutants show a maternal effect on elongation of the zygote and its asymmetric division (Ueda et al., 2017). These results suggest that both paternal (SSP) and maternal (HDG11/12) factors converge on the *WOX8* promoter to promote zygote asymmetry and embryo patterning (Fig. 2B).

Several other genes and pathways promote zygote elongation and the asymmetry of the first division, though their relationship to the *YDA* pathway is not entirely clear. Mutations in the *GROUNDED (GRD)* gene, encoding an RKD transcription factor, affect zygote elongation, reduce suspensor cell identity, and increase apical cell identity, reminiscent of *yda* mutants (Jeong, Palmer, & Lukowitz, 2011; Waki, Hiki, Watanabe, Hashimoto, & Nakajima, 2011). In addition, *grd* mutants have a strong synergistic interaction with *wox8/wox9* mutants, causing zygote arrest. Forced expression of *GRD* converts basal to apical cell fates, and dominant expression of *YDA* requires *GRD* to convert apical to basal cell fates, suggesting that *GRD* acts downstream of *YDA*. However, *GRD* is apparently not regulated by the *YDA* kinase cascade (Jeong et al., 2011). Thus, GRD is required for *YDA* signaling, perhaps as a co-factor for downstream targets of YDA, such as *WRKY2*. Polar transport of the phytohormone auxin, and transcriptional responses to this hormone, is also required for the asymmetry of the first division of the zygote, and for correct divisions in the derivatives of the apical cell (Berleth & Jürgens, 1993; Friml et al., 2003; Hamann, Mayer, & Jürgens, 1999; Hardtke & Berleth, 1998; Mayer & Jürgens, 1993; Yoshida et al., 2014). Many *AUXIN RESPONSIVE FACTOR (ARF)* transcription factor genes with cell type specific expression patterns exist in *A. thaliana*, but the exact *ARF* genes that promote the asymmetry of the first division are currently not known (Rademacher et al., 2011). microRNAs have also been shown to be required for polarization of the zygote and its first asymmetric division; *WOX2* and *WOX8* expression; and establishment of early cell fates in the embryo (Armenta-Medina et al., 2017; Nodine & Bartel, 2010; Seefried, Willmann, Clausen, & Jenik, 2014). However, the exact miRNAs that promote these functions remain to be determined.

5. Specification of the epidermis, vascular, and ground tissue

After the asymmetric division of the zygote establishes the main axis of the embryo, the apical and basal cell lineages adopt different patterns of cell division and growth, and new cell fates are generated with almost every cell division. The apical cell undergoes two longitudinal divisions to produce the four-celled proembryo, which swells isodiametrically. All cells divide anticlinally (perpendicular to the surface), giving rise to an eight-celled embryo containing an upper tier of four cells, which will produce the shoot structures of the plant, and a lower tier of four cells, which will produce the hypocotyl (embryonic stem). Each of these 8 cells then divides periclinally (parallel to the surface of the embryo) to produce the proto-epidermis. This dermatogen stage embryo has acquired an axis of radial polarity, with internal and external tissues. The inner cells of the upper tier then divide longitudinally, producing internal cells which are the precursors of the shoot meristem, and a middle layer of cells which eventually generate the cotyledons (lateral organs of the embryo with a structure like leaves). The inner cells of the lower tier also divide longitudinally, producing internal cells that are the primordia of the vascular tissue, and a middle layer of cells that are the primordia of the ground tissue (structural tissue) of the stem and root. Thus, the early globular (32 cell) embryo contains precursors of the epidermis, vascular tissue, ground tissue, and shoot meristem. By the mid heart stage, the embryo has initiated cotyledon primordia, and contains shoot and root meristems, several files of vascular primordia, two layers of ground tissue primordia, and the epidermis (Barton & Poethig, 1993; Jürgens & Mayer, 1994; see Jenik, Gillmor, & Lukowitz, 2007 for a comprehensive review of the first 10 years of molecular studies on early *A. thaliana* embryogenesis). Meanwhile, the basal cell lineage undergoes simple anticlinal divisions to produce the suspensor, an extra-embryonic tissue which pushes the embryo into the nutrient rich endosperm, and which serves as a conduit for nutrients from the mother plant (Kawashima & Goldberg, 2010). The uppermost cell of the suspensor, called the hypophysis, generates the quiescent center of the root meristem, and the root cap (Fig. 3).

The phytohormone indole acetic acid, commonly known as auxin, is required for most aspects of early embryogenesis, including specification of the epidermis, vascular tissue, and root meristem. Indole acetic acid is synthesized from the amino acid tryptophan in a two-step process, by the enzymes TRYPTOPHANE AMINOTRANSFERASE OF ARABIDOPSIS 1 (TAA1) and the YUCCA monoxygenases (YUC1-YUC11) (reviewed in Zhao, 2012). Auxin regulates gene transcription by degrading the INDOLE ACETIC ACID (IAA) proteins *via* the ubiquitin pathway (Dharmasiri, Dharmasiri, & Estelle, 2005; Kepinski & Leyser, 2005). IAA proteins (not to be confused with indole acetic acid itself) are repressors of ARF transcription factors. In the presence of auxin, the auxin receptor TRANSPORT INHIBITOR RESPONSE 1 (TIR1) ubiquitinates IAA proteins, leading to their destruction and the derepression of ARF proteins, which can then promote (or repress) auxin target genes (reviewed in Parry & Estelle, 2006). Auxin transcriptional responses can be monitored by use of reporters containing synthetic ARF binding elements, such as DR5, placed in front of a reporter gene encoding a green fluorescent protein (DR5::GFP) (Friml et al., 2003). Polar transport of auxin is essential for early embryogenesis, and is accomplished primarily by PIN FORMED (PIN) auxin efflux proteins, most of which are localized in the plasma membrane, and transport auxin out of the cell into the extracellular space. By contrast, AUXIN RESISTANT (AUX) and LIKE AUXIN RESISTANT (LAX) proteins promote uptake of auxin into the cell, increasing the efficiency by which PIN proteins can create auxin gradients within tissues (Robert et al., 2015).

Evidence from DR5-driven GFP expression and the localization of the PIN7 protein suggests that the zygote and early embryo are initially a sink for auxin. Recent experiments have strongly suggested that the integuments of the seed coat are the source of this auxin (Robert et al., 2018). After the first division of the zygote, PIN7 is localized at the apical end of the basal cell, and a weak DR5 signal can be detected in the apical cell. PIN1 is observed in the embryo beginning at the octant stage, and is initially found in the plasma membrane in the interior part of all cells of the embryo (Friml et al., 2003). At the dermatogen stage, PIN1 is localized at the apical sides of the lower epidermal cells, and *TAA1* is expressed in the epidermal cells of the upper tier (Robert et al., 2013). Two rounds of cell division later, at the globular stage, an auxin gradient is established where auxin is transported apically in the epidermis, and basipitally in the internal provascular tissue. This is accomplished by localization of PIN1 proteins at the apical end of the lower tier epidermal cells and at the basal end of the lower tier internal cells, along with synthesis of auxin in the epidermis of the upper tier by TAA1, YUC1 and YUC4 (Friml et al., 2003; Robert et al., 2013). The internal, basipetal transport of auxin creates a strong maximum of auxin transcriptional response in the hypophysis, which will give rise to the quiescent center of the root meristem. Consistent with this new basipetal auxin transport pattern, PIN7 proteins change their polarity to basal, transporting auxin into the suspensor (reviewed in Jeong et al., 2016) (Fig. 4).

The most compelling evidence for the importance of auxin transport and transcriptional response in early embryogenesis comes from the analysis of mutations affecting the *PIN*, *ARF* and *IAA* genes. *pin7* embryos have cell division orientation defects in the apical cell, and affect the definition of the boundary between the proembryo and the suspensor, with these defects becoming more extreme when the function of additional PIN proteins is compromised. Combinations of *pin1*, *pin3*, *pin4* and *pin7* mutants affect the oriented cell divisions creating the protodermis, and result in lack of

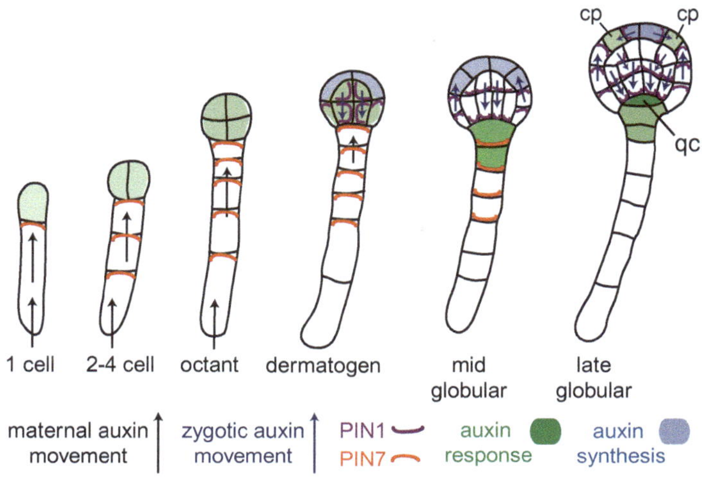

Fig. 4 Polar auxin transport, synthesis and transcriptional responses in early *A. thaliana* embryogenesis. The early embryo is initially a sink for maternally supplied auxin, produced in the integuments of the ovule and transported through the suspensor to the embryo by PIN7, where a weak auxin transcriptional response is induced. Beginning at the dermatogen stage, the embryo begins to synthesize its own auxin, and PIN1 is expressed in the interior plasma membranes of the interior cells of the embryo, as well as on the apical end of the lower tier protodermal cells. By the globular stage, PIN1 proteins in the lower tier protodermal cells continue to transport auxin apically, while a basal gradient of auxin is established in the interior of the embryo, resulting in a strong auxin transcriptional response (as monitored by DR5::GFP expression) in the hypophysis and the upper cells of the suspensor. At this stage, PIN7 proteins have changed their polarity from apical to basal, and are now transporting auxin away from the hypophysis. In late globular embryos, the basal transport of auxin in the embryo has triggered a high auxin transcriptional response in the hypophyseal cell, triggering an asymmetric division which produces a lens-shaped cell which will form the quiescent center (qc) of the root meristem. Apical transport of auxin in the lower epidermal cells, as well as synthesis of auxin in the upper epidermis, results in formation of a weak auxin response on the upper flanks of the embryo, where the cotyledon primordia (cp) will form. Movement of this auxin from the epidermis to the center of the embryo, and subsequently to the lens-shaped cell, will establish the pattern of provascular tissue from the cotyledon primordia to the qc. Once cotyledon primordia are established, a minimum of auxin and maximum of cytokinin responses is required to establish the shoot apical meristem between the vascular strands leading to the two cotyledons. *Figure design based on Jenik, P. D., Gillmor, C. S., & Lukowitz, W. (2007). Embryonic patterning in* Arabidopsis thaliana. Annual Review of Cell and Developmental Biology, 23, *207–236.*

vascularization along the apical-basal axis. Loss of *MONOPTEROS (MP)* (also known as *ARF5*), or gain of function of *BODENLOS* (also known as *IAA12*), causes similar defects, demonstrating that this ARF/IAA pair plays important roles in pattern formation in early embryogenesis (Berleth & Jürgens, 1993; Friml et al., 2003; Hamann et al., 1999).

3D reconstructions of embryo cell division and cell expansion patterns have shown in more detail the exact cellular geometries created by oriented cell divisions in early embryogenesis. Using mutants affecting auxin transcriptional responses, these reconstructions have demonstrated that auxin is required for the periclinal divisions in the octant stage embryo that create the proto-epidermis (Yoshida et al., 2014). These divisions are unusual because they deviate from the default mode of cell division in plants, which is to form new cell walls of the shortest distance possible (Besson & Dumais, 2011). Recent modeling of cell division planes and microtubule orientation in early embryos has provided evidence that the first few oriented divisions of the *A. thaliana* zygote can be explained by the effect of cell shape on cortical microtubule orientation, the regulation of microtubule stability at the edges of cells, and the effect of auxin on stabilizing microtubule orientation along newly formed cell walls (Chakrabortty et al., 2018). Oriented cell divisions are also essential for the specification of precursors for vascular and ground tissue, which are generated by longitudinal division of the inner cells of the dermatogen stage embryo. The inner daughter cell of this division is the vascular precursor, and the outer daughter cell is the precursor of ground tissue. The transcription factor *MP* is essential for establishing both these cell fates, through *TARGET OF MONOPTEROS (TMO)* genes, transcription factors that are direct targets of MP (Schlereth et al., 2010). This periclinal division and specification of the vascular precursors is regulated by TMO5 and its binding partner LONESOME HIGHWAY (De Rybel et al., 2013). After promoting the division that leads to the initiation of vascular and ground tissue cells, *MP* promotes the asymmetric division of the ground cell precursor causing the proliferation of ground tissue cells (Möller et al., 2017).

Development of the epidermis depends on the homeodomain transcription factors *A. thaliana MERISTEM LAYER 1 (AtML1)* and *PROTODERMAL FACTOR 2 (PDF2)*, both of which are expressed in the protoderm beginning at the dermatogen stage (Abe, Katsumata, Komeda, & Takahashi, 2003; Lu, Porat, Nadeau, & O'Neill, 1996). In the absence of expression of both *AtML1* and *PDF2*, embryos do not develop a functional epidermis (Abe et al., 2003). WOX transcription factors also promote differentiation of epidermal, vascular and ground tissue, though their downstream targets are not defined. *WOX8* and *WOX9* are required for specification of epidermal, vascular and ground tissues, as *wox8/wox9* double mutants develop only double files of cells in the embryo proper, with ectopic protrusions (Breuninger, Rikirsch, Hermann, Ueda, & Laux, 2008). *wox2* embryos show defects in epidermal cell divisions beginning at the dermatogen stage (Haecker et al., 2004). *WOX8* and *WOX9* act upstream of auxin transcription responses, as *wox8/wox9* double mutants have an expanded DR5-driven GFP expression from the hypophysis to throughout the embryo, and lack *PIN1* expression (Breuninger et al., 2008). However, *wox2/wox8* mutants enhance the phenotype of *mp* embryos (Breuninger et al., 2008), such that the exact relationship between WOX transcription factors and auxin in early patterning of the epidermis, vascular and ground tissue remains to be discovered.

6. Initiation of shoot and root meristems

The shoot apical meristem (SAM) and root apical meristem (RAM) can be divided into several zones with distinct properties: the organizing center (OC) cells that divide slowly to provide cells for the surrounding stem cells; the stem cells that divide often; and a third region with proliferative cells that are not able to self-renew and differentiate after dividing several times (Weigel & Jürgens, 2002). The initiation and maintenance of the different zones within meristems involve different transcriptional networks, hormone signaling pathways, and cell-cell communication. Here, we focus on the establishment of the SAM and RAM during embryogenesis, as the maintenance of the meristems has been discussed in past reviews (Soyars, James, & Nimchuk, 2016; Wendrich & Weijers, 2013).

The shoot meristem of *A. thaliana* embryos is morphologically distinguishable at the late heart stage of embryogenesis, as the small cytoplasmic cells of the epidermis (the L1 cells) located between the cotyledon primordia, as well as the two underlying cell layers (the L2 and L3) (Barton & Poethig, 1993) (Fig. 5A). These three cell layers constitute the stem cells of the shoot meristem, and are marked by expression of the *CLAVATA3 (CLV3)* gene, which encodes a small peptide that promotes the stem cell niche (Reddy & Meyerowitz, 2005). The organizing center of the stem cell niche is located directly underneath the L3 layer, and is marked by expression of the WUSCHEL (WUS) transcription factor. The size of the shoot meristem is controlled by a regulatory loop between *WUS* and *CLV3*, where *WUS* promotes *CLV3* expression, and *CLV3* represses *WUS* expression (Schoof et al., 2000). At the same time, *WUS* maintains the organizing center itself by repressing differentiation

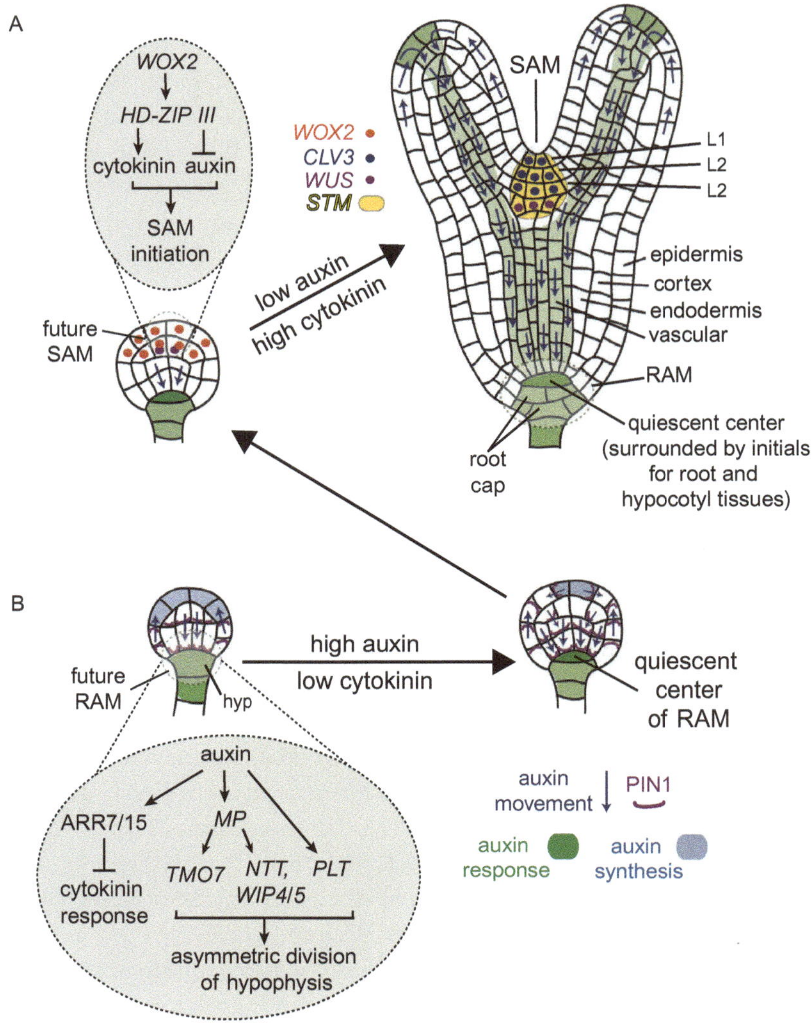

Fig. 5 Specification of shoot and root apical meristems. (A) The first molecular marker of the shoot apical meristem is *WUS*, which is expressed in the inner two cells of the upper tier at the dermatogen stage. In late globular embryos, *WOX2* is expressed throughout the upper tier of the embryo. *WOX2* is essential for initiation of the SAM, where it promotes expression of HD-ZIP III transcription factors, which in turn promote cytokinin responses and repress auxin responses. This high cytokinin/low auxin response environment is required for initiation of the SAM, which develops into a three-layered structure, with L1, L2 and L3 layers expressing *CLV3*, and *WUS* expressed in the organizing center below the L3 layer. (B) In contrast to the SAM, RAM specification requires a high auxin, low cytokinin environment. Basal auxin transport creates high auxin levels in the hypophysis cell (the uppermost cell of the suspensor). Auxin promotes transcription of the cytokinin response regulators *ARR7* and *ARR15,* which repress cytokinin responses. Auxin also promotes *MP* expression, which triggers the asymmetric division of the hypophysis through its targets *TMO7*, *NTT*, *WIP4* and *WIP5*. Auxin directly promotes expression of the *PLT* family of transcription factors, which also promote asymmetric division of the hypophysis. By the late heart stage (shown on the right in (A)), root cap, vascular, endodermal, and cortex tissues have been specified, with their initials surrounding the quiescent center.

(Ikeda, Mitsuda, & Ohme-Takagi, 2009). Meanwhile, the transcription factor SHOOTMERISTEMLESS (STM) maintains the meristem in an undifferentiated state by preventing incorporation of cells of the SAM into lateral organ primordia (Fig. 5A). CUP SHAPED COTYLEDON (CUC) transcription factors initially promote *STM* expression, and then act to restrict its domain and to delineate the boundaries of organ primordia (Long & Barton, 1998).

Expression of *WUS* can first be detected at the dermatogen stage, in the inner cells of the upper tier. After the anticlinal division of the upper tier cells, *WUS* expression is restricted to the innermost daughter cells, and by the heart stage, *WUS* remains restricted to a small number of cells underlying the *CLV3* expression domain (Fig. 5A). *wus* mutants terminate the SAM at the seedling stage, indicating that *WUS* is absolutely required for meristem maintenance (Laux, Mayer, Berger, & Jürgens, 1996). However, *wus* mutants initiate *CLV3* expression normally at the heart stage, demonstrating that, despite its early and specific expression in precursors of the organizing center of the SAM, *WUS* is not required for initiation of the SAM (Zhang, Tucker, Tucker, Hermann, & Laux, 2017). Recent results demonstrate that the *WUS*-related gene *WOX2*, along with *WOX1*, *WOX3* and *WOX5*, are the key genes for initiation of the SAM (Zhang, Tucker, et al., 2017). *WOX2* is expressed in the egg cell and zygote, and marks the apical cell lineage after the first division of the zygote (Breuninger et al., 2008). *wox2* mutants affect divisions in the protoderm, and have a low penetrance of defects in cotyledon initiation. The latter is greatly enhanced by combining with mutations in the *WOX* subclade genes *WOX1*, *WOX3* and *WOX5*: *wox1/wox2/wox3/wox5* quadruple mutants have variable effects on seedlings, producing some seedlings which completely lack a shoot meristem (Breuninger et al., 2008).

Using the stem cell marker *CLV3*, Zhang, Tucker, et al. (2017) showed that *wox1/2/3/5* quadruple mutants cause initiation of the SAM to be delayed or eliminated, and that the orientation of cell divisions in the precursors of the SAM is aberrant in *wox1/2/3/5* mutants. Aberrant cell divisions and lack of *CLV3* expression in the SAM precursor region are due to elevated auxin transport and auxin transcriptional activity, and decreased synthesis of the phytohormone cytokinin. Zhang, Tucker, et al. (2017) showed that the HD-ZIP III transcription factors *PHABULOSA (PHB)*, *PHAVOLUTA (PHV)* and *REVOLUTA (REV)* act downstream of *WOX2* to repress auxin activity, and promote cytokinin activity, in the cells that form the SAM

(Fig. 5A). This result is very satisfying because it agrees with classic experiments in plant tissue culture showing that new shoot meristems can be induced by treatment of plant tissues with high cytokinin and low auxin (Skoog & Miller, 1957). Nonetheless, important questions remain, such as which factors act upstream of *WUS*, and which are responsible for maintaining the SAM once it has been initiated by *WOX2* activity.

Unlike the specification of the SAM, which requires cytokinin signaling and a minimum of auxin transcriptional responses, specification of the root apical meristem (RAM) requires an auxin maximum and a cytokinin minimum (Fig. 5B). This is achieved starting at the early globular (32 cell) stage, when basipetal transport of auxin creates an auxin maximum in the hypophysis, the uppermost cell of the suspensor (Friml et al., 2003). At the same time, auxin represses cytokinin responses by direct transcriptional regulation of the *ARABIDOPSIS RESPONSE REGULATOR (ARR)* genes *ARR7* and *ARR15*, which repress cytokinin signaling (Müller & Sheen, 2008). The hypophyseal cell undergoes an asymmetric division, where the lens-shaped upper daughter cell establishes the organizing center of the RAM, known as the quiescent center (QC). The cells surrounding the QC form the stem cells of the root meristem: the lower daughter cell of this asymmetric division forms the columella stem cells (responsible for producing the root cap), and the cells above the QC form the stem cells of the epidermis, ground and vascular tissue (reviewed in Palovaara et al., 2016).

The ARF transcription factor *MP* plays a major role in hypophysis specification, as *mp* mutant embryos show aberrant hypophysis divisions, eventually resulting in rootless seedlings (Berleth & Jürgens, 1993; Hardtke & Berleth, 1998) (Fig. 5B). *MP* directly promotes asymmetric division of the hypophysis through at least two different downstream pathways. One pathway involves the non-cell autonomous *TMO7* gene, encoding a small bHLH transcription factor which is expressed in the provascular cells adjacent to the hypophysis, and moves into the hypophysis to promote asymmetric division (Schlereth et al., 2010). *MP* also promotes the asymmetric division of the hypophysis in a cell-autonomous manner, through its direct targets *NO TRANSMITTING TRACT (NTT), WIP DOMAIN PROTEIN4 (WIP4)* and *WIP5,* which are expressed in the hypophysis. *ntt/wip4/wip5* triple mutants lack asymmetric divisions of the hypophysis, and do not form roots (Crawford et al., 2015). Auxin also promotes hypophysis divisions through *ARF9* (and by repressing *IAA10*) (Rademacher et al., 2012). The APETALA2-domain transcription factors *PLETHORA1*

(PLT1), PLT2, AINTEGUMENTA-LIKE6 (AIL6) and *BABYBOOM (BBM)* together promote the asymmetric division of the hypophysis. Expression of *PLT* genes is dependent on auxin (Aida et al., 2004), and these transcription factors show a gradient of expression that is highest in lens-shaped cell, and decreases distally along the maturing root (Galinha et al., 2007). After the asymmetric division of the hypophysis establishes the lens-shaped cell, its QC identity is maintained by *WOX5*, in part by restricting cell divisions through repression of *CYCLIN D* (Forzani et al., 2014; Sarkar et al., 2007).

7. Techniques and resources for molecular studies of embryogenesis

Recent technical advances in embryo culture and microscopy, and improved genomics resources, have led to improvements in our understanding of early embryogenesis in flowering plants. In this section, we review recent advances in techniques that are valuable for the study of early embryogenesis, and transcriptome and epigenome studies that are useful sources of information for hypothesis generation and interpretation of mutant phenotypes of previously studied genes. Table 1 lists papers describing these genomic resources, which the reader is encouraged to consult for additional information.

In vitro ovule cultivation methods have been developed for live cell imaging of developing zygotes and embryos of *A. thaliana* (Gooh et al., 2015; Kimata et al., 2016; Kurihara, Kimata, Higashiyama, & Ueda, 2017). Advances in microcopy, such as spinning-disk confocal and two-photon excitation, allow specimen penetration two or three times greater than traditional confocal microscopy, and cause less photobleaching and photodamage. These new methods and techniques have led to a fate map of early embryogenesis, to demonstration that the endosperm is not required for early embryo development (Gooh et al., 2015; Ngo et al., 2007), and have shown that laser ablation of the apical cell of the embryo leads to differentiation of a new apical cell from the suspensor lineage, indicating that the suspensor cells are totipotent in early embryogenesis (Liu et al., 2015). Combined with molecular markers for the cytoskeleton, *in vitro* embryo culture and advanced microscopy have uncovered the real-time dynamics of zygote polarization, such as nuclear migration, cytoskeleton rearrangement, and the timing of cell division (Kimata et al., 2016).

Table 1 Transcriptome and epigenome studies of embryos, endosperms and seeds.

Plant species	Tissue and developmental stage	Isolation method	Sequencing method	Major contributions	References
A. thaliana	– Globular – Heart – Torpedo	Laser-capture microdissection (LCM)	Microarrays	Reported that temporal differences in gene expression are more significant than spatial differences	Spencer, Casson, and Lindsey (2007)
A. thaliana	– Zygote – Quadrant – Globular – Heart – Torpedo – Bent – Mature	Manual	Microarrays	Differentiation of gene expression patterns between different stages of embryogenesis	Xiang et al. (2011)
A. thaliana	– 2–4 cell – Globular	Manual	RNAseq	Allele-specific profiling revealed maternal dominance at early stages of embryo development	Autran et al. (2011)
A. thaliana	– Embryo Proper (EP) – Suspensor (SUS) – Micropylar Endosperm (MCE) – Peripheral Endosperm (PEN) – Chalazal Endosperm(CZE) – Chalazal Seed Coat (CZSC) – Distal Seed Coat(SC) From fertilization through maturity	Laser-capture microdissection (LCM)	Microarrays	Most complete transcriptional profiles by embryo, endosperm and seed regions	Belmonte et al. (2013)

A. thaliana	– Proembryo – Suspensor – Whole embryo	FANS	Microarrays	Reported minimal contamination with cellular or nuclear material from embryo-surrounding cells	Slane et al. (2014)
A. thaliana	– Globular stage embryos	Manual dissection	RNA sequencing	Profiled pri-miRNAs (miRNA precursors) present in embryos	Armenta-Medina et al. (2017) and Lepe-Soltero et al. (2017)
A. thaliana	– Endosperm nuclei	INTACT	– CHIP – DNA bisulfite sequencing	Study of epigenetic dynamics of endosperm	Moreno-Romero, Santos-González, Hennig, and Köhler (2017)
A. thaliana	– Root stem cell niche	INTACT	Microarrays	Adapted, optimized and validated the INTACT method to efficiently isolate cell-type-specific nuclei from *A. thaliana* embryo	Palovaara et al. (2017)
A. thaliana	– Epidermal cells of torpedo stage	Laser Assisted Microdissection (LAM)	RNAseq	Used ultra-low quantity of RNA	Sakai et al. (2018)
Nicotiana tabacum	– Egg cells – Zygotes – Two-celled embryos	Manual	EST sequencing	Maternal-to-zygotic transition (MZT) is initiated prior to zygotic division in tobacco	Zhao et al. (2011)

Continued

Table 1 Transcriptome and epigenome studies of embryos, endosperms and seeds.—cont'd

Plant species	Tissue and developmental stage	Isolation method	Sequencing method	Major contributions	References
Zea mays	– Endosperm at 10 DAP	Manual	RNAseq	Using reciprocal crosses between B73 and Mo17 identified a total of 38 noncoding imprinted RNAs. 25 are maternally expressed whereas 13 are paternally expressed	Zhang et al. (2011)
Zea mays	– Endosperm 14 DAP – Embryo 14 DAP	Manual	RNAseq	Reciprocal crosses between B73 and Mo17 identify 100 putative imprinted genes in maize endosperm, some of them conserved with rice and *A. thaliana*	Waters et al. (2011)
Zea mays	– Endosperm 14 DAP from 5 reciprocal hybrids	Manual	RNAseq	Found 500 genes that exhibit statistically significant parent-of-origin effects in maize endosperm tissue	Waters et al. (2013)
Zea mays	– Embryo 9DAP – Endosperm 9DAP	Manual	RNAseq	Mapped the transcriptional differences between the embryo and endosperm and the transcriptional network governing maize seed development	Lu et al. (2013)

Zea mays	– Embryo at 15 DAP – Embryo at 21 DAP – Embryo at 27 DAP	Manual	RNAseq	Different patterns of expression of genes involved in embryo development and storage protein accumulation	Teoh et al. (2013)
Zea mays	– Whole seed at 0, 3, and 5 (DAP) – Endosperms at 7, 10, and 15 DAP	Manual	RNAseq	Using B73 and Mo17 reciprocal crosses, authors show that 10 DAP endosperm-specific Maternally Expressed Genes (MEGs) are involved in nutrient uptake and allocation and the auxin signaling pathway	Xin et al. (2013)
Zea mays	– Aleurone endosperm – Embryo-surrounding region endosperm – Basal endosperm transfer layer – Central starchy endosperm – Conducting zone endosperm All at 6DAP	LCM	RNAseq	Report a module that includes GRNs and might regulate maize endosperm cell differentiation	Zhan et al. (2015)
Zea mays	– Embryos (10–38 DAP) – Endosperm (6–38 DAP) – Whole seed (0–38 DAP)	Manual	RNAseq	Spatio-temporal transcriptome atlas of B73 maize seed	Chen et al. (2014)

Continued

Table 1 Transcriptome and epigenome studies of embryos, endosperms and seeds.—cont'd

Plant species	Tissue and developmental stage	Isolation method	Sequencing method	Major contributions	References
Zea mays	– 5 DAP embryo – 7 DAP embryo – 9 DAP embryo – 11 DAP embryo – 13 DAP embryo – 5 DAP endosperm From reciprocal crosses between B73 and Mo17	– Manual – Laser-capture microdissection (LCM)	RNAseq	Gene imprinting occurs in maize embryos	Meng et al. (2018)
Oryza sativa	– Endosperm 5DAP – Embryos 6DAP	Manual	RNAseq	Identification of 262 candidate imprinted loci in the endosperm, and only three in the embryo	Luo et al. (2011)
Oryza sativa	Spatial datasets of: – Early globular – Late globular – Coleoptiler	Laser Microdissection (LM)	Microarrays	Identified homologous genes from *A. thaliana* with known functions in embryogenesis and unique and uncharacterized genes with expression patterns during embryogenesis	Itoh et al. (2016)
Oryza sativa	– Isogenic zygotes at three time points: 2.5, 5, and 9 hap – Hybrid zygotes japonica × indica (2.5 hap) and indica × japonica (2.5 hap) – Hybrid zygotes japonica × indica (9 hap) and indica × japonica (9hap)	Manual	RNAseq	In rice, zygotic genome activation (ZGA) occurs in the unicellular zygote, with a maternal transcript bias	Anderson et al. (2017)

Solanum peruvianum and *Solanum chilense*	14 dap endosperms from reciprocal crosses of *S. peruvianum* × *S. chilense*	Laser-assisted microdissection	RNA sequencing	Inter-specific hybridization has drastic effects on imprinted gene expression in the endosperm	Florez-Rueda et al. (2016)
Triticum aestivum L	– Egg cell – 2-Cell proembryo	Manual	EST sequencing	Composition of two-celled proembryos was significantly distinct from egg cells, reflecting transcriptional activity in 2-cell proembryo	Sprunck et al. (2005)
Pinus pinaster	– Early embryo – Pre-cotyledonary embryos – Early cotyledonary embryos – Cotyledonary embryos – Mature embryos	Manual	Microarrays	Comparative analysis with *A. thaliana* embryogenesis highlighted genes involved in auxin responses and epigenetic regulation	de Vega-Bartol et al. (2013)
Brassica napus	– Embryo at 17 DAP – Embryo at 35 DAP – Embryo at 52 DAP	Manual	RNAseq	Reported differentially expressed genes involved in fatty acid metabolism	Deng, Yan, Zhang, Tang, and Yuan (2015)
Linum usitatissimum	– Globular – Heart – Torpedo – Cotyledon – Mature – Seed coats at Globular – Seed coat at Torpedo – Endosperm (pooled globular to torpedo stages)	Manual	Sanger sequencing of ESTs	Reported unigenes likely representing flax-specific genes	Venglat et al. (2011)

Due to the importance of auxin in embryo patterning, a collection of fluorescent transcriptional reporters for the *ARF* gene family has been used to generate a map of *ARF* expression during embryogenesis (Rademacher et al., 2011). Versatile dyes like Renaissance (Musielak, Schenkel, Kolb, Henschen, & Bayer, 2015), and fluorescent markers for subcellular structures, such as the plasma membrane, plasmodesma, endosomes, Golgi apparatus, nuclear pores, F-actin, have been developed and will allow more detailed investigations into cell biological aspects of early embryo development (Liao & Weijers, 2018; Wendrich et al., 2015). Computational analysis of detailed confocal reconstructions of early embryos has led to new insights as to the exact geometries of early cell divisions (Belcram, Palauqui, & Pastuglia, 2016; Yoshida et al., 2014). To study epigenetic regulation during early embryogenesis, protocols to immunolocalize chromatin marks in young whole-mount *A. thaliana* embryos have been developed (Garcia-Aguilar & Autran, 2018).

As discussed throughout this review, research on gene expression at the cellular level is essential for a mechanistic understanding of early embryo development. Many studies have carefully documented the expression and function of several genes simultaneously. However, because of the size and relative inaccessibility of plant embryos, genome-wide approaches to gene expression in embryos have been challenging. Nevertheless, in the last 10 years, embryo transcriptomes have been produced from many different species, and have led to important insights. Manual and laser capture microdissection have been used to isolate whole embryos at different developmental stages, and to isolate specific embryonic tissue domains for transcriptional analysis of species such as *A. thaliana*, flax, *Pinus*, maize, *Brassica*, rice, tomato and tobacco (Belmonte et al., 2013; de Vega-Bartol et al., 2013; Florez Rueda, Grossniklaus, & Schmidt, 2016; Itoh et al., 2016; Lu et al., 2013; Venglat et al., 2011) (Table 1).

The first embryo transcriptomes of *A. thaliana* used microarray analysis of RNA isolated by laser-capture microdissection of discrete tissues of globular, heart, and torpedo stage embryos. Spatial differences in gene expression were found to be less significant than temporal differences (Spencer et al., 2007). Xiang et al. (2011) isolated whole live *A. thaliana* embryos by hand dissection, and performed genome-wide profiling using microarrays, revealing genes with maternal or/and paternal contributions to zygote development, chromosomal level clustering of temporal expression in embryogenesis, and embryo-specific functions. Protocols for efficient isolation of young *A. thaliana* embryos, yielding up to 40 embryos in 1–4 h, were also developed (Autran et al., 2011; Raissig, Gagliardini, Jaenisch, Grossniklaus, & Baroux, 2013). Embryos isolated by this protocol were shown to be suitable for RNA sequencing, analysis of DNA methylation, fluorescence *in situ* hybridization, immunostaining, and reporter gene assays. The most complete transcriptional profiles of gene activity in regions and subregions of the *A. thaliana* seed have been described in Belmonte et al. (2013). Using laser-capture microdissection (LCM) and microarrays, they profiled gene expression in the embryo proper and suspensor; micropylar, peripheral, and chalazal subregions of the endosperm; and the chalazal and distal seed coat. Another study profiled parent-of-origin gene expression in the endosperm of inter-specific hybrids of tomato (Florez-Rueda et al., 2016). More recently, protocols for LCM have been optimized, allowing transcriptomic studies of single cells such as egg cells (Florez Rueda et al., 2016). Sakai et al. (2018) also developed a protocol for RNA sequencing (RNAseq) analysis of ultra-low quantities of template, combining laser-assisted microdissection and RNAseq for transcriptomic analysis of epidermal cells of torpedo stage embryos.

Fluorescence-activated nuclear sorting (FANS) of embryos with specific domains marked by GFP expression has been used to analyze cell type-specific transcriptomes in the early *A. thaliana* embryos (Slane et al., 2014). INTACT (Isolation of Nuclei Tagged in specific Cell Types), another method that relies on cell-specific markers to purify nuclei of discrete cell types, has been used to generate a transcriptome atlas of early *A. thaliana* embryo development with a focus on the root stem cell niche (Palovaara et al., 2017). A modification of the INTACT method was used to purify endosperm nuclei and generate parent-specific epigenome profiles. Nuclei obtained by this method can be used for ChIP and DNA bisulfite treatment followed by next-generation sequencing to study histone modifications and DNA methylation profiles (Moreno-Romero et al., 2017), suggesting that this protocol could be adapted for epigenetic studies of early embryos. Due to the difficulty of isolating such tiny amounts of target cells or nuclei from embryos that are surrounded by thousands of cells of maternal tissue, adequate controls must be performed to determine the degree of maternal contamination (Schon & Nodine, 2017).

RNAseq experiments have also been performed on seeds of monocots such as rice and maize, as well as other species of economic importance. Xin et al. (2013) sequenced RNA from hand dissected 3 and 5 DAP maize seed (embryo and endosperm), and Lu et al. (2013) sequenced RNA from 9 DAP seeds of maize (embryo and endosperm). Laser-capture microdissection and microarray analysis were used to study spatial and temporal gene expression in rice (Itoh et al., 2016). Crops with seeds of industrial importance like canola (*Brassica napus*), and flax (*Linum usitatissimum*) have been studied to generate molecular resources that can be utilized toward improving seed traits. Venglat et al. (2011) reported EST libraries of the developing seed partitioned into embryo, endosperm, and seed coat compartments relative to vegetative tissues. RNA from developing canola embryos at 17, 35 and 52 DAP has been sequenced (Deng et al., 2015). Microarray analysis of several different stages of embryo development in the gymnosperm loblolly pine (*Pinus pinaster*) has also been conducted (de Vega-Bartol et al., 2013).

8. Conclusions and perspectives

In the last 10 years, tremendous progress has been made on understanding the genetic and molecular mechanisms that drive early embryogenesis in flowering plants. Much of this knowledge stems from studies of *A. thaliana*, an excellent model for embryogenesis due to its tremendous genetic and genomic resources, and relative ease of observation of fixed and live embryos. However, important questions on embryogenesis remain to be answered. These include determining the relative importance of maternal and zygotic regulation of early embryogenesis, which factors establish the initial polarity of the zygote, and how the principle tissues of the embryo are established. More broadly, it will be important to elucidate the mechanisms that control early embryogenesis in monocots, which have important differences with dicots such as *A. thaliana*. These differences include a lack of regular cell divisions in the early embryo, and the formation of a persistent and massive endosperm in the seed. It will also be of great interest to determine which mechanisms of development from higher plants have been conserved during evolution, by studying embryogenesis in more basal plants such as gymnosperms, liverworts and ferns.

Acknowledgments

We apologize to colleagues whose work was not mentioned due to space limitations. Research on embryogenesis in the Gillmor laboratory is funded by CONACyT Ciencia Básica Grant 237480, and by Grant CN-17-64 from the University of California Institute for Mexico and the United States (UC MEXUS) and the Consejo Nacional de Ciencia y Tecnología de México (CONACyT).

References

Abe, M., Katsumata, H., Komeda, Y., & Takahashi, T. (2003). Regulation of shoot epidermal cell differentiation by a pair of homeodomain proteins in *Arabidopsis*. *Development (Cambridge, England)*, *130*(4), 635–643.

Aida, M., Beis, D., Heidstra, R., Willemsen, V., Blilou, I., Galinha, C., et al. (2004). The PLETHORA genes mediate patterning of the *Arabidopsis* root stem cell niche. *Cell*, *119*(1), 109–120.

Anderson, S. N., Johnson, C. S., Chesnut, J., Jones, D. S., Khanday, I., Woodhouse, M., et al. (2017). The zygotic transition is initiated in unicellular plant zygotes with asymmetric activation of parental genomes. *Developmental Cell*, *43*(3), 349–358.e4.

Anderson, S. N., Johnson, C. S., Jones, D. S., Conrad, L. J., Gou, X., Russell, S. D., et al. (2013). Transcriptomes of isolated Oryza sativa gametes characterized by deep sequencing: Evidence for distinct sex-dependent chromatin and epigenetic states before fertilization. *The Plant Journal for Cell and Molecular Biology*, *76*(5), 729–741.

Armenta-Medina, A., Lepe-Soltero, D., Xiang, D., Datla, R., Abreu-Goodger, C., & Gillmor, C. S. (2017). *Arabidopsis thaliana* miRNAs promote embryo pattern formation beginning in the zygote. *Developmental Biology*, *431*(2), 145–151.

Autran, D., Baroux, C., Raissig, M. T., Lenormand, T., Wittig, M., Grob, S., et al. (2011). Maternal epigenetic pathways control parental contributions to *Arabidopsis* early embryogenesis. *Cell*, *145*(5), 707–719.

Aw, S. J., Hamamura, Y., Chen, Z., Schnittger, A., & Berger, F. (2010). Sperm entry is sufficient to trigger division of the central cell but the paternal genome is required for endosperm development in *Arabidopsis*. *Development (Cambridge, England)*, *137*(16), 2683–2690. https://doi.org/10.1242/dev.052928.

Babu, Y., Musielak, T. J., Henschen, A., & Bayer, M. (2013). Suspensor length determines developmental progression of the embryo in *Arabidopsis thaliana*. *Plant Physiology*, *162*(3), 1448–1458.

Bai, F., Daliberti, M., Bagadion, A., Xu, M., Li, Y., Baier, J., et al. (2016). Parent-of-origin-effect rough endosperm mutants in maize. *Genetics*, *204*(1), 221–231.

Baroux, C., Autran, D., Gillmor, C. S., Grimanelli, D., & Grossniklaus, U. (2008). The maternal to zygotic transition in animals and plants. *Cold Spring Harbor Symposia on Quantitative Biology*, *73*(0), 89–100.

Baroux, C., Autran, D., Raissig, M. T., Grimanelli, D., & Grossniklaus, U. (2013). Parental contributions to the transcriptome of early plant embryos. *Current Opinion in Genetics & Development*, *23*(1), 72–74.

Baroux, C., Blanvillain, R., & Gallois, P. (2001). Paternally inherited transgenes are down-regulated but retain low activity during early embryogenesis in *Arabidopsis*. *FEBS Letters*, *509*(1), 11–16.

Baroux, C., & Grossniklaus, U. (2015). The maternal-to-zygotic transition in flowering plants: Evidence, mechanisms, and plasticity. *Current Topics in Developmental Biology*, *113*, 351–371.

Barton, M., & Poethig, R. (1993). Formation of the shoot apical meristem in *Arabidopsis thaliana*: An analysis of development in the wild type and in the *shoot meristemless* mutant. *Development (Cambridge, England)*, *119*(3), 823–831.

Bayer, M., Nawy, T., Giglione, C., Galli, M., Meinnel, T., & Lukowitz, W. (2009). Paternal control of embryonic patterning in *Arabidopsis thaliana*. *Science (New York, N.Y.)*, *323*(5920), 1485–1488.

Belcram, K., Palauqui, J.-C., & Pastuglia, M. (2016). Studying cell division plane positioning in early-stage embryos. *Methods in Molecular Biology (Clifton, NJ)*, *1370*, 183–195 [chapter 14].

Belmonte, M. F., Kirkbride, R. C., Stone, S. L., Pelletier, J. M., Bui, A. Q., Yeung, E. C., et al. (2013). Comprehensive developmental profiles of gene activity in regions and

subregions of the *Arabidopsis* seed. *Proceedings of the National Academy of Sciences of the United States of America, 110*(5), E435–E444.
Berger, F. (2011). Imaging fertilization in flowering plants, not so abominable after all. *Journal of Experimental Botany, 62*(5), 1651–1658.
Berleth, T., & Jürgens, G. (1993). The role of the *monopteros* gene in organising the basal body region of the *Arabidopsis* embryo. *Development (Cambridge, England), 118*, 575–587.
Besson, S., & Dumais, J. (2011). Universal rule for the symmetric division of plant cells. *Proceedings of the National Academy of Sciences of the United States of America, 108*(15), 6294–6299.
Breuninger, H., Rikirsch, E., Hermann, M., Ueda, M., & Laux, T. (2008). Differential expression of WOX genes mediates apical-basal axis formation in the *Arabidopsis* embryo. *Developmental Cell, 14*(6), 867–876.
Chakrabortty, B., Willemsen, V., de Zeeuw, T., Liao, C.-Y., Weijers, D., Mulder, B., et al. (2018). A plausible microtubule-based mechanism for cell division orientation in plant embryogenesis. *Current Biology, 28*(19), 3031–3043.e2.
Chaudhury, A. M., Ming, L., Miller, C., Craig, S., Dennis, E. S., & Peacock, W. J. (1997). Fertilization-independent seed development in *Arabidopsis thaliana. Proceedings of the National Academy of Sciences of the United States of America, 94*(8), 4223–4228.
Chen, L.-Q., Lin, I. W., Qu, X.-Q., Sosso, D., McFarlane, H. E., Londoño, A., et al. (2015). A cascade of sequentially expressed sucrose transporters in the seed coat and endosperm provides nutrition for the *Arabidopsis* embryo. *The Plant Cell, 27*(3), 607–619.
Chen, J., Strieder, N., Krohn, N. G., Cyprys, P., Sprunck, S., Engelmann, J. C., et al. (2017). Zygotic genome activation occurs shortly after fertilization in maize. *The Plant Cell, 29*(9), 2106–2125.
Chen, J., Zeng, B., Zhang, M., Xie, S., Wang, G., Hauck, A., et al. (2014). Dynamic transcriptome landscape of maize embryo and endosperm development. *Plant Physiology, 166*(1), 252–264.
Chettoor, A. M., Phillips, A. R., Coker, C. T., Dilkes, B., & Evans, M. M. S. (2016). Maternal gametophyte effects on seed development in maize. *Genetics, 204*(1), 233–248.
Clark, J. K., & Sheridan, W. F. (1988). Characterization of the two maize embryo-lethal defective kernel mutants rgh*-1210 and fl*-1253b: Effects on embryo and gametophyte development. *Genetics, 120*(1), 279–290.
Costa, L. M., Marshall, E., Tesfaye, M., Silverstein, K. A. T., Mori, M., Umetsu, Y., et al. (2014). Central cell-derived peptides regulate early embryo patterning in flowering plants. *Science (New York, N.Y.), 344*(6180), 168–172.
Crawford, B. C. W., Sewell, J., Golembeski, G., Roshan, C., Long, J. A., & Yanofsky, M. F. (2015). Plant development. Genetic control of distal stem cell fate within root and embryonic meristems. *Science, 347*(6222), 655–659.
De Rybel, B., Möller, B., Yoshida, S., Grabowicz, I., Barbier de Reuille, P., Boeren, S., et al. (2013). A bHLH complex controls embryonic vascular tissue establishment and indeterminate growth in *Arabidopsis. Developmental Cell, 24*(4), 426–437.
de Vega-Bartol, J. J., Simões, M., Lorenz, W. W., Rodrigues, A. S., Alba, R., Dean, J. F. D., et al. (2013). Transcriptomic analysis highlights epigenetic and transcriptional regulation during zygotic embryo development of Pinus pinaster. *BMC Plant Biology, 13*(1), 123.
Del Toro-De León, G., Garcia-Aguilar, M., & Gillmor, C. S. (2014). Non-equivalent contributions of maternal and paternal genomes to early plant embryogenesis. *Nature, 514*(7524), 624–627.
Del Toro-De León, G., Lepe-Soltero, D., & Gillmor, C. S. (2016). Zygotic genome activation in isogenic and hybrid plant embryos. *Current Opinion in Plant Biology, 29*, 148–153.

Deng, W., Yan, F., Zhang, X., Tang, Y., & Yuan, Y. (2015). Transcriptional profiling of canola developing embryo and identification of the important roles of BnDof5.6 in embryo development and fatty acids synthesis. *Plant and Cell Physiology*, *56*(8), 1624–1640.

Denninger, P., Bleckmann, A., Lausser, A., Vogler, F., Ott, T., Ehrhardt, D. W., et al. (2014). Male–female communication triggers calcium signatures during fertilization in *Arabidopsis*. *Nature Communications*, *5*, 4645.

Dharmasiri, N., Dharmasiri, S., & Estelle, M. (2005). The F-box protein TIR1 is an auxin receptor. *Nature*, *435*(7041), 441–445.

Dong, J., MacAlister, C. A., & Bergmann, D. C. (2009). BASL controls asymmetric cell division in *Arabidopsis*. *Cell*, *137*(7), 1320–1330.

Dresselhaus, T., Sprunck, S., & Wessel, G. M. (2016). Fertilization mechanisms in flowering plants. *Current Biology*, *26*(3), R125–R139.

Evans, M. M., & Kermicle, J. L. (2001). Interaction between maternal effect and zygotic effect mutations during maize seed development. *Genetics*, *159*(1), 303–315.

Faure, J.-E., Rotman, N., Fortuné, P., & Dumas, C. (2002). Fertilization in *Arabidopsis thaliana* wild type: Developmental stages and time course. *The Plant Journal for Cell and Molecular Biology*, *30*(4), 481–488.

Fitch, K. R., Yasuda, G. K., Owens, K. N., & Wakimoto, B. T. (1997). 1 Paternal effects in Drosophila: Implications for mechanisms of early development. *Current Topics in Developmental Biology*, *38*, 1–34.

Florez Rueda, A. M., Grossniklaus, U., & Schmidt, A. (2016). Laser-assisted microdissection (LAM) as a tool for transcriptional profiling of individual cell types. *Journal of Visualized Experiments*, *111*, e53916.

Florez-Rueda, A. M., Paris, M., Schmidt, A., Widmer, A., Grossniklaus, U., & Städler, T. (2016). Genomic imprinting in the endosperm is systematically perturbed in abortive hybrid tomato seeds. *Molecular Biology and Evolution*, *33*(11), 2935–2946.

Forzani, C., Aichinger, E., Sornay, E., Willemsen, V., Laux, T., Dewitte, W., et al. (2014). WOX5 suppresses CYCLIN D activity to establish quiescence at the center of the root stem cell niche. *Current Biology*, *24*(16), 1939–1944.

Friedman, W. E. (1999). Expression of the cell cycle in sperm of *Arabidopsis*: Implications for understanding patterns of gametogenesis and fertilization in plants and other eukaryotes. *Development (Cambridge, England)*, *126*(5), 1065–1075.

Friml, J., Vieten, A., Sauer, M., Weijers, D., Schwarz, H., Hamann, T., et al. (2003). Efflux-dependent auxin gradients establish the apical-basal axis of *Arabidopsis*. *Nature*, *426*(6963), 147–153.

Galinha, C., Hofhuis, H., Luijten, M., Willemsen, V., Blilou, I., Heidstra, R., et al. (2007). PLETHORA proteins as dose-dependent master regulators of *Arabidopsis* root development. *Nature*, *449*(7165), 1053–1057.

Garcia-Aguilar, M., & Autran, D. (2018). Localization of chromatin Marks in *Arabidopsis* early embryos. *Methods in Molecular Biology (Clifton, NJ)*, *1675*(5), 419–441.

Garcia-Aguilar, M., & Gillmor, C. S. (2015). Zygotic genome activation and imprinting: Parent-of-origin gene regulation in plant embryogenesis. *Current Opinion in Plant Biology*, *27*, 29–35.

Golden, T. A., Schauer, S. E., Lang, J. D., Pien, S., Mushegian, A. R., Grossniklaus, U., et al. (2002). *SHORT INTEGUMENTS1/SUSPENSOR1/CARPEL FACTORY*, a dicer homolog, is a maternal effect gene required for embryo development in *Arabidopsis*. *Plant Physiology*, *130*(2), 808–822.

Gooh, K., Ueda, M., Aruga, K., Park, J., Arata, H., Higashiyama, T., et al. (2015). Live-cell imaging and optical manipulation of *Arabidopsis* early embryogenesis. *Developmental Cell*, *34*(2), 242–251.

Grossniklaus, U. (2017). Polyspermy produces tri-parental seeds in maize. *Current Biology*, *27*(24), R1300–R1302.
Grossniklaus, U., & Schneitz, K. (1998). The molecular and genetic basis of ovule and megagametophyte development. *Seminars in Cell & Developmental Biology*, *9*(2), 227–238.
Grossniklaus, U., Vielle-Calzada, J. P., Hoeppner, M. A., & Gagliano, W. B. (1998). Maternal control of embryogenesis by *MEDEA*, a polycomb group gene in *Arabidopsis*. *Science (New York, N.Y.)*, *280*(5362), 446–450.
Grover, J. W., Kendall, T., Baten, A., Burgess, D., Freeling, M., King, G. J., et al. (2018). Maternal components of RNA-directed DNA methylation are required for seed development in Brassica rapa. *The Plant Journal for Cell and Molecular Biology*, *94*(4), 575–582.
Haecker, A., Gross-Hardt, R., Geiges, B., Sarkar, A., Breuninger, H., Herrmann, M., et al. (2004). Expression dynamics of WOX genes mark cell fate decisions during early embryonic patterning in *Arabidopsis thaliana*. *Development (Cambridge, England)*, *131*(3), 657–668.
Hamamura, Y., Nishimaki, M., Takeuchi, H., Geitmann, A., Kurihara, D., & Higashiyama, T. (2014). Live imaging of calcium spikes during double fertilization in *Arabidopsis*. *Nature Communications*, *5*, 4722.
Hamamura, Y., Saito, C., Awai, C., Kurihara, D., Miyawaki, A., Nakagawa, T., et al. (2011). Live-cell imaging reveals the dynamics of two sperm cells during double fertilization in *Arabidopsis thaliana*. *Current Biology*, *21*(6), 497–502.
Hamann, T., Mayer, U., & Jürgens, G. (1999). The auxin-insensitive bodenlos mutation affects primary root formation and apical-basal patterning in the *Arabidopsis* embryo. *Development (Cambridge, England)*, *126*(7), 1387–1395.
Hardtke, C. S., & Berleth, T. (1998). The *Arabidopsis* gene *MONOPTEROS* encodes a transcription factor mediating embryo axis formation and vascular development. *The EMBO Journal*, *17*(5), 1405–1411.
He, S., He, S., Sun, Y., Sun, Y., Yang, Q., Yang, Q., et al. (2017). A novel imprinted gene NUWA controls mitochondrial function in early seed development in *Arabidopsis*. *PLoS Genetics*, *13*(1), e1006553.
Higashiyama, T., Kuroiwa, H., Kawano, S., & Kuroiwa, T. (1997). Kinetics of double fertilization in Torenia fournieri based on direct observations of the naked embryo sac. *Planta*, *203*(1), 101–110.
Huang, J.-B., Liu, H., Chen, M., Li, X., Wang, M., Yang, Y., et al. (2014). ROP3 GTPase contributes to polar auxin transport and auxin responses and is important for embryogenesis and seedling growth in *Arabidopsis*. *The Plant Cell*, *26*(9), 3501–3518.
Ikeda, M., Mitsuda, N., & Ohme-Takagi, M. (2009). *Arabidopsis* WUSCHEL is a bifunctional transcription factor that acts as a repressor in stem cell regulation and as an activator in floral patterning. *The Plant Cell*, *21*(11), 3493–3505.
Ingouff, M., Hamamura, Y., Gourgues, M., Higashiyama, T., & Berger, F. (2007). Distinct dynamics of HISTONE3 variants between the two fertilization products in plants. *Current Biology*, *17*(12), 1032–1037.
Ingouff, M., Sakata, T., Li, J., Sprunck, S., Dresselhaus, T., & Berger, F. (2009). The two male gametes share equal ability to fertilize the egg cell in *Arabidopsis thaliana*. *Current Biology*, *19*(1), R19–R20.
Itoh, J.-I., Sato, Y., Hibara, K.-I., Shimizu-Sato, S., Kobayashi, H., Takehisa, H., et al. (2016). Genome-wide analysis of spatiotemporal gene expression patterns during early embryogenesis in rice. *Development (Cambridge, England)*, *143*(7), 1217–1227.
Jahnke, S., & Scholten, S. (2009). Epigenetic resetting of a gene imprinted in plant embryos. *Current Biology*, *19*(19), 1677–1681.
Jenik, P. D., Gillmor, C. S., & Lukowitz, W. (2007). Embryonic patterning in *Arabidopsis thaliana*. *Annual Review of Cell and Developmental Biology*, *23*, 207–236.

Jeong, S., Eilbert, E., Bolbol, A., & Lukowitz, W. (2016). Going mainstream: How is the body axis of plants first initiated in the embryo? *Developmental Biology*, *419*(1), 78–84.

Jeong, S., Palmer, T. M., & Lukowitz, W. (2011). The RWP-RK factor GROUNDED promotes embryonic polarity by facilitating YODA MAP kinase signaling. *Current Biology*, *21*(15), 1268–1276.

Jürgens, G., & Mayer, U. (1994). Arabidopsis. In J. Bard (Ed.), *EMBRYOS color atlas of development* (pp. 7–22). London: Wolf Publishing.

Jürgens, G., Mayer, U., Torres Ruíz, R., Berleth, T., & Miséra, S. (1991). Genetic analysis of pattern formation in the *Arabidopsis* embryo. *Development (Cambridge)*, *113*(Suppl. 1), 27–38.

Kawashima, T., & Goldberg, R. B. (2010). The suspensor: Not just suspending the embryo. *Trends in Plant Science*, *15*(1), 23–30.

Kawashima, T., Maruyama, D., Shagirov, M., Li, J., Hamamura, Y., Yelagandula, R., et al. (2014). Dynamic F-actin movement is essential for fertilization in *Arabidopsis thaliana*. *eLife*, *3*, e04501.

Kepinski, S., & Leyser, O. (2005). The *Arabidopsis* F-box protein TIR1 is an auxin receptor. *Nature*, *435*(7041), 446–451.

Kimata, Y., Higaki, T., Kawashima, T., Kurihara, D., Sato, Y., Yamada, T., et al. (2016). Cytoskeleton dynamics control the first asymmetric cell division in *Arabidopsis* zygote. *Proceedings of the National Academy of Sciences of the United States of America*, *113*(49), 14157–14162.

Köhler, C., Hennig, L., Spillane, C., Pien, S., Gruissem, W., & Grossniklaus, U. (2003). The Polycomb-group protein MEDEA regulates seed development by controlling expression of the MADS-box gene *PHERES1*. *Genes & Development*, *17*(12), 1540–1553.

Kong, J., Lau, S., & Jürgens, G. (2015). Twin plants from supernumerary egg cells in *Arabidopsis*. *Current Biology*, *25*(2), 225–230.

Kurihara, D., Kimata, Y., Higashiyama, T., & Ueda, M. (2017). In vitro ovule cultivation for live-cell imaging of zygote polarization and embryo patterning in *Arabidopsis thaliana*. *Journal of Visualized Experiments*, *127*, e55975.

Laux, T., Mayer, K. F., Berger, J., & Jürgens, G. (1996). The *WUSCHEL* gene is required for shoot and floral meristem integrity in *Arabidopsis*. *Development (Cambridge, England)*, *122*(1), 87–96.

Lee, M. T., Bonneau, A. R., & Giraldez, A. J. (2014). Zygotic genome activation during the maternal-to-zygotic transition. *Annual Review of Cell and Developmental Biology*, *30*, 581–613.

Lepe-Soltero, D., Armenta-Medina, A., Xiang, D., Datla, R., Gillmor, C. S., & Abreu-Goodger, C. (2017). Annotating and quantifying pri-miRNA transcripts using RNA-Seq data of wild type and serrate-1 globular stage embryos of *Arabidopsis thaliana*. *Data in Brief*, *15*, 642–647.

Li, N., & Li, Y. (2015). Maternal control of seed size in plants. *Journal of Experimental Botany*, *66*(4), 1087–1097.

Liao, C.-Y., & Weijers, D. (2018). A toolkit for studying cellular reorganization during early embryogenesis in *Arabidopsis thaliana*. *The Plant Journal for Cell and Molecular Biology*, *93*(6), 963–976.

Liu, Y., Li, X., Zhao, J., Tang, X., Tian, S., Chen, J., et al. (2015). Direct evidence that suspensor cells have embryogenic potential that is suppressed by the embryo proper during normal embryogenesis. *Proceedings of the National Academy of Sciences*, *112*(40), 12432–12437.

Long, J. A., & Barton, M. K. (1998). The development of apical embryonic pattern in *Arabidopsis*. *Development (Cambridge, England)*, *125*(16), 3027–3035.

Loppin, B., Lepetit, D., Dorus, S., Couble, P., & Karr, T. L. (2005). Origin and neofunctionalization of a Drosophila paternal effect gene essential for zygote viability. *Current Biology*, *15*(2), 87–93.
Lord, E. M., & Russell, S. D. (2002). The mechanisms of pollination and fertilization in plants. *Annual Review of Cell and Developmental Biology*, *18*(1), 81–105.
Lu, X., Chen, D., Shu, D., Zhang, Z., Wang, W., Klukas, C., et al. (2013). The differential transcription network between embryo and endosperm in the early developing maize seed. *Plant Physiology*, *162*(1), 440–455.
Lu, P., Porat, R., Nadeau, J. A., & O'Neill, S. D. (1996). Identification of a meristem L1 layer-specific gene in *Arabidopsis* that is expressed during embryonic pattern formation and defines a new class of Homeobox genes. *The Plant Cell*, *8*(12), 2155.
Lukowitz, W., Roeder, A., Parmenter, D., & Somerville, C. (2004). A MAPKK kinase gene regulates extra-embryonic cell fate in *Arabidopsis*. *Cell*, *116*(1), 109–119.
Luo, M., Bilodeau, P., Koltunow, A., Dennis, E. S., Peacock, W. J., & Chaudhury, A. M. (1999). Genes controlling fertilization-independent seed development in *Arabidopsis thaliana*. *Proceedings of the National Academy of Sciences of the United States of America*, *96*(1), 296–301.
Luo, M., Taylor, J. M., Spriggs, A., Zhang, H., Wu, X., Russell, S., et al. (2011). A genome-wide survey of imprinted genes in rice seeds reveals imprinting primarily occurs in the endosperm. *PLoS Genetics*, 7(6), e1002125.
Mansfield, S. G., & Briarty, L. G. (1991a). Early embryogenesis in Arabidopsis thaliana. I. The mature embryo sac. *Canadian Journal of Botany*, *69*, 447–460.
Mansfield, S. G., & Briarty, L. G. (1991b). Early embryogenesis in Arabidopsis thaliana. II. The developing embryo. *Canadian Journal of Botany*, *69*, 461–476.
Mayer, U., & Jürgens, G. (1993). Apical-basal pattern formation in the *Arabidopsis* embryo: Studies on the role of the gnom gene. *Development (Cambridge, England)*, *117*, 149–162.
Mayer, U., Torrez Ruíz, R., Berleth, T., Miséra, S., & Jürgens, G. (1991). Mutations affecting body organization in the *Arabidopsis* embryo. *Nature*, *353*, 402–407.
McElver, J., Tzafrir, I., Aux, G., Rogers, R., Ashby, C., Smith, K., et al. (2001). Insertional mutagenesis of genes required for seed development in *Arabidopsis thaliana*. *Genetics*, *159*(4), 1751–1763.
Meinke, D. W., & Sussex, I. M. (1979a). Embryo-lethal mutants of *Arabidopsis thaliana*. A model system for genetic analysis of plant embryo development. *Developmental Biology*, *72*(1), 50–61.
Meinke, D. W., & Sussex, I. M. (1979b). Isolation and characterization of six embryo-lethal mutants of *Arabidopsis thaliana*. *Developmental Biology*, *72*(1), 62–72.
Meng, D., Zhao, J., Zhao, C., Luo, H., Xie, M., Liu, R., et al. (2018). Sequential gene activation and gene imprinting during early embryo development in maize. *The Plant Journal*, *93*(3), 445–459.
Mòl, R., Dumas, C., & Matthys-Rochon, E. (1994). The kinetics of cytological events during double fertilization in *Zea mays* L. *Plant Journal*, *5*(2), 197–206.
Möller, B. K., Hove Ten, C. A., Xiang, D., Williams, N., Williams, N., López, L. G., et al. (2017). Auxin response cell-autonomously controls ground tissue initiation in the early *Arabidopsis* embryo. *Proceedings of the National Academy of Sciences of the United States of America*, *114*(12), E2533–E2539.
Moreno-Romero, J., Santos-González, J., Hennig, L., & Köhler, C. (2017). Applying the INTACT method to purify endosperm nuclei and to generate parental-specific epigenome profiles. *Nature Protocols*, *12*(2), 238–254.
Mori, T., Kuroiwa, H., Higashiyama, T., & Kuroiwa, T. (2006). GENERATIVE CELL SPECIFIC 1 is essential for angiosperm fertilization. *Nature Cell Biology*, *8*(1), 64–71.
Müller, B., & Sheen, J. (2008). Cytokinin and auxin interaction in root stem-cell specification during early embryogenesis. *Nature*, *453*(7198), 1094–1097.

Muralla, R., Lloyd, J., & Meinke, D. (2011). Molecular foundations of reproductive lethality in *Arabidopsis thaliana*. *PLoS One*, *6*(12), e28398.

Musielak, T. J., Schenkel, L., Kolb, M., Henschen, A., & Bayer, M. (2015). A simple and versatile cell wall staining protocol to study plant reproduction. *Plant Reproduction*, *28*(3–4), 161–169.

Nakel, T., Tekleyohans, D. G., Mao, Y., Fuchert, G., Vo, D., & Gross-Hardt, R. (2017). Triparental plants provide direct evidence for polyspermy induced polyploidy. *Nature Communications*, *8*(1), 1033.

Neuffer, M. G., & Sheridan, W. F. (1980). Defective kernel mutants of maize. I. Genetic and lethality studies. *Genetics*, *95*(4), 929–944.

Ngo, Q. A., Baroux, C., Guthörl, D., Mozerov, P., Collinge, M. A., Sundaresan, V., et al. (2012). The Armadillo repeat gene ZAK IXIK promotes *Arabidopsis* early embryo and endosperm development through a distinctive gametophytic maternal effect. *The Plant Cell*, *24*(10), 4026–4043.

Ngo, Q. A., Moore, J. M., Baskar, R., Grossniklaus, U., & Sundaresan, V. (2007). *Arabidopsis* GLAUCE promotes fertilization-independent endosperm development and expression of paternally inherited alleles. *Development (Cambridge, England)*, *134*(22), 4107–4117.

Nodine, M. D., & Bartel, D. P. (2010). MicroRNAs prevent precocious gene expression and enable pattern formation during plant embryogenesis. *Genes & Development*, *24*(23), 2678–2692.

Nodine, M. D., & Bartel, D. P. (2012). Maternal and paternal genomes contribute equally to the transcriptome of early plant embryos. *Nature*, *482*(7383), 94–97.

Ohad, N., Margossian, L., Hsu, Y. C., Williams, C., Repetti, P., & Fischer, R. L. (1996). A mutation that allows endosperm development without fertilization. *Proceedings of the National Academy of Sciences of the United States of America*, *93*(11), 5319–5324.

Ohad, N., Yadegari, R., Margossian, L., Hannon, M., Michaeli, D., Harada, J. J., et al. (1999). Mutations in FIE, a WD polycomb group gene, allow endosperm development without fertilization. *The Plant Cell*, *11*(3), 407–416.

Olsen, L. I., Hansen, T. H., Larue, C., Østerberg, J. T., Hoffmann, R. D., Liesche, J., et al. (2016). Mother-plant-mediated pumping of zinc into the developing seed. *Nature Plants*, *2*(5), 16036.

Pagnussat, G. C., Yu, H.-J., Ngo, Q. A., Rajani, S., Mayalagu, S., Johnson, C. S., et al. (2005). Genetic and molecular identification of genes required for female gametophyte development and function in *Arabidopsis*. *Development (Cambridge, England)*, *132*(3), 603–614.

Palovaara, J., de Zeeuw, T., & Weijers, D. (2016). Tissue and organ initiation in the plant embryo: A first time for everything. *Annual Review of Cell and Developmental Biology*, *32*(1), 47–75.

Palovaara, J., Saiga, S., Wendrich, J. R., Van't Wout Hofland, N., Van Schayck, J. P., Hater, F., et al. (2017). Transcriptome dynamics revealed by a gene expression atlas of the early *Arabidopsis* embryo. *Nature Plants*, *3*(11), 894–904.

Parry, G., & Estelle, M. (2006). Auxin receptors: A new role for F-box proteins. *Current Opinion in Cell Biology*, *18*(2), 152–156.

Pignatta, D., Erdmann, R. M., Scheer, E., Picard, C. L., Bell, G. W., & Gehring, M. (2014). Natural epigenetic polymorphisms lead to intraspecific variation in *Arabidopsis* gene imprinting. *eLife*, *3*, e03198.

Pillot, M., Baroux, C., Vazquez, M. A., Autran, D., Leblanc, O., Vielle-Calzada, J.-P., et al. (2010). Embryo and endosperm inherit distinct chromatin and transcriptional states from the female gametes in *Arabidopsis*. *The Plant Cell*, *22*(2), 307–320.

Pires, N. D., Bemer, M., Müller, L. M., Baroux, C., Spillane, C., & Grossniklaus, U. (2016). Quantitative genetics identifies cryptic genetic variation involved in the paternal regulation of seed development. *PLoS Genetics*, *12*(1), e1005806.

Prigge, M. J., & Wagner, D. R. (2001). The *Arabidopsis SERRATE* gene encodes a zinc-finger protein required for normal shoot development. *The Plant Cell, 13*(6), 1263–1279.

Rademacher, E. H., Lokerse, A. S., Schlereth, A., Llavata-Peris, C. I., Bayer, M., Kientz, M., et al. (2012). Different auxin response machineries control distinct cell fates in the early plant embryo. *Developmental Cell, 22*(1), 211–222.

Rademacher, E. H., Möller, B., Lokerse, A. S., Llavata-Peris, C. I., van den Berg, W., & Weijers, D. (2011). A cellular expression map of the *Arabidopsis* AUXIN RESPONSE FACTOR gene family. *The Plant Journal for Cell and Molecular Biology, 68*(4), 597–606.

Raissig, M. T., Bemer, M., Baroux, C., & Grossniklaus, U. (2013). Genomic imprinting in the *Arabidopsis* embryo is partly regulated by PRC2. *PLoS Genetics, 9*(12), e1003862.

Raissig, M. T., Gagliardini, V., Jaenisch, J., Grossniklaus, U., & Baroux, C. (2013). Efficient and rapid isolation of early-stage embryos from *Arabidopsis thaliana* seeds. *Journal of Visualized Experiments, 76*, e50371.

Ray, S., Golden, T., & Ray, A. (1996). Maternal effects of the short integument mutation on embryo development in *Arabidopsis. Developmental Biology, 180*(1), 365–369.

Reddy, G. V., & Meyerowitz, E. M. (2005). Stem-cell homeostasis and growth dynamics can be uncoupled in the *Arabidopsis* shoot apex. *Science (New York, N.Y.), 310*(5748), 663–667.

Robert, H. S., Grones, P., Stepanova, A. N., Robles, L. M., Lokerse, A. S., Alonso, J. M., et. al. (2013). Local auxin sources orient the apical-basal axis in *Arabidopsis* embryos. *Current Biology, 23*(24), 2506–2512.

Robert, H. S., Grunewald, W., Sauer, M., Cannoot, B., Soriano, M., Swarup, R., et al. (2015). Plant embryogenesis requires AUX/LAX-mediated auxin influx. *Development (Cambridge, England), 142*(4), 702–711.

Robert, H. S., Park, C., Gutièrrez, C. L., Wójcikowska, B., Pěnčík, A., Novák, O., et al. (2018). Maternal auxin supply contributes to early embryo patterning in *Arabidopsis. Nature Plants, 4*(8), 548–553.

Ronceret, A., Gadea-Vacas, J., Guilleminot, J., & Devic, M. (2008). The alpha-N-acetyl-glucosaminidase gene is transcriptionally activated in male and female gametes prior to fertilization and is essential for seed development in *Arabidopsis. Journal of Experimental Botany, 59*(13), 3649–3659.

Ronceret, A., Gadea-Vacas, J., Guilleminot, J., Lincker, F., Delorme, V., Lahmy, S., et al. (2008). The first zygotic division in *Arabidopsis* requires de novo transcription of thymidylate kinase. *The Plant Journal for Cell and Molecular Biology, 53*(5), 776–789.

Ronceret, A., Guilleminot, J., Lincker, F., Gadea-Vacas, J., Delorme, V., Bechtold, N., et al. (2005). Genetic analysis of two *Arabidopsis* DNA polymerase epsilon subunits during early embryogenesis. *The Plant Journal for Cell and Molecular Biology, 44*(2), 223–236.

Sakai, K., Taconnat, L., Borrega, N., Yansouni, J., Brunaud, V., Paysant-Le Roux, C., et al. (2018). Combining laser-assisted microdissection (LAM) and RNA-seq allows to perform a comprehensive transcriptomic analysis of epidermal cells of *Arabidopsis* embryo. *Plant Methods, 14*(1), 635.

Sarkar, A. K., Luijten, M., Miyashima, S., Lenhard, M., Hashimoto, T., Nakajima, K., et al. (2007). Conserved factors regulate signalling in *Arabidopsis thaliana* shoot and root stem cell organizers. *Nature, 446*(7137), 811–814.

Schlereth, A., Möller, B., Liu, W., Kientz, M., Flipse, J., Rademacher, E. H., et al. (2010). *MONOPTEROS* controls embryonic root initiation by regulating a mobile transcription factor. *Nature, 464*(7290), 913–916.

Schmitz, R. J., Schultz, M. D., Urich, M. A., Nery, J. R., Pelizzola, M., Libiger, O., et al. (2013). Patterns of population epigenomic diversity. *Nature, 495*(7440), 193–198.

Schon, M. A., & Nodine, M. D. (2017). Widespread contamination of *Arabidopsis* embryo and endosperm transcriptome data sets. *The Plant Cell, 29*(4), 608–617.

Schoof, H., Lenhard, M., Haecker, A., Mayer, K. F., Jürgens, G., & Laux, T. (2000). The stem cell population of *Arabidopsis* shoot meristems is maintained by a regulatory loop between the *CLAVATA* and *WUSCHEL* genes. *Cell, 100*(6), 635–644.

Seefried, W. F., Willmann, M. R., Clausen, R. L., & Jenik, P. D. (2014). Global regulation of embryonic patterning in *Arabidopsis* by MicroRNAs. *Plant Physiology, 165*(2), 670–687.

Skoog, F., & Miller, C. O. (1957). Chemical regulation of growth and organ formation in plant tissues cultured *in vitro*. *Symposia of the Society for Experimental Biology, 11*, 118–131.

Slane, D., Kong, J., Berendzen, K. W., Kilian, J., Henschen, A., Kolb, M., et al. (2014). Cell type-specific transcriptome analysis in the early *Arabidopsis thaliana* embryo. *Development (Cambridge, England), 141*(24), 4831–4840.

Soyars, C. L., James, S. R., & Nimchuk, Z. L. (2016). Ready, aim, shoot: Stem cell regulation of the shoot apical meristem. *Current Opinion in Plant Biology, 29*, 163–168.

Spencer, M. W. B., Casson, S. A., & Lindsey, K. (2007). Transcriptional profiling of the *Arabidopsis* embryo. *Plant Physiology, 143*(2), 924–940.

Springer, P. S., Holding, D. R., Groover, A., Yordan, C., & Martienssen, R. A. (2000). The essential Mcm7 protein PROLIFERA is localized to the nucleus of dividing cells during the G(1) phase and is required maternally for early *Arabidopsis* development. *Development (Cambridge, England), 127*(9), 1815–1822.

Sprunck, S., Baumann, U., Edwards, K., Langridge, P., & Dresselhaus, T. (2005). The transcript composition of egg cells changes significantly following fertilization in wheat (*Triticum aestivum* L.). *The Plant Journal for Cell and Molecular Biology, 41*(5), 660–672.

Sprunck, S., Rademacher, S., Vogler, F., Gheyselinck, J., Grossniklaus, U., & Dresselhaus, T. (2012). Egg cell-secreted EC1 triggers sperm cell activation during double fertilization. *Science (New York, N.Y.), 338*(6110), 1093–1097.

St Johnston, D. (2002). The art and design of genetic screens: *Drosophila melanogaster*. *Nature Reviews. Genetics, 3*(3), 176–188.

Tadros, W., & Lipshitz, H. D. (2009). The maternal-to-zygotic transition: A play in two acts. *Development (Cambridge, England), 136*(18), 3033–3042.

Tekleyohans, D. G., Mao, Y., Kägi, C., Stierhof, Y.-D., & Gross-Hardt, R. (2017). Polyspermy barriers: A plant perspective. *Current Opinion in Plant Biology, 35*, 131–137.

Teoh, K. T., Requesens, D. V., Devaiah, S. P., Johnson, D., Huang, X., Howard, J. A., et al. (2013). Transcriptome analysis of embryo maturation in maize. *BMC Plant Biology, 13*(1), 19.

Tessadori, F., van Zanten, M., Pavlova, P., Clifton, R., Pontvianne, F., Snoek, L. B., et al. (2009). *PHYTOCHROME B* and *HISTONE DEACETYLASE 6* control light-induced chromatin compaction in *Arabidopsis thaliana*. *PLoS Genetics, 5*(9), e1000638.

Ueda, M., Aichinger, E., Gong, W., Groot, E., Verstraeten, I., Vu, L. D., et al. (2017). Transcriptional integration of paternal and maternal factors in the *Arabidopsis* zygote. *Genes & Development, 31*(6), 617–627.

Ueda, M., Zhang, Z., & Laux, T. (2011). Transcriptional activation of *Arabidopsis* axis patterning genes *WOX8/9* links zygote polarity to embryo development. *Developmental Cell, 20*(2), 264–270.

Venglat, P., Xiang, D., Qiu, S., Stone, S. L., Tibiche, C., Cram, D., et al. (2011). Gene expression analysis of flax seed development. *BMC Plant Biology, 11*, 74.

Vielle-Calzada, J. P., Baskar, R., & Grossniklaus, U. (2000). Delayed activation of the paternal genome during seed development. *Nature, 404*(6773), 91–94.

Vogiatzaki, E., Baroux, C., Jung, J.-Y., & Poirier, Y. (2017). PHO1 exports phosphate from the Chalazal seed coat to the embryo in developing *Arabidopsis* seeds. *Current Biology, 27*(19), 2893–2900.e3.

Waki, T., Hiki, T., Watanabe, R., Hashimoto, T., & Nakajima, K. (2011). The *Arabidopsis* RWP-RK protein RKD4 triggers gene expression and pattern formation in early embryogenesis. *Current Biology, 21*(15), 1277–1281.

Wang, H., Ngwenyama, N., Liu, Y., Walker, J. C., & Zhang, S. (2007). Stomatal development and patterning are regulated by environmentally responsive mitogen-activated protein kinases in *Arabidopsis*. *The Plant Cell*, *19*(1), 63–73.

Wang, Y., Tsukamoto, T., Noble, J. A., Liu, X., Mosher, R. A., & Palanivelu, R. (2017). *Arabidopsis* LORELEI, a maternally expressed imprinted gene, promotes early seed development. *Plant Physiology*, *175*(2), 758–773.

Waters, A. J., Bilinski, P., Eichten, S. R., Vaughn, M. W., Ross-Ibarra, J., Gehring, M., et al. (2013). Comprehensive analysis of imprinted genes in maize reveals allelic variation for imprinting and limited conservation with other species. *Proceedings of the National Academy of Sciences of the United States of America*, *110*(48), 19639–19644.

Waters, A. J., Makarevitch, I., Eichten, S. R., Swanson-Wagner, R. A., Yeh, C.-T., Xu, W., et al. (2011). Parent-of-origin effects on gene expression and DNA methylation in the maize endosperm. *The Plant Cell*, *23*(12), 4221–4233.

Webb, M. C., & Gunning, B. E. (1991). The microtubular cytoskeleton during development of the zygote, proembryo and free-nuclear endosperm in *Arabidopsis thaliana* (L.) Heynh. *Planta*, *184*(2), 187–195.

Weigel, D., & Jürgens, G. (2002). Stem cells that make stems. *Nature*, *415*, 751–754.

Weijers, D., Geldner, N., Offringa, R., & Jürgens, G. (2001). Seed development: Early paternal gene activity in *Arabidopsis*. *Nature*, *414*(6865), 709–710.

Wendrich, J. R., Möller, B. K., Uddin, B., Radoeva, T., Lokerse, A. S., De Rybel, B., et al. (2015). A set of domain-specific markers in the *Arabidopsis* embryo. *Plant Reproduction*, *28*(3–4), 153–160.

Wendrich, J. R., & Weijers, D. (2013). The *Arabidopsis* embryo as a miniature morphogenesis model. *The New Phytologist*, *199*(1), 14–25.

Xiang, D., Venglat, P., Tibiche, C., Yang, H., Risseeuw, E., Cao, Y., et al. (2011). Genome-wide analysis reveals gene expression and metabolic network dynamics during embryo development in *Arabidopsis*. *Plant Physiology*, *156*(1), 346–356.

Xin, H.-P., Peng, X.-B., Ning, J., Yan, T.-T., Ma, L.-G., & Sun, M.-X. (2011). Expressed sequence-tag analysis of tobacco sperm cells reveals a unique transcriptional profile and selective persistence of paternal transcripts after fertilization. *Sexual Plant Reproduction*, *24*(1), 37–46.

Xin, M., Yang, R., Li, G., Chen, H., Laurie, J., Ma, C., et al. (2013). Dynamic expression of imprinted genes associates with maternally controlled nutrient allocation during maize endosperm development. *The Plant Cell*, *25*(9), 3212–3227.

Xu, J., Zhang, H.-Y., Xie, C.-H., Xue, H.-W., Dijkhuis, P., & Liu, C.-M. (2005). EMBRYONIC FACTOR 1 encodes an AMP deaminase and is essential for the zygote to embryo transition in *Arabidopsis*. *The Plant Journal for Cell and Molecular Biology*, *42*(5), 743–756.

Yoshida, S., Barbier de Reuille, P., Lane, B., Bassel, G. W., Prusinkiewicz, P., Smith, R. S., et al. (2014). Genetic control of plant development by overriding a geometric division rule. *Developmental Cell*, *29*(1), 75–87.

Yu, T.-Y., Shi, D.-Q., Jia, P.-F., Tang, J., Li, H.-J., Liu, J., et al. (2016). The *Arabidopsis* receptor kinase ZAR1 is required for zygote asymmetric division and its daughter cell fate. *PLoS Genetics*, *12*(3), e1005933–23.

Yuan, G.-L., Li, H.-J., & Yang, W.-C. (2017). The integration of Gβ and MAPK signaling cascade in zygote development. *Scientific Reports*, 7(1), 8732.

Zhan, J., Thakare, D., Ma, C., Lloyd, A., Nixon, N. M., Arakaki, A. M., et al. (2015). RNA sequencing of laser-capture microdissected compartments of the maize kernel identifies regulatory modules associated with endosperm cell differentiation. *The Plant Cell*, *27*(3), 513–531.

Zhang, Z., Tucker, E., Tucker, E., Hermann, M., & Laux, T. (2017). A molecular framework for the embryonic initiation of shoot meristem stem cells. *Developmental Cell*, *40*(3), 264–277.e4.

Zhang, Y., Wang, P., Shao, W., Zhu, J.-K., & Dong, J. (2015). The BASL polarity protein controls a MAPK signaling feedback loop in asymmetric cell division. *Developmental Cell*, *33*(2), 136–149.

Zhang, M., Wu, H., Su, J., Wang, H., Zhu, Q., Liu, Y., et al. (2017). Maternal control of embryogenesis by MPK6 and its upstream MKK4/MKK5 in *Arabidopsis*. *The Plant Journal for Cell and Molecular Biology*, *92*(6), 1005–1019.

Zhang, M., Zhao, H., Xie, S., Chen, J., Xu, Y., Wang, K., et al. (2011). Extensive, clustered parental imprinting of protein-coding and noncoding RNAs in developing maize endosperm. *Proceedings of the National Academy of Sciences of the United States of America*, *108*(50), 20042–20047.

Zhao, Y. (2012). Auxin biosynthesis: A simple two-step pathway converts tryptophan to indole-3-acetic acid in plants. *Molecular Plant*, *5*(2), 334–338.

Zhao, J., Xin, H., Qu, L., Ning, J., Peng, X., Yan, T., et al. (2011). Dynamic changes of transcript profiles after fertilization are associated with de novo transcription and maternal elimination in tobacco zygote, and mark the onset of the maternal-to-zygotic transition. *The Plant Journal for Cell and Molecular Biology*, *65*(1), 131–145.

CHAPTER NINETEEN

Histology versus phylogeny: Viewing plant embryogenesis from an evo-devo perspective

John W. Chandler, Wolfgang Werr*
Department of Biology, Biocenter, University of Cologne, Cologne, Germany
*Corresponding author: e-mail address: werr@uni-koeln.de

Contents

Abstract

The goal of evolutionary developmental (evo-devo) biology compares inter-organism developmental processes to infer ancestral relationships and evolutionary adaptations. Frameworks to address macroevolutionary traits such as plant embryogenesis commonly involve two complementary approaches. Historically, focus has been placed on comparative morphology and histology, but more recently, accumulating genome data from diverse taxa have elicited the construction of molecular phylogenies, which aid the identification of gene homologies and orthologies that have been adaptive and that underlie differences in form. Distinguishing between ancestral or derived traits in phyletic or cladistic-driven approaches is challenging, but relates to the broader applicability of existing developmental models such as *Arabidopsis thaliana*.

Current Topics in Developmental Biology, Volume 131
ISSN 0070-2153
https://doi.org/10.1016/bs.ctdb.2018.11.009

1. Plant embryogenesis relates to the dominance of a diplontic lifestyle

Land plants (embryophytes) are monophyletic and evolved approximately 470 million years ago from aquatic green algae (Plackett, Di Stilio, & Langdale, 2015). Extant charophytes comprise a sister group to the embryophytes and produce a unicellular diploid zygote that quickly undergoes meiosis to produce gametes. It is accepted that delayed meiotic division of the zygote was a prerequisite for the intercalation of mitotic divisions into the diploid phase (Bower, 1908). A defining characteristic of embryophyte evolution was the change from a haplontic to a diplontic lifestyle. The former persists in extant bryophytes, in which a large haploid gametophyte is dominant and the diploid phase is transient and the sporophyte remains small. In contrast, the diplontic cycle of higher plants is characterized by a dominant diploid phase with a large sporophyte and short-lived, small haploid gametophytes. In lower plant radiations, the sporophyte often remains parasitically attached to the gametophyte, a phenomenon termed zygote retention. A rapid increase in sporophyte dominance and size occurred in the Devonian period. This coincided with an explosion in the diversity of sporophyte body plans that maximized fitness to a dry terrestrial environment, and included branching and the acquisition of indeterminate growth.

These adaptations had consequences for embryogenesis: first, the reduction of the haploid phase altered the maternal environment of the egg cell, zygote and early embryo, and second, the multicellularity and increasing dominance of the diploid phase required a three-dimensional sporophyte body with outer, protodermal or epidermal cells and specialized inner cell types, i.e., the innovation of a centro-radial axis in addition to an apical–basal axis. This increasing body plan complexity is directly associated with embryo morphology, since post-embryonic growth depends on multicellular embryonic meristems. The sporophyte of extant mosses and hornworts mainly elongates via a basal meristem at an intermediate position between the sporangium and foot (Ligrone, Duckett, & Renzaglia, 2012a, 2012b). A polysporangiophyte invention was a shift in the meristematic region to create a well-defined apical meristem, which contains stem-cell niches that divide asymmetrically and release recruitable cells for diverse cell types or tissues. However, the potential to generate three-dimensional tissue depends on multiplanar cell divisions within apical meristems, in contrast to apical cells in charophytes or some bryophytes, where one or two cutting faces give rise to filamentous or planar cell arrays. In lower plant radiations, a transient sporophyte remains attached to the gametophyte and contains a single embryonic meristem, but an increase in sporophyte autonomy necessitated the evolution of rooting systems that were initially rhizoid-based before becoming roots in vascular plants. Conserved molecular mechanisms coordinate stem-cell niches in the *Arabidopsis thaliana* apical shoot or basal root meristems, and suggest that the latter might have evolved via a duplication of the ancestral apical trait, a possibility compatible with whole-genome duplications in seed plant lineages (Sarkar et al., 2007).

The use of comparative approaches to infer evolutionary trajectories of embryonic development from the orthology of morphological characteristics or genes between divergent taxa must consider the key features of embryo evolution (Fig. 1A): (i) the fusion of two germ cells to form a diploid zygote within a gametophytic context (Fig. 1B), (ii) zygotic cell division and resulting apical–basal polarity, (iii) multiplanar apical meristems to elaborate a three-dimensional higher plant sporophyte, which relates to an inner–outer or radial organization; subsequent acquisitions were (iv) a root system and (v) increasing cell-type complexity, which probably involved cell-to-cell communication and hormones. This review aims to contextualize the diversity of embryonic patterning programs in relation to functional studies mainly driven by the study of model species and to outline the challenges inherent in reaching robust conclusions concerning embryo evolution.

2. Embryonic phases and patterns

Embryogenesis generally begins with the syngamy of germ cell nuclei, i.e., egg cells and sperms in higher plants. Embryo or diploid sporophyte development starts within haploid gametophytic tissue: the archegonium in lower radiations or the embryo sac in angiosperms. Comparative embryo morphology has a long-standing tradition in land plants (Wardlaw, 1955) and special emphasis has focussed on angiosperms (Maheshwari, 1950), where embryogenesis consists of three phases. First, multiple rounds of mitosis initially generate a globular embryo from the zygote, which possesses apical–basal and radial axes. Despite the monophyletic classification of embryophytes, the choreography of embryonic ontogeny differs enormously among and between seed plants and lower plant radiations (Fig. 1C). However, the discrete patterns described in eudicots (Johansen, 1950) do not correlate with taxa phylogeny (Cridge, Dearden, & Brownfield, 2016). Only few model

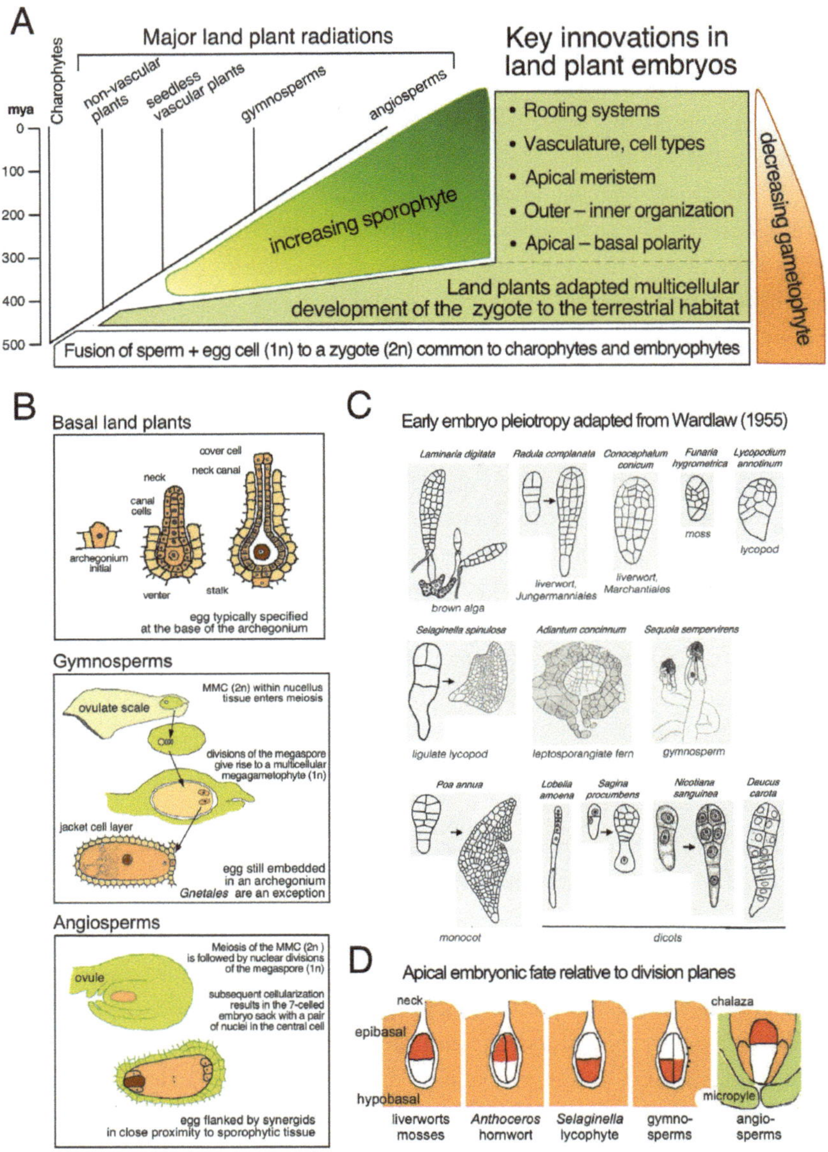

Fig. 1 Overview of embryophyte embryonic diversity. (A) A schematic illustration of key innovations during embryogenesis evolution; (B) the differing context of egg cell development in basal land plants, gymnosperms and angiosperms. MMC = megasporophyte mother cell; (C) Representative histological variation in early embryogenesis cell divisions across the plant kingdom; (D) zygote polarity and initial cell division planes in different plant radiations: exoscopic (liverworts, mosses and hornworts), or endoscopic (lycophytes, gymnosperms, angiosperms).

angiosperm species remain studied today, of which the very early model *Capsella bursa-pastoris* (Hanstein, 1870) and *A. thaliana* (ten Hove, Lu, & Weijers, 2015) show a stereotypic onograd/crucifer cell division pattern. Although this early embryo developmental paradigm is atypical among angiosperms, its stereotypy outlines how embryonic apical–basal polarity and radial organization are established and relates to cell fates and cell type-specific gene expression patterns, which are functionally and evolutionarily relevant. In seed plants, the middle embryogenesis phase consists of the elaboration of apical and basal meristems, the vasculature and body plan. The third phase involves metabolic changes and morphological adaptations associated with seed development, such as cotyledons to store nutrients for germination.

3. Apical–basal polarity: A unifying feature with exceptions

A general feature of initial embryonic zygotic divisions is a filamentous, axial structure or "primitive spindle" (Bower, 1922). The first division plane is mostly transversal relative to the archegonium neck/venter or the micropyle/chalaza axis in seed plants and is often symmetrical in lower plant radiations, resulting in two equally sized daughter cells: a hypobasal cell at the base of the venter and an epibasal cell toward the archegonium neck. Exoscopic polarity, shown by extant bryophytes, defines when the embryonic apical pole develops from the epibasal cell (Fig. 1D); in contrast, endoscopic polarity in most vascular plants is when the hypobasal cell gives rise to the embryo apical pole (Fig. 1D). An axis inversion apparently occurred evolutionarily after the basal bryophyte lineage, although some ferns also show exoscopic polarity. In species of hornwort, i.e., anthocerotophyta, the first cell division is longitudinal, which indicates that apical embryonic fate can be established independently of the initial cell division plane (Wardlaw, 1955; Fig. 1D). Endoscopic polarity in angiosperms results from the first transversal or oblique zygotic division, to give a small apical cell oriented to the chalazal pole of the megagametophyte or embryo sac, and a large basal cell toward the micropyle (Wardlaw, 1955).

A general feature of eudicots, exemplified by *A. thaliana*, is that the chalazal–micropylar polarity of the egg prepatterns the apical–basal polarity of the zygote and the prospective primary shoot–root axis. However, in *Zea mays* (maize) as a well-characterized monocot, all cell division planes following the first asymmetric division at the apical pole are random and create a club-shaped transition stage embryo that consists apically of a small-celled embryo proper, subtended by a larger-celled suspensor (Chandler, Nardmann, & Werr, 2008), The shoot–root axis develops obliquely to the apical–basal polarity of the zygote, with the shoot apical meristem (SAM) at the adaxial face of the embryo proper directing away from the future endosperm, and the root meristem at the center of the multicellular boundary with the subtending suspensor.

An additional embryogenesis feature shown by some gymnosperms is a syncytial phase of several cycles of nuclear duplication without cytokinesis before cellularization, which results in a cell mass, or tiers of cells, with the upper cells forming a proembryo and the basal tiers dividing and elongating to form the suspensor (Cairney & Pullman, 2007). Discrete embryonic and basal suspensor cell fates at the apical or basal pole, respectively, are thus established after the syncytial phase (Fig. 1D). Many gymnosperm taxa display either simple polyembryony, where egg cells within different archegonia are independently fertilized to generate multiple zygotes within the seed, or cleavage polyembryony, in which proembryos divide, but only one embryo usually survives to maturity. Cotyledon numbers are also variable, with *Ginkgo biloba* and *Ephedra trifurca* initiating two, but many taxa have more.

The high diversity of embryogenesis programs within individual taxa and the poor correlation between embryo type and taxon impede the identification of morphological embryonic plesiomorphies (ancestral traits). This is exemplified by debates concerning the orthology of the monocot scutellum and dicot cotyledons, or of bryophyte rhizoids and higher plant roots. Given the diversity of initial cell division patterns and deeply branched land plant phylogenies over temporally long trajectories, the early axial polarity of land plant embryos appears to represent a common principle and a highly adaptive trait. This raises questions concerning its common molecular basis, and how evolutionarily informative genetic models such as *A. thaliana* are?

4. Embryonic transcriptional regulation

In animals, maternally encoded proteins drive early embryo development prior to the activation of zygotic gene expression, which is known as the maternal-to-zygotic transition. During early embryo development, parental genomes contribute non-equivalently to early plant embryogenesis and at least for a subset of functionally relevant embryo-expressed genes, the paternal genome is effectively silenced and becomes activated within a few days following fertilization (Del Toro-De León, García-Aguilar, &

Gillmor, 2014; Vielle-Calzada, Baskar, & Grossniklaus, 2000). The differential regulation of parental alleles during zygotic gene activation is regulated mainly epigenetically (Autran et al., 2011; reviewed in García-Aguilar & Gillmor, 2015).

By analogy to the morphological development of animal embryos, the so-called development "hourglass model" has been applied to plant embryos, in which early embryos from the same phylum show broad morphological variation, but converge to a similar form in mid-embryogenesis, which is termed the phylotypic stage, before subsequently diverging once again morphologically. Similarly, a transcriptional hourglass model has been proposed for the mid-phase of plant embryogenesis, which is characterized by little transcriptional variation and the preferential activity of ancient genes (Drost, Gabel, Grosse, & Quint, 2015; Drost, Janitza, Grosse, & Quint, 2017; Quint et al., 2012). However, this might not be a general feature of plant embryos.

Transcriptome studies have revealed that in a bryophyte model, gene expression in sporophyte and gametophyte generations is more similar than in *A. thaliana*, which shows greater generation-biased gene activity (Szövényi, Rensing, Lang, Wray, & Shaw, 2011). The preferential enrichment of transcription factors expressed in bryophyte gametophytes in *A. thaliana* sporophytes suggests that gametophytic genetic networks have been recruited to the sporophyte. In contrast, genes with sporophyte-biased expression in both moss and *A. thaliana* are enriched for biological pathways involved in adaptation to the terrestrial versus aquatic environment, such as osmotic stress, hormonal intercellular communication and UV tolerance (Szövényi et al., 2011). The paralogous bHLH transcription factors ROOT HAIR DEFECTIVE6 (RHD6) and RHD SIX-LIKE1 positively regulate root hair development in the *A. thaliana* sporophyte as well as rhizoids and caulonema cells of *Physcomitrella patens* gametophytes (Menand et al., 2007). Their phylogeny is informative in highlighting molecular similarities between rhizoids or roots, and gene recruitment from a bryophyte ancestor as a mechanism to explain the increasing cellular diversity of land plant sporophytes.

Other transcription factors probably acquired from gametophytes and exploited for sporophyte functions include the MIKC* class of type II MADS-box genes, which have retained similar functions in bryophyte and angiosperm gamete differentiation (Zobell, Faigl, Saedler, & Münster, 2010). However, the presence of a single MIKC^c type II representative in the gametophytes of several charophycean algae and the large expansion of this family in angiosperms (Tanabe et al., 2005) suggests that neo-functionalizations accompanied recruitment for sporophyte or floral development. In contrast, the evolution of *LEAFY* exemplifies the opposite scenario; although two *LEAFY* paralogues regulate the first zygotic cell divisions in *P. patens* (Tanahashi, Sumikawa, Kato, & Hasebe, 2005), angiosperm *LEAFY* orthologues regulate the floral transition and have no functions in embryogenesis. Inferring the ancestral state within the deeply branched land plant phylogeny is challenging, as extant species are the result of long independent evolutionary trajectories: gymnosperms and angiosperms diverged 300 million years ago (mya) (Bowe, Coat, & dePamphilis, 2000), and the last common ancestor of moss (*P. patens*) or the eudicot *A. thaliana* possibly existed 400 mya, during which time, genes have potentially been recruited, amplified and neofunctionalized, or lost in specific lineages.

Transcription factors are traceable across wide phylogenetic distances and are classified according to conserved protein domains. One example is the DNA-binding homeodomain (HD), first identified from homeotic gene mutations in *Drosophila melanogaster*. The HD is a eukaryotic invention and defines a large protein superfamily that contains many discrete sub-branches, including the plant-specific WUSCHEL-related homeobox (WOX) family or the three amino acid length extension (TALE) HD protein family, which is widespread among plant and animal kingdoms. Evo-devo interest in WOX genes has been motivated by the function of its founding member *WUSCHEL* (*WUS*), which promotes stem-cell fate in the *A. thaliana* SAM, and because WOX members are important transcriptional determinants in the early embryo. However, moss WOX genes are of an ancestral *WOX13* type, and despite their conservation among all analyzed embryophytes, are not orthologous to *WUS* or its relatives that pattern the early *A. thaliana* embryo, which are mostly unique to seed plants or angiosperms (Nardmann & Werr, 2012). Therefore, evo-devo comparisons based on WOX gene transcription patterns in the early *A. thaliana* embryo are phylogenetically limited in depth.

5. *WOX8* is a major target of zygote polarity determinants in *A. thaliana*

The first zygotic division in *A. thaliana* results in a small apical cell at the chalazal pole and a larger basal cell at the micropyle. *WOX2* and *WOX8* are maternally transcribed in the egg cell, but are differentially expressed after the first zygotic cell division: *WOX2* in the small apical cell and *WOX8* and its paralogue *WOX9* in the large basal cell (Haecker et al., 2004). Basal cell fate is deterministic and relies on a complex interplay between intrinsic pollen or egg cell proteins and extrinsic signals from the megagametophyte to the zygote. The WRKY2 transcription factor is expressed in egg cell and zygote, but requires phosphorylation to actively regulate *WOX8* and *WOX9* transcription. This activation involves the YODA mitogen-activated protein kinase kinase kinase, whose activity depends on the sperm-derived *SHORT SUSPENSOR* (*SSP*) mRNA (Bayer et al., 2009), which is only translated to a membrane-associated pseudokinase in the zygote (Ueda et al., 2017; Ueda, Zhang, & Laux, 2011).

The central cell and subsequently the endosperm provide extrinsic signals; namely, EMBRYO-SURROUNDING FACTOR1 (ESF1) and CLAVATA3/ESR1-RELATED8 (CLE8), a potential receptor kinase ligand (Fig. 2A). Abnormalities in basal cell fate and altered cell division planes are common to *esf1* and *cle8* mutants, and to mutants in the plasma membrane receptor kinase ZYGOTE ARRESTED1 (ZAR1). ZAR1 is

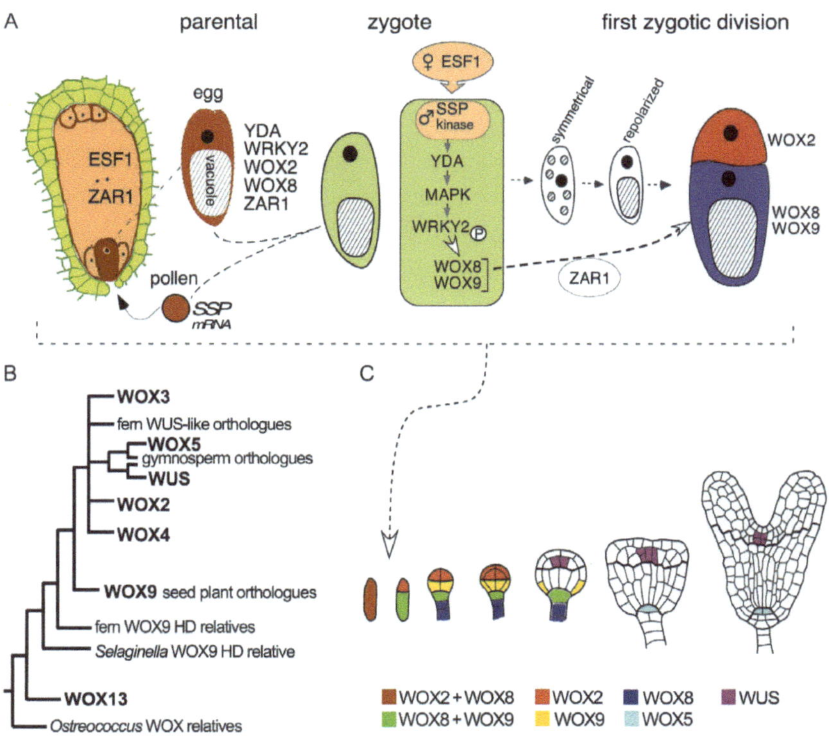

Fig. 2 Zygote polarity and cell fates in the *A. thaliana* embryo. (A) Known molecular determinants in zygote, egg or after the first asymmetric cell division. (B) Phylogenetic limitations within the WOX gene family. (C) WOX gene expression patterns and cell fates in the early *A. thaliana* embryo until the heart stage.

maternally expressed at the micropylar pole of the eight-nucleate embryo sac and after cellularisation in the egg, its flanking synergids and the central cell (Yu et al., 2016). The master regulator of polar zygote development, WOX8, is integral to these phenotypes and it should be recalled that the large vacuole at the micropylar pole of the *A. thaliana* egg is fragmented and the zygote undergoes a histologically symmetrical stage, before active vacuolar repolarization precedes the initial zygotic division. The plane of division and the restriction of *WOX2* and *WOX8/9* gene transcription to the small apical or large basal cell, respectively, thus depends on maternal gene products deposited in the egg, paternal inputs via SSP, and external cues from surrounding maternal gametophytic tissue.

This sophisticated network is difficult to monitor phylogenetically; for example, neither *WOX8/9* nor *WOX2* belongs to the ancestral *WOX13* group (Deveaux et al., 2008). Similar limitations in inferring phylogenetic relationships between ancestral and derived gene family members exist for many developmental marker genes, such as the MIKC* class of type II MADS-box genes, TALE-HD gene family members that specify meristem identity, or the *ASYMMETRIC LEAF1/ROUGH SHEATH/PHANTASICA* gene family of MYB-transcription factors that specify lateral leaf primordia. All these gene families expanded during vascular plant evolution, often in association with whole-genome duplications. The paternally derived SSP pseudokinase is also a recent evolutionary addition within the BRASSINOSTEROID KINASE family that is limited to the Brassicaceae (Shao-Lun & Adams, 2010). It possibly derives from a similar paleopolyploidization to that which created the *WOX8* and *WOX9* paralogues (Vanneste, Maere, & Van de Peer, 2014). Limited phylogenetic support exists for WOX8 prior to seed plants: fern WOX8-type sequences share differences in the DNA-binding homeodomain (HD) relative to the ancestral WOX13 type, but the flanking protein sequences differ substantially and, accordingly, they form an outgroup relative to seed plant members within the WOX8/9 clade (Fig. 2B; Nardmann & Werr, 2013). According to HD-swaps within the WUS protein, amino acid signatures discriminate the ancestral WOX13 type from the derived WOX8/9 and WUS proteins (Dolzblasz et al., 2016). Thus, DNA target-site specificity potentially began to diverge from the ancestral type in lycophytes such as *Selaginella moellendorffii*, but the different domain structures among WOX8/9 orthologues in ferns and seed plants hinder establishing functional equivalence of the respective proteins.

6. From zygote polarity to multicellular meristems

In contrast to WOX8, firm phylogenetic support exists for WOX2 below seed plants. A distant WOX2 relative in the leptosporangiate fern *Ceratopteris richardii* (Nardmann & Werr, 2012), *CrWUL*, promotes stem-cell identity (Zhang, Jiao, Jiao, Zhao, & Zhu, 2017) and marks immediate descendants of the root apical initial, so-called merophytes, which are pluripotent and undergo stereotypic cell divisions (Nardmann & Werr, 2012). Cell proliferation and pluripotency are essential prerequisites for the elaboration of cell types in the embryo proper, and relate to multicellular stem-cell niches in the shoot and root meristems. The stereotypic cell division patterns of the *A. thaliana* zygote and early embryo as well as WOX gene expression are illustrative in this context: the basal *WOX8*-expressing cell divides transversally to form a filamentous suspensor, whereas the apical *WOX2*-positive cell divides longitudinally to generate a four-celled proembryo, before transversal divisions result in apical and basal cell tiers of four cells each. Subsequent periclinal divisions generate a radial organization with eight outer protodermal cells and eight inner cells, which continue to divide without stereotypy to establish the vascular ground cells, before isodiametric growth enlarges the spherical embryo. This cellular histology is accompanied by specific patterns of *WOX2*, *WOX8*, and *WOX9* activity: *WOX2* is expressed throughout the four-cell proembryo, but becomes restricted to the upper cell tier at the eight-cell stage and *WOX8* expression remains suspensor-specific. In contrast, *WOX9* is present in the lower cell tier of the eight-cell embryo and in the uppermost suspensor cell or hypophysis, which divides transversally to generate precursors of the root quiescent center (QC) and the columella root cap (Fig. 2C). The upper lens-shaped QC precursor cell then expresses *WOX5* (Haecker et al., 2004). At the apical (chalazal) pole, WOX1, WOX2, WOX3 and WOX5 redundantly activate *WUS* in the organizing centre of the SAM, which explains the subtle *wox2* single mutant phenotype of periclinal divisions at the eight-cell stage (Zhang, Tucker, Hermann, & Laux, 2017). The initial stereotypic cell division pattern in *A. thaliana* terminates with the elaboration of the primary shoot and root meristems and the subtle single WOX gene mutant phenotypes indicate substantial redundancy within or outside the WOX gene family and illustrate the robustness of the regulatory network.

7. Plesiomorphies in seed plants

In maize, *ZmWOX2A* expression is confined to the apical embryo proper and prepatterns the SAM position, as in *A. thaliana* (Nardmann, Zimmermann, Durantini, Kranz, & Werr, 2007). The specific expression of maize *WOX8/9* orthologues in cells of the basal lineage was supported by transcriptome analyses of the maize zygote after the first asymmetric cell division (Chen et al., 2017). However, the apical specificity of *ZmWOX2A* was not confirmed, suggesting it is activated later, consistent with the assumption that maize embryogenesis involves an initial cell proliferation phase before specific cell types are established (Zhao, Begcy, Dresselhaus, & Sun, 2017). Alternatively, *ZmWOX2A* transcripts are transient and are dependent on WOX8/9 functions (Breuninger, Rikirsch, Hermann, Ueda, & Laux, 2008) that are potentially lost by microdissection of the apical and basal cells after the first zygotic division. However, the lateral expression domain of *ZmWOX2A* in the transition stage maize embryo prepatterns the SAM, in functional homology to its specific expression in the apical domain of the *A. thaliana* 16-cell embryo (Nardmann et al., 2007). Similarly, in *Pinus sylvestris*, *PsWOX2* transcription was demonstrated in apical cells of young gymnosperm embryos during the polyembryonic stage, suggesting that *WOX2* association with apical embryonic fate is plesiomorphic from the base of seed plants. Although cellular expression data for WOX8/9 orthologues in *P. sylvestris* are lacking, the downregulation of *Picea abies WOX2* via RNAi (Zhu, Moschou, Alvarez, Sohlberg, & von Arnold, 2016) suggests that *WOX2* function is conserved in embryonic protoderm development among land plants, and has acquired a potentially gymnosperm-specific function in suspensor cell expansion.

The early egg of the fruit fly *D. melanogaster* illustrates that nuclei within a common cellular environment can respond in concert or individually to incoming sperm-derived or extrinsic signals. However, during the syncytial phase of nuclear divisions in some gymnosperms, including *P. sylvestris*, cellular contacts or secreted ligands analogous to *A. thaliana* ESF1 or CLE8 would have to originate from the surrounding megagametophytic tissue, which differs from the triploid endosperm of angiosperms. The *P. sylvestris* megagametophyte develops over a 15-month period, starting with a slow free nuclear division phase, which is stimulated by pollination (13 months), followed by cellularization (2 months), before two to three archegonia and egg cells are specified and fertilized. Despite these differences in the origin and environment of the gymnosperm egg or zygote, RNAi data from somatic embryos suggest that *WOX8* regulates the basal fate of zygote development in seed plants.

The conservation of apical *WOX2* expression in young embryos is striking, as gene orthologues exist in all available seed plant genomes (W. Werr, Unpublished Data). The long independent evolutionary trajectories of *A. thaliana*, maize and *P. sylvestris* suggest that *WOX2* recruitment to embryonic apical cell fate possibly occurred in a common ancestor of seed plants, although convergent evolution cannot formally be excluded. The *WOX2* plesiomorphy is similar to the *WOX3* association with marginal and plate meristems and *WOX4* recruitment to the vascular cambium (Nardmann & Werr, 2013), and also relates to *WUS* activation in the organizing centre of the prospective SAM (Zhang, Tucker, et al., 2017). WUS-like relatives in the leptosporangiate ferns *C. richardii* and *Cyathea australis* share the WUS-box, EAR-like functional motifs and HD adaptations with WUS and can partially substitute for the WUS stem cell-promoting function in *A. thaliana* when appropriately expressed in the SAM organizing centre (Zhang, Jiao, et al., 2017). Phylogenetically, *WOX2* thus relates to stem cell-promoting functions, independent of whether the megagametophyte size is reduced in angiosperms, intermediate in gymnosperms or dominant in ferns, and is a prime candidate for an apical zygotic fate determinant in evo-devo comparisons.

8. The auxin connection

The role of *WOX2* in *A. thaliana* stem cell maintenance via mediating auxin and cytokinin concentrations, and the connection between polar auxin transport (Zhang, Tucker, et al., 2017) and *WOX2* expression in conifers (Palovaara & Hakman, 2009), links transcriptional and hormonal regulation in apical–basal polarity. The intercalation of a vegetative growth phase early in embryo evolution to establish a longitudinal axis of polarity for sporophyte elongation was potentially driven by auxin, a determinant of embryonic patterning in *A. thaliana* (Friml et al., 2003). The components of auxin biology, including auxin synthesis and homeostasis via conjugation, perception and signal transduction, are conserved among embryophytes and the auxin response machinery is encoded by the *P. patens* and *S. moellendorffii* genomes (Finet & Jaillais, 2012). Polar auxin transport is present in the charophyte alga, *Chara corallina*, and also plays a role in apical–basal polarity formation in embryos of the brown alga *Fucus distichus* (Basu, Sun, Brian, Quatrano, & Muday, 2002), which is phylogenetically distinct from the red/green algae and green plant lineage. This suggests that polar auxin transport is an ancestral trait in zygote development. However, we cannot differentiate whether it represents a plesiomorphy present in a common ancestor or has been independently acquired in the brown algal and green land plant lineages. Evidence that auxin has been independently recruited several times during evolution is provided by its differential integration into the developmental networks of roots and rhizoids, which are non-homologous structures. Despite caveats in establishing auxin response as a plesiomorphy, the ubiquity of a complete auxin signaling module across gymnosperm, monocot, and eudicot embryos supports the coevolution of individual auxin response components over a huge evolutionary time-scale. It is conceivable that changes within individual machinery constituents have been adaptive and have resulted in micro- and macro-evolutionary changes in embryonic patterning. These adaptations include AUXIN RESPONSE FACTOR (ARF) truncation, the expansion of ARF and AUXIN/INDOLE ACETIC ACID (AUX/IAA) transcription factor families and their interaction specificities, endoplasmic reticulum-associated auxin responses, and the manipulation of auxin concentrations via local biosynthesis.

9. Functional tests and limitations

Another studied transcription factor family with respect to embryophyte evolution is that of TALE-HD master regulator proteins of the diploid zygotic phase in unicellular green soil algae. *Chlamydomonas reinhardtii* expresses the GAMETOPHYTE-SPECIFIC MINUS1 (GSM1) or GAMETOPHYTE-SPECIFIC PLUS1 (GSP1) proteins in "minus" and "plus" type gametes, respectively. Following syngamy, GSM1 and GSP1 dimerize and enter the nucleus (Lee, Lin, Joo, & Goodenough, 2008) to control zygotic genes. GSM1 is a KNOTTED-LIKE HOMEOBOX (KNOX) protein, whereas GSP1 is a BELL-like HD protein and both protein families have expanded independently among land plants. The KNOX family bifurcated in charophyte algae (Frangedakis, Saint-Marcoux, Moody, Rabbinowitsch, & Langdale, 2016) and class I KNOX proteins regulate cell division and differentiation in determinate and indeterminate meristems in diploid tissues, but are not associated with haploid gametophytic growth (Sakakibara, Nishiyama, Deguchi, & Hasebe, 2008). Class I KNOX genes diverged substantially in the non-vascular bryophytes (liverworts, mosses and hornworts), with extant moss sequences being most similar to those in vascular plants. In reciprocal complementation experiments, Class I KNOX proteins from either *A. thaliana*, the moss *P. patens*, the fern *C. richardii* or the lycophyte *Selaginella kraussiana*

(Frangedakis et al., 2016) have been used to rescue either the *brevipedicellus* (*bp*) mutant of *A. thaliana*, or a *P. patens* triple *knox* mutant. The *P. patens* triple mutant could not be complemented by any of the vascular plant KNOX proteins; however, the *bp* mutant was complemented by KNOX proteins from *P. patens* and *S. kraussiana*, but was only partially complemented by those from *C. richardii*. This suggests that moss KNOX genes have retained a broad developmental function, which has been refined via gene neo- and subfunctionalization in vascular plants. In contrast, close GSP1-like BELL relatives appear to have been lost from land plant genomes, and it is hypothesized that the extant species-specific BELL and KNOX gene complements evolved from independent gene duplications. In contrast, to their KNOX partners, BELL proteins are key genetic triggers of the gametophyte-to-sporophyte transition, based on ectopic overexpression of the native BELL1 protein in *P. patens*, which leads to the formation of embryos and diploid sporophytes from gametophytic cells in the absence of fertilization (Horst et al., 2016). All KNOX and BELL proteins function interdependently via dimerization, an ancestral trait that is conserved in animals and plants (Bürglin, 1998). KNOX protein motifs have remained evolutionarily conserved, despite the emergence of phylum- or species-specific anomalies, whereas much greater structural variation has been tolerated in the BELL family (Frangedakis et al., 2016). Obligate heterodimerization partners apparently diverged at different rates, although their developmental potentials remain unresolved in a context where major macroevolutionary adaptations can result from adaptations in a few genes (Doebley & Lukens, 1998). BELL1 overexpression in moss (Horst et al., 2016) exemplifies its recruitment during somatic embryogenesis, but the identity of its KNOX partner remains unknown. In addition to differences in heterodimerization potentials between family members, different KNOX/BELL dimers might recognize different *cis*-regulatory elements in target gene promoters. The availability of protein partners can potentially be addressed by transcriptomic approaches, but identifying their target gene promoters within a few cells remains a challenge, although *cis*-regulatory adaptations are evolutionarily potentially as important as *trans*-regulatory changes in the generation of phenotypic diversity (Alonso-Blanco, Mendez-Vigo, & Koornneef, 2005).

10. Conclusions and perspectives

The huge morphological diversity in embryonic ontogeny across the plant kingdom reflects long independent evolutionary trajectories with high extinction rates (Forest, 2009). Existing embryophyte phylogenies are constantly subject to revision, which impacts the interpretation of character evolution. One example is the anthophyte hypothesis that, based on morphological cladistics, suggested that angiosperms and the Gnetales belonged to the same clade. However, molecular phylogenies have since reached a consensus that Gnetales are not the sister group to the angiosperms, but are most closely related to conifers (Soltis, Bell, Kim, & Soltis, 2008). Molecular systematics is cladistic-driven and presupposes that taxa are monophyletic and that molecular relationships correlate with phylogenetic descent. The increasing number of available whole-genome sequences, especially those of key bryophytes, polysporangiophyte mosses and gymnosperms, or fern models such as *C. richardii* (Plackett et al., 2015), will improve the robustness of taxonomic relationships. The emerging use of charophytes to identify molecular genetic components present at the base of embryophytes will further substantiate molecular phylogenies (Domozych, Popper, & Sørensen, 2016), and may help to elaborate potential KNOX/BELL partners involved in the haploid/diploid transition or the WOX2–WOX8 antagonism in apical and basal cells after the first zygotic division. The *C. richardii* WOX8/9 clade member CrWOXA shares homology in the DNA-binding homeodomain with seed plant members and is specifically expressed in stem cells, such as root apical mother cells, whereas CrWUL resides in the WUS-clade that includes WOX2 and is active in descending proliferative merophytes (Nardmann & Werr, 2012). A robust evolutionary association between fern and seed plants would be highly informative with respect to zygote polarity and cell fate.

Instead of the *de novo* evolutionary invention of the entire integrated gametophytic network of sperm and maternal signals, it is more likely that individual gene functions were recruited into existing networks. The delivery of SSP via sperm into the *Arabidopsis* egg illustrates one important aspect of syngamy, but the GSM1/GSP1 module in algae or the obligate KNOX/BELL partnership, which regulates the haploid/diploid phase transition in moss, probably represent more ancestral innovations. Comparative transcriptome analyses of sperm and egg cells would address potential differences between key bryophyte and fern species and be highly supported by robust genome data and phylogenies. Another paradigm of embryogenesis evolution is zygote polarity, an adaptive trait that is either exoscopic or endoscopic among basal land plant radiations, and transcriptomic approaches in species that differ in polar zygote development should reveal common principles. The atypical longitudinal first zygotic division in species of the hornwort *Anthoceros* is redolent of the transient symmetry and repolarization of the *A. thaliana* zygote, where extrinsic signals from the cellular environment have been functionally demonstrated. Accordingly, existing transcriptome data for the egg, zygote or its polar cellular descendants could benefit from corresponding data that depict gene expression in the surrounding cellular environment, which is megagametophytic in radiations up to gymnosperms, but includes the endosperm in angiosperms.

The creation of robust molecular phylogenies remains by far the most powerful method to generalize functional data from model species such as *A. thaliana* on the evolutionary scale. However, several limitations remain: first, gene homologies become increasingly unreliable or undetectable, concomitant with increasing evolutionary distance; second, identifying conserved gene functions depends on establishing functional orthology among species, which in turn, relies on heterologous transgenic approaches that are currently unfeasible for many taxa. Third, the combinatorial coevolution of gene functions cannot be detected. However, the resolution of RNA-sequencing, the increasing number of genome sequences, phylogenetic methods and bioinformatic resources available today intersect a long-standing histological study of embryo development. Current functional data and genetic resources in the bryophyte *P. patens* and the angiosperm *A. thaliana* define extant anchors for evo-devo comparisons. Given these advantages, the choice and depth of analysis in appropriate model species is more important than genetic resources alone, to address common principles of early zygotic development in land plants, a macro-evolutionary question that reflects adaptations within the diplontic life cycle of extant species.

Acknowledgments

We acknowledge the following sources of funding: the Deutsche Forschungsgemeinschaft via Grant WE1262/11-2 and the Sonderforschungsbereich 680 (SFB 680 A5), and the Cluster of Excellence on Plant Sciences (CEPLAS).

References

Alonso-Blanco, C., Mendez-Vigo, B., & Koornneef, M. (2005). From phenotypic to molecular polymorphisms involved in naturally occurring variation for plant development. *The International Journal of Developmental Biology*, *49*, 717–732.

Autran, D., Baroux, C., Raissig, M. T., Lenormand, T., Wittig, M., Grob, S., et al. (2011). Maternal epigenetic pathways control parental contributions to *Arabidopsis* early embryogenesis. *Cell*, *145*, 707–719.

Basu, S., Sun, H., Brian, L., Quatrano, R. L., & Muday, G. K. (2002). Early embryo development in *Fucus distichus* is auxin sensitive. *Plant Physiology*, *130*, 292–302.

Bayer, M., Nawy, T., Giglione, C., Galli, M., Meinnel, T., & Lukowitz, W. (2009). Paternal control of embryonic patterning in *Arabidopsis thaliana*. *Science*, *323*, 1485–1488.

Bowe, L. M., Coat, G., & dePamphilis, C. W. (2000). Phylogeny of seed plants based on all three genomic compartments: Extant gymnosperms are monophyletic and Gnetales' closest relatives are conifers. *Proceedings of the National Academy of Sciences of the United States of America*, *97*, 4092–4097.

Bower, F. O. (1908). *The origin of a land flora. A theory based upon the facts of alternation*. London: Macmillan & Co. Ltd.

Bower, F. O. (1922). The primitive spindle as a fundamental feature in the embryology of plants. *Proceedings of the Royal Society of Edinburgh*, *43*, 1–36.

Breuninger, H., Rikirsch, E., Hermann, M., Ueda, M., & Laux, T. (2008). Differential expression of *WOX* genes mediates apical-basal axis formation in the *Arabidopsis* embryo. *Developmental Cell*, *14*, 867–876.

Bürglin, T. R. (1998). The PBC domain contains a MEINOX domain: Coevolution of Hox and TALE homeobox genes? *Development Genes and Evolution*, *208*, 113–116.

Cairney, J., & Pullman, G. S. (2007). The cellular and molecular biology of conifer embryogenesis. *New Phytologist*, *176*, 511–536.

Chandler, J., Nardmann, J., & Werr, W. (2008). Plant development revolves around axes. *Trends in Plant Science*, *13*, 78–84.

Chen, J., Strieder, N., Krohn, N. G., Cyprys, P., Sprunck, S., Engelmann, J. C., et al. (2017). Zygotic genome activation occurs shortly after fertilization in maize. *Plant Cell*, *29*, 2106–2125.

Cridge, A. G., Dearden, P. K., & Brownfield, L. R. (2016). Convergent occurrence of the developmental hourglass in plant and animal embryogenesis? *Annals of Botany*, *117*, 833–843.

Del Toro-De León, G., García-Aguilar, M., & Gillmor, C. S. (2014). Non-equivalent contributions of maternal and paternal genomes to early plant embryogenesis. *Nature*, *514*, 624–627.

Deveaux, Y., Toffano-Nioche, C., Claisse, G., Thareau, V., Morin, H., Laufs, P., et al. (2008). Genes of the most conserved WOX clade in plants affect root and flower development in Arabidopsis. *BMC Evolutionary Biology*, *8*, 291.

Doebley, J., & Lukens, L. (1998). Transcriptional regulators and the evolution of plant form. *The Plant Cell*, *10*, 1075–1082.

Dolzblasz, A., Nardmann, J., Clerici, E., Causier, B., van der Graaff, E., Chen, J., et al. (2016). Stem cell regulation by *Arabidopsis WOX* genes. *Molecular Plant*, *9*, 1028–1039.

Domozych, D. S., Popper, Z. A., & Sørensen, I. (2016). Charophytes: Evolutionary giants and emerging model organisms. *Frontiers in Plant Science*, *7*, 1470.

Drost, H. G., Gabel, A., Grosse, I., & Quint, M. (2015). Evidence for active maintenance of phylotranscriptomic hourglass patterns in animal and plant embryogenesis. *Molecular Biology and Evolution*, *32*, 1221–1231.

Drost, H. G., Janitza, P., Grosse, I., & Quint, M. (2017). Cross-kingdom comparison of the developmental hourglass. *Current Opinion in Genetics & Development*, *45*, 69–75.

Finet, C., & Jaillais, Y. (2012). AUXOLOGY: When auxin meets plant evo-devo. *Developmental Biology*, *369*, 19–31.

Forest, F. (2009). Calibrating the tree of life: Fossils, molecules and evolutionary timescales. *Annals of Botany*, *104*, 789–794.

Frangedakis, E., Saint-Marcoux, D., Moody, L. A., Rabbinowitsch, E., & Langdale, J. A. (2016). Nonreciprocal complementation of KNOX gene function in land plants. *New Phytologist*, *216*, 591–604.

Friml, J., Vieten, A., Sauer, M., Weijers, D., Schwarz, H., Hamann, T., et al. (2003). Efflux-dependent auxin gradients establish the apical-basal axis of *Arabidopsis*. *Nature*, *426*, 147–153.

García-Aguilar, M., & Gillmor, C. S. (2015). Zygotic genome activation and imprinting: Parent-of-origin gene regulation in plant embryogenesis. *Current Opinion in Plant Biology*, *27*, 29–35.

Haecker, A., Gross-Hardt, R., Geiges, B., Sarkar, A., Breuninger, H., Herrmann, M., et al. (2004). Expression dynamics of *WOX* genes mark cell fate decisions during early embryonic patterning in *Arabidopsis thaliana*. *Development*, *131*, 657–668.

Hanstein, J. (1870). Die Entwicklung des Keimes der Monokotylen und Dikotylen. *Botanische Abhandlungen aus dem Gebiete der Morphologie und Physiologie*, *1*, 1–112.

Horst, N. A., Katz, A., Pereman, I., Decker, E. L., Ohad, N., & Reski, R. (2016). A single homeobox gene triggers phase transition, embryogenesis and asexual reproduction. *Nature Plants*, *2*, 15209.

Johansen, D. A. (1950). *Plant embryology. Embryogeny of the spermatophyta*. Waltham, MA: Chronica Botanica Co. William Dawson & Sons, Ltd., London.

Lee, J. H., Lin, H., Joo, S., & Goodenough, U. (2008). Early sexual origins of homeoprotein heterodimerization and evolution of the plant KNOX/BELL family. *Cell*, *133*, 829–840.

Ligrone, R., Duckett, J. G., & Renzaglia, K. S. (2012a). Major transitions in the evolution of early land plants: A bryological perspective. *Annals of Botany*, *109*, 851–871.

Ligrone, R., Duckett, J. G., & Renzaglia, K. S. (2012b). The origin of the sporophyte shoot in land plants: A bryological perspective. *Annals of Botany*, *110*, 935–941.

Maheshwari, P. (1950). *An introduction to the embryology of angiosperms*. New York: McGraw Hill.

Menand, B., Yi, K., Jouannic, S., Hoffmann, L., Ryan, E., Linstead, P., et al. (2007). An ancient mechanism controls the development of cells with a rooting function in land plants. *Science*, *316*, 1477–1480.

Nardmann, J., & Werr, W. (2012). The invention of WUS-like stem cell-promoting functions in plants predates leptosporangiate ferns. *Plant Molecular Biology*, *78*, 123–134.

Nardmann, J., & Werr, W. (2013). Symplesiomorphies in the WUSCHEL clade suggest that the last common ancestor of seed plants contained at least four independent stem cell niches. *New Phytologist*, *199*, 1081–1092.

Nardmann, J., Zimmermann, R., Durantini, D., Kranz, E., & Werr, W. (2007). WOX gene phylogeny in Poaceae: A comparative approach addressing leaf and embryo development. *Molecular Biology and Evolution*, *24*, 2474–2484.

Palovaara, J., & Hakman, I. (2009). *WOX2* and polar auxin transport during spruce embryo pattern formation. *Plant Signaling & Behavior*, *4*, 153–155.

Plackett, A. R., Di Stilio, V. S., & Langdale, J. A. (2015). Ferns: The missing link in shoot evolution and development. *Frontiers in Plant Science*, *6*, 972.

Quint, M., Drost, H. G., Gabel, A., Ullrich, K. K., Bönn, M., & Grosse, I. (2012). A transcriptomic hourglass in plant embryogenesis. *Nature*, *490*, 98–101.

Sakakibara, K., Nishiyama, T., Deguchi, H., & Hasebe, M. (2008). Class 1 KNOX genes are not involved in shoot development in the moss *Physcomitrella patens* but do function in sporophyte development. *Evolution & Development*, *10*, 555–566.

Sarkar, A. K., Luijten, M., Miyashima, S., Lenhard, M., Hashimoto, T., Nakajima, K., et al. (2007). Conserved factors regulate signalling in *Arabidopsis thaliana* shoot and root stem cell organizers. *Nature*, *446*, 811–814.

Shao-Lun, L., & Adams, K. L. (2010). Dramatic change in function and expression pattern of a gene duplicated by polyploidy created a paternal effect gene in the Brassicaceae. *Molecular Biology and Evolution*, *27*, 2817–2828.

Soltis, D. E., Bell, C. D., Kim, S., & Soltis, P. S. (2008). Origin and early evolution of angiosperms. *Annals of the New York Academy of Sciences*, *1133*, 3–25.

Szövényi, P., Rensing, S. A., Lang, D., Wray, G. A., & Shaw, A. J. (2011). Generation-biased gene expression in a bryophyte model system. *Molecular Biology and Evolution*, *28*, 803–812.

Tanabe, Y., Hasebe, M., Sekimoto, H., Nishiyama, T., Kitani, M., Henschel, K., et al. (2005). Characterization of MADS-box genes in charophycean green algae and its implication for the evolution of MADS-box genes. *Proceedings of the National Academy of Sciences of the United States of America*, *102*, 2436–2441.

Tanahashi, T., Sumikawa, N., Kato, M., & Hasebe, M. (2005). Diversification of gene function: Homologs of the floral regulator FLO/LFY control the first zygotic cell division in the moss *Physcomitrella patens*. *Development*, *132*, 1727–1736.

ten Hove, C. A., Lu, K. J., & Weijers, D. (2015). Building a plant: Cell fate specification in the early *Arabidopsis* embryo. *Development*, *142*, 420–430.

Ueda, M., Aichinger, E., Gong, W., Groot, E., Verstraeten, I., Vu, L. D., et al. (2017). Transcriptional integration of paternal and maternal factors in the *Arabidopsis* zygote. *Genes & Development*, *31*, 617–627.

Ueda, M., Zhang, Z., & Laux, T. (2011). Transcriptional activation of *Arabidopsis* axis patterning genes WOX8/9 links zygote polarity to embryo development. *Developmental Cell*, *20*, 264–270.

Vanneste, K., Maere, S., & Van de Peer, Y. (2014). Tangled up in two: A burst of genome duplications at the end of the cretaceous and the consequences for plant evolution. *Philosophical Transactions of the Royal Society of London Series B: Biological Sciences*, *369*(1648), pii: 20130353.

Vielle-Calzada, J. P., Baskar, R., & Grossniklaus, U. (2000). Delayed activation of the paternal genome during seed development. *Nature*, *404*, 91–94.

Wardlaw, C. W. (1955). *Embryogenesis in plants*. London: Methuen & Co Ltd. New York: John Wiley and Sons Inc.

Yu, T.-Y., Shi, D.-Q., Jia, P.-F., Tang, J., Li, H.-J., Liu, J., et al. (2016). The Arabidopsis receptor kinase ZAR1 is required for zygote asymmetric division and its daughter cell fate. *PLoS Genetics*, *12*, e1005933.

Zhang, Y., Jiao, Y., Jiao, H., Zhao, H., & Zhu, Y. X. (2017). Two-step functional innovation of the stem-cell factors WUS/WOX5 during plant evolution. *Molecular Biology and Evolution*, *34*, 640–653.

Zhang, Z., Tucker, E., Hermann, M., & Laux, T. (2017). A molecular framework for the embryonic initiation of shoot meristem stem cells. *Developmental Cell*, *40*, 264–277.

Zhao, P., Begcy, K., Dresselhaus, T., & Sun, M.-S. (2017). Does early embryogenesis in eudicots and monocots involve the same mechanism and molecular players? *Plant Physiology*, *173*, 130–142.

Zhu, T., Moschou, P. N., Alvarez, J. M., Sohlberg, J. J., & von Arnold, S. (2016). *WUSCHEL-RELATED HOMEOBOX 2* is important for protoderm and suspensor development in the gymnosperm Norway spruce. *BMC Plant Biology*, *16*, 19.

Zobell, O., Faigl, W., Saedler, H., & Münster, T. (2010). MIKC* MADS-box proteins: Conserved regulators of the gametophytic generation of land plants. *Molecular Biology and Evolution*, *27*, 1201–1211.

CHAPTER TWENTY

Apomixis in flowering plants: Developmental and evolutionary considerations

Gloria León-Martínez, Jean-Philippe Vielle-Calzada*
Group of Reproductive Development and Apomixis, UGA Laboratorio Nacional de Genómica para la Biodiversidad (Langebio), Cinvestav Irapuato, Irapuato, Mexico
*Corresponding author: e-mail address: vielle@cinvestav.mx

Contents

Abstract

Apomixis refers to a set of reproductive mechanisms that invariably rely on avoiding meiotic reduction and fertilization of the egg cell to generate clonal seeds. After having long been considered a strictly asexual oddity leading to extinction, the integration of more than 100 years of embryological, genetic, molecular, and ecological research has revealed apomixis as a widely spread component of the dynamic processes that shape flowering plant evolution. Apomixis involves several flexible and versatile developmental pathways that can be combined within the ovule to produce offspring. Here we review the large body of classic and contemporaneous contributions that have addressed unreduced gamete formation, haploid induction, and parthenogenesis in flowering plants. We emphasize similarities and differences between sexual and apomictic reproduction, and highlight their implications for the evolutionary emergence of asexual reproduction through seeds. On the basis of these comparisons, we propose a model that associates the developmental origin of apomixis to a dynamic epigenetic landscape, in which environmental fluctuations reversibly influence female reproductive development through mechanisms of hybridization and polyploidization.

Current Topics in Developmental Biology, Volume 131
ISSN 0070-2153
https://doi.org/10.1016/bs.ctdb.2018.11.014

1. Apomixis: Confronting the evolutionary prevalence of sex

Understanding the origins of asexual reproduction in multicellular organisms is one of the most challenging problems faced by modern evolutionary biology. Perhaps owing to Darwin's failure to shed light in the central explanation of why sex is so prevalent in all branches of life (Darwin, 1862), several decades of theoretical studies have yet to provide a reasonable explanation for the evolutionary emergence of mechanisms that can give rise to a viable embryo from a cell which is not formed by the fusion of a male and female gamete. In the plant kingdom, a major consequence of species survival and adaptation is the emergence of a multitude of asexual reproductive alternatives that take advantage of vegetative propagation and seed formation. Since genetic variation is interpreted as essential for adaptation, it is often thought that sex is necessary for the perpetuation of a species. However, many flowering plants are able to produce seeds without undergoing meiosis and gamete fusion. Apomixis refers a set of heritable developmental mechanisms that represent a direct challenge to the main theories behind the evolutionary success of sexual reproduction, by consistently giving rise to viable clonal progeny through seeds. The phenomenon was first described as an intriguing abnormality found in *Alchornea ilicifolia* (Smith, 1841), 28 years before Mendel attempted to recapitulate his classic findings on inheritance by attempting crosses with *Hieracium* subgenus *Pilosella*, unaware that apomixis is common in this phenotypically diverse taxon (Bicknell, Catanach, Hand, & Koltunow, 2016; Nogler, 2006). From the 1920s to the 1960s, classic plant embryologists characterized and named the different forms of apomixis by comparing the cytological organization of apomictic and sexual female gametophytes in a wide group of species, documenting its taxonomic distribution and describing its prevalent developmental variants (Gustafsson, 1946). Paraffin embedding, microtome sectioning, and chromosome smearing were the standard techniques for this type of embryological studies. Their large collective surveys showed that apomixis is not a rare exception but rather a reproductive strategy that depends on conserved developmental mechanisms widely spread among flowering plants (Carman, 1997). In the 1970s, the introduction of optical clearing of whole-mounted ovules (in replacement of microtome sectioning) resulted in the rapid and efficient observation of thousands of specimens, completely transforming the aims and scope of studies focusing on megagametogenesis and early seed formation. Complementing approaches that allowed the assessment of genetic variation, these methodologies opened the possibility for large-scale genetic and cytological studies of plants resulting from intra- and interspecific hybridization between mixed populations of sexual and apomict individuals. Such studies were often part of breeding attempts for transferring apomixis to economically important cereals, such as *Zea mays* (maize), *Oryza sativa* (rice), *Pennisetum glaucum* (pearl millet), from their wild apomict relatives *Tripsacum dactyloides, Elymus rectisetus,* and *Pennisteum squamulatum*, respectively (Hannah & Bashaw, 1987). Although unsuccessful in generating agronomically valuable apomictic crops, analysis of the resulting germplasm provided crucial insights into structural, functional, and developmental aspects of apomixis. In particular, several detailed characterizations of genomic regions co-segregating with apomictic traits allowed the identification of candidate regions involved in the control of apomixis, either as genetic determinants, or as downstream factors acting during female reproductive development.

Apomixis invariably circumvents meiotic reduction (apomeiosis) and fertilization of the egg cell (parthenogenesis), avoiding the mechanisms that create genetic variation during sexual reproduction, but maintaining the benefits of seed dispersal. Gametophytic apomixis entails the formation of an unreduced female gametophyte, in which the egg cell is the precursor of the clonal embryo. In contrast, sporophytic apomixis relies on a somatic (sporophytic) cell of the ovule to directly give rise to a viable embryo. Male gametogenesis is usually not affected, and functional pollen is required for viable endosperm formation in more than 90% of apomictic species. Sexual reproduction and apomixis are not mutually exclusive. Whereas all forms of apomixis are heritable, most individual plants exhibit the trait as a facultative alternative to sexual reproduction. Over the past three decades, developmental research has mainly focused on attempting to elucidate the genetic basis and molecular mechanisms that regulate apomixis, either through the elucidation of its genetic control in comparison to sexuality, or through the identification of genes that can induce components of apomixis in sexual model species. In parallel, the integration of ecological and population genetic data has provided a fresh view on the mechanisms of adaptation, distribution, evolutionary advantages, and possible origins of apomixis, expanding the taxonomic and biogeographic distribution of apomictic genera, and discussing their implications within a phylogenetic context. The importance endosperm formation and its bearing for apomixis and genomic imprinting have been reviewed elsewhere (Grossniklaus, 2001; Haig & Westoby, 1991; Hands, Rabiger, & Koltunow, 2016; Kotlunow & Grossniklaus, 2003). Here we extensively review the main developmental components that relate apomixis to sexual reproduction in flowering plants. We also discuss the evolutionary implications that emerge from a detailed comparison of the genetic basis and molecular mechanisms controlling unreduced gamete formation and haploid induction in sexual species to those regulating apomeiosis and parthenogenesis during apomixis.

2. Female gametogenesis in sexual and apomictic species

In sexual flowering plants, floral organogenesis gives rise to male (anther) and female (ovule) reproductive organs that harbor the formation of meiotic precursors (see chapters "Pre-meiotic anther development" by van der Linde and Walbot, this issue and "Development and evolution of the unique ovules of flowering plants" by Gasser and Skinner, this issue). Whereas many male meiocytes differentiate in the anther, usually a single female meiocyte (or megaspore mother cell, MMC) differentiates in the developing ovule. The MMC undergoes meiosis to generate four reduced cells (megaspores). In the majority of sexual species, a single functional megaspore gives rise to the female gametophyte, whereas the three additional megaspores die without further expansion or division. After substantial cellular enlargement, the nucleus of the haploid functional megaspore undergoes three endomitotic divisions before cellularizing and giving rise to a differentiated female gametophyte composed of seven cells: two companion synergids, the egg cell, a binucleated central cell, and three antipodals (see chapter "Development and function of the flowering plant female gametophyte" by Serbes et al., this issue). Whereas this developmental pathway is the most prevalent among sexual flowering plants, no less 88 families are polysporic and contain species that avoid cytokinesis during female meiosis to incorporate either two (bispory) or all four (tetraspory) haploid nuclei into subsequent gametogenesis (Carman, 1997), giving rise to female gametophytes containing a reduced egg and central cell that require fertilization to initiate embryo and endosperm formation, respectively. The resulting patterns of endomitosis and cellularization are variable among species, leading in some cases to genetically distinct cells in the gametophyte, especially in the case of tetraspory. Some sexual species can also form seeds containing multiple embryos, most often by embryonic cleavage following fertilization of the egg cell, or by fertilization of synergids that acquire an egg cell identity (Johri, Ambegaokar, & Srivastava, 1992).

In contrast, apomictic species can develop embryos from unreduced female gametophytes in which the egg cell develops by parthenogenesis (gametophytic apomixis), or directly from a sporophytic cell in a differentiating ovule (sporophytic apomixis through adventitious or nucellar embryony). Gametophytic apomixis relies on mechanisms that ensure the formation of gametes from an apomeiotically derived cell. In the case of diplospory, the gametophytic precursor is the MMC; as a consequence, sexual and apomictic reproduction are mutually exclusive in a single ovule, but not in different ovules of an individual. In the case of apospory, the gametophytic precursor is a sporophytic cell (apoporous initial cell, AI), often positioned in the vicinity of the MMC. As a consequence, sexually derived and apomeiotic female gametophytes can coexist within the same ovule. As compared to sexual development, unreduced female gametophytes develop precociously in several species, suggesting that temporal asynchronies are inherent to apomixis pathways (Carman, Jamison, Elliott, Dwivedi, & Naumova, 2011). In contrast, sporophytic apomicts are able to form adventitious embryos from almost any non-gametophytic cell in the developing ovule. These embryos develop in parallel to meiotically derived female gametogenesis within the same ovule, and rely on fertilization of the central cell for endosperm development and seed viability. The seeds of sporophytic apomicts contain multiple embryos, one of which is usually the result of fertilization of a meiotically derived egg cell, and the rest being derived from sporophytic cells. As a consequence of all these alternatives allowing the coexistence of sexuality and apomixis at the species, population, or individual levels, most natural apomicts are classified as facultative, confirming the regular identification of recombinant genotypes and ploidy variants in their natural habitat. Also, and despite the classification of apomixis appears quite restrictive, it is not uncommon to find populations or individual plants exhibiting a combination of diplospory, apospory, and adventitious embryony (Aliyu, Schranz, & Sharbel, 2010; Bicknell, Borst, & Koltunow, 2000; Hojsgaard, Greilhuber, et al., 2014; Hojsgaard, Klatt, Baier, Carman, & Hörandl, 2014; Hojsgaard, Martínez, & Quarin, 2013). This narrow reproductive association between sexual and asexual reproduction in natural apomicts suggests that meiosis, unreduced gamete formation, double fertilization, parthenogenesis, and early embryo development must be controlled by developmental mechanisms that share a close evolutionary history in the context of plant adaptation.

3. Distribution of apomixis and interrelation with polyploidy and hybridization

Apomixis has been reported in at least 78 of 460 families of flowering plants (Carman, 1997; Hojsgaard, Greilhuber, et al., 2014; Hojsgaard, Klatt, et al., 2014; Hörandl & Hojsgaard, 2012), including members of basal orders such as the Magnolids (Piperales, Canellales, Magnoliales, and Laurales), as well as families in 25 out of 43 dicotyledoneous, and 5 out of 11 monocotyledoneous orders. It is particularly prevalent in the *Rosaceae*, *Asteraceae*, and *Poaceae* (Fig. 1). Although initially thought to represent an ancient evolutionary predisposition manifested by certain clades (Richards, 2003),

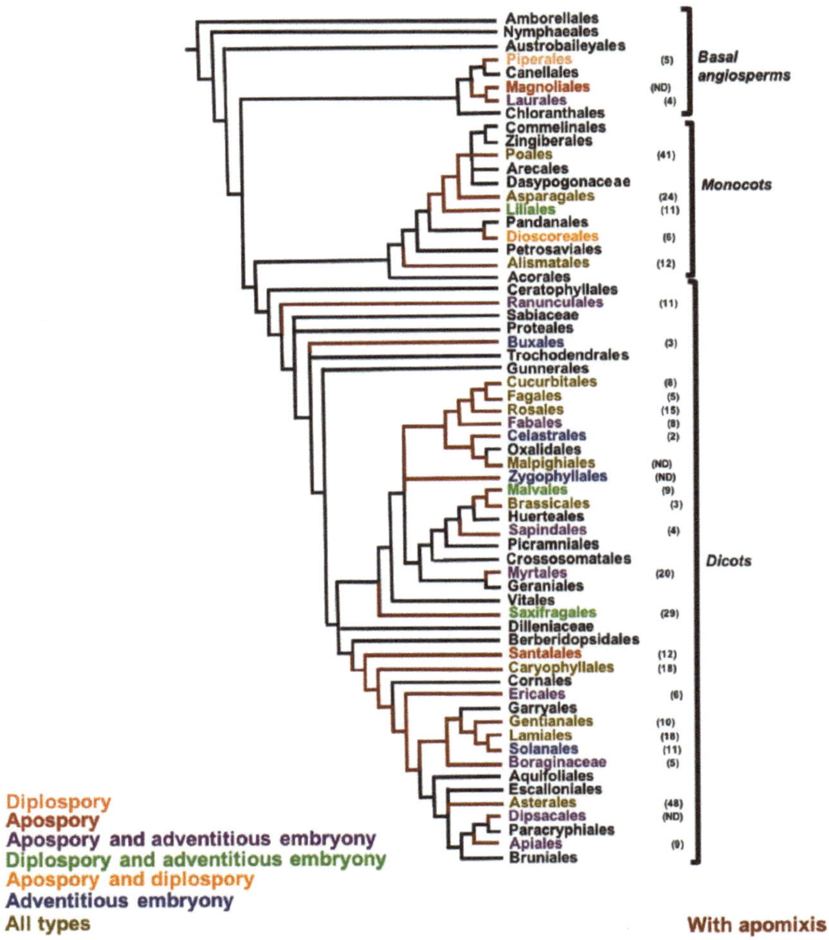

Fig. 1 Distribution of the main types of apomixis among the orders and genera of flowering plants. Phylogenetic lineages unequivocally exhibiting apomixis are marked in red. *Modified from Hojsgaard, D., Klatt, S., Baier, R., Carman, J. G., & Hörandl E. (2014). Taxonomy and Biogeography of Apomixis in Angiosperms and Associated Biodiversity Characteristics.* CRC Critical Reviews in Plant Sciences, *33, 414–427.*

current phylogenetic trees superimposed with apomixis distribution support a model of multiple independent emergence and rapid spreading among large families, coupled to frequent reversion from facultative apomixis to obligate sexuality (Fig. 1). Considering that the reproductive characterization of female gametogenesis is not available for 112 families, the total number of apomictic species is expected to increase. The three main types of apomixis occur in all major clades of flowering plants, adventitious embryony being the most frequent form (148 genera), followed by apospory (110 genera), and diplospory (68 genera). However, the majority of historical records only reported the most prevalent type, and not the frequent coexistence of at least two if not all three main types. Apomixis is not only absent in *Amborella trichopoda*, the most basal angiosperm, but also from other non-flowering seed plant lineages, including gymnosperms. After conducting a large taxonomic comparison of female reproductive anomalies recorded in flowering plants, Carman (1997) concluded that apomictic, polysporic, and polyembryonic species have a similar phylogenetic distribution, suggesting that the genetic basis controlling apomixis, polyspory, and polyembryony are developmentally related from an evolutionary perspective (Carman, 1997; Carman et al., 2011).

Current biogeographical records show that some apomicts have wider distributions than their sexual relatives, with a tendency for some genera to abundantly occupy regions that were affected by the last glacial period (Bierzychudek, 1985; Hörandl, Cosendal, & Temsch, 2008; Kearney, 2005; Paun, Stuessy, & Hörandl, 2006). Almost all gametophytic apomicts are polyploids, mostly allopolyploids presumed to derive from hybridization between diploid sexual parents (Bierzychudek, 1985). Gametophytic apomicts usually form populations composed of individuals harboring different levels of ploidy, in which occasional fertilization of an unreduced egg cell, or parthenogenesis of a reduced one, can give rise to individuals having new ploidy levels, generating what is referred to as an "agamic complex" (Asker & Jerling, 1992; Stebbins, 1980). However, many polyploid species reproduce without showing any form of apomixis. The natural occurrence of scarce diploid populations that retain apomictic reproduction, and the frequent recovery of diploid individuals in several apomictic species, indicates that polyploidy is not an obligatory requirement for functional apomixis (Bicknell, 1997; Nogler, 1984; Voigt-Zielinski, Piwczyński, & Sharbel, 2012). In contrast, diploid apomicts are quite common in genera showing adventitious embryony (Asker & Jerling, 1992; Gustafsson, 1946), suggesting that the trait might be dependent on a less restrictive genetic dosage control. Since sporophytic apomixis does not depend on the formation of unreduced gametes, they do not form agamic complexes.

Gametophytic apomixis is also tightly correlated with hybridization. As in the case of polyploidy, hybridization in itself does not induce apomixis *de novo*. Since first generation hybrids (F1s) can often show poor fertility, the perception that hybridization can result in meiotic aberrations led to the hypothesis that unreduced gamete formation could represent the only mechanism to generate functional female gametes in apomictic hybrids (Arnold, 1997). However, polyploidization of hybrids can also lead to meiotic stability through homologous pairing of the duplicated chromosomes (Comai, 2005). Recent results in the *Ranunculus auricomus* agamic complex suggest that synthetic hybrids generated by crossing sexual diploid and tetraploid taxa representing phylogenetically distant species, result in reduced fertility, apomeiosis, and developmental asynchronicity during female gametogenesis, suggesting that hybridization rather than polyploidy can lead to developmental alterations in the ovule of facultative apomictic species (Hojsgaard, Greilhuber, et al., 2014; Hojsgaard, Klatt, et al., 2014). However, despite polyploidization and hybridization being recognized as important preconditions for apomixis emergence (Asker & Jerling, 1992; De Storme & Geelen, 2013; Grossniklaus, 2001; Spillane, Steimer, & Grossniklaus, 2001), there is no evidence suggesting that their combined presence is sufficient to generate reproductive innovations leading to apomeiosis or autonomous embryogenesis on the basis of a previously established and strictly sexual developmental pathway.

4. Genetic control of apomixis

Apomixis is genetically controlled and inherited as a dominant trait (Grossniklaus, Nogler, & van Dijk, 2001; Hand & Koltunow, 2014; Nogler, 1984; Ozias-Akins & van Dijk, 2007). Early genetic evidence suggested that a single locus was responsible for its control, and multiple studies have revealed that loci segregating with apomixis exhibit suppressed recombination. The identification of rare recombinants or the genetic dissection of large genomic deletions co-segregating with apomixis revealed that apomeiosis, parthenogenesis, and exceptional cases of fertilization-independent endosperm formation, are controlled by independent loci (Albertini et al., 2001; Catanach, Erasmuson, Podivinsky, Jordan, & Bicknell, 2006; Conner, Gunawan, & Ozias-Akins, 2013; Noyes & Rieseberg, 2000; Ogawa, Johnson, Henderson, & Koltunow, 2013; Schallau et al., 2010; van

Dijk, Tas, Falque, & Bakx-Schotman, 1999). In *Hieracium praealtum*, a member of the subgenus *Pilosella*, apomixis is controlled independently by two unlinked loci: *LOSS OF APOMEIOSIS* (*LOA*) and *LOSS OF PARTHENOGENESIS* (*LOP*). Whereas *LOA* is required for apomeiosis and suppression of all sexually derived megaspores, the gametophytic activity of *LOP* is necessary for autonomous development of both embryo and endosperm (Catanach et al., 2006: Koltunow et al., 2011). Deletion of *LOA* or *LOP* results in partial reversion to sexuality (Catanach et al., 2006; Koltunow et al., 2011), suggesting that apomixis is superimposed on a sexual pathway that remains functional in the absence of apomixis (Koltunow, Ozias-Akins, & Siddiqi, 2013). However, several regulatory pathways differ between sexual and apomictic gametophytes, including cell cycle control, hormonal pathways, epigenetic, and transcriptional regulation (Schmidt et al., 2014). *LOA* function requires the initiation of megaspore formation in the early ovule, suggesting that in *Hieracium* spp. the specification of AIs is dependent on the initiation of sexual development, supporting the view that apomictic development may cause a spatio-temporal deregulation of the mechanisms that control sexual development, triggering the ectopic specification of AIs (Hand & Koltunow, 2014).

In several species, the suppression of recombination around apomixis loci is associated to heterochromatic or highly repetitive genomic regions with strong tendencies to allelic divergence. In several species, long and laborious efforts of genetic and physical mapping have narrowed down the genomic location of loci involved in the initiation of apomixis to a few hundred kilobases (Akiyama et al., 2011; Ochogavia et al., 2011; Ozias-Akins & van Dijk, 2007). In *P. squamulatum*, for example, the apospory-specific genomic region (ASGR) maps to a telomeric location containing abundant repetitive sequences (Akiyama et al., 2004). The hemizygous chromosomal region containing *LOA* in *Hieracium* spp. is structurally reminiscent of the hemizygous ASGR region in *Pennisetum* spp., suggesting a convergent evolution that might be necessary for function and maintenance of the trait, or eventually the parallel accumulation of genomic structural traits derived from apomictic reproduction (Okada et al., 2011). Repetitive sequences in *LOA* are not required for functional apomeiosis (Kotani et al., 2013), suggesting that conserved structural features around apomict loci are a consequence of asexual reproduction and suppressed recombination, perhaps as a means to maintain the elements necessary for apomixis (Hand & Koltunow, 2014; Kotlunow & Grossniklaus, 2003). In *Paspalum* spp., the hemizygous apomictic controlling locus (ACL) also shows suppression of recombination and abundant local rearrangements, likely caused by the activity of transposable elements (Calderini et al., 2006). In *Taraxacum* spp., refinement of the genetic linkage map showed a bias toward apomictic plants among recombinants between the two most closely linked molecular markers to the *DIPLOSPOROUS* (*DIP*) locus that controls diplospory, suggesting a possible duplication of a presumed *DIP* gene (Vijverberg, Milanovic-Ivanovic, Bakx-Schotman, & van Dijk, 2010). In the case of sexual diploid *Erigeron strigosus*, the absence of apomixis in diploids is supported by evidence suggesting univalent inheritance of the locus co-segregating with diplospory, associated with negative selection against recessive gametophytic lethality linked to the locus controlling parthenogenesis (Noyes & Rieseberg, 2000). Recently, shotgun sequencing of bacterial artificial chromosomes (BACs) that physically mapped to the *P. squamulatum* ASGR allowed the identification of the first gene that is directly involved in the control of parthenogenesis in a natural apomict (see Section 6), opening the possibility for the identification of other genes controlling apomixis by pursing ongoing equivalent approaches in several species.

5. Unreduced female gamete formation and apomeiosis

The generation of gametes that maintain the sporophytic chromosome number (unreduced gametes or 2n-gametes) is likely the most important evolutionary force involved in the origin of plant polyploids (Bretagnolle & Thompson, 1995; De Storme & Geelen, 2013; Otto & Whitton, 2000; Ramsey & Schmeske, 1998; Tayalé & Parisod, 2013). Although largely unexplored at the experimental level (Schinkel et al., 2017), it is generally thought that polyploids originate from unreduced gametes that fuse with reduced ones, resulting in unstable triploid offspring that serves as intermediate to generate a new polyploid through a second fusion of mixed-ploidy gametes (Harlan & De Wet, 1975; Hojsgaard, 2018). After Clausen, Keck, and Heisey (1945), Harlan and De Wet (1975) documented no less than 106 genera, in which the consistent generation of viable unreduced gametes likely gave rise to progeny with higher ploidy levels. Additional reviews summarized studies documenting the production of unreduced gametes in 20 species from 17 different genera (Bretagnolle & Thompson, 1995), establishing that unreduced gamete formation and not sporophytic chromosome doubling is the main driving force behind polyploidy.

While the formation of female unreduced gametes is intrinsic to gametophytic apomixis (diplospory and apospory), many sexual species can form unreduced gametes from meiotic abnormalities that affect chromosomal segregation or cell division (Bretagnolle & Thompson, 1995; Brownfield & Köhler, 2011; D'Amato, 1989; De Storme & Geelen, 2013; Hermsen, 1984; Rammana, 1992; Veilleux, 1985). The mechanisms leading to the formation of female unreduced gametes operate during the meiotic phase of megaspore formation or in pre-meiotic precursor cells. Although they may result from a wide variety of meiotic deviations, two prevalent mechanisms are generally distinguished: first division restitution (FDR) and second division restitution (SDR). The term "restitution" refers to the formation of a single nucleus with an unreduced chromosome number, instead of two nuclei with reduced chromosome numbers (Ramanna, 1979). Two successive rounds of division occur during normal meiosis (Fig. 2). Whereas the first division (meiosis I) leads to the separation of paired homologous chromosomes, the second division (meiosis II) leads to the separation of sister chromatids of an individual chromosome. FDR is the result of an

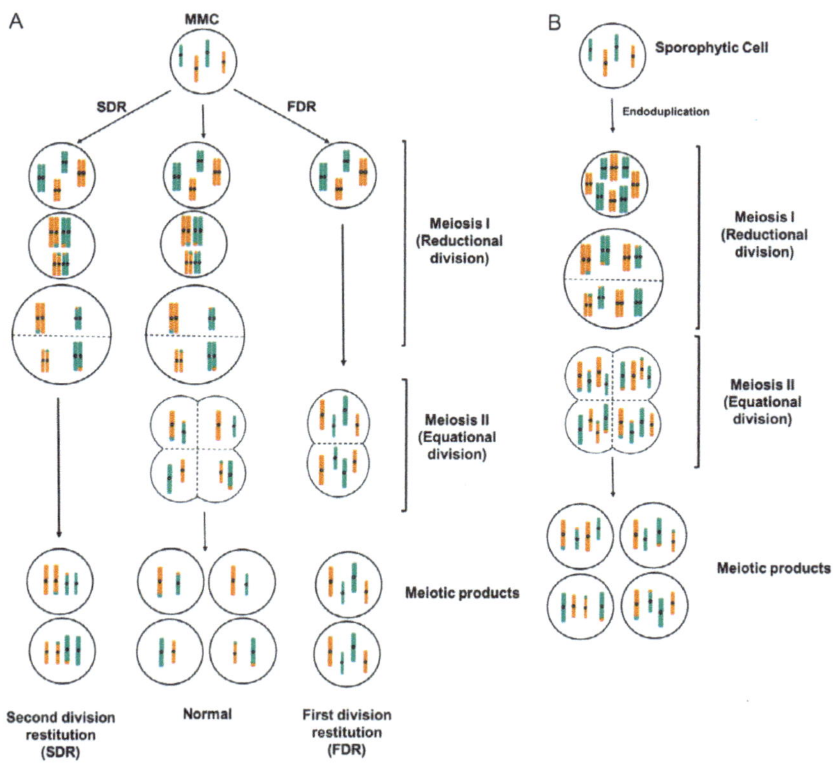

Fig. 2 The three main categories of unreduced female gamete formation during meiosis in a hypothetical flowering plant containing 2n = 4 chromosomes in sporophytic cells.

abnormal meiosis I giving rise to a single nucleus containing all non-sister chromatids. In contrast, unreduced gametes derived from SDR result from the omission of sister chromatid separation during meiosis II (Fig. 2A). In both cases, two of the four chromatids of a homologous set of chromosomes are recovered in the gametes. When compared to apomixis, these two mechanisms have different developmental consequences. Unreduced female gametes derived from FDR preserve the parental chromosome constitution, including heterozygosity in non-sister chromatids from the centromere to the first cross-over location, and therefore can be considered to be similar to those formed during gametophytic apomixis. Documented cytological defects leading to FDR involve failure of spindle formation during metaphase I (Nassar, 1992), abnormal cytokinesis (Simpson & Davis, 1983; Tavoletti, 1991; Werner & Peloquin, 1991a), or synaptic defects impeding chromosome segregation (Ramanna, 1983; Werner & Peloquin, 1991b). In contrast, unreduced female gametes derived from SDR undergo normal meiotic recombination and do not transmit the intact chromosome constitution of the maternal genotype. A third but less frequent mechanism of unreduced gamete formation depends on endopolyploidization of precursor cells prior to meiosis (Lelley, Mahmoud, & Lein, 1987; Wagenvoort, De Vries-Van Hulten, Winkelhorst, & Den Nijs, 1990; Fig. 2B). In this case, chromosome doubling can be dependent on endoreplication or spindle disturbance or the absence of cytokinesis during mitosis (D'Amato, 1989). The subsequent meiotic division occurs without anomalies, giving rise to unreduced gametes in which independent chromosome assortment and recombination modify their genetic constitution as compared to their precursors.

Nowadays the understanding of the mechanisms leading to unreduced gamete formation is gender-biased, as most assessments have been conducted during male rather than female gametogenesis. Male meiosis is invariably more frequent and cytologically more easily accessible than female meiosis. Studies of unreduced gamete frequencies in male organs are predominantly based on measures of pollen size and DNA content, whereas the assessment of DNA content in the egg cell nucleus using flow cytometry is complicated by the presence of unreduced sporophytic tissue from the ovule. Unreduced female gamete estimates have been invariably based on progeny frequencies combined with flow cytometric seed screening that determine parental gamete types based on the relative predicted ploidy of the resulting embryo and endosperm. For these types of indirect measurements, one must consider that preferential genetic transmission through reduced rather than unreduced gametes, or poor viability of progeny resulting from inter-ploidy fertilization, will cause an underestimation of the frequency of unreduced female gametes. As a consequence, it is still difficult to assess the evolutionary impact of unreduced female gamete formation in the emergence of apomictic polyploids. Current estimations—mainly conducted in cultivated plants—show that the phenomenon is variable among species, ecotypes, and even reproductive organs of a single individual (McCoy, 1982; Veilleux, 1985; Veilleux & Lauer, 1981). Population studies indicate that only a few individuals produce high frequencies of unreduced gametes (Armenta-Medina, Demesa-Arevalo, & Vielle-Calzada, 2011; Carman, 1997; Vijverberg et al., 2010). However, it is not clear whether natural populations conform to these tendencies, as it is possible that past artificial selection during domestication might have influenced the prevalence of unreduced gamete formation, and only few studies—dominated by male assessments—have attempted to estimate the frequency of unreduced gamete formation in populations of wild species (Kreiner, Kron, & Husband, 2017a, 2017b).

Monogenetic recessive inheritance of unreduced gamete formation has been reported in several species. In maize, for example, Rhoades and Dempsey (1966) demonstrated that unreduced female gamete formation is under the genetic control of the *elongate1* (*el1*) mutant. A detailed cytological analysis revealed that the defect by which diploid female gametophytes arise in *el1* is the absence of meiosis II, leading to dyad cells directly initiating female gametogenesis (Barrell & Grossniklaus, 2005). Comparative mapping before the advent of full genomic sequencing indicated that the genomic region containing a locus related to diplospory in highly apomictic accessions of *T. dactyloides* was homologous to the region where *el1* maps in the maize genome (Grimanelli et al., 1998). However, the molecular nature of the corresponding gene was never revealed, and it remains unclear if an *el1* homolog is involved in the initiation of at least certain types of diplospory. In *Solanum tuberosum* (potato), abnormal synapsis of homologous chromosomes is considered to be under the control of *desynapsis1* (*ds1*), which forms unreduced gametes during both male and female meiosis (Jongedijk & Ramanna, 1988; Jongedijk, Ramanna, Sawor, & Hermsen, 1991). Additional loci appear to regulate the expression of *ds1* as well as the frequency of cells affected during meiotic development (Jongedijk & Ramanna, 1989). Monogenetic inheritance of unreduced gamete formation is also found in *Trifolium pretense*, *Petunia hybrida* (Maizonnier, 1976), and *Medicago* spp. (Veronesi, Tavoletti, & Mariani, 1990), in all cases with phenotypic penetrance modulated by several genetic modifiers. As many of the mutants affecting meiosis exhibit male or female specificity, there is little genetic correlation between unreduced female and male gamete formation (De Haan, Maceira, Lumaret, & Delay, 1992; Veronesi, Mariani, & Bingham, 1986; Werner & Peloquin, 1987). Mutant alleles responsible for unreduced gamete phenotypes often show a variable degree of penetrance, suggesting that defects leading to their emergence might result from alleles with variable expressivity (Veilleux & Lauer, 1981).

More recently, three genes in which single mutations give rise to variable frequencies of unreduced female gametes were identified in *Arabidopsis thaliana* (d'Erfurth et al., 2010, 2008, 2009; Erilova et al., 2009; Ravi, Marimuthu, & Siddiqi, 2008; Wang, Jha, Chen, Doonan, & Yang, 2010). However, defects in only one of them give rise to a FDR phenotype, whereas the two others cause SDR by abolishing meiosis II. Whereas several mutations in the *SWITCH1(SWI1)/DYAD* gene cause defects in both male and female meiosis, specific mutant alleles such as *dyad* lead only to unreduced female gamete formation by mechanisms reminiscent of diplospory (Mercier et al., 2001; Siddiqi, Ganesh, Grossniklaus, & Subbiah, 2000). Homozygous *dyad* individuals are mostly female sterile. However, they usually produce 10–20 sexually derived seeds that are triploid and give rise to progeny that fully retains parental heterozygosity across the genome, a trait that is also found in unreduced female gametes derived from apomixis (Ravi et al., 2008). *SWI1/DYAD* is fundamental for sister chromatid cohesion and recombination during prophase of meiosis I. Homozygous individuals defective in *CYCLIN A1;2* (*CYCA1;2*), also known as *TARDY ASYNCHRONOUS MEIOSIS* (*TAM*), produce about 30% of viable unreduced female gametes that fail to enter meiosis II, giving rise to recombined cells that contain both sister chromatids of each chromosome (d'Erfurth et al., 2010; Magnard, Yang, Chen, Leary, & McCormick, 2001; Wang et al., 2010). CYCA1;2 is a cyclin A protein necessary for meiotic cell cycle progression. Recessive alleles of *OMISSION OF SECOND DIVISION1* (*OSD1*) also complete meiosis I but fail to enter meiosis II, producing close to 85% of recombined unreduced female gametes that can give rise to polyploid progeny. Although OSD1 has no obvious conserved domains or precise function, it belongs to the UV14 family of plant-specific proteins thought to have a role in modifying cyclin-dependent kinase activity (Hase, Trung, Matsunaga, & Tanaka, 2006). Other proteins involved in the production of male but not female unreduced gametes include ARABIDOPSIS PARALLEL SPINDLE1 (ATPS1), involved in meiosis II spindle orientation (d'Erfurth et al., 2008), JASON, a protein of unknown function (Erilova et al., 2009), and TETRASPORE/STUD (TES/STUD), a predicted kinesin that affects meiosis II spindle formation and subsequent cytokinesis (Hülskamp et al., 1997; Spielman et al., 1997; Yang et al., 2003). Finally, d'Erfurth et al. (2010) were able to generate nearly 100% functional, unreduced female gametophytes by generating a triple *A. thaliana* mutant that combines suppression of recombination with absence of meiosis II. The combination of two mutants, *recombination8* (*rec8*) with *sporulation11* (*spo11*), abolished recombination without chromosome damage. The resulting double mutant was crossed with *osd1* to abolish meiosis II, obtaining viable, unreduced female gametophytes that maintain heterozygosity when fertilized by sexually derived sperm cells. Megaspore formation in this triple *Mitosis instead of Meiosis* (*MiMe*) mutant background is reminiscent of diplospory in apomicts, since the resulting unreduced female gametophytes are genetically identical to the MMC.

Do natural populations reproducing by gametophytic apomixis exhibit a higher frequency of unreduced gamete formation (uncoupled from parthenogenesis) than their sexual counterparts? The question has barely been addressed, as apomictic species have rarely been included in studies aiming at estimating unreduced gamete frequency. The broadest survey for a specific family was conducted across 22 species of *Brassicaceae*, generally showing 0.1–2% of unreduced male gametes, a range that is equivalent to those found for cultivated species (Kreiner et al., 2017a, 2017b). The study included populations of *Cardamine bulbosa*, *C. concatenata*, and *C. diphylla*, all reproducing asexually by leaf bulbils. However, it did not include species of the *Boechera* genus reproducing by apomixis, or sexual ecotypes of *A. thaliana*. Populations of *Cardamine* spp. produced significantly more unreduced male gametes than populations from mixed-mating and outcrossing species, suggesting that asexual vegetative propagation might favor their maintenance by limiting opportunities for selection exerted during seed-dependent reproduction. But in the case of apomixis, the evolutionary relationship between asexuality and unreduced gamete formation is not fully clear. To this date, only three natural populations have been reported as having unreduced female gamete frequencies exceeding 2%. Interestingly, one of them has significant rates of apomictic seed formation (*Malus coronaria*; Kron & Husband, 2009), and two others mostly reproduce through asexual propagation (*Cardamine* spp.; Kreiner et al., 2017b), in agreement with the possibility that asexuality can act as an evolutionary stabilizing agent of unreduced gamete formation. It is also well known that fertilization of unreduced eggs by meiotically derived sperm cells (also referred as 2n + n or BIII fertilization) occurs in apomictic species at frequencies ranging from 1% to 3% (Bashaw & Hignight, 1990; Bashaw, Hussey, & Hignight, 1992; Leblanc & Mazzucato, 2001). As in the case of apomictic *Poaceae* (Bashaw & Hanna, 1990), hybrids derived from BIII fertilization in the alpine species *Ranunculus kuepferi* arise from crosses between facultative aposporous tetraploids and sexual diploid individuals, suggesting that the formation of unreduced egg cells is the major pathway toward polyploidization in *R. kuepferi*, and its occurrence in diploids perhaps the first evolutionary step toward apomixis (Schinkel et al., 2017). In the case of aposporous species such as *Ranunculus* spp., the production of unreduced female gametes might not directly depend on a meiotic malfunction, but rather on the ectopic differentiation of sporophytic cells into gametophytic precursors. Representing the deleterious products of meiotic aberrations, unreduced gametes in sexual populations should tend to be affected by the action of a selective balance that results in their maintenance at relatively low frequencies, with the emergence of *de novo* polyploidy (neopolyploidy), depending on the contribution of individuals that exhibit unusually high frequencies of unreduced gamete formation. In contrast, by conferring a fitness advantage in combination with parthenogenesis, unreduced gametes might turn to be adaptive and positively selected in populations reproducing by apomixis (Bicknell & Koltunow, 2004; Lovell et al., 2013), giving rise to higher frequencies of female unreduced gametes.

6. Haploid induction and parthenogenesis

The autonomous development of an embryo through parthenogenesis, a key component of gametophytic apomixis, can also occur spontaneously in many sexual species, giving rise to progeny carrying the gametic chromosome number (n instead of 2n). The recovery of haploid individuals can nowadays be induced through *in vitro* cultured anthers or microspores (androgenesis), ovaries or ovules (gynogenesis), or by distant hybridization using irradiated pollen (Dunwell, 2010; Forster, Heberle-Bors, Kasha, & Touraev, 2007). The spontaneous formation of embryos from meiotically derived female gametes is the main cause of natural haploid induction in flowering plants, a phenomenon that, as apomixis, is not restricted to any particular taxonomic group. Haploids were initially reported in at least 71 species belonging to 16 families (Kimber & Riley, 1963), including the *Brassicaceae* (*Brassica campestris*, *B. olaracea*, and *B. napus*, among others),

Malvaceae (several *Gossypium* spp.), *Rosaceae*, *Poaceae*, *Solanaceae*, and *Compositeae*. Additional reports include the identification of spontaneous haploid formation in *Agropyron tsukushiense* (Sakamoto, 1964), *Medicaco sativa* (cultivated alfalfa; Bingham, 1969), *Prunus persica* (peach; Toyama, 1974), *Citrus* spp. (Karasawa, 1971; Yahata et al., 2005) and *Trillium smallii* (Uchino, 1973). Haploid embryos can arise from spontaneous development of the egg cell associated with single fertilization of the central cell, or by parental genome elimination following double fertilization. However, in most cases the cytological mechanisms by which gametic cells give rise to an embryo are unknown.

Many haploids are members of polyploidy series in which species with higher chromosome numbers have multiples of the basic haploid number. It is therefore common to find cases of haploid induction giving rise to autopolyhaploid or allopolyhaploid individuals. Meiosis in these haploids often shows anomalies caused by the absence of homologous chromosomes, revealing pairing affinities that usually remain undetected during diploid meiosis (Kimber & Riley, 1963). Since the observation of chiasmata is routinely accepted as evidence of segmental homology, haploid meiosis has been used to determine the chromosome constitution and its evolutionary implications in relation to a presumed diploid or polyploid status of extinct ancestors. Whereas allopolyhaploids will have a strong tendency to form a low level of bivalents, with frequent aneuploidies and unbalanced gametes that result in sterility, autopolyhaploids will show frequent formation of partial or sometimes complete synapses in fertile individuals. Interestingly, when McClintock analyzed maize meiosis in strict haploid plants carrying a single chromosome set (monoploids), she surprisingly found a high level of pairing during prophase of meiosis I (McClintock, 1933). Despite the rarity of chiasma formation, she noticed that the association of some heterologous chromosomes at pachytene was highly reminiscent of homologous pairing. Later, Brown (1958) reported equivalent results in some *Gossypium* spp. (cotton) hybrids, suggesting that the nature of non-homologous chromosome pairing in some monoploids might yield novel insights into the evolutionary understanding of meiosis and its reproductive alternatives, including ancestors of current apomictic species.

In several cases, haploid embryos can be generated by interspecific pollination with a distantly related species (Kasha & Kao, 1970; Laurie, O'Donoughue, & Bennett, 1990). Although detailed cytological analysis documenting the kinetics of double fertilization and early seed development is lacking, it is likely that the egg and central cell are able to fuse with two sperm cells following pollen tube entry into the female gametophyte. However, subsequent division of the zygote results in the elimination of paternally inherited chromosomes, giving rise to viable haploid embryos. Since endosperm development is also defective following gamete fusion, in most cases haploid embryos have to be rescued by *in vitro* culture. Interestingly, several environmental factors, such as temperature and light intensity, influence the frequency of haploid recovery (Bitsch, Gröger, & Lelley, 2000; Campbell, Griffin, Burritt, & Conner, 2001; Garcia-Llamas, Ramirez, & Ballesteros, 2004; Sanei, Pickering, Kumke, Nasuda, & Houben, 2011), a characteristic similar to the modifier effect exerted by environmental factors over apomixis (Gounaris, Sherwood, Gounaris, Hamilton, & Gustine, 1991; Hjelmqvist & Grazi, 1964; Knox, 1967). Interspecific pollination is routinely used to generate haploids in *Hordeum vulgare* (barley) by using *H. bulbosum* as a pollen donor (Devaux, 2003). The method is particularly successful if combined with homozygous barley plants carrying the *haploid initiator* (*hap*) locus that generates up to 30% haploid embryos (Hagberg & Hagberg, 1980). In the case of *Triticum durum* (durum wheat), parental lines can produce more than 90% of parthenogenetic haploids when carrying a homozygous wheat-rye translocation from the "Veery" cultivar and cytoplasm from *Aegilops kotschyi* that was introduced using the "Salmon" cultivar as maternal parent (Hsam & Zeller, 1993; Matzk, 1996). Neither the translocation nor the *Aegilops* cytoplasm alone triggers parthenogenesis, indicating that the system is dependent on nuclear-cytoplasmic interactions (Matzk, 1996; Matzk, Meyer, Bäumlein, Balzer, & Schubert, 1995). Haploid embryos of *Triticum aestivum* (wheat), *Avena Sativa* (oat), *Secale cereale* (rye), and triticale (a hybrid of wheat and rye) can also be obtained at variable frequencies by pollinating with maize (Immonen & Tenhola-Roininen, 2004; Inagaki, 2003; Laurie & Bennett, 1986; Wedzony, 2003). The same approach is used in *Pyrus communis* (pear) by crossing with *Malus pumila* (apple; Inoue, Sakuma, Kasumi, Hara, & Tsukihashi, 2004). Maternal dihaploids are also generated in tetraploid *S. tuberosum* by crosses to diploid *S. phureja*. Interestingly, in this particular case the resulting endosperm is functional and supports the parthenogenetic development of the embryo, avoiding the need of rescue through *in vitro* culture (De Maine, 2003). Haploid induction by chromosome elimination is to some degree reminiscent of semigamy, an unusual form of haploid embryogenesis, in which fusion of male and female nuclei does not occur. Instead both gametes divide independently, giving rise to a chimeric embryo containing cellular sectors derived from either the male or female haploid genomes. Turcotte and Feaster (1963) isolated a mutant of the cotton *G. barbadense*, which exhibits high frequencies of semigamy; they also demonstrated that the defect was caused by the allele *Se* (Stelly & Rooney, 1989). Although semigamy has also been recorded in *Coix aquatica* (Rao & Narayana, 1980) and *Theobroma cacao* (cocoa; Lanaud, 1988), none of the genes responsible for its occurrence have been isolated.

Haploid embryogenesis is crucial in the generation of so-called double haploids (DHs), completely homozygous parental lines that are obtained by artificially doubling the chromosome number of haploid progenitors. Despite the fact that DHs are essential for fast commercial hybrid production, the genetic basis of haploid induction remains poorly understood. It is only lately that a few genes causing haploid embryo formation through chromosome elimination have been identified in sexual model species. In maize, several alleles of the recessive mutation *indeterminate gametophyte1* (*ig1*) have been used as haploid inducers (Chase, 1963, 1964; Kermicle, 1969, 1971). In addition to abnormal nuclear proliferation in the female gametophyte, homozygous *ig1* individuals show simultaneous production of maternal and paternal (androgenetic) haploids at frequencies ranging from 0.6% to 2% (Kermicle, 1969). The *Ig1* gene encodes a LATERAL ORGAN BOUNDARIES (LOB) domain protein similar to the ASSYMETRIC LEAVES2 (AS2) protein of *A. thaliana* (Evans, 2007). Several other lines producing maternal haploids at unusually high frequencies became widely exploited for breeding purposes, including Stock6 and its derived germplasms (Coe, 1959a, 1959b; Chalyk, 1994; Chalyk & Chebotar, 2000; Liu & Song, 2000; Eder & Chalyk, 2002). Subsequent breeding resulted in higher frequencies of haploid induction at the expense of increasing rates of seed abortion. In Stock6-derived inducer lines such as HZI or RWK, maternal haploid frequencies can reach more than 10% and 13%, respectively (Kelliher et al., 2017; Zhang et al., 2008). Zhao, Xu, Xie, Chen, and Jin (2013) generated two additional Stock6-derived lines—CAU^{B} containing B chromosomes and CAU^{YFP} containing the H3 histone variant CENH3 coupled to a reporter gene—to follow molecular marker inheritance during haploid induction. CENH3 is known to replace the standard histone H3 in centromeric nucleosomes. A detailed cytogenetic analysis confirmed that chromosome elimination often starts at initial stages of embryogenesis. Recently, the gene responsible for the Stock6 phenotype was named *MATRILINEAL* (*MTL*). It encodes a pollen-specific phospholipase that is highly conserved in the *Poaceae* and exclusively expressed in the sperm cell's cytoplasm (Kelliher et al., 2017), confirming the importance of nuclear-cytoplasmic interactions for male genome transmittance. These findings indicate that maternal haploids in Stock6 may also depend on paternal chromosome elimination following double fertilization. However, careful cytological observations and quantitative assessment of paternal DNA in haploids will be necessary to precisely identify the mechanism underlying maize haploid induction.

In *A. thaliana*, haploid maternal embryos were also recovered by parental-specific genome elimination through a transgenic manipulation of CENH3. To construct a chimeric CENH3 protein that could complement a *cenh3* mutant, the N-terminal tail sequence of *CENH3* was replaced by the tail sequence of a conventional *A. thaliana* histone H3, fused with a GFP reporter to construct the *GFP-tailswap* transgene (Ravi & Chan, 2010). When crossed to wild-type plants as the female parent, *GFP-tailswap* plants produced 25–45% maternal haploids and close to 50% aneuploids. If crossed as a male to wild-type females, haploid frequencies were reduced to 4–5% for paternal haploids and to less than 11% for aneuploids. Haploids invariably contained the chromosomes from the wild-type parent and the cytoplasm from the maternal parent. As all crop species have CENH3, it is likely that haploids induced by CENH3 modification may be extended to other plant species (Comai, 2014; Ishii, Karimi-Ashtiyani, & Houben, 2016). Interestingly, in *H. vulgare* × *H. bulbosum* hybrids, both *CENH3* genes are normally transcribed, and the CENH3 protein of *H. vulgare* is properly localized to its native centromeres. In contrast, the *H. bulbosum* centromere has no CENH3 protein during mitotic anaphase, leading to chromosome elimination (Watts, Kumar, & Bhat, 2016). These results suggest that the loss of protein translation rather than repression of *CENH3* transcription is involved in chromosome elimination. The evidence showing that CENH3 is involved in haploid induction based on either intra- (maize and *A. thaliana*) or interspecific (barley) pollination suggests that defective centromeric functions involved in kinetochore assembly during the first division of the zygote might be the most common cause of chromosome elimination in sexual flowering plants. A synthetic strategy capable of producing viable clonal seeds was designed in *A. thaliana* by crossing the *GFP-tailswap* line, which causes parental chromosome elimination, with the *MiMe* triple mutant that generates high frequencies of unreduced gametes (Marimuthu et al., 2011). When *GFP-tailswap* was crossed as a male, 34% of the resulting F1 plants were maternal clones. When *GFP-tailswap* was crossed as a female, 42% of paternal clones were recovered, suggesting that the strategy is more efficient when female chromosomes are eliminated. Although successful in the establishment of a synthetic method for clonal reproduction through seeds, the strategy does not recapitulate the mechanisms that prevail in natural apomicts.

Are the frequencies of haploid parthenogenesis equivalent in apomictic individuals as compared to their sexual counterparts? The question has not been experimentally addressed. Because autonomous apomicts such as *Hieracium* spp. can generate viable seeds in complete absence of pollination (Hand & Koltunow, 2014) and developing embryos are routinely observed prior to pollen tube entry in every *Poaceae* apomict examined (Naumova & Vielle-Calzada, 2001; Vielle, Burson, Bashaw, & Hussey, 1995), it is unlikely that paternal-specific genome elimination represents an important mechanism of autonomous embryo formation in apomicts. Although no gene capable of inducing parthenogenetic development has been discovered in dicots, a first gene causing parthenogenesis of the egg cell in a natural apomict was isolated from *P. squamulatum*. The *PsASGR-BABYBOOM-like* (*PsASGR-BBML*) gene is expressed in apomictically derived egg cells before double fertilization, and its transgenic activity in sexual individuals of tetraploid pearl millet lines can induce the production of viable maternal haploids (Conner, Mookkan, Huo, Chae, & Azias-Akins, 2015; Gualtieri et al., 2006). RNAi-dependent silencing of *PsASGR-BBML* expression in F1 apomictic hybrids between *P. squamulatum* and *P. glaucum* resulted in a significant reduction of parthenogenetic frequencies. *PsASGR-BBML* encodes a transcription factor member of the BBM-like clade of the APETALA2/ ETHYLENE RESPONSE FACTOR (AP2/ERF) DNA-binding domain family, which is highly conserved in the plant kingdom. In *A. thaliana*, ectopic expression of a member of this family (*BBM*) leads to the formation of somatic embryos in seedling cotyledons (Boutilier et al., 2002). The expression of *PsASGR-BBML* can also induce parthenogenesis in maize and rice, but not in *A. thaliana* (Conner, Podio, & Ozias-Akins, 2017), suggesting that repression of autonomous egg cell division in dicots might involve mechanisms different from those prevailing in monocots. These differences might be caused by unexplored redundancy of genetic factors crucial for avoiding parthenogenesis, or divergent evolution during the establishment of molecular mechanisms necessary to ensure egg cell quiescence prior to double fertilization. For example, the only gene encoding a RWP-RK domain (RKD)-containing transcription factor found in the basal liverwort *Marchantia polymorpha* functions as a regulator of the gametophyte-sporophyte transition by establishing and/or maintaining the quiescent state of the egg cell in the absence of fertilization (Rövekamp, Bowman, & Grossniklaus, 2016).

But in *A. thaliana*, although the ectopic expression of *RKD1* or *RKD2* from the same transcription factor gene family induces an egg cell-like transcriptome in sporophytic cells (Koszegi et al., 2011), single loss-of-function mutations in the corresponding genes do not cause defects in the egg cell. This finding suggests high genetic redundancy among the RKD genes in *A. thaliana*, or the subsequent emergence of genetic and perhaps biochemical interactions between *BBM-like* and *RKD* activities in dicotyledonous but not monocotyledonous plants. Overall, the current molecular distinctions between the mechanisms responsible for maintenance of egg cell quiescence in the main clades of flowering plants support the view of a flexible sexual gamete evolving different developmental mechanisms to prevent parthogenogenesis.

7. Environmental and epigenetic regulation of female gametogenesis

Whereas temperature changes are known to affect sexual reproduction (Barnabás, Jäger, & Fehér, 2008; Hedhly, 2011; Hedhly, Hormaza, & Herrero, 2009; Peng et al., 2004), the balance between female meiosis and apomeiosis in facultative gametophytic apomicts is influenced by several environmental factors, such as temperature, photoperiod, soil salinity, and drought stress (Amasino, 2010; Gounaris et al., 1991; Hjelmqvist & Grazi, 1964; Keller & Körner, 2003; Klatt et al., 2016; Knox, 1967; Kurepin, Walton, Reid, Chinnappa, & Pharis, 2007; Lokhande, Ogawa, Tanaka, & Hara, 2003; Sakai et al., 2006). In several apomict grasses, prolonged photoperiod often leads to a decrease in apomeiosis frequencies (Evans & Knox, 1969; Gupta, Roy, & Singh, 1969; Quarin, 1986; Saran & Dewet, 1976). This abiotic response is in agreement with large-scale transcriptional studies showing significant changes in the expression patterns of several hundred metabolic genes when sexual development is replaced by apomixis (Schmidt et al., 2014; Sharbel et al., 2010), including a significant enrichment of genes involved in the biosynthesis of polyamines during aposporous initiation in *Boechera gunnisoniana* (Schmidt et al., 2014). Low levels of reactive oxygen species (ROS) promote the acquisition of meiotic fate in maize anthers (Kelliher & Walbot, 2012), and phenylpropanoid polyamine conjugates that abundantly accumulate in developing flowers are known to act as inhibitors of ROS, suggesting that the enhanced accumulation of polyamines during apomixis might result in better protection against oxidative stress (Schmidt et al., 2014). Although a role for ROS regulating sexual or apomict development remains to be elucidated, the previous hypothesis supports the view that oxidative stress—perhaps caused by extended UV exposure through long photoperiods (Klatt et al., 2016)—may be responsible for activating meiosis-specific proteins that initiate double-strand break formation and DNA repair, increasing recombination frequency (Hörandl & Hadacek, 2013).

Epigenetic regulation is likely responsible of the reproductive response to abiotic factors in flowering plants by controlling developmental and cell fate decisions through DNA or chromatin modifications. Several studies have provided evidence revealing that small RNA (sRNA)-dependent mechanisms play an important role in cell specification within the ovule (Martinez & Köhler, 2017; Schmidt, Schmid, & Grossniklaus, 2015; Wang & Köhler, 2017), a discovery with relevant implications for our understanding of how the female gametophytic lineage is established during both sexual and apomictic reproduction. The genome of *A. thaliana* contains 10 ARGONAUTE (AGO) proteins involved in genome regulation mediated by a large variety of small RNAs, such as microRNAs (miRNA) and small interfering RNAs (siRNA; Mallory & Vaucheret, 2010). Among them, the AGO4 clade (composed of AGO4, AGO6, AGO8 and AGO9) is active in the RNA-dependent DNA methylation (RdDM) pathway that regulates the transcriptional silencing of transposons and repeats by mediating DNA methylation and heterochromatin formation (Law & Jacobsen, 2010). Mutations in all members of the *A. thaliana AGO4* clade are dominant, and lead to aposporous-like phenotypes by which somatic sporophytic cells give rise to apomeiotically derived female gametophytes (Olmedo-Monfil et al., 2010; Hernandez-Lagana, Rodriguez-Leal, Lúa, & Vielle-Calzada, 2016). The same phenotype was found in mutants affecting additional members of the RdDM pathway, including *polymeraseIV* (*polIV*), *dicer-like3* (*dcl3*), *rna-dependent rna polymerase6* (*rdr6*), and *supressor of gene silencing3* (*sgs3*), but not in *ago1* or *dcl1* that are required for miRNA biogenesis and function (Hernandez-Lagana et al., 2016; Olmedo-Monfil et al., 2010). These findings suggest that silencing of heterochromatic repetitive regions is crucial to distinguish sexual from aposporous development. Similar phenotypic defects were also reported in several alleles of *MNEME* (*MEM*), a RNA helicase that is preferentially expressed in the MMC (Schmidt et al., 2011). Whereas a possible *MEM* function related to RdDM has not been investigated, AGO9 preferentially interacts with 24-nt sRNAs derived from transposable elements (TEs), mainly belonging to ancient families of retrotransposons. Its function is necessary for silencing TEs in female gametes (Olmedo-Monfil et al., 2010), a mechanism reminiscent of PIWI-dependent pathways in animals (Brennecke et al., 2007; Juliano, Wang, & Lin, 2011). The majority of targeted TEs are located in pericentromeric regions, suggesting a link between gametophytic cell fate and heterochromatic silencing (Duran-Figueroa & Vielle-Calzada, 2010). The action of an AGO9-dependent pathway has been suggested to represent a defense response against potential transpositionally induced damage during meiosis or female gametogenesis, a mechanism that could ensure the maintenance of genome integrity during the sporophyte-to-gametophyte transition in the ovule (Armenta-Medina et al., 2011). Similar epigenetic effects were discovered in maize, where mutations in *AGO104*—a homolog of *AGO4* in *A. thaliana*—also give rise to apomeiotic gametes (Garcia-Aguilar, Michaud, Leblanc, & Grimanelli, 2010). Phenotypes of mutants affecting the DNA methyltransferases *DMT102* and *DMT103* (closely related to *CHROMETHYLTRANSFERASE3* and *DOMAINS REARRANGED METHYLTRANSFERASE2* of *A. thaliana*) are also reminiscent of apospory (Singh et al., 2011), supporting an essential role for RdDM mechanisms in gametic cell specification. In addition to RdDM, large-scale chromatin reprogramming establishes an epigenetic and transcriptional state in the *A. thaliana* MMC that is distinct from the rest of the developing ovule (She et al., 2013). Consistent with methylation being involved in reproductive fate, significant changes in genomic methylation patterns occur during the formation of triploid diplosporous dandelions produced from a diploid sexual individual fertilized by polyploid pollen donors (Verhoeven, van Dijk, & Biere, 2010; Verhoeven, Jansen, van Dijk, & Biere, 2010). On the basis of this growing body of evidence, suggesting continuous action of differential methylation through female gametogenesis and early embryo formation (Belmonte et al., 2013; Jullien, Susaki, Yelagandula, Higashiyama, & Berger, 2012; Pillot et al., 2010; Schmid et al., 2012), the discovery of epigenetic components acting in the ovule could have important implications for understanding the evolutionary trends that shape innovations in reproductive development and adaptation.

8. Evolutionary considerations on the origins of apomixis

The prevalence of an independent gametophyte over the sporophyte in the life cycle of non-vascular plants and many tracheophytes complicates their developmental comparison to angiosperms in terms of the genetic basis and molecular mechanisms that control their reproductive habits. Asexual reproduction is rare in non-vascular plants such as mosses (*Bryophyta*), hornworts (*Anthocerotophyta*), and liverworts (*Marchantiophyta*), but frequent in ferns (*Polypodiopsida*), where it occurs in almost 10% of species (Burt, 2000; Mogie, 1992). Although production of unreduced gametophytes through diplospory has been observed in liverworts (Smith, 1979), parthenogenesis has never been observed in nature, the only reported case occurring under *in vitro* conditions (Lal, 1984). And although apospory is common in ferns, it is never found in association with parthenogenesis, which is extremely rare in more than 12,000 species (Mogie, 1992). Parthenogenesis also appears to have minimal evolutionary consequences for the genetic structure of approximately 730 species of gymnosperms, being rare and in most cases unsuccessful (Droga, 1966; Durzan et al., 1994).

The origin of apomixis in flowering plants must be related to the selective forces responsible for the evolution of sexual reproduction through meiosis and fertilization. Extant eukaryotes share many genes essential for meiosis, confirming it was already present in their last common ancestor, probably as a precondition already existing in bacteria (Bernstein & Bernstein, 2013; Cavalier-Smith, 2010). Although a form of reductional division could appear in asexual unicellular eukaryotes, e.g., in response to a need of alternating between haploidy and diploidy due to accidental replications of the nuclear genome in the absence of cytokinesis (Mable & Otto, 1998; Rescan, Lenormand, & Roze, 2016), the origin of meiosis remains unsolved. The evolutionary benefits and adaptive functions of meiosis in the context of sexual costs are still under discussion (Lenormand, Engelstädter, Johnston, Wijnker, & Haag, 2016; Mirzaghaderi & Hörandl, 2018). As a consequence, understanding the origin of unreduced gamete formation as a transgenerationally stable component of gametophytic apomixis is still at an early stage. An attractive hypothesis suggests that, from an evolutionary point of view, the main function of meiosis could be related to DNA repair (perhaps in response to oxidative damage) rather than crossing-over or recombination (Hörandl, 2009; Hörandl & Hadacek, 2013), implying that homologous chromosome pairing and double-strand break formation during prophase of meiosis I should be among the most ancestral and conserved components of a primitive reductional division. In agreement with this hypothesis, facultative apomicts could maintain DNA repair functions and purifying selection through sexually reduced male and female gametes (Lovell, Williamson, Wright, McKay, & Sharbel, 2017; Mirzaghaderi & Hörandl, 2018). Theoretical considerations suggest that unreduced gamete formation should be detrimental to fitness, both through a reduction in the potential contribution to random fertilization, and through the generation of triploid (2n + n) embryos with low fertility and viability (Husband & Sabara, 2003; Ramsey & Schmeske, 1998). The mechanisms of unreduced gamete formation are likely to emerge under unstable climatic conditions or stressful environments, in which polyploidy could provide a selective advantage (Vanneste, Maere, & Van de Peer, 2014). Perhaps the only selective force that can provide an evolutionary advantage for the production of unreduced gametes is the acquisition of reproductive stability through gametophytic apomixis (Bicknell & Koltunow, 2004, Kotlunow & Grossniklaus, 2003; Kreiner et al., 2017a, 2017b). Genotyping-by-sequencing approaches to estimate the frequency of unreduced egg cell fertilization in gametophytic apomicts could provide important evidence to confirm or discard our current views on this evolutionary problem.

Our current genetic, molecular, and phylogenetic evidence suggests that apomixis has emerged multiple independent times from a deregulation of the sexual developmental pathway (Grossniklaus, 2001; Grossniklaus, Moore, & Gagliano, 1998; Hojsgaard, Greilhuber, et al., 2014; Hojsgaard, Klatt, et al., 2014; Hörandl & Hojsgaard, 2012; Koltunow, 1993; Kotlunow & Grossniklaus, 2003; Sharbel et al., 2010; Vielle-Calzada, Crane, & Stelly, 1996). The phylogenetic distribution illustrated in Fig. 1 confirms that frequent evolutionary reversions occur from facultative apomixis to obligate sexuality (Ortiz et al., 2013; Taliaferro & Bashaw, 1966). Based on a significant correlation between the occurrence of sexual (polyspory) and asexual (apomixis) abnormalities during female gametogenesis, Carman (1997) suggested that apomixis could result from the hybridization-derived confrontation of divergent alleles that cause a temporal deregulation of cell specification and fate during early ovule development. Hörandl and Hojsgaard (2012) extended this hypothesis by suggesting the existence of an ancestral predisposition in all flowering plants that could be favored by polyploidization and hybridization. If such a predisposition is confirmed, triploid evolutionary transitions associated with genomic disparities caused by hybridization could preclude developmental alterations to sexuality that transiently induce apomixis as a requirement to enhance the evolutionary establishment of new polyploids (Hojsgaard, 2018). Their eventual reversion to sexuality would cause the re-establishment of meiosis under developmental circumstances that favor polyspory through precocious megaspore formation in the absence of cytokinesis (Hörandl & Hojsgaard, 2012).

Polyploidization and hybridization could also influence the natural sources of epigenetic variation that shape the developmental versatility exhibited by the ovule of flowering plants, probably as an adaptive response to changing environmental factors. Waddington proposed that epigenetic adaptation could be a dynamic mechanism to stabilize phenotypic variation through "canalization," the natural tendency of a trait to resist mutation or environmental modifications (Waddington, 1942). Following his hypothesis, canalization of mechanisms leading to the survival of a single meiotic product (monospory) could depend on the epigenetic landscape that prevails in the developing ovule through the action of non-coding RNA (ncRNA)-dependent regulatory pathways (Fig. 3). Obligate monospory would require the establishment and maintenance of consistent methylation patterns and/or chromatin epigenetic marks through compatible recognition between ncRNAs and their targets. Following hybridization, a lack of sequence recognition involving ncRNAs in divergent genotypes could result in epigenetic changes through DNA methylation or chromatin modification, leading to a deregulation of female meiosis and cell specification in the ovule. The epigenetic landmarks involved in the regulation of transcriptionally active or silent elements would not be targeted by ncRNAs produced from divergent parental genomes, as sequence complementarity would tend to diverge in recently formed hybrids. Polyploidization could also lead to this type of deregulation through chromosome duplications

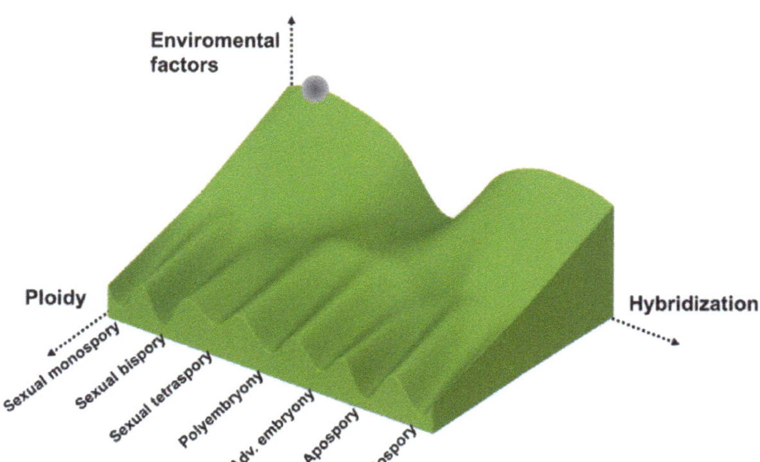

Fig. 3 Sexual reproduction and apomixis as reversible alternatives in the evolutionary epigenetic landscape that guides female gametogenesis. In this epigenetic landscape, adaptation is flexibly modulated by shaping forces that depend on ploidy, hybridization, and the influence of environmental factors.

and re-arrangements that quantitatively affect the epigenetic mechanisms controlling genomic methylation through the action of ncRNAs. Environmental factors affecting female meiosis constitute the third evolutionary force shaping the epigenetic mechanisms that control the reproductive outcome. Following this model, the formation of unreduced female gametes could depend on sources of epigenetic variation that do not contribute to the stable range of phenotypes manifested under monospory. These sources of epigenetic variation would only appear after the occurrence of hybridization and polyploidy (Grossniklaus, 2001), under the influence of a wide variety of environmental perturbations.

Smith used to say that from an evolutionary perspective, sexual reproduction could be more important than life itself (Smith, 1979). By incorporating apomixis as part of the epigenetic landscape required for versatile adaptation, flowering plants might have developed one of the most evolutionary robust alternatives to sexuality. The continuous investigation of natural forms of epigenetic variation in multicellular organisms should offer new and exciting insights into the mechanisms of reproductive innovation and evolution that prevail in the plant kingdom.

Acknowledgments

Research in our laboratory is funded by Consejo Nacional de Ciencia y Tecnología (Conacyt, México), and Dupont Pioneer regional initiatives in favor of local smallhold farmers. G.L.M. was the recipient of a postdoctoral fellowship from Conacyt.

References

Akiyama, Y., Conner, J. A., Goel, S., Morishige, D. T., Mullet, J. E., Hanna, W. W., et al. (2004). High-resolution physical mapping in *Pennisetum squamulatum* reveals extensive chromosomal heteromorphism of the genomic region associated with apomixis. *Plant Physiology, 134*, 1733–1741.

Akiyama, Y., Goel, S., Conner, J. A., Hanna, W. W., Yamada-Akiyama, H., & Ozias-Akins, P. (2011). Evolution of the apomixis transmitting chromosome in Pennisetum. *BMC Evolutionary Biology, 11*, 289.

Albertini, E., Porceddu, A., Ferranti, F., Reale, L., Barcaccia, G., Romano, B., et al. (2001). Apospory and parthenogenesis may be uncoupled in Poa pratensis: A cytological investigation. *Sexual Plant Reproduction, 14*, 213–217.

Aliyu, O. M., Schranz, M. E., & Sharbel, T. F. (2010). Quantitative variation for apomictic reproduction in the genus Boechera (Brassicaceae). *American Journal of Botany, 97*, 1719–1731.

Amasino, R. (2010). Seasonal and developmental timing of flowering. *The Plant Journal, 61*, 1001–1013.

Armenta-Medina, A., Demesa-Arevalo, E., & Vielle-Calzada, J. P. (2011). Epigenetic control of cell specification during female gametogenesis. *Sexual Plant Reproduction, 24*, 137–147.

Arnold, M. L. (1997). *Natural hybridization and evolution* (p. 232). New York, NY, USA: Oxford University Press.
Asker, S. E., & Jerling, L. (1992). *Apomixis in plants* (p. 298). Boca Raton, FL: CRC press.
Barnabás, B., Jäger, K., & Fehér, A. (2008). The effect of drought and heat stress on reproductive processes in cereals. *Plant, Cell and Environment*, *31*, 11–38.
Barrell, P. J., & Grossniklaus, U. (2005). Confocal microscopy of whole ovules for analysis of reproductive development: The *elongate1* mutant affects meiosis II. *The Plant Journal*, *43*, 309–320.
Bashaw, E. C., & Hanna, W. W. (1990). Apomictic reproduction. In G. P. Chapman (Ed.), *Reproductive verstatility in the grasses* (p. 298). Cambridge UK: Cambridge University Press.
Bashaw, E. C., & Hignight, K. W. (1990). Gene transfer in apomictic buffel grass through fertilization of an unreduced egg. *Crop Science*, *30*, 571–575.
Bashaw, E. C., Hussey, M. A., & Hignight, K. W. (1992). Hybridization (n + n and 2n + n) of facultative apomictic species in the Pennisetum agamic complex. *International Journal of Plant Sciences*, *153*, 466–470.
Belmonte, M. F., Kirkbride, R. C., Stone, S. L., Pelletier, J. M., Bui, A. Q., Yeung, E. C., et al. (2013). Comprehensive developmental profiles of gene activity in regions and subregions of the Arabidopsis seed. *Proceedings of the National Academy of Sciences of the United States of America*, *10*, E435–E444.
Bernstein, C., & Bernstein, H. (2013). Evolutionary origin and adaptative function of meiosis. In C. Bernstein & H. Bernstein (Eds.), *Meiosis* (pp. 41–45). Rijeka, Croatia: InTech.
Bicknell, R. A. (1997). Isolation of a diploid, apomictic plant of *Hieracium aurantiacum*. *Sexual Plant Reproduction*, *10*, 168–172.
Bicknell, R. A., Borst, N. K., & Koltunow, A. (2000). Monogenic inheritance of apomixes in two *Hieracium* species with distinct developmental mechanisms. *Heredity*, *84*, 228–237.
Bicknell, R., Catanach, A., Hand, M., & Koltunow, A. (2016). Seeds of doubt: Mendel's choice of *Hieracium* to study inheritance, a case of right plant, wrong trait. *Theoretical and Applied Genetics*, *129*, 2253–2266.
Bicknell, R. A., & Koltunow, A. M. (2004). Understanding apomixis: Recent advances and remaining conundrums. *The Plant Cell*, *16*, S228–S245.
Bierzychudek, P. (1985). Patterns in plant parthenogenesis. *Experientia*, *41*, 1255–1264.
Bingham, E. (1969). Haploids from cultivated Alfalfa, Medicago sativa L. *Nature*, *221*, 865–866.
Bitsch, C., Gröger, S., & Lelley, T. (2000). Effect of parental genotypes on haploid embryo and plantlet formation in wheat × maize crosses. *Euphytica*, *103*, 319–323.
Boutilier, K., Offringa, R., Sharma, V. K., Kieft, H., Ouellet, T., Zhang, L., et al. (2002). Ectopic expression of *BABY BOOM* triggers a conversion from vegetative to embryonic growth. *The Plant Cell*, *14*, 1737–1749.
Brennecke, J., Aravin, A. A., Stark, A., Dus, M., Kellis, M., Sachidanandam, R., et al. (2007). Discrete small RNA-generating loci as master regulators of transposon activity in *Drosophila*. *Cell*, *128*, 1089–1103.
Bretagnolle, F., & Thompson, J. D. (1995). Gametes with the somatic chromosome number: Mechanisms of their formation and role in the evolution of autopolyploid plants. *New Phytologist*, *129*, 1–22.
Brown, M. S. (1958). The implications of pachytene and metaphase pairing for species differentiation. Proceedings of the X international congress of genetics. *Vol. 2* (p. 36).
Brownfield, L., & Köhler, C. (2011). Unreduced gamete formation in plants: Mechanisms and prospects. *Journal of Experimental Botany*, *62*, 1659–1668.
Burt, A. (2000). Perspective: Sex, recombination and the efficacy of selection—Was Weismann right? *Evolution*, *54*, 337–351.

Calderini, O., Chang, S. B., De Jong, H., Busti, A., Paolocci, F., Arcioni, S., et al. (2006). Molecular cytogenetics and DNA sequence analysis of an apomixis-linked BAC in Paspalum simplex reveal a non pericentromere location and partial microcolinearity with rice. *Theoretical and Applied Genetics*, *112*, 1179–1191.

Campbell, A. W., Griffin, W. B., Burritt, D. J., & Conner, A. J. (2001). The importance of light intensity for pollen tube growth and embryo survival in wheat X maize crosses. *Annals of Botany*, *87*, 517–522.

Carman, J. G., Jamison, M., Elliott, E., Dwivedi, K. K., & Naumova, T. N. (2011). Apospory appears to accelerate onset of meiosis and sexual embryo sac formation in sorghum ovules. *BMC Plant Biology*, *11*, 9.

Carman, J. G. (1997). Asynchronous expression of duplicate genes in angiosperms may cause apomixis, bispory, tetraspory, and polyembryony. *Biological Journal of the Linnean Society*, *61*, 51–94.

Catanach, A. S., Erasmuson, S. K., Podivinsky, E., Jordan, B. R., & Bicknell, R. (2006). Deletion mapping of genetic regions associated with apomixis in Hieracium. *Proceedings of the National Academy of Sciences of the United States of America*, *103*, 18650–18655.

Cavalier-Smith, T. (2010). Origin of the cell nucleus, mitosis and sex: Roles of intracellular coevolution. *Biology Direct*, *4*, 5–7.

Clausen, J. D., Keck, D., & Heisey, W. M. (1945). Experimental studies on the nature of species. II. Plant evolution through amphiploidy and autoplyploidy, with examples from the Madiinae. *Vol. 564* (pp. 1–174). Carnegie Institute of Washington Publisher.

Coe, E. H. (1959a). A line of maize with high haploid frequency. *The American Naturalist*, *93*, 381–382.

Coe, E. H. (1959b). Mutation of CI-a line with 2–3% haploids. *Vol. 30* (pp. 96–99). Maize Genetics Cooperation Newsletter.

Comai, L. (2005). The advantages and disadvantages of being polyploid. *Nature Reviews Genetics*, *6*, 836–846.

Comai, L. (2014). Genome elimination: Translating basic research into a future tool for plant breeding. *PLoS Biology*, *12*, e1001876.

Conner, J. A., Mookkan, M., Huo, H., Chae, K., & Azias-Akins, P. (2015). A parthenogenesis gene of apomict origin elicits embryo formation from unfertilized eggs in a sexual plant. *Proceedings of the National Academy of Sciences of the United States of America*, *112*, 11205–11210.

Conner, J. A., Podio, M., & Ozias-Akins, P. (2017). Haploid embryo production in rice and maize induced by *PsASGR-BBML* transgenes. *Plant Reproduction*, *30*, 41–52.

Conner, J. A., Gunawan, G., & Ozias-Akins, P. (2013). Recombination within the apospory specific genomic region leads to the uncoupling of apomixis components in *Cenchrus ciliaris*. *Planta*, *238*, 51–63.

Chalyk, S. T. (1994). Properties of maternal haploid maize plants and potential application to maize breeding. *Euphytica*, *79*, 13–18.

Chalyk, S. T., & Chebotar, O. D. (2000). Regular segregation of four recessive marker genes among maternal haploids in maize. *Plant Breeding*, *119*, 363–364.

Chase, S. S. (1963). Androgenesis its use for transfer of maize cytoplasm. *Journal of Heredity*, *54*, 152–158.

Chase, S. S. (1964). Monoploids and diploids of maize: A comparison of genotypic equivalents. *American Journal of Botany*, *51*, 928–933.

D'Amato, F. (1989). Polyploidy in cell differentiation. *Caryologia*, *42*, 183–211.

d'Erfurth, I., Cromer, L., Jolivet, S., Girard, C., Horlow, C., Sun, Y., et al. (2010). The CYCLIN-A CYCA1;2/TAM is required for the meiosis I to meiosis II transition and cooperates with OSD1 for the prophase to first meiotic division transition. *PLoS Genetics*, *6*, e1000989.

d'Erfurth, I., Jolivet, S., Froger, N., Catrice, O., Novatchkova, M., Simon, M., et al. (2008). Mutations in *AtPS1* (*Arabidopsis thaliana Parallel Spindle 1*) lead to the production of diploid pollen grains. *PLoS Genetics*, *4*, e1000274.
d'Erfurth, I., Jolivet, S., Froger, N., Catrice, O., Novatchkova, M., & Mercier, R. (2009). Turning meiosis into mitosis. *PLoS Biology*, 7, e1000124.
Darwin, C. R. (1862). On the two forms, or dimorphic-condition, in the species of Primula and on their remarkable sexual relations. *Journal of the Proceedings of the Linnean Society of Botany*, *6*, 77–96. London.
De Haan, A., Maceira, N. O., Lumaret, R., & Delay, J. (1992). Production of 2n gametes in diploid subspecies of *Dactylis gomerata* L. 2. Occurrence and frequency of 2n eggs. *Annals of Botany*, *69*, 345–350.
De Maine, M. J. (2003). In M. Maluszynski, et al. (Ed.)*Potato haploid technologies. In doubled haploid production in crop plants: A manual* (pp. 241–247). Kluwer Academic Publishers.
De Storme, N., & Geelen, D. (2013). Sexual polyploidization in plants—Cytological mechanisms and molecular regulation. *New Phytologist*, *198*, 670–684.
Devaux, P. (2003). The *Hordeum bulbosum* (L.) method. In M. Maluszynski, et al. (Ed.), *Doubled haploid production in crop plants: A manual* (pp. 15–19). London: Kluwer Academic Publishers.
Droga, P. D. (1966). Observations on Abies pindrow with a discussion on the question of occurrence of apomixis in gymnosperms. *Silvae Genetica,15,* 11–20.
Dunwell, J. M. (2010). Haploids in flowering plants: Origins and exploitation. *Plant Biotechnology Journal*, *8*, 377–424.
Duran-Figueroa, N., & Vielle-Calzada, J. P. (2010). ARGONAUTE9-dependent silencing of transposable elements in pericentromeric regions of Arabidopsis. *Plant Signaling & Behavior*, *5*, 1476–1479.
Durzan, D. J., Jokinen, K., Guerra, M. P., Santerre, A., Chalupa, V., & Havel, L. (1994). Latent diploid parthenogenesis and parthenote cleavage in egg-equivalents of Norway spruce. *International Journal of Plant Sciences*, *155*, 677–688.
Eder, J., & Chalyk, S. (2002). In vivo haploid induction in maize. *Theoretical and Applied Genetics*, *104*, 703–708.
Erilova, A., Brownfield, L., Exner, V., Rosa, M., Twell, D., Mittelsten Scheid, O., et al. (2009). Imprinting of the Polycomb proup gene MEDEA serves as a ploidy sensor in Arabidopsis. *PLoS Genetics*, *5*, e1000663.
Evans, M. M. (2007). The indeterminate gametophyte1 gene of maize encodes a LOB domain protein required for embryo sac and leaf development. *Plant Cell*, *19*, 46–62.
Evans, L. T., & Knox, R. B. (1969). Environmental control of reproduction in *Themeda* australis. *Australian Journal of Botany*, *17*, 375–389.
Forster, B. P., Heberle-Bors, E., Kasha, K. J., & Touraev, A. (2007). The resurgence of haploids in higher plants. *Trends in Plant Science*, *12*, 368–375.
Garcia-Aguilar, M., Michaud, C., Leblanc, O., & Grimanelli, D. (2010). Inactivation of a DNA methylation pathway in maize reproductive organs results in apomixis-like phenotypes. *The Plant Cell*, *22*, 3249–3267.
Garcia-Llamas, C., Ramirez, M. C., & Ballesteros, J. (2004). Effect of pollinator on haploid production in durum wheat crossed with maize and pearl millet. *Plant Breeding*, *123*, 201–203.
Gounaris, E. K., Sherwood, R. T., Gounaris, I., Hamilton, R. H., & Gustine, D. L. (1991). Inorganic salts modify embryo sac development in sexual and aposporous *Cenchrus ciliaris*. *Sexual Plant Reproduction*, *4*, 188–192.

Grimanelli, D., Leblanc, O., Espinosa, E., Perotti, E., Gonzalez de Leon, D., & Savidan, Y. (1998). Mapping diplosporous apomixis in tetraploid Tripsacum: One gene or several genes. *Heredity, 80*, 33–39.

Grossniklaus, U. (2001). From sexuality to apomixis: molecular and genetic approaches. In Y. Savidan, J. G. Carman, & T. Dresselhaus (Eds.), *The flowering of apomixis: From mechanisms to genetic engineering* (pp. 168–211). CIMMYT IRD EU Press.

Grossniklaus, U., Moore, J. M., & Gagliano, W. B. (1998). Molecular and genetic approaches to understanding and engineering apomixis: *Arabidopsis* as a powerful tool. In S. S. Virmani, E. A. Siddiq, & K. Muralidharan (Eds.), *Vol. 1996. Advances in hybrid rice technology proceedings of the 3rd international symposium on hybrid rice* (pp. 187–211). Manila, Philippines: International Rice Research Institute.

Grossniklaus, U., Nogler, G. A., & van Dijk, P. J. (2001). How to avoid sex: The genetic control of gametophytic apomixis. *The Plant Cell, 13*, 1491–1498.

Gualtieri, G., Conner, J. A., Morishige, D. T., Moore, L. D., Muller, J. E., & Ozias-Akins, P. (2006). A segment of the apospory-specific genomic region is highly microsyntenic not only between the apomicts *Pennisetum squamulatum* and buffelgrass, but also with a rice chromosome 11 centromeric-proximal genomic region. *Plant Physiology, 140*, 963–971.

Gupta, P. K., Roy, R. P., & Singh, A. P. (1969). Aposporous apomixis: Seasonal variation in tetraploid *Dichanthium annulatum* (Forssk.) Stapf. *Portugaliae Acta Biologica, 11*, 253–260.

Gustafsson, A. (1946). Apomixis in higher plants. I. The mechanisms of apomixes. Lunds Universitets Årskrift N.F. Avd. 2 Bd. 42. Nr 3In *Kungl Fysiografiska Sållskapets Hadlingar NF Bd 57 Nr 3* (p 370). Lund C.W.K. Gleerup.

Hagberg, A., & Hagberg, G. (1980). High frequency of spontaneous haploids in the progeny of an induced mutation in barley. *Hereditas, 93*, 341–343.

Haig, D., & Westoby, M. (1991). Genomic imprinting in endosperm: Its effect or development in crosses between species, and between different ploidies of the same species, and its implications for the evolution of apomixis. *Philosophical Transactions of The Royal Society B Biological Science, 33*, 1–13.

Hand, M. L., & Koltunow, A. M. (2014). The control of Apomixis: Asexual seed formation. *Genetics, 197*, 441–450.

Hands, P., Rabiger, D. S., & Koltunow, A. (2016). Mechanisms of endosperm initiation. *Plant Reproduction, 29*, 215–225.

Hannah, W. W., & Bashaw, E. C. (1987). Apomixis: Its identification and use in plant breeding. *Crop Science, 27*, 1136–1139.

Harlan, J. R., & De Wet, J. M. J. (1975). On Ö Winge and a prayer: The origins of polyploidy. *The Botanical Review, 41*, 361–390.

Hase, Y., Trung, K. H., Matsunaga, T., & Tanaka, A. (2006). A mutation in the uvi4 gene promotes progression of endo-reduplication and confers increased tolerance towards ultraviolet B light. *The Plant Journal, 46*, 317–326.

Hedhly, A. (2011). Sensitivity of flowering plant gametophytes to temperature fluctuations. *Environmental and Experimental Botany, 74*, 9–16.

Hedhly, A., Hormaza, J. I., & Herrero, M. (2009). Global warming and sexual plant reproduction. *Trends in Plant Science, 14*, 30–36.

Hermsen, J. G. T. (1984). Nature, evolution, and breeding of polyploids. *Iowa State Journal of Research, 58*, 411–420.

Hernandez-Lagana, E., Rodriguez-Leal, D., Lúa, J., & Vielle-Calzada, J.-P. (2016). A multigenic network of ARGONAUTE4 clade members controls early megaspore formation in Arabidopsis. *Genetics, 204*, 1045–1056.

Hjelmqvist, H., & Grazi, F. (1964). Studies on variation in embryo sac development. *Botanical Notes, 117*, 141.

Hojsgaard, D. (2018). Transient activation of apomixis in sexual neotriploids may retain genomically altered states and enhance polyploid establishment. *Frontiers in Plant Science, 9*, 230.

Hojsgaard, D., Greilhuber, J., Pellino, M., Paun, O., Sharbel, T. F., & Hörandl, E. (2014). Emergence of apospory and bypass of meiosis via apomixis after sexual hybridisation and polyploidisation. *New Phytologist, 204*, 1000–1012.

Hojsgaard, D., Klatt, S., Baier, R., Carman, J. G., & Hörandl, E. (2014). Taxonomy and biogeography of Apomixis in angiosperms and associated biodiversity characteristics. *CRC Critical Reviews in Plant Sciences, 33*, 414–427.

Hojsgaard, D., Martínez, E., & Quarin, C. (2013). Competition between meiotic and apomictic pathways during ovule and seed development results in clonality. *New Phytologist, 197*, 336–347.

Hörandl, E. (2009). A combinational theory for maintenance of sex. *Heredity, 103*, 445–457.

Hörandl, E., & Hojsgaard, D. (2012). The evolution of apomixis in angiosperms: A reappraisal. *Plant Biosystems, 146*, 681–693.

Hörandl, E., Cosendal, A.-C., & Temsch, E. M. (2008). Undertsanding the geographic distributions of apomictic plants: A case for pluralistic approach. *Plant Ecology and Diversity, 1*, 309–320.

Hörandl, E., & Hadacek, F. (2013). The oxidative damage initiation hypothesis for meiosis. *Plant Reproduction, 26*, 351–367.

Hsam, S. L., & Zeller, F. J. (1993). Haploid induction in durum wheat by the interaction of Aegilops kotschyi cytoplasm and 1BL/1RS chromosomal interchange. *Theoretical and Applied Genetics, 86*, 951–954.

Hülskamp, M., Parekh, N. S., Grini, P., Schneitz, K., Zimmermann, I., Lolle, S. J., et al. (1997). The STUD gene is required for male-specific cytokinesis after telophase II of meiosis in Arabidopsis thaliana. *Developmental Biology, 187*, 114–124.

Husband, B. C., & Sabara, H. A. (2003). Reproductive isolation between autotetraploids and their diploid progenitors in fireweed *Chamerion angustifolium* (Onagraceae). *New Phytologist, 161*, 703–713.

Immonen, S., Tenhola-Roininen, T., Maluszynski, M., et al. (2004). Protocol for rye anther culture. In *Doubled haploid production in crop plants: A manual* (pp. 141–149). Kluwer Academic Publishers.

Inagaki, M. N., Maluszynski, M., et al. (2003). Doubled haploid production in wheat through wide hybridization. In *Doubled haploid production in crop plants: A manual* (pp. 53–58). Kluwer Academic Publishers.

Inoue, E., Sakuma, F., Kasumi, M., Hara, H., & Tsukihashi, T. (2004). Maternal haploidization of Japanese pear through inter-generic hybridization with apple. *Acta Horticulturae, 663*, 815–818.

Ishii, T., Karimi-Ashtiyani, R., & Houben, A. (2016). Haploidization via chromosome elimination: Means and mechanisms. *Annual Review of Plant Biology, 67*, 421–438.

Johri, B. M., Ambegaokar, K. B., & Srivastava, P. S. (1992). *Comparative embryology of angiosperms: Vol. 1* (pp. 1171–1174). London: Springer-Verlag.

Jongedijk, E., & Ramanna, M. S. (1988). Synaptic mutants in potato, *Solanum tuberosum* L. I. Expression and identity of genes for desynapsis. *Genome, 30*, 664–670.

Jongedijk, E., & Ramanna, M. S. (1989). Synaptic mutants in potato, *Solanum tuberosum* L. II. Concurrent reduction of chiasma frequencies in male and female meiosis of ds-1 (desynapsis) mutants. *Genome, 32*, 1054–1062.

Jongedijk, E., Ramanna, M. S., Sawor, Z., & Hermsen, J. G. (1991). Formation of first division restitution (FDR) 2n-megaspores through pseudohomotypic division in ds-1 (desynapsis) mutants of diploid potato: Routine production of tetraploid progeny from 2xFDR × 2xFDR crosses. *Theoretical and Applied Genetics, 5*, 645–656.

Juliano, C., Wang, J., & Lin, H. (2011). Uniting germline and stem cells: The function of Piwi proteins and the piRNA pathway in diverse organisms. *Annual Review of Genetics*, *45*, 447–469.

Jullien, P. E., Susaki, D., Yelagandula, R., Higashiyama, T., & Berger, F. (2012). DNA methylation dynamics during sexual reproduction in Arabidopsis thaliana. *Current Biology*, *22*, 1825–1830.

Karasawa, K. (1971). On the occurrence of haploide seedlings in Citrus natsudaidai Hayata. *Bull Sakushingakuin Junior College for Woman*, 1–2.

Kasha, K. J., & Kao, K. N. (1970). High frequency haploid production in barley (Hordeum vulgare L.). *Nature*, *225*, 874–876.

Kearney, M. (2005). Hybridization, glaciation and geographical parthenogenesis. *Trends in Ecology & Evolution*, *20*, 495–502.

Keller, F., & Körner, C. (2003). The role of photoperiodism in alpine plant development. *Antarctic and Alpine Research*, *35*, 361–368.

Kelliher, T., & Walbot, V. (2012). Hypoxia triggers meiotic fate acquisition in maize. *Science*, *337*, 345–348.

Kelliher, T., Starr, D., Richbourg, L., Chintamanani, S., Delzer, B., Nuccio, M. L., et al. (2017). MATRILINEAL, a sperm-specific phospholipase, triggers maize haploid induction. *Nature*, *542*, 105–109.

Kermicle, J. L. (1969). Androgenesis conditioned by a mutation in maize. *Science*, *166*, 1422–1424.

Kermicle, J. L. (1971). Pleiotropic effects on seed development of the indeterminate gametophyte gene in maize. *American Journal of Botany*, *58*, 1–7.

Kimber, G., & Riley, R. (1963). Haploid angiosperms. *Botanical Review*, *29*, 480–531.

Klatt, S., Hadacek, F., Hodac, L., Brinkmann, G., Eilerts, M., Hojsgaard, D., et al. (2016). Photoperiod extension enhances sexual megaspore formation and triggers metabolic reprogramming in facultative apomictic *Ranunculus auricomus*. *Frontiers in Plant Science*, *7*, 278.

Knox, B. (1967). Apomixis: Seasonal and population differences in a grass. *Science*, *157*, 325–326.

Koltunow, A. M. (1993). Apomixis: Embryo sacs and embryos formed without meiosis or fertilization in ovules. *The Plant Cell*, *5*, 1425–1437.

Koltunow, A. M., Ozias-Akins, P., & Siddiqi, I. (2013). Apomixis. In P. W. Becraft (Ed.), *Seed genomics* (pp. 83–110). New York: Wiley.

Koltunow, A. M. G., Johnson, S. D., Rodrigues, J. C., Okada, T., Hu, Y., Tsuchiya, T., et al. (2011). Sexual reproduction is the default mode in apomictic Hieracium subgenus Pilosella, in which two dominant loci function to enable apomixis. *The Plant Journal*, *66*, 890–902.

Koszegi, D., Johnston, A. J., Rutten, T., Czihal, A., Altschmied, L., Kumlehn, J., et al. (2011). Members of the RKD transcription factor family induce an egg cell-like gene expression program. *The Plant Journal*, *67*, 280–291.

Kotani, Y., Henderson, S. T., Suzuki, G., Johnson, S. D., Okada, T., Siddons, H., et al. (2013). The LOSS OF APOMEIOSIS (LOA) locus in *Hieracium praealtum* can function independently of the associated large-scale repetitive chromosomal structure. *New Phytologist*, *201*, 973–981.

Kotlunow, A. M., & Grossniklaus, U. (2003). Apomixis: A developmental perspective. *Annual Review of Plant Biology*, *54*, 547–574.

Kreiner, J. M., Kron, P., & Husband, B. C. (2017a). Evolutionary dynamics of unreduced gametes. *Trends in Genetics*, *33*, 583–593.

Kreiner, J. M., Kron, P., & Husband, B. C. (2017b). Frequency and maintenance of unreduced gametes in natural plant populations: Associations with reproductive mode, life history and genome size. *New Phytologist*, *214*, 879–889.

Kron, P., & Husband, B. C. (2009). Hybridization and the reproductive pathways. *Botany, 87*, 864–874.

Kurepin, L. V., Walton, L. J., Reid, D. M., Chinnappa, C. C., & Pharis, R. P. (2007). Photoperiod, light quality, and irradiance effects on flowering in the alpine and prairie ecotypes of *Stellaria longipes*. *Canadian Journal of Botany, 85*, 538–544.

Lal, M. (1984). The culture of bryophytes including apogamy, apospory, parthenogenesis and protoplasts. In A. F. Dyer & J. G. Duckett (Eds.), *The experimental biology of bryophytes* (pp. 97–115). London: Academic Press.

Lanaud, C. (1988). Origin of haploids and semigamy in *Theobroma cacao* L. *Euphytica, 38*, 221–228.

Laurie, D. A., & Bennett, M. D. (1986). Wheat x maize hybridization. *Canadian Journal of Genetics and Cytology, 28*, 313–316.

Laurie, D. A., O'Donoughue, L. S., & Bennett, M. D. (1990). Wheat X maize and other wide sexual hybrids: Their potential for genetic manipulation and crop improvement. In J. P. Gustafson (Ed.), *Gene manipulation in plant improvement II* (pp. 95–126). Boston, MA: Stadler genetics symposia series, Springer.

Law, J. A., & Jacobsen, S. E. (2010). Establishing, maintaining and modifying DNA methylation patterns in plants and animals. *Nature Reviews Genetics, 11*, 204–220.

Leblanc, O., & Mazzucato, A. (2001). Screening procedures to identify and quantify apomixis. In E. Y. Savidan, J. G. Carman, & T. Dresselhaus (Eds.), *The flowering of Apomixis: From mechanisms to genetic* (pp. 121–136). CIMMYT IRD EU Press.

Lelley, T., Mahmoud, A. A., & Lein, V. (1987). Genetics and cytology of unreduced gametes in cultivated rye (*Secale cereale* L.). *Genome, 29*, 635–638.

Lenormand, T., Engelstädter, J., Johnston, S. E., Wijnker, E., & Haag, C. R. (2016). Evolutionary mysteries in meiosis. *Philosophical Transactions of the Royal Society B, 371*, 20160001.

Liu, Z.-Z., & Song, T.-M. (2000). The breeding and identification of haploid inducer with high frequency parthenogenesis in maize. *Acta Automatica Sinica, 26*, 570–574.

Lokhande, S. D., Ogawa, K., Tanaka, A., & Hara, T. (2003). Effect of temperature on ascorbate peroxidase activity and flowering of *Arabidopsis thaliana* ecotypes under different light conditions. *Journal of Plant Physiology, 160*, 57–64.

Lovell, J. T., Aliyu, O. M., Mau, M., Schranz, M. E., Koch, M., Kiefer, C., et al. (2013). On the origin and evolution of apomixis in *Boechera*. *Plant Reproduction, 26*, 309–315.

Lovell, J. T., Williamson, R. J., Wright, S. I., McKay, J. K., & Sharbel, T. F. (2017). Mutation accumulation in an asexual relative of Arabidopsis. *PLoS Genetics, 13*, e1006550.

Mable, B. K., & Otto, S. P. (1998). The evolution of life cycles with haploid and diploid phases. *BioEssays, 20*, 453–462.

Magnard, J. L., Yang, M., Chen, Y. C., Leary, M., & McCormick, S. (2001). The Arabidopsis gene tardy asynchromous meiosis is required for the normal pace and synchrony of cell division during male meiosis. *Plant Physiology, 127*, 1157–1166.

Maizonnier, D. (1976). Production de tetraploïdes et de trisomiques naturels chez le Pétunia. *Annales d'Amelioration des Plantes, 26*, 305–318.

Mallory, A., & Vaucheret, H. (2010). Form, function, and lation of ARGONAUTE proteins. *The Plant Cell, 22*, 3879–3889.

Marimuthu, M. P., Jolivet, S., Ravi, M., Pereira, L., Davda, J. N., Cromer, L., et al. (2011). Synthetic clonal reproduction through seeds. *Science, 331*, 876.

Martinez, G., & Köhler, C. (2017). Role of small RNAs in epigenetic reprogramming during plant sexual reproduction. *Current Opinion in Plant Biology, 36*, 22–28.

Matzk, F. (1996). The "Salmon system" of wheat—A suitable model for apomixis research. *Hereditas, 125*, 299–301.

Matzk, F., Meyer, H. M., Bäumlein, H., Balzer, C. J. H., & Schubert, I. (1995). A novel approach to the analysis of the initiation of embryo development in graminae. *Sexual Plant Peproduction, 8*, 266–272.

McClintock, B. (1933). The association of non-homologous parts of chromosomes in the mid-prophase of meiosis in Zea mays. *Zeitschrift Fur Zellforschung Und Mikroskopische Anatomie, 19*, 191–237.

McCoy, T. J. (1982). Inheritance of 2n pollen formation in diploid alfalfa (*Medicago sativa* L.). *Canadian Journal of Genetics and Cytology, 24*, 315–323.

Mercier, R., Vezon, D., Bullier, E., Motamayor, J. C., Sellier, A., Lefèvre, F., et al. (2001). SWITCH1 (SWI1): A novel protein required for the establishment of sister chromatid cohesion and for bivalent formation at meiosis. *Genes & Development, 15*, 1859–1871.

Mirzaghaderi, G., & Hörandl, E. (2018). The evolution of meiotic sex and its alternatives. *Proceedings of the Royal Society of London B, 283*, 20161221.

Mogie, M. (1992). *The evolution of asexual reproduction in plants*. London, UK: Chapman and Hall.

Nassar, N. M. A. (1992). Production of triploid cassava, *Manihot esculenta* Crantz by hybrid diploid gametes. *Field Crops Research, 30*, 173–192.

Naumova, T. N., & Vielle-Calzada, J. P. (2001). Ultraestructural analysis of apomictic development. In Y. Savidan, J. G. Carman, & T. Dresselhaus (Eds.), *The flowering of apomixis: From mechanisms to genetic engineering* (p. 243). CIMMYT IRD EU Press.

Nogler, G. A. (2006). The lesser-known Mendel: His experiments on Hieracium. *Genetics, 172*, 1–6.

Nogler, G. A. (1984). In B. M. Johri (Ed.), *Gametophytic apomixes. Embryology of angiosperms (eBS-V)* (pp. 475–518). Berlin: Springer-Verlag.

Noyes, R. D., & Rieseberg, L. H. (2000). Two independent loci control agamospermy (Apomixis) in the triploid flowering plant Erigeron annuus. *Genetics, 155*, 379–390.

Ochogavia, A. C., Seijo, J. G., Gonzalez, M., Podio, M., Duarte Silveira, E., Machado Lacerda, A. L., et al. (2011). Characterization of retrotransposon sequences expressed in inflorescences of apomictic and sexual Paspalum notatum plants. *Sexual Plant Reproduction, 24*, 231–246.

Ogawa, D., Johnson, S. D., Henderson, S. T., & Koltunow, A. M. (2013). Genetic separation of autonomous endosperm formation (AutE) from two other components of apomixis in Hieracium. *Plant Reproduction, 26*, 113–123.

Okada, T., Ito, K., Johnson, S. D., Oelkers, K., Suzuki, G., Houben, A., et al. (2011). Chromosomes carrying meiotic avoidance loci in three apomictic eudicot Hieracium subgenus Pilosella species share structural features with two monocot apomicts. *Plant Physiology, 157*, 1327–1341.

Olmedo-Monfil, V., Duran-Figueroa, N., Arteaga-Vazquez, M., Demesa-Arevalo, E., Autran, D., Grimanelli, D., et al. (2010). Control of female gamete formation by a small RNA pathway in Arabidopsis. *Nature, 464*, 628–632.

Ortiz, J. P. A., Quarin, C., Pessino, S., Acuña, C., Martínez, E. J., Espinoza, F., et al. (2013). Harnessing apomictic reproduction in grasses: What we have learned from Paspalum. *Annals of Botany, 112*, 767–787.

Otto, S. P., & Whitton, J. (2000). Polyploid incidence and evolution. *Annual Review of Genetics, 34*, 401–437.

Ozias-Akins, P., & van Dijk, P. J. (2007). Mendelian genetics of apomixis in plants. *Annual Review of Genetics, 41*, 509–537.

Paun, O., Stuessy, T. F., & Hörandl, E. (2006). The role of hybridization, polyploidization and glaciation in the origin and evolution of the apomictic Ranunculus cassubicus complex. *New Phytologist, 171*, 223–236.

Peng, S., Huang, J., Sheehy, J. E., Laza, R. C., Visperas, R. M., Zhong, X., et al. (2004). Rice yields decline with higher night temperature from global warming. *Proceedings of the National Academy of Sciences of the United States of America, 101*, 9971–9975.

Pillot, M., Baroux, C., Vazquez, M. A., Autran, D., Leblanc, O., Vielle-Calzada, J. P., et al. (2010). Embryo and endosperm inherit distinct chromatin and transcriptional states from the female gametes in Arabidopsis. *The Plant Cell, 22*, 307–320.

Quarin, C. L. (1986). Seasonal changes in the incidence of apomixis of diploid, triploid, and tetraploid plants of *Paspalum cromyorrhizon. Euphytica, 35*, 515–522.

Ramanna, M. S. (1979). A re-examination of the mechanisms of 2n gametes formation in potato and its implications for breeding. *Euphytica, 28*, 537–561.

Ramanna, M. S. (1983). First division restitution gametes through fertile desynaptic mutants of potato. *Euphytica, 32*, 337–350.

Rammana, M. S. (1992). The use of 2n-gamtes in breeding polysomic polyploid species: Some achievements and perspectives. In A. Mariani & S. Tavoletti (Eds.), *Gametes with somatic chromosome number in the evolution and breeding of polyploid polysomic species* (pp. 91–99). Perugia, Italy: Achievements and perspectives.

Ramsey, J., & Schmeske, D. W. (1998). Pathways, mechanisms, and rates of polyploid formation in flowering plants. *Annual Review of Ecology and Systematics, 29*, 467–501.

Rao, P. N., & Narayana, D. S. (1980). Occurrence and identification of semigamy in *Coix aquatica* (tribe Maydeae). *Journal of Heredity, 71*, 117–120.

Ravi, M., & Chan, S. W. (2010). Haploid plants produced by centromere mediated genome elimination. *Nature, 464*, 615–618.

Ravi, M., Marimuthu, M. P. A., & Siddiqi, I. (2008). Gamete formation without meiosis in Arabidopsis. *Nature, 451*, 1121–1124.

Rescan, M., Lenormand, T., & Roze, D. (2016). Interactions between genetic and ecological effects on the evolution of life cycles. *The* American Naturalist, *187*, 19–34.

Rhoades, M. M., & Dempsey, E. (1966). Induction of chromosome doubling at meiosis by the elongate gene in maize. *Genetics, 54*, 505–522.

Richards, J. A. (2003). Apomixis in flowering plants: An overview. *Transactions of the Rosyal Society of London B, 358*, 1085–1093.

Rövekamp, M., Bowman, J. L., & Grossniklaus, U. (2016). Marchantia MpRKD regulates the gametophyte-sporophyte transition by keeping egg cells quiescent in the absence of fertilization. *Current Biology, 26*, 1782–1789.

Sakai, S., Harrison, R. D., Momose, K., Kuraji, K., Nagamasu, H., Yasunari, T., et al. (2006). Irregular droughts trigger mass flowering in aseasonal tropical forests in Asia. *American Journal of Botany, 93*, 1134–1139.

Sakamoto, S. (1964). Cytogenetic studies in the tribe triticeae. I. A polyhaploid plant of Agropyron tsukushiense var. transiens ohwi found in a state of nature. *Japanese Journal of Genetics, 39*, 393–400.

Sanei, M., Pickering, R., Kumke, K., Nasuda, S., & Houben, A. (2011). Loss of centromeric histone H3 (CENH3) from centromeres precedes uniparental chromosome elimination in interspecific barley hybrids. *Proceedings of the National Academy of Sciences of the United States of America, 108*, E498–E505.

Saran, S., & Dewet, J. M. J. (1976). Environmental control of reproduction in *Dichanthium* intermedium. *Journal of Cytology and Genetics, 11*, 22–28.

Schallau, A., Arzenton, F., Johnston, A. J., Hähnel, U., Koszegi, D., Blattner, F. R., et al. (2010). Identification and genetic analysis of the APOSPORY locus in Hypericum perforatum L. *The Plant Journal, 62*, 773–784.

Schinkel, C. C. F., Kirchheimer, B., Dullinger, S., Geelen, D., De Storme, N., & Hörandl, E. (2017). Pathways to polyploidy: Indications of a female triploid bridge in the alpine species *Ranunculus kuepferi* (Ranunculaceae). *Plant Systematics and Evolution, 303*, 1093–1108.

Schmid, M. W., Schmidt, A., Klostermeier, U. C., Barann, M., Rosenstiel, P., & Grossniklaus, U. (2012). A powerful method for transcriptional profiling of specific cell types in eukaryotes: Laser-assisted microdissection and RNA sequencing. *PLoS One*, 7, pe29685.

Schmidt, A., Schmid, M. W., & Grossniklaus, U. (2015). Plant germline formation: Commun concepts and developmental flexibility in sexual and asexual reproduction. *Development*, *142*, 229–241.

Schmidt, A., Schmid, M. W., Klostermeier, U. C., Qi, W., Guthörl, D., Sailer, C., et al. (2014). Apomictic and sexual germline development differ with respect to cell cycle, transcriptional, hormonal and epigenetic regulation. *PLoS Genetics*, *10*, e1004476.

Schmidt, A., Wuest, S. E., Vijverberg, K., Baroux, C., Kleen, D., & Grossniklaus, U. (2011). Transcriptome analysis of the Arabidopsis megaspore mother cell uncovers the importance of RNA helicases for plant germline development. *PLoS Biology*, *9*, e1001155.

Sharbel, T. F., Voigt, M. L., Corral, J. M., Galla, G., Kumlehn, J., Klukas, C., et al. (2010). Apomictic and sexual ovules of Boechera display heterochronic global gene expression patterns. *The Plant Cell*, *22*, 655–671.

She, W., Grimanelli, D., Rutowicz, K., Whitehead, M. W., Puzio, M., Kotlinski, M., et al. (2013). Chromatin reprogramming during the somatic-to-reproductive cell fate transition in plants. *Development*, *140*, 4008–4019.

Siddiqi, I., Ganesh, G., Grossniklaus, U., & Subbiah, V. (2000). The dyad gene is required for progression through female meiosis in Arabidopsis. *Development*, *127*, 197–207.

Simpson, C. E., & Davis, K. S. (1983). Meiotic behavior of a male fertile triploid Arachis L. hybrid. *Crop Science*, *11*, 325–327.

Singh, M., Goel, S., Meeley, R. B., Cantec, C., Parrinello, H., Michaud, C., et al. (2011). Production of viable gametes without meiosis in maize deficient for an ARGONAUTE protein. *The Plant Cell*, *23*, 443–458.

Smith, A. J. E. (1979). Cytogenetics, biosystematics and evolution in Bryophyta. *Advances in Botanical Research*, *6*, 195–276.

Smith, J. (1841). *Notice of a plant which produces seeds without any apparent action of pollen* (p. 18). Transactions of the Linenan Society.

Spielman, M., Preuss, D., Li, F.-L., Browne, W. E., Scott, R. J., & Dickinson, H. G. (1997). TETRASPORE is required for male meiotic cytokinesis in Arabidopsis thaliana. *Development*, *124*, 2645–2657.

Spillane, C., Steimer, A., & Grossniklaus, U. (2001). Apomixis in agriculture: The quest for clonal seeds. *Sexual Plant Reproduction*, *14*, 179–187.

Stebbins, G. L. (1980). Polyploidy in plants: Unsolved problems and prospects. In W. H. Lewis (Ed.), *Polyploidy. biological relevance* (pp. 495–520). New York: Plenum press.

Stelly, D. M., & Rooney, W. L. (1989). Delimitation of the $Le2^{dav}$complementary lethality system of *Gossypium* to intracellular interaction. *Journal of Heredity*, *80*, 100–103.

Taliaferro, C. M., & Bashaw, E. C. (1966). Inheritance and control of obligate apomixis in breeding buffelgrass, *Pennisetum ciliare*. *Crop Science*, *5*, 473–476.

Tavoletti, S. (1991). Cytology of 2n egg production in *Medicago sativa ssp. falcata*. In F. Veronesi, S. Bullita, & S. Carelda (Eds.), *Ploidy and chromosome manipulation in forage breeding. Proceedings of the 17th meeting of the fodder crops section of Eucarpia, Alghero, Italy* (pp. 162–164).

Tayalé, A., & Parisod, C. (2013). Natural pathways to polyploidy in plants and consequences for genome reorganization. *Cytogenetic and Genome Research*, *140*, 79–96.

Toyama, T. K. (1974). Haploidy in peach. *Horticultural Science*, *9*, 187–188.

Turcotte, E. L., & Feaster, C. V. (1963). Haploids: High frequency production from single-embryo seeds in a line of Pima cotton. *Science*, *140*, 1407–1408.

Uchino, A. (1973). A spontaneous haploid plant of *Trillium smallii*. *Japanese Journal of Genetics*, *48*, 65–67.

van Dijk, P. J., Tas, I. C. Q., Falque, M., & Bakx-Schotman, T. (1999). Crosses between sexual and apomictic dandelions (Taraxacum). II. The breakdown of apomixis. *Heredity*, *83*, 715–721.

Vanneste, K., Maere, S., & Van de Peer, Y. (2014). Tangled up in two: A burst of genome duplications at the end of the cretaceous and the consequences for plant evolution. *Philosophy Transactions of the Royal Society London B Biological Sciences*, *369*, 1648.
Veilleux, R. E., & Lauer, F. I. (1981). Variation for 2n pollen production in clones of *Solanum phureja* Juz. And Buk. *Theoretical and Applied Genetics*, *59*, 95–100.
Veilleux, R. (1985). Diploid and polyploid gametes in crop plants: Mechanisms of formation and utilization in plant breeding. *Plant Breeding Review*, *3*, 252–288.
Verhoeven, K. J., Van Dijk, P. J., & Biere, A. (2010). Changes in genomic methylation patterns during the formation of triploid asexual dandelion lineages. *Molecular Ecology*, *19*, 315–324.
Verhoeven, K. J. F., Jansen, J. J., van Dijk, P. J., & Biere, A. (2010). Stress-induced DNA methylation changes and their heritability in asexual dandelions. *New Phytologist*, *185*, 1108–1118.
Veronesi, F., Mariani, A., & Bingham, E. T. (1986). Unreduced gametes in diploid Medicago and their importance in alfalfa breeding. *Theoretical and Applied Genetics*, *72*, 37–41.
Veronesi, F., Tavoletti, S., & Mariani, A. (1990). Identification of 2n and 4n gamete producers in an experimental population of diploid Medicago. *Journal of Genetics an Breeding*, *44*, 143–148.
Vielle, J.-P., Burson, B. L., Bashaw, E. C., & Hussey, M. A. (1995). Early fertilizations events in the sexual and aposporous egg apparatus of Pennisetum ciliare (L) link. *The Plant Journal*, *9*, 309–316.
Vielle-Calzada, J.-P., Crane, C. F., & Stelly, D. M. (1996). Apomixis: The asexual revolution. *Science*, *274*, 1322–1323.
Vijverberg, K., Milanovic-Ivanovic, S., Bakx-Schotman, T., & van Dijk, P. J. (2010). Genetic fine-mapping of DIPLOSPOROUS in Taraxacum (dandelion; Asteraceae) indicates a duplicated DIP-gene. *BMC Plant Biology*, *10*, 154.
Voigt-Zielinski, M. L., Piwczyński, M., & Sharbel, T. F. (2012). Differential effects of polyploidy and diploidy on fitness of apomictic Boechera. *Sexual Plant Reproduction*, *25*, 97–109.
Waddington, C. H. (1942). Canalization of development and the inheritance of acquired characters. *Nature*, *150*, 563–565.
Wagenvoort, M., De Vries-Van Hulten, M., Winkelhorst, G. D., & Den Nijs, A. P. M. (1990). In *Cytological mechanisms responsible for 2n pollen production and unilaterla sexual polyploidization in perrenial ryegrass (Lolium perenne). Proceedings of the 16th meeting of the fodder crops section of Eucarpia, Wageningen, The Netherlands.*
Wang, G., & Köhler, C. (2017). Epigenetic processes in flowering plant reproduction. *Journal of Experimental Botany*, *68*, 797–807.
Wang, Y., Jha, A. K., Chen, R., Doonan, J. H., & Yang, M. (2010). Polyploidy-associated genomic instability in Arabidopsis thaliana. *Genesis*, *48*, 254–263.
Watts, A., Kumar, V., & Bhat, S. R. (2016). Centromeric histone H3 protein: From basic study to plant breeding applications. *Journal of Plant Biochemistry and Biotechnology*, *25*, 339–348.
Wedzony, M. (2003). Protocol for doubled haploid production in hexaploid Triticale (*x Triticosecale* Wittm.) by crosses with maize. In M. Maluszynskiet al. (Eds.), *Doubled haploid production in crop plants: A manual* (pp. 135–140). Kluwer Academic Publishers.
Werner, J. E., & Peloquin, S. J. (1991a). Occurrence and mechanism of 2n egg formation in 2x potato. *Genome*, *34*, 975–982.
Werner, J. E., & Peloquin, S. J. (1991b). Yield and tuber characteristics of 4x progeny from 2x × 2x crosses. *Potato Research*, *34*, 261–267.
Werner, J. E., & Peloquin, S. J. (1987). Frequency and mechanism of 2n egg formation in haploid tuberosum-wild species F1 hybrids. *American Potato Journal*, *64*, 641–654.

Yahata, M., Kurogi, H., Kunitake, H., Nagano, K., Yabuya, T., Yamashita, K., et al. (2005). Evaluation of reproductive functions in a haploid pummelo by crossing with several diploid citrus cultivars. *Journal of the Japanese Society of Horticultural Science, 74*, 281–288.
Yang, C.-Y., Spielman, M., Coles, J. P., Li, Y., Ghelani, S., Bourdon, V., et al. (2003). TETRASPORE encodes a kinesin required for male meiotic cytokinesis in Arabidopsis. *The Plant Journal, 34*, 229–240.
Zhang, Z., Qiu, F., Liu, Y., Ma, K., Li, Z., & Xu, S. (2008). Chromosome elimination and in vivo haploid production induced by stock 6-derived inducer line in maize (*Zea mays* L.). *Plant Cell Reports, 27*, 1851–1860.
Zhao, X., Xu, X., Xie, H., Chen, S., & Jin, W. (2013). Fertilization and uniparental chromosome elimination during crosses with maize haploid inducers. *Plant Physiology, 163*, 721–731.

Further reading

Bingham, T., & Binek, A. (1969). Comparative morphology of haploids from cultivated alfalfa, *Medicago sativa* L. *Crop Science, 9*, 749–751.
Cavalier-Smith, T. (2002). Origins of the machinery of recombination and sex. *Heredity, 88*, 125–141.
Nakajima, K. (2018). Be my baby: Patterning toward plant germ cells. *Current Opinion in Plant Biology, 41*, 110–115.
Ren, J., Wu, P., Trampe, B., Tian, X., Lübberstedt, T., & Chen, S. (2017). Novel technologies in doubled haploid line development. *Plant Biotechnology Journal, 15*, 1361–1370.
Rines, H. W. (2003). Oat haploids from wide hybridization. In M. Maluszynskiet al. (Eds.), *Doubled haploid production in crop plants: A manual* (pp. 155–159). Kluwer Academic Publishers.
Rodriguez-Leal, D., & Vielle-Calzada, J.-P. (2012). Regulation of apomixis: Learning from sexual experience. *Current Opinion in Plant Biology, 5*, 549–555.

CHAPTER TWENTY-ONE

Seeds—An evolutionary innovation underlying reproductive success in flowering plants

Célia Baroux*, Ueli Grossniklaus
Department of Plant and Microbial Biology & Zurich-Basel Plant Science Center, University of Zurich, Zurich, Switzerland
*Corresponding author: e-mail address: cbaroux@botinst.uzh.ch

Contents

Abstract

"Seeds nourish, seeds unite, seeds endure, seeds defend, seeds travel," explains the science writer Thor Hanson in his book *The Triumph of Seeds* (2015). The seed is an ultimate product of land plant evolution. The nursing and protective organization of the seed enable a unique parental care of the progeny that has fueled seed plant radiation. Seeds promote dispersal and optimize offspring production and thus reproductive fitness through biological adaptations that integrate environmental and developmental cues. The composite structure of seeds, uniting tissues that originate from three distinct organisms, enables the partitioning of tasks during development, maturation,

Current Topics in Developmental Biology, Volume 131
ISSN 0070-2153
https://doi.org/10.1016/bs.ctdb.2018.11.017

and storage, while a sophisticated interplay between the compartments allows the fine-tuning of embryonic growth, as well as seed maturation, dormancy, and germination. In this review, we will highlight peculiarities in the development and evolution of the different seed compartments and focus on the molecular mechanisms underlying the interactions between them.

1. Introduction

The Triumph of Seeds (Hanson, 2015) is an ode to the diversity and exquisite functionalities of seeds that "conquered the plant kingdom and shaped human history" as the author writes. The extraordinary resilience and vital energy of seeds are illustrated by the germination of a Judean date palm tree from a 2000-year-old seed excavated from archaeological remnants of Masada's siege (73–74 CE, Israel) (Sallon et al., 2008). Another example is the unexpected sprouting of several-hundred-year-old seeds from ruins of the British Museum's Botany Department blazed by bombardments in the Second World War, or the basil seeds returning from 1 year on the international space station exposed to cosmic rays, yet germinating successfully (Hanson, 2015). While the fact that DNA starts to degrade after a few decades of seed storage (Walters, Reilley, Reeves, Baszczak, & Richards, 2007) puts some of these reports in serious doubt, they do illustrate that seeds are long-lived propagules ideally suited for species dispersal. Certainly, seeds are a remarkable evolutionary innovation. In seed plants (Spermatophyta), reproduction is no longer left to chance at every step, from gamete union to development of the fragile progeny, its dispersal and immediate germination, sometimes in an adverse environment. With the appearance of seeds, the female gametophyte is retained on and nourished by the mother plant, male gametes are protected and transported within pollen, and the embryo is encapsulated in a protective and nourishing structure and released only when favorable conditions are met. Clearly benefitting reproductive success, the appearance of seeds coincides with the rapid radiation and extraordinary diversification of land plants, particularly the flowering plants (angiosperms) (Harris & Davies, 2016; Linkies, Graeber, Knight, & Leubner-Metzger, 2010; Willis et al., 2014).

This book chapter provides an overview of the structure of the seed, presents selected evolutionary and functional aspects of the seed compartments, and discusses recent findings on the interplay between them that fine-tunes seed biology. Whenever appropriate, the reader is referred to other reviews that discuss specific topics in a more exhaustive manner than possible here.

Detailed descriptions principally focus on the eudicot *Arabidopsis thaliana* and occasionally on the monocots *Zea mays* (maize) and *Oryza sativa* (rice). This choice has the virtue to showcase selected molecular mechanisms and genetic regulatory pathways but does not render justice to the remarkable diversity of seed properties, both with respect to physiology and morphology, that is found among the >300,000 species of the seed plants (Harris & Davies, 2016).

From a botanical viewpoint, the seed is a fertilized ovule. An extension to this basic definition includes caryopses (kernels) where the seed coat is fused to the ovary wall (pericarp) of the fruit. Ovules are the female reproductive units of seed plants and comprise the female gametophyte (embryo sac) surrounded by the nucellus and integuments (Gasser & Skinner, 2019). In gymnosperms, the female gametophyte comprises one or several egg cells surrounded by a dense multicellular tissue that will nurse the embryo following fertilization. In angiosperms, the female gametophyte is reduced to a small number of cells and the embryo-nourishing tissue is not preformed. About 70% of angiosperms contain an eight-nucleated, seven-celled embryo sac of the *Polygonum* type with two female gametes, the egg and central cell, two synergid cells assisting in double fertilization, and three antipodal cells (Maheshwari, 1950; Serbes, Palovaara, & Groß-Hardt, 2019). Fertilization of the egg and central cell with one sperm cell each produces the zygote and the endosperm, respectively, which jointly develop within the seed integuments. At maturity, the seed is usually dormant and comprises a juvenile plant embryo and stored nutrients sufficient to support its initial growth after germination. The tissue dedicated to nutrient storage varies: in many species with a non-persistent endosperm, nutrients are stored in the cotyledons, while other seeds contain a persistent tissue surrounding the embryo, which can be the endosperm or the nucellus (perisperm) (Cocucci, 2005; Lu & Magnani, 2018). Dormancy is established during seed maturation and exerts control over germination until environmental conditions are favorable (Finch-Savage & Footitt, 2017). Dormancy is thus a trait that likely brought considerable benefits to the reproductive success of seed plants.

Seeds evolved in the late Devonian (~370 Mya) with an extension of maternal tissues surrounding the egg chamber and the proliferation of tissue surrounding the embryo. Seed shape, function, and diversity as we know it today were largely established in the Cretaceous (~140 Mya) following the expansion of angiosperms (Linkies et al., 2010). The evolutionary success of seeds lies in their composite structure integrating multiple functions. With a protective seed coat, a nourishing embryo-surrounding tissue, and the embryo itself, anatomy recapitulates the functional attributes of the seed, fine-tuned by an interplay between these different compartments that is based on a variety of signaling pathways.

2. The seed is composed of three compartments of distinct developmental origin

The seed is a composite structure in which three genetically distinct compartments cohabitate and cooperate in a coordinated manner to form the mature seed.

The seed coat is purely of maternal origin. It is composed of several tissue layers derived from the maternal sporophyte, differing in structure and function depending on the species (Radchuk & Borisjuk, 2014). In most eudicots, the seed coat is bitegmic, with an inner and an outer integument. In *A. thaliana*, the seed coat is histologically well described and its development has been genetically dissected (Golz et al., 2018; Haughn & Chaudhury, 2005). Development and structure of the three-layered inner and the two-layered outer integument are distinct, showing tissue specializations serving specific functions. The innermost endothelial layers produce proanthocyanidins (PAs), oligomeric flavonoid compounds, which upon oxidation form toxic quinones as well as other active secondary metabolites, which protect the seed. The outer layers store starch during development until they differentiate, with epidermal cells turning into a mucilage-producing, water-impermeable layer, while subepidermal cells produce a thick cell wall at their basis, isolating the outer from the inner integument. At maturity, the seed coat desiccates and becomes a physiologically inert testa, impregnated with PAs upon release from the endothelial layer (Haughn & Chaudhury, 2005). In legumes, the inner integuments retain photosynthetically active parenchyma cells, and the outer layers are composed of thickened and water-impermeable sclerenchyma and palisade cells. In cereal grains, the ovule integuments are fused with the pericarp, overtaking the function of the seed coat (Kovalchuk et al., 2010).

The seed coat program is initiated at fertilization by release of a developmental block exerted by the *Polycomb* group protein FERTILIZATION INDEPENDENT ENDOSPERM (FIE) in both *A. thaliana* and rice (Huang et al., 2016; Roszak & Köhler, 2011). Coordination of seed coat development then requires a hormonal cross talk with the endosperm (Figueiredo, Batista, Roszak, Hennig, & Köhler, 2016), for which the endothelium plays a critical role (see below). Endothelial differentiation is regulated by the

TRANSPARENT TESTA (*TT*) class of genes. *TT* genes were uncovered in genetic screens for pigmentless or translucid seeds. Their analysis revealed that testa formation is under the control of two classes of molecular factors: MYB- and MADS-domain transcription factors on the one hand and enzymes involved in PA biogenesis on the other. The two MADS-box genes *SEEDSTICK* (*STK*) and *TT16* act in concert to regulate the formative divisions of the endothelium as *stk tt16* double mutants completely lack this innermost layer of the maternal seed coat (Mizzotti et al., 2012). The closely related MADS-box genes *SHATTERPROOF1* (*SHP1*) and *SHP2* also work together with *TT6* in endothelium differentiation but have partially antagonistic functions (Ehlers et al., 2016). Triple loss-of-function mutants of these genes have a malformed seed coat, impairing also endosperm and embryo development. Similarly, epidermal seed coat development is primarily controlled by a dual set of molecular factors: transcription factors similar to those governing trichome, root hair, and endothelium development and enzymes of the mucilage biosynthesis pathway (Haughn & Chaudhury, 2005). The analysis of mutant seeds with incomplete epidermal seed coat differentiation indicates an important role for the mucilage in germination and seed viability (Golz et al., 2018; Haughn & Chaudhury, 2005).

The embryo-nourishing tissue has a distinct developmental origin and genetic composition in gymnosperms and angiosperms. In gymnosperms, a specialized extension of the female gametophyte forms prior to fertilization and provides nutrients to the developing embryo after fertilization. In contrast, the embryo-nourishing tissue of angiosperms is a product of fertilization and hence carries genetic contributions from both parents. During double fertilization, not only the egg but also the central cell of the female gametophyte gets fertilized by a sperm and subsequently gives rise to the endosperm. In endospermic species, the endosperm develops at the expense of the nucellus, which partially or totally disappears. The endosperm proliferates in a syncytium, cellularizes, and is later consumed by the developing embryo or the germinated seedling, depending on the species (Floyd & Friedman, 2000; Lopes & Larkins, 1993; Olsen, 2004; Fig. 1). In perispermic species, such as sugar beet, coffee and quinoa, the endosperm poorly develops and the embryo nurturing function is largely overtaken by the nucellus, of maternal origin, which proliferates following fertilization (Lu & Magnani, 2018).

Flowering plants show a diversity of endosperm developmental patterns: nuclear, cellular, and helobial (Geeta, 2003; Fig. 1). The nuclear type of endosperm is very common in angiosperms, including most rosids, the Magnoliales, Laurales, Piperales, and Caryophyllales, as well as the monocot

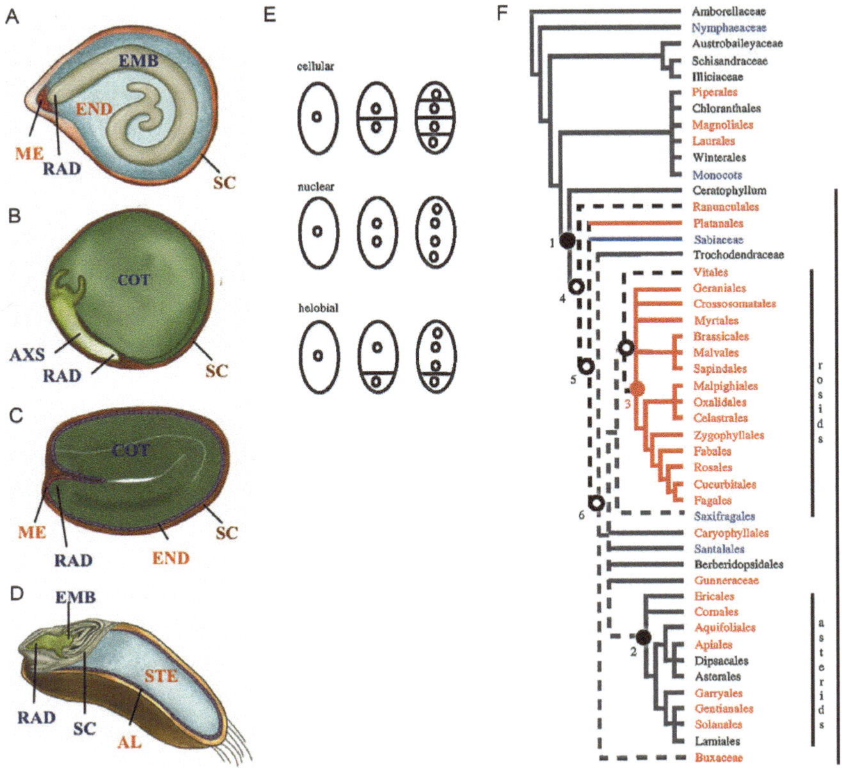

Fig. 1 Different endosperm types and developmental patterns. (A-D) Persistent and non-persistent endosperm in eudicots and monocot seeds. Illustrations of the embryo and endosperm in (A) *Solanum lycopersicum* (tomato), (B) *Pisum sativum* (pea), (C) *Arabidopsis thaliana*, (D) *Triticum aestivum* (wheat). The endospermic and embryonic tissues are labeled in orange and blue, respectively. END, endosperm; EMB, embryo; ME, micropylar endosperm; RAD, radicle; SC, seed coat; COT, cotyledons; AXS, embryonic axis; STE, starchy endosperm; AL, aleurone layer; SC, scutellum. The ME region is colored in red (A and C). (E and F) Nuclear, cellular, and helobial types of endosperm development (E) and their occurrence in angiosperms (F). Red, nuclear; black, cellular; blue, helobial. *, monocots have a further diversification of helobial (e.g., in Alismatales) and nuclear (e.g., in cereals). *Panels (A)-(D): After Yan, D., Duermeyer, L., Leoveanu, C., & Nambara, E. (2014). The functions of the endosperm during seed germination.* Plant and Cell Physiology, 55, *1521–1533. Panels (E) and (F): Modified after Geeta, R. (2003). The origin and maintenance of nuclear endosperms: Viewing development through a phylogenetic lens.* Proceedings. Biological Sciences, 270, *29–35.*

cereals. In a first phase, proliferation takes place in a syncytium without cytokinesis, producing a liquid endosperm. After a critical number of nuclei have been produced, cellularization occurs and the tissue develops through regular cell divisions in a second phase. Cellularization corresponds to a critical physiological transition in nutrient storage and embryo provisioning, which can profoundly impact seed viability (Hehenberger, Kradolfer, & Köhler, 2012; Lafon-Placette & Köhler, 2014). In species with cellular endosperm, present in the Asterales, Lamiales and the basal-most angiosperm *Amborella trichopoda*, cell wall formation is concomitant with cell division. Helobial endosperm is a mixture of the first two types, where after the first cellular endosperm division, one nucleus will follow the cellular type while the other engages in the nuclear type. Depending on the species, the nuclear or cellular part of the endosperm is emphasized. The helobial type is less frequent, found for instance in the Saxifragales and the monocot order Alistamales. The cellular type is supposed to be ancestral with multiple emergences of the nuclear type during angiosperm evolution (Geeta, 2003). But reconstructing the evolution of endosperm development based on the different types is complicated by the further partitioning of the endosperm into distinct domains with regard to their progression through proliferation and cellularization (Boisnard-Lorig et al., 2001; Olsen, 2004). In *A. thaliana*, three domains are distinguished: the peripheral endosperm, comprising a large central vacuole and a peripheral multinucleate cytoplasm during the first phase; the chalazal endosperm, corresponding to a multinucleate mass that rarely cellularizes and comprises giant, polyploid nuclei (Baroux, Fransz, & Grossniklaus, 2004; Boisnard-Lorig et al., 2001); and the micropylar endosperm, forming a dense tissue that rapidly cellularizes around the embryo and is thought to serve as a nutrient transfer interface (Lafon-Placette & Köhler, 2014; Lopes & Larkins, 1993; Olsen, 2004). In cereals, the three domains correspond to the starchy endosperm, the basal endosperm transfer layer (BETL), and the embryo surrounding region (ESR), respectively (Olsen, 2004).

As an embryo-nourishing tissue, the endosperm is the functional equivalent of the female gametophytic tissue in gymnosperm seeds. But in contrast to the latter, it develops only upon fertilization. Our current understanding is that the unfertilized central cell has the potential to proliferate into endosperm but is developmentally suppressed until fertilization. In *A. thaliana,* several layers of control involving *Polycomb* Repressive Complex 2 (PRC2), various transcription factors, and hormonal regulators act iteratively to activate endosperm development (reviewed in Hands, Rabiger, & Koltunow, 2016). PRC2 confers histone 3 lysine 27 trimethylation (H3K27me3), a repressive epigenetic mark that controls various developmental processes in plants and animals (Bemer & Grossniklaus, 2012; Mozgova, Köhler, & Hennig, 2015). The FIS-PRC2 comprises the C2H2 zinc-finger protein

FERTILIZATION-INDEPENDENT SEED2 (FIS2), the SET-domain containing histone methyltransferase MEDEA (MEA), the WD-40 protein FIE, and the p55-like MULTICOPYSUPPRESSOR OF IRA1 (MSI1) protein (Bemer & Grossniklaus, 2012; Mozgova et al., 2015). Several members of this complex are named after a property of their loss-of-function alleles, which enable development of endosperm and seed-like structures in the absence of fertilization (reviewed in Grossniklaus & Paro, 2014; Grossniklaus, Spillane, Page, & Köhler, 2001; Hands et al., 2016; Henning & Derkacheva, 2009). Possibly, in apomictic species where seeds develop in the absence of fertilization (León-Martínez & Vielle-Calzada, 2019), PRC2 repression in the central cell may be relaxed (Grossniklaus et al., 2001; Rodrigues, Tucker, Johnson, Hrmova, & Koltunow, 2008). In *A. thaliana*, while abolishing FIS-PRC2 function is sufficient to trigger auxin signaling that activates mitosis in the endosperm (Figueiredo & Köhler, 2018), full development, notably cellularization, is not recapitulated in mutants of the *fis* class, demonstrating the existence of additional checkpoints. Sustained auxin signaling, possibly in interplay with signals from other seed compartments, is proposed to coordinate endosperm cellularization and differentiation (reviewed in Figueiredo, Batista, & Köhler, 2018; Figueiredo & Köhler, 2018). Reviewing the influence of auxin on seed development in species beyond *A. thaliana,* including cereals, fruit trees and other woody species (Figueiredo et al., 2018), and considering the role of cytokinin in endosperm proliferation (Li, Nie, Tan, & Berger, 2013), it is safe to predict that hormonal pathways play a prominent role in the control of endosperm development, which will be unraveled soon. In addition, endosperm development is under the control of cell cycle regulators that coordinate the mitotic cycles until cellularization and consecutive endoreduplication (reviewed in Dante, Larkins, & Sabelli, 2014). In *A. thaliana*, endosperm proliferation is coordinated by a key transcription factor, SHORT HYPOCOTYL BLUE1 (SHB1), which regulates *MINISEED3* (*MINI3*) and *HAIKU2* (*IKU2*), genes encoding a WRKY transcription factor and an LEUCINE-RICH REPEAT (LRR) receptor-like kinase (RLK), respectively (Zhou et al., 2009). The *IKU* pathway, which in addition to *MINI3* and *IKU2* also includes *IKU1* coding for a VQ motif protein (Wang et al., 2010), is thought to coordinate cytokinin-mediated control of endosperm proliferation by establishing a gradient through the different proliferation domains (Li et al., 2013). Future investigations will refine the interplay between transcriptional, hormonal, signaling, and epigenetic controls operating in the regulation of endosperm growth.

The seed ultimately encapsulates the plant's offspring, the embryo. A seed generally hosts only one embryo, except in some gymnosperms where fertilization of multiple eggs occasionally leads to several embryos, although all but one often degenerate (Maheshwari, 1950; reviewed in Baroux, Spillane, & Grossniklaus, 2002). The embryo projects the basic body plan of the plant, where tissue types are organized along an apical-basal and a radial axis. The mature embryo comprises a radicle, embryonic leaves (cotyledons), and apical meristems that will generate shoot and root organs after germination (Jenik, Gillmor, & Lukowitz, 2007). The molecular control of cell fate determination and embryo morphogenesis, differentiation, and maturation has been described in several excellent reviews (e.g., De Smet, Lau, Mayer, & Jürgens, 2010; Jenik et al., 2007; ten Hove, Lu, & Weijers, 2015). The embryo of eudicots and monocots shows notable differences, including a less predictive patterning, an additional dorsoventral axis, and specialized embryonic extensions in the latter (Zhao, Begcy, Dresselhaus, & Sun, 2017). Nevertheless, in both clades, embryogenesis starts with (i) an asymmetric division of the zygote, defining the apical and basal lineages that lead to the formation of the embryo proper and a suspensor or suspensor-like structure, respectively, (ii) the embryos undergo a transition from radial to bilateral symmetry, and (iii) they share apical-basal and radial symmetries along which the meristems and tissue layers are organized, respectively. This points toward a set of conserved molecular circuits in monocot and dicot embryogenesis, involving the WUSCHEL-RELATED HOMEOBOX (WOX) family of transcription factors and polar auxin transport by the PIN-FORMED (PIN) family of auxin efflux carriers (Zhao et al., 2017). Other chapters of this book provide an in-depth review of the genetic control and evolution of plant embryogenesis (Armenta-Medina & Gillmor, 2019; Chandler & Werr, 2019).

3. Seed evolution is intimately linked to double fertilization

Seeds are acknowledged as an ultimate product of evolution in the plant kingdom, offering competitive benefits for the propagation of the next generation, and hence reproductive success. Seed evolution required several key steps (Linkies et al., 2010): (i) the appearance of heterospory; (ii) the retention of the megaspore, female gametophyte, and embryo within the megasporangium/nucellus; (iii) the formation of an endosporic embryo sac within the nucellus; and (iv) the formation of integuments delimiting an aperture (micropyle), enabling the entry of the male gametophyte (pollen tube). Seed-like structures are already found in fossils of extinct seed ferns in the late Devonian, thus predating the (paraphyletic) group of seed plant ancestors (progymnosperms). In these fossil records, the megasporangium is enveloped by sterile branches of the sporophyte, sometimes surrounded again by another cupule, thus resembling the modern integuments (see Gasser & Skinner, 2019). Recent discussions also emphasize the importance of megasporangium transformation into the modern nucellus of seed plants. Notably, its evolution into a non-dehiscent structure retaining the megaspore until embryogenesis is thought to be a key transition step, enabling both endomegagametogenesis and encapsulation. Another step was the acquisition of a transient fate to accommodate the embryo-nourishing tissue, the female gametophyte itself as in gymnosperms, or the endosperm as in most angiosperms (Magnani, 2018). Nucellus elimination, however, did not evolve as a consistent trait. In perispermic seeds it proliferates into a storage tissue at the expense of the endosperm (Cocucci, 2005). The MADS-box gene *TT16* is proposed to have been a key molecular component toward the evolution of such a transient fate (Magnani, 2018). In *A. thaliana*, this transcription factor controls the nucellus cell death program, enabling endosperm proliferation (Xu et al., 2016). *TT16* seems to be epigenetically regulated by PRC2 (Xu et al., 2016). This additional, versatile level of control of a key factor for nucellar fate possibly allowed for evolutionary variations in the nucellus elimination program among plant species.

The endosperm of angiosperms is, in contrast to the nursing tissue of gametophytic origin in gymnosperms, a product of fertilization. The evolutionary acquisition of contributions from both parents is thought to have benefited the control of resource allocation to the embryo, shifting from maternal to biparental control (reviewed in Baroux et al., 2002). In most but the basal lineages, the central cell is homo-diploid, thus generating a triploid endosperm with a two-to-one maternal-to-paternal (2m:1p) genome ratio. Basal angiosperms, however, also comprise species, for instance the water lily *Nuphar polysepalum*, with a 1m:1p diploid endosperm derived from a four-nucleate embryo sac with a haploid central cell (Williams & Friedman, 2002; Fig. 2). This observation suggests that a diploid, 1m:1p genetic composition might represent the ancestral state of the endosperm; in this scenario, the most common 2m:1p triploid state may have evolved later. This hypothesis, whereby double fertilization evolved before endosperm triploidy, is supported by the observation that in some distantly related, non-flowering plant species two egg cells in a female gametophyte

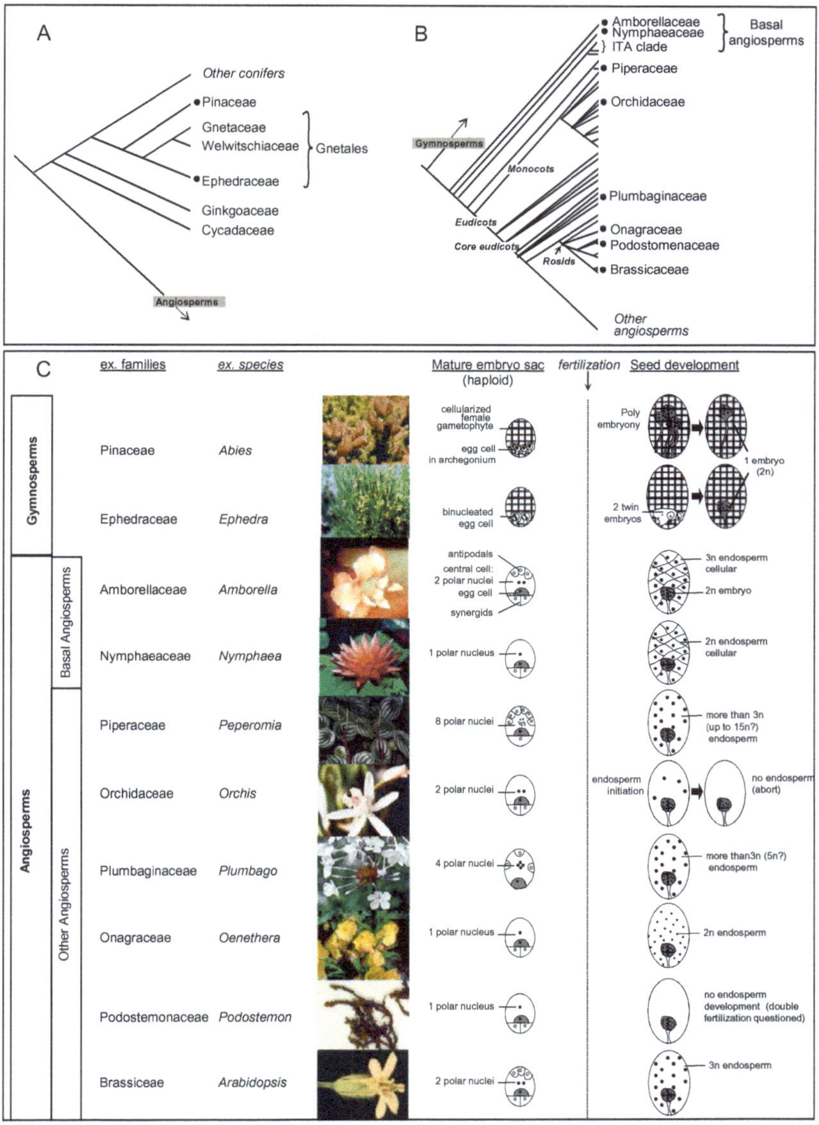

Fig. 2 Reconstruction of the evolutionary ontogeny of the endosperm by phylogenetically anchored comparative embryology. Schematic representation of the phylogenetic trees of (A) gymnosperms and (B) angiosperms. Only species relevant for the text or the illustration in (C) are indicated. (C) A series of examplary species illustrating the diversity of embryo sac structures among extant plants and the subsequent diversity of endosperm ploidy. The embryo sac depicted for *Arabidopsis thaliana* is of the *Polygonum* type. Note that since the first publication of this figure, *Amborella trichopoda* has been shown to have three synergid cells and not two as depicted here (Friedman, 2006).

(Continued)

can get fertilized by the two sperm from one pollen, giving rise to two embryos (Carmichael & Friedman, 1995; Friedman, 1992); interestingly supernumerary embryos usually degenerate, at the possible benefit of the surviving one. Such "altruistic" twin embryos might well be evolutionary precursors of a 1m:1p endosperm, which later gained a second maternal contribution (reviewed in Baroux et al., 2002; Friedman, 2006). Yet, the frequent occurrence of basal lineages with a triploid endosperm challenges this hypothesis. In particular, also the basal-most extant angiosperm, *A. trichopoda*, produces a triploid endosperm (Williams, 2009), and it is currently unclear whether triploid endosperm is an ancestral or derived character.

The evolution of triploidy is often interpreted as a result of, but also a means for, parental conflicts. Parental conflicts arise in situations where alleles from the two parents have different effects on the relative fitness of the progeny. Such conflicts unfold largely, but not exclusively, at the level of gene expression controlling resource allocation from the mother to its offspring (Patten et al., 2014). In angiosperms, the maternal parent has an interest to provide equal resources to all its progeny, irrespective of the father. In contrast, alleles derived from the paternal parent would benefit if more resources were provided to the seeds carrying them, in particular in cases with mixed fatherhood (Haig & Westoby, 1989). In this context a higher maternal genome dosage in the triploid endosperm could have evolved to reinforce equal resource distribution by the maternal parent, opposing maximized provisioning promoted by the paternal parent

Fig. 2—Cont'd The variations in endosperm ploidy for Piperaceae and Plumbaginaceae, the transient presence of endosperm in Orchidaceae and its absence from Podostemonaceae are all only predicted, on the basis of cytological studies. The ploidy number of the endosperm has been precisely measured in some basal angiosperms and is known for *A. thaliana*. For comparison to the angiosperm female gametophyte (embryo sac), the female gametophyte of two gymnosperm species is shown with copious cellularized haploid tissue and one to five archegonia, each harboring one egg cell. In the case of *Ephedra nevadensis*, egg cells are binucleate with a normal egg nucleus and a ventral canal nucleus. In gymnosperms, "simple complex polyembryony" occurs, whereby several egg cells can be fertilized (by distinct sperm nuclei), and each zygote then generates four embryo clones. The extent of this polyembryony varies between species, and a simplified form is depicted here. Ultimately, only one embryo will survive while the others degenerate (gray dashed lines in the example of *Abies alba*). In the case of *E. nevadensis*, both nuclei in the egg cell are fertilized by two sperm nuclei discharged by a single pollen tube (double fertilization), and polyembryony also applies to both fertilization products. *After Baroux, C., Spillane, C., & Grossniklaus, U. (2002). Evolutionary origins of the endosperm in flowering plants.* Genome Biology, 3, *reviews1026.1–reviews1026.5.*

(Haig & Westoby, 1989). The characterization of genetic mutants conveying a maternal effect on endosperm development and seed viability has provided strong support to this hypothesis (reviewed in Grossniklaus, 2005; Grossniklaus et al., 2001; Köhler & Weinhofer-Molisch, 2010; Pires & Grossniklaus, 2014). One of them, the *mea* mutant of *A. thaliana*, nicely illustrates maternal influence on endosperm development: irrespective of the paternal contribution, seeds inheriting a defective *mea* maternal allele prove inviable with a malformed endosperm and overgrown embryo (Grossniklaus, Vielle-Calzada, Hoeppner, & Gagliano, 1998). Several mechanisms can contribute to a dominant representation of maternal products (Dilkes & Comai, 2004; Messing & Grossniklaus, 1999): (i) stoichiometric expression of dosage-sensitive factors; (ii) selective silencing of paternal alleles, an epigenetic phenomenon referred to as genomic imprinting (see below); and (iii) maternally stored products inherited from the egg or central cell.

However, triploidy has also predictable evolutionary benefits in masking deleterious mutations altering endosperm function (Cailleau, Cheptou, & Lenormand, 2010). Modeling the evolution of ploidy in the endosperm suggests a stronger influence of masking effects than provisioning conflicts. The latter are indeed expected to lead to an escalation, which is not seen in the parental contribution and genomic composition of the endosperm (Cailleau et al., 2010). A conciliating view is to consider a combined impact of heterosis, masking deleterious effects, and allocation conflicts in the evolution of double fertilization and the doubling of the maternal dosage in the endosperm.

The seed coat also shows great evolutionary diversity. In angiosperms, three main classes are distinguished based on the number of integument layers: unitegmic, bitegmic and the hybrid bifid class. In the latter, the basal part of the seed coat has an unclear origin while the apical part consists of two integuments (Coen & Magnani, 2018). The bitegmic situation is thought to be the ancestral one, whereas unitegmic and bifid configurations might have emerged from integument fusion or developmental suppression (Coen & Magnani, 2018; Gasser & Skinner, 2019). Possibly, the repeated emergence of unitegmic species in evolution (Coen & Magnani, 2018) may have benefited from a certain flexibility in the regulation of seed coat development conveyed by epigenetic regulators, notably the PRC2 components FIE and MSI1 (Hennig, Taranto, Walser, Schönrock, & Gruissem, 2003; Roszak & Köhler, 2011). The conversion of the outer integument into a leaf-like structure in specific loss-of-function mutants of *A. thaliana,* and the expression of outer integument molecular markers in arils of ancestral plants, provide support to the hypothesis that the seed coat evolved from enveloping leafy structures (Coen & Magnani, 2018; Gasser & Skinner, 2019). The evolution of seed coat layers has not yet been elucidated but the identification of numerous master regulators in *A. thaliana* should enable evo-devo approaches that will shed light onto this issue in the near future. The observation that seed coat initiation is controlled by an epigenetic regulator commonly used in developmental transitions (Xiao, Jin, & Wagner, 2017) suggests that recruitment of FIE as an epigenetic switch may have been key to the evolution of this seed compartment.

4. The seed coat—A polyvalent capsule

The seed coat acts as a physical barrier against microscopic invaders (bacteria, fungi, and possibly viruses), a property reinforced by the secretion of defensive molecules (peptides and secondary metabolites such as cyanogenic glycosides, terpenoids, and flavonoids, e.g., PAs) (Kovalchuk et al., 2010; Radchuk & Borisjuk, 2014). Being water impermeable in most cases, the seed coat physically protects the embryo from precocious imbibition and water-driven damages at low temperatures. Yet the seed coat permits gas exchange, regulating seed metabolism necessary to activate nutrient transport from maternal tissues to the endosperm. The selective properties of the seed coat, however, limit oxidation in the interior of the seed, a condition thought to favor seed longevity (Kovalchuk et al., 2010). Some species retained photosynthetic activity in the seed coat. In addition to contributing to the carbon budget (CO_2 fixation and redistribution as photoassimilates), this likely enables the seed to integrate light cues as environmental signals for the control of seed dormancy and germination (Kovalchuk et al., 2010). In *A. thaliana*, and likely in many other species, the seed coat also acts as a nutrient provider at early stages of embryo development. It does so by uploading phloem-derived compounds at the base of the seed (chalaza) connected to the maternal tissue (reviewed in Coen & Magnani, 2018; Radchuk & Borisjuk, 2014). In particular, sucrose transporters, such as SWEET12, SWEET15, and SWEET11 in *A. thaliana*, orchestrate the apoplastic transport of sucrose from maternal tissues toward the seed coat and endosperm (Chen et al., 2015; Hedhly et al., 2016). In cereals, such as maize and rice, seed filling is essentially regulated by the BETL of the endosperm. SWEET transporters in the BETL are essential for hexose transport and particularly the variant SWEET4 seems to have been selected during domestication of the cereals (Sosso et al., 2015). In addition, loss-of-function of *SWEET11* and *SWEET15* in rice leads to hyperaccumulation of starch in the pericarp and deficient export of sucrose in the endosperm (Yang, Luo, Yang, Frommer, & Eom, 2018). This illustrates the complex routes of sugar transport from maternal tissues to the seed coat/pericarp of caryopses, and endosperm nutrition largely relying on the exquisite specialization and cell-specific expression of SWEET transporters.

Another role of the seed coat is to provide physical constraints that modulate embryo and endosperm growth (Creff, Brocard, & Ingram, 2015). At maturity, the seed coat of *A. thaliana* undergoes programmed cell death (Ingram, 2017) and becomes a physiologically inert testa in dried mature seeds (Radchuk & Borisjuk, 2014). The role of the *A. thaliana* seed coat becomes restricted to a mechanical barrier preventing germination and providing physical protection. In mature seeds, the seed coat plays a central role in seed dormancy, exerting control over germination for the benefit of successful growth and survival in a suitable environment. Dormancy is also regulated physiologically and light, temperature, moisture, and oxidation cues are integrated to fine-tune seed germination (Finch-Savage & Footitt, 2017). In some species, seed coat scarification, i.e., physical damage, is sufficient to break dormancy (Nee, Xiang, & Soppe, 2017). Seed coat integrity controls the level of oxidation in the seed, through direct O_2 uptake or through the oxidation of flavonoids that accumulated in the seed coat and are released upon physical damage (El-Maarouf-Bouteau & Bailly, 2008; Nee et al., 2017). Sensing oxidation levels triggers oxidative signaling and promotes germination (El-Maarouf-Bouteau & Bailly, 2008). In other species, such as *A. thaliana* and many legumes, physiological and hormonal regulation dominates over seed coat integrity in the control of germination (Willis et al., 2014). Yet, oxidative signaling is thought to cross talk with the hormonal control of germination at the level of endosperm rupture. Abscisic acid (ABA) and the *DELAY IN GERMINATION1* (*DOG1*) gene are key regulators repressing seed germination (Finch-Savage & Footitt, 2017). However, ABA is mostly contributed by the single endosperm cell layer remaining in mature *A. thaliana* seeds and not the inert testa itself (Lee, Piskurewicz, Tureckova, Strnad, & Lopez-Molina, 2010). In *Pinus taede* (loblolly pine), the seed coat expresses biochemical determinants of germination that are mitigated by chilling moisture and not by physical rupture of the seed coat alone (Cooke, Cooke, & Gifford, 2002). Those likely involve the phytohormones ABA and gibberellic acid (GA) as in flowering plants (Zhao & Jiang, 2014). Similar to angiosperms, pine seed germination is also aided by the remaining nucellar layer below the seed coat (Guo, Shen, & Shi, 2018). This points toward a conserved role of the seed coat and the cell layers underneath it to control embryo germination via physical and physiological cues.

Finally, the seed coat also plays an important role in seed dispersal, a function recognized early by botanists reporting on the properties of diverse seed morphologies contributed by seed coat and fruit extensions (Hanson, 2015). Finally, the mucilage of the seed coat has multiple functions in the preservation of seeds against adverse conditions, in dispersal, storage, hydration, and facilitating germination (Golz et al., 2018; Yang, Baskin, Baskin, & Huang, 2012). Interestingly, the dead seed coat is also a storage compartment for enzymes with antimicrobial and detoxifying activities, helping in long-term survival and the preservation of germination efficiency as shown in *Sinapis alba* (white mustard) (Raviv et al., 2017).

5. The endosperm—A nurse, a referee, and a check-point

The primary role of the endosperm is that of a nutrient supplier sustaining embryo development in species with non-persistent endosperm (e.g., *A. thaliana*, *Phaseolus vulgaris*) and the germinating seedling in cereals and legumes with persistent endosperm. The analysis of mutants showing embryo arrest or defective endosperm, and which were affected in nutrient accumulation and transport, revealed a critical stage with respect to provisioning. Until the late globular stage, the embryo is able to develop with no or very limited endosperm tissue. This has been well demonstrated by mutants lacking fertilization of the central cell, such as in *glauce* or in *cdka;1* mutants, where the embryo develops until the globular stage in the absence of a developing endosperm (Gusti et al., 2009; Iwakawa, Shinmyo, & Sekine, 2006; Ngo, Moore, Baskar, Grossniklaus, & Sundaresan, 2007; Xu, Ren, Song, & Liu, 2015). Up to that stage, the young embryo is fed via the suspensor, importing nutrients from the seed coat (reviewed in Lafon-Placette & Köhler, 2014). But in heart-stage embryos, the endosperm becomes the main source of carbon and nitrogen for embryonic organogenesis. Endosperm cellularization coincides with lipid catabolism and sucrose remobilization, which is linked to sucrose and hexose transport to the embryo by the SWEET11 transporter (Baud et al., 2005; Chen et al., 2015; Kondou et al., 2008). In addition, nitrogen supply to the embryo is secured by amino acid (AA) transporters, notably USUALLY MULTIPLE ACIDS MOVE IN AN OUT TRANSPORTER25 (UMAMIT25) exporting AAs from the endosperm and AMINO ACID PERMEASE1 (AAP1) at the embryonic surface, importing AAs toward the embryo (reviewed in Dinkeloo, Boyd, & Pilot, 2018; Karmann, Müller, & Hammes, 2018). The endosperm also supplies biotin (vitamin B7) to the embryo via SUCROSE TRANSPORTER5 (SUC5), which is essential for fatty acid metabolism in the embryo (Pommerrenig et al., 2013). In legumes, where provisioning also switches from hexose to sucrose during organogenesis and cotyledons are important for starch storage, the embryo epidermis develops a specialized nutrient transfer layer (Borisjuk, Wang, Rolletschek, Wobus, & Weber, 2002).

A further role of the endosperm is to control germination, a process enabling adaption to seasonal changes of the environment, which is thought to have greatly contributed to the successful radiation of flowering plants (Finch-Savage & Footitt, 2017). Germination of the embryo is suppressed by ABA, a phytohormone produced in the endosperm. While the ABC TRANSPORTERS ABCG25 and ABCG31 export ABA to the apoplastic compartment, ABCG30 is an active importer, translocating this phytohormone to the embryo (Kang et al., 2015). ABA levels are modulated by transcription factors of the DELLA family, integrating light and temperature signals as well as GA levels to release the germination block and induce endosperm and testa rupture (reviewed in Chahtane, Kim, & Lopez-Molina, 2017).

The tissues surrounding the embryo are largely maternally dominated in both gymnosperms and angiosperms. This provides an effective maternal control to provisioning and embryo germination. The evolutionary acquisition of a paternal contribution in the endosperm of flowering plants, however, has opened the possibility for parental conflicts to develop over resource allocation and the developmental control of the progeny. This situation is manifested by the evolution of epigenetic mechanisms such as genomic imprinting, which results in the sole maternal or paternal expression of specific loci involved in the above-mentioned processes.

6. Genomic imprinting—The seed as a battle ground of parental conflicts

Genomic imprinting is an epigenetic phenomenon leading to the differential expression of maternal and paternal alleles. Imprinting is thought to have evolved with the emergence of reproductive modes relying on maternal provisioning in higher plants and mammals (Haig & Westoby, 1989; Pires & Grossniklaus, 2014). In flowering plants where it has been well studied (*Arabidopsis* spp., *Capsella* spp., *Ricinus communis* (castor bean), *Solanum lycopersicum* (tomato), *Sorghum bicolor* (sorghum), rice, and maize), imprinting is prevalent in the endosperm (reviewed in Gehring & Satyaki, 2017) but it also occurs in young embryos of maize, rice, and *A. thaliana* (Chen et al., 2018; Jahnke & Scholten, 2009; Luo et al., 2011; Raissig, Bemer, Baroux, & Grossniklaus, 2013; Scholten, 2010). Several comprehensive reviews on the mechanisms of genomic imprinting have been published (Grossniklaus, 2005; Köhler, Wolff, & Spillane, 2012; Ohnishi, Sekine, & Kinoshita, 2014; Raissig, Baroux, & Grossniklaus, 2011; Rodrigues & Zilberman, 2015; Satyaki & Gehring, 2017) and only selected aspects and examples are discussed here.

6.1 Imprinting mechanisms in the endosperm

Early studies aiming at explaining the maternal expression bias at numerous loci in the endosperm identified DNA methylation as a major epigenetic modification that differentiates maternal and paternal genomes (reviewed in Messing & Grossniklaus, 1999). Differential DNA methylation was then measured for a number of loci with robust, but low-resolution molecular methods (Lauria et al., 2004). The advent of high-throughput profiling of transcriptomes and methylomes in a tissue- and stage-specific manner has provided a copious list of imprinted candidate genes with allele-specific expression that often correlates with differential DNA methylation. The central question remains, however, whether DNA methylation constitutes a primary imprint inherited by the gametes or represents an epigenetic mark for maintenance during endosperm development. Given the difficulty to assess the gametic methylomes, the question has proven arduous to answer for many years and most models are inferred from data on DNA methylation in endosperm harvested several days after fertilization.

Current models favor DNA methylation as a primary imprint for many, but not all, maternally expressed genes (MEGs; reviewed in Gehring & Satyaki, 2017; Raissig et al., 2011; Rodrigues & Zilberman, 2015; Fig. 3). For some MEG loci, for instance *FIS2*, cytosine methylation is established in the promoter region prior to gametogenesis in both sexes but is specifically removed by the DNA glycosylase DEMETER (DME) (Choi et al., 2002; Gehring et al., 2006) in the central cell prior to fertilization (Jullien, Katz, Oliva, Ohad, & Berger, 2006). This results in paternally methylated but

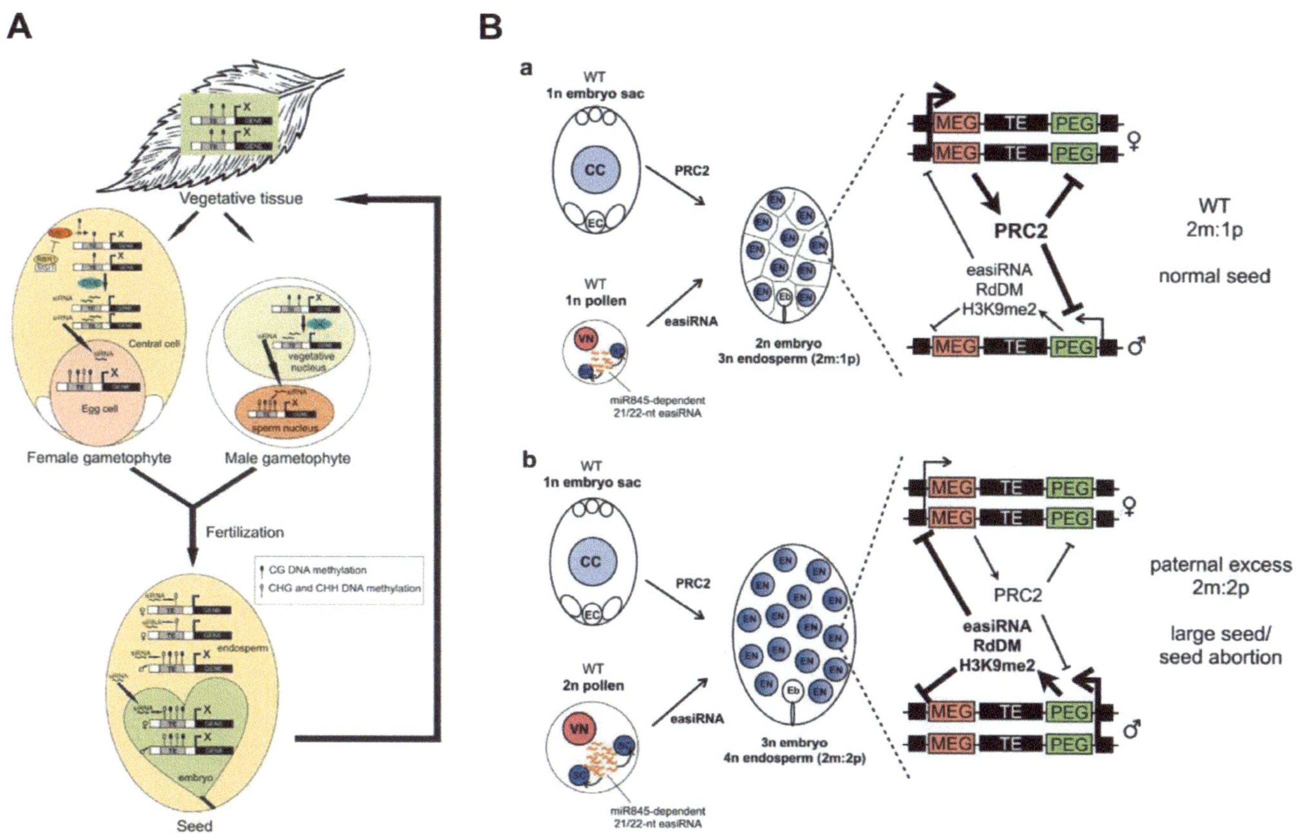

Fig. 3 See legend on next page.

Fig. 3 Genomic imprinting in the endosperm and role of imprinting in triploid block. (A) Reprogramming of epigenetic marks in gametes (Köhler & Weinhofer-Molisch, 2010). During female gametogenesis, expression of the DNA methyltransferase *MET1* is repressed by the Retinoblastoma pathway involving RETINOBLASTOMA-RELATED1 (RBR1) and its interacting partner MSI1, causing passively reduced DNA methylation. In the mature central cell, the DNA glycosylase DEMETER (DME) is expressed and actively removes DNA methylation from maternal alleles. *DME* is not expressed in egg or sperm cells; therefore, maternal and paternal alleles remain methylated. In the embryo, an initial demethylation of maternal alleles can occur by an as yet unknown mechanism (not shown in model); however, maternal alleles are then remethylated and silenced in the embryo. In the vegetative nucleus of the male gametophyte, lack of the chromatin remodeling factor DEFICIENT IN DNA METHYLATION1 (DDM1) causes a reduction in DNA methylation levels and transposon reactivation that is accompanied by production of small interfering RNAs (siRNAs). These siRNAs are possibly transported into sperm cells to induce asymmetric CHG and CHH (H is A, C or T) DNA methylation preventing transposon reactivation. Similarly, in the central cell and in the endosperm, the widespread loss of DNA methylation is likely to cause genome-wide transcriptional reactivation of transposons and repeat sequences, resulting in massive production of siRNAs that possibly trigger asymmetric DNA methylation in egg cell and embryo. However, transport of siRNAs from accessory cells to the gametes awaits experimental demonstration. (B) A model for genome dosage responses in the endosperm of *Arabidopsis thaliana* (Borges et al., 2018). Biogenesis of 21–22-nt easiRNA is triggered by miR845, resulting from the activation of retrotransposons during meiosis and in the vegetative nucleus (VN). easiRNAs accumulate in the sperm cells (SC) that are transported to the embryo sac to perform double fertilization of the haploid egg (EC) and diploid central cell (CC), giving rise to the embryo (Eb) and endosperm (EN), respectively. (a) In wild-type endosperm, maternally expressed genes (MEGs), such as *MEA* and *FIS2*, encode components of the *Polycomb* Repressive Complex 2 (PRC2) that regulates expression of paternally expressed genes (PEGs) via deposition of H3K27me3. (b) In interploid hybrids with paternal excess, the endosperm over-proliferates and fails to cellularize, leading to seed abortion ("triploid block"). The phenotype resulting from this interploidy hybridization barrier resembles that of maternal PRC2 mutants, where PEGs are also over-expressed and the endosperm over-proliferates leading to seed abortion phenotypes. Similar to mutations in PRC2 components, the triploid block can be suppressed by mutations in particular PEGs. miR845b-dependent easiRNAs are an important component of this pathway, perhaps by targeting TEs that flank imprinted genes. For example, increased paternal dosage and easiRNA activity, RNA-directed DNA methylation (RdDM) and H3K9me, may contribute to down-regulation of MEGs and PRC2 activity, which in turn leads to up-regulation of PEGs and seed abortion. *Panel (A): Figure and modified legend after Köhler, C., & Weinhofer-Molisch, I. (2010). Mechanisms and evolution of genomic imprinting in plants.* Heredity, 105, *57–63. Panel (B): Figure and modified legend after Borges, F., Parent, J.-S., van Ex, F., Wolff, P., Martínez, G., Köhler, C., & Martienssen, R. A. (2018). Transposon-derived small RNAs triggered by miR845 mediate genome dosage response in Arabidopsis.* Nature Genetics, 50, *186–192.*

maternally demethylated, and thus expressed, alleles in the endosperm (Choi et al., 2002; Gehring et al., 2006). The same model was proposed for the imprinted loci *MEA* and *FWA* (Gehring et al., 2006; Jullien, Katz, et al., 2006, Jullien, Kinoshita, Ohad, & Berger, 2006; Kinoshita et al., 2004; Vielle-Calzada et al., 1999; Xiao et al., 2003), the maternal alleles of which are indeed demethylated in the endosperm several days after fertilization. However, a detailed analysis of the imprinting control region of *MEA* showed that a short promoter fragment devoid of any DNA methylation can confer imprinted expression (Wöhrmann et al., 2012). Models that involve higher order chromatin structure, e.g., chromatin loops, which are regulated by DNA methylation and work in combination with yet unknown factors setting the primary imprints reconcile these results (Grossniklaus, 2015; Wöhrmann et al., 2012). Moreover, analysis of DNA methylation of the *FWA* promoter in the central cell showed that it was fully methylated, indicating that demethylation occurs after fertilization and that the primary imprint is not DNA methylation (Wöhrmann et al., 2012). Indeed, several other MEG loci are not marked by DNA methylation in the gametes and establish differential methylation only after fertilization (e.g., *maternally expressed in embryo1* (*mee1*) and *Fie2* in maize; Gutierrez-Marcos et al., 2006; Jahnke & Scholten, 2009) or are found differentially methylated on H3K27 in the endosperm (e.g., *VARIANT IN METHYLATION 5* (*VIM5*) in *A. thaliana*; Hsieh et al., 2011). The quest for alternative imprints is still ongoing but histone methylation, particularly H3K27me3, is an attractive candidate. Genome-wide profiling showed that H3K27me3 is preferentially enriched at DME target loci in the endosperm and at some pericentromeric paternal alleles where, however, it has no influence on the imprinting status of MEGs (Moreno-Romero, Jiang, Santos-Gonzalez, & Köhler, 2016). In addition to maintaining imprinting at DME target loci, H3K27me3 may also serve as primary imprint for unmethylated MEGs in *A. thaliana* but this remains to be demonstrated. Yet, the function of DNA methylation is ambiguous as it is not always associated with transcriptional silencing of the imprinted allele. Indeed, paternal expression of the rice *OsYUCCA11* gene requires gene body methylation while silencing of the maternal allele is associated with absence of DNA methylation but presence of H3K27me3 (Du, Luo, Zhang, Finnegan, & Koltunow, 2014). This is reminiscent of imprinting regulation at the *PHERES1* (*PHE1*) locus in *A. thaliana* (Köhler et al., 2003; Köhler, Page, Gagliardini, & Grossniklaus, 2005; Makarevich, Villar, Erilova, & Köhler, 2008). In addition, the discovery that allele-specific expression is largely controlled by the RNA-dependent DNA methylation (RdDM) machinery in maize (Erdmann, Satyaki, Klosinska, & Gehring, 2017) suggests that small RNAs may be involved in reinforcing MEG imprinting. However, how these siRNA would act in *cis* and not in *trans* on the other allele remains to be elucidated. The answer may lay in the peculiar cytological organization and histone composition of the endosperm's chromatin, which is largely inherited from the central cell (Baroux, Pien, & Grossniklaus, 2007; Erdmann et al., 2017) and could provide structural requirements protecting maternal loci from siRNA targeting. In support of this hypothesis, imprinted loci are associated with differences in nucleosome coverage and distribution in maize (Dong et al., 2018). Thus, while some common thread in imprinting control arose from the analysis of individual model loci such as *MEA*, *FIS2* and *FWA*, genome-wide studies brought up many questions and suggest diverse and complex regulatory mechanisms of genomic imprinting. In addition, transposable elements (TEs) seem to make an important contribution to the imprinting epigenetic landscape (Anderson & Springer, 2018). Possibly, *de novo* insertion of TEs may have generated an epigenetically sensitized region hijacked by nearby genes that became differentially marked by virtue of having distinct epigenetic machineries in male and female gametes (Anderson & Springer, 2018). Consistent with such a scenario, TEs seem to have been co-opted as an siRNA source inherited by one parent to regulate imprinted loci in the endosperm (Fig. 3) (Anderson & Springer, 2018). The involvement of TEs in imprinting mechanisms is very exciting as they are expected to be powerful modifiers of the imprinting landscape and, hence, in fostering evolutionary diversity.

6.2 Are imprinted genes conserved across species?

Some imprinted loci and particularly imprinted genes in similar functional categories are shared between dicots and monocots as revealed by comparative profiling studies in *Arabidopsis* spp., *Capsella* spp., tomato, castor bean, sorghum, rice, and maize (reviewed in Chen et al., 2018; Florez-Rueda et al., 2016; Gehring & Satyaki, 2017). This finding suggests evolutionary conservation for key cellular functions, such as auxin biosynthesis, RdDM, and the control of DNA and histone methylation (see below). Nevertheless, species-specific imprinting is observed (Pires & Grossniklaus, 2014) and would deserve further attention: this approach may reveal specializations pertinent to the different modes of seed development in monocots and dicots, and notably the distinct inter-relationships between the three seed compartments. In addition, defining common sets of imprinted genes within and between species is hampered by variable technical and experimental approaches and strongly differing analytical pipelines for which, however, improvements have been proposed (Wyder, Raissig, & Grossniklaus, 2018). A thorough comparative approach between closely related Brassicaceae concluded that only 14% and 29% of imprinted MEGs and PEGs, respectively, are conserved between *A. thaliana* and *Capsella rubella*, a situation that may be attributed in part to differences in mating systems but also to the different distribution of TEs that are notably associated with PEGs in *A. thaliana* (Hatorangan, Laenen, Steige, Slotte, & Köhler, 2016).

6.3 Function of imprinted genes in the endosperm

Imprinted loci are enriched in genes coding for chromatin modifiers that promote DNA and histone methylation (PCR2 components and members of the RdDM pathway in *A. thaliana*, maize, and rice), MADS-box genes (such as *PHE1*), and auxin biosynthesis genes (such as the endosperm-specific *YUCCAs*, reviewed in Gehring & Satyaki, 2017). The fact that some imprinting regulators are themselves imprinted may support the notion that the endosperm is a hotspot of parental conflicts, which contributed to the evolution of genomic imprinting. The function of the vast majority of imprinted genes or candidate genes has not yet been elucidated. Thorough functional studies are necessary to disentangle relevant imprinted genes from inconsequential side effects of differential epigenetic landscapes inherited by the endosperm (Berger, Vu, Li, & Chen, 2012). Nevertheless, functional analyses of a few imprinted genes, including those whose discovery was instrumental to developing the concept of imprinting and its underlying mechanisms in plants, illuminate their role in seed development (reviewed in Gehring & Satyaki, 2017; Raissig et al., 2011). For instance, the imprinted maternally expressed *MEA*locus in *A. thaliana* restricts endosperm and embryo proliferation (Grossniklaus et al., 1998) and the imprinted *maternally expressed gene1 (meg1)* locus in maize controls BETL development and, hence, seed provisioning (Costa et al., 2012). Imprinting in the endosperm may influence more than nutritive functions: the *CYSTEINE PROTEINASE1* (*CP1*) and the *ALLANTOINASE* (*ALN*) loci are MEGs controlling dormancy. Loss of their preferentially maternal expression in the endosperm releases dormancy during imbibition (Piskurewicz et al., 2016).

6.4 Imprinting in the control of interploidy and interspecific barriers

Endosperm abortion is a common cause of hybrid seed failure upon certain interspecific or interploidy crosses (Schatlowski & Köhler, 2012). Studies on the underlying molecular mechanisms were long disconnected from imprinting. However, a decade ago, the observation that such hybridization failures were linked with parental genome dosage (Dilkes & Comai, 2004) and the emerging view that genomic imprinting widely affected the dosage of parental products converged. Key to the emergence of this concept was the identification of the *jason* (*jas*) mutant in *A. thaliana*, which leads to the formation of unreduced pollen, thus changing the dosage of the paternal genome, including that of the imprinted PEG *PHE1*. When *jas* pollen is used in crosses with wild-type diploid plants viable "triploid" seeds, usually lethal in other accessions, are produced. The triploid 1m:2p embryo and the tetraploid 2m:2p endosperm are mostly viable, allowing the investigation of unbalanced parental contributions to the seeds. The paternal inheritance of the *jas* mutation leads to upregulation of PRC2 target genes, such as *PHE1*, in the seed reminiscent of seeds lacking maternal PRC2 function as, for instance, in the *mea* mutant (Erilova et al., 2009). Based on these findings, the idea that the imprinted *MEA* gene serves as a "ploidy sensor" in the *A. thaliana* endosperm was proposed (Erilova et al., 2009). In this scenario, a ploidy sensor consists of a macromolecular complex whose activity requires a stoichiometric balance of components contributed by each parent (e.g., see Veitia, Bottani, & Birchler, 2008). A stoichiometric imbalance occurs in interploidy crosses but also in interspecific crosses where the dosage of parental genomes is no longer appropriate (Burkart-Waco, Ngo, Lieberman, & Comai, 2015; Dilkes & Comai, 2004). Endosperm failure in interploidy and interspecific crosses share delayed cellularization of the endosperm which then fails to undergo the physiological transition to provisioning the embryo (Schatlowski & Köhler, 2012; von Wangenheim & Peterson, 2004), leading to seed abortion as observed in loss-of-function mutants affecting PRC2 components (Hehenberger et al., 2012). Changes in the relative parental contributions to gene expression is also associated with interspecific seed failure in tomato (Florez-Rueda et al., 2016). In rice, manipulating dosage in one of the parental species allowed the recovery of a cellular hybrid endosperm with restored expression levels of the key imprinted genes *MADS87* and *YUCCA11* (Tonosaki et al., 2018). The endosperm thus provides dosage-sensitive control on seed development, exerted in part by genomic imprinting as well as small RNAs (see Fig. 3B and legend), that plays a central function in interspecific hybridization barriers, thereby influencing plant evolution (Schatlowski & Köhler, 2012).

7. Signaling interplay between seed compartments

The seed is a composite structure comprising three genetically and functionally distinct compartments, which develop in a coordinated manner. Recent studies have revealed an exquisite communication between these compartments that not only contributes developmental coordination but also establishes a tripartite regulatory circuit. Because of their interplay, it is difficult to disentangle effects of nutrient allocation (from the seed coat to the endosperm and from the endosperm to the embryo) from inter-compartment signaling. Here, we will focus mainly on hormonal and peptide-based signaling (see Fig. 4).

7.1 Interplay between the endosperm and the embryo

Recent mutant analyses have uncovered reciprocal effects between the endosperm and the embryo during seed development. The development of the endosperm and embryo starts at fertilization. To a certain extent, both fertilization products can develop autonomously for the first few cell cycles, as shown by *A. thaliana* mutants producing embryos that develop up to the globular stage in the absence of any endosperm (e.g., Iwakawa et al., 2006) and embryoless seeds (or seeds with embryos arrested at the zygote stage) with a proliferative endosperm (Ronceret et al., 2008). A signaling-based dialog between both fertilization products is already established prior to fertilization. The central cell produces a set of three CLAVATA3/EMRBYO SURROUNDING REGION-RELATED (CLE) peptides that will later promote the development of the embryo's suspensor (Costa et al., 2014). If ESR1 peptides are not produced, embryo patterning is impaired (Costa et al., 2014) Malformed suspensors are likely unable to transport nutrients to the embryo and to regulate cell fate in the basal region of the embryo (Kawashima & Goldberg, 2010). The maternally produced ESR1 induces signaling through the paternally expressed YODA MAP kinase kinase kinase, featuring an unsuspected and intriguing case of inter-parental, intertissue dialog (Costa et al., 2014).

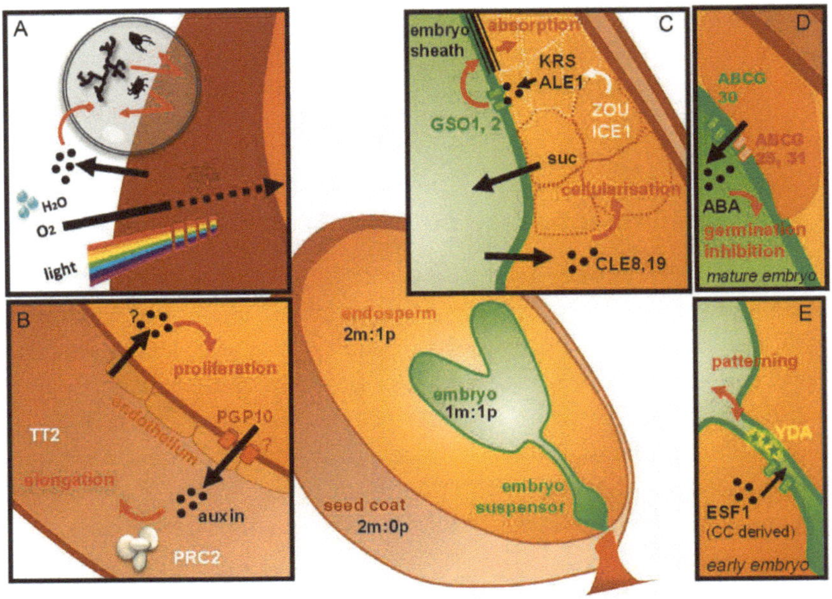

Fig. 4 Interplay between the seed compartments and the seed and its environment. The seed of angiosperms is a composite structure made of a seed coat (maternal integuments), encapsulating the embryo and the endosperm, an embryo-nourishing tissue. Embryo and endosperm are fertilization products with contributions from both maternal (m) and paternal (p) parents. They are genetically identical except for their ploidy and relative parental genome dosage (1m:1p and 2m:1p, respectively; see main text), while the seed coat is of maternal origin (2m:0p). (A) The seed coat is a major interface integrating environmental signals, such as moisture, light, and oxidation. Proanthocyanidins in the endothelium contribute to oxidative signaling influencing seed biology. The seed coat also secretes antifungal and antibacterial peptides and secondary metabolites (black circles), reinforcing physical protection exerted by the sclereids and mucilage of the outermost layer against pathogens (magnifying lens). (B-E) Selected examples of interactions between the seed compartments known in *Arabidopsis thaliana*. (B) Interactions between the seed coat and the endosperm. (C-E) Interactions between the embryo and endosperm during endosperm development (C), at the mature embryo stage (D), and at the early embryo stage (E). The interactions displayed involve peptide and hormone signaling. Signaling molecules are represented as black dots (CLE peptides; auxin and ABA phytohormones; suc, sucrose; ?, unknown signal), transporters as cylinders colored according to the tissue where they are expressed (green, embryo; orange, endosperm; brown, seed coat) and labeled if their identity is known. In some instances, transcriptional regulators (white lettering) and signal transduction factors (yellow lettering) have been identified that regulate signal production (e.g., in C, ZOU and ICE1 transcription factors), downstream signaling (e.g., in B, the IKU2 LRR receptor-like kinase, in E, the YODA MAP kinase kinase kinase), or downstream transcriptional regulation (e.g., in B, by PRC2 and the TT2 transcription factor). Black arrows indicate the transport/signaling axis at the core of the interaction. Red arrows indicate a process triggered by the signals or downstream components (B: seed coat elongation, endosperm proliferation; C: endosperm cellularization, endosperm absorption, D: germination, E: embryo patterning).

We have previously discussed that the endosperm, as nutrient supplier, exerts a profound influence on embryonic growth. Particularly, provisioning routes are redrawn at endosperm cellularization and this developmental transition is essential for embryo maturation. Interestingly, the embryo itself controls endosperm cellularization and the process involves small secreted peptides of the CLE family. CLEs are 12 AA long secreted peptides that are sensed by LRR-RLKs (Yamaguchi, Ishida, & Sawa, 2016). CLE/LRR-RLK complexes provide non-cell autonomous intercellular signaling modules, which are reused throughout plant development and in environmental responses (Yamaguchi et al., 2016). The CLE8 and CLE19 peptides are specifically produced in the embryo and control endosperm differentiation (Fiume & Fletcher, 2012; Xu et al., 2015). In their absence, the endosperm fails to complete cellularization, and the physiological transition in provisioning that is necessary to form a mature embryo.

At a later stage, when the seed of eudicotsenters the maturation phase, the endosperm is absorbed by the embryo. Endosperm breakdown is cooperatively controlled by the embryo and the endosperm. Central to this process is KERBEROS (KRS), a signaling peptide belonging to the cysteine-rich peptide (CRP) family. KRS is specifically produced in the endosperm under the control of a heterodimer between the basic Helix-Loop-Helix (bHLH) nuclear factors ZHOUPI (ZOU) and INDUCER OF CBP1 EXPRESSION1 (ICE1). KRS is perceived by the embryonic transmembrane RLKs GASSHO1 (GSO1) and GSO2, which promote the formation of a cuticle surrounding the embryo. The embryo sheath is composed of extensin-rich material derived from the endosperm and enables physical separation prior to endosperm breakdown (Moussu et al., 2017; Yang et al., 2008). The conservation of ZOU in monocots and gymnosperms suggests that the ZOU/KRS/GSO signaling module is an evolutionary conserved module (Dou, Zhang, Yang, & Feng, 2018; Yang et al., 2008). CRPs are commonly found in eudicots and monocots, and future research will provide additional insights into the combinatorial association and function of CRP signal/receptor modules in development and environmental responses (Marshall, Costa, & Gutierrez-Marcos, 2011). In the maize kernel, the endosperm-specific CRP MEG1 regulates the differentiation of the important BETL tissue involved in nutrient provisioning. Additional CRPs belonging to the MEG, BETL1–4, and BAP families were identified and may also contribute to signaling during kernel development. Many CRPs show antifungal activity, but the functional relevance of most of these CRPs for seed development awaits demonstration (Doll, Depege-Fargeix, Rogowsky, & Widiez, 2017).

In addition to peptide-based signaling, hormonal cross talk takes place between the endosperm and the embryo. Recent work has highlighted the importance of auxin signaling in the endosperm for embryo patterning in both maize and *A. thaliana*, suggesting an evolutionary conserved process. Auxin fluxes are contributed both by transport at the epidermis of the embryo and through the suspensor (reviewed in Locascio, Roig-Villanova, Bernardi, & Varotto, 2014). However, an interplay with local auxin production in the embryo is to be expected (Moller & Weijers, 2009). This situation complicates the genetic and physiological dissection of inter-compartment signaling in embryo patterning.

Unraveling the specific and reciprocal contributions of the embryo and endosperm on each other's development is challenged by the genetic relatedness, apart from ploidy, between both fertilization products. Heterofertilization where two genetically different sperm cells fuse with the egg and the central cell, respectively, may provide an avenue for genetic dissection of the embryo and endosperm contribution. This was recently achieved using an *A. thaliana* mutant that produces a single sperm cell, followed by a second pollination with wild-type pollen (Maruyama et al., 2013). In maize, similiar studies based on the formation of kernels with embyos and endosperm of distinct genotypes, can be performed using B-A translocation stocks or spontaneous heterofertilization (Grossniklaus, 2017; Neuffer & Sheridan, 1980).

7.2 Signaling interplay between the seed coat and the endosperm

We have previously discussed that seed coat differentiation is initiated at fertilization (Roszak & Köhler, 2011). This developmental coordination is accomplished by a signaling process between the central cell and the integuments in which the phytohormone auxin again plays a central role. Produced by the central cell and later by the developing endosperm, auxin is rapidly exported toward the seed coat, where it acts in concert with PRC2 derepression to promote cell elongation. In this cross talk, the endothelium plays a central role. Notably P-GLYCOPROTEIN10 (PGP10) is thought to assist in auxin transport (Figueiredo et al., 2016). Similarly, in maize, auxin is produced in the endosperm, and defective biosynthesis or signaling leads to small, unviable kernels. Although it is difficult to disentangle effects of auxin and sugar metabolism on the development of the different compartments, auxin likely also regulates the expansion of maternal tissues in the maize kernel (Doll et al., 2017).

Genetic screens for mutations affecting seed size have revealed a handful of factors controlling the coordinated development of endosperm and seed coat in a non-autonomous manner. The transcription factor MIN3, for instance, modulates cell division in the endosperm and promotes cell elongation in the seed coat (Garcia et al., 2003). However, *MIN3* is expressed only in the endosperm and embryo, and thus acts non-cell autonomously on the seed integuments. Reciprocally, the endothelium expressed transcription factor TT2, which promotes seed coat cell elongation, also promotes endosperm growth (Garcia, Fitz Gerald, & Berger, 2005). MIN3 and TT2 control transcriptional programs that likely involve the production of mobile signals of yet unknown nature, which fine-tune the coordinated development of both compartments.

These examples illustrate the complex coordination of seed compartment development involving molecular signaling. Mobile signals may be hormones, small peptides (CRPs or CLEs), but also small mobile RNAs, for which new roles in plant development are continuously being uncovered (D'Ario, Griffiths-Jones, & Kim, 2017).

8. Conclusions and outlook

The seed is a highly successful evolutionary innovation. It provides a protective and nourishing environment for the next generation, promoting its survival and dispersal. In this chapter, we reviewed selected aspects of the molecular mechanisms and regulatory control circuits controlling seed formation. Over the past decade, research has highlighted intricate interactions between the seed compartments that allow the fine-tuning of seed formation. In angiosperms, endosperm and embryo deploy an astonishing diversity of communication modes, involving phytohormones, signaling peptides, small RNAs, metabolic signaling, and more at distinct developmental stages. This interplay between the two fertilization products provides developmental checkpoints that ensure the success of seed development and are affected in interspecific hybridization, which has shaped angiosperm evolution. Refining our understanding of the dialog between embryo and endosperm will also provide answers to long-held problems in breeding strategies impaired by interspecific crossing barriers. Similarly, unraveling the reciprocal interactions between the maternal seed coat and the two fertilization products influence seed size and quality, which are of great agronomic interest.

The judicious choice of model plants amenable to molecular genetic approaches, genetic screens, and high-throughput genome-wide profiling approaches has clearly fueled our knowledge on the molecular mechanisms controlling seed development. It is time, however, to develop approaches capturing the extraordinary evolutionary diversity of seed biology among spermatophytes. This requires the identification of new model organisms positioned at key branching points of the phylogenetic tree of seed plants, and translating knowledge of target genes and pathways to these species. Future progress will also require the investigation of species displaying unique trajectories of seed development in unbiased ways, e.g., using forward genetic approaches, to shed light onto evolutionary novelties. The recent advances in sequencing techniques and methods for functional studies using targeted mutation and genome editing, open the way for studies in non-model systems in order to harness the astounding diversity in seed formation. Comprehensive analyses at the genomic, transcriptomic, and epigenomic level will allow the investigation of diversity within and among species, which will lead to new insights into seed development. Looking at the molecular consequences of unbalanced parental contributions in seeds will provide insights that may allow plant breeders to overcome crossing barriers, giving them access to novel genepools. Future efforts should be devoted to identify key genes that have influenced the evolution of seed structures and their developmental control. Interesting questions, among others, are the evolutionary transition from a maternal to a biparental embryo-nourishing tissue. How and when in evolution signaling modules serving inter-compartment interactions were recruited to the seed of angiosperms is another exciting question. It can now be addressed using gene editing techniques that will accelerate functional evo-devo analyses, which will shed light onto the evolution of seeds.

Acknowledgments

We apologize to our colleagues whose work was not cited in the review due to space constraints. We thank Hanspeter Schöb and Nina Chumak for corrections and help with the bibliography. Our work in the area of plant reproduction is supported by the University of Zürich and grants of the Swiss National Science Foundation and SystemsX.ch, the Swiss Initiative for Systems Biology.

References

Anderson, S. N., & Springer, N. M. (2018). Potential roles for transposable elements in creating imprinted expression. *Current Opinion in Genetics & Development*, *49*, 8–14.

Armenta-Medina, A., & Gillmor, C. S. (2019). Genetic, molecular and parent-of-origin regulation of early embryogenesis in flowering plants. In U. Grossniklaus (Ed.), *Current topics in developmental biology, Vol. 131*. Plant development and evolution (pp. 497–543).

Baroux, C., Fransz, P., & Grossniklaus, U. (2004). Nuclear fusions contribute to polyploidization of the gigantic nuclei in the chalazal endosperm of *Arabidopsis*. *Planta*, *220*, 38–46.

Baroux, C., Pien, S., & Grossniklaus, U. (2007). Chromatin modification and remodeling during early seed development. *Current Opinion in Genetics & Development*, *17*, 473–479.

Baroux, C., Spillane, C., & Grossniklaus, U. (2002). Evolutionary origins of the endosperm in flowering plants. *Genome Biology*, *3*. reviews1026.1–reviews1026.5.

Baud, S., Wuilleme, S., Lemoine, R., Kronenberger, J., Caboche, M., Lepiniec, L., et al. (2005). The *At*SUC5 sucrose transporter specifically expressed in the endosperm is involved in early seed development in *Arabidopsis*. *The Plant Journal*, *43*, 824–836.

Bemer, M., & Grossniklaus, U. (2012). Dynamic regulation of *Polycomb* group activity during plant development. *Current Opinion in Plant Biology*, *15*, 523–529.

Berger, F., Vu, T. M., Li, J., & Chen, B. (2012). Hypothesis: Selection of imprinted genes is driven by silencing deleterious gene activity in somatic tissues. *Cold Spring Harbor Symposia on Quantitative Biology*, *77*, 23–29.

Boisnard-Lorig, C., Colon-Carmona, A., Bauch, M., Hodge, S., Doerner, P., Bancharel, E., et al. (2001). Dynamic analyses of the expression of the HISTONE::YFP fusion protein in *Arabidopsis* show that syncytial endosperm is divided in mitotic domains. *The Plant Cell*, *13*, 495–509.

Borges, F., Parent, J.-S., van Ex, F., Wolff, P., Martínez, G., Köhler, C., et al. (2018). Transposon-derived small RNAs triggered by *miR845* mediate genome dosage response in *Arabidopsis*. *Nature Genetics*, *50*, 186–192.

Borisjuk, L., Wang, T. L., Rolletschek, H., Wobus, U., & Weber, H. (2002). A pea seed mutant affected in the differentiation of the embryonic epidermis is impaired in embryo growth and seed maturation. *Development*, *129*, 1595–1607.

Burkart-Waco, D., Ngo, K., Lieberman, M., & Comai, L. (2015). Perturbation of parentally biased gene expression during interspecific hybridization. *PLoS One*, *10*, e0117293.

Cailleau, A., Cheptou, P. O., & Lenormand, T. (2010). Ploidy and the evolution of endosperm of flowering plants. *Genetics*, *184*, 439–453.

Carmichael, J. S., & Friedman, W. E. (1995). Double fertilization in *Gnetum gnemon*: The relationship between the cell cycle and sexual reproduction. *The Plant Cell*, 7(12), 1975–1988.

Chahtane, H., Kim, W., & Lopez-Molina, L. (2017). Primary seed dormancy: A temporally multilayered riddle waiting to be unlocked. *Journal of Experimental Botany*, *68*, 857–869.

Chandler, J. W., & Werr, W. (2019). Histology versus phylogeny: Viewing plant embryogenesis from an evo-devo perspective. In U. Grossniklaus (Ed.), *Current topics in developmental biology, Vol. 131*. Plant development and evolution (pp. 545–564). Elsevier.

Chen, C., Li, T., Zhu, S., Liu, Z., Shi, Z., Zheng, X., et al. (2018). Characterization of imprinted genes in rice reveals conservation of regulation and imprinting with other plant species. *Plant Physiology*, *177*, 1754–1771.

Chen, L. Q., Lin, I. W., Qu, X. Q., Sosso, D., McFarlane, H. E., Londono, A., et al. (2015). A cascade of sequentially expressed sucrose transporters in the seed coat and endosperm provides nutrition for the *Arabidopsis* embryo. *The Plant Cell*, *27*, 607–619.

Choi, Y., Gehring, M., Johnson, L., Hannon, M., Harada, J. J., Goldberg, R. B., et al. (2002). DEMETER, a DNA glycosylase domain protein, is required for endosperm gene imprinting and seed viability in arabidopsis. *Cell*, *110*, 33–42.

Cocucci, A. E. (2005). Morphogenetic seed types of spermatophyta. *Plant Systematics and Evolution*, *250*, 1–6.

Coen, O., & Magnani, E. (2018). Seed coat thickness in the evolution of angiosperms. *Cellular and Molecular Life Sciences*, *75*, 2509–2518.

Cooke, J., Cooke, B., & Gifford, D. (2002). Loblolly pine seed dormancy: Constraints to germination. *New Forests*, *23*, 239–256.

Costa, L. M., Marshall, E., Tesfaye, M., Silverstein, K. A., Mori, M., Umetsu, Y., et al. (2014). Central cell-derived peptides regulate early embryo patterning in flowering plants. *Science, 344*, 168–172.

Costa, L. M., Yuan, J., Rouster, J., Paul, W., Dickinson, H., & Gutierrez-Marcos, J. F. (2012). Maternal control of nutrient allocation in plant seeds by genomic imprinting. *Current Biology, 22*, 160–165.

Creff, A., Brocard, L., & Ingram, G. (2015). A mechanically sensitive cell layer regulates the physical properties of the *Arabidopsis* seed coat. *Nature Communications, 6*, 6382.

Dante, R. A., Larkins, B. A., & Sabelli, P. A. (2014). Cell cycle control and seed development. *Frontiers in Plant Science, 5*, 493.

D'Ario, M., Griffiths-Jones, S., & Kim, M. (2017). Small RNAs: Big impact on plant development. *Trends in Plant Science, 22*, 1056–1068.

De Smet, I., Lau, S., Mayer, U., & Jürgens, G. (2010). Embryogenesis—The humble beginnings of plant life. *The Plant Journal, 61*, 959–970.

Dilkes, B. P., & Comai, L. (2004). A differential dosage hypothesis for parental effects in seed development. *The Plant Cell, 16*, 3174–3180.

Dinkeloo, K., Boyd, S., & Pilot, G. (2018). Update on amino acid transporter functions and on possible amino acid sensing mechanisms in plants. *Seminars in Cell & Developmental Biology, 74*, 105–113.

Doll, N. M., Depege-Fargeix, N., Rogowsky, P. M., & Widiez, T. (2017). Signaling in early maize kernel development. *Molecular Plant, 10*, 375–388.

Dong, X., Chen, J., Li, T., Li, E., Zhang, X., Zhang, M., et al. (2018). Parent-of-origin-dependent nucleosome organization correlates with genomic imprinting in maize. *Genome Research, 28*, 1020–1028.

Dou, M., Zhang, Y., Yang, S., & Feng, X. (2018). Identification of *ZHOUPI* orthologs in rice involved in endosperm development and cuticle formation. *Frontiers in Plant Science, 9*, 223.

Du, M., Luo, M., Zhang, R., Finnegan, E. J., & Koltunow, A. M. (2014). Imprinting in rice: The role of DNA and histone methylation in modulating parent-of-origin specific expression and determining transcript start sites. *The Plant Journal, 79*, 232–242.

Ehlers, K., Bhide, A. S., Tekleyohans, D. G., Wittkop, B., Snowdon, R. J., & Becker, A. (2016). The MADS box genes *ABS, SHP1*, and *SHP2* are essential for the coordination of cell divisions in ovule and seed coat development and for endosperm formation in *Arabidopsis thaliana*. *PLoS One, 11*, e0165075.

El-Maarouf-Bouteau, H., & Bailly, C. (2008). Oxidative signaling in seed germination and dormancy. *Plant Signaling & Behavior, 3*, 175–182.

Erdmann, R. M., Satyaki, P. R. V., Klosinska, M., & Gehring, M. (2017). A small RNA pathway mediates allelic dosage in endosperm. *Cell Reports, 21*, 3364–3372.

Erilova, A., Brownfield, L., Exner, V., Rosa, M., Twell, D., Mittelsten Scheid, O., et al. (2009). Imprinting of the *Polycomb* group gene *MEDEA* serves as a ploidy sensor in *Arabidopsis*. *PLoS Genetics, 5*, e1000663.

Figueiredo, D. D., Batista, R. A., & Köhler, C. (2018). Auxin regulates endosperm cellularization in *Arabidopsis*. *bioRxiv*, 239301.

Figueiredo, D. D., Batista, R. A., Roszak, P. J., Hennig, L., & Köhler, C. (2016). Auxin production in the endosperm drives seed coat development in *Arabidopsis*. *eLife, 5*, e20542.

Figueiredo, D. D., & Köhler, C. (2018). Auxin: A molecular trigger of seed development. *Genes & Development, 32*, 479–490.

Finch-Savage, W. E., & Footitt, S. (2017). Seed dormancy cycling and the regulation of dormancy mechanisms to time germination in variable field environments. *Journal of Experimental Botany, 68*, 843–856.

Fiume, E., & Fletcher, J. C. (2012). Regulation of *Arabidopsis* embryo and endosperm development by the polypeptide signaling molecule CLE8. *The Plant Cell, 24*, 1000–1012.
Florez-Rueda, A. M., Paris, M., Schmidt, A., Widmer, A., Grossniklaus, U., & Stadler, T. (2016). Genomic imprinting in the endosperm is systematically perturbed in abortive hybrid tomato seeds. *Molecular Biology and Evolution, 33*, 2935–2946.
Floyd, S. K., & Friedman, W. E. (2000). Evolution of endosperm developmental patterns among basal flowering plants. *International Journal of Plant Sciences, 161*, S57–S81.
Friedman, W. E. (1992). Evidence of a pre-angiosperm origin of endosperm: Implications for the evolution of flowering plants. *Science, 255*, 336–339.
Friedman, W. E. (2006). Embryological evidence for developmental lability during early angiosperm evolution. *Nature, 441*, 337–340.
Garcia, D., Fitz Gerald, J. N., & Berger, F. (2005). Maternal control of integument cell elongation and zygotic control of endosperm growth are coordinated to determine seed size in *Arabidopsis*. *The Plant Cell, 17*, 52–60.
Garcia, D., Saingery, V., Chambrier, P., Mayer, U., Jürgens, G., & Berger, F. (2003). *Arabidopsis haiku* mutants reveal new controls of seed size by endosperm. *Plant Physiology, 131*, 1661–1670.
Gasser, C. S., & Skinner, D. J. (2019). Development and evolution of the unique ovules of flowering plants. In U. Grossniklaus (Ed.), *Current topics in developmental biology, Vol. 131*. Plant development and evolution (pp. 373–399). Elsevier.
Geeta, R. (2003). The origin and maintenance of nuclear endosperms: Viewing development through a phylogenetic lens. *Proceedings. Biological Sciences, 270*, 29–35.
Gehring, M., Huh, J. H., Hsieh, T.-F., Penterman, J., Choi, Y., Harada, J. J., et al. (2006). DEMETER DNA glycosylase establishes *MEDEA Polycomb* gene self-imprinting by allele-specific demethylation. *Cell, 24*, 495–506.
Gehring, M., & Satyaki, P. R. (2017). Endosperm and imprinting, inextricably linked. *Plant Physiology, 173*, 143–154.
Golz, J. F., Allen, P. J., Li, S. F., Parish, R. W., Jayawardana, N. U., Bacic, A., et al. (2018). Layers of regulation—Insights into the role of transcription factors controlling mucilage production in the *Arabidopsis* seed coat. *Plant Science, 272*, 179–192.
Grossniklaus, U. (2017). Polyspermy produces tri-parental seeds in maize. *Current Biology, 27*, R1300–R1302.
Grossniklaus, U. (2005). Genomic imprinting in plants: A predominantly maternal affair. In P. Meyer (Ed.), *Annual plant reviews: Plant epigenetics* (pp. 174–200). UK: Blackwell, Sheffield.
Grossniklaus, U. (2015). Molecular regulation of reproductive development. In B. B. Buchanan, W. Gruissem, & R. L. Jones (Eds.), *Biochemistry and molecular biology of plants* (2nd ed., pp. 872–924). USA: Wiley Blackwell.
Grossniklaus, U., & Paro, R. (2014). Transcriptional silencing by *Polycomb*-group proteins. *Cold Spring Harbor Perspectives in Biology, 6*, 019331.
Grossniklaus, U., Spillane, C., Page, D. R., & Köhler, C. (2001). Genomic imprinting and seed development: Endosperm formation with and without sex. *Current Opinion in Plant Biology, 4*, 21–27.
Grossniklaus, U., Vielle-Calzada, J. P., Hoeppner, M. A., & Gagliano, W. B. (1998). Maternal control of embryogenesis by *MEDEA*, a *Polycomb* group gene in *Arabidopsis*. *Science, 280*, 446–450.
Guo, C. C., Shen, Y. B., & Shi, F. H. (2018). Investigating seed dormancy in *Pinus bungeana* Zucc. ex Endl.: Understanding the contributions of enclosing tissues and temperature on germination. *Forests, 9*, 401.
Gusti, A., Baumberger, N., Nowack, M., Pusch, S., Eisler, H., Potuschak, T., et al. (2009). The *Arabidopsis thaliana* F-box protein FBL17 is essential for progression through the second mitosis during pollen development. *PLoS One, 4*, e4780.

Gutierrez-Marcos, J. F., Costa, L. M., Dal Pra, M., Scholten, S., Kranz, E., Perez, P., et al. (2006). Epigenetic asymmetry of imprinted genes in plant gametes. *Nature Genetics, 38*, 876–878.

Haig, D., & Westoby, M. (1989). Parent-specific gene-expression and the triploid endosperm. *The American Naturalist, 134*, 147–155.

Hands, P., Rabiger, D. S., & Koltunow, A. M. (2016). Mechanisms of endosperm initiation. *Plant Reproduction, 29*, 215–225.

Hanson, T. (2015). *The triumph of seeds: How grains, nuts, kernels, pulses & pips conquered the plant kingdom and shaped human history*. New York, NY: Basic Books.

Harris, L. W., & Davies, T. J. (2016). A complete fossil-calibrated phylogeny of seed plant families as a tool for comparative analyses: Testing the 'time for speciation' hypothesis. *PLoS One, 11*, e0162907.

Hatorangan, M. R., Laenen, B., Steige, K. A., Slotte, T., & Köhler, C. (2016). Rapid evolution of genomic imprinting in two species of the *Brassicaceae*. *The Plant Cell, 28*, 1815–1827.

Haughn, G., & Chaudhury, A. (2005). Genetic analysis of seed coat development in *Arabidopsis*. *Trends in Plant Science, 10*, 472–477.

Hedhly, A., Vogler, H., Schmid, M. W., Pazmino, D., Gagliardini, V., Santelia, D., et al. (2016). Starch turnover and metabolism during flower and early embryo development. *Plant Physiology, 172*, 2388–2402.

Hehenberger, E., Kradolfer, D., & Köhler, C. (2012). Endosperm cellularization defines an important developmental transition for embryo development. *Development, 139*, 2031–2039.

Henning, L., & Derkacheva, M. (2009). Diversity of *Polycomb* group complexes in plants: same rules, different players? *Trends on Genetics, 25*, 414–423.

Hennig, L., Taranto, P., Walser, M., Schönrock, N., & Gruissem, W. (2003). *Arabidopsis MSI1* is required for epigenetic maintenance of reproductive development. *Development, 130*, 2555–2565.

Hsieh, T. F., Shin, J., Uzawa, R., Silva, P., Cohen, S., Bauer, M. J., et al. (2011). Regulation of imprinted gene expression in *Arabidopsis* endosperm. *Proceedings of the National Academy of Sciences of the United States of America, 108*, 1755–1762.

Huang, X., Lu, Z., Wang, X., Ouyang, Y., Chen, W., Xie, K., et al. (2016). Imprinted gene *OsFIE1* modulates rice seed development by influencing nutrient metabolism and modifying genome H3K27me3. *The Plant Journal, 87*, 305–317.

Ingram, G. C. (2017). Dying to live: Cell elimination as a developmental strategy in angiosperm seeds. *Journal of Experimental Botany, 68*, 785–796.

Iwakawa, H., Shinmyo, A., & Sekine, M. (2006). *Arabidopsis CDKA;1*, a *cdc2* homologue, controls proliferation of generative cells in male gametogenesis. *The Plant Journal, 45*, 819–831.

Jahnke, S., & Scholten, S. (2009). Epigenetic resetting of a gene imprinted in plant embryos. *Current Biology, 19*, 1677–1681.

Jenik, P. D., Gillmor, C. S., & Lukowitz, W. (2007). Embryonic patterning in *Arabidopsis thaliana*. *Annual Review of Cell and Developmental Biology, 23*, 207–236.

Jullien, P. E., Katz, A., Oliva, M., Ohad, N., & Berger, F. (2006). *Polycomb* group complexes self-regulate imprinting of the *Polycomb* group gene *MEDEA* in *Arabidopsis*. *Current Biology, 16*, 486–492.

Jullien, P. E., Kinoshita, T., Ohad, N., & Berger, F. (2006). Maintenance of DNA methylation during the *Arabidopsis* life cycle is essential for parental imprinting. *The Plant Cell, 18*, 1360–1372.

Kang, J., Yim, S., Choi, H., Kim, A., Lee, K. P., Lopez-Molina, L., et al. (2015). Abscisic acid transporters cooperate to control seed germination. *Nature Communications, 6*, 8113.

Karmann, J., Müller, B., & Hammes, U. Z. (2018). The long and winding road: Transport pathways for amino acids in *Arabidopsis* seeds. *Plant Reproduction, 31*, 253–261.
Kawashima, T., & Goldberg, R. B. (2010). The suspensor: Not just suspending the embryo. *Trends in Plant Science, 15*, 23–30.
Köhler, C., Hennig, L., Spillane, C., Pien, S., Gruissem, W., & Grossniklaus, U. (2003). The *Polycomb* group protein MEDEA regulates seed development by controlling expression of the MADS-box gene *PHERES1*. *Genes & Development, 17*, 1540–1553.
Kinoshita, T., Miura, A., Choi, Y., Kinoshita, Y., Cao, X., Jacobsen, S. E., et al. (2004). One-way control of *FWA* imprinting in *Arabidopsis* endosperm by DNA methylation. *Science, 303*, 521–523.
Köhler, C., Page, D. R., Gagliardini, V., & Grossniklaus, U. (2005). The *Arabidopsis thaliana* MEDEA *Polycomb* group protein controls expression of *PHERES1* by parental imprinting. *Nature Genetics, 37*, 28–30.
Köhler, C., & Weinhofer-Molisch, I. (2010). Mechanisms and evolution of genomic imprinting in plants. *Heredity, 105*, 57–63.
Köhler, C., Wolff, P., & Spillane, C. (2012). Epigenetic mechanisms underlying genomic imprinting in plants. *Annual Review of Plant Biology, 63*, 331–352.
Kondou, Y., Nakazawa, M., Kawashima, M., Ichikawa, T., Yoshizumi, T., Suzuki, K., et al. (2008). RETARDED GROWTH OF EMBRYO1, a new basic helix-loop-helix protein, expresses in endosperm to control embryo growth. *The Plant Physiology, 147*, 1924–1935.
Kovalchuk, N., Li, M., Wittek, F., Reid, N., Singh, R., Shirley, N., et al. (2010). Defensin promoters as potential tools for engineering disease resistance in cereal grains. *Plant Biotechnology Journal, 8*, 47–64.
Lafon-Placette, C., & Köhler, C. (2014). Embryo and endosperm, partners in seed development. *Current Opinion in Plant Biology, 17*, 64–69.
Lauria, M., Rupe, M., Guo, M., Kranz, E., Pirona, R., Viotti, A., et al. (2004). Extensive maternal DNA hypomethylation in the endosperm of *Zea mays*. *The Plant Cell, 16*, 510–522.
Lee, K. P., Piskurewicz, U., Tureckova, V., Strnad, M., & Lopez-Molina, L. (2010). A seed coat bedding assay shows that *RGL2*-dependent release of abscisic acid by the endosperm controls embryo growth in *Arabidopsis* dormant seeds. *Proceedings of the National Academy of Sciences of the United States of America, 107*, 19108–19113.
León-Martínez, G., & Vielle-Calzada, J.-P. (2019). Apomixis in flowering plants: Developmental and evolutionary considerations. In U. Grossniklaus (Ed.), *Current topics in developmental biology, Vol. 131*. Plant development and evolution (pp. 565–604). Elsevier.
Li, J., Nie, X., Tan, J. L., & Berger, F. (2013). Integration of epigenetic and genetic controls of seed size by cytokinin in *Arabidopsis*. *Proceedings of the National Academy of Sciences of the United States of America, 110*, 15479–15484.
Linkies, A., Graeber, K., Knight, C., & Leubner-Metzger, G. (2010). The evolution of seeds. *The New Phytologist, 186*, 817–831.
Locascio, A., Roig-Villanova, I., Bernardi, J., & Varotto, S. (2014). Current perspectives on the hormonal control of seed development in *Arabidopsis* and maize: A focus on auxin. *Frontiers in Plant Science, 5*, 412.
Lopes, M. A., & Larkins, B. A. (1993). Endosperm origin, development, and function. *The Plant Cell, 5*, 1383–1399.
Lu, J., & Magnani, E. (2018). Seed tissue and nutrient partitioning, a case for the nucellus. *Plant Reproduction, 31*, 309–317.
Luo, M., Taylor, J. M., Spriggs, A., Zhang, H., Wu, X., Russell, S., et al. (2011). A genome-wide survey of imprinted genes in rice seeds reveals imprinting primarily occurs in the endosperm. *PLoS Genetics*, 7, e1002125.
Magnani, E. (2018). Seed evolution, a 'simpler' story. *Trends in Plant Science, 23*, 654–656.

Maheshwari, P. (1950). *An introduction to the embryology of angiosperms.* New York, NY: McGraw-Hill Book Company.

Makarevich, G., Villar, C. B., Erilova, A., & Köhler, C. (2008). Mechanism of *PHERES1* imprinting in *Arabidopsis. Journal of Cell Science, 121,* 906–912.

Marshall, E., Costa, L. M., & Gutierrez-Marcos, J. (2011). Cysteine-rich peptides (CRPs) mediate diverse aspects of cell-cell communication in plant reproduction and development. *Journal of Experimental Botany, 62,* 1677–1686.

Maruyama, D., Hamamura, Y., Takeuchi, H., Susaki, D., Nishimaki, M., Kurihara, D., et al. (2013). Independent control by each female gamete prevents the attraction of multiple pollen tubes. *Developmental Cell, 25,* 317–323.

Messing, J., & Grossniklaus, U. (1999). Genomic imprinting in plants. In R. Ohlsson (Ed.), *Genomic imprinting: An interdisciplinary approach* (pp. 23–40). Berlin, Heidelberg: Springer Berlin Heidelberg.

Mizzotti, C., Mendes, M. A., Caporali, E., Schnittger, A., Kater, M. M., Battaglia, R., et al. (2012). The MADS box genes *SEEDSTICK* and *ARABIDOPSIS* B_{sister} play a maternal role in fertilization and seed development. *The Plant Journal, 70,* 409–420.

Moller, B., & Weijers, D. (2009). Auxin control of embryo patterning. *Cold Spring Harbor Perspectives in Biology, 1,* a001545.

Moreno-Romero, J., Jiang, H., Santos-Gonzalez, J., & Köhler, C. (2016). Parental epigenetic asymmetry of PRC2-mediated histone modifications in the *Arabidopsis* endosperm. *The EMBO Journal, 35,* 1298–1311.

Moussu, S., Doll, N. M., Chamot, S., Brocard, L., Creff, A., Fourquin, C., et al. (2017). *ZHOUPI* and *KERBEROS* mediate embryo/endosperm separation by promoting the formation of an extracuticular sheath at the embryo surface. *The Plant Cell, 29,* 1642–1656.

Mozgova, I., Köhler, C., & Hennig, L. (2015). Keeping the gate closed: Functions of the polycomb repressive complex PRC2 in development. *The Plant Journal, 83,* 121–132.

Nee, G., Xiang, Y., & Soppe, W. J. (2017). The release of dormancy, a wake-up call for seeds to germinate. *Current Opinion in Plant Biology, 35,* 8–14.

Neuffer, M. G., & Sheridan, W. F. (1980). Defective kernel mutants of maize. I. Genetic and lethality studies. *Genetics, 95,* 929–944.

Ngo, Q. A., Moore, J. M., Baskar, R., Grossniklaus, U., & Sundaresan, V. (2007). *Arabidopsis GLAUCE* promotes fertilization-independent endosperm development and expression of paternally inherited alleles. *Development, 134,* 4107–4117.

Ohnishi, T., Sekine, D., & Kinoshita, T. (2014). Genomic imprinting in plants: What makes the functions of paternal and maternal genes different in endosperm formation? *Advances in Genetics, 86,* 1–25.

Olsen, O. A. (2004). Nuclear endosperm development in cereals and *Arabidopsis thaliana. The Plant Cell, 16,* S214–S227.

Patten, M. M., Ross, L., Curley, J. P., Queller, D. C., Bonduriansky, R., & Wolf, J. B. (2014). The evolution of genomic imprinting: Theories, predictions and empirical tests. *Heredity, 113,* 119–128.

Pires, N. D., & Grossniklaus, U. (2014). Different yet similar: Evolution of imprinting in flowering plants and mammals. *F1000prime Reports, 6,* 63.

Piskurewicz, U., Iwasaki, M., Susaki, D., Megies, C., Kinoshita, T., & Lopez-Molina, L. (2016). Dormancy-specific imprinting underlies maternal inheritance of seed dormancy in *Arabidopsis thaliana. eLife, 5,* e19573.

Pommerrenig, B., Popko, J., Heilmann, M., Schulmeister, S., Dietel, K., Schmitt, B., et al. (2013). SUCROSE TRANSPORTER5 supplies *Arabidopsis* embryos with biotin and affects triacylglycerol accumulation. *The Plant Journal, 73,* 392–404.

Radchuk, V., & Borisjuk, L. (2014). Physical, metabolic and developmental functions of the seed coat. *Frontiers in Plant Science, 5,* 510.

Raissig, M. T., Baroux, C., & Grossniklaus, U. (2011). Regulation and flexibility of genomic imprinting during seed development. *The Plant Cell*, *23*, 16–26.

Raissig, M. T., Bemer, M., Baroux, C., & Grossniklaus, U. (2013). Genomic imprinting in the *Arabidopsis* embryo is partly regulated by PRC2. *PLoS Genetics*, *9*, e1003862.

Raviv, B., Aghajanyan, L., Granot, G., Makover, V., Frenkel, O., Gutterman, Y., et al. (2017). The dead seed coat functions as a long-term storage for active hydrolytic enzymes. *PLoS One*, *12*, e0181102.

Rodrigues, J. C., Tucker, M. R., Johnson, S. D., Hrmova, M., & Koltunow, A. M. (2008). Sexual and apomictic seed formation in *Hieracium* requires the plant *Polycomb*-group gene *FERTILIZATION INDEPENDENT ENDOSPERM*. *The Plant Cell*, *20*, 2372–2386.

Rodrigues, J. A., & Zilberman, D. (2015). Evolution and function of genomic imprinting in plants. *Genes & Development*, *29*, 2517–2531.

Ronceret, A., Gadea-Vacas, J., Guilleminot, J., Lincker, F., Delorme, V., Lahmy, S., et al. (2008). The first zygotic division in *Arabidopsis* requires *de novo* transcription of thymidylate kinase. *The Plant Journal*, *53*, 776–789.

Roszak, P., & Köhler, C. (2011). *Polycomb* group proteins are required to couple seed coat initiation to fertilization. *Proceedings of the National Academy of Sciences of the United States of America*, *108*, 20826–20831.

Sallon, S., Solowey, E., Cohen, Y., Korchinsky, R., Egli, M., Woodhatch, I., et al. (2008). Germination, genetics, and growth of an ancient date seed. *Science*, *320*, 1464.

Satyaki, P. R., & Gehring, M. (2017). DNA methylation and imprinting in plants: Machinery and mechanisms. *Critical Reviews in Biochemistry and Molecular Biology*, *52*, 163–175.

Serbes, I. E., Palovaara, J., & Groß-Hardt, R. (2019). Development and function of the flowering plant female gametophyte. In U. Grossniklaus (Ed.), *Current topics in developmental biology, Vol. 131*. Plant development and evolution (pp. 401–434). Elsevier.

Schatlowski, N., & Köhler, C. (2012). Tearing down barriers: Understanding the molecular mechanisms of interploidy hybridizations. *Journal of Experimental Botany*, *63*, 6059–6067.

Scholten, S. (2010). Genomic imprinting in plant embryos. *Epigenetics*, *5*, 455–459.

Sosso, D., Luo, D., Li, Q. B., Sasse, J., Yang, J., Gendrot, G., et al. (2015). Seed filling in domesticated maize and rice depends on SWEET-mediated hexose transport. *Nature Genetics*, *47*, 1489–1493.

ten Hove, C. A., Lu, K. J., & Weijers, D. (2015). Building a plant: Cell fate specification in the early *Arabidopsis* embryo. *Development*, *142*, 420–430.

Tonosaki, K., Sekine, D., Ohnishi, T., Ono, A., Furuumi, H., Kurata, N., et al. (2018). Overcoming the species hybridization barrier by ploidy manipulation in the genus *Oryza*. *The Plant Journal*, *93*, 534–544.

Veitia, R. A., Bottani, S., & Birchler, J. A. (2008). Cellular reactions to gene dosage imbalance: Genomic, transcriptomic and proteomic effects. *Trends in Genetics*, *24*, 390–397.

Vielle-Calzada, J.-P., Thomas, J., Spillane, C., Collucio, A., Hoeppner, M. A., & Grossniklaus, U. (1999). Maintenance of genomic imprinting at the *Arabidopsis MEDEA* locus requires zygotic *DDM1* activity. *Genes & Development*, *13*, 2971–2982.

von Wangenheim, K. H., & Peterson, H. P. (2004). Aberrant endosperm development in interploidy crosses reveals a timer of differentiation. *Developmental Biology*, *270*, 277–289.

Walters, C., Reilley, A. A., Reeves, P. A., Baszczak, J., & Richards, C. M. (2007). The utility of aged seeds in DNA banks. *Seed Science Research*, *16*, 169–178.

Wang, A., Garcia, D., Zhang, H., Feng, K., Chaudhury, A., Berger, F., et al. (2010). The VQ motif protein IKU1 regulates endosperm growth and seed size in *Arabidopsis*. *The Plant Journal*, *63*, 670–679.

Williams, J. H. (2009). *Amborella trichopoda* (Amborellaceae) and the evolutionary developmental origins of the angiosperm progamic phase. *American Journal of Botany*, *96*, 144–165.

Williams, J. H., & Friedman, W. E. (2002). Identification of diploid endosperm in an early angiosperm lineage. *Nature*, *415*, 522–526.

Willis, C. G., Baskin, C. C., Baskin, J. M., Auld, J. R., Venable, D. L., Cavender-Bares, J., et al. (2014). The evolution of seed dormancy: Environmental cues, evolutionary hubs, and diversification of the seed plants. *The New Phytologist*, *203*, 300–309.

Wöhrmann, H. J., Gagliardini, V., Raissig, M. T., Wehrle, W., Arand, J., Schmidt, A., et al. (2012). Identification of a DNA methylation-independent imprinting control region at the *Arabidopsis MEDEA* locus. *Genes & Development*, *26*, 1837–1850.

Wyder, S., Raissig, M. T., & Grossniklaus, U. (2018). Consistent reanalysis of genome-wide imprinting studies in plants using generalized linear models increases concordance across datasets. *Scientific Reports*. in press.

Xiao, W., Gehring, M., Choi, Y., Margossian, L., Pu, H., Harada, J. J., et al. (2003). Imprinting of the *MEA Polycomb* gene is controlled by antagonism between MET1 methyltransferase and DME glycosylase. *Developmental Cell*, *5*, 891–901.

Xiao, J., Jin, R., & Wagner, D. (2017). Developmental transitions: Integrating environmental cues with hormonal signaling in the chromatin landscape in plants. *Genome Biology*, *18*, 88.

Xu, W., Fiume, E., Coen, O., Pechoux, C., Lepiniec, L., & Magnani, E. (2016). Endosperm and nucellus develop antagonistically in *Arabidopsis* seeds. *The Plant Cell*, *28*, 1343–1360.

Xu, T. T., Ren, S. C., Song, X. F., & Liu, C. M. (2015). *CLE19* expressed in the embryo regulates both cotyledon establishment and endosperm development in *Arabidopsis*. *Journal of Experimental Botany*, *66*, 5217–5227.

Yamaguchi, Y. L., Ishida, T., & Sawa, S. (2016). CLE peptides and their signaling pathways in plant development. *Journal of Experimental Botany*, *67*, 4813–4826.

Yang, X. J., Baskin, J. M., Baskin, C. C., & Huang, Z. Y. (2012). More than just a coating: Ecological importance, taxonomic occurrence and phylogenetic relationships of seed coat mucilage. *Perspectives in Plant Ecology, Evolution and Systematics*, *14*, 434–442.

Yang, S., Johnston, N., Talideh, E., Mitchell, S., Jeffree, C., Goodrich, J., et al. (2008). The endosperm-specific *ZHOUPI* gene of *Arabidopsis thaliana* regulates endosperm breakdown and embryonic epidermal development. *Development*, *135*, 3501–3509.

Yang, J., Luo, D., Yang, B., Frommer, W. B., & Eom, J. S. (2018). *SWEET11* and *15* as key players in seed filling in rice. *The New Phytologist*, *218*, 604–615.

Zhao, P., Begcy, K., Dresselhaus, T., & Sun, M. X. (2017). Does early embryogenesis in eudicots and monocots involve the same mechanism and molecular players? *Plant Physiology*, *173*, 130–142.

Zhao, G. W., & Jiang, X. W. (2014). Roles of gibberellin and auxin in promoting seed germination and seedling vigor in *Pinus massoniana*. *Forest Science*, *60*, 367–373.

Zhou, Y., Zhang, X., Kang, X., Zhao, X., Zhang, X., & Ni, M. (2009). SHORT HYPOCOTYL UNDER BLUE1 associates with *MINISEED3* and *HAIKU2* promoters *in vivo* to regulate *Arabidopsis* seed development. *The Plant Cell*, *21*, 106–117.

Kolibri 360 GmbH
Ferdinand-Jühlke-Straße 7
99095 Erfurt, Deutschland
produktsicherheit@kolibri360.de